T0215693

Lecture Notes in Computer Science 12837

De-Shuang Huang · Kang-Hyun Jo ·
Jianqiang Li · Valeriya Gribova ·
Abir Hussain (Eds.)

Intelligent Computing Theories and Application

17th International Conference, ICIC 2021
Shenzhen, China, August 12–15, 2021
Proceedings, Part II

 Springer

Editors
De-Shuang Huang
Tongji University
Shanghai, China

Kang-Hyun Jo
University of Ulsan
Ulsan, Korea (Republic of)

Jianqiang Li
Shenzhen University
Shenzhen, China

Valeriya Gribova
Far Eastern Branch of the Russian Academy
of Sciences
Vladivostok, Russia

Abir Hussain
Department of Computer Science
Liverpool John Moores University
Liverpool, UK

ISSN 0302-9743 ISSN 1611-3349 (electronic)
Lecture Notes in Computer Science
ISBN 978-3-030-84528-5 ISBN 978-3-030-84529-2 (eBook)
https://doi.org/10.1007/978-3-030-84529-2

LNCS Sublibrary: SL3 – Information Systems and Applications, incl. Internet/Web, and HCI

This Springer imprint is published by the registered company Springer Nature Switzerland AG
The registered company address is: Gewerbestrasse 11, 6330 Cham, Switzerland

Preface

The International Conference on Intelligent Computing (ICIC) was started to provide an annual forum dedicated to the emerging and challenging topics in artificial intelligence, machine learning, pattern recognition, bioinformatics, and computational biology. It aims to bring together researchers and practitioners from both academia and industry to share ideas, problems, and solutions related to the multifaceted aspects of intelligent computing.

ICIC 2021, held in Shenzhen, China, during August 12–15, 2021, constituted the 17th International Conference on Intelligent Computing. It built upon the success of ICIC 2020 (Bari, Italy), ICIC 2019 (Nanchang, China), ICIC 2018 (Wuhan, China), ICIC 2017 (Liverpool, UK), ICIC 2016 (Lanzhou, China), ICIC 2015 (Fuzhou, China), ICIC 2014 (Taiyuan, China), ICIC 2013 (Nanning, China), ICIC 2012 (Huangshan, China), ICIC 2011 (Zhengzhou, China), ICIC 2010 (Changsha, China), ICIC 2009 (Ulsan, South Korea), ICIC 2008 (Shanghai, China), ICIC 2007 (Qingdao, China), ICIC 2006 (Kunming, China), and ICIC 2005 (Hefei, China).

This year, the conference concentrated mainly on the theories and methodologies as well as the emerging applications of intelligent computing. Its aim was to unify the picture of contemporary intelligent computing techniques as an integral concept that highlights the trends in advanced computational intelligence and bridges theoretical research with applications. Therefore, the theme for this conference was "Advanced Intelligent Computing Technology and Applications". Papers that focused on this theme were solicited, addressing theories, methodologies, and applications in science and technology.

ICIC 2021 received 458 submissions from authors in 21 countries and regions. All papers went through a rigorous peer-review procedure and each paper received at least three review reports. Based on the review reports, the Program Committee finally selected 192 high-quality papers for presentation at ICIC 2021, which are included in three volumes of proceedings published by Springer: two volumes of *Lecture Notes in Computer Science* (LNCS) and one volume of *Lecture Notes in Artificial Intelligence* (LNAI).

This volume of LNCS includes 62 papers.

The organizers of ICIC 2021, including Tongji University and Shenzhen University, China, made an enormous effort to ensure the success of the conference. We hereby would like to thank all the ICIC 2021 organizers, the members of the Program Committee, and the referees for their collective effort in reviewing and soliciting the papers. We would like to thank Ronan Nugent, executive editor from Springer, for his frank and helpful advice and guidance throughout as well as his continuous support in publishing the proceedings. In particular, we would like to thank all the authors for contributing their papers. Without the high-quality submissions from the authors, the success of the conference would not have been possible. Finally, we are especially grateful to the International Neural Network Society and the National Science Foundation of China for their sponsorship.

August 2021

De-Shuang Huang
Kang-Hyun Jo
Jianqiang Li
Valeriya Gribova
Abir Hussain

Organization

General Co-chairs

De-Shuang Huang   Tongji University, China
Zhong Ming   Shenzhen University, China

Program Committee Co-chairs

Kang-Hyun Jo   University of Ulsan, South Korea
Jianqiang Li   Shenzhen University, China
Valeriya Gribova   Far Eastern Branch of Russian Academy of Sciences, Russia

Organizing Committee Co-chairs

Qiuzhen Lin   Shenzhen University, China
Cheng Wen Luo   Shenzhen University, China

Organizing Committee Members

Lijia Ma   Shenzhen University, China
Jie Chen   Shenzhen University, China
Jia Wang   Shenzhen University, China
Changkun Jiang   Shenzhen University, China
Junkai Ji   Shenzhen University, China
Zun Liu   Shenzhen University, China

Award Committee Co-chairs

Ling Wang   Tsinghua University, China
Abir Hussain   Liverpool John Moores University, UK

Tutorial Co-chairs

Kyungsook Han   Inha University, South Korea
Prashan Premaratne   University of Wollongong, Australia

Publication Co-chairs

Vitoantonio Bevilacqua   Polytechnic of Bari, Italy
Phalguni Gupta   Indian Institute of Technology Kanpur, India

Special Session Co-chairs

Michal Choras	University of Science and Technology in Bydgoszcz, Poland
Hong-Hee Lee	University of Ulsan, South Korea

Special Issue Co-chairs

M. Michael Gromiha	Indian Institute of Technology Madras, India
Laurent Heutte	Université de Rouen, France
Hee-Jun Kang	University of Ulsan, South Korea

International Liaison Co-chair

Prashan Premaratne	University of Wollongong, Australia

Workshop Co-chairs

Yoshinori Kuno	Saitama University, Japan
Jair Cervantes Canales	Autonomous University of Mexico State, Mexico

Publicity Co-chairs

Chun-Hou Zheng	Anhui University, China
Dhiya Al-Jumeily	Liverpool John Moores University, UK

Exhibition Contact Co-chairs

Qiuzhen Lin	Shenzhen University, China

Program Committee

Mohd Helmy Abd Wahab	Universiti Tun Hussein Onn Malaysia, Malaysia
Nicola Altini	Polytechnic University of Bari, Italy
Waqas Bangyal	University of Gujrat, Pakistan
Wenzheng Bao	Xuzhou University of Technology, China
Antonio Brunetti	Polytechnic University of Bari, Italy
Domenico Buongiorno	Politecnico di Bari, Italy
Hongmin Cai	South China University of Technology, China
Nicholas Caporusso	Northern Kentucky University, USA
Jair Cervantes	Autonomous University of Mexico State, Mexico
Chin-Chih Chang	Chung Hua University, Taiwan, China
Zhanheng Chen	Shenzhen University, China
Wen-Sheng Chen	Shenzhen University, China
Xiyuan Chen	Southeast University, China

Wei Chen	Chengdu University of Traditional Chinese Medicine, China
Michal Choras	University of Science and Technology in Bydgoszcz, Poland
Angelo Ciaramella	Università di Napoli, Italy
Guojun Dai	Hangzhou Dianzi University, China
Weihong Deng	Beijing University of Posts and Telecommunications, China
YanRui Ding	Jiangnan University, China
Pu-Feng Du	Tianjing University, China
Jianbo Fan	Ningbo University of Technology, China
Zhiqiang Geng	Beijing University of Chemical Technology, China
Lejun Gong	Nanjing University of Posts and Telecommunications, China
Dunwei Gong	China University of Mining and Technology, China
Wenyin Gong	China University of Geosciences, China
Valeriya Gribova	Far Eastern Branch of Russian Academy of Sciences, Russia
Michael Gromiha	Indian Institute of Technology Madras, India
Zhi-Hong Guan	Huazhong University of Science and Technology, China
Ping Guo	Beijing Normal University, China
Fei Guo	Tianjin University, China
Phalguni Gupta	Indian Institute of Technology Kanpur, India
Kyungsook Han	Inha University, South Korea
Fei Han	Jiangsu University, China
Laurent Heutte	Université de Rouen Normandie, France
Jian Huang	University of Electronic Science and Technology of China, China
Chenxi Huang	Xiamen University, China
Abir Hussain	Liverpool John Moores University, UK
Qinghua Jiang	Harbin Institute of Technology, China
Kanghyun Jo	University of Ulsan, South Korea
Dah-Jing Jwo	National Taiwan Ocean University, Taiwan, China
Seeja K R	Indira Gandhi Delhi Technical University for Women, India
Weiwei Kong	Xi'an University of Posts and Telecommunications, China
Yoshinori Kuno	Saitama University, Japan
Takashi Kuremoto	Nippon Institute of Technology, Japan
Hong-Hee Lee	University of Ulsan, South Korea
Zhen Lei	Institute of Automation, CAS, China
Chunquan Li	Harbin Medical University, China
Bo Li	Wuhan University of Science and Technology, China
Xiangtao Li	Jilin University, China

Hao Lin	University of Electronic Science and Technology of China, China
Juan Liu	Wuhan University, China
Chunmei Liu	Howard University, USA
Bingqiang Liu	Shandong University, China
Bo Liu	Academy of Mathematics and Systems Science, CAS, China
Bin Liu	Beijing Institute of Technology, China
Zhi-Ping Liu	Shandong University, China
Xiwei Liu	Tongji University, China
Haibin Liu	Beijing University of Technology, China
Jin-Xing Liu	Qufu Normal University, China
Jungang Lou	Huzhou University, China
Xinguo Lu	Hunan University, China
Xiaoke Ma	Xidian University, China
Yue Ming	Beijing University of Posts and Telecommunications, China
Liqiang Nie	Shandong University, China
Ben Niu	Shenzhen University, China
Marzio Pennisi	University of Eastern Piedmont Amedeo Avogadro, Italy
Surya Prakash	IIT Indore, India
Prashan Premaratne	University of Wollongong, Australia
Bin Qian	Kunming University of Science and Technology, China
Daowen Qiu	Sun Yat-sen University, China
Mine Sarac	Stanford University, USA
Xuequn Shang	Northwestern Polytechnical University, China
Evi Sjukur	Monash University, Australia
Jiangning Song	Monash University, Australia
Chao Song	Harbin Medical University, China
Antonino Staiano	Parthenope University of Naples, Italy
Fabio Stroppa	Stanford University, USA
Zhan-Li Sun	Anhui University, China
Xu-Qing Tang	Jiangnan University, China
Binhua Tang	Hohai University, China
Joaquin Torres-Sospedra	UBIK Geospatial Solutions S.L., Spain
Shikui Tu	Shanghai Jiao Tong University, China
Jian Wang	China University of Petroleum, China
Ling Wang	Tsinghua University, China
Ruiping Wang	Institute of Computing Technology, CAS, China
Xuesong Wang	China University of Mining and Technology, China
Rui Wang	National University of Defense Technology, China
Xiao-Feng Wang	Hefei University, China
Shitong Wang	Jiangnan University, China
Bing Wang	Anhui University of Technology, China
Jing-Yan Wang	New York University Abu Dhabi, Abu Dhabi

Dong Wang	University of Jinan, China
Gai-Ge Wang	Ocean University of China, China
Yunhai Wang	Shandong University, China
Ka-Chun Wong	City University of Hong Kong, Hong Kong, China
Hongjie Wu	Suzhou University of Science and Technology, China
Junfeng Xia	Anhui University, China
Shunren Xia	Zhejiang University, China
Yi Xiong	Shanghai Jiao Tong University, China
Zhenyu Xuan	University of Texas at Dallas, USA
Bai Xue	Institute of Software, CAS, China
Shen Yin	Harbin Institute of Technology, China
Xiao-Hua Yu	California Polytechnic State University, USA
Naijun Zhan	Institute of Software, CAS, China
Bohua Zhan	Institute of Software, CAS, China
Fa Zhang	Institute of Computing Technology, CAS, China
JunQi Zhang	Tongji University, China
Le Zhang	Sichuan University, China
Wen Zhang	Huazhong Agricultural University, China
Zhihua Zhang	Beijing Institute of Genomics, CAS, China
Shixiong Zhang	Xidian University, China
Qi Zhao	University of Science and Technology of Liaoning, China
Yongquan Zhou	Guangxi University for Nationalities, China
Fengfeng Zhou	Jilin University, China
Shanfeng Zhu	Fudan University, China
Quan Zou	University of Electronic Science and Technology of China, China

Additional Reviewers

Nureize Arbaiy	Shutao Mei	Na Cheng
Shingo Mabu	Jing Jiang	Menglu Li
Farid Garcia Lamont	Yuelin Sun	Zhenhao Guo
Lianming Zhang	Haicheng Yi	Limin Jiang
Xiao Yu	Suwen Zhao	Kun Zhan
Shaohua Li	Xin Hong	Cheng-Hsiung Chiang
Yuntao Wei	Ziyi Chen	Yuqi Wang
Jinglong Wu	Hailin Chen	Bahattin Karakaya
Weichiang Hong	Xiwei Tang	Tejaswini Mallavarapu
Sungshin Kim	Shulin Wang	Jun Li
Chen Li	Di Zhang	Sheng Yang
Tianhua Guan	Sijia Zhang	Laurent Heutte

Pufeng Du
Atif Mehmood
Jonggeun Kim
Eun Kyeong Kim
Hansoo Lee
Yiqiao Cai
Wuritu Yang
Weitao Sun
Guihua Tao
Jinzhong Zhang
Wenjie Yi
Lingyun Huang
Chao Chen
Jiangping He
Wei Wang
Jin Ma
Liang Xu
Vitoantonio Bevilacqua
Huan Liu
Lei Deng
Di Liu
Zhongrui Zhang
Qinhu Zhang
Yanyun Qu
Jinxing Liu
Shravan Sukumar
Long Gao
Yifei Wu
Tianhua Jiang
Lixiang Hong
Tingzhong Tian
Yijie Ding
Junwei Wang
Zhe Yan
Rui Song
S. A. K. Bangyal
Giansalvo Cirrincione
Xiancui Xiao
X. Zheng
Vincenzo Randazzo
Huijuan Zhu
Dongyuan Li
Jingbo Xia
Boya Ji
Manilo Monaco
Xiaohua Yu

Zuguo Yu
Jun Yuan
Punam Kumari
Bowei Zhao
X. J. Chen
Takashi Kurmeoto
Pallavi Pandey
Yan Zhou
Mascot Wang
Chenhui Qiu
Haizhou Wu
Lulu Zuo
Juan Wang
Rafal Kozik
Wenyan Gu
Shiyin Tan
Yaping Fang
Alexander Moopenn
Xiuxiu Ren
Aniello Castiglione
Qiong Wu
Junyi Chen
Meineng Wang
Xiaorui Su
Jianping Yu
Lizhi Liu
Junwei Luo
Yuanyuan Wang
Xiaolei Zhu
Jiafan Zhu
Yongle Li
Xiaoyin Xu
Shiwei Sun
Hongxuan Hua
Shiping Zhang
Xiangtian Yu
Angelo Riccio
Yuanpeng Xiong
Jing Xu
Chienyuan Lai
Guo-Feng Fan
Zheng Chen
Renzhi Cao
Ronggen Yang
Zhongming Zhao
Yongna Yuan

Chuanxing Liu
Panpan Song
Joao Sousa
Wenying He
Ming Chen
Puneet Gupta
Ziqi Zhang
Davide Nardone
Liangxu Liu
Huijian Han
Qingjun Zhu
Hongluan Zhao
Rey-Sern Lin
Hung-Chi Su
Conghua Xie
Caitong Yue
Li Yan
Tuozhong Yao
Xuzhao Chai
Zhenhu Liang
Yu Lu
Jing Sun
Hua Tang
Liang Cheng
Puneet Rawat
Kulandaisamy A.
Jun Zhang
Egidio Falotico
Peng Chen
Cheng Wang
Jing Li
He Chen
Giacomo Donato Cascarano
Shaohua Wan
Cheng Chen
Jie Li
Ruxin Zhao
Jiazhou Chen
Guoliang Xu
Congxu Zhu
Deng Li
Piyush Joshi
Syed Sadaf Ali
Kuan Li
Teng Wan
Hao Liu

Yexian Zhang
Xu Qiao
Lingchong Zhong
Wenyan Wang
Xiaoyu Ji
Weifeng Guo
Yuchen Jiang
Van-Dung Hoang
Yuanyuan Huang
Zaixing Sun
Honglin Zhang
Yu-Jie He
Rong Hu
Youjie Yao
Naikang Yu
Giulia Russo
Dian Liu
Cheng Liang
Iyyakutti Iyappan Ganapathi
Mingon Kang
Xuefeng Cui
Hao Dai
Geethan Mendiz
Brendan Halloran
Yue Li
Qianqian Shi
Zhiqiang Tian
Ce Li
Yang Yang
Jun Wang
Ke Yan
Hang Wei
Yuyan Han
Hisato Fukuda
Yaning Yang
Lixiang Xu
Yuanke Zhou
Shihui Ying
Wenqiang Fan
Zhao Li
Zhe Zhang
Xiaoying Guo
Zhuoqun Xia
Na Geng
Xin Ding
Balachandran Manavalan

Lianrong Pu
Di Wang
Fangping Wan
Renmeng Liu
Jiancheng Zhong
Yinan Guo
Lujie Fang
Ying Zhang
Yinghao Cao
Xhize Wu
Chao Wu
Ambuj Srivastava
Prabakaran R.
Xingquan Zuo
Jiabin Huang
Jingwen Yang
Qianying Liu
Tongchi Zhou
Xinyan Liang
Xiaopeng Jin
Yumeng Liu
Junliang Shang
Shanghan Li
Jianhua Zhang
Wei Zhang
Han-Jing Jiang
Kunikazu Kobayashi
Shenglin Mu
Jing Liang
Jialing Li
Zhe Sun
Wentao Fan
Wei Lan
Josue Espejel Cabrera
José Sergio Ruiz Castilla
Rencai Zhou
Moli Huang
Yong Zhang
Joaquín Torres-Sospedra
Xingjian Chen
Saifur Rahaman
Olutomilayo Petinrin
Xiaoming Liu
Lei Wang
Xin Xu
Najme Zehra

Zhenqing Ye
Zijing Wang
Lida Zhu
Xionghui Zhou
Jia-Xiang Wang
Gongxin Peng
Junbo Liang
Linjing Liu
Xiangeng Wang
Y. M. Nie
Sheng Ding
Laksono Kurnianggoro
Minxia Cheng
Meiyi Li
Qizhi Zhu
Pengchao Li
Ming Xiao
Guangdi Liu
Jing Meng
Kang Xu
Cong Feng
Arturo Yee
Kazunori Onoguchi
Hotaka Takizawa
Suhang Gu
Zhang Yu
Bin Qin
Yang Gu
Zhibin Jiang
Chuanyan Wu
Wahyono Wahyono
Kaushik Deb
Alexander Filonenko
Van-Thanh Hoang
Ning Guo
Deng Chao
Jian Liu
Sen Zhang
Nagarajan Raju
Kumar Yugandhar
Anoosha Paruchuri
Lei Che
Yujia Xi
Ma Haiying
Huanqiang Zeng
Hong-Bo Zhang

Yewang Chen
Sama Ukyo
Akash Tayal
Ru Yang
Junning Gao
Jianqing Zhu
Haizhou Liu
Nobutaka Shimada
Yuan Xu
Shuo Jiang
Minghua Zhao
Jiulong Zhang
Shui-Hua Wang
Sandesh Gupta
Nadia Siddiqui
Syeda Shira Moin
Ruidong Li
Mauro Castelli
Ivanoe De Falco
Antonio Della Cioppa
Kamlesh Tiwari
Luca Tiseni
Ruizhi Fan
Grigorios Skaltsas
Mario Selvaggio
Xiang Yu
Huajuan Huang
Vasily Aristarkhov
Zhonghao Liu
Lichuan Pan
Zhongying Zhao
Atsushi Yamashita
Ying Xu
Wei Peng
Haodi Feng
Jin Zhao
Shunheng Zhou
Changlong Gu
Xiangwen Wang
Zhe Liu
Pi-Jing Wei
Haozhen Situ
Xiangtao Chen
Hui Tang
Akio Nakamura
Antony Lam

Weilin Deng
Xu Zhou
Shuyuan Wang
Rabia Shakir
Haotian Xu
Zekang Bian
Shuguang Ge
Hong Peng
Thar Baker
Siguo Wang
Jianqing Chen
Chunhui Wang
Xiaoshu Zhu
Yongchun Zuo
Hyunsoo Kim
Areesha Anjum
Shaojin Geng
He Yongqiang
Mario Camana
Long Chen
Jialin Lyu
Zhenyang Li
Tian Rui
Duygun Erol Barkana
Huiyu Zhou
Yichuan Wang
Eray A. Baran
Jiakai Ding
Dehua Zhang
Insoo Koo
Yudong Zhang
Zafaryab Haider
Vladimir Shakhov
Daniele Leonardis
Byungkyu Park
Elena Battini
Radzi Ambar
Noraziah Chepa
Liang Liang
Ling-Yun Dai
Xiongtao Zhang
Sobia Pervaiz Iqbal
Fang Yang
Si Liu
Natsa Kleanthous
Zhen Shen

Chunyan Fan
Jie Zhao
Yuchen Zhang
Jianwei Yang
Wenrui Zhao
Di Wu
Chao Wang
Fuyi Li
Guangsheng Wu
Yuchong Gong
Weitai Yang
Yanan Wang
Bo Chen
Binbin Pan
Chunhou Zheng
Bowen Song
Guojing Wu
Weiping Liu
Laura Jalili
Xing Chen
Xiujuan Lei
Marek Pawlicki
Hao Zhu
Wang Zhanjun
Mohamed Alloghani
Yu Hu
Baohua Wang
Hanfu Wang
Hongle Xie
Guangming Wang
Fuchun Liu
Farid Garcia-Lamont
Hengyue Shi
Po Yang
Wen Zheng Ma
Jianxun Mi
Michele Scarpiniti
Yasushi Mae
Haoran Mo
Gaoyuan Liang
Pengfei Cui
Yoshinori Kobayashi
Kongtao Chen
Feng Feng
Wenli Yan
Zhibo Wang

Ying Qiao
Qiyue Lu
Dong Li
Heqi Wang
Tony Hao
Chenglong Wei
My Ha Le
Yu Chen
Naida Fetic
Bing Sun
Zhenzhong Chu
Meijing Li
Wentao Chen
Mingpeng Zheng
Zhihao Tang
Li Keng Liang
Alberto Mazzoni
Liang Chen
Meng-Meng Yin
Yannan Bin
Wasiq Khan
Yong Wu
Juanjuan Shi
Shiting Sun
Xujing Yao
Wenming Wu
Na Zhang
Anteneh Birga
Yipeng Lv
Qiuye Wang
Adrian Trueba
Ao Liu
Bifang He
Jun Pang
Jie Ding
Shixuan Guan
Boheng Cao
Bingxiang Xu
Lin Zhang
Mengya Liu
Xueping Lv
Hee-Jun Kang
Yuanyuan Zhang
Jin Zhang
Lin Chen
Runshan Xie

Zichang Tan
Fengcui Qian
Xianming Li
Jing Wang
Yuexin Zhang
Fan Wang
Yanyu Li
Qi Pan
Jiaxin Chen
Yuhan Hao
Xiaokang Wang
Jiekai Tang
Wen Jiang
Nan Li
Zhengwen Li
Yuanyuan Yang
Wenbo Chen
Wenchong Luo
Jiang Xue
Xuanying Zhang
Lianlian Zhong
Liu Xiaolin
Difei Liu
Bowen Zhao
Bowen Xue
Churong Zhang
Xing Xing Zhang
Yang Guo
Lu Yang
Jinbao Teng
Yupei Zhang
Keyu Zhong
Mingming Jiang
Chen Yong
Haidong Shao
Weizhong Lin
Leyi Wei
Ravi Kant Kumar
Jogendra Garain
Teressa Longjam
Zhaochun Xu
Zhirui Liao
Qifeng Wu
Nanxuan Zhou
Song Gu
Bin Li

Xiang Li
Yuanpeng Zhang
Dewu Ding
Jiaxuan Liu
Zhenyu Tang
Zhize Wu
Zhihao Huang
Yu Feng
Chen Zhang
Min Liu
Baiying Lei
Jiaming Liu
Xiaochuan Jing
Francesco Berloco
Shaofei Zang
Shenghua Feng
Xiaoqing Gu
Jing Xue
Junqing Zhu
Wenqiang Ji
Muhamad Dwisnanto Putro
Li-Hua Wen
Zhiwen Qiang
Chenchen Liu
Juntao Liu
Yang Miao
Yan Chen
Xiangyu Wang
Cristina Juárez
Ziheng Rong
Jing Lu
Lisbeth Rodriguez Mazahua
Rui Yan
Yuhang Zhou
Huiming Song
Li Ding
Alma Delia Cuevas
Zixiao Pan
Yuchae Jung
Chunfeng Mi
Guixin Zhao
Yuqian Pu
Hongpeng Ynag
Yan Pan
Rinku Datta Rakshit
Ming-Feng Ge

Mingliang Xue	Jiatong Li	Francesco Fontanella
Fahai Zhong	Enda Jiang	Rahul Kumar
Shan Li	Yichen Sun	Alessandra Scotto di Freca
Qingwen Wu	Yanyuan Qin	Nicole Cilia
Tao Li	Chengwei Ai	Annunziata Paviglianiti
Liwen Xie	Kang Li	Jacopo Ferretti
Daiwei Li	Jhony Heriberto Giraldo Zuluaga	Pietro Barbiero
Yuzhen Han	Waqas Haider Bangyal	Seong-Jae Kim
Fengqiang Li	Tingting Dan	Jing Yang
Chenggang Lai	Haiyan Wang	Dan Yang
Shuai Liu	Dandan Lu	Dongxue Peng
Cuiling Huang	Bin Zhang	Wenting Cui
Wenqiang Gu	Cuco Cristanno	Wenhao Chi
Haitao Du	Antonio Junior Spoleto	Ruobing Liang
Bingbo Cui	Zhenghao Shi	Feixiang Zhou
Yang Lei	Ya Wang	Jijia Kang
Xiaohan Sun	Shuyi Zhang	Huawei Huang
Inas Kadhim	Xiaoqing Li	Peng Li
Jing Feng	Yajun Zou	Yunfeng Zhao
Xin Juan	Chuanlei Zhang	Xiaoyan Hu
Hongguo Zhao	Berardino Prencipe	Li Guo
Masoomeh Mirrashid	Feng Liu	Lei Du
Jialiang Li	Yongsheng Dong	Xia-An Bi
Yaping Hu	Rong Fei	Xiuquan Du
Xiangzhen Kong	Zhen Wang	Ping Zhu
Mixiao Hou	Jun Sang	Young-Seob Jeong
Zhen Cui	Jun Wu	Han-Gyu Kim
Na Yu	Xiaowen Chen	Dongkun Lee
Meiyu Duan	Hong Wang	Jonghwan Hyeon
Baoping Yuan	Daniele Malitesta	Chae-Gyun Lim
Umarani Jayaraman	Fenqiang Zhao	Dingna Duan
Guanghui Li	Xinghuo Ye	Shiqiang Ma
Lihong Peng	Hongyi Zhang	Mingliang Dou
Fabio Bellavia	Xuexin Yu	Jansen Woo
Giosue' Lo Bosco	Xujun Duan	Shanshan Hu
Zhen Chen	Xing-Ming Zhao	Hai-Tao Li
Jiajie Xiao	Jiayan Han	Francescomaria Marino
Chunyan Liu	Weizhong Lu	Jiayi Ji
Yue Zhao	Frederic Comby	Jun Peng
Yuwen Tao	Taemoon Seo	Shirley Meng
Nuo Yu	Sergio Cannata	Lucia Ballerini
Liguang Huang	Yong-Wan Kwon	Haifeng Hu
Duy-Linh Nguyen	Heng Chen	Jingyu Hou
Kai Shang	Min Chen	
Wu Hao	Qing Lei	

Contents – Part II

Intelligent Modeling Technologies for Smart Cities

Knowledge Discovery and Data Mining

Machine Learning

Theoretical Computational Intelligence and Applications

Intelligent Computing in Computer Vision

BIDGAN: Blind Image Deblurring with Improved CycleGAN and Frequency Filtering

Yina Zhou, Caiwang Zhang, and Xiaoyong Ji[⊠]

School of Electronic Science and Engineering, Nanjing University,
Nanjing, China
jxy@nju.edu.cn

Abstract. The details of images can be severely impaired when they are captured in motion. To overcome such a blind image deblurring problem, we propose a deblurring method based on improved CycleGAN. This method can remarkably improve the definition of unknown blur kernel images and provide instant higher quality images for the application of many image processing technologies concerned with object detection. Unpaired samples are adopted for training due to the limited sample size. We use random rotation to amplify training effect on distorted images, and it indeed improves the accuracy of irregular image recognition. To minimize the problems of pattern collapse and gradient vanishing, the network adopts Wasserstein distance instead of Jensen-Shannon divergence. Switchable Normalization is also employed because it is more robust to mint-batch size than the former Instance Normalization. The test results show that the deblurring model has achieved better performance in terms of image quality on several popular benchmarks, such as Peak Signal to Noise Ratio (PSNR), Structural Similarity (SSIM) and Kernel MMD; running time has been reduced by 80% compared with conventional deblurring methods based on CNN; mAP on GOPRO Large street view has been improved within the framework of YOLOv3 to overcome the object detection challenge of blurred images. Ultimately, the bright spots have been filtered in the frequency domain and the problem of fake recognition has been solved, so we can retain more image details to provide better pre-processing quality for later object detection.

Keywords: CycleGAN · Blind image deblurring · Motion blur · Frequency filtering

1 Introduction

According to the World Report on Vision released by World Health Organization in 2019 [1], 2.2 billion people worldwide are visually impaired or blind. They comprise about 30% of the world population, especially in low and medium income countries. With the progress of social civilization, travel demand of the blind has gradually attracted the attention all over the world. Though guide dogs are the best auxiliary partners, they require high training cost and a long training cycle, so they can hardly meet the needs of the huge number of blind people. Wearing wearable devices may be

© Springer Nature Switzerland AG 2021
D.-S. Huang et al. (Eds.): ICIC 2021, LNCS 12837, pp. 3–17, 2021.
https://doi.org/10.1007/978-3-030-84529-2_1

another choice to solve the navigation problem for the blind, but the high price discourages most of them. Smartphones and Location Based Services (LBS) technology have become widespread, and opened up a new way to assist the blind. Google [2] proposed a lightweight deep neural network based on embedded devices such as mobile phones, enabling image recognition to be completed on mobile phone CPUs. Under the premise of mobile phone recognition and navigation, blurs caused by relative motion between the blind and the scene greatly affect the recognition results. Aiming at reducing motion blur, we designed a method for solving the problems of object detection based on the blur images in a mobile terminal. By removing motion blur, image features with unclear edge details in the original image can transmit richer semantic information, making the recognition result more accurate and reliable. Thus, it is possible to replace high-speed cameras and processing systems with relatively inexpensive mobile phones, which improves the real-time efficiency of recognition task on mobile terminals.

Motion blur refers to the blur caused by the relative movement between the camera and the object. Image demotion blur is an important research direction in the field of image restoration. According to whether the blur kernel is known, the image deblurring problem is mainly divided into Blind Image Deblurring (BID) and Non-Blind Image Deblurring (NBID). For NBID, the known blur kernel is used for deconvolution to complete image restoration. In 2011, Zhao et al. [3] proposed a new image deblurring algorithm based on Richardson-Lucy (RL), which is an image restoration algorithm in the time domain. It can better suppress ringing response and image noise problems, and retain certain details of the image. In 2016, Luo et al. [4] proposed an image deblurring method combining Total Variation (TV) and Fractional Order Total Variation (FOTV) model to establish a convex optimization model for image deblurring, employing a variable split and alternative direction method to solve the optimization model quickly. The results of experiments show that the PSNR and the SSIM are provided simultaneously. On the other hand, for BID, the blur kernel needs to be estimated first, transformed from BID to NBID, and then deconvolution is carried out for image restoration. For the unknown blur kernel, the normalized sparse prior regularization is used to estimate the complex blur kernel [5], and then the image diffusion function is derived by the geometric relationship model of binocular camera. In the actual situation, due to the great random motion and the complex and unpredictable relative movement, images from handheld shooting conform to the relevant theories of BID, which is the focus of this paper. As for BID, the conventional method uses rich image physical parameters and complex mathematical calculation to realize motion blur restoration, and it has the disadvantages of many iterations, large amount of calculation, high requirements on camera and enormous influence of environment noise.

Machine learning is used to replace the conventional algorithm in this paper, and a new BID method with unknown blur kernel is proposed. Inspired by Image Super Resolution (ISR) [6] and Style Transfer [7] problem, blur images and sharp images are employed as the training set for CycleGAN, creating a better generator to complete motion blur removing. Moreover, Wasserstein distance is used to replace Jensen–Shannon (JS) divergence distance to improve network robustness. Switchable Normalization, which removes image noise information better, is also employed instead of the original Instance Normalization. The quality of the generated image, the similarity between the generated image and the real image and the running time are all improved. In the experiment, the random distortion factor is added into the training set, and the performance in time domain and frequency domain is considered as well, so a more comprehensive blur image preprocessing method can be proposed by this paper.

2   Blind Image Deblurring Based on CycleGAN

2.1   Generative Adversarial Network

Generative Adversarial Network [8], proposed by Ian J. Goodfellow et al. in 2014, is a generation model frame trained by mutual adversarial processes. As one of the most popular unsupervised learning methods in deep learning models, generative adversarial network is widely used in image generation, data augmentation, style transfer, etc. Inspired by zero-sum game, the generator network and the discriminator network are trained simultaneous. The generator generates a new data distribution G(z) from a given noise z, and the discriminator D(x) distinguishes the output of the generator (D = 0) from the real data distribution (D = 1). The generator tries to generate data closer to the true distribution, while the discriminator tries to distinguish the real data from the distributed data more accurately. The two networks tune the training process by back propagation so as to obtain the Nash equilibrium, D (G (z)) = 0.5.

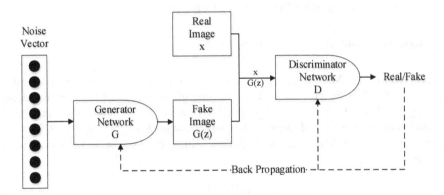

Fig. 1. The network structure of GAN

The basic structure of generative adversarial network is shown in Fig. 1, where G is the generator, for generating expected data, and D is the discriminator, for determining the authenticity of the generated data to optimize the performance of the generator. The loss function of the whole network is:

$$L_{GAN}(D, G, x, z) = E_{x \sim P_{data}(x)}[\log D(x)] + E_{z \sim P_z(z)}[\log(1 - D(G(z)))] \qquad (1)$$

where $P_{data}(x)$ is the data distribution, $P_z(z)$ is the model distribution, and the input z is a sample from a simple noise distribution. The two opposite networks, generator network and discriminator network, conduct the training collaboratively. They promote the update of two network models by restraining each other in training, in order to achieve the best result. This means the discriminant ability of the discriminator reaches the maximum, and meanwhile the difference between the real data distribution and the model distribution reaches the minimum. The optimization objective of the final function is: $\min_{G}(\max_{D}(L_{GAN}(D, G, x, z))$.

GAN has received extensive attention in the field of image generation, however, training of conventional GAN faces many problems, such as mode collapse, vanishing gradient, etc. Several GAN variant structure models are proposed based on the conventional structure. In the field of style transfer, Zhu et al. [7] proposed CycleGAN in 2017 to solve image2image problems. Under the condition of unpaired training set, CycleGAN adopts cycle consistency loss to form a cycle network, in order to complete the unsupervised style transfer problem. In the field of image generation, Isola et al. [9] proposed conditional GAN in 2017, adding certain labels to constrain the generation, and to produce network output images with distinctive features. Radford et al. [10] proposed a deep convolutional generative network in 2015, and introduced it into GAN. The combination of CNN and GAN stabilizes GAN initially. Martin et al. [11] proposed Wasserstein GAN in 2017, to replace JS divergence distance with Wasserstein Loss, and alleviate the potential problem of pattern collapse. Ishaan et al. [12] proposed WGAN-GP based on WGAN, to improve the condition of continuity limitation, and solve the gradient vanishing problem. The generated samples are also of higher quality, thus improving the stability of training.

2.2   CycleGAN Based on Wasserstein Loss

Image Deblurring Model. Suppose $I_B \in R^{h \times w}$ represents original blurred input image, $I_S \in R^{h \times w}$ is the corresponding sharp output image, where h and w represent the height and width of the image respectively. $K(M)$ is the unknown blur kernel determined by motion field M, degenerating sharp image IS into blur image IB. N is the random system noise. The common formulation of unknown blur kernel model is as follows:

$$I_B = k(M) * I_S + N \tag{2}$$

The goal of deblurring is to convert the blurring image IB into corresponding sharp image IS. In the reconstruction process, the motion of the blind holding the mobile phone is quite random, and the relative motion with the shooting object is complex and unpredictable, making the random noise in the system unable to be ignored. Yet, compared with traditional deblurring methods, deep learning can model with any type of data distribution [13], reproducing solving sharp images from blur images accurately.

Network Structure. Compared with the conventional GAN network, CycleGAN consists of two mirrored GAN networks to form a ring network, which ensures that the generator with unsupervised training by unpaired data sets can output images with high similarity to the original ones, without losing the features of the original image. Each one-way GAN has two generators and one discriminator.

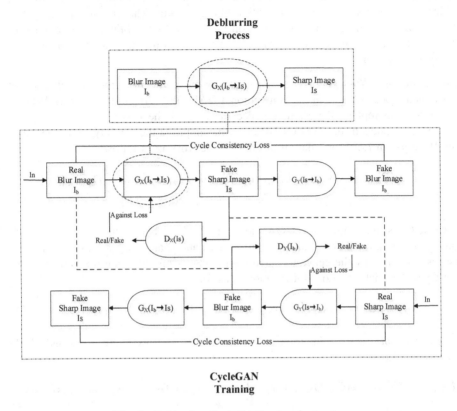

Fig. 2. Deblurring CycleGAN network structure

The structure of CycleGAN used for blind deblurring is shown in Fig. 2. The upper part shows the use of the trained deblurring generator to remove image motion blur, and the lower part shows the training process of CycleGAN. The two mirrored ring GANs cooperate with each other to ensure that the input and output of the whole network are of high similarity. For each one-way GAN:

1. Input real and blur images into the network to train the discriminator;
2. The real input blur image is transformed by the first generator into a fake sharp image, which is sent to the discriminator, giving feedback to the generator to improve the authenticity of the generated image;
3. The generated fake sharp image is transformed into a fake blur image by the second generator. Meanwhile, the cycle consistency loss between the generated fake blur image and the real blur image is calculated to ensure the features in the original image are not lost during the deblurring process; and
4. Repeat until the network converges, and reserve the first deblurring generator.

Similar to the above process, input real and sharp images into the other one-way network. The two GANs are trained alternately with each other to form a closed-loop CycleGAN. The convergence CycleGAN consists of two available generator models: deblurring generator and simulated motion blur generator. The first deblurring generator is retained for removing image motion blur taken by the blind. The second simulated motion blur generator can also be used to expand the training set, which is not the focus in this paper.

Loss Function. Wasserstein Loss: In order to overcome the problems of gradient vanish and mode collapse in the conventional GAN, Wasserstein loss used in WGAN is proposed to replace the JS divergence distance of the original model. Gradient penalty is introduced into WGAN-GP to further improve the stability of the network. The Wasserstein Loss formulation of one-way network is as follow:

$$L_{WGAN-GP}(G_X, D_X, x, y, z) = E_{x \sim P_{I_s}}[\log D_X(x)] + E_{y \sim P_{I_b}}[\log(1 - D_X(G_X(y)))]$$
$$- \lambda E_{z \sim P_{penalty}}[(\|\nabla_z D_X(z)\| - 1)^2] \tag{3}$$

Cycle Consistency Loss: CycleGAN learns both GX and GY at the same time, which can realize the mutual conversion of blur images and sharp images, ensuring that the images generated by the generator have high feature similarity with the input images. The Cycle Consistency Loss formulation is as follow:

$$L_{CYC}(G_X, G_Y) = E_{x \sim P_{I_s}}[\|G_Y(G_X(x)) - x\|_1] + E_{y \sim P_{I_b}}[\|G_X(G_Y(y)) - y\|_1] \tag{4}$$

We formulate the loss function as a combination of Wasserstein Loss and Cycle Consistency Loss:

$$L_{TOTAL} = L_{WGAN-GP}(G_X, D_X, x, y, z) + L_{WGAN-GP}(G_Y, D_Y, x, y, z) + \lambda L_{CYC}(G_X, G_Y) \tag{5}$$

where the λ is the weight to balance the Wasserstein Loss and the Cycle Consistency Loss.

2.3 Generator Network

The generator network mainly completes the conversion of $Ib \rightarrow Is$ and $Is \rightarrow Ib$. Considering the structure of auto-encoder, the generator consists of an encoder, a feature transformation and a decoder, shown in Fig. 3.

As shown in Fig. 3, the convolution layers composed of three convolutional neural networks on the left side of the generator completes the down-sampling and extracts features from the input blurred images. The image is compressed into several feature vectors to complete encoding. Next, blur images are transformed into sharp images by 9 ResNet Blocks to remove image motion blur. ResNet applies identity mapping of the redundant layer in the deep network to ensure that the errors of the overall network will not increase, thus preventing feature loss of input images.

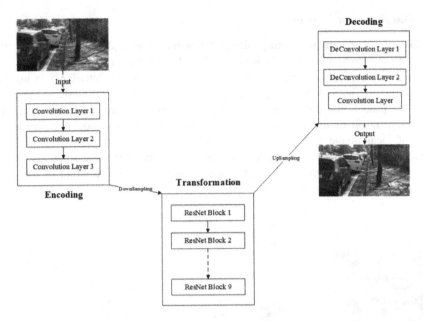

Fig. 3. Generator network

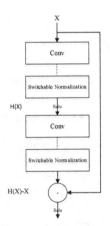

Fig. 4. ResNet structure

The structure of ResNet is shown in Fig. 4. X represents the input of ResNet model, and H represents convolution function. The network learns the difference between input and output, the residual error H(X) – X [14]. Finally, in decoding, the decoder receives Feature Map from the feature converter and restores the high-dimensional features to low-dimensional features by two reverse convolution layers, completes up-sampling, and sends it to a convolution layer to generate a sharp image. Being symmetrical to each other, the encoder and the decoder complete the image2image style transfer task.

2.4   Discriminator Network

The discriminator network mainly distinguishes the generated image from the real image. As the conventional GAN is not applicable to image style transfer with high resolution and high-definition, the structure of discriminator in PatchGAN is used in this paper. PatchGAN maps the input image to an N × N matrix X and the value of X_{ij} resents the probability of the receptive field patch is true. The average value of X_{ij} the final output result of the discriminator, which realizes the extraction and characterization of local image features. PatchGAN is conducive to the generation of high-resolution images. Compared with conventional GAN, it achieves a better loss representation for partial images with large feature differences. The structure of discriminator in PatchGAN is shown in Fig. 5.

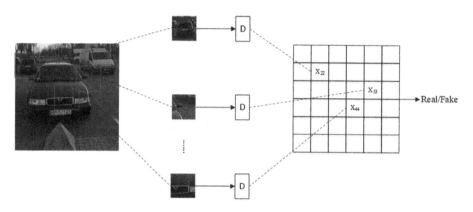

Fig. 5.   Discriminator network

2.5   Switchable Normalization for Motion Blur Removing

Normalization is an essential part in deep learning. The dimensions used in different feature evaluation indicators are often different. In order to reduce the impact of data distribution change on model convergence, normalization is required to make data in the same order of magnitude, solving the comparability problem among feature indexes.

There are several types of normalization algorithms. In addition to conventional BatchNorm algorithm [15] and InstanceNorm algorithm [16], there are ReNorm algorithm [17], LayerNorm algorithm [18], GroupNorm algorithm [19], etc. BatchNorm (BN) is used to calculate the mean and standard deviation of all pixels in a batch of images. InstanceNorm (IN) is used to normalize a single image. Batch Size is usually small or even equals to 1 when training CycleGAN in style transfer task of a single image, calculating mean and standard deviation. The samples in each task have independent styles, so InstanceNorm is often used to solve the problem of individual-based sample distribution problem. However, when removing image motion blur, it is found that IN pays too much attention to the independent features of images, which affects the overall convergence and generalization of the model. It is also very unfavorable to the processing of details in images, which affects the overall deblurring effect.

Thus, SwitchableNorm, a more adaptable algorithm, is considered for the motion blur removing task. It combines BN, LN and IN to determine the appropriate normalization algorithm for each layer of the network by differentiable learning and gives weights on each algorithm to obtain the best normalization of the network itself. The common formulation of SN is as follow:

$$\widehat{h}_{ncij} = \gamma \frac{h_{ncij} - \sum_{k \in \Omega} \omega_k \mu_k}{\sqrt{\sum_{k \in \Omega} \omega'_k \sigma_k^2 + \varepsilon}} + \beta \tag{6}$$

The mean and variance of three normalization algorithms, BN, LN and IN, are calculated together in SN. The weights of these 6 parameters are also calculated and normalized using SOFTMAX. In the formulation, λ and β represent scaling and translation coefficients respectively. ω represents the set of BN, LN and IN. μ and σ^2 represent the mean and the variance. ω_k and ω'_k respectively represent the weight coefficients of the mean and variance parameters corresponding to different normalization algorithms.

3   Experiments

3.1   GOPRO Dataset

The experiments are conducted on GOPRO Large, whose dataset consists of street scene images (including sharp and blurred) taken by high-frame cameras and synthetic blurred images. There are 2103 pairs of blurred and sharp images in 720 pixel $\times$ 1280 pixel quality, captured from moving videos in various scenes. Sharp images record the traffic vehicle in the street scene, from which the outline, texture and color can be clearly observed. They contain various vehicles, pedestrians, zebra crossings, signal lights and other possible objects.

3.2   Model Training

In order to further improve the generalization ability and recognition accuracy of the model for hand-held objects with motion blur including those from non-vertical angles. The data enhancement method is adopted to train this network. The input training samples are random adjusted, defined in Eq. 7:

$$(\alpha, \beta, \gamma) = (\frac{\pi}{180} \times Random(\lambda_1), \frac{\pi}{180} \times Random(\lambda_2), \frac{\pi}{180} \times Random(\lambda_3)) \tag{7}$$

where α, β and γ are the angle difference between the camera coordinate system and the world coordinate system on the x-axis, y-axis and z-axis, respectively. λ is a random angle, which is simulated hand-held shooting angle, $\lambda_i \in (0, 30)$. β is a random deformation factor. Random rotation distortion is performed between the three axes to simulate the real scene shot by the blind, so as to achieve a better training effect on the sample. Due to the limit of network and machine performance, Batch_Size is set at 1, Adam optimizer is adopted and Relu function is used as an activation function.

3.3   Experiment Results

To evaluate the motion blur removing model, subjective visual effect and objective data index are considered to comprehensively evaluate deblurring effect. From subjective visual effect, the details on outline, texture and color of the measured objects that sampled in motion-blurred images, deblurring images and sharp images are compared. Moreover, YOLOv3 on images with different resolution is also introduced to prove the deblurring effect. From objective data index, peak signal-to-noise ratio (PSNR), structural similarity index (SSIM) and kernel maximum mean discrepancy (Kernel MMD) are respectively calculated. Image quality evaluation score, distortion degree between deblurring images and sharp images and difference between real distribution and generated distribution, which means the retention of the features in original images, are comprehensively considered to evaluate the deblurring effect.

Subjective Visual Effect. Figure 6 includes a group of blur images, deblurring images and sharp images, and shows contract result of object detection on different images. It can be intuitively sensed that the blur images have been improved in details such as outline, texture and color after deblurring. YOLOv3 [20], a relatively mature recognition network in object detection domain, can identify most common objects. However, its training set is limited, as it does not consider some rare condition, including motion blur. In the background of guidance by hand-held device, blur caused by relative motion is an important factor affecting the recognition accuracy, and determining the security of the guidance system. From the comparison of the mAP results of images with three different blurring degrees, it can be concluded that the recognition accuracy of the deblurring images is significantly improved, moreover, the target object in the far end of view can also be recognized.

| (a) blur images | (b) deblurring images | (c) sharp images |

Fig. 6. Discriminator network

Objective Data Index. Considering the image2image network structure, this paper selects peak signal-to-noise ratio (PSNR), structural similarity index (SSIM) and kernel maximum mean discrepancy (Kernel MMD) as the evaluation standard to evaluate blur images and deblurring images and calculate the running time of each algorithm on a single GPU. In addition, in order to verify the advantage of deblurring method in this paper, some conventional methods are selected for comparative experiments, including a deblurring method based on convolutional neural networks proposed by Nah et al. in 2017 [21], and a common deep generative model pix2pix [9]. Conventional CycleGAN that uses IN as normalization [7] are also added in the comparative experiment.

PSNR is selected to calculate the ratio of the maximum possible power of images and the destructive power involved with accuracy, which can be used to quantify the quality of images after deblurring. SSIM index of the input and output images, a kind of full reference image quality evaluation index, is calculated to measure the similarity of the images from brightness, contrast and structure respectively, so as to evaluate the distortion degree of images after deblurring compared with the original images. Kernel MMD calculates the difference between real distribution and generative distribution from a limited number of samples to evaluate the generative adversarial network.

Table 1. Evaluation indexes

	Nah et al.	Pix2pix	IN-CycleGAN	SN-CycleGAN
PSNR	30.79	25.13	29.43	**30.45**
SSIM	0.88	0.85	0.91	**0.92**
Kernel MMD	–	0.21	0.19	**0.16**
Time	4.33 s	0.93 s	**0.85 s**	

The evaluation results of the four models under four evaluation indexes are shown in Table 1. In this paper, the proposed deblurring network based on SN normalization algorithm has good performance in terms of SSIM, kernel maximum mean difference and running time. As for the PSNR index, it is also very close to the convolutional neural network using L2 divergence, but it avoids the estimation of motion blur kernel by traditional convolutional neural network method, reduces several calculations of parameters, and cuts the running time by 80%. It is shown that the algorithm proposed in this paper can deal with the blurring problem caused by camera shake or object movement in the process of hand-held shooting. Compared with other deblurring networks, it can better balance the quality and clarity of the generated images and reduce computational complexity of network. Under the premise of reserving image features after deblurring, the noise is safely removed and the detail information is retained to the maximum. Obviously, it is a better choice for hand-held shooting motion blur removing.

4   Image Frequency Domain Analysis

We run YOLO on deblurring images, and the result shows that 'person' is recognized next to 'motorbike' and 'trunk' respectively in groups 2 and 3, and it is the false recognition that does not appear in sharp images. Then, we consider their frequency domain, and carry out two-dimension Fourier transform on the images. The result is shown in Fig. 7.

(a) blur images

(b) deblurring images

(c) sharp images

Fig. 7. Two-dimension Fourier transform on the images

As can be seen from the figure, the high-frequency information of deblurring images is enhanced compared with blur images, which explains why the details of the image after deblurring are enhanced in the time domain to achieve the effect of deblurring. But at the same time, compared with original sharp images, weak or non-existent high-frequency spectral components appeared in blur images, and located in the center of the four quadrants of the magnitude spectrum. It is judged that this is a kind of fake enhancement after deblurring. Spectrum information of the bright spots in deblurring images is extracted and filtered, then the two-dimensional Fourier transform results are shown in Fig. 8.

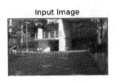
Input Image

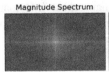
Magnitude Spectrum

Input Image

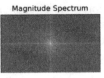
Magnitude Spectrum

Fig. 8. Two-dimension Fourier transform results after bright spots information is extracted

After filtering, the two-dimensional spectral highlight information is removed. We run the inverse Fourier transform and do object detection based on YOLO on these images. The results are shown in Fig. 9.

Fig. 9. Object detection results after bright spots information is extracted

As can be seen from Fig. 9, recognition result on the filtered images has a slight decline compared with the original deblurring images, but the recognition rate is significantly increased in comparison blurred images. That means the filtering is helpful in removing some false recognition. 'Person', which does not appear in the original sharp images, is recognized in original deblurring images next to 'motorbike' and 'trunk'. This is a false judgement result. After filtering the bright spot spectral information in the images, there is no false recognition result as in the original sharp images. In conclusion, we have enhanced the accuracy of the deblurring and improved the safety of the whole guide system.

5 Conclusion

This paper proposed a kernel-free blind image deblurring method and introduced CycleGAN, which is a Generative Adversarial Network optimized by using a combination of Wasserstein loss and cycle consistency loss. In addition, SN is employed as a normalization algorithm instead of the original IN normalization algorithm. It is more suitable for deblurring tasks and enhances image operation details. The method has improved performance under SSIM, NIQE and Kernel MMD indexes, and increased recognition accuracy of traffic obstacle targets under the framework of YOLOV3.

Moreover, through removing the false enhancement information in the frequency domain, we have eliminated the false recognition of blurring images. The recognition accuracy of deblurring network and the security of the whole guidance system have been considerably improved.

References

1. World Health Organization.: World report on vision (2019)
2. Howard, A.G., Zhu, M., Chen, B., et al.: Mobilenets: efficient convolutional neural networks for mobile vision applications. arXiv preprint arXiv:1704.04861 (2017)
3. Zhao, B., Zhang, W., Ding, H., et al.: Noval image deblurring algorithm based on Richardson-Lucy. PhD thesis (2011)
4. Luo, G., Yang, X.: Combining total variation and fractionalorder tatal variation for image deblurring. Comput. Eng. Des. **37**(7), 857–1861 (2016)
5. Tang, M., Peng, G., Zheng, H.: Blind image deblurring based on regularization method. Appl. Res. Comput. **31**(2), 596–599 (2014)
6. Ledig, C., Theis, L., Huszár, F., et al.: Photo-realistic single image super-resolution using a generative adversarial network. In: Proceedings of the IEEE Conference on Computer Vision and Pattern Recognition, pp. 4681–4690 (2017)
7. Zhu, J.Y., Park, T., Isola, P., et al.: Unpaired imageto-image translation using cycle-consistent adversarial networks. In: Proceedings of the IEEE International Conference on Computer Vision, pp. 2223–2232 (2017)
8. Goodfellow, I., Pouget-Abadie, J., Mirza, M., et al.: Generative adversarial nets. In: Advances in Neural Information Processing Systems, pp. 2672–2680 (2014)
9. Isola, P., Zhu, J.Y., Zhou, T., et al.: Image-to-image translation with conditional adversarial networks. In: Proceedings of the IEEE Conference on Computer Vision and Pattern Recognition, pp. 1125–1134 (2017)
10. Radford, A., Metz, L., Chintala, S.: Unsupervised representation learning with deep convolutional generative adversarial networks. arXiv preprint arXiv:1511.06434 (2015)
11. Arjovsky, M., Chintala, S., Bottou, L.: Wasserstein gan. arXiv preprint arXiv:1701.07875 (2017)
12. Gulrajani, I., Ahmed, F., Arjovsky, M., et al.: Improved training of wasserstein gans. In: Advances in Neural Information Processing Systems, pp. 5767–5777 (2017)
13. Huang, G., Liu, Z., Van Der Maaten, L., et al.: Densely connected convolutional networks. In: Proceedings of the IEEE Conference on Computer Vision and Pattern Recognition, pp. 4700–4708 (2017)
14. He, K., Zhang, X., Ren, S., et al.: Deep residual learning for image recognition. In: Proceedings of the IEEE Conference on Computer Vision and Pattern Recognition, pp. 770–778 (2016)
15. Ioffe, S., Szegedy, C.: Batch normalization: accelerating deep network training by reducing internal covariate shift. arXiv preprint arXiv:1502.03167 (2015)
16. Huang, X., Belongie, S.: Arbitrary style transfer in real-time with adaptive instance normalization. In: Proceedings of the IEEE International Conference on Computer Vision, pp. 1501–1510 (2017)
17. Ioffe, S.: Batch renormalization: Towards reducing minibatch dependence in batch-normalized models. In: Advances in Neural Information Processing Systems, pp. 1945–1953, (2017)

18. Ba, J.L., Kiros, J.R., Hinton, G.E.: Layer normalization. arXiv preprint arXiv:1607.06450 (2016)
19. Wu, Y., He, K.: Group normalization. In: Ferrari, V., Hebert, M., Sminchisescu, C., Weiss, Y. (eds.) ECCV 2018. LNCS, vol. 11217, pp. 3–19. Springer, Cham (2018). https://doi.org/10.1007/978-3-030-01261-8_1
20. Redmon, J., Farhadi, A.: Yolov3: an incremental improvement. arXiv preprint arXiv:1804.02767 (2018)
21. Nah, S., Hyun Kim, T., Mu Lee, K.: Deep multi-scale convolutional neural network for dynamic scene deblurring. In Proceedings of the IEEE Conference on Computer Vision and Pattern Recognition, pp. 3883–3891 (2017)

Emotional Interaction Computing of Actors in the Mass Incidents

Yi-yi Wang[1,2] and Fan-liang Bu[1(✉)]

[1] Police Information Engineering and Cyber Security College,
People's Public Security University of China, Beijing 100038, China
[2] Academic Affairs Office, Beijing Police College, Beijing 102202, China

Abstract. The occurrence of mass incidents has an important impact on social and public security. The inspiring and polarization of people's emotions is an important factor affecting the generation and deterioration of mass incidents. Put forward an emotional interaction computation model of actors in the mass incidents combining the group emotional contagion theory with the agent-based modeling method. In the modeling process, pick up some major factors such as emotional value, emotional expression degree, sensitivity, and simulate the emotional interaction of actors. A large number of simulation results show that in the progress of emotional interaction of actors in the mass incidents there is individual emotional spiral phenomenon. It directly leads to group emotional polarization, and affects the actors' violent behavior. Group emotional contagion is an important factor to promote the development of the incident.

Keywords: The mass incidents · Agent model · Emotion · Affective computing

Frequent mass incidents have brought a huge test to the stability of social security. In addition, mass incidents can easily cause emotional infection and polarization among the people. At the same time, the stimulation and polarization of public emotions are also important factors that affect the generation and deterioration of mass incidents. Studying the psychological characteristics of participants in mass incidents can help the public security organs more accurately grasp the law of occurrence and development of mass incidents.

Many scholars have used computer modeling and simulation methods to study mass incidents. Mark Granovetter constructed a group behavior threshold model with the concept of threshold as the core. Starting from the frequency distribution of the threshold, the decision-making process of individual behavior in the group was calculated through the reasonable selection of parameters. The model was used to simulate riot behavior, the spread of rumors, etc., and was used for reference by many scholars afterwards [1]. Wander Jager used multi-agent-based modeling method to simulate the gathering and conflict of two types of people, and simulated the tendency and avoidance process in conflict [2]. The research team at the University of Groningen, the Netherlands represented by Nanda Wijermans, from the perspective of the individual, summarized the factors affecting the individual into three categories: physical environment, social environment and individual factors, and extracted relevant parameters

© Springer Nature Switzerland AG 2021
D.-S. Huang et al. (Eds.): ICIC 2021, LNCS 12837, pp. 18–30, 2021.
https://doi.org/10.1007/978-3-030-84529-2_2

from each category to construct an analysis group Model of factors affecting the probability of violent behavior. The CROSS model they established points out the misunderstanding of group behavior in traditional research [3, 4].

With the deepening of research, more scholars have realized that the group behavior model lacking psychological theory is too rudimentary. Many scholars have conducted in-depth research on the emotions of participants in group events. Wang Lei verified the simulation results of the emotional power model of group events through psychological experiments [5]. Ji Hao analyzed the manifestation of group psychological demands in social event communication cases [6]. Li Congdong conducts theoretical research on group negative emotions by identifying and measuring the communication mechanism of group emotions [7].

Although the current research on the emotions of participants in mass events has achieved some results, but there are few research results on the mechanism of participants' emotion changes. First of all, this article draws the emotional characteristics of participants in group events based on the theory of group emotional infection. Secondly, define the parameters of individual emotional expression degree, tendency and sensitivity, and propose a calculation model of emotional interaction among participants in group events. Finally, based on the Agent-based modeling method, combined with actual cases to simulate the individual emotional evolution process during the infection process, to find out the basic law of the emotional evolution of the participants in the event.

1 Theoretical Basis of Group Psychology Research

Emotions are people's attitude experiences and corresponding behavioral responses to objective things. It has two special forms of existence: internal experience and external expression [8]. Emotion has a significant impact on the individual's psychological and behavioral processes. The earliest research on emotions can be traced back to the era of Plato and Aristotle. Later, James-Lange theory, Cannon's thalamus theory of emotion, behavioral school's emotion theory, emotion activation theory, etc. Appeared. The more influential in contemporary emotional theories are the biological viewpoints, cognitive viewpoints, organizational viewpoints, sociocultural viewpoints, etc. [9].

1.1 Emotional Infection Mechanism

The theory of emotional infection was first proposed by Hatfield, which believed that individuals would automatically and continuously imitate other people's facial expressions, actions, sounds, postures and behaviors in the process of interaction, and tend to capture other people's emotions [10]. Subsequent research results of many scholars have proved that emotional infections are real. Barsade pointed out: The reason why a group is not an independent collection of individuals is precisely because of the strong emotional interaction within the group. This process is more complex than the emotional interaction between only two individuals [11].

In the study of emotional infection mechanism, scholars have proposed a variety of theories, including imitation-feedback mechanism, association-learning mechanism, cognitive mechanism, direct induction mechanism, etc. The most influential one is the imitation-feedback mechanism. The imitation-feedback mechanism points out that emotional infection includes two processes: imitation and feedback. Scholars have found that people tend to imitate the emotional expressions of others, including facial expressions, intonation, posture, and movements. In 1976, Smith pointed out that "when we see someone else's leg or arm being hit, we will also unconsciously contract our leg or arm." This is an active imitation. Lundqvist found that when people watch facial pictures containing different emotions, their facial muscle groups will show different reactions. For example, when watching happy facial expressions, people's cheek muscle groups show more muscle activity; and when watching angry facial expressions, people's brow muscles have more muscular activity. This study shows that people subconsciously imitate the facial expressions of other people's emotions, and they are synchronized. The imitation of people's expressions, sounds, postures and actions will continue to produce feedback and stimulation to people's emotional experience, which is the feedback process. People perceive their own facial movements or other physiological reactions. At this moment, people's emotions will be affected by people's perception. On the one hand, emotions can be strengthened due to their externalization performance, and on the other hand it can also be weakened due to the control of the degree of externalization performance.

1.2   Emotional Characteristics of Participants in Mass Incidents

Based on the above theory, the emotions of participants in mass events are mainly composed of the following characteristics:

First, emotions are related to cognition. Emotions come from highly positive or negative evaluations, that is, the organism evaluates the meaning of the triggering event or situation in a certain number of dimensions, and the result of the evaluation determines the emotions that will occur. It is generally considered that the evaluation is highly cognitive [12]. Participants will generate corresponding emotions based on their own social attitudes and previous damage to their interests.

Second, emotions are constantly changing, and they tend to be transient and sudden. In the process of the formation of mass events, participants experienced the process of emotional accumulation, emotional stimulation, emotional infection and emotional burning. The dramatic changes in emotions seriously affected the behavioral tendencies of the participants.

Finally, the interaction between individuals has a great influence on mood changes. In the case of crowds gathering, due to close contact, psychological cues and behavioral imitation are prone to occur. The probability of emotional infection is very high.

In mass events, most participants have negative emotions. Barsade's experimental results confirm that: compared with positive emotions, negative emotions are more likely to be contagious in a group [11]. Due to the heterogeneity of individuals, the degree of negative emotions they have has certain differences. Negative emotions flow between individuals, and finally reach the result of emotional aggregation. And the anger of the participants will be superimposed due to the gathering of the crowd.

Barsade described the emotional analogy attraction theory that: I feel the same as you, I have to continue to strengthen, which will make me feel comfortable and also make me attractive to you. Therefore, the result of this superposition is that the negative emotions of each participant are improved.

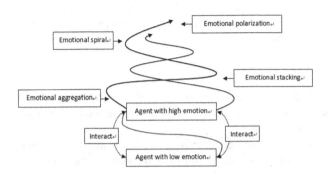

Fig. 1. Process diagram of emotion transmission

In a group event, after the participant enters the group, he quickly senses the emotions of other members of the group, and is affected by the emotion sharing mechanism and the similar emotional attraction. The negative emotions of the participants will soon be generated with the other people's, and it is strengthened through emotional resonance and continues to pass on. Due to multiple emotional interactions between participants, the emotions showed a spiraling result. The whole process is shown in Fig. 1.

2   Computational Model of Emotional Interaction Among Participants in Mass Events

In the model, the emotion receiver is defined as agent$r$, and the emotion sender is defined as agent s, and agent is both the receiver and the sender. The parameters in the model are defined as follows (as shown in Table 1).

The sentiment value of agent, denoted by e, ranges from 0 to 1. In mass events, participants mostly have negative emotions, and the larger the value of e, the greater the negative emotion value.

The degree of emotional expression of agent s, denoted by α, ranges from 0 to 1. Due to the differences in individual personalities, some people are extroverted, lively, and good at expressing their emotions, and some people are introverted, calm, and not good at expressing their emotions, so people have different degrees of transforming their inner emotions into external expressions. The sensitivity of agent r, denoted by β, ranges from 0 to 1. People with high sensitivity are more likely to accept the influence of others.

Table 1. The parameters of the agent in the model

Parameter code	Parameter meaning	value range
e	The sentiment value of agent	[0,1]
α	Emotional expression degree of agent s	[0,1]
β	The sensitivity of agent r	[0,1]
h	The degree of contact between agent r and agent s	[0,1]
δ	The agent's tendency to accept emotions	[0,1]
r	The sight radius of agent r	1

The degree of contact between agent r and agent s, represented by h, with a value between 0 and 1. The larger the value of h, the more likely the emotional infection between the two will occur.

Which emotional valence agent tends to accept, denoted by δ, ranges from 0 to 1. When δ is 0, it means only accept the influence of positive emotions, and when δ is 1, it means only accept the influence of negative emotions.

We set the sight radius of agent r to be r that all agent in his sight can interact with him. The larger the crowd, the denser the crowd, and the greater the number of other agent in his sight. When there is only one agent s, define the received emotion intensity of agent r as:

$$E_r^* = \alpha\beta h \tag{1}$$

The larger the value of E_r^*, the greater the degree of influence on agent r, and the more likely it is for emotional infection to occur.

Then the emotional intensity of all agent s in the line of sight received by agent r is:

$$E_r = \sum_s E_r^* \tag{2}$$

The positive influence received by agent r is expressed as follows:

$$P = e_r e_{sp} \tag{3}$$

Where e_{sp} represents the weighted sum of the intensity of the positive sentiment sent by the sender.

$$e_{sp} = \sum_{sp} \frac{\alpha h}{\sum_s \alpha h} e_s \tag{4}$$

In the same way, the negative influence received by agent r is expressed as follows:

$$N = e_r e_{sn} \tag{5}$$

Where e_{sn} represents the weighted sum of the negative emotion intensity sent by the sender.

$$e_{sn} = \sum_{sn} \frac{\alpha h}{\sum_s \alpha h} e_s \tag{6}$$

The dynamics of agent r's emotional value is defined as:

$$e_r(t+1) = e_r(t) + E_r[\delta N - (1-\delta)P] \tag{7}$$

3   Model Simulation and Empirical Analysis

3.1   Review of the Death of a Woman in Jingwen Mall

In the early morning of May 3, 2013, a 22-year-old woman from Anhui fell to her death on her own in the Jingwen Mall. Later, her boyfriend Peng used the Internet to spread information about her bizarre death and incited his fellow villagers to help talk to the mall. On May 7th, "Beijing Jingwen 22-year-old beauty who jumped from a building in a strange secret room" was posted on a forum and received nearly 2 million views. Since then, a series of rumors have spread wildly.On the morning of May 8, nearly a hundred people gathered at the gate of Jingwen Mall, seriously disrupting social order and traffic.

3.2   Model Simulation

Taking the previous case as a background, the individual emotional characteristics are analyzed. In the simulation, set the network population size to 500 people, the field population size to 100 people, and 13 people who mainly instigated the incident. Using Netlogo simulation software, randomly select 5 interacting agent from the crowd as samples, and observe when configuring different parameters, The changes in their emotional values. For the two population sizes, we designed two simulations to study the influence of individual sensitivity and emotional expression on emotional infection. The simulation experiment settings are shown in Table 2.

Table 2. Simulation experiment setting

Simulation number	Simulation purpose
Simulation 1	Testing the effect of sensitivity β on emotional infections in the real population
Simulation 2	Testing the effect of emotional expression degree α on emotional infections in the real population
Simulation 3	Testing the effect of sensitivity β on emotional infections in the internet crowd
Simulation 4	Testing the effect of emotional expression degree α on emotional infections in the internet crowd
Simulation 5	Test the emotional spiral in emotional infection

3.2.1   The Real Crowd Simulation

In simulation 1, set the tendency of agent to 0.6, and the degree of emotional expression to 0.6. When the sensitivity of agent is 0.1, 0.5, 0.7 and 0.9, their emotional value changes are shown in Fig. 2, the group average emotional value changes are shown in Fig. 3, and the number of emotional values greater than 0.85 changes is shown in Fig. 4.

It can be seen from the simulation results that when the sensitivity is 0.1, the emotional value of agent basically remains unchanged. It is seldom affected by the emotions of others, and emotional infections will not occur. When the sensitivity increases, emotional infections begin to appear. Increasing the value of the individual's sensitivity will increase the speed and amplitude of the rise of the individual's emotional value.

In the second simulation, set the sensitivity to 0.6 and the tendency to 0.6. When the degree of emotional expression is 0.1, 0.5, 0.7 and 0.9, their emotional value changes are shown in Fig. 5, the group average emotional value changes are shown in Fig. 6, and the number of emotional values greater than 0.85 changes is shown in Fig. 7.

It can be seen from the simulation results that when the degree of emotional expression is 0.1, agent has less impact on other individuals, and the phenomenon of emotional infection is not obvious, and each agent keeps their initial emotional value unchanged. When the degree of emotional expression increases, the phenomenon of emotional infection among individuals becomes obvious.

3.2.2   Network Crowd Simulation

In simulation three, set the propensity to 0.6 and the emotional expression level to 0.6. When the sensitivity is 0.1, 0.5, 0.7 and 0.9, their emotional value changes are shown in Fig. 8, the group average emotional value changes are shown in Fig. 9, and the number of emotional values greater than 0.85 changes is shown in Fig. 10. It can be seen that when the sensitivity increases, the rising speed and amplitude of the individual emotional value also increase, and the emotional infection phenomenon is obvious.

In simulation four, set the sensitivity to 0.6 and the tendency to 0.6. When their emotional expression levels are 0.1, 0.5, 0.7, and 0.9, their emotional value changes are shown in Fig. 11, the group's average emotional value changes are shown in Fig. 12, and the number of emotional values greater than 0.85 changes are shown in Fig. 13. It can be seen that when the degree of emotional expression increases, the phenomenon of emotional infection between individuals obviously occurs.

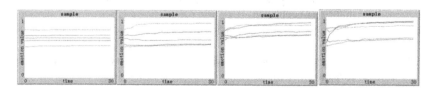

Fig. 2. The evolution of sentiment value when the sensitivity is 0.1, 0.5, 0.7 and 0.9

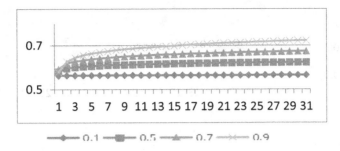

Fig. 3. Changes in the average sentiment value of the group with different sensitivity

3.2.3 Emotional Spiral in Emotional Infection

In simulation five, first set the sensitivity to 0.6 and the emotional expression level to 0.6. When the inclination of 5 samples is set to 1, 0.5, 0.6, 0.2, and 0.8, the three sets of simulation results are shown in Fig. 14. It can be seen that when the initial emotional difference of Agents is large, the emotional infection between them is small.

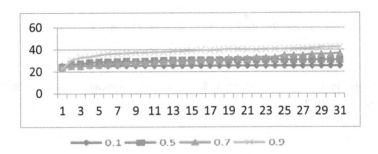

Fig. 4. Changes in the number of Agents with sentiment values greater than 0.85 with different sensitivity

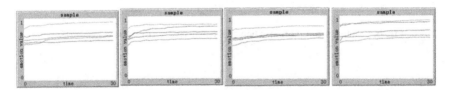

Fig. 5. Changes in emotion when the emotion expression level is 0.1, 0.5, 0.7 and 0.9

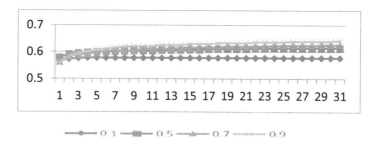

Fig. 6. Changes in the average sentiment value of the group with different emotion expression level

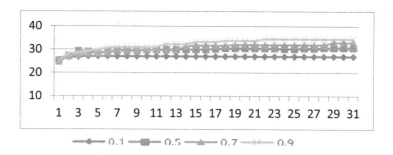

Fig. 7. Changes in the number of Agents with sentiment values greater than 0.85 with different emotion expression level

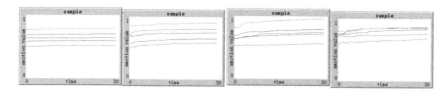

Fig. 8. Changes in emotion when the sensitivity is 0.1, 0.5, 0.7 and 0.9

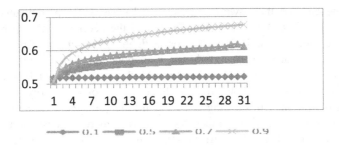

Fig. 9. Changes in the average sentiment value of the group when the sensitivity is 0.1, 0.5, 0.7 and 0.9

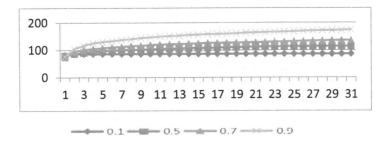

Fig. 10. Changes in the number of Agents with sentiment values greater than 0.85 when the sensitivity is 0.1, 0.5, 0.7 and 0.9

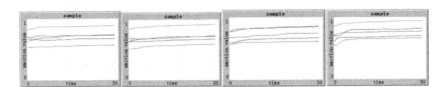

Fig. 11. Changes in emotion value when the emotion expression level is 0.1, 0.5, 0.7 and 0.9

From the development of individual emotions under different sensitivities, it can be seen that the individual's cognitive concepts and experiences have a significant influence on the development of group emotions and the entire event state. When the sensitivity increases, the speed and amplitude of the increase in emotional value will also increase; the speed and amplitude of the increase in the number of over-emotional individuals in the group will also increase. There has been rapid changes in individual emotional values, as well as emotional polarization and emotional aggregation. It shows that the higher the sensitivity, the more influence individuals accept from others, and the more obvious the emotional contagion. In mass incidents, the sensitivity of participants depends on their own experiences and ideas. If the participants are in an environment where social conflicts accumulate and the people have deep grievances, they will have prejudice and dissatisfaction with the government or related departments. After the incident, these individuals with potential dissatisfaction are highly sensitive. They quickly resonate emotionally with the same individuals in the crowd. The mutual infection of emotions produces huge psychological energy, which makes the participants irrational. In the Jingwen incident, the spreaders of rumors used keywords such as "Jingwen girl was raped by 7 security guards", "suspicious", "bizarre death" to attract a lot of attention. At the same time, they used "the police refused to open the case" and "dark insider", "Police refused to check the surveillance requirements" to incite dissatisfaction. Although the Beijing police have repeatedly voiced their doubts, facing the existing evidence, the public still has doubts.

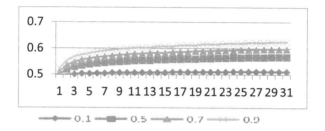

Fig. 12. Changes in the average sentiment value of the group when the emotion expression level is 0.1, 0.5, 0.7 and 0.9

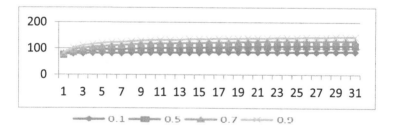

Fig. 13. Changes in the number of Agents with sentiment values greater than 0.85 when the emotion expression level is 0.1, 0.5, 0.7 and 0.9

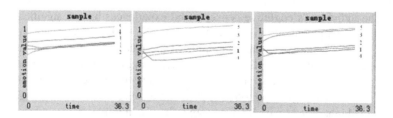

Fig. 14. Changes in sentiment value when sample tendencies are 1, 0.5, 0.6, 0.2 and 0.8

Mass incidents do not happen for no reason. Behind these incidents, far-reaching contradictions and problems must be hidden. If we can take effective measures to release the accumulated interests and emotions in the stage of accumulation of contradictions, we believe that we can greatly reduce the probability of mass incidents.

From the individual emotional development situation under different emotional expression levels, it can be seen that individual incitement has a significant impact on the development of group emotions and the entire event. When the degree of expression of other emotions increases, the speed and amplitude of the increase in emotional value also increase. It can be seen that the higher the degree of emotional expression of other individuals affected, the more influences they receive from others, and the more obvious the phenomenon of emotional infection. In a group event, the more emotional the individual is, the higher the degree of emotional expression. When the emotional value is the same, the degree of emotional expression depends on the participant's personality and other factors. It can be seen that the greater the number of radicals in the incident, the greater the impact on other individuals. Controlling militants in time after the incident and reducing the number of militants can reduce the scale of emotional infections to a certain extent. In the Beijing-Wen incident, after 13 criminal suspects including Peng were arrested, the attention of the masses was cooled, dissatisfaction was calmed, and the incident was calmed down.

When the initial emotional difference is larger, the emotional infection between them is smaller. They are more inclined to communicate with each other with high emotional similarity, and the aggregation and superimposition of emotions will occur if the emotional similarity is high. For example, the sample No. 3 in Fig. 13 has a tendency of 0.6, but in the right figure of Fig. 13, because the initial emotional difference between him and the sample No. 1 and No. 2 is small, he is more affected by the samples No. 1 and No. 2 Influence, so the sentiment value decreases slightly at the beginning. In the left panel of Fig. 13, the initial sentiment difference between sample No. 3 and sample No. 5 is small, so the sentiment value of sample No. 3 rises rapidly and aggregates with the sentiment value of sample No. 5. After the aggregation took place, there was a superposition of emotions, and the emotions of each individual were improved. This is exactly consistent with the emotional attraction theory described by Basade.

Let's observe sample No. 4 again, and his propensity is 0.2. In the right panel of Fig. 13, since his emotional similarity with other samples is low, his emotional value does not change significantly. In the middle the left graph of Fig. 13, the sentiment value of sample No. 4 dropped first and then increased. Individuals with a tendency of 0.2 tend to receive positive influences, which means that their sentiment value should be positive. Shows a downward trend. It can be seen that if the mood of the group is raised, it is possible that the spiral of mood will occur. In mass incidents, the higher the group's negative emotions, the more easily individuals will be attracted by the group. Even if there is a short-term decline in negative emotions, they will rebound back under the influence of other individuals, forming a spiral of emotions.

4 Conclusion

Combining group emotional infection theory with agent-based modeling and simulation research methods, a computational model of the emotional interaction process of participants in group events is proposed. Through the analysis of the experimental and simulation results, we conclude that when the sensitivity increases, the speed and amplitude of the rise of the individual's emotional value also increase. It can be seen that the higher the individual's sensitivity, the more they accept the influence of others. The more obvious the infection. When the emotional expression degree of other individuals increases, the speed and amplitude of the individual emotional value increase accordingly. It can be seen that the higher the emotional expression degree of other individuals that affect the individual, the more the individual accepts the influence of others, and the more obvious the emotional infection phenomenon. In the process of emotional infection, there will be a phenomenon of aggregation and spiraling of emotions. The higher the negative sentiment of the group in the event, the more easily the individual will be attracted by the group. Even if there is a short-term decline in negative sentiment, it will rebound back under the influence of other individuals, forming an emotional spiral. For mass incidents, if we can take effective measures to release the accumulated interests and emotions during the accumulation stage of contradictions, we believe that the probability of mass incidents can be reduced to a large extent. Controlling militants in time after the incident and reducing the number of militants can reduce the scale of emotional infections to a certain extent.

References

1. Granovetter, M.: Threshold models of collective behavior. Am. J. Sociol. **86**(6), 1420–1443 (1978)
2. Jager, W.: Roel popping, hanswan de sande clustering and fighting party crowds: simulating the approach-avoidance conflict. J. Artif. Soc. Soc. Simul. **4**(3), 1–10 (2001)
3. Nanda, W., Rene, J., Wander, J., et al.: Modelling crowd dynamics: influence factors related to the probability of a riot. In: Frederic, A. (eds.) Proceedings of ESSA 4th Conference of the European Social Simulation Association 2007, pp.1–13 Groningen: ESSA (2007)
4. Wijermans, N.: Understanding Crowd Behaviour: [Doctoral]. University of Groningen, Groningen (2011)
5. Lei, W., Ping, F.: A system-of-dynamic model of intergroup emotional transmission. J. Psychol.l Sci. **37**(3), 678–682 (2014)
6. Ji Hao, S., Bing, L.M.: Intention of internet rumor information emotional spreading behavior: a study based on hot-events. J. Intelligence **33**(11), 36–39 (2014)
7. Congdong, L., Xiangyu, H.: Study on rural group negative emotional spreading mechanism based on social network theory. Appl. Res. Comput.s **32**(1), 85–88 (2015)
8. Wang, Z.: Artificial Emotion, pp.1–10. China Machinery Industry Press, Beijing (2009)
9. Hatfield, E., Cacioppo, J.T., Rapson, R.L.: Primitive Emotional Contagion. Emotion and Social Behavior, pp.151–177. Sage, New York (1992)
10. Barsade, S.G.: The ripple effect: emotional contagion and its influence on group behavior. Adm. Sci. Q. **47**(1), 644–675 (2002)
11. Du, J.: Research on Emotional Infection and Face in Service Recovery, p.53. Nankai University Press, Tianjin (2010)

Multi Spatial Convolution Block for Lane Lines Semantic Segmentation

Yan Wu[✉], Feilin Liu, Wei Jiang, and Xinneng Yang

College of Electronics and Information Engineering, Tongji University,
Shanghai 201804, China
{yanwu,1933048,1631730,1830836}@tongji.edu.cn

Abstract. As semantic segmentation models can accurately distinguish object category and contour from the background of images, its application in autonomous driving has been widely studied in recent years. As one of the most important perception modules in autonomous driving, many modern lane line detection models adopt the method of semantic segmentation or instance segmentation. Compared to other objects in traffic scenes, lane lines have a linear structure characteristic of regular shape. Based on that, we design a semantic segmentation model to take advantage of the special structure of lane lines. Combining VH-stage's two-branches horizontal and vertical one-dimensional convolution with SCNN's spatial convolution, we propose a multi spatial convolution block (MSCB). We evaluate our method on PSV and TSD datasets and find MSCB substantially improves the accuracy of lane lines semantic segmentation by up to 4%.

Keywords: Semantic segmentation · Lane line detection · Autonomous driving

1 Introduction

For autonomous vehicles, the lane detection module provides traffic information to judge the status of the vehicle on the road, so as to assist the vehicle in deviation correction or path planning. Because the semantic segmentation [1] method can accurately distinguish the class and contour position of objects in the image, its application in visual perception has been paid more and more attention. Especially with the development of deep learning [2], the accuracy and speed of the semantic segmentation model are constantly improved, making it more suitable for practical application. As an important module of the perception system in autonomous driving, most of the modern lane line detection models adopt the method of semantic segmentation or instance segmentation to obtain accurate lane information.

In real traffic scenarios, there can be many vehicles on different road lanes, which makes the driving condition potentially quite complex. This makes it not possible to simply keep driving in the self lane all the time. When a vehicle decides to move out of the self lanes, it is necessary to know the types of lane lines on both sides before changing lanes or overtaking without violating traffic rules. Therefore, the identification of lane lines is of great significance to the decision-making system of autonomous

© Springer Nature Switzerland AG 2021
D.-S. Huang et al. (Eds.): ICIC 2021, LNCS 12837, pp. 31–41, 2021.
https://doi.org/10.1007/978-3-030-84529-2_3

vehicles. However, most of the existing semantic segmentation-based lane line detection models just divided or segmented the lane area according to the lane sequence and didn't pay much attention to its classes. So in this paper, we focus on the multi-classification and semantic segmentation of lane lines at the same time.

Compared with other objects in traffic scenes, lane lines have the characteristics of linear structure with regular shape. Combining the advantages of VH-stage [3] and SCNN [4], we propose a multi-branch spatial convolution block (MSCB) and add it to our previously proposed lane semantic segmentation model DFNet [5]. We verify the effect of MSCB in two datasets of look-around PSV and forward-looking TSD and find it significantly improves the classification accuracy and segmentation details.

2   Related Work

The traditional lane detection method mainly extracts features such as color [6], structure [7], and then combines Hough transform, Kalman filter, and other methods for recognition, which is greatly affected by scene transformation, occlusion, wear, and other factors. Compared with traditional manual feature extraction methods, the ability of automatic feature extraction by deep learning makes lane detection more accurate and robust.

The current deep learning-based lane detection methods, according to the focus of detection work, can be roughly divided into the following categories: 1. The output form of lane lines, such as point set or segmentation mask; 2. The category of lane lines, which is classified according to the color or shape of lines; 3. The instantiation of the lane line, which distinguishes each lane line according to different lanes.

According to the different output forms, the methods used to output point sets are mostly based on classification or object detection, while the segmentation mask output comes from semantic segmentation or instance segmentation-based methods. E2E-LMD [8] takes lane detection as a column-level classification task. The end of the network divides the feature maps by column and then makes two sequential classifications to get the point sets of lane lines. VPGNet [9] established a dataset of lane and road markings under rainy and poor illumination environments, then used grid cells and multi-label classification to represent targets.

Different from the above methods, the semantic segmentation-based model classifies and labels the pixels of the whole image, so as to obtain the segmentation mask of lane lines. However, few studies in this category can distinguish different types of lanes. VH-HFCN [3] collected and produced a look-around dataset for lane lines and parking slots detection. Lane lines are divided into four classes: white solid line, white dashed line, yellow solid line and yellow dashed line. Based on the linear structure of lane line, they proposed a double branch module of transverse convolution branch and longitudinal convolution branch. In [10], lane lines are divided into double yellow solid lines, single yellow dashed lines, single red solid lines and single white solid lines. Inspired by the characteristics of limited size and certain width of lane lines, they proposed feature size selection and decreasing expansion module.

On the other hand, the common instance segmentation-based method consists of two branches. Lane segmentation branch first divides the image into lane line and background, then lane embedding branch separates the segmented lane line according to different lane instances [11, 12]. TPSeg [13] proposed a view angle conversion convolution layer to transform the middle feature map into a top view for feature extraction, taking the characteristics of the near wide and far narrow lane line from the forward view and the consistent width of the lower lane line from the top view into account.

Considering the shape of lane lines, many methods have been proposed to take advantage of its structural characteristics. Lanenet [11] used the transformation matrix to transform the feature map to the top view angle after semantic segmentation, and then used curve fitting to extract lane lines. Li [14] used a multi-objective depth convolution network to extract geometric structure attribute features of lane lines, and combined with RNN to detect lane lines. VH-stage uses two branches to extract horizontal and vertical structural features respectively. SCNN [4] proposed spatial convolution to introduce temporal spatial information. Compared to using post-processing modules, it is more convenient for training and testing to improve the network structure directly.

In this paper, we take the linear structure of lane as the breakthrough point, and propose the MSCB module. We combine it with the basic model DFNet to get further improvement.

3   Method

In this paper, we propose a spatial feature rearrangement module that takes advantage of the shape prior of lane line to enhance the linear feature extraction capability of segmentation models. The proposed module, which is called Multi Spatial Convolution Block (MSCB), is then combined with the DFNet framework [5] to further improve the lane line segmentation accuracy. The overview of the proposed model is shown in Fig. 1.

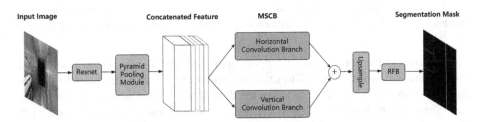

Fig. 1. Overview of the proposed model which combines MSCB in the middle of the DFNet framework.

In view of the linear characteristics of keeping horizontal and vertical in most lane lines, VH-stage [3] adopts one-dimensional convolution and transverse longitudinal branches. Through the one-dimensional convolution, the vertical branch mainly extracts the features in the vertical direction, and the horizontal branch mainly extracts the features in the horizontal direction. Then, the feature maps are fused by pixel-level adding to get the feature extraction of all directions. On the other hand, SCNN [4] adopts a novel spatial convolution that cuts feature maps into rows or columns and transmits information between them, so as to fully extract the spatial relationship. It is especially suitable for objects with continuous shapes or strong spatial relationships such as lane lines. The obtained spatial information is then processed in multiple single directions, which ensures the sequential transmission and structural integrity of the lane line features.

Inspire by the characteristics of the horizontal and vertical branches in VH-stage and the spatial convolution in SCNN, we propose the MSCB which replaces the one-dimensional convolution in VH-stage with the spatial convolution in SCNN. In this way, MSCB possesses the abundant feature extraction ability of dual-branch structure and the linear shape-specific advantage of spatial convolution at the same time, which is especially suitable for lane lines. The structure of MSCB is shown in Fig. 2.

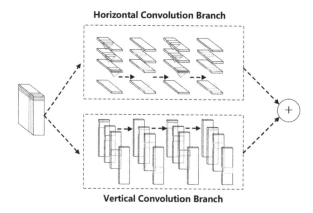

Fig. 2. The proposed Multi Spatial Convolution Block (MSCB) used to enhance the lane line features with a horizontal convolution branch (upper) and a vertical convolution branch (lower).

As shown in the figure above, the upper branch of MSCB only contains horizontal spatial convolution with 9×1 filters that extract lane line features in the order of top-down, bottom-up and alternate twice in the direction of spatial information transmission. Similarly, the lower branch only contains the vertical spatial convolution with 1×9 filters and forms a symmetrical structure with the upper branch. These two branches are then fused by a pixel-level adding operation to integrate the spatial features of different directions.

We choose to put the MSCB between the pyramid pooling module and the upper sampling operation in DFNet for computational cost purposes. The number of feature map channels after the pyramid pooling module is relatively small compared to the previous stage, which is consistent with the number of categories to be classified. While the feature map size needs to be magnified by 8 times after upsampling, the position between the two modules allows it to have the minimum number of channels and size, which reduces the cost of convolutional calculation in MSCB.

4   Experiment

In our experiments, the proposed method is evaluated on two datasets, the look-around dataset PSV [3] and the forward-looking dataset TSD [15]. Lane lines are labeled in both datasets by category (dashed or solid) and color (white or yellow).

PSV is built by The Tongji Intelligent Electric Vehicle (TiEV) team, which contains a total of 4249 panoramic RGB images (2975 images for training and 1274 for testing) with labeled ground truth of 6 object classes, i.e. background, parking slots, white solid line, white dashed line, yellow solid line and yellow dashed line. Some example images with their labels are shown in Fig. 3.

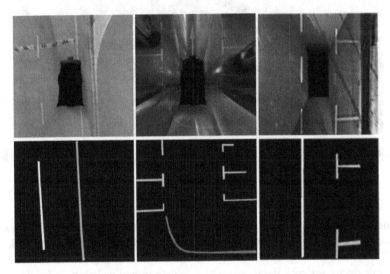

Fig. 3. Example images from the PSV dataset with its lane line segmentation labels.

The TSD dataset is collected and constructed by Xi'an Jiaotong University. It is also used as the training and test set for the off-line competition of China's smart car future challenge, which includes lane line detection, traffic signal detection, vehicle detection and other tasks. We select its subset of the lane detection task as the second dataset in this paper, which contains 4721 images for training and 877 images for testing. The original label of the TSD dataset is set up to mark the vertex position of the lane boundary, which is not suitable for the segmentation task. We transform it according to the coordinates provided by the labels so that it can be used for the lane line semantic segmentation task. In TSD dataset, lane lines are divided into five categories, i.e. background, white solid line, white dashed line, yellow solid line and yellow dashed line. Since the label map is converted from the vertex coordinates, the labeling of the dashed line is consistent with that of the PSV dataset, and the gap part is labeled as the dashed line category. Sample images with their labels of the TSD dataset is shown in Fig. 4.

Fig. 4. Example images from the TSD dataset with its lane line segmentation labels.

Our experiments are implemented on PyTorch, and trained on NVIDIA GeForce GTX TITAN X graphic card with 12G memory. The input images are cropped to a unified size of 600 × 600. According to the memory of the graphic card, we set the batch size to 3. For other training parameters, the momentum is 0.9, the weight decay is 1×10^{-4}, the max training epoch is 50, and the initial learning rate is 1×10^{-3}. The learning rate decreases with iterations, which is determined as below in Eq. (1), where lr_original, current_iter, max_iter and lr_decay refer to the original learning rate, current and max iteration number and the learning rate decay parameter respectively. We use the mean intersection over union(mIoU) and IoU of each class to evaluate our method.

$$lr = lr\_original * \left(1 - \frac{current\_iter}{max\_iter}\right)^{lr\_decay} \tag{1}$$

In order to verify the effectiveness of MSCB module in improving the accuracy of lane semantic segmentation, we insert MSCB, VH-stage and SCNN modules in the same position of DFNet to compare the IoU of each class and mIoU. Meanwhile, we take PSPNet [16] and DFNet as the reference.

Table 1 shows the experimental results on PSV dataset. It can be seen that compared to the DFNet baseline, models with VH-stage, SCNN and MSCB all have a certain improvement effect on the accuracy. Among them, MSCB brings the highest gain, increasing mIoU by 3%. Almost every class of lanes, which includes parking signs, white solid lines, yellow solid lines and yellow dashed lines, is increased by nearly 5%. Different from other classes, the accuracy of the white dashed line has little change compared with DFNet. We assume this happened because the lanes are also white discontinuous lines, which caused certain interclass interference.

Figure 5 shows the segmentation results after adding several modules to DFNet on PSV dataset. For the panoramic image, the width of lane line is consistent, the boundary of the segmented lane line is clear, and there is no bump in the segmented area of the dashed line. As can be seen from the third column, when the space gap of the dashed line is large, a part of the space in VH-stage is mistakenly classified as background. SCNN and MSCB with spatial convolution can better identify the dashed line as a whole.

Table 1. The segmentation results on PSV dataset. Mean IoU for the accuracy metric (%).

Model	Background	Parking	White solid	White dashed	Yellow solid	Yellow dashed	mIoU
PSPNet [16]	98.71	51.88	57.77	37.28	57.70	42.25	57.60
DFNet [5]	98.45	58.40	59.80	49.90	68.32	64.32	66.53
DFNet + VH-stage	98.56	61.08	61.36	49.45	68.35	66.28	67.51
DFNet + SCNN	98.60	61.23	63.13	49.60	68.63	67.58	68.13
DFNet + MSCB	**98.70**	**62.81**	**64.38**	49.19	**72.63**	**69.66**	**69.56**

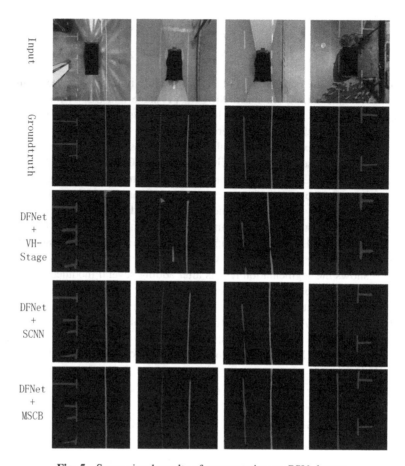

Fig. 5. Some visual results of segmentation on PSV dataset.

Table 2 shows the experimental results on TSD dataset. In terms of the mIoU, all of the three modules have a small improvement, while the MSCB brings a higher gain of 4%. The improvement effect of the dashed line is significantly higher than that of the solid line. Compare to the 2% accuracy improvement for the white solid line, 1% for the yellow solid line, nearly 7% for the white dashed line and 9% for the yellow dashed line are reached. We believe this happened because the forward-looking perspective brings a visual effect of near large and far small. The near part of the lane line with the same length and width occupies a longer and wider area than the far lane in the image, which results in a larger gap between the dashed line. This make the two segments of the dashed line easily be mistaken as two independent solid lines rather than a complete dashed line segment.

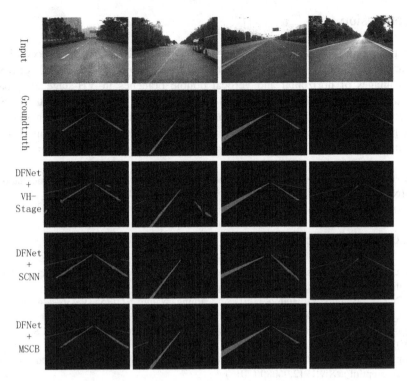

Fig. 6. Some visual results of segmentation on TSD dataset.

Table 2. The segmentation result on TSD dataset. Mean IoU for the accuracy metric (%).

Model	Background	White solid	White dashed	Yellow solid	Yellow dashed	mIoU
PSPNet [16]	99.22	31.61	47.09	64.95	8.53	50.28
DFNet [5]	99.22	42.99	51.03	74.78	45.16	62.64
DFNet + VH-stage	99.12	42.62	53.68	72.74	54.91	64.61
DFNet + SCNN	99.18	42.73	53.71	76.22	53.30	65.03
DFNet + MSCB	99.21	**44.93**	**57.95**	75.61	54.40	**66.42**

The segmentation results on TSD dataset are present in Fig. 6. Compared to other models, the result of MSCB is more complete in lane line boundary. Especially for the dashed line, the bump in the gap part and the boundary between the end of lines are obviously more rough and fuzzy in other methods. At the far end of the dashed line, it is easy to ignore the small gap, resulting in the dashed line segments being mistakenly identified as solid lines. In general, the model with MSCB module performs better in terms of overall detail processing of lane lines.

5 Conclusion

In this paper, we propose the Multi Spatial Convolution Block (MSCB) according to the characteristics of the lane line as linear strip regular objects. This module draws on the advantages of horizontal and vertical double branches in VH-stage and spatial convolution in SCNN. The experimental results on PSV and TSD show that MSCB improves the accuracy of the segmentation model, and is slightly better than VH-stage and SCNN. In our future work, we will further improve the segmentation completeness of the dashed line, and improve the recognition accuracy of the empty segment parts.

Acknowledgments. This work was supported by the National Natural Science Foundation of China (No.U19A2069).

References

1. Garcia-Garcia, A., Orts-Escolano,. S, Oprea, S., et al: A review on deep learning techniques applied to semantic segmentation. arXiv preprint arXiv:1704.06857 (2017)
2. LeCun, Y., Bengio, Y., Hinton, G.: Deep learning. Nature **521**(7553), 436–444 (2015)
3. Wu, Y., Yang, T., Zhao, J., et al.: VH-HFCN based parking slot and lane markings segmentation on panoramic surround view. In: 2018 IEEE Intelligent Vehicles Symposium (IV), pp. 1767–1772. IEEE (2018)
4. Pan, X., Shi, J., Luo, P., et al.: Spatial as deep: spatial CNN for traffic scene understanding. arXiv preprint arXiv:1712.06080 (2017)
5. Jiang, W., Wu, Y., Guan, L., Zhao, J.: DFNet: semantic segmentation on panoramic images with dynamic loss weights and residual fusion block. In: 2019 International Conference on Robotics and Automation (ICRA), pp. 5887–5892. IEEE (2019)
6. Chiu, K.Y., Lin, S.F.: Lane detection using color-based segmentation. In: Intelligent Vehicles Symposium, pp. 706–711. IEEE (2005)
7. Loose, H., Franke, U., Stiller, C.: Kalman particle filter for lane recognition on rural roads. In: Intelligent Vehicles Symposium, pp. 60–65 IEEE (2009)
8. Seungwoo, Y., Heeseok, L., Heesoo, M., et al.: End-to-end lane marker detection via row-wise classification. arXiv preprint arXiv: 2005.08630 (2020)
9. Lee, S., Kim, J., Yoon, J.S., et al.: VPGNet: vanishing point guided network for lane and road marking detection and recognition. In: Proceedings of the IEEE International Conference on Computer Vision, pp. 1965–1973. IEEE (2017)
10. Lo, S.Y., Huang, H.M., Chan, S.W., et al.: Multi-class lane semantic segmentation using efficient convolutional networks. arXiv preprint arXiv:1907.09438 (2019)
11. Neven, D., Brabandere, B., Georgoulis, S., et al.: Towards end-to-end lane detection: an instance segmentation approach. In: 2018 IEEE Intelligent Vehicles Symposium (IV). pp. 286–291. IEEE (2018)
12. Chang, D., Chirakkal, V., Goswami, S., et al.: Multi-lane detection using instance segmentation and attentive voting. arXiv preprint arXiv:2001.00236 (2020)
13. Yu, Z., Ren, X., Huang, Y., et al.: Detecting lane and road markings at a distance with perspective transformer layers. arXiv preprint arXiv: 2003.08550 (2020)

14. Li, J., Mei, X., Prokhorov, D., et al.: Deep neural network for structural prediction and lane detection in traffic scene. In: IEEE Transactions on Neural Networks & Learning Systems, vol. 28, no. 3, pp. 690–703. IEEE (2017)
15. Institute of artificial intelligence and robotics, XJTU. TSD-max: Real traffic scene data set. http://trafficdata.xjtu.edu.cn/. Accessed 1 Mar 2021
16. Zhao, H., Shi, J., Qi, X., et al.: Pyramid scene parsing network. In: Proceedings of the IEEE Conference on Computer Vision and Pattern Recognition, pp. 2881–2890. IEEE (2017)

VISFF: An Approach for Video Summarization Based on Feature Fusion

Wei-Dong Tian[✉], Xiao-Yu Cheng, Bin He, and Zhong-Qiu Zhao

School of Computer Science and Information Engineering,
Hefei University of Technology, Hefei 230009, China
wdtian@hfut.edu.cn

Abstract. In recent years, the amount of video data has been increasing explosively, and the requirements for video summarization technology have also increased. Video summarization is a summary of the video. By browsing the video summarization, users can quickly understand the content of the video. The traditional video summarization algorithms extract the global features of the video frames to form video summarization. However, these algorithms have obvious disadvantages. Therefore, we propose a method to generate video summarization by fusing the global and local features of video frames, and clustering video frames by DBSCAN algorithm. By comparing with the video summarization manually selected by multiple users, we achieve better results on OVP and YouTube datasets than previous algorithms.

Keywords: Video summarization · Global feature · Local feature · Feature fusion · Cluster

1 Introduction

Due to the widespread use of digital cameras and mobile phones, and the significant improvement of the speed in network information transmission, the amount of video data has been increasing explosively in recent years [1]. Video platform applications such as YouTube, Tik Tok are playing an increasingly important role in our daily life. Most people spend a significant amount of time every day watching videos in order to study, entertain, socialize, watch the news, etc.

In addition, social security issues have become increasingly prominent in recent years. The demand for surveillance camera systems is also increasing [2]. However, browsing or processing huge amounts of video data can be time-consuming. A mechanism for summarizing video data is necessary.

Video summarization mechanism is to compress a complete video and remove redundant and unimportant information on the premise of retaining the main information. The main purpose is to help users quickly and conveniently understand the content of the whole video.

Video summarization mechanism includes static video summarization and dynamic video summarization. Static video summaries are representative keyframes selected from the original videos [3]. Through these key frames, users can understand the overall content of the video. Dynamic video summaries are new videos composed of

© Springer Nature Switzerland AG 2021
D.-S. Huang et al. (Eds.): ICIC 2021, LNCS 12837, pp. 42–53, 2021.
https://doi.org/10.1007/978-3-030-84529-2_4

wonderful video clips selected from the original videos, which are more suitable for the highlights of sports events.

Users can learn more video content efficiently in less time through static video summaries. Shot-based algorithm [4], significant content change-based algorithm [5] and clustering-based algorithm [6] are the most popular traditional key frame extraction methods. In the general video summary algorithm, the main process of key frame extraction is to extract the global features of the frame firstly, such as color, texture and motion information, then to analyze the changes of visual features by local differentiation between adjacent frames in the whole video, and finally to select representative frames to form a video summary [7].

However, only extracting the global features of frames for video summarization often ignores the following situations: (1) The contents of the two frames are similar, but the extracted global features are much different (see Fig. 1). (2) The two images are similar on the whole, but differ in some local regions which users may concern about. As a result, the global features of the two images are similar, but the meanings are different (see Fig. 2).

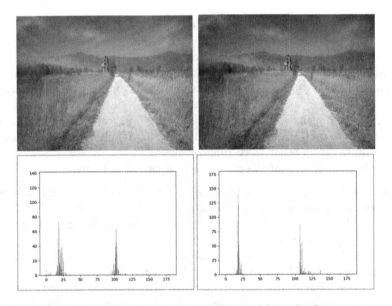

Fig. 1. Two similar images have different global color feature.

Therefore, the global color feature cannot represent the content of the original image well, and it is not enough only to select the global color feature for video summarization.

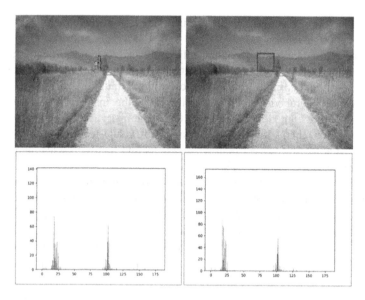

Fig. 2. Two images are similar on the whole while they differ in some obvious local regions.

Based on the above discussion, we propose a novel method of image feature extraction. The method first extracts the global color features of the image in the HSV color space, and then extracts the object features of each frame with object detection network. The extracted object features actually reflect the local information of the image. Finally, the global color features and local object features are fused, and the fused features are used in video frame clustering to complete static video summarization, so as to obtain better video summarizations.

The main contributions of this paper are summarized as follows:

(1) We propose a static video summarization method based on feature fusion, which takes into account both global and local features of video frames.
(2) We use DBSCAN for video frame clustering instead of K-means algorithm and improve the quality of static video summaries.
(3) We evaluate our static video summarization mechanism with reference to the evaluation method CUS (Comparison of User Summaries) which is proposed in VSSUM, which reduce the subjectivity of the evaluation task, quantify the quality of video summarization, and allow more objective comparisons between different technologies.

2   Related Works

Video summarization is a summary of the video. By browsing the video summarization, users can quickly understand the content of the video. The traditional video summarization algorithms extract the global features of the video frames to form video summarization.

Avila et al. proposed a video summarization method based on color feature extraction and K-means clustering algorithm [8]. The method divided and sampled the original video in time sequence, applied the color histogram to the HSV color space, and eliminated the meaningless frames by calculating the standard deviation of the frame feature vectors. Then K-means algorithm was used for frame clustering. Since K-means algorithm has the defect of requiring preset K value, and the size of video summarization was related to the length of video and the intensity of the change of video content, different videos have different K value. Therefore, this algorithm adopted Euclidean distance to calculate the distance between successive frames taken. If the distance was greater than the threshold τ, the value of k was increased by 1. Through the number of peak values, a reasonable estimate of k value was made. Key frame extraction, firstly, identified the cluster with more than the average number of elements from all the clusters as the key cluster, and then selected the video frame closest to the clustering center from the key cluster as the key frame. Then, similar frames were eliminated based on Euclidean distance, and a group of key frames obtained finally form a video summarization.

In recent years, with the rapid development of deep learning, some deep learning ideas have been applied to the generation of video summaries [9]. The deep learning method firstly used convolutional neural network to obtain the key information of each video frame. Then, the feature information was input into the network, and the score of each frame and the similarity between frames were obtained through the recurrent neural network, and the time division of the whole video was performed by using the similarity between frames to avoid the repetition of key frames. Finally, according to the score of each frame and the number of required key frames, the key frames were obtained to generate video summaries.

3   Video Summarization Based on Feature Fusion

Figure 3 illustrates the steps of our method to produce static video summaries.

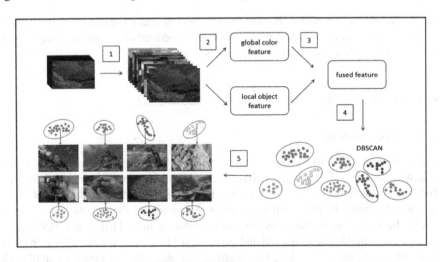

Fig. 3. VISFF approach

Initially, we sample the original video and select the video frames according to the preset sampling rate (step 1). In the next step (step 2), we will calculate the color histogram of each frame in the HSV color space to form the color feature of each frame. Meanwhile, YOLOv4 convolutional neural network is used for object detection. The object features of each frame are formed according to the type and number of objects. After that (step 3), we will normalize the global color feature and the local object feature, and fuse the two features in series to form the fusion feature of each frame. Then (step 4), DBSCAN clustering algorithm is used to cluster each frame. Finally, (step 5), the clustering center of each cluster is selected as the key frame, and then these frames are arranged according to their order in the original video to generate the final video summaries. Each step is detailed in next subsections.

3.1 Video Frame Preprocessing

The first step in producing a video summary is to pre-sample the video frames. Since the original video has a large number of frames and the information difference of successive frames is very small, there is a large degree of information redundancy. When extracting the video summary, if every frame is taken into account, the calculation amount will be particularly huge. Therefore, we use the method of pre-sampling, according to the preset sampling rate, to sample the video. Referring to existing work in this field, the sampling rate is often set at one frame per second [8]. Note that fps, i.e., the number of frames per second in a video, is reciprocal to the preset sampling rate. In other words, the sampling rate will vary depending on the video, rather than a fixed number. In our method, we select the frames in the middle position in one second as the pre-sampled frames. For example, if the fps of the video is 30, we will select the 15th frame as the pre-sampled frame.

$$sampling\_rate = \frac{1}{fps} \tag{1}$$

After pre-sampling, we will eliminate meaningless frames. Meaningless frames are often defined as monochromatic frames, which are generated due to scene changes. The method we use to eliminate meaningless frames is to calculate the standard deviation of the hue value between pixels. If the standard deviation is close to zero, it can be considered as a monochromatic frame. We will do this in parallel with color feature extraction in the next step to make it easier to calculate.

3.2 Feature Extraction

As mentioned in the introduction, we believe that the global color feature cannot represent the content of the original image well, and it is not enough only to select the global color feature for video summarization. We will extract both the global color feature and the local object feature to represent the content of the original video frame.

Color histogram is usually used to extract color features [10]. Since HSV color space is more suitable for human visual perception [11], we convert video frames in RGB color space into HSV color space firstly. After obtaining the pixels value of each frame in HSV color space, we need to calculate the standard deviation of hue value of each pixel. If the standard deviation is close to zero, the frame can be considered as a monochromatic frame, which is meaningless to video summarization, so it can be removed.

Next, we calculate the color histogram of video frames. The hue component is selected as the color feature of the video frame in the end. Since the color histogram calculates the global information of the video frame, the color feature is a kind of global feature. The color histogram is represented as a 180-dimensional vector formally (Fig. 4).

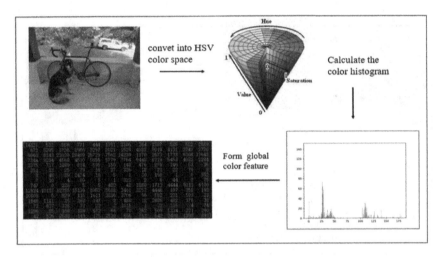

Fig. 4. Global color feature extraction

As an important research direction of computer vision, object detection has been widely used in face detection, pedestrian detection and unmanned driving. With the breakthrough of big data, computer hardware technology and deep learning algorithm in image classification, object detection algorithm based on deep learning has become the mainstream. At present, the mainstream object detection algorithm based on deep learning was divided into two-stage object detection algorithm and one-stage object detection algorithm. Compared with the two-stage object detection algorithm, the one-stage object detection algorithm has the advantage of fast speed and can learn the generalized features of the object. The latest one-stage object detection algorithms, such as YOLOv3 [12] and YOLOv4 [13], have been greatly improved in accuracy and real-time performance.

We use the latest one-stage object detection model YOLOv4 to detect the objects which appear in video frames. On the basis of the original YOLO object detection architecture, YOLOv4 adopts the best optimization strategy in the field of convolutional neural network in recent years. It has been optimized in data processing, backbone network, network training, activation function, loss function and other aspects, and has greatly improved detection accuracy and real-time performance [13].

We train the YOLOv4 object detection network on the MS COCO dataset. When extracting the object features of video frames, we will consider both the category and quantity of objects. The strategy we adopted is that for each video frame, the initial object feature is an 80-dimensional zero vector, and each component of the vector represented the number of the corresponding object in the video frame. Then, each video frame is taken as the input of the YOLOv4 network, and each inferred bounding box indicates that an object is detected. Then, the vector component corresponding to the object is added by one, then we obtain the object feature containing the number and type of objects in each video frame. Since the object feature reflects the local information of the video frame, the object feature is a kind of local feature (Fig. 5).

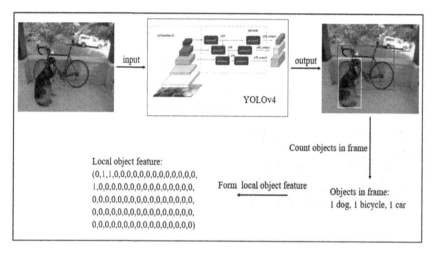

Fig. 5. Local object feature extraction

3.3   Feature Fusion

We have obtained the global color feature and local object feature of the video frame through feature extraction. Both features are formally vectors, but have different dimensions. In order to avoid the shortcoming that the original image cannot be well represented only by color features, the two features need to be fused. Since the order of magnitude between the two characteristic data is of large difference, it is necessary to normalize the two vectors respectively before fusion. Besides, considering that the dimensions of the two features are different, and in each feature, the components of the features are independent of each other, so the common dimensionality reduction methods such as principal component analysis (PCA) are not appropriate [14]. So our strategy is to combine two vectors into one vector in series. The dimensions of the combined vector are equal to the sum of the dimensions of the two vectors.

$$feature_{fused} = Concat\left(feature_{color}, feature_{object}\right) \tag{2}$$

3.4 DBSCAN Clustering

In the field of video summarization, the commonly used clustering method is K-means algorithm. However, K-means has an obvious disadvantage, that is, the number of clusters (i.e., the value of K) needs to be set in advance before using this method [15]. For videos of different lengths, the number of key frames contained in the final generated video summarization is also different. Some approaches will first calculate the difference of features between successive frames. If the difference exceeds a certain threshold, the value of K will be added by one. However, the number of clusters thus determined and the number of frames in the video summarization cannot prove to be correlated.

In our method we will use DBSCAN cluster algorithm instead of K-means. We visualize our fusion features, which represent the distribution of video frame data. Since our fusion features are high-dimensional, we need to use some dimension-reduction methods to display these high-dimensional data points in low-dimensional space (two-dimensional space or three-dimensional space). Therefore, we adopted the latest dimension-reduction method T-SNE. We randomly select some videos from the dataset and analyze the distribution patterns of video frames. It can be seen that the data distribution patterns of different videos are very different. K-means clustering method is more suitable for convex data distribution [16]. For such arbitrary data distribution, density-based clustering algorithm is often more appropriate (Fig. 6).

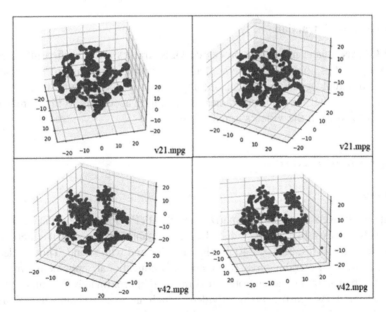

Fig. 6. Fusion features distribution of v21.mpg and v42.mpg in OVP dataset

DBSCAN defines a cluster as the largest collection of density-connected points, which can divide regions with sufficient density into clusters. DBSCAN algorithm needs to determine two parameters: (1) ε: the radius of the adjacent region around a point; (2) minPts: the number of points at least in the adjacent region. In our method, this value is set at 3. DBSCAN divides the sample points into three categories: (1) Core point, i.e., within the adjacent region of the radius ε, the number of sample points is no less than minPts; (2) Border points, i.e., within the adjacent region of the radius ε, the number of sample points is less than minPts, but there are core points in these points. (3) Outliers, which are neither core points nor border points, can be considered as noise in clustering.

The specific clustering process of DBSCAN algorithm is as follows:

(1) Check the unexamined object p in the sample set. If p is not processed (classified into a cluster or marked as noise), then check its adjacent region of the radius ε. If the number of sample points contained is not less than minPts, establish a new cluster C and add all points in it to the candidate set N;

(2) For all the sample points q which have not been processed in the candidate set N, check their adjacent region of the radius ε, and add these sample points to N if the number of sample points contained is not less than minPts; If q doesn't belong to any cluster, then q is added to C;

(3) Repeat step (2), continue to check the unprocessed objects in N, until the current candidate set N is empty;

(4) Repeat steps (1) to steps (3) until all sample points belong to a certain cluster or are marked as noise.

3.5 Keyframe Extraction

After DBSCAN clustering, all the processed video frames are classified into a certain cluster or marked as noise. In our method, all noise frames can be retained or discarded according to actual needs. We select one representative video frame in each cluster randomly. Finally, all the selected frames are arranged according to their order in the original video to form the video summarization.

4 Experimental Results

4.1 Datasets

We used two datasets for video summarization: the OVP (Open Video) dataset [17] and the YouTube dataset [8]. Open Video dataset contain 50 videos. All videos are in MPEG-1 format (30 fps, 352 × 240 pixels), in color and with sound. These videos are distributed among several genres (documentary, educational, ephemeral, historical, lecture) and their duration varies from 1 to 4 min and approximately 75 min of video in total. YouTube dataset also contain 50 videos which are from websites like YouTube. These videos are distributed among several genres (cartoons, news, sports, commercials, TV-shows and home videos) and their duration varies from 1 to 10 min. Many researches in the field of video summarization have been done on these two video datasets.

4.2   Evaluation Method

In the field of video summarization, the general evaluation method is the subjective evaluation by researchers themselves. Some evaluation metrics, such as compression ratio, cannot reflect the quality of video summarization well. Therefore, the method we use is CUS, i.e., comparison of user summaries. The CUS evaluation method will compare the video summaries automatically formed by the algorithm with the video summaries selected by users [8]. For the OVP and YouTube datasets, each video corresponds to five video summaries selected by different users. In this way, the opinions of each user are considered comprehensively, so as to reduce the subjectivity of video summarization evaluation. In the CUS evaluation process, the similarity extent of two frames is measured by comparing the Manhattan distance of fusion features between two frames. If this value is less than 0.5, the two frames are considered to have matched successfully. Finally, two objective evaluation metrics are used to reflect the quality of video summaries. The evaluation metrics are accuracy rate and error rate, which are defined as follows:

$$CUS_A = \frac{n_{mAS}}{n_{US}} \qquad (3)$$

$$CUS_E = \frac{n_{\overline{m}AS}}{n_{US}} \qquad (4)$$

n_{US} means the number of frames in user summary. n_{mAS} means the number of matching keyframes from automatic summary (AS). And $n_{\overline{m}AS}$ means the number of non-matching keyframes from AS.

4.3   Evaluation on OVP Dataset

As shown in Table 1, our method achieves high accuracy and low error rate simultaneously on the OVP data set, which proves the effectiveness of our method.

Table 1. CUS evaluation method results on OVP dataset

Methods	CUS_A	CUS_E
OV [17]	0.70	0.57
DT [18]	0.53	0.29
STIMO [19]	0.72	0.58
VSUMM$_1$ [8]	0.85	0.38
VSUMM$_2$ [8]	0.70	0.27
VISFF (our method)	**0.82**	**0.29**

4.4   Evaluation on YouTube Dataset

We divided videos in the YouTube dataset into six categories based on its type: cartoons, news, sports, commercials, TV-shows and home video. As shown in Table 2, our method also achieves high accuracy and low error rate on the YouTube data set.

Table 2. CUS evaluation method results on YouTube dataset

Video categories	Number of videos	VSUMM	Our method
Cartoons	10	0.87/0.22	0.81/0.26
News	15	0.88/0.32	**0.92/0.28**
Sports	17	0.76/0.65	**0.83/0.48**
Commercials	2	0.93/0.06	0.93/0.06
TV-shows	5	0.91/0.33	0.89/**0.27**
Home	1	0.85/0.23	**0.87/0.21**
Weighted average	50	0.84/0.40	**0.86/0.33**

It can be seen that in the YouTube dataset, except cartoon videos, the effect of other types of videos has been improved. The reason might be that the YOLOv4 network is more suitable for the detection of objects in real scenes. Using fusion feature for video summarization will reduce the quality of video summarization when the effect of object detection is poor. It can be inferred that our method is more suitable for the application of video summarization in real scenes.

5   Conclusion

In this paper, we propose VISFF, an approach for video summarization based on feature fusion. We first extract the global color feature and local object feature of video frames, and then fuse the two features. Next, we use DBSCAN clustering algorithm to cluster the fused features, and finally select one representative video frame in each cluster to form static video summarization. According to the experimental results of CUS evaluation method, VISFF has different extent of improvement in OVP and YouTube dataset compared with the methods that only uses color features for video summarization, which can be used for practical application of video summarization.

Acknowledgment. This work was supported in part by the National Natural Science Foundation of China under Grants 61976079 & 61672203, in part by Anhui Natural Science Funds for Distinguished Young Scholar under Grant 170808J08, and in part by Anhui Key Research and Development Program under Grant 202004a05020039.

References

1. Basavarajaiah, M., Sharma, P.: GVSUM: generic video summarization using deep visual features. Multimed. Tools Appl. 1–18 (2021)
2. Chamasemani, F.F., Affendey, L.S., Mustapha, N., Khalid, F.: Video abstraction using density-based clustering algorithm. Vis. Comput. **34**(10), 1299–1314 (2017). https://doi.org/10.1007/s00371-017-1432-3
3. Kannappan, S., Liu, Y., Tiddeman, B.: Human consistency evaluation of static video summaries. Multimed. Tools Appl. **78**(9), 12281–12306 (2018). https://doi.org/10.1007/s11042-018-6772-0
4. Fu, Y., Liu, H., Cheng, Y., et al.: Key-frame selection in WCE video based on shot detection. In: IEEE Intelligent Control and Automation, pp. 5030–5034 (2012)
5. Jiang, M., Sadka, A., Crookes, D.: Advances in video summarization and skimming. In: Grgic, M., Delac, K., Ghanbari, M., (eds.) Recent Advances in Multimedia Signal Processing and Communications, pp. 27–50. Springer, Berlin (2009). https://doi.org/10.1007/978-3-642-02900-4_2
6. Zhang, Q., Yu, S.-P., Zhou, D.-S., Wei, X.-P.: An efficient method of key-frame extraction based on a cluster algorithm. J. Hum. Kinet. **39**, 5–13 (2013)
7. Mei, S., Guan, G., Wang, Z., Wan, S., He, M., Feng, D.D.: Video summarization via minimum sparse reconstruction. Pattern Recogn. **48**, 522–533 (2015)
8. de Avila, S.E.F., Lopes, A.P.B., da Luz, A., de Albuquerque Araújo, A.: VSUMM: a mechanism designed to produce static video summaries and a novel evaluation method. Pattern Recogn. Lett. **32**, 56–68 (2011)
9. Zhang, K., Chao, W.-L., Sha, F., Grauman, K.: Video summarization with long short-term memory. In: Leibe, B., Matas, J., Sebe, N., Welling, M. (eds.) ECCV 2016. LNCS, vol. 9911, pp. 766–782. Springer, Cham (2016). https://doi.org/10.1007/978-3-319-46478-7_47
10. Trémeau, A., Tominaga, S., Plataniotis, K.N.: Color in image and video processing: most recent trends and future research directions. EURASIP J. Image Video Process. **2008**, 1–26 (2008)
11. James, I.S.P., Angeline, D.M.D.: HSV color histogram based content based image retrieval. Digit. Image Process. **4**, 440–443 (2012)
12. Redmon, J., Farhadi, A.: YOLOv3: an incremental improvement. arXiv preprint arXiv:1804.02767 (2018)
13. Bochkovskiy, A., Wang, C.-Y., Liao, H.-Y.M.: YOLOv4: optimal speed and accuracy of object detection. arXiv preprint (2020)
14. Suykens, J.K., Van Gestel, T., Vandewalle, J., De Moor, B.: A support vector machine formulation to PCA analysis and its kernel version. IEEE Trans. Neural Netw. Learn. Syst. **14**, 447–450 (2003)
15. MacQueen, J.: Some methods for classification and analysis of multivariate observations. University of California Press (1967)
16. Aljarah, I., Faris, H., Mirjalili, S.: Evolutionary Data Clustering: Algorithms and Applications. The University of Jordan, Amman (2021)
17. DeMenthon, D., Kobla, V., Doermann, D.: Video summarization by curve simplification. In: Multimedia, pp. 211–218 (1998)
18. Mundur, P., Rao, Y., Yesha, Y.: Keyframe-based video summarization using Delaunay clustering. Int. J. Digit. Libr. **6**, 219–232 (2006)
19. Furini, M., Geraci, F., Montangero, M., Pellegrini, M.: STIMO: STIll and MOving video storyboard for the web scenario. Multimed. Tools Appl. **46**, 47–69 (2010)

Understanding Safety Based on Urban Perception

Felipe Moreno-Vera$^{(\boxtimes)}$ (iD)

Universidad Católica San Pablo, Arequipa, Perú
felipe.moreno@ucsp.edu.pe

Abstract. Currently, one important field on machine learning is Urban Perception Computing is to model the way in which humans can interact and understand the environment that surrounds them. This process is performed using convolutional models to learn and identify some insights which define the concept of perception of a place (e.g. a street image). One approach of this field is urban perception of street images, we will focus on this approach to study the safety perception of a city and try to explain why and how the perception can be predicted by a mathematical model. As result, we present an analysis about the influence and impact of the visual components on the safety criteria and also an explanation about why a certain decision on the perception of the safety of the streets, such as safe or unsafe.

Keywords: Urban perception · Urban computing · Interpretability · LIME · Computer vision · Perception computing · Deep learning · Street-level imagery · Visual processing · Street View · Cityscape · Perception learning · Grad-CAM

1 Introduction

"Cities are designed to shape and influence the lives of their inhabitants" [13]. Various studies have shown that the visual appearance of cities plays a central role in human perception and reaction to said environment such as "The image of the city" [15]. A notable example is the Broken Window Theory [39] which suggests that visual signs of environmental disruption, such as broken windows, abandoned cars, trash, and graffiti, can induce social outcomes like increase crime levels. This theory has had a great influence on public policy strategies that lead to aggressive police tactics to control the manifestations of social and physical disorder. For example, in social experiments and studies on the perceived quality of life in the streets of New York, comparing impeccable places such as shopping malls (clean walls, orderly, quiet) with other places in which graffiti or garbage is presented [10, 15, 28, 37] concluding that in places where "the rules are violated" it means that in the long term, none of the rules will be fulfilled in that place negatively influenced by the environment (e.g. graffiti, garbage). In addition, other studies have shown that the visual aspect of the spaces of a city affect the psychological state of its inhabitants [9, 13]; Other studies show that the impact of green areas in urban cities [27, 38] has a positives relation to safety perception. In this study, we present a deep learning-based methodology and a model to

© Springer Nature Switzerland AG 2021
D.-S. Huang et al. (Eds.): ICIC 2021, LNCS 12837, pp. 54–64, 2021.
https://doi.org/10.1007/978-3-030-84529-2_5

predict and understand human perceptions of the physical setting of a place. The approach is able to predict, understand and explain predictions of the security perception accurately for a new urban region. Second, we studied the relationship between urban visual elements and perceptions, and tried to determine "the importance of visual elements and their influence over a specific perception". This result helps urban planners and researchers to understand the positive or negative impact of various visual components by exploring urban patterns. The present document is organized by the following: Sect. 2 is about Related works; Sect. 3 we present our Methodology; Sect. 4 Discussions and results; and finally, Conclusions of this work. Our main contributions is a methodology to train and explain urban perceptions from street-images.

2   Related Works

Previous works have a difficulty to explain the direct relation between visual appearance of a city and their corresponding non-visual attributes. These works made an study focus on find a relation between datasets like reports like crimes statistic, robbery rate, house prices, population density, graffiti presence (local reports), and a danger perception survey; with visual appearance of one city.

2.1   Urban Perception

There is a selected works based on urban perception and how to determine using computational methods. The main goal of these works is to correlate a visual appearance of a city with their non-visual attributes like crimes, house prices, perception surveys, etc. These works are solving questions like "What makes Paris look like Paris?" [6] to compare, differentiate and correlate the visual appearance (features) between 12 cities. A similar approach was proposed to answer "What Makes London Look Beautiful, Quiet, and Happy?" [22] exploring 700,000 street-images through a online web survey. [4] studied the correlation between visual non-attributes from city and their visual appearance using several dataset like crimes statistic, robbery rate, house prices, population density, graffiti presence (local reports), and a danger perception survey. In addition, MIT Media Lab releases the Place Pulse dataset [25] which is compose by a street images from difference main cities like New York, Boston, Linz, and Salzburg; and a corresponding perceptual score associated. This work was born from the attempt to relate people's perception of a street through an online survey. This dataset conduced new studies like urban mapping [20] which performs a classification/regression task using and comparing the performance of features extractors like Gist, SIFT + Fisher Vectors, and DeCAF [7]. StreetScore [17] compares GIST, Geometric Probability Map, Text on Histograms, Color Histograms, Geometric Color Histograms, HOG 2x2, Dense SIFT, LBP, Sparse.

Understanding Safety Based on Urban Perception3SIFT histograms, and SSIM features extractors doing a similar research. Following this methodology, a similar study was performed over the city Bogotá, Colombia called Wmodi [1]. In summary, these works have difficulty in extracting information about the natural image because they use classical image features including Hog+Color descriptor, Locality-Sensitive Hashing, Gist, [4], SIFT Fisher Vectors, DeCAF features [20], geometric classification map, color Histograms, HOG2x2, and DenseSIFT [17]. Other works use non-linear models to predict images like SVM [5] and Linear Regression [20], Support Vector Regression was used in [17], RankingSVM was used in [21], SVR was implemented in [4], Multi Task Learning, Transfer Learning based models on ImageNet, and pre-trained networks in [1, 8, 11, 16, 18, 40].

2.2   Model Interpretation.

Model interpretation methods helps us to get insights and understand our learning process and the behavior of a model. In Interpretable Machine Learning, there are several works whose purpose is to understand and explain predictions. Usually, models like CNN are called "black-box" due to they have a large number of parameters distributed in hidden layers with unknown information shared through each layer. Previous works such as LIME [23], SHAP [14], and Anchor [24] explain a model based on their local and global level features components. Other approach based on gradient attribution methods to generate feature maps of an input to provide a visual idea about the explanation like Saliency Maps [32], Gradient [31], Integrated Gradients [36], DeepLIFT [30], CAM [41], Grad-CAM [29], Guided Back Propagation [35], Guided gradCAM [29], and SmoothGrad [3]. These methods are useful to explain simple or complex black box models identifying the dependence of variables and determine if one of them can be isolated or not, in addition to which one has a better representation for prediction depending on the input type. In this work, our approach is to understand the behavior of the urban perception trained on a convolutional network based model using the Place Pulse dataset, composed by images and associate perceptual scores. We want to understand which features impact positive and negative in the perception of safety in street images.

3   Methodology

Our methodology was divided into three parts: (i) dataset pre-processing, (ii) Model training and evaluation, and (iii) Model Interpretation.

3.1   Dataset

Place Pulse has two versions, the first one is Place pulse 1.0 is a dataset composed by a set of images and their correspond perceptual scores. The second one, 4 F. Moreno-Vera et al. Place Pulse 2.0 [8] is a dataset composed of a set of comparisons between 2 images, containing the latitude and longitude for each image. In addition, each comparison has the respective winner (or draw).

Place Pulse 1.0. At the end of 2013, Place Pulse 1.0 contains a total of 73,806 comparisons of 4,109 images from 4 cities: New York City (including Manhattan and parts of Queens, Brooklyn and The Bronx), Boston (including parts of Cambridge), Linz and Salzburg of two countries (US and Austria) and three types of comparisons: safe, wealth, y unique. This dataset has been pre-processed for quick use, containing information on the position of each image (latitude and longitude), perception score for each category, an image identifier and the city to which said image belongs (Table 1).

Table 1. Data summary about Place Pulse 1.0 and their respective category mean.

Place Pulse 1.0

City	# images	Safe mean	Wealth mean	Unique mean
Linz	650	4.85	5.01	4.83
Boston	1237	4.93	4.97	4.76
New York	1705	4.47	4.31	4.46
Salzburg	544	4.75	4.89	5.04
Total	*4136*			

Place Pulse 2.0. In 2016, Place Pulse 2.0 already contained around 1.22 million comparisons of 111,390 images of 56 cities in 28 countries across the 5 continents and six types of comparisons: safe, wealth, depress, beautiful, boring, and lively. This dataset contain 8 columns: image ID (left and right), latitude and longitude (of each image), the result of the comparison, and the respective evaluated category. We perform an algorithm proposed by [26] to pre-process all comparisons in the dataset: for each compared image i with other images j many times in different categories, we define as the intensity of perception of any image i as the percentage of times that the image was selected. Besides, the intensity of j affects i intensity. Due to this, we define the positive rate Wi (1) and the negative rate Li (2) of an image i corresponding to a specific category:

$$W_i = \frac{w_i}{w_i + d_i + l_i} \tag{1}$$

$$L_i = \frac{l_i}{w_i + d_i + l_i} \tag{2}$$

Where, wi is the number of wins, li number of loses, and di draws; From the Eqs. 1 and 2 we can calculate the perceptual score associated for each an image i called Q-score with notation qi, k in a category k:

$$q_{i,k} = \frac{10}{3}\left(W_{i,k} + \frac{1}{n_{i,k}^w}\left(\sum_{j_1} W_{j_1,k}\right) - \frac{1}{n_{i,k}^l}\left(\sum_{j_2} L_{j_2,k}\right) + 1\right) \tag{3}$$

The Eq. 3 is the perceptual score of the image i to be ranked, where j is an image compared to i, n w i is equal to the total number of images i beat and n l i is equal to the total number of images to which i lost. Besides, j1 is the set of images that loses against the image i and j2 is the set of images that wins against the image i. Finally, Q is normalized to fit the range 0 to 10, this scale is a standard when you evaluate perceptions [19]. In this scores, 10 represents the highest possible score for a given question. As an example, if an image receives a calculated score of 0 for the question "Which place looks safer?" that means that specific image is perceived as the least safe image in the dataset (Table 2).

Table 2. Statistics obtained after process all comparisons from Place Pulse, containing information about images per cities in each continent and the mean score for each category asked.

Place Pulse 2.0		
Continent	# cities	# images
America	22	50,028
Europe	22	38,747
Asia	7	11,417
Oceania	2	6,097
Africa	3	5,101
Total	56	111,390

(a)

Place Pulse 2.0		
Category	# comparisons	mean
Safety	368,926	5.188
Lively	267,292	5.085
Beautiful	175,361	4.920
Wealthy	152,241	4.890
Depressing	132,467	4.816
Boring	127,362	4.810
Total	1,223,649	

(b)

3.2   Experiments

In this work, we adapted the VGG16 [33] architecture and our adapted a modification to GAP [12] called VGG16-GAP, we modify the last layer of the blockconv5, taking the Max-Pooling layer and replacing for a GAP layer. This modification aims to extract more informative and high-level features from input images through Global Average Pooling. Once we extract features with this architecture presented in Fig. 1. Then we remove last 2 Fully Connected layer from original model architecture after layer 13, we call the output of this layer as the features extracted from VGG-GAP. We train our Place Pulse dataset focus on two main cities: Boston and New York with perceptual metric safety (we study other metric as well). We select both cities because these cities have the most images quantity and comparisons between images.

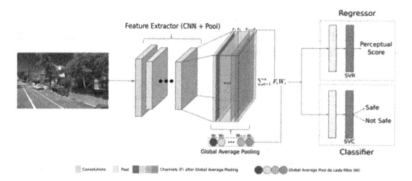

Fig. 1. Statistics obtained after process all comparisons from Place Pulse, containing information about images per cities in each continent and the mean score for each category asked.

We will focus our study comparing three main feature extractors: VGG16, VGG16-GAP, and GIST. Due to the reported results in several previous works in street images described above, we select GIST as baseline feature extractor. To train our VGG16-GAP model, we used a Transfer Learning fine-tuned strategy in the classification and regression task. We use the pre-trained weights of ImageNet dataset, which contains millions of images across 1,000 classes. Then, we freeze our 10 first layers (from input until block4conv3), training only the Block-Conv5. To train the dataset, we make two experiments. The first one is the classification task: To perform this task we divided our images into two labels for each category in the dataset: e.g. safe and not safe. To select which set of images will be safe or unsafe, we define a parameter called δ with a value between 0,05 - 0,5. This delta will creates a subset using the binary labels yi,k $\in \{1, -1\}$ for both training and testing as:

$$y_{i,k} = \begin{cases} 1 & \text{if } (q_{i,k}) \text{ in the top } \delta\% \\ -1 & \text{if } (q_{i,k}) \text{ in the bottom } \delta\% \end{cases} \tag{4}$$

We parameterize the classification problem by a variable δ and calculate performance as we adjust δ. As we move the value of our parameter δ the problem becomes more difficult since the visual appearance of the positive and negative images starts to become less evident up to the point when $\delta = 0,5$. At the same time when δ has smaller values the positive and negative images are easier to classify but we have access to less data. We learn models to predict yi,k from input image representations xi using the following methods to extract features: VGG16 based-line, VGG16-GAP, and GIST. We train and compare the behavior of linear and non-linear models regularized with l2 like Logistic Regression: L(y, f(x)) = Pn i = 1 log(e (−yi ∗ f(xi)) + 1), Linear SVC: L (y, f(x)) = Pn i = 1 max(0, 1 − yif(xi)), Ridge Classifier: L(y, f(x)) = sgn(‖y − f(x)‖2 2 + ‖w‖2 2), VGG16-Softmax, and VGG16-GAP -Softmax both with loss function L (y, f(x)) = e (yi − f(xi)) Pn k = 1 e (yk − f(xk)) + ‖w‖2 2.

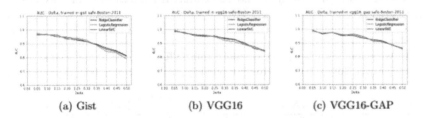

(a) Gist   (b) VGG16   (c) VGG16-GAP

Fig. 2. Test classification results: 10 KFolds cross-validation Avg AUC over the city Boston trained using GIST, VGG16, and VGG16-GAP as feature extractor. VGG16-GAP achieves higher metric values than Gist and VGG16 along the different values of our parameter δ.

We evaluated our binary classifier model behavior using the Area Under the Curve (AUC) metric which depends on Precision-Recall as we report in Fig. 2. We use a regularization l2 to avoid overfit our model. We set the regularization parameter C using held-out data and learn wk using training data {xi,yi,k}.

Table 3. We report Test classification for $\delta = 0,5$ in two scenarios: (a) training and testing perceptual prediction models on images from the same city, and (b) training models on images from one city and testing images from another city. We present that our VGG16-GAP has a better performance except in the perceptual category uniquely.

Training	Métrica	Test Boston			Test New York		
		VGG16-GAP	*VGG16*	*Gist*	*VGG16-GAP*	*VGG16*	*Gist*
Boston	*safety*	**71.428**	70.322	71.064	**67.741**	59.354	64.721
	wealthy	**67.741**	63.88	66.334	**65.897**	64.183	61.458
	uniquely	**63.354**	61.935	62.486	63.773	**63.858**	63.564
New York	*safety*	**66.459**	65.512	65.842	**66.968**	64.741	66.874
	wealthy	**64.748**	62.111	63.265	**63.8032**	60.001	62.997
	uniquely	**68.322**	64.748	66.349	62.895	62.468	**62.968**

The second one is the regression task: To perform this task we divided our images in the same way as we divided before. In this case, we want to predict not the category but the perceptual score associate with an image which we calculated before (Eq. 3). Here, our ground truth labels are yi,k = qi,k for image i and perceptual measure k. Therefore, we make predictions ŷi, k as a linear combination of the extracted features xi corresponding to image i as follows:

$$\hat{y}_k = q_{i,k} \tag{5}$$

To perform our experiments, we train our set (xi, yi, k) using linear and nonlinear methods regularized with l2 like: Ridge: L(y, f(x)) = ||y − f(x)||2 2 + ||w||2 2, Lasso: L (y, f(x)) = 1 2 ∗ n ||y − f(x)||2 P 2 + ||w||1, Linear SVR: L(y, f(x)) = n i = 1 max(0, |yi − f(xi)|), and a simple Linear Regression: L(y, f(x)) = ||y − f(x)||2 2. We choose the Pearson coefficient as a metric of regression models, we select this metric because we want to achieve a high correlation between extracted features from images with their correspond yi, k:

Table 4. We report Test regression task trained with SVR-l_2. We note that regression task has a best behavior over Boston and VGG16-GAP provides the best result in both cities.

City	Feature Extractor	Methods			
		LinearSVR	LinearRegress	Lasso	Ridge
Boston	VGG16-GAP	**0.7095**	0.5717	**0.71342**	**0.7462**
	VGG16	0.6832	**0.6354**	0.70001	0.7163
	GIST	0.66643	0.41612	0.6569	0.6658
New York	VGG16-GAP	0.5649	**0.6196**	**0.6503**	**0.7209**
	VGG16	**0.6062**	0.60487	0.64531	0.70986
	GIST	0.59157	0.5734	0.61991	0.68732

3.3   Model Explanation

In this work, we want to understand why our street images that are predicted as "safe" or "not safe". To do this, we compare two explainers: The first one is LIME, a local interpretable model-agnostic technique. LIME explains a blackbox model by simulating local candidates close to the original prediction. Using these predictions, LIME generates a random distribution set of possible predictions based on L2 distance called "local fidelity" taken as reference the original prediction. Then, LIME select which possibles random noises could be a good samples to evaluate using its Submodular Pick Algorithm (SP-LME) trained by a SVM. The second one is Grad-CAM, this method presents a strong behavior in interpret convolutional networks [2]. In this work, a comparison of robustness against adversarial attack was performed. This work shows that Grad-CAM is strong against adversarial attacks, unlike CAM [41], Gradient Input [31], Integrated Gradients [36], GBP [35], Smoothed Gradients [34], Grad-CAM and Guided Grad-CAM [29], and DeepLIFT [30]. As we can see on Fig. 3, both methods highlight different regions for the same prediction sample. We can easily visualize which part of an input image was learned by the model and which regions are relevant to the prediction.

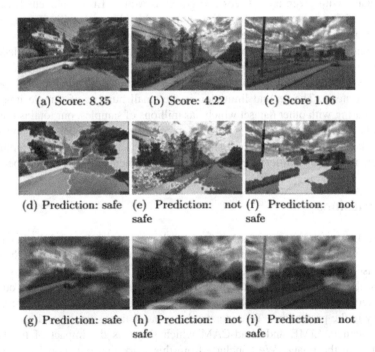

(a) Score: 8.35   (b) Score: 4.22   (c) Score 1.06

(d) Prediction: safe   (e) Prediction: not safe   (f) Prediction: not safe

(g) Prediction: safe   (h) Prediction: not safe   (i) Prediction: not safe

Fig. 3. Image from Boston (first row) with their respective predicted scores and class predicted. Besides, we present LIME outputs (second row) in which green regions mean positive and red ones mean negatives impacts of the features of a prediction. Furthermore, Grad-CAM results (third row) only shows the highlighted positive regions with more importance for the prediction. (Color figure online)

4 Discussions

This work presents a methodology to teach a machine how to learn features to differentiate perceptions using the Place Pulse dataset, and explain predictions about urban perception. This work was focused on safety perception processing and calculate the safe perceptual scores of street images. We adapt the VGG16 model modifying the MaxPool for a GAP operation layer. We compare VGG16, VGG16-GAP, and GIST performance in regression and classification task varying the quantify of images depending on our parameter δ varying from 0,05 to 0,5. For evaluations, we calculate the Area Under the Curve (AUC) for classification task. For regression, we trust in the Pearson Correlation Coefficient which report the correlation between a image and their associated perceptual score (see Tables 3 and 4). To understand our model predictions, we use and compare two model explainer like LIME and grad-CAM. With these both methods, we analyze the resulting highlighted regions about safety perception predicted per image and visualize the impact of important features as you can see in Fig. 3. For unsafe predictions, Grad-CAM highlights asphalt, fence, and walls. Instead of LIME, which presents a random behavior over the regions, sometimes highlight sky, asphalt, trees, grass, cars, earth or fences. For safe predictions, Grad-CAM highlighted regions are associated with green areas (trees and grass) as well as LIME. Nonetheless, LIME has a lack of importance due to the main features which have a positive impact on safe prediction usually are shadows, clouds, or asphalt as well.

We found three main limitations in this work. The first one is about the Place Pulse dataset that was constructed using a online survey. Here each volunteer chooses between two images that are the most "safe" depending on their biased personal perception criteria. The second limitation is the small number of sample images per city. Comparing with other dataset which has millions of samples, our total is not above of 100,000 generating a lack of robustness when training a model with few sample data. The last limitation is the impossibility of creating a general city perceptual predictor, due to the large difference between cities and their unique visual appearance.

5 Conclusions

In this work, we propose a methodology that allows us to understand the behavior of the urban perception of safety from street images. To do this, we pre-process the dataset Place Pulse 2.0 analyzing the 110 thousand images obtained by comparisons and calculated their corresponding perception scores in six categories. We focus our study on the safety scores to analyze which parts of the images are impacting positively and negatively in the predictions. To understand this predictions, we use and compare two model explainers LIME and Grad-CAM which show us the impact of the features extracted from the image. We conclude from this work that our model is capable to predict the safety perception from street image. Besides, we show the correlation between high safety perception with the presence of trees or green areas.

References

1. Acosta, S.F., Camargo, J.E.: Predicting city safety perception based on visual image content. In: Vera-Rodriguez, R., Fierrez, J., Morales, A. (eds.) CIARP 2018. LNCS, vol. 11401, pp. 177–185. Springer, Cham (2019). https://doi.org/10.1007/978-3-030-13469-3_21
2. Adebayo, J., Gilmer, J., Muelly, M., Goodfellow, I., Hardt, M., Kim, B.: Sanity checks for saliency maps (2018)
3. Ancona, M., Ceolini, E., Oztireli, C., Gross, M.: A unified view of gradient-based attribution methods for deep neural networks. ETH Zurich (2017)
4. Arietta, S.M., Efros, A.A., Ramamoorthi, R., Agrawala, M.: City forensics: using visual elements to predict non-visual city attributes. IEEE Trans. Vis. Comput. Graph. **20**(12), 2624–2633 (2014)
5. Boser, B.E., Guyon, I.M., Vapnik, V.N.: A training algorithm for optimal margin classifiers. In: Proceedings of the Fifth Annual Workshop on Computational Learning Theory, pp. 144–152. ACM (1992)
6. Doersch, C., Singh, S., Gupta, A., Sivic, J., Efros, A.: What makes paris look like paris? (2012)
7. Donahue, J., et al.: DeCAF: a deep convolutional activation feature for generic visual recognition. In: International Conference on Machine Learning, pp. 647–655 (2014)
8. Dubey, A., Naik, N., Parikh, D., Raskar, R., Hidalgo, C.A.: Deep learning the city: quantifying urban perception at a global scale. CoRR abs/1608.01769 (2016). http://arxiv.org/abs/1608.01769
9. Kaplan, R., Kaplan, S.: The Experience of Nature: A Psychological Perspective. Cambridge University Press, Cambridge (1989)
10. Keizer, K., Lindenberg, S., Steg, L.: The spreading of disorder. Science (New York, N.Y.) **322**, 1681–1685 (2008). https://doi.org/10.1126/science.1161405
11. León-Vera, L., Moreno-Vera, F.: Car monitoring system in apartments' garages by small autonomous car using deep learning. In: Lossio-Ventura, J.A., Muñante, D., Alatrista-Salas, H. (eds.) SIMBig 2018. CCIS, vol. 898, pp. 174–181. Springer, Cham (2019). https://doi.org/10.1007/978-3-030-11680-4_18
12. Lin, M., Chen, Q., Yan, S.: Network in network. arXiv preprint arXiv:1312.4400 (2013)
13. Lindal, P.J., Hartig, T.: Architectural variation, building height, and the restorative quality of urban residential streetscapes. J. Environ. Psychol. **33**, 26–36 (2013)
14. Lundberg, S.M., Lee, S.I.: A unified approach to interpreting model predictions. In: Guyon, I., et al. (eds.) Advances in Neural Information Processing Systems 30, pp. 4765–4774. Curran Associates, Inc. (2017). http://papers.nips.cc/paper/7062-aunified-approach-to-interpreting-model-predictions.pdf
15. Lynch, K.: Reconsidering the image of the city. In: Rodwin, L., Hollister, R.M., (eds.) Cities of the Mind, pp. 151–161. Springer, Boston (1984). https://doi.org/10.1007/978-1-4757-9697-1_9
16. Moreno-Vera, F.: Performing deep recurrent double Q-learning for Atari games. In: 2019 IEEE Latin American Conference on Computational Intelligence (LA-CCI), pp. 1–4 (2019). https://doi.org/10.1109/LA-CCI47412.2019.9036763
17. Naik, N., Philipoom, J., Raskar, R., Hidalgo, C.: StreetScore: predicting the perceived safety of one million streetscapes. In: 2014 IEEE Conference on Computer Vision and Pattern Recognition Workshops (2014)
18. Naik, N., Raskar, R., Hidalgo, C.A.: Cities are physical too: using computer vision to measure the quality and impact of urban appearance. Am. Econ. Rev. **106**(5), 128–132 (2016)

19. Nasar, J.L.: The Evaluative Image of the City. , Thousand Oaks (1998)
20. Ordonez, V., Berg, T.L.: Learning high-level judgments of urban perception. In: Fleet, D., Pajdla, T., Schiele, B., Tuytelaars, T. (eds.) ECCV 2014. LNCS, vol. 8694, pp. 494–510. Springer, Cham (2014). https://doi.org/10.1007/978-3-319-10599-4_32
21. Porzi, L., Rota Bulò, S., Lepri, B., Ricci, E.: Predicting and understanding urban perception with convolutional neural networks (2015). https://doi.org/10.1145/2733373.2806273
22. Quercia, D., O'Hare, N.K., Cramer, H.: Aesthetic capital: what makes London look beautiful, quiet, and happy? In: Proceedings of the 17th ACM Conference on Computer Supported Cooperative Work & Social Computing, pp. 945–955. ACM (2014)
23. Ribeiro, M.T., Singh, S., Guestrin, C.: Why should I trust you?: Explaining the predictions of any classifier. In: Proceedings of the 22nd ACM SIGKDD International Conference on Knowledge Discovery and Data Mining, pp. 1135–1144. ACM (2016)
24. Ribeiro, M.T., Singh, S., Guestrin, C.: Anchors: high-precision model-agnostic explanations. In: Thirty-Second AAAI Conference on Artificial Intelligence (2018)
25. Salesses, M.P.: Place Pulse: measuring the collaborative image of the city. Ph.D. thesis. Massachusetts Institute of Technology (2012)
26. Salesses, P., Schechtner, K., Hidalgo, C.A.: The collaborative image of the city: mapping the inequality of urban perception. PLOS ONE 10, e0119352 (2013)
27. Sampson, R.J., Morenoff, J.D., Gannon-Rowley, T.: Assessing "neighborhood effects": social processes and new directions in research. Annu. Rev. Sociol. 28(1), 443–478 (2002)
28. Schroeder, H.W., Anderson, L.M.: Perception of personal safety in urban recreation sites. J. Leis. Res. 16(2), 178–194 (1984)
29. Selvaraju, R.R., Cogswell, M., Das, A., Vedantam, R., Parikh, D., Batra, D.: GradCAM: visual explanations from deep networks via gradient-based localization. In: Proceedings of the IEEE International Conference on Computer Vision, pp. 618–626 (2017)
30. Shrikumar, A., Greenside, P., Kundaje, A.: Learning important features through propagating activation differences. arXiv preprint arXiv:1704.02685 (2017)
31. Shrikumar, A., Greenside, P., Shcherbina, A., Kundaje, A.: Not just a black box: learning important features through propagating activation differences (2016)
32. Simonyan, K., Vedaldi, A., Zisserman, A.: Deep inside convolutional networks: Visualising image classification models and saliency maps. arXiv preprint arXiv:1312.6034 (2013)
33. Simonyan, K., Zisserman, A.: Very deep convolutional networks for large-scale image recognition. In: International Conference on Learning Representations (ICLR) (2014)
34. Smilkov, D., Thorat, N., Kim, B., Vi´egas, F., Wattenberg, M.: SmoothGrad: removing noise by adding noise (2017)
35. Springenberg, J.T., Dosovitskiy, A., Brox, T., Riedmiller, M.: Striving for simplicity: the all convolutional net. arXiv preprint arXiv:1412.6806 (2014)
36. Sundararajan, M., Taly, A., Yan, Q.: Axiomatic attribution for deep networks (2017)
37. Tokuda, E.K., Silva, C.T., Cesar Jr, R.M.: Quantifying the presence of graffiti in urban environments. CoRR abs/1904.04336 (2019). http://arxiv.org/abs/1904.04336
38. Ulrich, R.S.: Visual landscapes and psychological well-being. Landsc. Res. 4(1), 17–23 (1979)
39. Wilson, J.Q., Kelling, G.L.: Broken windows. Atl. Mon. 249(3), 29–38 (1982)
40. Zhang, F., et al.: Measuring human perceptions of a large-scale urban region using machine learning. Landsc. Urban Plann. 180, 148–160 (2018)
41. Zhou, B., Khosla, A., Lapedriza, A., Oliva, A., Torralba, A.: Learning deep features for discriminative localization. In: Proceedings of the IEEE Conference on Computer Vision and Pattern Recognition, pp. 2921–2929 (2016)

Recognition of Multiple Panamanian Watermelon Varieties Based on Feature Extraction Analysis

Javier E. Sánchez-Galán[1,4], Anel Henry[2], Fatima Rangel[2],
Emmy Sáez[2], Kang-Hyun Jo[3], and Danilo Cáceres-Hernández[2,4(✉)]

[1] Facultad de Ingeniería de Sistemas Computacionales,
Universidad Tecnológica de Panamá, Panama, Panama
javier.sanchezgalan@utp.ac.pa

[2] Facultad de Ingeniería Eléctrica, Universidad Tecnológica de Panamá,
Panama, Panama
{anel.henry, fatimam.rangel, emmy.saez,
danilo.caceres}@utp.ac.pa

[3] Intelligent Systems Laboratory, Graduate School of Electrical Engineering,
University of Ulsan, Ulsan, South Korea
acejo@ulsan.ac.kr

[4] Sistema Nacional de Investigación (SNI), SENACYT, Panama, Panama

Abstract. In this paper we present a multi-watermelon varieties recognition analysis using the color and texture information. The goal is to classify the watermelon taking into account its features, despite its variety. To improve the efficacy of the proposed method the images were preprocessed using morphological and adaptive threshold methods. Also, to extract the candidate region for feature extraction, vertical and horizontal histograms were used. Finally, the model was tested on a group of watermelons to prove the effectiveness of the implementation; the result shows it is capable of classifying watermelon varieties by just considering the presence or not of stripes, as well as the color information. The importance of this study is that it uses as test subjects local varieties (Panamanian) of watermelons, thus helping the knowledge of national export products with computer vision.

Keywords: Clustering · Feature extraction · Model fitting · Recognition · Watermelon

1 Introduction

Watermelon (Citrullus lanatus) is a species of flowering plant, that makes part of the Cucurbitaceae family. Is a popularly grown fruit in both subsistence farming and commercial settings, with more than a 1,000 varieties. It can easily grown in tropical and temperate weathers [1], with Asian countries taking over 70% of the commercial production. Nowadays, watermelon constitutes one of the main export activities of Panama [2]. The export watermelons usually comply to external characteristics, that vary market to market such as: size (small, also called uni-personal or medium sized),

© Springer Nature Switzerland AG 2021
D.-S. Huang et al. (Eds.): ICIC 2021, LNCS 12837, pp. 65–75, 2021.
https://doi.org/10.1007/978-3-030-84529-2_6

weight (between 2.5 and 6.0 kg/unit), shape (symmetrical, uniform, waxy and a shiny surface), pulp internal color (live red) and pulp sugar content (often measured as Brix degrees (∘Bx) [3]. Determination of external watermelon characteristics have been commonly via RGB images. Several authors have been focusing in using supervised and unsupervised classification methods, such as: K-Means [4, 5], Fuzzy Logic [6], Artificial Neural Networks [7, 8]. However, their solutions are variety-specific, and sometimes not re-usable for local market varieties. The objectives of this study are twofold:

- Developing a vision strategy for the task of watermelon recognition that can be used in a belt conveyor.
- Define the best set of features that allows the recognition of watermelon between Panamanian varieties.

The rest of this article is structured as follows: Sect. 2, presents the proposed method, Sect. 3, shows the experimental results obtained, Sect. 4, provides overall conclusions of the article and focuses suggesting future work.

2   Proposed Method

The ongoing result of the proposed method to recognize watermelons based on feature extraction is presented. The multi-watermelon recognition proposed algorithm has six steps, which are shown in Fig. 2 (A) preprocessing, (B) finding the region of interest, (C) candidate detection, (D) shape estimation, (E) classification, and (F) clustering analysis. Once the input image is capture, the algorithm start by extracting the edge information from the input image. To do that, a prepossessing task includes the use of morphological operators, the difference of Gaussian filter, Otsu and the Canny operator. Then, the region of interest (ROI) is extracted by using a vertical and horizontal histogram analysis. Lastly, the classification is done by extracting a bounding box region from the ROI that is split in their respective RGB channels. Hence, the green channel is used to separate the image into two classes. Once the classes are extracted, the mean value from both classes are computed, the distance between this values is used to separate the watermelon. Finally, a clustering method is used to determine the level of the watermelon maturity. For this study different watermelon varieties found in local market, and export varieties were used. As it can be seen if Fig. 1, the selected varieties (Anna Hybrid, Quetzali and Joya Hybrid) have very distinctive markings, also are very unique in terms of terms shape, and depth of colours. Which makes for an interesting feature based analysis.

Fig. 1. Image samples from different types of watermelon used in the proposed method.

Taking the notable characteristics of each watermelon variety into account. An analysis focusing on the exploitation of color and geometry information for visual watermelon classification was devised. Figure 2 shows the flow diagram of the proposed approach.

The proposed algorithm can be briefly explained as follows:

- For an input image a preprocessing task was performed. This stage includes color image smoothing, feature enhancement algorithm, global adaptive binarization threshold, and an edge detection operator.
- The frequency distribution along both axis of the previous preprocessed image were extracted as well as an analysis based on the number of detected edge is proposed to find the candidate detection step (Region).
- After extracting the candidate region, an strategy to classify the type of watermelon was developed. In this process, shape, color information, as well as the orientation of the watermelon must be taken into account.
- Finally, an strategy to determine the watermelon maturity level was previously presented in [5].

Each of these steps will be explained in the following subsections.

2.1   Image Preprocessing

In this step, the algorithm works with a set of watermelon images as is input. It is best if the images show notable as is our case (see Fig. 1), for example: difference between image in size, as well as, color, texture, geometry and shape of the watermelon image used in this stage. Watermelon is a fruit that can be usually describe computationally using an elliptical or circular shape, but it depends greatly on the variety. By focusing on those issues and considering that the proposed algorithm will be used to develop an strategy for an autonomous belt conveyor system another set of sub-steps were implemented. First, for uneven illumination and noise removal problem a morphological approach was proposed. Second, to remove problems given by changes in frequency, as well as, homogeneous areas within the image a difference of Gaussian (DoG) filter was implemented [9, 10]. Due to the fact that watermelon differ in texture and color an adaptive binarization algorithm was proposed, more precisely the Otsu method [11]. Finally, in order to extract the edges from the image, Canny operator was implemented [12]. Figure 3 shows the result of each mentioned sub-step.

2.2   Candidate Detection

Once the edges are disclosed, the next step consist in determining the size in pixel in both direction of the candidate object. The aim of this step is to estimate the localization of the watermelon within the image. To do that, an intensity-statistics analysis was applied [13]. In order to achieve this, a vertical and horizontal distribution analysis along the axis of the edge detected image is used. By applying this process the Region of Interest (ROI) is determined by locating the boundaries of the watermelon within the edge images and its max and min of x-coordinate and y-coordinate. Figure 4 shows the process of detecting the candidate regions in both vertical and horizontal directions.

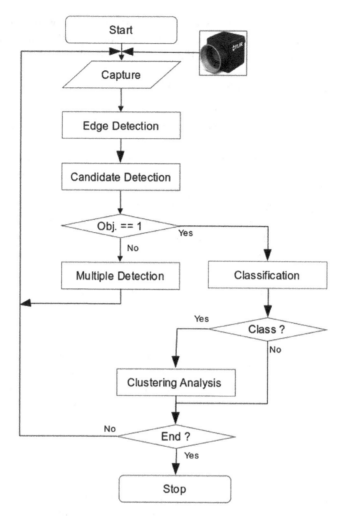

Fig. 2. Algorithm scheme for watermelon recognition based feature extraction analysis using single camera.

2.3   Classification

As noted beforehand, it is important to emphasize that this was done considering the algorithm will be fully implemented in a belt conveyor, in which watermelon will move along a predefined trajectory with in an arbitrary orientation, according to the shape of the watermelon. Taking into consideration the aforementioned, from the previous step the center of the watermelon and the radius candidate was computed. To reduce the processing time, by using the computed information (xc, yc, rc) the region where the watermelon appears was cropped, see Fig. 5.

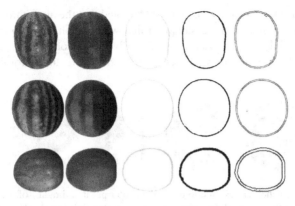

Fig. 3. Algorithm scheme for the edge detection step. The first columns show the input image. The second column shows the dilate results. The third column shows the Difference of Gaussian results. The fourth column shows the Otsu results. The fifth column shows the edge detection result by using Canny operator.

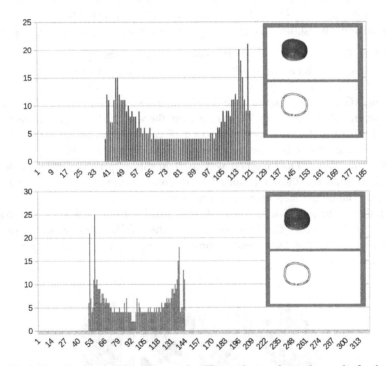

Fig. 4. Vertical and horizontal histogram results. The top image shows the results for the vertical histogram. The bottom image shows the results for the horizontal histogram. x_{min}, x_{max}, y_{min}, and y_{max}, define the width and height in pixels of the watermelon.

From the four images in Fig. 5 can be easily noted that the algorithm might crop the watermelon region based on the assumption that watermelon has a round shape. This problem arises mainly due to orientation. To solve this problem, the algorithm esti-mates the orientation using the watermelon green striped texture information.

Fig. 5. Watermelon candidate extraction using the center point and radius of the candidate. The images show the results for two different types of watermelon that vary in shape and orientation. (Color figure online)

Therefore, the idea consists of fully exploiting the texture and color information to estimate the best fitting line candidate from the watermelon strips. To this end, in order to achieve, this step uses a set of four (4) sub-steps: bounding box extraction, splitting image channels, adaptive threshold, and line fitting estimation, as is exemplified in Fig. 6.

- Bounding box extraction. From the cropped image, see (Fig. 5), the algorithm extracts a region with a size of 30% of the cropped image. The extracted image was centered at (x_c, y_c).
- Splitting RGB image channels. The color image was split into its 3 RGB channels. By observing the green image histogram, from the watermelon data set, was noticed that this channel shows a good separation of color classes between the strips. In order to reduce the computing time the green channels was used.
- Adaptive threshold. Otsu method was used in this step due to the fact that the watermelons do not show the same feature color information among watermelon varieties. The output step gives a binary image.
- Line fitting estimation. The step consists in estimating from the watermelon strip the best fitting line. To that, from the binary image the coordinate points were extracted. By applying the Random sample consensus (RANSAC) the best-fit model is computed.

Fig. 6. Best fitting estimation. The first image shows the watermelon bounding box extraction, centered at (xc, yc). The second image shows the green channel color. The third image shows the binary images by using Otsu method. The fourth images shows the fitting line step, points in blue are randomly selected by the algorithm to compute a set of fitting linear models (lines in green) to estimate the best model that has the best fit (line in magenta). (Color figure online)

Then, given the estimated intercept and slope, the algorithm estimates the angle of rotation. Once the algorithm rotates the image, the next step entails identifying the variety of watermelons. In that sense, in order to test the algorithm watermelon with and without stripes had been tested. Hence, the algorithm computed the distance between the given the foreground and background computed means values using the Otsu method fused with the green color channel using the AND logic operations, (third image in Fig. 6). To this end, the algorithm was able to watermelon with strip from those without strip. Although, the proposed idea focuses on the analysis of watermelon with strip (Anna Hybrid, Quetzali and Joya Hybrid watermelon types, respectively), the current algorithm have proved effective in the task of watermelon recognition based on color and texture information.

2.4 Classification

Finally, an strategy to determine the watermelon maturity level was previously presented in [5]. The main idea, from the watermelon data set a group of images were selected for the training process. The RGB color images were divided in groups. Then, the R, G, and B channels were used to determine the best cluster groups that can be used for the recognition task; consequently the group of cluster with the nearest mean for each R, G, and B that represent the maturity level classification were computed. Finally, each new captured image was analyzed using the resulting trained classifier.

3 Experimental Results

All the experiments were done on Intel® Core™ i7-9750H CPU @ 2.60 GHz × 12, 15 GB RAM. The implementation was done in C++ under Ubuntu 18.04. The algorithm used a group of two variety of watermelon in texture and color, the first one is a watermelon with lime solid colors and the second one with strips. For watermelon with stripes a set of three types were used, Anna Hybrid, Quetzali and Joya Hybrid watermelon types. To test the algorithm a set of 10 images with a different image size were taken. Figure 7 shows the result for the proposed algorithm tested in a our available data sets (https://sistemasinteligentes.utp.ac.pa/watermelon/). For the case of watermelon with strip the proposed strategy shows a good performance. Although the algorithm was able to work with watermelon with/without stripes, it can be observed that for the case of without stripes additional considerations need to be made, see Fig. 7 column 6 and 7, showing the last 3 steps.

Table 1 shows the results for the watermelon recognition step for the main process shown in Fig. 2. The first columns show the set of image used in the experiment. The second column indicates that the watermelons have or not stripes, S means the presence of stripes while wS mean the absence of stripes. The next two columns (3 and 4) indicate the pixel size of the image. The next two columns (5 and 6) show the size in the real world of the watermelon. These two columns indicate the difference in size between the watermelon varieties. It should be stated that the information of the last two image was not acquired at the moment of the capture. The seventh column indicates the type of the variety, Quetzali, Hoya and Ana The last column indicate if the watermelon fulfill the requirement for the export nor not, E stands for export while L stands for local consumption.

Fig. 7. Experimental Results for a set of 4 different types of watermelon with different geometric, scale and position within the image samples. The first row shows the data set used in this study. The second shows the edge detection results. Third row shows the ROI and the bounding box region used to estimate the orientation, ellipsoid and rectangle in red. The four rows show the adaptive thresholding results by using the Otsu Method. The fifth show the best line fitting result after using RANSAC, line in magenta. The last row shows the ongoing results for a set of different variety of watermelon (Color figure online)

Table 1. Experimental results for multi-watermelon recognition.

I	F	Iw [p]	Ih [p]	Ww [cm]	Wh [cm]	V	T
F1	S	360	270	16.5	15.5	Q	E
F2	S	720	540	16.5	15.5	Q	E
F3	S	360	270	17.5	17.0	H	E
F4	S	720	540	17.5	17.0	H	E
F5	S	360	270	10.5	11.0	A	E
F6	S	720	540	10.5	11.0	A	E
F7	S	360	270	23.0	18.0	Q	L
F8	S	360	270	21.0	19.5	Q	L
F9	wS	645	370	–	–	–	L
F10	wS	360	270	–	–	–	L

In Table 1 note that I stands for image sample, F stands for feature information, Iw stands for image width, Ih stands for images height, Ww stands for watermelon width in cm, Wh stands for watermelon height in cm, V stands for variety while Q stands for Quetzali, H stands for Hoya, A stands for Ana, T stands for Type while E stands for export and L stands for local type.

Table 2 shows the computing time for all main process shown in Fig. 2. It can be noticed that the computing time was related to the size and texture of the image. The higher computing time was given by the set of image in which watermelon have at least the half size of the image, for example from image 4 to 7. The first row shows the time for capturing the image. The second row show the time to extract the edge information. The third row shows the processing time for the detection task. The fourth row shows the computing time for the classification step. The last row shows the total computing time. F1 to F10 are the input image used in the experiment, while the last two columns show the average and standard deviation values for the main process. To evaluate the algorithm a set of quantitative and qualitative analysis were performed in order to verify robustness of recognition. Note: Fx stands for image sample (x stands for the index position in Table 1), C stands for image capture, E stands for edge detection, D stands for Detection, Cl stands for Classification, T stands for Computational time. A stands for average, S stands for standard deviation.

Table 2. Computing time in milliseconds

	F1	F2	F3	F4	F5	F6	F7	F8	F9	F10	A	S
C	3.4	3.5	0.9	8.6	7.6	16.5	15.4	1.54	3.3	3.33	6.43	5.56
E	2.4	2.4	2.8	6.3	4.6	7.8	7.33	5.9	2.4	2.3	4.4	2.2
D	7.4	7.4	11.9	21.4	17.5	30.3	35.4	19.4	7.5	7.4	16.6	10.8
Cl	0.07	0.08	0.08	0.08	0.08	0.09	0.09	0.09	0.07	0.07	0.08	0.00
T	13.3	13.4	15.8	36.4	29.8	54.7	58.2	26.9	13.4	13.0	27.5	17.3

3.1   Edge Detection Evaluation

The algorithm was tested in a set of 10 images with difference in type, size, color and texture; all these image edges were correctly extracted. The strategies used to extract the watermelon edge produce good results. Images in Figs. 7 second file clearly showed the watermelon edge, ignoring unwanted edge due to the presence of strip or illumination condition. This result reflected in decreasing errors and processing time.

3.2   Ellipsoid Detection

To test this section, the area and perimeter of the watermelon were computed and compared with respect to the existing ground truth watermelon information which was manually labeled. Table 1 shows the results for a data set with include watermelon that does not have the characteristic to export, for example shape, size or watermelon that were not cut well at the harvest period. It can be noticed, this error is around 10%.

However, the project is intended to work with watermelon for export purpose, a set of watermelon that can be exported was used for the testing task. Taken this into account, Table 2 shows the result for this group of watermelons. From Tables 3 y 4 it can see the effect of a selection which does not fulfill the export requirement, errors drops from approximately 10% to 2%.

Table 3. Average computing error in percentage for the full data set.

	Ground truth	Computed error	[%]
Area [pixel^2]	19,229.0	21,260.0	10.56
Perimeter [pixel]	444.0	462.0	3.76

Table 4. Average computing error in percentage for the data set of watermelons for export

	Ground truth	Computed error	[%]
Area [pixel^2]	13,372.0	13,618.0	1.83
Perimeter [pixel]	369.0	370.0	0.42

3.3   Fitting Estimation

As mentioned, the algorithm has been tested in a set of different type of watermelon with or without strips. For the set of image in the fifth column Fig. 7, the line fitting results extract for all those cases is shown. It should be highlighted that the experiments were carried out under controlled field conditions using watermelons with difference in size. Similarly, strips pattern does not align perfectly as well there are not similar between watermelons. Although this differences between the strips, the algorithm was able to estimate the line fitting. However, for the cases of watermelon with strips, as the size of the watermelon decreases, the estimation error increases. On the other hand, the line fitting was computed on watermelon without strip due to the presence of a blurry reflection on the watermelon surface with brightness covering. The error can be seen in the last column in Fig. 7, further strategies will be implemented in order to improve the result of this step.

4   Conclusions and Future Work

This article describes the preliminary results of an ongoing experiment of putting together a classification system that has the possibility to detect and recognize three local watermelon varieties in a real time. This approach focuses on pattern analysis looking into the color and texture information. By combining the color and texture information the algorithm was able to recognize two different types of watermelon, with or without stripes. Figure 7 show that there is a relationship between the bounding box regions used in the orientation task. The lower the quality of the image decreases the orientation performance task. Finally, the presented strategy has been demonstrated to be feasible for the task of watermelon recognition. For the future works will be improvements by using supervised learning algorithms, as well as the design of real-time embedded systems for a belt conveyor. At the same time, in order to develop an robust image based machine vision system an spectral analysis will be implemented.

Acknowledgment. This work was supported by the Secretaría Nacional de Ciencia, Tecnología e Innovación de Panamá (SENACYT) Grant funded by the Panamanian Goverment (Project 165-2019-FID18-060). The authors would like to thank the Sistema Nacional de Investigación of Panamá, the Universidad Tecnológica de Panamá for their administrative support and contribution.

References

1. Wehner, T.C.: Watermelon. In: Prohens, J., Nuez, F. (eds.) Vegetables I, pp. 381–418. Springer, New York (2008). https://doi.org/10.1007/978-0-387-30443-4_12
2. Barba, A., Espinosa, J., Suris, M.: Adopción de prácticas para el manejo agroecológico de plagas en la sandía (citrullus lanatus thumb.) en Azuero, Panamá. Revista de Protección Vegetal **30**(2), 104–114 (2015)
3. Osorio, N., González, R., Guerra, J., Aguilera, C.: Manual Técnico: Manejo Integral del Cultivo de Sandía (Citrullus lanatus). Instituto de Investigación Agropecuaria de Panamá, Panamá (2012)
4. Liantoni, F., et al.: Watermelon classification using k-nearest neighbours based on first order statistics extraction. J. Phys. Conf.: Ser. **1175**(1), 012114 (2019)
5. Royo, A.H., Kung, K., Jo, K., Hernández, D.C.: Design and implementation of a smart system for watermelon recognition. In: 2019 12th International Conference on Human System Interaction (HSI), pp. 82–86 (2019)
6. Rahman, F.Y.A., Baki, S.R.M.S., Yassin, A.I.M., Tahir, N.M., Ishak, W.I.W.: Monitoring of watermelon ripeness based on fuzzy logic. In: 2009 WRI World Congress on Computer Science and Information Engineering, vol. 6, pp. 67–70. IEEE (2009)
7. Kutty, S.B., et al.: Classification of watermelon leaf diseases using neural network analysis. In: 2013 IEEE Business Engineering and Industrial Applications Colloquium (BEIAC), pp. 459–464. IEEE (2013)
8. Ahmad Syazwan, N., Shah Rizam, M.S.B., Nooritawati, M.T.: Categorization of watermelon maturity level based on rind features. Proc. Eng. **41**, 1398–1404 (2012)
9. Wu, Y.-T., Yeh, J.-S., Wu, F.-C., Chuang, Y.-Y.: Tangent-based binary image abstraction. J. Imag. **3**(2), 16 (2017)
10. Huang, S., Li, M., Wang, X., Zhao, X., Yang, L., Peng, Z.: Infrared small target detection with directional difference of Gaussian filter. In: 2017 3rd IEEE International Conference on Computer and Communications (ICCC), pp. 1698–1701 (2017)
11. Otsu, N.: A threshold selection method from gray-level histograms. IEEE Trans. Syst. Man Cybern. **9**(1), 62–66 (1979)
12. Canny, J.: A computational approach to edge detection. IEEE Trans. Pattern Anal. Mach. Intell. PAMI **8**(6), 679–698 (1986)
13. Hernández, D.C., Jo, K.-H.: Outdoor stairway segmentation using vertical vanishing point and directional filter. In: International Forum on Strategic Technology 2010, pp. 82–86 (2010)

STDA-inf: Style Transfer for Data Augmentation Through In-data Training and Fusion Inference

Tao Hong, Yajun Zou, and Jinwen Ma[✉]

Department of Information and Computational Sciences, School of Mathematical Sciences and LMAM, Peking University, Beijing 100871, China
{paul.ht,zouyj}@pku.edu.cn, jwma@math.pku.edu.cn

Abstract. Style transfer has been effectively applied to data augmentation. However, previous work requires careful selection of style images out of the concerned datasets, and neglects the impact of style transfer on the inference procedure. In this paper, we propose a novel method **STDA-inf** for image classification: Style Transfer for **D**ata **A**ugmentation through **in**-data training and **f**usion inference. Firstly, we acquire the transferred training data in an adaptive way of in-data, in which style images are extracted from the training data itself. An online end-to-end training strategy is utilized to create an adversarial training effect, thereby alleviating the overfitting on textures when identifying different classes. Moreover, we fuse the outputs of the original and transferred images from the trained network, obtaining a more accurate classification. It is demonstrated by the experiments that our proposed method outperforms the previous style augmentation method with 7% improvement of classification accuracy on STL-10 and 3% on Caltech-256 dataset, respectively. Its superiority is also demonstrated over the other data augmentation methods.

Keywords: Style transfer · Data augmentation · In-data style · Fusion inference

1 Introduction

In recent years, deep neural networks have performed superiorly in many computer vision tasks such as classification, object detection, and so on. Driven by deep learning research, more effective data are needed under the challenges of lack of data, expensive tags, imbalanced categories, *etc*. Data augmentation is a powerful tool to solve this problem. In general, generating new samples via label-preserving transformations [1] can expand the training dataset, resulting in a better performance on the relative models.

On the other hand, the data learning mechanism of networks is also worthy of exploration. In terms of human perception, it is naturally believed that we classify objects majorly by shapes. Nevertheless in [2], Geirhos *et al.* found out that the ImageNet trained Convolutional Neural Networks (CNNs) are strongly biased towards recognizing textures rather than shapes. Since texture is considered to be closely related to image style, this discovery leads researchers to utilize the style to implement

© Springer Nature Switzerland AG 2021
D.-S. Huang et al. (Eds.): ICIC 2021, LNCS 12837, pp. 76–90, 2021.
https://doi.org/10.1007/978-3-030-84529-2_7

classification strategies. Style transfer [3, 4] appeals to be a promising efficient tool to achieve advanced data augmentation, by utilizing the effect of textures to divert more attention of networks to shapes.

Jackson *et al.* adopted an arbitrary style transfer network to perform style randomization [5]. Yet the style images are specially chosen from the Office dataset [6], which lacks a relationship with the concerned dataset of the given task. We refer to these style images as *out-of-data*. In [7], it reveals that styles adding too many colors and shapes on original images lead to bad performance, which reflects the difficulty of locating proper out-of-data styles. And in [7], the *offline* style transfer policy is adopted, such that the transferred dataset is given in advance and stored locally. Certainly, this non-end-to-end policy desires extra space storage, especially when trying many styles one time. Besides, all the previous work only concentrates on the style augmentation during training, neglecting to explore its effect on inference.

In this paper, we apply the style transfer based data augmentation on classification tasks with improvement and innovation (named **STDA-inf**). We extract the style images from the given dataset in the way of so-called *in-data*, which is more adaptive and controllable. In-data style augmentation harvests more robust models about style features at a higher level, which is orthogonal with other augmentation methods. Note that our proposed online training is end-to-end. Moreover, we adopt the style augmentation into the inference course in a way of *fusion inference*. The original image and the transferred images are fed into the trained network, and the final classification result is obtained via the weighted sum of their classification results. Our experiments strongly verify the effectiveness of STDA-inf. The main contributions of our work are as follows:

- We improve the selection source of styles from *out-of-data* to *in-data*, and get the *state-of-the-art* performance of style augmentation. The latter fully utilizes texture information of the datasets themselves, to create an adversarial training effect, thereby avoiding further overfitting, especially overfitting textures. And random selection can already reach good enough behavior.
- We adopt the *fusion inference* which gets a remarkable improvement of accuracy in comparison with the universe inference. This operation makes full use of style transferred information, because not only the original distribution is learned during training, but also the style transferred distribution.
- We present a comprehensive exploration on style augmentation in detail, such as style selection strategy, transfer policy, *etc.* And we present the superiority over interpolation-based data augmentation such as Mixup [8].

The rest of the paper is organized as follows. In Sect. 2, we introduce and review the related work. Our proposed method is presented in Sect. 3. In Sect. 4, experiments and analyses are given to illustrate the efficiency and effectiveness of our method. Finally, a brief conclusion is made in Sect. 5.

2 Related Work

2.1 Style Transfer

Style transfer means adopting a new style of an image into another image. Style is thought related to the variance or eigenvalue or gradient of the pixel tensor. The first attempted work adopts Gram matrix to encode the deep features of style representation [3, 4]. In [9], Huang *et al.* proposed an adaptive instance normalization (AdaIN) layer, achieving faster speed. Different from artistic effects, Luan *et al.* introduced a photo-realistic loss term to optimize towards photorealistic visual effects [10]. To alleviate the time consumption, PhotoWCT [11, 12] adopts a non-end-to-end architecture to insert whitening and coloring transform (WCT) modules in auto-encoders. Further, Yoo *et al.* [13] proposed Wavelet Corrected Transfer (WCT^2) aiming at eliminating post-processing steps while preserving fine details. And neural architecture search is adopted in StyleNAS [14].

In a word, the current mainstream framework adopts CNNs to encode content images (denoted as c) and style images (denoted as s) into feature maps. After encoding, we try to transfer the feature maps with style relative modules and decode them into style transferred images (denoted as cs). We adopt the AdaIN [9] model as the style transfer module in this paper. AdaIN simply aligns the channel-wise mean and variance of the content to match the style's, without learnable affine parameters. The input is simply shifted with mean μ and scaled with variance σ:

$$\text{AdaIN}(c, s) = \sigma(s)\left(\frac{c - \mu(c)}{\sigma(c)}\right) + \mu(s) \tag{1}$$

After getting cs, we adopt weighted interpolation between cs and s to control the degree of style transfer and content reservation. Note that we all set the weight of cs as 1.0.

2.2 Data Augmentation

Data augmentation has been a standard part of training deep neural networks ever since the work of Krizhevsky *et al.* [15]. Data augmentation is plausible to avoid overfitting so that the trained models will acquire stronger generalization capacity. Certain traditional augmentation techniques are very popular and effective, including horizontal flipping, random rotation, random cropping. They can make the model get more robust training in position, angle, *etc.* [15].

On the other hand, the convex combinations of pairs of inputs and their labels (Mixup) [8] also generate new samples, which inspires a series of augmentation methods such as MixMatch [16]. Cutout [17] refers to randomly masking out square regions of inputs, while CutMix [18] means that patches are cut and pasted among training images. Further, Manifold Mixup [19] interpolates between feature maps rather than just inputs to get better representations of hidden states.

3 Proposed Approach

In this section, we propose a systematic style augmentation approach STDA-inf, whose overview is shown in Fig. 1.

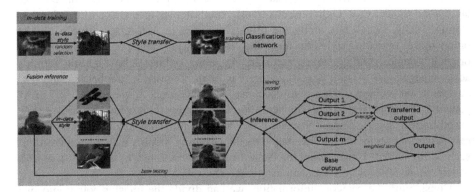

Fig. 1. The overview of our proposed approach STDA-inf: in-data training & fusion inference.

3.1 In-data Training

Before feeding training samples into a classification network, we usually apply some transformations Φ on them (with a certain probability p) such as horizontal flipping, which is just the traditional augmentation. Style augmentation also works in this stage, then the network is optimized towards

$$\min \mathcal{L}(f(\Phi(x)), y) \tag{2}$$

where $\mathcal{L}, f, x$ and y denote the loss function, network, sample and label.

During training, content images are the whole training samples generally. And style images in the way of in-data are from a subset of the training samples. Denoting the training and test set as Tr and Te, and the set of content images and style images as C and S respectively. Then for every paired (c, s), we have.

$$\text{Train: } c \in C = Tr, s \in S \subset Tr \tag{3}$$

$$\text{Inference: } c \in C = Te, s \in S \subset Tr \tag{4}$$

where the explanation of inference will be given in Sect. 3.2.

It's worth noting that Φ_{in} (in-data style transfer) could be utilized to force the network to train towards an adversarial direction. This is superior to Φ_{out} (out-of-data style transfer) adopted in the previous work [5, 7]. As for the chosen style number $|S|$, we can adjust it according to the whole training number $|C|$. For example, choosing 10 images per class from STL-10 (total 10 class) or choosing 100 images from Caltech-256 is a good choice.

In addition to random choice, we have explored different choice strategies called *min-loss* and *max-loss* choice. Feeding the original training samples into the trained models, we can sort them according to the classification loss $\mathcal{L}(f(x), y)$. Min-loss means choosing top images with the smallest loss as styles, while max-loss is corresponding to top images with the largest loss.

Moreover, we can divide the transfer procedure $\Phi(x)$ into two modes, called *offline* and *online* respectively. Offline means transferring samples in advance of training and storing locally, *i.e.* all raw inputs are original samples or transferred samples. While online means operating style transfer in the pre-processing stage, *i.e.* all raw inputs are original samples. Offline is space-consuming while online is time-consuming. It's acknowledged that to some extent, the more patterns we provide, the better result data augmentation will get. Offline will consume too much space if we want to get many patterns. Therefore, we mainly adopt end-to-end online operation mode in our work. As introduced in Sect. 2.1, AdaIN module is very fast, relatively speaking. This is one reason why we prefer to combine online mode with AdaIN rather than other algorithms such as StyleNAS. Under online mode, we set the style transfer proportion p as 0.3, which means 30% of training samples get transferred.

Considering the category information, we can divide the transfer policy into two modes: *inter-class transfer* and *intra-class transfer*. During training, if we transfer the images of one class with the style images of the corresponding class, we call this operation *inter-class training*. In other words, every paired (c, s) comes from the same class. If we transfer regardless of the class match between contents and styles, we call the operation *intra-class training*. Intra-class training defeats inter-class training at model performance, since the former retards the overfitting while the latter exacerbates the overfitting on the textures. We adopt intra-class training mode unless otherwise specified.

3.2 Fusion Inference

After training, the universal approach is testing original test samples ($x \in Te$, called *base test*) via the trained model directly. Supposing there are n classes, then the last layer of a network is an n-dimension vector v_{base}. The index of the largest value in the vector represents the divided category.

Apart from base test samples Te, we transfer them with style images (chosen from $S \subset Tr$, the same as training) to get transferred test samples $\Phi(Te)$. After every base sample is transferred once, we call it a round. So, we can get one complete test data in every round. Something different from the training case, *inter-class test* means style images are all from one class in one round, while *intra-class test* means transferring without the limit of class in each round. We adopt intra-class test mode unless otherwise specified. For a specific sample $\Phi(x)$, it has the same content but different styles in different rounds. Denoting the vector of the last layer in round i as v_i, then we get the average vector

$$v_{avg} = \frac{1}{m} \sum_{i=1}^{m} v_i \tag{5}$$

where m denotes the total rounds. Furthermore, we can interpolate between v_{base} and v_{avg} with the weight coefficient $\beta \in [0, 1]$, then the final vector is

$$v_{\text{final}} = \beta \cdot v_{\text{base}} + (1 - \beta) \cdot v_{\text{avg}} \tag{6}$$

Classifying according to v_{final} rather than v_{base} will get a considerable promotion in accuracy since training also happens on style transferred distribution rather than just on original distribution. Although classifying just according to a certain round's vector v_i gets lower accuracy than v_{base}, every round's vector contains its judgment for classes. (We infer that since $|S|$ is far smaller than $|C|$, the classifier tries to grasp style patterns naturally. After all, it is much easier to remember a few patterns than many.) Therefore, the test accuracy increases along with the fusion of vectors within a certain range. We call this inference method *fusion inference*.

Examining the two hyperparameters, m and β, in the above algorithm, how shall we determine them? Within a certain range, increasing of m brings cumulative improvement on accuracy. A reasonable m should bring great improvement while not consume too much time, which can be fine-tuned in different datasets. As for β, we adopt a simple grid search strategy to search for the proximate optimal β_{opt}, with steps of 0.1 and 0.01. β improves the accuracy over a wide range of right intervals in $[0, 1]$, which reflects the domination of v_{base}.

4 Experiments

4.1 Dataset

We evaluate our proposed STDA-inf on several datasets for classification task, *i.e.* STL-10 and Caltech-256, inherited from our mainly compared work [5, 7].

- STL-10[1] has 10 classes such as airplane, bird, monkey, with 500 training images and 800 test images per class. Besides, STL-10 has 100000 unlabeled images for unsupervised learning. Every image is 96×96 pixels. Because the number of test images exceeds training images, data augmentation plays a big role.
- Caltech-256[2] consists of 257 classes and the number of images per class is not equal. We randomly choose 60 images per class as the training dataset, the remaining images as the test dataset. Every image is about 200–300 pixels, so it is resized to 224×224 firstly.

In the out-of-data way, the style images are mainly chosen from the Office dataset [6] in the same way as [5], with 7 inherited from [7]. These oil paintings are so colorful that their style transfer effect is more intense than general style images of in-data way. On the other hand, AdaIN is more intense than StyleNAS. Taking an image from STL-10 as an example, we can observe the different degrees of style transfer in Fig. 2. *AdaIN module*

---

[1] https://cs.stanford.edu/~acoates/stl10/.

[2] http://www.vision.caltech.edu/Image_Datasets/Caltech256/.

cooperated with style images of in-data way is plausible for style augmentation, neither too intense like out-of-data way nor too soft and slow like StyleNAS.

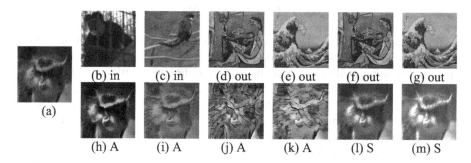

Fig. 2. Different effects of style transfer. (a) is an original image of the monkey class in STL-10. The 1st row is the style image and the 2nd row is the corresponding transferred image. *In* and *out* respectively represent in-data way and out-of-data way. *A* and *S* represent AdaIN and StyleNAS, respectively.

4.2   Implementation Details

We finish all the experiments on 2 NVIDIA Tesla P100 GPUs by PyTorch framework. The classification network is the widely applied ResNet50 [20]. Different datasets just need a little change in the stride and kernel size of convolution. Our adopted style transfer model AdaIN is released from Github[3]. AdaIN model adopts the pre-trained VGG19 as the encoder, and then only needs to train the decoder. Apart from style augmentation, we also exploit two traditional augmentation methods (*i.e.* horizontal flipping and random cropping, abbreviated as *tra*) and Mixup, *etc.*

- Training STL-10: total epoch is 150, training batch is 256, optimizer is Adam with momentum $\beta_1 = 0.9$, $\beta_2 = 0.999$, initial learning rate is 0.001 and it decays by 0.1 at the epoch of 80, 120, weight decay is 5×10^{-4}.
- Training Caltech-256: total epoch is 150, training batch is 32, optimizer is SGD with default parameters, initial learning rate is 0.01 and it decays by 0.2 at the epoch of 60, 120, weight decay is 5×10^{-4}.

4.3   Experiment Results

We focus on STL-10 dataset firstly and systematically illustrate the experiment results on it. Then there is much similarity on Caltech-256.

STL-10. We first present comprehensive results which verify the strength of our STDA-inf over out-data style augmentation and other augmentation methods like Mixup. Next, the presentation will follow the order as: style source, style number, hyperparameter optimization, robustness exploration and time analysis.

---

[3] https://github.com/xunhuang1995/AdaIN-style.

Table 1. Test accuracy (%) of our STDA-inf on STL-10 with and without traditional augmentation. Out style with number 10 and in style with number 10 × 10. † are the reproduce results of compared work.

+Tra	Style type	Base test	Rounds m			
			1	3	10	15
w/o	out random	67.18†	71.08	71.63	72.31	72.54
60.17	in random	68.95	71.73	71.89	72.83	72.88
	in min-loss	68.48	71.05	72.09	72.85	72.94
	in max-loss	**70.33**	71.90	72.79	73.89	**74.05**
w/	out random	79.56†	81.76	81.84	81.88	81.89
78.13	in random	**82.26**	83.43	83.70	**83.78**	83.69
	in min-loss	81.42	81.75	81.95	82.03	82.14
	in max-loss	81.54	82.04	82.31	82.41	82.59

As Table 1 shows, style augmentation of in-data way behaves much better than out-of-data way. Out-of-data is not as stable as in-data, getting bigger variance in the repeated experiments. From out-of-data to in-data, the base test accuracy gets around 3% improvement. As for fusion inference, 1 round improves the accuracy a lot and enlarging to 15 rounds still harvests some improvement. Without or with traditional augmentation, the test accuracy gets beyond 4% or 1% improvement further, after 15-rounds fusion inference. As we can see, our method STDA-inf could at most improve 14% without traditional augmentation and 5% with traditional augmentation compared to the baseline.

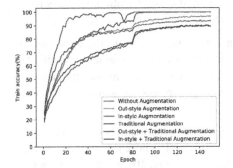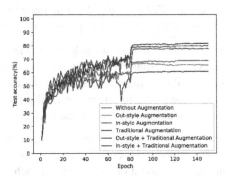

Fig. 3. The comparison of training and test accuracy on STL-10 under different augmentation methods.

Moreover, we compare the training accuracy and test accuracy in Fig. 3. The jump points exactly correspond to the changing points of learning rate. Note that the corresponding configuration is: *in random* with 100 styles; *out random* with 10 styles. 100 and 10 are chosen according to the search of proper number (shown in Table 3). In-data way would manifest greater superiority if we both choose the same style number. As we can see, training without augmentation or only with traditional augmentation causes severe overfitting. While training with style augmentation is harder to converge to a very high accuracy due to the attack of styles' uncertainty. And training with both traditional and style augmentation brings a better cumulative effect. Also, in-data way behaves much better than out-of-data, which verifies our first contribution powerfully.

Table 2. Test accuracy (%) of composite augmentation methods on STL-10. Column 2&4 shows the superiority of our style augmentation over interpolation-based augmentation methods. *Style* corresponds to our *in random* way in Table 1.

	Baseline	+style	+tra	+tra+style
Baseline	60.17	–	78.13	–
+Mixup [19]	63.74	68.98	81.14	83.34
+Cutout [3]	61.10	**71.45**	80.13	82.06
+CutMix [18]	63.39	71.21	80.90	**83.63**
+Manifold [15]	60.10	68.35	76.45	77.21
+style (ours)	**68.95**	–	**82.26**	–

In addition, we compare style augmentation with Mixup, Cutout, CutMix and Manifold Mixup. The specific parameter configuration is explained in the Appendix. MixUp *etc.* are simply linear interpolations while style augmentation utilizes semantic information (nonlinear). As Table 2 shows, our style augmentation performs better (see Column 2&4), and it can be used in conjunction with existing forms of data augmentation to further improve model performance (see Column 3&5). It needs to be emphasized that these augmentation methods don't work during inference, yet style augmentation performs better along with fusion inference. And the Test Time Augmentation (TTA) is carried out on the basis of traditional augmentation, which is trivial and doesn't get significant promotion, compared to our fusion inference.

Style Source. Except for selecting styles randomly, we investigate the cross-entropy loss of every training sample and sort them. The minimal loss is in the magnitude of 10^{-6} while the maximal loss is in the magnitude of 10^{-3}. It seems that the max-loss images are harder to classify while the min-loss images are easier, somewhere related to the semantic information of their styles.

Style Number. As shown in Table 3, we investigate the proper number of style images. In this experiment, out-of-data style images are randomly chosen, while in-data style images are chosen on average from each class (for example, 100 means 10 images per class). We speculate that the variation trend is: as the number of style images $|S|$ increases, the test accuracy increases first and then decreases. And the optimal $|S|$ of in-data way seems bigger than out-of-data way. A reasonable explanation is that the intensity of style augmentation should be in an appropriate range. To put it in another way, style images of out-of-date way are so colorful that too many intense styles make training models hard to catch dominant patterns and converge. In turn, too few soft styles can't give full play to the role of data augmentation. It's worth noting that in *out random*, 7 images are inherited from [7]. [7] choose 8 different styles that look different from each other and only 7 styles bring about positive performance. The test accuracy is lower than 67.18% as we substitute the 7 images with other random images.

Table 3. Searching proper number of style images: test accuracy (%) on STL-10 without traditional augmentation. The baseline is 60.17%.

Style type	Style number		
	10	30	100
out random	**67.18**	66.98	66.29
in random	69.23	**69.30**	68.95
in min-loss	67.19	67.48	**68.48**
in max-loss	69.63	**70.75**	70.33

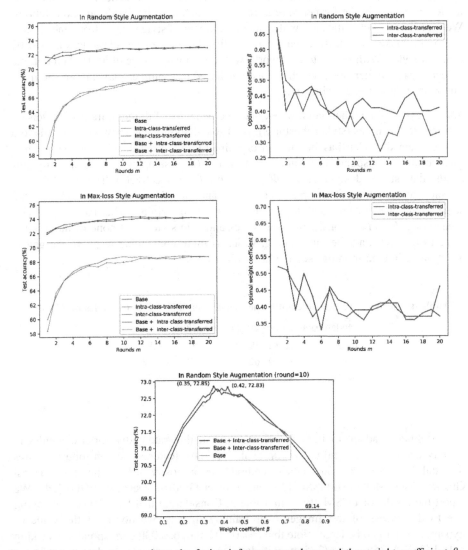

Fig. 4. Hyperparameter search on the fusion inference rounds m and the weight coefficient β. The horizontal line represents the accuracy of base test. Transferred without base means classifying only via the style transferred samples.

Hyperparameter Optimization. Fusion inference is effective no matter which kind of style type. Figure 4 shows a complete hyperparameter search on the fusion inference rounds m and the weight coefficient β on STL-10. Note that the presented figure corresponds to one result of the repeated experiment. Since the base test dominates in fusion inference, the accuracy only with transferred samples can't exceed the base test. Within a certain range, the accuracy increases as rounds m increases until steady. And the accuracy after only 1 round gets remarkable improvement. On STL-10, around 15-rounds harvests optimal test accuracy. Considering the accuracy and the time consumption together, we can fix $m = 5$ or less. Besides *intra-class test*, we compare *inter-class test* with it, and find that the test accuracy gets somewhat higher in the *inter-class* mode, especially in the former several rounds, which is in line with our expectations. We speculate that classification with a smaller number of style images from one class in one round makes the classifier easier to identify the patterns per class, less misled by different styles. With regard to β, it works well over a wide range of $[0, 1]$. And optimal β decreases as m increases, which means transferred test samples play a more and more important role in the classification.

Robustness. On the other hand, we explore the behavior of different trained models on different test samples (style robustness). As Table 4 shows, each row corresponds to a kind of trained model (baseline model or trained through out-data/in-data way) while each column corresponds to a kind of test dataset (base samples or transferred by out-data/in-data styles, denoted as Te, $\Phi_{\mathrm{out}}(Te)$, $\Phi_{\mathrm{in}}(Te)$ respectively). The out styles are the same as the 10 images adopted in out-data random training and the in styles are the same as the 100 images adopted in in-data random training. The 10 mixed styles (generating $\Phi_{\mathrm{mix}}(Te)$) consist of 5 out styles and 5 in styles, and none of them is the same as the styles adopted in training, which is more convincing to prove the generalization of in-data training (see the last Column).

Table 4. Accuracy (%) of different trained models on different test samples of STL-10.

Model\Data	Te	$\Phi_{\mathrm{out}}(Te)$	$\Phi_{\mathrm{in}}(Te)$	$\Phi_{\mathrm{mix}}(Te)$
base	78.13	19.71	23.86	16.84
out	80.79	**62.60**	46.81	36.58
in (ours)	**82.06**	55.05	**61.89**	**46.45**

Speaking of adversarial attacks, Kurakin *et al.* illustrated adversarial examples in the physical world [21] and Ilyas *et al.* explored robust features and non-robust features in detail [22]. Besides, many attack methods are proposed such as the classical Fast Gradient Sign Method (FGSM) [23] and Project Gradient Descent (PGD) [24]. We report the results of FGSM attack in Table 5. Considering that STL-10 is challenging due to the excess of test images over training images, we don't set the value of hyperparameter ϵ to be large. Note that $\epsilon = 0.004, 0.016, 0.030$ correspond to attacking only 1/255, 4/255, and 8/255 magnitude of pixels, respectively. In-data way of style augmentation defeats out-of-data way. Style augmentation performs better when the attack is not very intense. And Mixup and CutMix perform well. We speculate that on one hand, it's harder to get strong robustness on STL-10 since it has a bigger proportion of test samples. On the other hand, style augmentation is not as intense as augmentation methods like Mixup. But when involving with the disturbance of styles, attacking with other classes' styles will make it harder for the classifier to classify samples during training. Thus we can avoid the overfitting of textures to grasp essential patterns, and thereby obtain stronger (style) robustness.

Table 5. Accuracy (%) after FGSM white-box attack with different intensity on STL-10. Note that except for the baseline, all other rows adopt the traditional augmentation. And the last 3 rows belong to ours.

ϵ	0	0.004	0.016	0.030
Baseline	60.91	25.28	1.90	0.23
Traditional	78.13	57.50	16.34	7.00
Mixup	81.14	57.20	24.43	17.19
Cutout	80.13	56.56	13.71	6.39
CutMix	80.90	54.74	**24.81**	**20.10**
Manifold	76.45	54.15	17.64	10.93
out random	80.79	60.14	15.68	6.15
in random	**83.15**	**62.10**	17.76	9.06
in min-loss	81.53	58.40	16.26	7.68
in max-loss	81.74	59.51	15.14	6.35

Time Analysis. Finally, we illustrate the time consumption briefly, as shown in Table 6. On the basis of training without any data augmentation, traditional augmentation takes about another 0.03 h (hour) while style augmentation takes about another 0.30 h. As for the inference time, one-round fusion inference takes about 120 s (second), compared to 82 s of the base test. As for Mixup *etc.*, they are not very time consuming. The time consumption of style transfer is also relative to the resized size of content and style images. In future work, the function and mechanism of style transfer can be explored further, especially reducing the inference time.

Table 6. Training (2 GPU) and inference (1 GPU) time on STL-10. 120 s is the inference time in one round of fusion inference.

Augmentation	Base	+tra	+style
Training time (h)	1.30	+0.03	+0.30
Inference time (s)	82	–	120

Caltech-256. The main experiments on Caltech-256 dataset are the same as STL-10. However, Caltech-256 contains 257 classes so it's much more challenging. The exploration of the number of style images $|S|$ can be seen in Table 7. The variation trend is much similar to STL-10: the test accuracy increases first and then decreases as $|S|$ increases. We set $|S|$ as 100 and randomly choose style images from the whole training images. There is no need to explore the optimal $|S|$ since 100 is efficient enough. With style augmentation, it's observed that the training accuracy still has improvement space after the last epoch. So we extend the training epoch from 150 to 200 and change the learning rate by multiplying 0.1 at the 190th epoch, getting higher accuracy.

Table 7. Test accuracy (%) on Caltech-256 with traditional augmentation. The baseline is 60.86%. Note that 771 means choosing 3 style images per class for *in random*.

Style category	Style number			
	10	100	250	771
out random	61.48	**63.17**	62.74	63.13
in random	61.63	**64.44**	63.65	63.98

We both set $|S|$ as 100, and randomly choose style images from the whole training images for in-data way. Table 8 shows the fusion inference accuracy on Caltech-256 with traditional data augmentation: style images of in-data way perform better than out-of-data way. And the base test accuracy without traditional augmentation is 39.98%. We get 10+ improvement easily via style augmentation.

Table 8. Test accuracy (%) of our STDA-inf on Caltech-256 with traditional augmentation. The baseline is 60.85%. + means extending the training epoch to 200 and † is the reproduce result of compared work.

Style type	Style number	Base test	Rounds m			
			1	5	10	15
out random$^+$	100	63.36†	64.31	64.79	64.79	64.75
in random$^+$	100	**64.92**	65.79	66.18	66.35	**66.41**

5   Conclusion

In this paper, we have proposed a novel method of style augmentation named STDA-inf, which consists of in-data training and fusion inference. Style images of *in-data* way are more proper and targeted than *out-of-data* way during training since classifiers may overfit textures. In addition, the learned transfer distribution can be utilized during inference. Current methods of data augmentation harvest better performance combined with our method.

Improvements of style augmentation vary along with different transfer degrees, intense or soft. *Online* transfer mode creates much richer samples and is more flexible to control the transfer proportion compared to *offline* mode. As for the strategy of style choice, *intra-class* mode is superior to *inter-class* mode during training since it creates an adversarial effect. In future work, it's worth studying to reduce the time consumption of style augmentation (for example, using a unified network to accomplish style transfer and classification simultaneously), and apply it to other tasks such as object detection.

6 Appendix

Configuration of Compared Augmentation Methods
We illustrate the specific parameter configuration of compared data augmentation methods here. If the reference paper provides the experiments of STL-10, we adopt the parameters directly. If not, we try some simple tuning of parameter rs based on the reported datasets. Note that the traditional augmentation refers to horizontal flipping and random cropping.

- Mixup: No regulated parameters.
- Cutout: The mask area of square region is 24×24 without traditional augmentation and 32×32 with traditional augmentation.
- CutMix: The CutMix probability is 0.5 and distributional parameter $\beta = 1$.
- Manifold Mixup: We adopt the mixed layers as $[0, 1, 2]$. Then we try to take distributional parameter α as $0.2, 1, 2$, and get the best result when applying 1. But regrettably, this method still can't defeat the baseline. Maybe training for more epochs is necessary than the vanilla training, since Manifold Mixup is a strong regularizer.

Acknowledgment. This work was supported by the National Key Research and Development Program of China under grant 2018AAA0100205.

References

1. Yaeger, L.S., Lyon, R.F., Webb, B.J.: Effective training of a neural network character classifier for word recognition. In: Advances in Neural Information Processing Systems, pp. 807–816 (1997)
2. Geirhos, R., Rubisch, P., Michaelis, C., Bethge, M., Wichmann, F.A., Brendel, W.: ImageNet-trained CNNs are biased towards texture; increasing shape bias improves accuracy and robustness. In: The International Conference on Learning Representations (ICLR) (2019)
3. Gatys, L.A., Ecker, A.S., Bethge, M.: A neural algorithm of artistic style. Nature Communications (2015)
4. Gatys, L.A., Ecker, A.S., Bethge, M.: Image style transfer using convolutional neural networks. In: Proceedings of the IEEE Conference on Computer Vision and Pattern Recognition, pp. 2414–2423 (2016)
5. Jackson, P.T., Atapour-Abarghouei, A., Bonner, S., Breckon, T.P., Obara, B.: Style augmentation: data augmentation via style randomization. In: Proceedings of the IEEE Conference on Computer Vision and Pattern Recognition Workshops, pp. 83–92 (2019)

6. Saenko, K., Kulis, B., Fritz, M., Darrell, T.: Adapting visual category models to new domains. In: Daniilidis, K., Maragos, P., Paragios, N. (eds.) Computer Vision – ECCV 2010: 11th European Conference on Computer Vision, Heraklion, Crete, Greece, September 5-11, 2010, Proceedings, Part IV, pp. 213–226. Springer, Heidelberg (2010). https://doi.org/10.1007/978-3-642-15561-1_16

7. Zheng, X., Chalasani, T., Ghosal, K., Lutz, S., Smolic, A.: STaDA: style transfer as data augmentation. arXiv preprint arXiv:1909.01056 (2019)

8. Zhang, H., Cisse, M., Dauphin, Y.N., Lopez-Paz, D.: mixup: beyond empirical risk minimization. In: The International Conference on Learning Representations (ICLR) (2018)

9. Huang, X., Belongie, S.: Arbitrary style transfer in real-time with adaptive instance normalization. In: Proceedings of the IEEE International Conference on Computer Vision, pp. 1501–1510 (2017)

10. Luan, F., Paris, S., Shechtman, E., Bala, K.: Deep photo style transfer. In: Proceedings of the IEEE Conference on Computer Vision and Pattern Recognition, pp. 4990–4998 (2017)

11. Li, Y., Fang, C., Yang, J., Wang, Z., Lu, X., Yang, M.H.: Universal style transfer via feature transforms. In: Advances in Neural Information Processing Systems, pp. 386–396 (2017)

12. Li, Y., Liu, M.-Y., Li, X., Yang, M.-H., Kautz, J.: A closed-form solution to photorealistic image stylization. In: Ferrari, V., Hebert, M., Sminchisescu, C., Weiss, Y. (eds.) ECCV 2018. LNCS, vol. 11207, pp. 468–483. Springer, Cham (2018). https://doi.org/10.1007/978-3-030-01219-9_28

13. Yoo, J., Uh, Y., Chun, S., Kang, B., Ha, J.W.: Photorealistic style transfer via wavelet transforms. In: Proceedings of the IEEE International Conference on Computer Vision, pp. 9036–9045 (2019)

14. An, J., Xiong, H., Huan, J., Luo, J.: Ultrafast photorealistic style transfer via neural architecture search. In: AAAI, pp. 10443–10450 (2020)

15. Krizhevsky, A., Sutskever, I., Hinton, G.E.: ImageNet classification with deep convolutional neural networks. In: Advances in Neural Information Processing Systems, pp. 1097–1105 (2012)

16. Berthelot, D., Carlini, N., Goodfellow, I., Papernot, N., Oliver, A., Raffel, C.A.: MixMatch: a holistic approach to semi-supervised learning. In: Advances in Neural Information Processing Systems, pp. 5050–5060 (2019)

17. DeVries, T., Taylor, G.W.: Improved regularization of convolutional neural networks with cutout. arXiv preprint arXiv:1708.04552 (2017)

18. Yun, S., Han, D., Oh, S.J., Chun, S., Choe, J., Yoo, Y.: CutMix: regularization strategy to train strong classifiers with localizable features. In: Proceedings of the IEEE International Conference on Computer Vision, pp. 6023–6032 (2019)

19. Verma, V., et al.: Manifold mixup: better representations by interpolating hidden states. In: International Conference on Machine Learning, pp. 6438–6447. PMLR (2019)

20. He, K., Zhang, X., Ren, S., Jian, S.: Deep residual learning for image recognition. In: 2016 IEEE Conference on Computer Vision and Pattern Recognition (CVPR) (2016)

21. Kurakin, A., Goodfellow, I., Bengio, S.: Adversarial examples in the physical world. arXiv preprint arXiv:1607.02533 (2016)

22. Ilyas, A., Santurkar, S., Tsipras, D., Engstrom, L., Tran, B., Madry, A.: Adversarial examples are not bugs, they are features. In: Advances in Neural Information Processing Systems, pp. 125–136 (2019)

23. Goodfellow, I.J., Shlens, J., Szegedy, C.: Explaining and harnessing adversarial examples. arXiv preprint arXiv:1412.6572 (2014)

24. Madry, A., Makelov, A., Schmidt, L., Tsipras, D., Vladu, A.: Towards deep learning models resistant to adversarial attacks. arXiv preprint arXiv:1706.06083 (2017)

Abnormal Driving Detection Based on Human Pose Estimation and Facial Key Points Detection

Zihao Ye[1,2], Qize Wu[1,2], Xinxin Zhao[1,2], Jiajun Zhang[1,2], Wei Yu[1,2], and Chao Fan[1,2(✉)] (iD)

[1] The School of Artificial Intelligence and Computer Science,
Jiangnan University, Wuxi 214122, China
fanchao@jiangnan.edu.cn
[2] Jiangsu Key Laboratory of Media Design and Software Technology,
Jiangnan University, Wuxi 214122, China

Abstract. In recent years, there have been an increasing number of traffic accidents which are caused by the abnormal driving behaviors, including dangerous driving and fatigue driving. This research aims to identify these abnormal driving behaviors immediately and warn the drivers by proposing a hybrid approach based on human pose estimation and facial key points detection. Initially, we utilize OpenPose neural network model to detect the key points of body, face and hand in real time. These data are employed to judge the dangerous driving behaviors such as smoking, phoning, etc. Furthermore, Dlib model are selected to detect facial key points, which are passed to fatigue detection algorithm PERCLOS. Facial features are extracted as a supplement to human pose data. Finally, our proposed hybrid method for abnormal driving detection can effectively reduce the risk of traffic accidents. Hence, it has a great social significance and application value in the transportation field.

Keywords: Abnormal driving detection · Human pose estimation · Facial key points detection

1 Introduction

According to statistics from the Ministry of Public Security of the People's Republic of China in 2019, the number of motor vehicles in China reached 340 million. At the same time, the number of motor vehicle drivers reached 422 million [1]. What should never escape our attention is that the number of car accidents in China reached 200,114, among which the death toll reached 52,388 in 2019.

As abnormal driving behaviors, dangerous and fatigue driving are the most important causes of traffic accidents. Nevertheless, there is a lack of a method to detect both dangerous and fatigue driving simultaneously in the field of traffic safety. Especially, the last decade witnessed an upward trend in the application of deep learning and computer vision technology in driving behavior recognition [2]. Exploring the abnormal driving behaviors detection in real time is beneficial to the individual, family

© Springer Nature Switzerland AG 2021
D.-S. Huang et al. (Eds.): ICIC 2021, LNCS 12837, pp. 91–102, 2021.
https://doi.org/10.1007/978-3-030-84529-2_8

and society. Moreover, in the research of related fields, the detection method proposed in this paper is also innovative.

Under this premise, we make an effort to study the abnormal driving behavior detection based on human pose estimation and facial key points detection. OpenPose network model [9] is adopted to obtain the key point data of human pose, which will be used in the dangerous driving behavior detection algorithm. Further, Dlib library [3] is exploited to extract facial features. It will be used to calculate fatigue physical quantity PERCLOS to achieve fatigue detection. From the experimental results, we discovered that the proposed hybrid method has a good effect on the dataset.

The rest of this article is presented in the following steps. Related work is briefly discussed in Sect. 2. The proposed method is illuminated in Sect. 3. Section 4 shows the experimental results and gives some discussions about the results. Section 5 draws the conclusion of this paper.

2 Related Work

Considering the work of human key point detection, several real-time and high-accuracy detection algorithms are developed to meet the requirements of the real-life applications. Zhe et al. from CMU Perceptual Computing Lab propose a method called Part Affinity Fields (PAFs) [9], which can effectively estimate the attitude of multiple people and achieve real-time results while ensuring accuracy. They adopt the bottom-up method to detect a large number of people in real-time. However, their work demands the global context information [10]. In addition, Cheng et al. [11] proposed a bottom-up detection method called HigherHRNet. The method for human posture estimation applied the high-resolution feature pyramid to learn scale-aware representation. It has the advantages of training multi-resolution monitoring and reasoning multi-resolution aggregation, which can more accurately solve the problem of multi-person attitude estimation and scale change of key points.

Numerous scholars studied the face detection in previous work. Kazemi et al. [4] addressed the problem of face alignment for a single image. Their experiments showed how an ensemble of regression trees can be used to estimate the face's landmark positions directly from a sparse subset of pixel intensities. The MTCNN network algorithm [5] is another method for face detection. With a good detection performance, it is eminently suitable for face detection of drivers. This algorithm can be divided into four parts. Multi-scale image gold is generated from the original image at first. Then the P-Net network and R-Net network are utilized. Further, the O-Net network, a more powerful network, is used for improvement. Finally, the algorithm outputs a final face window and marks five feature points.

Many facial features are chosen for the determination of fatigue driving behaviors. A physical PERCLOS [6, 7] was implemented to measure fatigue based on the facial information of eyes. Three different criteria were given for measuring the fatigue: EM criterion, P70 criterion and P80 criterion. Feng et al. [8] analyzed the fatigue detection of vehicle drivers based on facial features of mouth since the state of mouth may change when the driver yawns.

In addition, in the aspect of fatigue driving detection, there is also a method based on physiological electrical signals. With the popularization of BCI (brain-computer interface) and development of signal processing technology, fatigue driving detection based on EEG signals has become possible. It basically completes the detection function by identifying the EEG signals of fatigue driving and performing classification [12, 13]. However, EEG-based detection approach is not the research object of this paper, so we are not going into further in this paper.

3   Methodology

3.1   Dangerous Driving Detection

In the part of human key point detection, OpenPose network model [9] can meet our requirements for real-time and high-accuracy detection. OpenPose is developed based on convolutional neural network and supervised learning with caffe as the framework. It can track people's facial expressions, torso, limbs and even fingers.

Openpose network is trained by CoCo human posture data set, which contains the coordinates and type information of human joint points. There are totally 17 joint points in CoCo training set, and the number of joint points in the algorithm is 18 (see Fig. 1). When we study the distribution law of the driver's limbs and design the posture discrimination algorithm, the joints of the human upper body need to be used (see Fig. 2). Hence, it is enough to detect upper body's joints. In order to study the distribution law of limb position under abnormal posture of drivers, it is necessary to output and record the coordinate data of joint points under abnormal posture. In the detection process, some joint points are not detected, which is indicated by (0, 0).

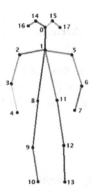

Fig. 1. Joint points of whole body

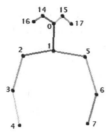

Fig. 2. The points of upper body

When the state of a driver changes from a normal driving posture to a calling posture, some features will change obviously, including the arm bending angle and distance between wrist joint and eye joint. As can be seen from Fig. 3, the features change when a driver changes from normal driving posture to calling posture. Both of these characteristics will have an obvious step [14] when the driver's driving attitude changes.

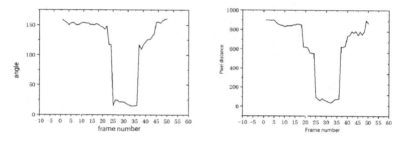

Fig. 3. Feature differences in different poses

For instance, when a driver is in an abnormal driving posture, the bending angle of both arms will be between 0 and 80°, and the pixel distance between the wrist key point and the eye key point will be between 0 and 457.4 [14]. Therefore, these data can be used for dangerous driving detection.

3.2 Fatigue Driving Detection

The state of face can be detected by collecting facial features, such as opening and closing of eyes and the opening degree of the mouth. On the one hand, the frequency and duration of eye closure can reflect the state of fatigue to some extent. On the other hand, the opening degree of the mouth changes when the driver yawns. In a driver yawns, the distance between the upper and lower lips of the mouth becomes larger, whereas the width of the mouth becomes smaller since the left and right ends of the mouth shrink toward the center of the mouth. Although the opening height of the mouth will increase when laughing and talking, the width will not decrease obviously.

The width of the mouth will even increase when laughing, so the opening degree of the mouth can be applied to judge yawning [8]. In this paper, we propose a method to detect fatigue driving behaviors by utilizing features of both eyes and mouth.

Researchers from Carnegie Mellon Institute put forward a physical PERCLOS to measure fatigue [6]. At present, PERCLOS method has three different criteria for judging fatigue, namely EM criterion, P70 criterion and P80 criterion. Its specific meaning [7] is listed as follows.

(1) EM criterion: If the pupil is covered by eyelid by more than 50%, the eye is considered closed;
(2) P70 criterion: If the pupil is covered by eyelid over 70%, the eye is considered closed;
(3) P80 criterion: If the pupil is covered by eyelid more than 80%, the eye is considered closed.

The subject shows signs of fatigue if the number of blinks reaches the threshold within the specified time. In this paper, it is considered that the eyelid may cover more than 50% or even 70% of the pupil when people pay special attention or are in meditation, so the P80 criterion is adopted in the system [6]. Statistics show that people blink about ten times in one minute, each time it takes about 0.3 or 0.4 s, and the interval between two blinks is about 2.8–4.0 s. However, due to the different nature of work, some professional workers need to pay close attention to their work, so they blink a little less, about 5–10 times. The frequency and duration of eye closure are closely related to fatigue.

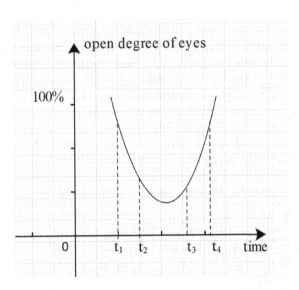

Fig. 4. The graph explaining the principle of PERCLOS

The principle of PERCLOS can be illustrate by Fig. 4. The parameter value P of PERCLOS is defined by formula (1). t1 represents the time when the eyes are fully opened until the eyes and face are closed by 20%. t2 represents the time when the eyes are fully opened until the eyelids are closed 80%. t3 indicates that the eyes are completely opened, completely closed and then opened for 20%. t4 indicates the time from the whole opening of the eyes to the 80% closing and then opening.

$$P = \frac{t_3 - t_2}{t_4 - t_1} \times 100\% \tag{1}$$

Carrying out calculation processing above on all image to be measured, we calculate the circumscribed rectangles of all feature images and record the height and width lx. This ratio is used as the opening and closing value of the eyes, and all the opening degrees are normalized to the range of [0, 1]. If the range is less than 20% of the opening degree, the eyes are considered to be closed. Taking the experimental group as an example, an image frame is collected every 0.25 s and the collection time is 10 s, with a series of data totaling 40 pieces. Blinking frequency is 0.2 times per second. Eye closing time is 2.25 s. Average blinking time is 1.12 s and PERCLOS value is 27.6%. After a series of preprocessing and calculation, the eye features are calculated and extracted.

In the extraction and calculation of eye labor characteristics, p80 is selected as the main standard when the fatigue condition judgment feature is selected to supplement the calculation of other eye condition parameters. Some commonly used parameters that characterize eye condition incorporate blinking frequency, average eyelash time, eye closing time, etc. These parameters have been proved to be closely related to fatigue condition [14].

From opening to closing (calculated that opening degree is less than 20%, that is, closing) to opening, eyes are recorded as an act of blinking. Blinking frequency refers to blinking behavior of eyes in unit cycle. Average blinking time refers to the average time that the eyes are closed in every blinking action per unit time. The average duration of eye closure is usually in a period of time [14].

To sum up, the following features of eye information are extracted and calculated according to eye information, such as blink frequency, eye closing time, eye opening time, PERCLOS, average blink time, etc. According to Table 1, the extracted features are calculated based on the image information of human eyes.

Table 1. The feature names and their expressions

Feature name	Expression
Time when opening degree is less than 80%	t_{high}
Time when opening degree is less than 20%	t_{low}
Blinking times	k
Blinking frequency	$blink = \frac{k}{10}$
PERCLOS	$PERCLOS = \frac{0.25 \times (t_{low} - 1)}{10 - 0.25 \times (t_{high} - 1)}$
Opening time of eyes	$awake = 10 - 0.25 \times (t_{low} - 1)$
Eye closure time	$fatigue = 0.25 \times (t_{low} - 1)$
Average blink time	$perawake = \frac{awake}{k}$

In order to avoid the discrimination error caused by different lip thickness [15], this paper puts forward the method of mouth inner contour opening feature to judge fatigue. As depicted in Fig. 5, yawning can be recognized by calculating the opening degree of mouth utilizing the ordinate of 51, 59, 53, 57 and the abscissa of 49, 55.

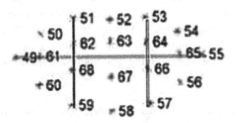

Fig. 5. Feature points of mouth

One example is adopting formula (2) to judge whether the mouth is open or not, so that we can determine whether people are yawning [8]. At the same time, the threshold should be reasonable and can distinguish yawning from normal speaking, eating or humming. Under normal circumstances, the driver's mouth is generally closed and the opening degree is a small value between 0 and 0.2. When yawning in fatigue state, the opening degree of mouth is large, generally more than 0.4. In this research, 0.4 is selected as the threshold for judging mouth opening.

$$opening\ degree = \frac{1}{2} \times \frac{(y_{51} + y_{53}) - (y_{59} + y_{57})}{x_{55} - x_{49}} \tag{2}$$

In order to judge fatigue accurately, it is necessary to distinguish yawning from other movement of mouth. It is universally acknowledged that the duration of mouth opening is different between yawning and speaking. When yawning, the duration of mouth opening is longer and the number of video frames occupied by mouth opening is more. On the contrary, the number of video frames occupied by mouth opening is less when speaking or eating. Yawning duration can also be approximated by the blink detection formula (1), which is expressed by the number of frames of mouth opening duration in a period of time.

4   Results and Discussion

The experimental data of body and face key points detection and fatigue detection come from the original papers [8, 16]. The data set involve following four parts. (1) CoCo: The data set contains the coordinates and type information of human joint points. (2) Pose estimation data set: The pose estimation data set of five drivers with six actions (drinking water, smoking, answering and calling, playing with mobile phones, scratching their heads and driving normally) [16]. (3) LFPW: Testset (224 rising face pictures marked with 68 feature points) and trainset (811 face pictures marked with 68 feature points). (4) Helen: Testset (330 rising face pictures marked with 68 feature points) and trainset (2000 face pictures marked with 68 feature points).

4.1 Results of Dangerous Driving Detection

The data set of driver's abnormal posture in the original paper is divided into five batches for evaluation. In the evaluation data, the driver's arm bending angle and pixel distance from wrist joint point to eye joint point are detected. Whether the driver is in an abnormal driving posture is judged by using abnormal posture judgment criterion. If the recognition result is consistent with the labeling result, the recognition is considered correct.

Table 2. Results of abnormal posture determination criteria

Accuracy test of abnormal attitude criterion	
Batch1(acc)	96.2%
Batch2(acc)	97.7%
Batch3(acc)	97.1%
Batch4(acc)	98.8%
Batch5(acc)	97.5%
Average(acc)	97.46%

According to Table 2, experimental results show that the average recognition accuracy of the proposed abnormal attitude criterion in the evaluation data set reaches 97.46%. The algorithm has achieved good results in judging whether the driver is in abnormal driving attitude. Some application instances are given in Figs. 6 and 7.

Fig. 6. An example of human pose estimation

Fig. 7. An example of hand detection

Figure 6 displays an instance of human pose estimation by using {7, 6, 20} or {4, 3, 2} to calculate the arm angle, and using {4, 0} or {7, 0} and {7, 17} to calculate the distance from hand to nose and ear. When these data reach the threshold, it can be inferred that dangerous driving behaviors have occurred, such as smoking and making phone calls. Hand detection can also be added in Fig. 7 in order to ensure more accurate measurement.

4.2 Results of Fatigue Driving Detection

In order to verify the accuracy of fatigue state judgment, fatigue judgment tests were carried out on eye state and mouth state respectively. Before the tests, the eigenvalues of the driver's eyes and mouth in fatigue state and normal state were collected respectively. Then we established the eigenvectors. The driver is in a dangerous state once he or she is tired when driving at high speed. This work only judges whether the driver is in a fatigue state or not without subdividing the fatigue state. Thus, we only build positive and negative samples [8] and give their corresponding labels.

Table 3. Analysis of fatigue state determination

Serial number	Times of eye fatigue	Successful detection times of eye fatigue	Times of mouth fatigue	Successful detection times of eye fatigue
1	5	5	5	5
2	10	10	7	7
3	7	6	2	2
4	15	13	3	3
5	8	8	3	3
Accuracy rate	93.3%		100%	

The experimental results and analysis are given in Table 3. The driver's mouth fatigue behavior can be detected. Further, the accuracy rate of eye fatigue behavior detection is as high as 93.3%. The reason why the accuracy does not reach 100% is that some drivers who participated in the test wore glasses. The reflection of lens leads to a missing detection.

In Fig. 8, fatigue driving detection is executed by counting the number of blinks and yawns. When blinking and yawning reach the threshold at a certain time, it is considered as a fatigue driving behavior (see Fig. 9).

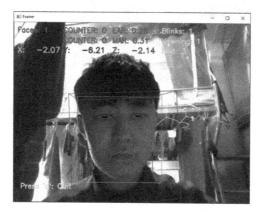

Fig. 8. An example of fatigue driving detection

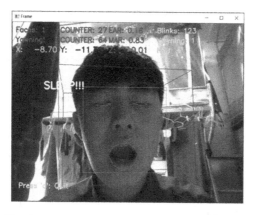

Fig. 9. The detection of a fatigue driving behavior

5 Conclusion

This paper proposed a hybrid approach to identify the abnormal driving behaviors of drivers based on human pose estimation and facial key points detection. We combined fatigue driving detection and dangerous driving behavior detection methods in our research. On one hand, human posture estimation was conducted to detect dangerous driving behaviors. We calculated the pixel distance between key points and simulated the bending angle of joints. On the other hand, we utilized the values of fatigue parameters PERCLOS, eye closing time and blink frequency to detect the fatigue driving behavior.

Dlib library and OpenPose neural network model are chosen to detect the key points of human body, face and hand in real time. Once these key points are obtained, we can calculate the distance between them. Setting the threshold, the fatigue driving and dangerous driving behaviors can be judged by combining a calculation formula and the parameters. Experimental results show that the accuracy of this detection method can reach 93.3% in fatigue driving and 97.46% in dangerous driving. Further, the detection speed is fast, which can meet the real-time requirements.

In future, enriched data set of driver posture will be provided to improve the precision of classification. Also, we will attempt to refine the detailed model and reduce the memory occupation in order to improve the real-time performance of algorithm.

Acknowledgement. This work was supported by the Foundation of Innovation and Entrepreneurship Training Program for College Students, 2021 (No. 2021186Z) and High-level Innovation and Entrepreneurship Talents Introduction Program of Jiangsu Province of China, 2019.

References

1. Mei, S.: Text data analysis and application of severe road traffic accidents in China. Henan Polytechnic University (2017)
2. Mu, G.: Study on dangerous driving behavior recognition based on deep learning. Hangzhou Dianzi University (2020)
3. Liu, Z.: Application of Dlib in face recognition technology. Pract. Electron. **2020**(21), 39–41 (2020)
4. Kazemi, V., Sullivan, J.: One millisecond face alignment with an ensemble of regression trees. In: Proceedings of the IEEE Conference on Computer Vision and Pattern Recognition (CVPR), pp. 1867–1874 (2014)
5. Zhang, K., Zhang, Z., Li, Z., Qiao, Y.: Joint face detection and alignment using multi-task cascaded convolutional networks. IEEE Signal Process. Lett. (SPL) **23**(10), 1499–1503 (2016)
6. Chen, Y.: Driver fatigue real-time monitoring system design and realize based on DSP technology. Shandong University (2010)
7. Dinges, D.F., Grace, R.: PERCLOS: a valid psychophysiological measure of alertness as assessed by psychomotor vigilance. Tech Brief (1998)
8. Feng, X., Fang, B.: Fatigue detection of vehicle drivers based on facial features, pp. 1–7. Mechanical Science and Technology for Aerospace Engineering (2020)

9. Zhe, C., Simon, T., Wei, S.E., Sheikh, Y.: Realtime multi-person 2D pose estimation using part affinity fields. In: Proceedings of the IEEE Conference on Computer Vision and Pattern Recognition (CVPR), pp. 1302–1310 (2017)
10. Huang, G., Li, Y.: A survey of human action and pose recognition. Comput. Knowl. Technol. 9(01), 133–135 (2013)
11. Cheng, B., et al.: HigherHRNet: scale-aware representation learning for bottom-up human pose estimation. In: Proceedings of the IEEE Conference on Computer Vision and Pattern Recognition (CVPR), pp. 5386–5395 (2020)
12. Liu, Y.: Driving behavior study based on Electroencephalography data analysis. Jiaotong University, Beijing (2019)
13. Xing, W.: Research on driving fatigue detection based on EEG signal recognition. Hangzhou Dianzi University (2019)
14. Yu, C.: Research on recognition method of driving fatigue characteristics based on physiological parameters. Yangzhou University (2020)
15. Wang, X., Tong, M., Wang, M.: Fatigue detection based on the inner profile characteristics of the mouth. Sci. Technol. Eng. 16(26), 240–244 (2016)
16. Chen, H.: Research on the method of driver abnormal posture recognition based on convolutional neural network. Dalian University of Technology (2020)

Uncertainty-Guided Pixel-Level Contrastive Learning for Biomarker Segmentation in OCT Images

Yingjie Bai[1,2], Xiaoming Liu[1,2(✉)], Bo Li[1,2], and Kejie Zhou[1,2]

[1] School of Computer Science and Technology, Wuhan University of Science and Technology, Wuhan 430065, China

[2] Hubei Province Key Laboratory of Intelligent Information Processing and Real-Time Industrial System, Wuhan 430065, China

Abstract. Optical coherence tomography (OCT) has been widely leveraged to assist doctors in clinical ophthalmic diagnosis, since it can show the hierarchical structure of the retina. The type and size of biomarkers are crucial in the classification and grading of diseases. Hence, automatic segmentation of biomarkers is important to quantitative analysis, which can reduce a heavy workload. In this paper, we propose a novel deep learning-based method for biomarker segmentation on OCT images. The contrastive learning is introduced to enhance the contextual relationship between pixels in the dataset instead of just in an image. In addition, uncertainty is used to weight the segmentation loss to prompt the network focus on the learning of hard pixels. At the same time, uncertainty is utilized to select hard pixels for guiding network to perform contrastive learning, which makes the segmentation result more accurate. The experiment results evaluated on a local dataset, demonstrate the effectiveness of the proposed biomarker segmentation framework.

Keywords: OCT · Biomarker · Deep learning · Contrastive learning · Uncertainty

1 Introduction

In clinical ophthalmology, biomarkers can help doctors diagnose and treat macular diseases since their size, shape and location information can reflect the type and grade of the disease. Optical coherence tomography (OCT) has become an indispensable tool for the diagnosis of eye diseases since it can provide important information for clinical decision-making [1]. Figure 1 shows the OCT images with subretinal fluid (SRF), intraretinal fluid (IRF), drusen and neovascularization.

Manual segmentation of these biomarkers requires professional knowledge and is time-consuming. During the past decades, many automatic segmentation methods have been proposed [2, 3]. Chen et al. [2] proposed a method for automatically segmenting drusen, which combined the morphological and structural features of the normal retina as the prior information. Machine learning has developed rapidly in recent years, which has been applied to medical image processing [4, 5]. Venhuizen et al. [4] segmented different levels of OCT lesion images based on the bag of words approach. In recent

© Springer Nature Switzerland AG 2021
D.-S. Huang et al. (Eds.): ICIC 2021, LNCS 12837, pp. 103–111, 2021.
https://doi.org/10.1007/978-3-030-84529-2_9

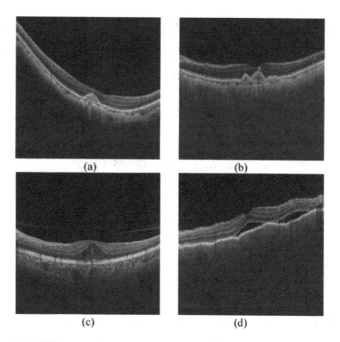

Fig. 1. Retinal OCT B-scan with neovascularization (a), drusen (b), IRF (c) and SRF (d).

years, deep learning-based methods have made great progress in the field of image processing [6–10]. Especially convolutional neural network (CNN) has made outstanding contributions. In [11], U-Net is used for lesion region segmentation. In [12], deep neural networks and ASPP are applied to segment SRF and pigment epithelium detachment lesions in OCT images. Fang et al. [13] applied a CNN which combines local and global information by a multi-resolution structure to detect lesion areas. These methods improve the performance of the biomarker segmentation task.

However, there are some shortcomings in the above methods. They only considered the relationship between pixels in an image, and ignored the relationship between pixels in the dataset. In addition, as shown in Fig. 1, the boundaries of some biomarkers are very complicated, and some biomarker regions are too small. In this paper, we propose an uncertainty-guided pixel-level contrastive learning architecture for biomarkers segmentation. Contrastive learning is introduced to make the pixel difference of similar biomarkers smaller at the level of the data set, and the pixel difference of different types of biomarkers becomes larger. Similar to some contrast learning methods [14, 15], which utilize large, external memories as a bank to store more negative samples. At the same time, uncertainty is introduced to mine the pixels that are hard to decide. And uncertainty-weighted segmentation loss and uncertainty-guided contrastive learning are used to improve the performance of the segmentation network.

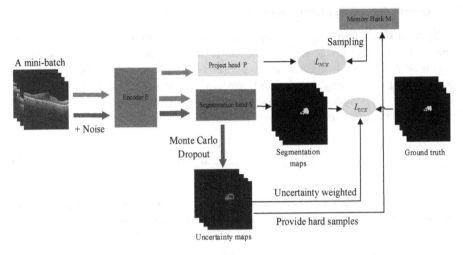

Fig. 2. An overview of the proposed method. (Color figure online)

Our contributions in this work are as follows:

(1) We use contrastive learning for retinal OCT image segmentation. To our best knowledge, this work is the first to use supervised contrastive learning into the field of OCT image analysis.
(2) The uncertainty is used to weight the segmentation loss to improve the segmentation performance of the network.
(3) The hard pixels of uncertainty mining to guide the network for performing contrastive learning.

2 Method

The network architecture of the proposed method is shown in Fig. 2, which consists of encoder E, segmentation head S, project head P and memory bank M.

2.1 The Workflow of the Proposed Method

As shown in Fig. 2. OCT B-scans are input into encoder E and the dense embeddings are obtained. The feature vector for the computation of the contrastive loss is obtained by project head P. The segmentation result is acquired by segmentation S, which is used to calculate the segmentation loss. Uncertainty map can reflect where pixel segmentation is more difficult. Therefore, we introduce it to the segmentation loss for guiding network to learn hard pixels. At the same time, the uncertainty map is used to select difficult samples and put them in the memory bank M to further guide comparative learning.

2.2 Uncertainty-Weighted Cross-Entropy Loss

As shown in Fig. 2 (red line), we added a noisy image based on Gaussian distribution sampling to the input image and input to the network. In the encoder E and segmentation head S, N stochastic forward are performed under the interference of dropout operation [16]. In this way, N predicted map can be obtained. Similar to method [17], the uncertainty of predicted map can be calculated. We average these N predicted maps $D_n^c(i,j)$ and obtain final predicted score D_{ave}^c, which can be defined as:

$$D_{ave}^c(i,j) = \frac{1}{N} \sum_{n=1}^{N} D_n^c(i,j) \tag{1}$$

Where C is the number of classes. Then, we can obtain the uncertainty map by computing the entropy of the prediction vector, which can be denoted as:

$$U(i,j) = - \sum_{c=1}^{C} D_{ave}^c(i,j) \log D_{ave}^c(i,j) \tag{2}$$

Where $U(i,j)$ is the uncertainty score of a pixel. Uncertainty shows how difficult it is for the network to classify pixels. When it is difficult for the network to distinguish the types of pixels, the uncertainty is higher. On the contrary, when the network is easy to distinguish the types of pixels, the uncertainty is lower. We weight it into the cross-entropy loss, so that the network can better learn those pixels that are difficult to distinguish, the definition is detailed as:

$$L_{UCE} = - \sum_{i=1}^{H} \sum_{j=1}^{W} U(i,j) Y(i,j) \log(T(i,j)) \tag{3}$$

Where $U(i,j)$ is the uncertainty value. $Y(i,j)$ is the ground truth label (in one-hot encoding). $T(i,j)$ is the predicted score.

2.3 Uncertainty-Guided Contrastive Learning

In this work, the pixel-level contrastive learning is introduced to the proposed method. As shown in Fig. 2, a mini-batch of OCT B-scans is input to the decoder E, and dense features D with shape of $H \times W \times C_{in}$ are obtained. At the same time, we get feature vector at each pixel position on D with shape of $1 \times 1 \times C_{in}$. Then, we can obtain the mapping feature vector with shape of $1 \times 1 \times C_{out}$ by the project head P. The P consists of two 1×1 convolution with ReLU. Note that, P is only used for training. At the same time, we distinguish between positive and negative samples based on the semantic category information of the pixels. Specifically, for a pixel i_c, the pixels with the same category are positive samples for i_c, the pixels with different categories are negative samples for i_c. In this way, we can divide the positive and negative samples into two sets ϕ_p and ϕ_n. In the pixels of the mini-batch, we perform the contrastive loss in [14]:

$$L_i^{NCE} = \frac{1}{|\phi_p^i|} \sum_{i^+ \in \phi_p^i} - \log \frac{\exp(i \cdot i^+ / \tau)}{\exp(i \cdot i^+ / \tau) + \sum_{i^- \in \phi_n^i} \exp(i \cdot i^- / \tau)} \tag{4}$$

where i^+ is a positive for i, i^- is a negative for i, $\tau > 0$ is a temperature hyperparameter. In this way, pixels of the same category are closer and pixels of different categories are further away. Visualization of features learned with pixel-level contrastive learning are shown in Fig. 3.

In order to learn representations, memory bank is important for contrastive learning to leverage massive data. If we use memory bank to store all samples, time of learning will increase greatly. Therefore, for every category, we build a pixel queue. In the most recent mini-batch, we randomly sample a small number of samples in each image. And add them into the pixel queue. However, in this way, sometimes it is impossible to capture the more characteristic pixels, since these pixels are few and randomly sampled. As mentioned in Sect. 2.2, uncertainty map shows the pixels that are hard to segment by the network, which can bring more gradient contributions. Therefore, we also randomly choose the pixels with large uncertainties and pull them into the queue in the memory bank M. In this way, it can guide the network to learn good representations.

2.4   Training and Testing

In the training phase, the segmentation network is trained by minimize a joint loss function L_{total}:

$$L_{total} = L_{UCE} + \lambda \sum_i L_{NCE} \tag{5}$$

The uncertainty-weighted cross-entropy loss is proposed in Eq. (3), the contrastive loss is proposed in Eq. (4). L_{total} can make the two losses complementary. The L_{UCE} makes network learn the difference between pixels on an image, L_{NCE} makes network learn the difference between pixels in the dataset. L_{total} makes network learn the context information in the image level and the context information in the dataset.

In the testing phase, we only used the trained encoder E and segmentation head S.

3   Experimentation and Results

3.1   Dataset

To evaluate the proposed method, we conduct experiments on a local dataset provided by Wuhan Aier Eye Hospital (denoted as WAEH), which contains 800 B-scans with neovascularization, drusen, SRF, IRF. Three quarters of them are randomly selected as the training set, the remaining quarter as the testing set. Each image was labeled at the pixel level by clinical experts. We resize all OCT B-scans to 512×512.

3.2   Comparative Methods and Metric

To investigate the segmentation performance of the proposed method, in this experiment, three several advanced methods for biomarker segmentation are compared: U-Net [11], ALNet [18] and LDN [13]. The U-Net is a widely used segmentation model. The ALNet introduced adversarial learning to U-Net. The LDN is applied to detect the diseased area with polygon supervision. To be fair, all methods are trained under pixel-level supervision.

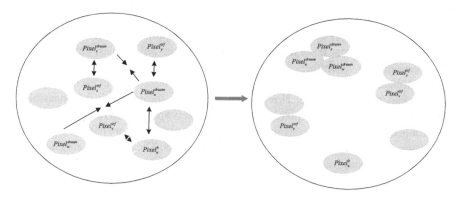

Fig. 3. Visualization of features learned with pixel-level contrastive learning. The superscript represents the category, and the subscript represents which image it comes from.

The Mean Intersection over Union (MIoU) is introduced to evaluate since it is used as metric in many segmentation papers, k represents the number of categories. Its formula is defined as follows:

$$MIoU = \frac{1}{k+1} \sum_{i=0}^{k} \frac{TP}{FN + FP + TP} \tag{6}$$

Where TP is true positive, FN is false negative and FP is false positive.

3.3   Results

In this section, we present the results of qualitative and quantitative comparison between the proposed and several comparison methods. Table 1 shows the quantitative comparison on MIoU between 4 methods for biomarker segmentation, the highest value is marked in bold. The IoU (n), IoU (d), IoU (i), IoU (s) and IoU (b) represent the Intersection over Union (IoU) of biomarker neovascularization, drusen, IRF, SRF and background respectively. The mIoU represents the overall segmentation performance. All segmentation results are shown in Fig. 4.

Table 1. Quantitative results of different methods on WAEH dataset (in percentage).

Methods	IoU (n)	IoU (d)	IoU (i)	IoU (s)	IoU (b)	mIoU
U-Net	74.5	73.2	67.7	70.8	80.5	73.3
ALNet	76.8	75.5	69.8	73.0	81.1	75.2
LDN	82.2	80.8	73.1	75.1	85.9	79.4
The proposed method	**85.1**	**83.4**	**76.4**	**78.2**	**88.3**	**82.3**

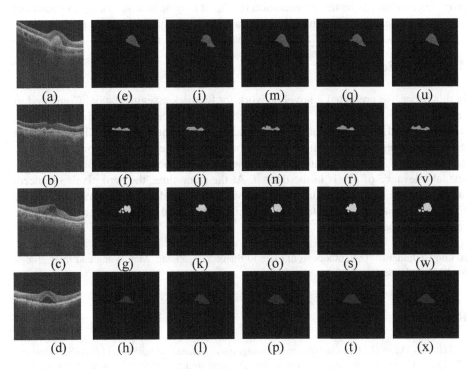

Fig. 4. Qualitative results on WAEH dataset. (a)–(d): original retinal OCT images of various biomarkers. (e)–(h): The regions annotated by ophthalmologists. (i)–(l): The segmentation maps generated by the U-Net. (m)–(p): The segmentation maps generated by the ALNet. (q)–(t): The segmentation maps generated by the LDN. (u)–(x): The segmentation maps generated by the proposed method.

As shown in the Table 1, U-Net achieves 73.3% in mIoU. Since ALNet added adversarial learning on the basis of U-Net, it increased by 1.9%. And LDN integrates the global and local information of the image, outperforming the ALNet by 4.2%. In addition, it can be seen that the proposed method achieves better segmentation results than other method, indicating that our network can accurately segment the biomarker area. The proposed method achieves the best performance with 82.3% in mIoU on WAEH datasets, outperforming the LDN by 2.9%. Specially, the reason may be that the introduced uncertainty-guided contrastive learning improves the segmentation performance of hard-to-segment areas.

The qualitative results of 4 methods are shown in Fig. 4. It can be clearly seen that the proposed method is closest to ground truth in overall. When multiple independent biomarkers are close to each other, it is easier for other methods to segment them into a whole. At the same time, when the biomarker is relatively small, it is more difficult for other methods to detect it. The proposed method uses the uncertainty to focus on the hard pixels such as small fluid (Fig. 4(w)). In addition, bottom boundary of the drusen and neovascularization is irregular and indistinguishable, we can see that the segmentation results of the other method are under-segmented in some areas, and the proposed method has better segmentation results. It may be that the contrastive learning reduces the distance within the class, increasing the distance between the classes.

4 Conclusion

In this paper, we presented a novel network for biomarker segmentation. Compared to several typical methods, the proposed method uses uncertainty to mine the region where it is hard to segment for network. And the uncertainty weighted CE loss to learn difficult samples for improving the performance of segmentation network. Moreover, the contrastive learning is introduced to the proposed method and enhance the contextual relevance of pixels in the dataset. At the same time, the hard pixels of uncertainty mining can help guide the network to conduct comparative learning. Experiments on the local dataset show that the proposed method can achieve better segmentation performance.

Acknowledgment. The authors would like to thank Ying Zhang, Man Wang and their groups for WAEH datasets.

References

1. Huang, D., et al.: Optical coherence tomography. Science **254**(5035), 1178-1181 (1991)
2. Chen, Q., et al.: Automated drusen segmentation and quantification in SD-OCT images. Med. Image Anal. **17**(8), 1058–1072 (2013)
3. Chen, Q., de Sisternes, L., Leng, T., Zheng, L., Kutzscher, L., Rubin, D.L.: Semi-automatic geographic atrophy segmentation for SD-OCT images. Biomed. Opt. Express **4**, 2729–2750 (2013)
4. Venhuizen, F.G., et al.: Automated staging of age-related macular degeneration using optical coherence tomography. Invest. Ophthalmol. Vis. Sci. **58**(4), 2318 (2017)
5. Liu, X., Wang, S., Zhang, Y., Liu, D., Hu, W.: Automatic fluid segmentation in retinal optical coherence tomography images using attention based deep learning. Neurocomputing **452**, 576–591 (2021)
6. Cao, G., Tang, Q., Jo, K.-H.: Aggregated deep saliency prediction by self-attention network. In: Huang, D.-S., Premaratne, P. (eds.) ICIC 2020. LNCS (LNAI), vol. 12465, pp. 87–97. Springer, Cham (2020). https://doi.org/10.1007/978-3-030-60796-8_8
7. Uddin, M.K., Lam, A., Fukuda, H., Kobayashi, Y., Kuno, Y.: Depth guided attention for person re-identification. In: Huang, D.-S., Premaratne, P. (eds.) ICIC 2020. LNCS (LNAI), vol. 12465, pp. 110–120. Springer, Cham (2020). https://doi.org/10.1007/978-3-030-60796-8_10

8. Liu, X., Yu, A., Wei, X., Pan, Z., Tang, J.: Multimodal MR image synthesis using gradient prior and adversarial learning. IEEE J. Sel. Top. Sign. Proces. **14**(6), 1176–1188 (2020)

9. Liu, X., et al.: Semi-supervised automatic segmentation of layer and fluid region in retinal optical coherence tomography images using adversarial learning. IEEE Access **7**, 3046–3061 (2019)

10. Altini, N., et al.: A Tversky loss-based convolutional neural network for liver vessels segmentation. In: Huang, D.-S., Bevilacqua, V., Hussain, A. (eds.) ICIC 2020. LNCS, vol. 12463, pp. 342–354. Springer International Publishing, Cham (2020). https://doi.org/10.1007/978-3-030-60799-9_30

11. Fauw, J.D., Ledsam, J.R., Romera-Paredes, B., Nikolov, S., Ronneberger, O.: Clinically applicable deep learning for diagnosis and referral in retinal disease. Nat. Med. **24**(9), 1342–1350 (2018)

12. Hu, J., Chen, Y., Yi, Z.: Automated segmentation of macular edema in OCT using deep neural networks. Med. Image Anal. **55**, 216 (2019)

13. Fang, L., Wang, C., Li, S., Rabbani, H., Chen, X., Liu, Z.: Attention to lesion: lesion-aware convolutional neural network for retinal optical coherence tomography image classification. IEEE Trans. Med. Imaging **38**(8), 1959–1970 (2019)

14. Wang, W., Zhou, T., Yu, F., Dai, J., Gool, L.V.: Exploring cross-image pixel contrast for semantic segmentation (2021)

15. Chen, X., Fan, H., Girshick, R., He, K.: Improved baselines with momentum contrastive learning (2020)

16. Tompson, J., Goroshin, R., Jain, A., LeCun, Y., Bregler, C.: Efficient object localization using convolutional networks. In: Proceedings of the IEEE Conference on Computer Vision and Pattern Recognition, pp. 648–656 (2015)

17. Gal, Y., Ghahramani, Z.: Dropout as a Bayesian approximation: representing model uncertainty in deep learning. In: International Conference on Machine Learning, pp. 1050–1059 (2016)

18. Tennakoon, R., Gostar, A.K., Hoseinnezhad, R., Babhadiashar, A.: Retinal fluid segmentation in OCT images using adversarial loss based convolutional neural networks. In: International Symposium on Biomedical Imaging, pp. 1436–1440 (2018)

Virtual Piano System Based on Monocular Camera

Yajing Wang and Liang Song[✉]

School of Informatics, Xiamen University, Xiamen 361005, China
songliang@xmu.edu.cn

Abstract. This paper studies the realization of the virtual piano system based on a monocular inverted camera. Compared to the traditional physical piano with the high price and large size, the virtual piano based on computer vision is economical and portable for novices to learn playing piano. However, while reducing learning cost, the virtual piano has to make compromises in terms of user experience and interactivity. On one hand, the software virtual piano only uses a computer keyboard, which cannot meet the players' cognition of the positions of physical piano keys. On the other hand, the virtual piano system which uses external binocular cameras is still at the theoretical stage. In practice, it is greatly limited by the problems of mutual coverage by left and right hands, large depth detection error, poor real-time performance, and so on. Therefore, in order to simulate the real experience of physical piano playing, this paper proposes the realization method of the virtual piano system based on a monocular inverted camera and a transparent plate with piano keys pattern. In detail, we use the offline modeling method to get the position of the piano keys and use the color threshold segmentation algorithm to realize the fingertips detection. Then, we detect the keys' position and depth according to fingertips pressing the keys, in turn, realize the virtual piano functions and complete the human-computer interaction. The experiments verify the practicality of the virtual piano system.

Keywords: Virtual piano · Monocular computer vision · Fingertip detection

1 Introduction

The primary goal of the virtual piano is to simulate the performance on the physical piano as real as possible. When the player presses the physical keys, the connecting rod inside the piano lifts the hammer [1], which strikes the steel strings to produce a wonderful sound. Thus, the internal structure of the piano is very complex, which inevitably leads to its huge volume, high cost, and difficult maintenance status. At the same time, more and more music lovers want to learn to play the piano but give up due to the above situations.

One implementation of virtual piano based on pure software is an important component of the digital audio workstation [2]. When software users press the computer keyboard keys which correspond to the piano keys, the computer automatically obtains the audio files of the piano keys, so that the computer sends out the sound

© Springer Nature Switzerland AG 2021
D.-S. Huang et al. (Eds.): ICIC 2021, LNCS 12837, pp. 112–121, 2021.
https://doi.org/10.1007/978-3-030-84529-2_10

corresponding to the piano keys and displays the visual result on the screen. In addition, the software user can adjust the volume, choose whether to record, play existing music and realize other functions according to the interface instructions. The software-only virtual piano has solved the problems of the traditional piano, such as huge volume, high cost, and difficult maintenance, and has certain audiences among piano lovers. However, its disadvantages are obvious. While solving the problem of portability, the user experience and authenticity are impacted. Users cannot memorize the actual position of the piano keys by hitting the keys on the computer keyboard, which is not beneficial for novices to learn to play the piano. Although the research of virtual piano is still in its initial stage, the promotion of handheld intelligent devices promotes its development. The key to the realization of the virtual piano is computer vision technology. Of course, we can combine a MIDI keyboard [3] with a digital audio workstation to provide a better virtual piano system for users. However, the MIDI keyboard is not only expensive but also cumbersome, and hence it is beyond the consideration in this paper.

At present, the use of different methods of image processing technology [4] can be divided into monocular vision technology and binocular vision technology. The existing monocular vision technology uses an infrared source to perceive the movements and depth of the fingers, so as to calculate the positions of the fingers. Binocular vision technology computes the movements and depth of fingers according to the difference of the positions of the fingers filmed by the two cameras, so as to detect whether the fingers press keys or not.

Monocular Vision. The most representative one is the virtual piano based on projection technology [5] with touch perception. In this technology, a projector is used to project a virtual piano keyboard on a horizontal desktop. When the user presses the keys on the keyboard, the camera will capture the infrared reflected by the fingers, and the computer will process the reflected information to calculate the pressing positions of fingers to realize the virtual piano. The projector dominates the quality of the whole virtual piano, which is generally expensive and difficult to popularize. Corso et al. [6] use an unsupervised learning technique to train a probabilistic graphical model with each node being a gesture word. Yeh et al. [7] propose a vision-based keystroke mechanism and the proposed keystroke detection algorithm can provide the function of a virtual piano application through a consumer-grade camera. Wyk et al. [8] propose a close to real-time and gesture-based virtual piano system which focuses on capturing a user's hand movements using sensors such as a camera and a motion controller. Begum et al. [9] present an Android application that allows users to point the camera of a handheld device towards the keyboard to play a virtual piano by using a keyboard drawn on a piece of paper. Qiao et al. [10] design a virtual piano system which allows users play piano in real-time by moving theirs hands above the Leap Motion controller. Guo et al. [11] present an AR-based individual training system for piano performance that uses only MIDI data as input.

Binocular Vision. When viewing the world with two eyes, the brain can use parallax to determine how far the objects are in the picture, thus understanding and recognizing the three-dimensional world. Binocular computer vision technology [12] is based on the principle of parallax. In this technology, the two identical performance cameras are fixed to the relative position of the target object. It gets the target location of the object distance (or depth) information by comparing and analyzing the two-camera images obtained at the same time. While monocular and binocular computer vision has advantages in depth detection, they can not guarantee the real-time requirement when applied in the virtual piano system. Du et al. [13] propose a virtual keyboard system based on true 3D optical distance camera. They analyze the hands' contour according to depth information acquired with the 3D range camera and fit the depth curve with different feature models to find fingertips. Malik et al. [14] propose a stereo hand tracking system that provides the 3D positions of a user's fingertips on and above the plane, and two downward-pointing cameras are attached above a planar surface. Hernanto et al. [15] use two ordinary webcams to realize the keys function of the virtual piano, and complete the segmentation of the hand region in the YCC color space.

The contribution of this paper consists of the following aspects. First, we use an inverted camera in the virtual piano system to solve the problem of not being able to detect fingertips when the left and right hands are covered by each other. Then, in the aspect of fingertip detection, we use the detection method based on the fingertip area, which can meet the real-time requirement on the premise of ensuring accuracy. In addition, this method can be used to further extend to identify the strength and duration of pressing keys.

2   System Implementation

We fixed a monocular inverted camera under the piano keyboard model to avoid the detection problem of mutual coverage by the left and right hands. At the same time, it provides an alternative method for fingertip depth detection. After piano keys offline modeling, based on computer vision technology, we adopt the image threshold segmentation algorithm and the contour recognition algorithm to realize the fingertip detection function in the image collected from the camera, so that we can figure out where our fingers are playing. In addition, we determine the depth of the fingers pressing by the size of the contour area of the fingertip pressing the key in relation to time.

2.1   Virtual Piano Keyboard

First, we model the piano keys offline [16] to get the coordinates of the key positions and save them in computer memory. The specific realization process is divided into the following three steps:

1) Turn on the camera to collect the target image.
2) Preprocess the image collected by the camera. First, the target color image is converted to a grayscale image. Then Gaussian filtering, edge detection, and expansion processing are performed on it. Finally, a binary image with closed edges and the contours of white keys is obtained.

3) Analyze the contour diagram and extract the position information of the target keys to prepare for the output of the key position after fingertip position recognition.

2.2   Fingertip Detection

The fingertip detection includes fingertip position detection and depth detection [17]. Each frame of image collected by the camera needs to be preprocessed in time when the virtual piano program is running, and the real-time requirements need to be met. This part is realized through the color threshold segmentation algorithm. The algorithm idea we adopt is that when the fingertips are played on the piano, the fingertips are squeezed on the transparent acrylic plate, which will produce obvious color changes and such color changes can be recognized through image processing. By analyzing the HSV image histogram, we extract the color range of the fingertips and their pressing area and then detect the target image to find out the area corresponding to the color range, and then extract the contour and calculate the area. The center of the area is taken as the fingertip position to determine its position on the piano keys. The timber and loudness of the piano keys are determined according to the changing relationship between the area and time. The virtual piano will make corresponding auditory feedback according to the above information.

HSV Color Threshold. Color space [18] is used to describe colors under certain standards, such as RGB, HSV, HIS, CMY, etc. In the HSV color space, the Hue uses an angled circle to indicate color. The Saturation indicates a value of 0 at the center of the circle, which increases along the radius of the circle, indicating that the color is becoming thicker. The Value represents the brightness of the color. At the bottom of the cone, the Value represents black, and at the top, white. In practical application, RGB color space representation is more susceptible to strong light, weak light, shadow, and other factors. By contrast, HSV space is more stable in the face of these light changes and can reflect the nature of color well. Therefore, we use HSV color space to detect the target image.

Object Segmentation Based on Histogram. The histogram is usually used in statistics to represent the distribution of data in the form of images, generally a two-dimensional statistical chart. In the color histogram, the abscissa represents the color, and the ordinate represents the number of pixels with the color value in the image. In the process of fingertip detection, we adopt the color threshold segmentation algorithm [19, 20] based on the histogram. We conduct image matting processing on the target area, analyze its histogram separately, and compare it with the color histogram of the target image. In this experiment, the extracted skin color range of the fingertip pressing area is H: 15.5°–28.1°, S: 0.004–0.5, V: 0.498–0.992. Considering the stability of the color space range of the target area, we try to extract and save the color space range of the target area, and then use the color histogram to find the target area in the target image, and finally form a binary image.

2.3   Performance Recognition

The same fingertip detection algorithm is used to extend the single-finger fingertip detection to the multi-finger fingertip detection, and the identified contour information of the multi-finger areas needs to be saved. In the experiment results, the fingertip position is not only accurately detected, but also the relationship between the fingertip pressing area and time is further processed, so as to realize the detection of fingertip depth. In the experiment, we test the area threshold of the fingertip pressing area, and the final selection range was 3000–5000. That is, when the detected fingertip area exceeds 3000, we consider the fingertip is pressing the piano key.

In the process of piano playing, when the fingertips press the piano and leave quickly, the keys sound is more rapid, otherwise, there is a longer delay. Through the analysis of the relationship between the area of fingertip pressing area and time, it is concluded that the shorter the time from fingertip pressing the key to fingertip leaving the key, the faster the audio playing speed of the key will be, and vice versa. The realization of fingertip depth detection means that the user experience and authenticity are better enhanced, which further enhances the piano's functionality. The above algorithm can satisfy the accuracy of performance recognition and the real-time requirement in the meanwhile, which are hard to both achieved by the depth detection based on binocular vision.

3   System Implementation

3.1   Experimental Settings

The virtual piano system is mainly composed of the following pieces of hardware equipment: a monocular camera, a personal computer, the transparent acrylic plate with a keyboard pattern. The Logitech C930e webcam is selected as the monocular camera, which is mainly used to collect images and is one of the core tools to realize the virtual piano system. Logitech uses Rightlight2 technology with high system stability, which can present clear images in different lighting environments, and the performance can still meet the requirements of the system under poor lighting conditions. The personal computer is also the Core tool of this virtual piano system. We choose a Dell G7 15 7590 notebook computer, which is equipped with an Intel Core i7 chip and 16G running memory. We made the piano keyboard model a sticker, and put it on the top of the acrylic display box in the center position, as the target image of the offline modeling of the piano keys, and at the same time facilitate the player to identify the position of the keys. Finally, we fix the monocular camera to the bottom of the display box and connect it to a personal computer, measure the distance between the camera and the keyboard so that its field of view is just the keyboard range, and fix the relative position of the camera and the keyboard to increase the portability of the system (see Fig. 1).

Fig. 1. The human computer interface of virtual piano system

The virtual piano program is written in C++ language, and the 4.1.2 version of the OpenCV library is called. The integrated development environment is Visual Studio 2017 X64 Debug/Release, the operating system is Windows10.

To our best knowledge, the present methods cannot satisfy both the real-time requirement and the depth detection accuracy. Therefore, we do not carry out comparative experiments but only describe our experimental results.

3.2 Experimental Results

In the keys offline modeling step [12], we obtain the contours of 14 white keys in the piano keyboard model (see Fig. 2), and the positions of the black keys can be located according to their position relationship with the white keys. In the process of contours extraction, the interference of small contours caused by the reflection of the camera light source on the transparent plate can be screened by contours area screening.

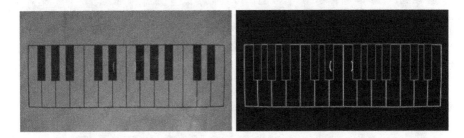

Fig. 2. The left and right figures are the physical and virtual keyboard respectively.

We use the designed fingertip detection algorithm based on the color threshold segmentation algorithm to test the fingertip detection accuracy. In Fig. 3, the left and right figures are the results of not pressing and pressing by single fingertips respectively. During the experiment, we have tested large numbers of target images, and the test result has satisfying accuracy. When processing the images stream (or video) collected by the camera, we set the frame rate to 30 FPS and the tested video time to

8 s, and we process 240 frames, including two frames whose detection effects are not desired. Although it does not affect the program operation and the implementation of the basic functions of the virtual piano, there is still room for improvement in the accuracy of fingertip detection.

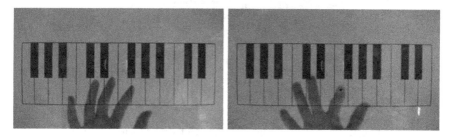

Fig. 3. Fingertip detection

Then, we carry out a fingertip detection test on multiple fingers (see Fig. 4). We save the contours of fingertip pressing areas of multiple fingers successively, so as to detect the fingertip positions when multiple fingers are playing simultaneously. Finally, we obtain a set of satisfying results. In Fig. 5, the image on the left shows an example of the virtual piano system with the camera above it, which can't tell if the keys are pressed by a covered finger. The image on the right shows the way we propose that the camera is located at the bottom, which is a good solution to the problem of fingertip detection when the left and right hands cover mutually.

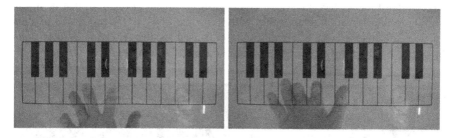

Fig. 4. Multiple fingertips detection

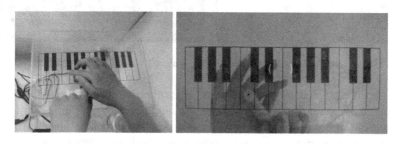

Fig. 5. The mutual cover situation and our solution

In the fingertip depth detection part, we calculate the area of the fingertip pressing areas and calculate the depth of the finger pressing key according to the relationship between its area change and time. Figure 6 shows the contour image of fingertip pressing area detected in single fingertip detection, double fingertips detection, multiple fingertips detection, and fingertip detection when the finger is covered by hands respectively. After extracting the contour of the target area, the area calculation function of OpenCV is used to calculate the area.

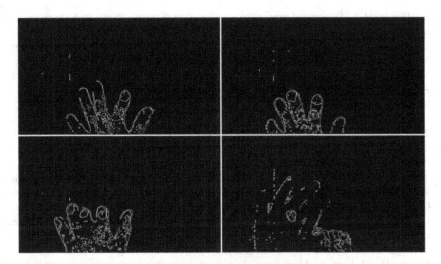

Fig. 6. Fingertips outline detection and pressing area

Finally, we test the real-time performance of the system. To start with, we set the frame rate to 30FPS and process 240 frames. The running time is about 210 s, which is slow. Therefore, on the premise of not affecting the fingertip detection effect, we reduce the frame rate to 12FPS and compress the running time to 100 s, and the processing speed is relatively ideal. In spite of this, we can still use program optimization, GPU acceleration, and other methods to further improve the real-time performance of the system.

4   Conclusion

In this paper, a virtual piano system based on a monocular inverted camera is implemented. The external equipment adopts a self-made piano model to enhance the user experience and authenticity. In addition, a color threshold segmentation algorithm based on HSV color space histogram is designed by using digital image processing technology to complete fingertip detection. After parameter adjustment and testing in experiments, the virtual piano system based on a monocular inverted camera designed in this paper has better portability, higher accuracy, and better interactivity. In spite of this, the system still has a lot of room for development. The external device can adopt an industrial way to embed the camera into the bottom of the piano model, thus further reducing its size and enhancing portability. At the same time, GPU acceleration mode can be used to further enhance the real-time image processing, so as to enhance user experience.

References

1. Conklin Jr., H.A.: Design and tone in the mechanoacoustic piano. Part I. Piano hammers and tonal effects. J. Acoust. Soc. Am. **99**(6), 3286–3296 (1996)
2. Marrington, M., et al.: Composing with the digital audio workstation. In: The Singer-Songwriter Handbook. Bloomsbury Publishing, London (2017)
3. Desnoyers-Stewart, J., Gerhard, D., Smith, M.: Mixed reality MIDI keyboard. In: Proceedings of the 13th International Symposium on CMMR, pp. 376–386 (2017)
4. Shapiro, L.: Computer Vision and Image Processing. Academic Press, Cambridge (1992)
5. DePra, Y., Fontana, F., Tao, L.: Infrared vs. ultrasonic finger detection on a virtual piano keyboard. In: International Computer Music Conference (2014)
6. Corso, J.J.: Techniques for vision-based human-computer interaction. Ph.D. thesis, Johns Hopkins University (2005)
7. Yeh, C.H., Tseng, W.Y., Bai, J.C., Yeh, R.N., Wang, S.C., Sung, P.Y.: Virtual piano design via single-view video based on multi finger actions recognition. In: 2010 3rd International Conference on Human-Centric Computing, pp. 1–5 (2010)
8. van Wyk, S., van der Haar, D.: A multimodal gesture-based virtual interactive piano system using computer vision and a motion controller. In: 2017 2nd Asia-Pacific Conference on Intelligent Robot Systems, pp. 1–6. IEEE (2017)
9. Begum, H., Shaheen, S., Moetesum, M., Siddiqi, I.: Digital beethoven—an android based virtual piano. In: 2017 13th International Conference on Emerging Technologies, pp. 1–5. IEEE (2017)
10. Qiao, W., Wei, R., Zhao, S., Huo, D., Li, F.: A real-time virtual piano based on gesture capture data. In: 2017 12th International Conference on Computer Science and Education, pp. 740–743. IEEE (2017)
11. Guo, R., Cui, J., Zhao, W., Li, S., Hao, A.: Hand-by-hand mentor: an AR based training system for piano performance. In: 2021 IEEE Conference on Virtual Reality and 3D User Interfaces Abstracts and Workshops, pp. 436–437. IEEE (2021)
12. Zhao, L.: Research on virtual piano based on computer binocular stereo vision. J. Phys.: Conf. Ser. **1533**, 032006 (2020)
13. Du, H., Oggier, T., Lustenberger, F., Charbon, E.: A virtual keyboard based on true-3d optical ranging. In: Proceedings of the British Machine Vision Conference, vol. 1, pp. 220-229 (2005)
14. Malik, S., Laszlo, J.: Visual touchpad: a two-handed gestural input device. In: Proceedings of the 6th International Conference on Multimodal Interfaces, pp. 289–296 (2004)
15. Hernanto, S., Suwardi, I.S.: Webcam virtual keyboard. In: Proceedings of the 2011 International Conference on Electrical Engineering and Informatics, pp. 1–5. IEEE (2011)
16. Goodwin, A., Green, R.: Key detection for a virtual piano teacher. In: 2013 28th International Conference on Image and Vision Computing New Zealand, pp. 282–287. IEEE (2013)
17. Torralba, A., Oliva, A.: Depth estimation from image structure. IEEE Trans. Pattern Anal. Mach. Intell. **24**(9), 1226–1238 (2002)

18. Kuehni, R.G.: Color space and its divisions. Color Res. Appl. **26**(3), 209–222 (2001). Endorsed by Inter-Society Color Council, The Colour Group (Great Britain), Canadian Society for Color, Color Science Association of Japan, Dutch Society for the Study of Color, The Swedish Colour Centre Foundation, Colour Society of Australia, Centre Francais de la Couleur

19. Lim, Y.W., Lee, S.U.: On the color image segmentation algorithm based on the thresholding and the fuzzy c-means techniques. Pattern Recogn. **23**(9), 935–952 (1990)

20. Gong, J., Jiang, Y., Xiong, G., Guan, C., Tao, G., Chen, H.: The recognition and tracking of traffic lights based on color segmentation and CAMSHIFT for intelligent vehicles. In: 2010 IEEE Intelligent Vehicles Symposium, pp. 431–435. IEEE (2010)

Wall-Following Navigation for Mobile Robot Based on Random Forest and Genetic Algorithm

Peipei Wu[(✉)], Menglin Fang, and Zuohua Ding

School of Information Science and Technology, Zhejiang Sci-Tech University, Hangzhou 310018, China

Abstract. Due to the high real-time requirements of robot navigation tasks, the nonlinearity of the robot system itself makes it difficult to model accurately, in addition, the existing "black box" navigation control model is difficult to verify, hard to achieve static and lack of interpretability. Therefore, in this paper, a navigation method for wall-following mobile robots based on random forest and genetic algorithm is proposed. In our proposed method, robots based on perceiving environment information need to autonomously decide the actions to be performed to implement subsequent navigation behaviors. On the UCI robot navigation dataset, the random forest model is first trained, and then control rules are extracted from the trained random forest, and genetic algorithms are used to optimize and filter these rules. Therefore, we have obtained a rule base with good generalization performance, that is, the controller. A comparative experiment on the same data set shows that the method proposed in this paper has a high accuracy rate of 89.19% and a recall rate of 89%. In 6 sets of comparative experiments, compared with the gravity search and particle swarm optimization assisted artificial neural network method and the gravity search and feedforward neural network method, the average accuracy of our proposed method is improved by 16.43% and 29.66%, respectively.

Keywords: Control rule · Random forest · Wall-following navigation · Genetic algorithm

1 Introduction

With the rapid development of robotics and artificial intelligence technology, robots have been widely used in industry, aerospace, national defense, transportation and other fields. Common types of robots include mobile robots, unmanned vehicles and robotic arms. Navigation is one of the core contents of robotics and intelligent control, which is the premise of robot performing tasks. This paper studies the wall-following navigation of the mobile robot, that is, during the navigation, the mobile robot moves along the edge of the object and keeps a safe distance from the wall [1]. Studying wall-following navigation is conducive to improving the robot's self-adaptability in complex and uncertain environments, and it is also helpful to improve the robot's ability to perform complex tasks.

© Springer Nature Switzerland AG 2021
D.-S. Huang et al. (Eds.): ICIC 2021, LNCS 12837, pp. 122–131, 2021.
https://doi.org/10.1007/978-3-030-84529-2_11

Wall-following navigation is a special type of reactive control, which requires the construction of a non-linear mapping from the sensor's perception space to the actuator's action space. Most of the existing wall-following navigation models come from robot motion planning models, and focus on model-free control, such as rule-based methods [2–4] and neural network-based methods [5, 6]. The rule-based method does not require accurate modeling of the controlled system, has the advantages of real-time response, strong practicability and good interpretability. However, the existing difficulty is how to obtain the rule base with good control performance [7, 8] (stability, smoothness of robot trajectory, etc.). Compared with the traditional control, the control based on neural network has the parallel distributed processing capacity, the control of nonlinear mapping ability and fault tolerance, but the "black box" nature of the neural network is hindered it in robot navigation, unmanned driving, health care and other fields of application, which determine the cause of the neural network to make a particular decision is a difficult task [9].

Dash et al. Proposed three different neural network methods to improve the accuracy of robot autonomous navigation along the wall: In [5], they proposed to use the gradient descent algorithm to train the neural network; In [10], they proposed the use of gravity search to set the optimal weight set of a feed forward neural network; In [6], they proposed a hybrid meta heuristic algorithm to train a multilayer artificial neural network. Neural networks can be trained to perform specific functions by adjusting the connection values between elements [11]. However, after the neural network trains the classifier, these weights do not give a clear explanation of the features. Therefore, it is difficult for people to understand after training [12] and a large number of parameters need to be adjusted. Eftekhary et al. [13] calculated the accuracy of Support Vector Machine (SVM) classification before and after using the normalization method, and finally used data envelopment analysis to sort the normalization method. These normalization methods have low processing capacity and improve the accuracy of SVM. However, SVM is complex and difficult to describe and understand [14]. Barakat et al. [15] pointed out that SVM and neural network are both "black box" models, which have inherent defects in the interpretation of models and results, and the learning model has no transparency and comprehensibility for humans. Beben et al. [16] proposed a mobile robot control system structure based on Erlang. This method creates virtual environment that can host an Erlang-based distributed learning library. The learning library uses a decision tree as a supervised learning algorithm. However, the decision tree is a single learner with lower learning performance than integrated learning.

Aiming at the above problems, this paper proposes a wall-following navigation method of mobile robot based on random forest and genetic algorithm, which extracts control rules from random forest and uses genetic algorithm to optimize the rules. Random forest [17, 18] is a representative model of integrated learning, which build a large number of base decision trees. Although each base decision tree naturally contains decision rules, the random forest is used directly for control problems, the large rule set may make the controller difficult to understand [19], even the conflict rules may cause "oscillation" problem in the trajectory of the robot. Therefore, this paper first extracts all the rules from the random forest, and then uses the heuristic optimization search algorithm to filter the rule set. Genetic algorithm (GA) is a global probabilistic search algorithm which simulates the process of biological evolution [20]. Compared with classical heuristic optimization algorithms, genetic algorithm has inherent implicit parallelism and better global optimization ability [21]. In this paper, the accuracy of rules is taken as the fitness function of the genetic algorithm, and the optimization is carried out in the rule set.

The method proposed in this paper combines the learning ability and generalization ability of random forest with the interpretability of rules to construct a rule-based controller [22, 23]. Optimizing rules through genetic algorithms to further ensure real-time control. The experiments on the UCI robot navigation data set show that the proposed method has a high accuracy rate of 89.19% compared with gravity search and particle swarm optimization assisted artificial neural network (GSPSO-ANN) method and gravity search and feedforward neural network (GS-FFNN) method. In 6 sets of comparative experiments, the average accuracy of the proposed method is 16.43% higher than that of GSPSO-ANN, and 29.66% higher than that of GS-FFNN.

The rest of the paper is structured as follows. Section 2 introduces the construction process of the navigation method. In Sect. 3, we present the experimental results and performance evaluation. The last section is a summary of the overall content of the paper.

2   Navigation Method

2.1   Control Rule Extraction

The decision tree is a decision analysis method to calculate the probability of the expected value of the net present value greater than or equal to zero, evaluate the project risk and judge its feasibility based on the probability of various situations. It is a graphical method to intuitively apply the probability analysis.

On the training set, this paper uses random forest to build a decision model RF, which contains 10 decision trees $T_1,...,T_{10}$. For each decision tree T_k, the rule R_k is extracted from it, and the rule set RS can be expressed as:

$$RS = \bigcup_{k=1}^{10} R_k \qquad (1)$$

Build a decision tree, first of all to all of the data as a node, and then choose from all of the data characteristics of a data characteristic of node split, generated a number of child nodes, and to determine each child node, if meet the stop dividing conditions, set the node to child nodes, the output for the node number of the largest category; Otherwise, re-select the data characteristics to segment the nodes.

According to the construction process of the decision tree, it can be known that for each T_k, a path P_{km} from the root node to the leaf node C_{km} represents a decision rule (assuming that the tree T_k contains m paths). Among them, the leaf node C_{km} where P_{km} is represents the follower of the regular P_{km} (conclusion). The n feature divisions X_{kmn} passed by P_{km} constitute the antecedent of the regular R_{km}. Therefore, R_{km} can be expressed as the combination of P_{km} and C_{km}:

$$R_{km} : \textbf{\textit{If}} \ X_{km1} \ and, \ \ldots, \ and \ X_{kmn} \ \textbf{\textit{Then}} \ C_{km}$$

Figure 1 shows a decision tree. For this decision tree, it can be known that $m = 5$, $P_1 = \{X[15] <= 1.159, X[18] <= 0.505, X[13] <= 1.561\}$, $C_1 = SLT$. From this path, a rule R_1 can be obtained for this paper: ***If*** $X[15] <= 1.159$ *and* $X[18] <= 0.505$ *and* X *[13] <= 1.561* ***Then*** *SLT*, that is, in this case the robot should turn slowly to the left.

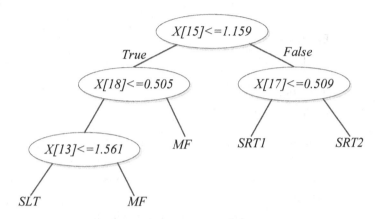

Fig. 1. Decision tree.

Similarly, you can get the other four rules:

R_2: ***If*** $X[15] <= 1.159$ *and* $X[18] <= 0.505$ *and* $X[13] > 1.561$ ***Then*** *MF*.
R_3: ***If*** $X[15] <= 1.159$ *and* $X[18] > 0.505$ ***Then*** *MF*.
R_4: ***If*** $X[15] > 1.159$ *and* $X[17] <= 0.509$ ***Then*** *SRT1*.
R_5: ***If*** $X[15] > 1.159$ *and* $X[17] > 0.509$ ***Then*** *SRT2*.

For ten decision trees $T_1, \ldots, T_{10}$. Extracting the rules according to the above steps, a total of 98 rules are obtained, and these 98 rules constitute *RS*.

2.2   Control Rule Optimization

Genetic algorithm is a heuristic search algorithm inspired by Darwin's evolution and based on the process of biological evolution. Its main characteristic is directly on the object structure, so different from other algorithms for solving the optimal solution, the genetic algorithm and continuity of function limit, there is no derivative method of probability optimization method, do not need to make sure the rules can automatically acquire and to guide the optimization of search space, adaptively adjust the search direction.

In this paper, genetic algorithm is used to optimize rule set RS and build a rule base *RB*. *RB* is a rule selected by the genetic algorithm after performing the operations of Fig. 2 under different parameters (crossover probability and mutation probability). Genetic algorithms can be expressed as an octuple, namely, $GA = \{r, n, pop, pf, ps, pc,$

pm, pmi}, among them *r* represents for rule, *n* stands for population size, *pop* stands for initial population, *pf* stands for fitness evaluation of each rule, *ps* stands for selection of rules, *pc* stands for crossover, *pm* stands for variation, and *pmi* stands for maximum iteration times.

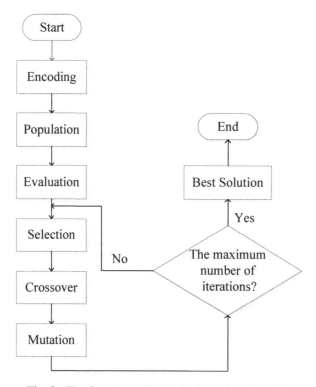

Fig. 2. The flow chart of optimization rules using GA.

The implementation of genetic algorithm optimization rules is as follows:

1) *Encoding*
   The rule set *RS* is encoded by binary encoding.
2) *Population initialization*
   The 98 rules were randomly assigned and initialized to form n populations.
3) *Fitness Function*
   Individual fitness was evaluated by calculating the accuracy of each rule in the test set.
4) *Selection*
   The genetic algorithm includes three selection methods, which are roulette selection method, random traversal sampling method and tournament selection method. The selection method adopted in this paper is roulette selection method, because random traversal sampling method and tournament selection method tend to fall into the local maximum, while roulette selection method can jump out of the local maximum. The roulette method first calculates the fitness of each rule in the

population. Then calculate the probability that each rule is inherited to the offspring population, formula is:

$$p(r_i) = \frac{f(r_i)}{\sum_{i=1}^{n} f(r_i)} \tag{2}$$

Then, the cumulative probability of each rule is calculated to construct a roulette wheel, formula is:

$$q_i = \sum_{j=1}^{i} p(r_j) \tag{3}$$

Finally, the rules with high probability are retained, and the rules with low probability are eliminated.

In this paper, the maximum number of iterations pmi is used as the condition for judging termination, and the rule base RB is output if the conditions are met; if the conditions are not met, after selection, crossing, and mutation, it is judged again whether the maximum number of iterations has been reached. If yes, output. If not, judge again through selection, crossover, and mutation until the maximum number of iterations is reached.

3   Experimental Results

This section first introduces the division of the data set and then gives the rules to extract from the trained random forest. Then, with the accuracy as the evaluation index, the optimization is carried out in the parameter space of the genetic algorithm. Finally, the performance of the model in the optimal parameter configuration is given.

3.1   Data Set

This paper uses UCI robot navigation dataset [24]. The dataset contains 5,456 samples, recording readings from 24 ultrasonic sensors (ranging in perception from 20cm to 300cm with a sampling frequency of 9 Hz) deployed on the SCITOS G5 robot and the actions performed by the robot. The robot has four discrete actions: Move-Forward (MF), Slight-Right-Turn (SRT1), Sharp-Right-Turn (SRT2) and Slight-Left-Turn (SLT). The category distribution of actions in the dataset is shown in Table 1.

Table 1. The distribution of actions in dataset.

Action	Samples	Proportion (%)
MF	2205	40.01
SRT1	826	15.03
SRT2	2097	38.43
SLT	328	6.01

Each row of data contains sensor readings X_{ij}, $i = 1, \ldots, 5456, j = 1, \ldots, 24$, robot action a_i, i.e. $D = \{(X_i, a_i) \mid i = 1, \ldots, 5456\}$. Examples of partial data samples such as those listed in Table 2. The dataset is divided into training set and testing set, i.e. $D = TrainSet + TestSet$.

Table 2. Part specification of sample data.

No.	X_1	X_2	X_3–X_{23}	X_{24}	Action
1	0.438	0.498	…	0.429	SRT1
73	0.800	0.855	…	0.878	MF
5259	1.839	2.215	…	1.828	SLT
5456	0.950	4.066	…	1.168	SRT2

The data set is divided into two mutually exclusive subsets using the hold-out method. 80% of the data is used for training and 20% of the data is used for testing.

4 Experimental Results

Will training on each ultrasonic sensor readings as a decision tree node, and then choose from all sensor readings out of a sensor readings to split decision tree node, generated a number of child nodes, and then to determine each child node, if meet the termination conditions of division, set up the node position, the output of the results for the node number of the largest category; If the termination condition is not met, the sensor reading is re-selected to segment the node until the splitting condition is met. Thus, multiple decision trees can be trained to form decision trees and forests, and 98 rules are extracted from the trained random forest. Some rules are as follows:

R_1: **If** X[13] <= 0.926 and X[14] <= 0.915 **Then** SRT1
R_7: **If** X[13] > 0.926 and X[14] > 0.915 and X[19] <= 0.497 and X[11] > 1.074 **Then** SRT2
R_{97}: **If** X[12] > 0.918 and X[14] > 0.902 and X[19] > 0.497 and X[17] <= 0.93 **Then** MF
R_{98}: **If** X[12] > 0.918 and X[14] > 0.902 and X[19] > 0.497 and X[17] > 0.93 **Then** SLT

In order to ensure the real-time performance of robot navigation control, genetic algorithm was used to screen 98 rules and select rules with good generalization performance.

The parameters of genetic algorithm, such as crossover probability and mutation probability will affect the performance of the model. In order to select rules with good decision effect and good generalization performance. The parameters of the genetic algorithm are set and the optimal parameters are selected. When the selectable values of crossover probability were 0.5, 0.6, 0.7 and 0.8, and the selectable values of variation probability were 0.001, 0.005, 0.01 and 0.05 respectively, the following 16 experiments were carried out. The accuracy of different parameter configurations was listed in Table 3, and the number of rules was listed in Table 4.

Table 3. The accuracy rates with different parameters.

pm	pc			
	0.5	0.6	0.7	0.8
0.001	85.52	87.90	86.49	84.97
0.005	85.89	88.76	86.87	85.71
0.01	85.52	86.68	85.28	85.09
0.05	83.08	87.72	83.14	**89.19**

Table 4. The number of rules with different parameters.

pm	pc			
	0.5	0.6	0.7	0.8
0.001	36	53	41	48
0.005	47	46	50	39
0.01	47	44	40	44
0.05	57	44	41	**44**

When the crossover probability and mutation probability are large, the damage probability to the existing population is large. When it is small, the original population cannot be updated well. Therefore, in this paper, in the given parameter space, the accuracy is used as a measure to optimize parameters. It can be known from Table 3 that when $pc = 0.8$ and $pm = 0.05$, the method in this paper achieves the highest accuracy, which is 89.19%. As can be seen from Table 4, the number of rules after optimization is 44, which is 54 fewer than the number before optimization. Fewer rules means less decision time. In this paper, the accuracy is 98%, the recall rate is 89%, and the F1 value is 93%.

In this paper, GSPSO-ANN model and GS-FFNN model are selected as comparison models. The results of the 6 comparative experiments of the three methods are listed in Table 5.

Table 5. Experimental comparison results.

No.	Our approach	GSPSO-ANN	GS-FFNN
1	86.49	66.87	47.51
2	86.68	70.40	53.57
3	87.72	71.36	53.57
4	87.90	71.89	54.71
5	88.76	73.66	69.72
6	89.19	73.99	69.72
Average	87.79	71.36	58.13

It can be known from Table 5 that among the three methods, the performance of the GS-FFNN method is the worst, and the performance of the GSPSO-ANN method is generally better than the GS-FFNN method. The decision-making method in this paper has the best performance under any parameters listed in the experiment. In 6 sets of comparative experiments, the highest accuracy of the method in this paper is 89.19%, and the average accuracy is 87.79%. 16.43% higher than the GSPSO-ANN method; 29.66% higher than the GS-FFNN method. Moreover, the precision, recall and F1 values of this paper are high.

The experimental results show that, of the three methods, the method proposed in this paper can more accurately implement the task of wall-following navigation of the mobile robot.

5 Conclusion

This paper studies the wall-following navigation of mobile robots, and proposes a navigation method based on random forest and genetic algorithm. The method proposed in this paper combines the powerful learning, generalization ability, and strong interpretability of control rules to build a rule-based controller. The rules are optimized by genetic algorithms to further reduce the rules and ensure the real-time control. The experimental results prove the effectiveness and feasibility of the method in this paper. In the future work, fuzzy rules or confidence rules will be used instead of existing control rules to enhance the ability to deal with uncertainty to further improve the robustness of the control system. And in order to better verify the good generalization performance and engineering application ability of the proposed method, we will consider more robot actions, such as Sharp-Left-Turn action. In addition, the method of this article will be implemented on a real mobile robot platform to verify and improve the practicability of the method in this paper.

References

1. Yershova, A., Tovar, B., Ghrist, R., et al.: Mapping and Pursuit-Evasion strategies for a simple wall-following robot. IEEE Trans. Rob. **27**(1), 113–128 (2011)
2. Juang, C.F., Chen, Y.H., Jhan, Y.H.: Wall-following control of a hexapod robot using a data-driven fuzzy controller learned through differential evolution. IEEE Trans. Industr. Electron. **62**(1), 611–619 (2014)
3. Hsu, C.H., Juang, C.F.: Evolutionary robot wall-following control using type-2 fuzzy controller with species-DE-activated continuous ACO. IEEE Trans. Fuzzy Syst. **21**(1), 100–112 (2012)
4. Juang, C.F., Hsu, C.H.: Reinforcement ant optimized fuzzy controller for mobile-robot wall-following control. IEEE Trans. Industr. Electron. **56**(10), 3931–3940 (2009)
5. Dash, T., Sahu, S.R., Nayak, T., et al.: Neural network approach to control wall-following robot navigation. In: IEEE International Conference on Advanced Communications, Control and Computing Technologies, pp. 1072–1076. IEEE, Piscataway (2014)
6. Dash, T., Swain, R.R., Nayak, T.: Automatic navigation of wall-following mobile robot using a hybrid metaheuristic assisted neural network. Data Sci. 1–17 (2017)

7. Chen, Y.Y.: Rules extraction for fuzzy control systems. In: Conference Proceedings, IEEE International Conference on Systems, Man and Cybernetics, pp. 526–527. IEEE (1989)
8. Chopra, S., Mitra, R., Kumar, V.: Fuzzy controller: choosing an appropriate and smallest rule set. Int. J. Comput. Cogn. **3**(4), 73–78 (2005)
9. Benitez, J.M., Castro, J.L.: Are artificial neural networks black boxes? IEEE Trans. Neural Netw. **8**(5), 1156–1164 (1997)
10. Dash, T., Nayak, T., Swain, R.R.: Controlling wall-following robot navigation based on gravitational search and feed forward neural network. In: Proceedings of the 2nd International Conference on Perception and Machine Intelligence, pp. 196–200. ACM, New York (2015)
11. Singh, M.K., Parhi, D.R.: Intelligent neuro controller for navigation of mobile robot. In: Proceedings of the International Conference on Advances in Computing, Communication and Control, pp. 123–128. ACM, New York (2009)
12. Craven, M., Shavlik, J.W.: Extracting tree structured representations of trained networks. In: Advances in Neural Information Processing Systems, pp. 24–30. MIT Press, Colorado (1996)
13. Eftekhary, M., Gholami, P., Safari, S., et al.: Rankin normalization methods for improving the accuracy of SVM algorithm by DEA method. Mod. Appl. Sci. **6**(10), 26–36 (2012)
14. Martens, D., Baesens, B., Gestel, T.V., et al.: Comprehensible credit scoring models using rule extraction from support vector machines. Eur. J. Oper. Res. **183**(3), 1466–1476 (2007)
15. Barakat, N.H., Bradley, A.P.: Rule extraction from support vector machines: a sequential covering approach. IEEE Trans. Knowl. Data Eng. **19**(6), 729–741 (2007)
16. Beben L, Sniezynski B, Turek W, et al.: Architecture of an Erlang-based learning system for mobile robot control. In: Proceedings of the 5th International Workshop on Evolutionary and Reinforcemen Learning for Autonomous Robot Systems, pp. 45–48 (2012)
17. Breiman, L.: Random forests. Mach. Learn. **45**(1), 5–32 (2001)
18. Chen, Z.H., You, Z.H., Guo, Z.H., et al.: Prediction of drug–target interactions from multi-molecular network based on deep walk embedding model. Front. Bioeng. Biotechnol. **8**, 338 (2020)
19. Han, L., Luo, S., Yu, J., et al.: Rule extraction from support vector machines using ensemble learning approach: an application for diagnosis of diabetes. IEEE J. Biomed. Health Inform. **19**(2), 728–734 (2014)
20. Wei, H., Tang, X.S.: A genetic-algorithm-based explicit description of object contour and its ability to facilitate recognition. IEEE Trans. Cybern. **45**(11), 2558–2571 (2014)
21. Talbi, E.G., Muntean, T.: Hill-climbing, simulated annealing and genetic algorithms: a comparative study and application to the mapping problem. In: Proceedings of the Twenty-Sixth Hawaii International Conference on System Sciences, pp. 565–573. IEEE, Piscataway (1993)
22. Mashayekhi M., Gras, R.: Rule extraction from random forest: the RF+HC methods. In: Barbosa, D., Milios, E. (eds.) Canadian AI 2015. LNCS, vol 9091, pp. 223–237. Springer, Cham (2015). https://doi.org/10.1007/978-3-319-18356-5_20
23. Jin, Y., Sendhoff, B.: Extracting interpretable fuzzy rules from RBF networks. Neural Process. Lett. **17**(2), 149–164 (2003)
24. Freire A L, Barreto G A, Veloso M, et al.: Short term memory mechanisms in neural network learning of robot navigation tasks: a case study. In: Latin American Robotics Symposium, pp. 1–6. IEEE, Piscataway (2009)

A Study of Algorithms for Controlling the Precision of Bandwidth in EMI Pre-testing

Shenglan Wu, Wenjing Hu, and Fang Zhang[(✉)]

Bohai University, 19, Keji Road, New Songshan District, Jinzhou City 121013, Liaoning Province, People's Republic of China
sensen@kw.ac.kr

Abstract. Introducing software control and automation in Electromagnetic Interference (EMI) testing can bring numerous benefits such as less testing cycle time and lower possibility of error caused by human, etc. With the development of the industrial automation, automated EMI testing has become a popular demand in the EMI field. However, compared with a traditional manual approach which can easily change the Resolution Bandwidth (RBW) of the spectrometer by a human operator when the testing is faced with a different situation, a software-driven, automated approach is relative lack of an intelligent adaptive method for frequency division. In this paper, we propose an algorithm that can easily fill the requirements mentioned above by parameterized some key value of spectrometers and related standards. After that, a comparative test was realized to verify the algorithm proposed could easily meet the actual requirement of frequency division in EMI automated test.

Keywords: Automated testing · Test standards · EMI testing

1 Introduction

With the use of various electronic equipment more and more intensively, the problem of electromagnetic interference is becoming more and more prominent. The deterioration of space electromagnetic environment is also getting more serious. In order to reduce the impact of performance of equipment and system in EMI, EMI tests should be carried out before electronic equipment is put into use [1–3].

According to many existing EMC (Electromagnetic Compatibility) test standards, each standard has its own specific requirement in bandwidth measurement. When a spectrometer is used in EMI pre-testing, it is difficult to meet the standard requirements due to its own characteristics or measuring precision. At the same time, EMI-receiver is expensive and the measurement time is too long. Therefore, a method is needed to make spectrometer work as an approximate substitute for EMI-receiver in EMI pre-testing. This paper proposes an algorithm for frequency division calculation. Using this algorithm, a spectrometer can make a reasonable frequency division of the whole working area and the measuring precision of spectrometer in each area can meet the requirements of the corresponding standards.

The EMI pre-testing system proposed in literature [4, 5] did not give too much description of the test accuracy. Literature [6] proposed a spectrum calculation method

© Springer Nature Switzerland AG 2021
D.-S. Huang et al. (Eds.): ICIC 2021, LNCS 12837, pp. 132–141, 2021.
https://doi.org/10.1007/978-3-030-84529-2_12

to optimize EMI receivers to reduce measuring time. Literature [7] points out that the traditional EMI test receiver is expensive and the measuring time is too long, so a kind of time-domain EMI measurement based on low-cost digital converter is involved. An EMI receiver simulation model is proposed in the literature [8], and the simulation model is verified through different tests to ensure that it has the functions of an EMI-receiver.

In order to save time and cost in EMI pre-testing, it is necessary to use spectrometer to conduct EMI test on the products. However, the mentioned litterateurs all use EMI-receivers as test instruments or do not have too many descriptions on the measuring precision. Based on the EMI automated testing [9–12], this paper proposes an algorithm that enables spectrometer to complete the automated pre-testing of EMI by parameterizing some key values.

2 Bandwidth Measurement Precision

The bandwidth measurement precision [13] of a spectrometer refers to the number of points that need to be captured for the spectrometer to scan the corresponding spectrum to form an image. It is related to the parameters of the spectrometer itself. The relevant tests require bandwidth limits in accordance with the relevant standards. This has led to some spectrometers cannot be suitable for the required precision [14]. This paper presents an example of a method that enables the measurement precision to meet the corresponding standard, using GJB151b-2013 as an example. One of the requirements for bandwidth in GJB151b-2013 is shown in Table 1.

Table 1. Bandwidth and measurement time in GJB151B

Frequency range	6 dB bandwidth (kHz)	dwell time (s)	Minimum measurement time
25 Hz–1 kHz	0.01	0.15	0.015 s/Hz
1 kHz–10 kHz	0.1	0.02	0.2 s/kHz
10 kHz–150 kHz	1	0.02	0.02 s/kHz
150 kHz–30 MHz	10	0.02	2 s/MHz
30 MHz–1 GHz	100	0.02	0.2 s/MHz
>1 GHz	1000	0.02	20 s/GHz

Based on the findings in [15], many characters of spectrometer, such as 3 dB IF bandwidth, quasi-peak, average detectors and pre-selectors, cannot fully meet the requirements of EMI-receiver and can be only used for factory pre-testing. The comparison between EMI-receiver and spectrometer is based on the expression in [15] as shown in Table 2. When conducting experiments for the general EMI receiver IF filter bandwidth of 6 dB bandwidth. Most of the instruments used in the general factory in pre-testing are spectrometers. For a general pre-testing in EMI, the error should be as small as possible with the normative test, so as to have a certain reference significance. Therefore, it needs a certain degree of conversion for spectrometer to simulate the work of EMI-receiver.

Table 2. Receiver versus Spectrometers

Comparative content Testing Instruments	Receiver	Spectrometers
Pre-selector	Generally immune to bandwidth signals and pre-selection	No pre-selector, low-pass filter used instead
Frequency sweep mode	Stepped, discrete point frequency scanning	This is achieved by a ramp or sawtooth signal controlling a swept signal source, where the change in frequency is continuous
IF filter bandwidth	6 dB bandwidth for amplitude and frequency characteristics	3 dB bandwidth for amplitude and frequency characteristics, also known as resolution bandwidth
Wave detection method	Average detector, peak detector, quasi-peak detector and effective time-dependent detector	Usually with peak and average detector
Scanning speed	Longer scanning times	Faster scanning times

3 Principle of the Algorithm

3.1 Analysis of the Problem

For a given starting and stopping frequency, the number of sampling points of the spectrometer is

$$\left(\frac{f_{stop} - f_{start}}{k} + 1\right) \times 2 \tag{1}$$

In the above equation, f_{stop} is the cut-off frequency of the input, f_{start} is the starting frequency of the input. The k in above formula means spacing between adjacent sampling points of a spectrometer. It can be interpreted to some extent as the standard bandwidth value. So taking the standard GJB151b-2013 as an example, the bandwidth here is the 6 dB bandwidth value shown in Table 1. When conducting the most basic conducted and radiated emission experiments, such as the CE102 test, the test frequency is 10 kHz–10 MHz, then according to Eq. (1) and Table 1, the basic number of points to be sampled by the spectrometer is 2254. Similarly, in the RE102 test, the test frequency is 10 kHz–18 GHz, at this time, according to formula (1) and GJB151b-2013, the basic number of points to be sampled by the spectrometer is 59656.

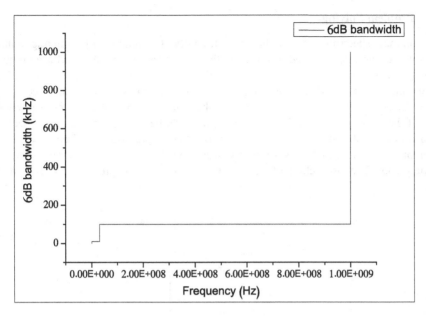

Fig. 1. 6 dB bandwidth sweep folding diagram in GJB151b-2013

A graph of the 6 dB bandwidth versus scan frequency under the GJB151b-2013 standard is shown as Fig. 1.The graph shows that it is a discontinuous step line segment. The bandwidth requirement is low at low frequencies and increases as the frequency increases. In the low frequency range, the precision requirements of the spectrometer are relatively simple to handle, while as the frequency increases, there is naturally a higher demand for its scanning capability, i.e. the scanning precision in a certain frequency band may be changed as required [16]. If the problem is known to be in an intrinsic band, the scanning precision can be relaxed for other bands that have been determined to be free of problems. The problematic bands can then be identified with greater precision [17]. This is where an algorithm needs to be designed that allows for easier and more efficient segmentation while complying with the standard, allowing the spectrometers to be scanned with maximum flexibility.

Therefore, according to the relevant standards, the majority of commercially available spectrometers are unable to meet the corresponding requirements for both the number of points acquired and the flexibility of the bandwidth when automated testing [18]. At the same time, when pre-processing, if you want to improve or reduce the precision of the test, you need to re-write the procedure according to the standard, which is more complicated. So we propose an algorithm which is able to perform a segmentation of the overall measurement range. At each segment, the spectrometer can sample with its own character and the precision of measurement can meet the requirements of relevant standards.

3.2   Problem Solving

For common spectrometers on the market, firstly, the number of sampling points generated on an interval is adjustable, i.e. there is a maximum value and a minimum value (the maximum and the minimum values may be equal to a constant value). This set of values can be used as a set of input parameters. At the same time, also according to the corresponding standard requirements, there are certain values for the bandwidth limit of the instrument, then this bandwidth limit can be used as an accuracy variable, i.e. a set of variable parameters, according to different precision requirements can be different precision segmentation of the spectrum, to achieve the corresponding standard requirements. The flow chart of the spectrum segmentation procedure is shown as Fig. 2.

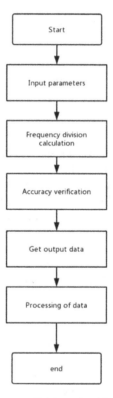

Fig. 2. Frequency division algorithms Flowchart

4   Program Implementation

4.1   Programming Principles

LabVIEW from National Instrument (NI) was chosen as the upper computer implementation software for the test verification. It is an ideal platform for the development of virtual instruments due to its excellent visualization interface and the many modules that have been packaged to make development very easy, as well as its good environmental support, excellent performance and scalability, and short development time.

The Spectrometer is directly connected to the PC. Due to the good packaging and comfortable environment of LabVIEW, the parameters of the spectrometer such as start/stop frequency, Resolution Bandwidth (RBW), Video Bandwidth (VBW) and reference level that need to be set up in order to design an automatic test software with it are all encapsulated in its existing packaging library. test conclusion data back to the test software, thus enabling the interactive processing of information [19].

After obtaining the input signal, i.e., the start and stop frequencies which we input, the frequency division is then calculated. The procedure of the calculation in the same segment is shown as pseudocode in Table 3.

Table 3. Pseudocode of frequency division.

Algorithm of frequency division calculation

Input StartFre,StopFre
Begin
If (StartFre >= StopFre)
Then return false
End If
If (StartFre < StopFre)
   Points ← ((StopFre - StartFre)/k)+1
If (Points < Min)
   Segments ← 1
   Segpoints ← Min
Else If (Points >= Min && points <= Max)
   Segments ← 1
   Segpoints ← Points
Else If (Points > Max)
   Segments ← Points/Max + 1
   SegPoints ←Points/Segments + 1
   /* k,Max,Min are variables that have already been parameterized*/
End If
End If
Output Segments,Segpoints
End

The implementation process is as follows,

1. The data to be parameterized is first entered, where each band includes the 6 dB bandwidth value k, the maximum and minimum acquisition frequency points of the spectrometer and the boundary frequency points of the relevant standard.
2. Enter the start and stop frequencies, determine if there is an error in the start and stop, and determine if the start and stop frequencies are in the same frequency band as the standard.
3. If within the same frequency band as the standard, the number of points to be acquired is calculated from the starting and ending frequencies and the k-value defined for each band and compared with the maximum and minimum number of points to be acquired by the machine; if not within the same frequency band as the standard, a comparison operation is performed for each band.
4. Calculate the total number of segments to be obtained and the number of points per segment based on the results of the comparison.
5. Output the number of points per segment and the number of segments as output parameters.

4.2   Specific Program Design

The frequency range of radiation harassment test for civil products in field environment is 30–1000 MHz. According to GJB151b-2013, the common test frequency range is 10 kHz–10 MHz for both CE102 and RE102 tests, so the start and stop frequencies are set to 1 MHz–10 MHz for the convenience of the test. The spectrometer used for the test is the domestic Rigol DSA815. It is small, lightweight, cost-effective and portable, and its full digital IF implementation ensures excellent performance and stable performance.

In the specific pre-testing experiment, the background noise is measured according to the relevant standards. A narrow-band square wave with a bandwidth of 60 Hz is now added as an interference signal at a frequency of 9 MHz. The background noise environment is measured according to the GJB151b-2013 standard as shown in Fig. 3.

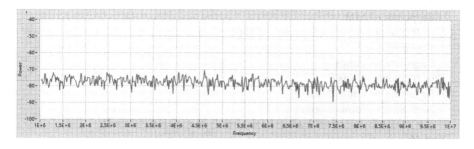

Fig. 3. Background noise scan after adding interference

At this point, the spectrometer does not capture the bottom noise image with the interference signal, so we can adjust the 6 dB bandwidth value (i.e. the parameterized k value) of each band according to the algorithm to make the spectrometer scan more accurate on top of the standard original. The results are shown in Fig. 4.

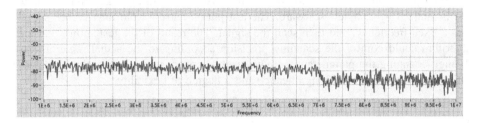

Fig. 4. Bottom noise scan after precision change

Since we already know that we have set the interference signal at 9 MHz, we can adjust the k value again for the band at 9 MHz to make the precision of the spectrometer more accurate for the target frequency while meeting the requirements. The results are shown in Fig. 5.

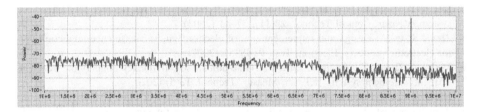

Fig. 5. Background noise scan after changing precision again

It is clear from the above figure that the initial standard was unable to detect the corresponding small interfering signals. And with a parametric change to the relevant parameters, a pre-set interference signal can be scanned. Experiments have shown that the algorithm is flexible in EMI pre-testing and can be adapted to the needs of automated EMI pre-testing in actual situation.

5 Conclusion

With the development of computer science, automated testing has become the mainstream way of industrial testing [20]. In EMI automated pre-testing, it is often necessary to change the frequency division rules to meet the requirements of different standards, so it can be cumbersome to develop automated test procedures based on these standards.

In this paper, an algorithm of frequency division in EMI pre-testing is proposed. Firstly, the algorithm proposed make the 6 dB bandwidth of the existing standard as a parameter, and then the limited value of the instrument about sampling points are similarly parameterized. After that, the frequency division is calculated based on the input parameters to get the corresponding number of points for each segment. And the numbers calculated are compared with the limited value of the instrument about sampling points which has been parameterized. According to the algorithm, a simple test is realized by Labview software. The test put an narrow-band interference signal into a fixed-frequency, and then scan the signal. The result of the test shows that the algorithm could control the scanning precision according to the actual requirements through changing the parameters we have parameterized, which shows that the algorithm has a certain practical significance and a degree of flexibility.

References

1. Liu, S., Wang, W., Hu, X., Wang, Y.: Study of resident narrowband signal rejection algorithm in EMI field test. J. Electron. Measur. Instrum. **29**(09), 1271–1277 (2015)
2. Zhang, L.Y., et al.: Adaptive noise spectrum estimation offset technique and its application. J. Harbin Inst. Technol. **047**(009), 31–35 (2015)
3. Ronald, P., Carrol, B., Praveen, A.: Electromagnetic environmental effects modeling, simulation & test validation for cosite mitigation an overview. IEEE Long Island Syst. Appl. Technol. Conf. **2008**, 1–6 (2008)
4. Liu, Y.Y., Guo, E.Q., et al.: Development of electromagnetic interference prediction test system. Measur. Control Technol. **22**(11), 37–41 (2003)
5. Zhu, J.S., et al.: Design and implementation of electromagnetic interference prediction test for electronic products. Electron. Test. **9**, 48–51 (2007)
6. Meng, J., Zhang, X., Zhang, L., et al.: A Time-domain low-frequency nonperiodic transient EMI measurement system. IET Sci. Measur. Technol. **13**(5) (2020)
7. Hartman, T., Grootjans, R., Moonen, N., Leferink, F.: Time-domain EMI measurements using a low cost digitizer to optimize the total measurement time for a test receiver. In: 2020 International Symposium on Electromagnetic Compatibility - EMC EUROPE, 2020, pp. 1–6 (2020)
8. Sun, M., Hu, J., He, J., Wu, C., Zhao, H., Kim, H.: EMI receiver model to evaluate conducted emissions from time-domain waveforms. In: 2020 IEEE International Conference on Computational Electromagnetics (ICCEM), pp. 166–168 (2020)
9. Mourougayane, K., Karunakaran, S., Satheesan, P., et al.: Automation of EMC tests–Software development and system integration approach. In: IEEE 2006 9th International Conference on Electromagnetic Interference and Compatibility (INCEMIC 2006), pp. 349–352 (2006)
10. Singh, D.K., Nagesha, S.P., Pande, D.C., Karthikeyan, S.: Design of an automated EMC test facility. Electromagnetic Interference and Compatibility, pp. 327–332 (2003)
11. Braun, S., Donauer, T., Russer, P.: A real-time time-domain EMI measurement system for full-compliance measurements according to CISPR 16–1–1. IEEE Trans. Electromagn. Compat. **50**(2), 259–267 (2008)
12. Peng, G.S.: EMC analysis of automation instruments. Telecom Power Technology (2019)
13. Qin, S.Y.: Research on the application of spectrometer to measure low-level signals. J. Electron. Measur. Instrum. Suppl. 51–55 (2009)
14. Chen, F.: The difference between receiver and spectrum analyzer - the choice of EMC measurement equipment. Safety EMC **02**, 23–24+32 (2001)

15. Zhao, W.H., Meng, X.R., Ma, H.Y., et al.: An adaptive algorithm for high-precision available bandwidth measurement. Comput. Measur. Control. **19**(006), 1297–1300 (2011)
16. Xing, K., Xue, M.L., Wang, Y.W.: Application of rapid EMC testing techniques for wireless receivers in complex electromagnetic environments. Aerosp. Measur. Technol. **035**(005), 21–26 (2015)
17. Singh, D.K., Nagesha, S.P., Pande, D.C., Karthikeyan, S.: Design of an automated EMC test facility. In: 8th International Conference on Electromagnetic Interference and Compatibility, INCEMIC 2003. 18–19 December 2003, pp.327–332 (2003)
18. Drinovsky, J., Svacina, J.: Automation measurement setups for EMC laboratory Course. In: 17th International Conference Radioelektronika, pp. 1–3 (2007)
19. Peter, J.S, Louis, L.S.: Higher-order spectral analysis of complex signals. Signal Process. **86** (11) (2006)
20. Craig, F.D.: Methods of performing computer controlled EMI compliance tests. International Symposium on Electromagnetic Compatibility **1990**, 305–309 (1990)

Intelligent Control and Automation

Flight Control for 6-DOF Quadrotor via Sliding Mode Integral Filter

Zinan Su, Aihua Zhang$^{(\boxtimes)}$, and Shaoshao Wang

College of Control Science and Engineering, Bohai University,
Jinzhou 121013, Liaoning, China

Abstract. This article mainly focuses on the ability of the quadrotor UAV to follow the desired trajectory when it is subjected to various external disturbances. In order to solve this problem, a closed-loop six-degree-of-freedom full control system will be proposed in the article. The sliding mode integral technology is introduced when designing the filter, which effectively solves the "computational explosion" problem in the traditional backstepping method by avoiding the analytical derivation process of the virtual control input. At the same time, an backstepping controller based on the above command filter is designed. After theoretical analysis, the stability of the closed-loop system is guaranteed, and the effectiveness of this method is further verified through simulation experiments.

Keywords: Trajectory tracking control · Sliding mode integral technology · Command filter backstepping · Quadrotor UAV

1 Introduction

In the past few decades, the use of unmanned aircraft systems (UAS) has continued to increase. The quadrotor unmanned aerial vehicle is an unmanned aerial vehicle with four propellers capable of vertical elevation [1]. The quadrotor is widely used in commercial and military applications due to its strong maneuverability, easy operation, better concealment, and strong stability. In recent years, quadrotor unmanned aerial vehicles have been widely used in aerial photography, power inspections, express transportation, agricultural plant protection, military investigations, geographic surveys, search and rescue and other fields [2–4]. As a result, the number of control algorithms designed for quadrotors is also increasing to meet the design requirements of UAVs under different missions [5].

All quadrotor UAV include translation and rotation, so it has six degrees of freedom (DOF) in mathematical model [6]. In order to better drive the quad-rotor UAV, many new methods are proposed to control or observe the position and attitude of the UAV [7]. The most classic is the linear control method. Among them, the Proportional-Integer-Derivative (PID)controller [8–11] has been widely used due to simple and easy control structure. In addition, Linear Quadratic Regulator (LQR) [12, 13] has been applied to UAVs to achieve control missions.

By comparing the above classic linear control methods, modern control methods have been widely developed in recent decades. More and more modern control

D.-S. Huang et al. (Eds.): ICIC 2021, LNCS 12837, pp. 145–160, 2021.
https://doi.org/10.1007/978-3-030-84529-2_13

methods are applied to quadrotor to complete various high-precision tasks based on complex nonlinear systems. Such as, sliding mode technology [14, 15], the adaptive control technique [16–18], the model predictive control technique [19, 20], the output feedback control [21], fuzzy logical technique [22], the neural networks [23], The robust performance of control [24]. In [14], a finite-time sliding-mode observer is designed to estimate the full state from the measurable output and identifies some types of disturbances. In [15], geometric attitude and position control laws are designed separately using a multi-input fast nonsingular terminal sliding mode control. In [16], introducing a new parameter uncertainty adaptive trajectory tracking control design algorithm. In [17], a reconfiguration control scheme is proposed for a quadrotor helicopter with loss of control effectiveness via simple adaptive control and quantum logic. In [18], an adaptive multivariable finite-time stabilizing control algorithm for second-order multivariable systems is developed. In [19], focuses on trajectory tracking of quadcopter using a Model Predictive control. In [20], the Model Predictive Controller used to achieve efficient and precise real-time feedback control. In [21], It develops a multivariable output feedback model reference adaptive control structure from a linearized multi-input multi-output quadrotor UAV model. In [22], focusing on the fuzzy neural networks (T2FNNs) for the trajectory tracking problem of quadrotor. In [23], proposed methods employ various forms of neural networks (NNs) to generate proper initial state values for RNNs. In [24] achieve the H tracking performance with a prescribed disturbance attenuation level. Although based on the various existing control methods, backstepping is still one of the most widely used and effective methods. But the backstepping method has a computational explosion problem should be solved. Motivated by the filter, this paper proposes a backstepping controller based on a sliding mode integral filter. Further, feedback and compensate the closed-loop controller to improve the trajectory tracking accuracy under the action of external disturbance, enhance the robustness of the system and shorten the response speed.

The main works are listed as follows:

1) The dual closed-loop control of 6-DOF of the quadrotor UAV is realized, in which the attitude loop and the position loop are respectively designed with a backstepping controller based on command filter (FBC). Due to the high coupling relationship between attitude and position, high-precision trajectory tracking can be achieved.
2) A sliding mode integral filter is designed, which effectively solves the "computational explosion" problem in the classical backstepping method by avoiding the analytical derivation process of the virtual control input.
3) Through a series of simulation experiments on the tracking effect, tracking error, and tracking accuracy under strong interference, it is further verified that FBC has superior anti-interference ability and control performance compared with CBC.

The remainder of this article is organized as follows: Sect. 2 describes the dynamic model of 6-DOF and the formulation of the UAV control square problem. Section 3 is presented the result of designing a backstepping controller based on a sliding mode integral filter and the stability analysis of the closed-loop system. Section 4 introduces experimental results. Section 5 concludes and provides direction for future work.

2 Mathematical Model and Problem Description

2.1 Mathematical Model of a Quadrotor

The quadrotor is an underdrive system with six degrees of freedom. In order to better show the mechanical structure of the quadrotor, Fig. 1 shows a 3D basic model of a quadrotor. The $E_e\{O_E X_E Y_E Z_E\}$ represents the earth fixed frame, the $B_b\{O_B X_B Y_B Z_B\}$ represents the body fixed frame, and O_B represents the mass center of the quadrotor.

In order to explain the method more simply and rigorously, the following assumptions are given [6, 25].

Assumption 1. The structure of the quadrotor UAV is rigid and strictly symmetric.

Assumption 2. The surface of the earth is a strict plane, which can be further defined as an inertial frame.

Assumption 3. The gravitational acceleration is constant and perpendicular to the surface of the earth.

Assumption 4. The disturbances due to the instability of the propeller blades can be ignored.

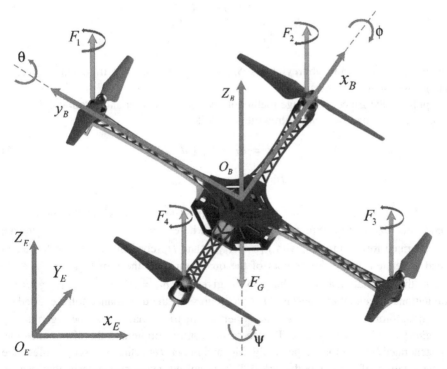

Fig. 1. The frame of quadrotor and its 3D Structure.

The three position information of the quadrotor is expressed as $\chi = [x, y, z]^T$, and the liner velocity is denoted as $v = [v_x, v_y, v_z]^T$ in the body fixed frame. Let $\delta = [\phi, \theta, \psi]^T$ be the three Euler angle, where $\phi \in \Re$ is the roll angle rotation around X_B - axis, $\theta \in \Re$ is the pitch angle rotation around Y_B - axis, $\psi \in \Re$ is the yaw angle rotation around Z_B - axis. The angular velocity as $\varpi = [\omega_\phi, \omega_\theta, \omega_\psi]^T$. The conversion relationship from E_e to B_b can be expressed by the following (1)–(2):

$$\dot{\chi} = Tv + \Delta f \tag{1}$$

$$\dot{\delta} = R\varpi + \Delta d \tag{2}$$

where $\Delta f, \Delta d$ is the uncertainty disturbance generated in system modeling and coupling, T represents the translation matrix of the quadrotor is given by

$$T = \begin{bmatrix} C_\psi C_\theta & C_\psi S_\theta S_\phi - S_\psi C_\phi & C_\psi S_\theta C_\phi + S_\psi S_\phi \\ S_\psi C_\theta & S_\psi S_\theta C_\phi + C_\psi C_\phi & S_\psi S_\theta C_\phi - C_\psi S_\phi \\ -S_\theta & C_\theta S_\phi & C_\theta C_\phi \end{bmatrix} \tag{3}$$

Further the rotation matrix of the quadrotor R is given by:

$$R = \begin{bmatrix} 1 & S_\phi T_\theta & C_\phi T_\theta \\ 0 & C_\phi & -S_\phi \\ 0 & S_\phi/C_\theta & S_\phi/C_\theta \end{bmatrix} \tag{4}$$

where $T, R \in \Re^{3\times3}$, the abbreviations $S_{(\cdot)}, C_{(\cdot)}, T_{(\cdot)}$ denote the triangular functions $\sin(\cdot), \cos(\cdot), \tan(\cdot)$, respectively.

Applying the Euler- Lagrange methodology, the translation and rotational dynamic equations of a quadrotor can be rewritten as follows:

$$m\dot{v} = F_U + F_G + d_\chi \tag{5}$$

$$I\dot{\varpi} + \varpi^\times I\varpi = \tau_f + \tau_d \tag{6}$$

where $m \in \Re$ indicates the mass of the entire quadrotor. $F_U \in \Re^3$ denote the translational force that can be expressed as $F_U = T\rho$, where $\rho = [0 \quad 0 \quad \sum_{i=1}^4 F_i]^T$ indicate the generated force in the B_b with $F_i = JH_i^2$, F_i and H_i represents the thrust force and speed generated by the four rotors of the quadrotor, J is the thrust factor. $F_G \in \Re^3$ denote the gravitational force that can be given by $F_G = [0 \quad 0 \quad -mg]^T$, g as the gravitational acceleration constant. $d_\chi \in \Re^3$ represents the disturbance force exerted on the quadrotor. $I \in \Re^{3\times3}$ is the inertia matrix that can be given by $I = diag\{ I_x \quad I_y \quad I_z \}$. $\tau_f, \tau_d \in \Re^3$ denote the control torque and the disturbance torque generated by rotation, respectively. Meanwhile, $\varpi^\times I\varpi$ is the Gyroscopic effect due to the rotation of a rigid body. $\varpi^\times \in \Re^{3\times3}$ is an anti-symmetric matrix that can be defined as follow:

$$\varpi^{\times} = \begin{bmatrix} 0 & -\varpi_3 & \varpi_2 \\ \varpi_3 & 0 & -\varpi_1 \\ -\varpi_2 & \varpi_1 & 0 \end{bmatrix} \tag{7}$$

In addition, the torque produced by the propellers of quadrotor can be expressed as

$$\tau_f = \begin{bmatrix} b_1 \left(H_3^2 - H_1^2 \right) \\ b_1 \left(H_4^2 - H_2^2 \right) \\ b_2 \left(H_3^2 + H_1^2 - H_4^2 - H_2^2 \right) \end{bmatrix} \tag{8}$$

where b_1, b_2 respectively represent the distance from each motor to the center of mass of the quadrotor and the disturbance torque.

2.2 Problem Statement

This article aims to enable the quadrotor UAV to accurately tracking the desired trajectory signal under the condition of interference. Therefore, the control problem can be described as any quadrotor with an initial yaw angle, there is a full 6-dof tracking control method for attitude and position is given to ensure that the quadrotor can still be controlled even under external disturbance. Follow the desired trajectory gradually. Further, the expected value of the position and attitude of the UAV can be given separately as $\chi_d = [x_d, y_d, z_d]^T \in \Re^3$ and $\delta_d = [\phi_d, \theta_d, \psi_d]^T \in \Re^3$, respectively. In order to further design the controller, the following formulas respectively define the position and attitude tracking errors: $\chi_e = \chi - \chi_d = [x_e, y_e, z_e]^T$ and $\delta_e = \delta - \delta_d = [\phi_e, \theta_e, \psi_e]^T$. In addition, linear velocity error and angular velocity error can be defined as $v_e = v - u_\chi, \varpi_e = \varpi - u_\delta$, respectively, where u_χ, u_δ represents the stable virtual control law of the position control and the attitude control system respectively.

Assumption 5. The given attitude desired signal and position desired signal satisfy second-order continuous derivability and bounded [6].

Notations: Throughout this paper, the following notations will be used. Given a vector $x = [x_1 \quad x_2 \quad \dots \quad x_n]^T \in \Re^n$ and a real number $o \in \Re$, define the multi-variable sign function $sign(x) = [sign(x_1) \quad sign(x_2) \quad \dots \quad sign(x_n)]^T$ and $\lfloor x \rfloor^o = \|x\|^P sign(x)$.

Lemma 1. The external disturbances in the quadrotor UAV system d_χ and τ_d is continuous and satisfying $\|d_\chi\| \le d_{n1}, \|\tau_d\| \le d_{n2}$ with d_{n1} and d_{n2} are unknown positive constants. Further, $\dot{d}_\chi(t)$ and $\dot{\tau}_d(t)$ be continuous functions.

Lemma 2. According to [26], the closed loop expressions for pitch and roll attitudes can be given by:

$$\begin{cases} \phi_d = \arcsin\left(\frac{U_x \sin\psi_d - U_y \cos\psi_d}{\|U_\chi\|} \right) \\ \theta_d = \arcsin\left(\frac{U_x \cos\psi_d + U_y \sin\psi_d}{U_z} \right) \end{cases} \tag{9}$$

where $U_\chi = [U_x \quad U_y \quad U_z]^T = U_\chi - \frac{F_G}{m}$.

3   Control Law Design and Stability Analysis

In order to achieve the goal of high-precision tracking of the quadrotor UAV proposed in the previous section under external disturbance, this chapter proposes a backstepping controller based on the design of sliding mode integral filter. As shown in the flowchart Fig. 2, the backstepping theory is used to design the position and attitude controllers with two sliding mode integral filter, respectively. Finally, this closed-loop system can accurately reconstruct the disturbance, so as to achieve accurate tracking of the desired trajectory in the presence of external interference, and to ensure that the closed-loop system is asymptotically stable.

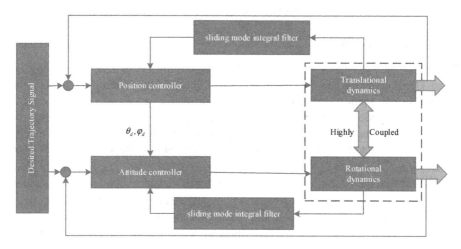

Fig. 2. Control flow chart of quadrotor

3.1   Attitude Control Law Design of Quadrotor

In this subsection, the attitude dynamics is given by (2) and (6). In order to designed attitude controller for the quadrotor which the command filter backstepping control method will be using. In the attitude angle subsystem, first design the feedback control $\bar{u}_\delta$, and design the sliding mode integral filter to approximate its derivative term, and further design the stable virtual control law u_δ. At the same time, design the real control law $\tau_d = u_\varpi$ in the attitude angular velocity subsystem.

Step1: In the attitude angle system, the tracking error can be converted into the following

$$\dot{\delta}_e = \dot{\delta} - \dot{\delta}_d = R\varpi + \Delta d - \dot{\delta}_d \tag{10}$$

Further design the feedback control law given below

$$\bar{u}_\delta = -R^{-1}\left(\gamma_{d1} - \dot{\delta}_d\right) \tag{11}$$

the given $\gamma_{d1} = \left(\delta_e \Lambda_{d1}^2\right)/\left(\delta_e^T \delta_e\right)$ is used to eliminate the influence of uncertain internal disturbances Δd.

Substituting (10) into (9) can get the following

$$\dot{\delta}_e = R(\varpi + u_\delta - \bar{u}_\delta) + \Delta d - \dot{\delta}_d = R(\varpi - \bar{u}_\delta) - \gamma_{d1} + \Delta d \tag{12}$$

Further stable virtual control can be obtained

$$u_\delta = -R^{-1}\lambda_\delta \delta_e + \hat{u}_\delta \tag{13}$$

where $\lambda_\delta \in \Re^{3\times3}$ is a positive definite matrix, design $\hat{u}_\delta$ to approximate feedback control $\bar{u}_\delta$.

Therefore, the attitude angle error can be expressed as follows

$$\dot{\delta}_e = R(\varpi_e + u_\delta - \bar{u}_\delta) - \gamma_{d1} + \Delta d = R\left(\varpi_e - R^{-1}\lambda_\delta \delta_e + \hat{u}_\delta - \bar{u}_\delta\right) - \gamma_{d1} + \Delta d \tag{14}$$

Step2: Further consideration of the attitude angular velocity subsystem, where a satisfies the following

$$\dot{\varpi}_e = \dot{\varpi} - \dot{u}_\delta = I^{-1}\left(\tau_f - \varpi^\times I\varpi + \tau_d\right) - \dot{u}_\delta \tag{15}$$

Further design the feedback control law given below

$$\bar{u}_\varpi = -I\left(-I^{-1}\varpi^\times I\varpi + \gamma_{d2} - \dot{u}_\delta + R\delta_e\right) \tag{16}$$

where $\gamma_{d2} = \left(\varpi_e \Lambda_{d2}^2\right)/\left(\varpi_e^T \varpi_e\right)$ is used to eliminate the influence of external disturbance τ_d.

Therefore, the attitude angular velocity error can be further given

$$\begin{aligned}
\dot{\varpi}_e &= I^{-1}\left(\tau_f - \bar{u}_\varpi + \bar{u}_\varpi - \varpi^\times I\varpi + \tau_d\right) - \dot{u}_\delta \\
&= I^{-1}\left(\tau_f - \bar{u}_\varpi + \tau_d\right) - \gamma_{d2} + R\delta_e
\end{aligned} \tag{17}$$

Define the control torque τ_f as

$$\tau_f = -I\lambda_\varpi \varpi_e + \hat{u}_\varpi \tag{18}$$

where $\lambda_\varpi \in \Re^{3\times3}$ is a positive definite matrix, design $\hat{u}_\varpi$ to approximate feedback control $\bar{u}_\varpi$.

The further attitude angular velocity error can be rewritten as

$$\dot{\varpi}_e = I^{-1}(-I\lambda_\varpi \varpi_e + \hat{u}_\varpi - \overline{u}_\varpi + \tau_d) - \gamma_{d2} + R\delta_e \tag{19}$$

In summary, the attitude angle and its angular velocity error equation can be described as

$$\begin{cases} \dot{\delta}_e = R(\varpi_e - R^{-1}\lambda_\delta \delta_e + \hat{u}_\delta - \overline{u}_\delta) - \gamma_{d1} + \varDelta d \\ \dot{\varpi}_e = I^{-1}(-I\lambda_\varpi \varpi_e + \hat{u}_\varpi - \overline{u}_\varpi + \tau_d) - \gamma_{d2} + R\delta_e \end{cases} \tag{20}$$

Step3: In the attitude angle subsystem, the Lyapunov-based filter is used to approximate the feedback control $\overline{u}_i, i = \delta, \varpi$ and its derivative, avoiding the analytical derivation process. Further define the system error vector $X = \begin{bmatrix} \delta_e^T, \varpi_e^T \end{bmatrix}$ and the filter error vector $\tilde{u} = \hat{u} - \overline{u} = \begin{bmatrix} (\hat{u}_\delta - \overline{u}_\delta)^T, (\hat{u}_\varpi - \overline{u}_\varpi)^T \end{bmatrix} := \begin{bmatrix} \tilde{u}_\delta^T, \tilde{u}_\varpi^T \end{bmatrix}$. At the same time, the filter subsystem based on attitude satisfies the stability of the input state relative to the state tracking error x. Design a sliding mode integral filter:

$$\dot{\hat{u}} = -A(\hat{u} - \overline{u}) - Bsign(\hat{u} - \overline{u}) - CX \tag{21}$$

where $A = diag\{A_\delta, A_\varpi\}, B = diag\{B_\delta, B_\varpi\}, C = diag\{C_\delta, C_\varpi\}, A_i, B_i, C_i \in \Re^{3\times3}, i = \delta, \varpi$ represents a positive definite matrix,

Further, it can be concluded that the filtering error a is dynamically satisfied as follows:

$$\dot{\tilde{u}} = \dot{\hat{u}} - \dot{\overline{u}} = -A\tilde{u} - Bsign(\tilde{u}) - CX - \dot{\overline{u}} \tag{22}$$

Please refer to Sect. 3.3 for the proof of the stability of the attitude system.

3.2   Position Control Law Design of Quadrotor

The position dynamic equation composed of (2) and (6) can be further designed the position controller and its filter with (23) and (24). This design process is similar to attitude control, please refer to the derivation process in Sect. 3.1 and the proof process of Theorem 1. First of all, the feedback control law of position should designed as

$$\overline{w}_\chi = -\frac{1}{m}(\gamma_{d3} - \dot{\chi}_d), \overline{w}_v = -m(\gamma_{d4} - \dot{w}_\chi + T\chi_e) \tag{23}$$

where $\gamma_{d3} = (\chi_e \varLambda_{d3}^2)/(\chi_e^T \chi_e), \gamma_{d4} = (v_e \varLambda_{d4}^2)/(v_e^T v_e)$ is used to eliminate uncertain internal disturbances $\varDelta f$ and the influence of external disturbance d_χ.

Further,

$$F_U = -m\lambda_\chi v_e + \hat{w}_v \tag{24}$$

where $\lambda_\chi \in \Re^{3\times3}$ is a positive definite matrix, $\hat{w}_v$ remain to approximate feedback control $\overline{w}_v$.

Define the position system error vector $Y = \left[\chi_e^T, v_e^T\right]$ and the filter error vector $\tilde{w} = \hat{w} - \overline{w} = \left[(\hat{w}_\chi - \overline{w}_\chi)^T, (\hat{w}_v - \overline{w}_v)^T\right] := \left[\tilde{w}_\chi^T, \tilde{w}_v^T\right]$, the filter subsystem based on position can be expressed as follow:

$$\dot{\hat{w}} = -P(\hat{w} - \overline{w}) - Q sign(\hat{w} - \overline{w}) - JY \tag{25}$$

where $P = diag\{P_\chi, P_v\}, Q = diag\{Q_\chi, Q_v\}, J = diag\{J_\chi, J_v\}.P_i, Q_i, J_i \in \Re^{3\times3}, i = \chi, v$ represents a positive definite matrix.

3.3   Stability Analysis

Theorem 1: The closed-loop system composed of the command backstepping controller (18) and the sliding mode integral filter (20) can ensure that the system output tracking error converges to a small neighborhood near zero for bounded initial conditions and external disturbances.

Proof: Choose a Lyapunov function $V = \frac{1}{2}X^TX + \frac{1}{2}\tilde{u}^T\tilde{u} = \frac{1}{2}\left(\delta_e^T\delta_e + \varpi_e^T\varpi_e\right) + \frac{1}{2}\tilde{u}^T\tilde{u}$ and substituting (11), (18) and (21) into $\dot{V}$ can get the following.

$$\begin{aligned}
\dot{V} &= \delta_e^T\dot{\delta}_e + \varpi_e^T\dot{\varpi}_e + \tilde{u}^T\dot{\tilde{u}} \\
&= \delta_e^T\left[R(\varpi_e - R^{-1}\lambda_\delta\delta_e + \hat{u}_\delta - \overline{u}_\delta) - \gamma_{d1} + \Delta d\right] \\
&+ \varpi_e^T\left[I^{-1}(-I\lambda_\varpi\varpi_e + \hat{u}_\varpi - \overline{u}_\varpi + \tau_d) - \gamma_{d2} + R\delta_e\right] + \tilde{u}^T\left(-A\tilde{u} - Bsign(\tilde{u}) - CX - \dot{\tilde{u}}\right) \\
&= -\delta_e^T\lambda_\delta\delta_e + \delta_e^TR\tilde{u}_\delta + \delta_e^T\Delta d - \Lambda_{d1}^2 - \varpi_e^T\lambda_\varpi\varpi_e + \varpi_e^TI^{-1}\tilde{u}_\varpi + \varpi_e^T\tau_d - \Lambda_{d2}^2 \\
&\quad -\tilde{u}^TA\tilde{u} - \tilde{u}^TBsign(\tilde{u}) - \tilde{u}^TCX - \tilde{u}^T\dot{\tilde{u}} \\
&\leq -\delta_e^T\lambda_\delta\delta_e + \delta_e^TR\tilde{u}_\delta + \|\delta_e\|\Lambda_{d1} - \Lambda_{d1}^2 - \varpi_e^T\lambda_\varpi\varpi_e + \varpi_e^TI^{-1}\tilde{u}_\varpi + \|\varpi_e\|\Lambda_{d2} - \Lambda_{d2}^2 \\
&\quad -K_{\min}(A)\|\tilde{u}\|^2 - K_{\min}(B)\|\tilde{u}\| + K_{\max}(C)\|\tilde{u}\|\|X\| + \|\tilde{u}\|\|\dot{\tilde{u}}\|
\end{aligned} \tag{26}$$

Further if $K_{\min}(B) > \left\|\dot{\tilde{u}}\right\|$, then

$$-K_{\min}(A)\|\tilde{u}\|^2 + K_{\max}(C)\|\tilde{u}\|\|X\| \leq -\frac{1}{2}K_{\min}(A)\|\tilde{u}\|^2 + \frac{1}{2}(K_{\max}(C))^2\|X\|^2/K_{\min}(A) \tag{27}$$

According to the perfect square formula:

$$\frac{1}{4}\delta_e^T\delta_e - \|\delta_e\|\Lambda_{d1} + \Lambda_{d1}^2 \geq 0, \frac{1}{4}\varpi_e^T\varpi_e - \|\varpi_e\|\Lambda_{d2} + \Lambda_{d2}^2 \geq 0 \tag{28}$$

$$-\frac{1}{2}K_{\min}(A)\|\tilde{u}\|^2 + \frac{1}{2}K_{\max}(C)\|\tilde{u}\|\|X\| \leq \frac{1}{2}(K_{\max}(C))^2\|X\|^2/K_{\min}(A) \tag{29}$$

Therefore, it can be concluded that Eq. (25) satisfies the following

$$
\begin{aligned}
\dot{V} \leq & - \delta_e^T \lambda_\delta \delta_e + \delta_e^T R \tilde{u}_\delta + \frac{1}{4}\delta_e^T \delta_e - \varpi_e^T \lambda_\varpi \varpi_e + \varpi_e^T I^{-1}\tilde{u}_\varpi + \frac{1}{4}\varpi_e^T \varpi_e \\
& - \frac{1}{2}K_{\min}(A)\|\tilde{u}\|^2 + \frac{1}{2}(K_{\max}(C))^2\|\mathrm{X}\|^2/K_{\min}(A) \\
\leq & - \delta_e^T \lambda_{\delta1}\delta_e - \delta_e^T \lambda_{\delta2}\delta_e + \delta_e^T R\tilde{u}_\delta + \frac{1}{4}\delta_e^T \delta_e - \varpi_e^T \lambda_{\varpi1}\varpi_e - \varpi_e^T \lambda_{\varpi2}\varpi_e + \varpi_e^T I^{-1}\tilde{u}_\varpi \\
& + \frac{1}{4}\varpi_e^T \varpi_e - \frac{1}{2}K_{\min}(A)\|\tilde{u}\|^2 + \frac{1}{2}(K_{\max}(C))^2\|\mathrm{X}\|^2/K_{\min}(A) \\
\leq & - \delta_e^T \lambda_{\delta1}\delta_e + \frac{1}{4}\|R\|^2\|\tilde{u}_\delta\|^2/K_{\min}(\lambda_{\delta2}) + \frac{1}{4}\delta_e^T \delta_e - \varpi_e^T \lambda_{\varpi1}\varpi_e + \frac{1}{4}\left\|I^{-1}\right\|^2\|\tilde{u}_\varpi\|^2/K_{\min}(\lambda_{\varpi2}) \\
& + \frac{1}{4}\varpi_e^T \varpi_e - \frac{1}{2}K_{\min}(A)\|\tilde{u}\|^2 + \frac{1}{2}(K_{\max}(C))^2\|\mathrm{X}\|^2/K_{\min}(A)
\end{aligned}
\tag{30}
$$

where $\lambda_\delta = \lambda_{\delta1} + \lambda_{\delta2}$, $\lambda_\varpi = \lambda_{\varpi1} + \lambda_{\varpi2}$ are positive definite matrices.
According to the perfect square formula:

$$
-\delta_e^T \lambda_{\delta2}\delta_e + \delta_e^T R\tilde{u}_\delta \leq -K_{\min}(\lambda_{\delta2})\delta_e^T \delta_e + \|R\|\|\delta_e\|\|\tilde{u}_\delta\| \leq \frac{1}{4}\|R\|^2\|\tilde{u}_\delta\|^2/K_{\min}(\lambda_{\delta2})
\tag{31}
$$

$$
\begin{aligned}
-\varpi_e^T \lambda_{\varpi2}\varpi_e + \varpi_e^T I^{-1}\tilde{u}_\varpi \leq & \\
& - K_{\min}(\lambda_{\varpi2})\varpi_e^T \varpi_e + \left\|I^{-1}\right\|\|\varpi_e\|\|\tilde{u}_\varpi\| \leq \frac{1}{4}\left\|I^{-1}\right\|^2\|\tilde{u}_\varpi\|^2/K_{\min}(\lambda_{\varpi2})
\end{aligned}
\tag{32}
$$

$$
-\frac{1}{2}K_{\min}(A)\|\tilde{u}\|^2 + \frac{1}{2}(K_{\max}(C))^2\|\mathrm{X}\|^2/K_{\min}(A) \leq -\frac{1}{4}K_{\min}(A)\|\tilde{u}\|^2
\tag{33}
$$

In order to solve further, the definition $\lambda_{\delta\tilde{u}}$, $\lambda_{\varpi\tilde{u}}$ is as follows:

$$
\lambda_{\delta\tilde{u}} = \frac{1}{4}\|R\|^2/K_{\min}(\lambda_{\delta2}), \quad \lambda_{\varpi\tilde{u}} = \frac{1}{4}\left\|I^{-1}\right\|^2/K_{\min}(\lambda_{\varpi2})
\tag{34}
$$

further

$$
\begin{aligned}
\dot{V} \leq & - \delta_e^T\left(\lambda_{\delta1} - \frac{1}{4}E\right)\delta_e + \lambda_{\delta\tilde{u}}\|\tilde{u}_\delta\|^2 - \varpi_e^T\left(\lambda_{\varpi1} - \frac{1}{4}E\right)\varpi_e + \lambda_{\varpi\tilde{u}}\|\tilde{u}_\varpi\|^2 - \frac{1}{4}K_{\min}(A)\|\tilde{u}\|^2 \\
\leq & - K_{\min}\left(\lambda_{\delta1} - \frac{1}{4}E\right)\|\delta_e\|^2 + \lambda_{\delta\tilde{u}}\|\tilde{u}_\delta\|^2 - K_{\min}\left(\lambda_{\varpi1} - \frac{1}{4}E\right) + \lambda_{\varpi\tilde{u}}\|\tilde{u}_\varpi\|^2 - \frac{1}{4}K_{\min}(A)\|\tilde{u}\|^2
\end{aligned}
\tag{35}
$$

where $E \in \Re^{3\times3}$ is an identity matrix. The problem can be further simplified. Choose parameters such that the following (35) holds, the Lyapunov function satisfies $\dot{V} \leq 0$.

$$\left(\sqrt{2}K_{\max}(C)/K_{\min}(A)\right)\|X\| \leq \|\tilde{u}\| \leq \|X\|/\sqrt{\alpha_{\tilde{u}}} \tag{36}$$

where $\alpha_{\tilde{u}}$ represents as following

$$\alpha_{\tilde{u}} = \min\left\{2\lambda_{\delta\tilde{u}}/K_{\min}\left(\lambda_{\delta 1} - \frac{1}{4}E\right), 2\lambda_{\varpi\tilde{u}}/K_{\min}\left(\lambda_{\varpi 1} - \frac{1}{4}E\right)\right\} \tag{37}$$

where $K_{\min}\left(\lambda_{\delta 1} - \frac{1}{4}E\right) > 0, K_{\min}\left(\lambda_{\varpi 1} - \frac{1}{4}E\right) > 0$.

Therefore, the system is asymptotically stable in the input state.

4 Experimental Results

In this section, simulation results are expressed to present the performance of the designed control method of the quadrotor. Meanwhile, simulations in the case that trajectory tracking with external random disturbances is carried out by MATLAB/SIMULINK, further, the experimental results of the sliding mode integral filtering backstepping method (FBC) is compared with the classical backstepping (CBC) in Chapter 3 in [25]. The experiment has verified its superiority. Simultaneously, the trajectory tracking diagram of the position will also be given.

The model parameters of the quadrotor [25] used in simulation are modified as: $m = 1.44\,kg$, $g = 9.8\,m/s^2$, $I_x = 0.03\,kgm^2$, $I_y = 0.03\,kgm^2$, $I_z = 0.04\,kgm^2$, $b_1 = 3.026 \times 10^{-5}Ns^2$, $b_2 = 3.122 \times 10^{-6}Ns^2$. The control parameters of FBC as follows $\gamma_{d1} = \gamma_{d2} = \gamma_{d3} = \gamma_{d4} = 1$; $\lambda_{\chi} = \lambda_v = diag\{1;1;1\}$; $A_i, B_i, C_i(i = \delta, \varpi) = diag\{0.1; 0.1;0.1\}$; $P_i, Q_i, J_i(i = \chi, v) = diag\{0.5;0.5;0.5\}$; $\lambda_{\delta} = \lambda_{\varpi} = diag\{0.85;1.1; 0.55\}$. The gains for the classical backstepping controller in Chapter 3 in [25] are set as $k_1 = k_2 = 3, k_3 = k_4 = 1, k_5 = k_6 = 2, k_7 = k_8 = k_9 = k_{10} = 10, k_{11} = k_{12} = 5$.

The purpose of the control is to allow the quadrotor to follow the desired path in the spiral path. The required trajectory is described as follows $\chi_d = [x_d \quad y_d \quad 1]^T$ where $x_d^2 + y_d^2 = 10^2$, and the desire yaw angle is $\psi_d = 0$. The initial position of the UAV model is $\chi = [0 \quad 0 \quad 0]^T$, and the initial attitude angles are $\delta = [0 \quad 0 \quad 0]^T$. Further, the initial translational and the initial angular velocity are given as $V = [0 \quad 0 \quad 0]^T$ and $\omega = [0 \quad 0 \quad 0]^T$ respectively.

The simulation results show that the quadrotor can tracking the trajectory with CBC and FBC strategy under random disturbance shown in Figs. 3 and 4, respectively. Further, the Fig. 5 shows the tracking of the position vectors, the Fig. 6 and 7 show the tracking error of the position and attitude respectively.

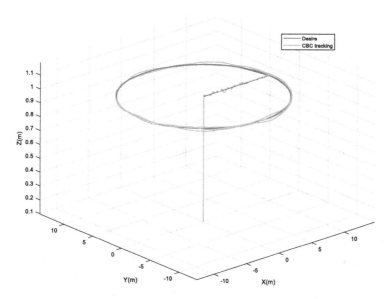

Fig. 3. The trajectory tracking based CBC under random disturbances

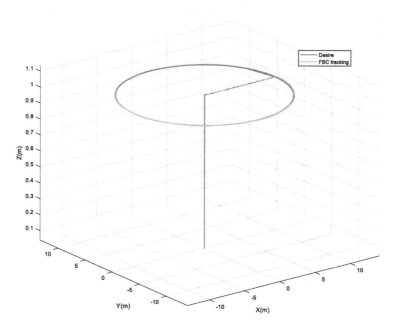

Fig. 4. The trajectory tracking based FBC under random disturbances

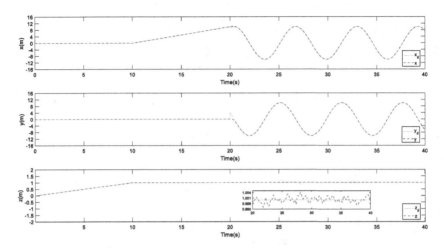

Fig. 5. The tracking situation based on FBC of the position vectors

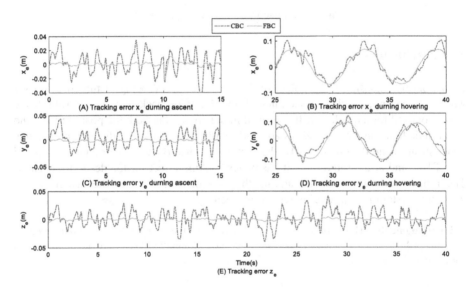

Fig. 6. Postition tracking error (x_e, y_e, z_e)

Through the simulation results of this part, we can clearly see that the FBC method based on the quadrotor UAV proposed in this paper can achieve accurate trajectory tracking. At the same time, through comparison with CBC, under strong external disturbance, FBC can obtain better High stability and anti-disturbance performance, the tracking error map given better illustrates this point.

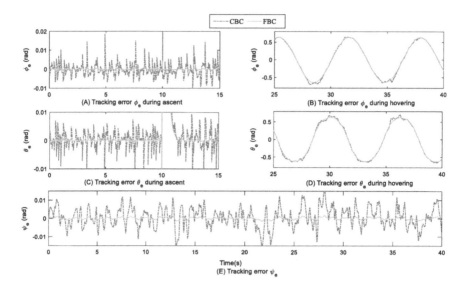

Fig. 7. Attitude tracking error $(\varphi_e, \theta_e, \psi_e)$

5 Conclusions

In this paper, regarding the problem of the quadrotor UAV being disturbed by strong winds in hovering, a sliding mode integral filter based on backstepping control scheme were proposed for a quadrotor. It is proved in the article that both the position loop and the attitude loop can be gradually stabilized. Further simulation verification proves that under strong wind disturbance, FBC has better anti-disturbance performance compared with CBC, and at the same time, it has better tracking effect for the desired trajectory. The future work will consider adding the idea of finite-time to the design of the quadrotor system to improve its response speed.

Acknowledgments. This works is partly supported by the Natural Science Foundation of Liaoning, China under Grant 2019MS008, Education Committee Project of Liaoning, China under Grant LJ2019003.

References

1. Liu, H., Bai, Y., Lu, G., Zhong, Y.: Robust attitude control of uncertain quadrotors. IET Control Theory Appl. **11**(2), 406–415 (2015)
2. Nonami, K.: Prospect and recent research & development for civil use autonomous unmanned aircraft as UAV and MAV. J. Syst. Des. Dyn. **1**(2), 120–128 (2007)
3. Lee, T., Leok, M., McClamroch, N.H.: Nonlinear robust tracking control of a quadrotor UAV on SE(3). Asian J. Control **15**(2), 391–408 (2013)

4. Yuan, C., Zhang, Y., Liu, Z.: A survey on technologies for automatic forest fire monitoring, detection, and fighting using unmanned aerial vehicles and remote sensing techniques. Canadian J. ForestbRes. **45**(7), 783–792 (2015)
5. Xu, R., Özgüner, Ü.: Sliding mode control of a class of underactuated systems. Automatica **44**(1), 233–241 (2008)
6. Xiao, B., Yin, S.: A new disturbance attenuation control scheme for quadrotor unmanned aerial vehicles. IEEE Trans. Industr. Inf. **13**(6), 2922–2932 (2017)
7. Castillo, P., Dzul, A., Lozano, R.: Real-time stabilization and tracking of a four-rotor mini rotorcraft. IEEE Trans. Control Syst. Technol. **12**(4), 510–516 (2004)
8. Bo, G., Xin, L., Hui, Z., Ling, W.: Quadrotor helicopter attitude control using cascade PID. In: 2016 Chinese Control and Decision Conference (CCDC), IEEE, Yinchuan, China (2016)
9. Zhang, Z.: Application of PID simulation control mode in quadrotor aircraft. In: 2020 International Conference on Computer Engineering and Application (ICCEA), IEEE, Guangzhou, China (2020)
10. Moreno-Valenzuela, J., Pérez-Alcocer, R., Guerrero-Medina, M., Dzul, A.: Nonlinear PID-type controller for quadrotor trajectory tracking. IEEE/ASME Trans. Mechatron. **23**(5), 2436–2447 (2018)
11. Karahan, M., Kasnakoglu, C.: Modeling and simulation of quadrotor UAV using PID controller. In: 11th International Conference on Electronics, Computers and Artificial Intelligence (ECAI), IEEE, Pitesti, Romania (2019)
12. Liu, C., Pan, J., Chang, Y.: PID and LQR trajectory tracking control for an unmanned quadrotor helicopter: experimental studies. In: 35th Chinese Control Conference (CCC), IEEE, Chengdu, China (2016)
13. Cohen, M.R., Abdulrahim, K., Forbes, J.R.: Finite-horizon LQR control of quadrotors on SE2(3) IEEE Robot. Autom. Lett. **5**(4), 5748–5755 (2020)
14. Ríos, H., Falcón, R., González, O.A., Dzul, A.: Continuous sliding-mode control strategies for quadrotor robust tracking: real-time application. IEEE Trans. Industr. Electron. **66**(2), 1264–1272 (2019)
15. Silva, A.L., Santos, D.A.: Fast nonsingular terminal sliding mode flight control for multirotor aerial vehicles. IEEE Trans. Aerosp. Electron. Syst. **56**(6), 4288–4299 (2020)
16. Zuo, Z.: Adaptive trajectory tracking control of a quadrotor unmanned aircraft. In: Proceedings of the 30th Chinese Control Conference, IEEE, Yantai, China (2011)
17. Chen, F., Wu, Q., Jiang, B., Tao, G.: A Reconfiguration scheme for quadrotor helicopter via simple adaptive control and quantum logic. IEEE Trans. Industr. Electron. **62**(7), 4328–4335 (2015)
18. Tian, B., Cui, J., Lu, H., Zuo, Z.: Adaptive finite-time attitude tracking of quadrotors with experiments and comparisons. IEEE Trans. Industr. Electron. **66**(12), 9428–9438 (2019)
19. Ganga, G., Dharmana, M.M.: MPC controller for trajectory tracking control of quadcopter. In: 2017 International Conference on Circuit, Power and Computing Technologies (ICCPCT), IEEE, Kollam, India (2017)
20. Torrente, G., Kaufmann, E., Föhn, P., Scaramuzza, D.: Data-driven MPC for Quadrotors. IEEE Robot. Autom. Lett. **6**(2), 3769–3776 (2021)
21. Selfridge, J.M., Tao, G.: Multivariable output feedback MRAC for a quadrotor UAV. In: 2016 American Control Conference (ACC), IEEE, Boston, MA, USA (2016)
22. Kayacan, E., Maslim, R.: Type-2 fuzzy logic trajectory tracking control of quadrotor VTOL aircraft with elliptic membership functions. IEEE/ASME Trans. Mechatron. **22**(1), 339–348 (2017)
23. Mohajerin, N., Waslander, S.L.: Multistep prediction of dynamic systems with recurrent neural networks. IEEE Trans. Neural Netw. Learn. Syst. **30**(11), 3370–3383 (2019)

24. Lin, X.L., Wu, C.F., Chen, B.S.: Robust adaptive fuzzy tracking control for MIMO nonlinear stochastic poisson jump diffusion systems. IEEE Trans. Cybern. **49**(8), 3116–3130 (2019)
25. Zhou, L., Zhang, J., She, H., Jin, H.: Quadrotor UAV flight control via a novel saturation integral backstepping controller. Automatika **60**, 193–206 (2019)
26. Zuo, Z., Wang, C.: Adaptive trajectory tracking control of output constrained multi-rotors systems. IET Control Theory Appl. **8**(13), 1163–1174 (2014)

An Enhanced Finite-Control-Set Model Predictive Control Strategy for PWM Rectifiers with Filter Inductance Mismatch

Van-Tien Le, Huu-Cong Vu, and Hong-Hee Lee$^{(\boxtimes)}$

Department of Electrical, Electronic and Computer Engineering,
University of Ulsan, Ulsan, South Korea
hhlee@mail.ulsan.ac.kr

Abstract. In finite-control-set model predictive control (FCS-MPC) for PWM rectifiers, the control performance is degraded seriously when the filter inductance is mismatched. To deal with this problem, this paper proposes a simple FCS-MPC strategy for PWM rectifiers to achieve high control performance regardless of the parameter mismatch. In the proposed FCS-MPC, the predictive model is modified by adding a correction term to compensate for the uncertainty, and high prediction accuracy is maintained irrespective of the parameter mismatch. The proposed FCS-MPC shows highly improved control performance compared to conventional FCS-MPC, and its effectiveness is experimentally verified.

Keywords: Pulse width modulation (PWM) rectifier · Finite-control-set model predictive control · Parameter mismatch

1 Introduction

Three-phase pulse width modulation (PWM) rectifiers known as AC/DC converters play important roles in various applications, such as renewable energy storage systems (ESSs) [1], power conversion, dc-motor drives, and distributed generation in micro-grids [2] due to its advantages of controllable power factor, sinusoidal input current, and adjustable output voltage [3]. To control the PWM rectifiers, many strategies have been proposed as proportional-integral (PI) controller [4], sliding mode control [5], deadbeat control [6], and model predictive control [7–9]. Among them, finite-control-set model predictive control (FCS-MPC) [9] has received wide attention because of its highlight features, such as very fast dynamic response, easy implementation with no modulation stage, and flexible control objectives [10].

In FCS-MPC, all possible switching vectors of a converter are used to predict future controlled variables based on system model and parameters. Among these switching vectors, the one that minimizes the error between the future controlled variables and their respective references is used to control the converter in the next sampling period [9]. As the principle shown, FCS-MPC directly uses the system model and parameters for determining the optimal switching vector, so the control performance is highly sensitive to system-parameter mismatch. Study in [11] has shown that even with a

© Springer Nature Switzerland AG 2021
D.-S. Huang et al. (Eds.): ICIC 2021, LNCS 12837, pp. 161–171, 2021.
https://doi.org/10.1007/978-3-030-84529-2_14

small underestimation of the system parameters, the control performance of FCS-MPC can be degraded significantly.

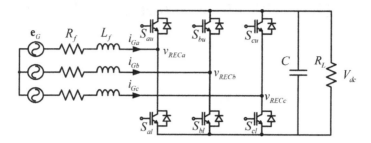

Fig. 1. Configuration of a two-level three-phase PWM rectifier.

To overcome performance degradation caused by system-parameter mismatch, some techniques have been proposed recently based on MPC [12–16]. In [12], an online disturbance estimator by Luenberger observer was proposed to remove the effects of the parameter mismatch. Nevertheless, designing observer gain in a closed loop system is not easy because of the trade-off between robustness and stability margin. In [13], model-free predictive control (MFPC) was proposed, where the system model and parameters are not required to predict the future controlled variables. Even though this method is robust against system-parameter mismatch, stagnant current-variation update is a critical drawback, which deteriorates the control performance. To avoid the stagnant current-variation update, an improved MFPC was proposed in [14]. However, this method is suffered from high computation burden because many low pass filters are required in implementation. In [14], another estimation technique based on Ohm's theory was proposed to remove filter-inductance mismatch for PWM rectifiers. But, this method is highly sensitive to grid-frequency variation. In [15], least-squares estimator was developed to online identify the system parameters. Even though parameter-mismatch problem is solved, this method is also suffered from high computation burden.

In this paper, we propose a simple FCS-MPC strategy for PWM rectifiers, which can deal with inductance mismatch effectively. In the proposed FCS-MPC, the parameter-mismatch problem is solved by modifying the predictive model with a correction term to compensate for the uncertainty. Regardless of the parameter mismatch, the proposed method provides high prediction accuracy, and its control performance is significantly improved compared to that of conventional FCS-MPC. The effectiveness and robustness of the proposed FCS-MPC have been verified by experiments.

2   Conventional FCS-MPC for PWM Rectifier

Figure 1 shows the configuration of two-level three-phase PWM rectifier consisting of six unidirectional power switches (IGBTs), an AC voltage source, a filter inductor with inductance L_f and resistance R_f, an output capacitor, and a DC load (resistance). For

Table 1. Switching states and voltage vectors of the PWM rectifier

SS & VV	S_a	S_b	S_c	$v_{REC\alpha}$	$v_{REC\beta}$
S_0, V_0	0	0	0	0	0
S_1, V_1	1	0	0	$2V_{dc}/3$	0
S_2, V_2	1	1	0	$V_{dc}/3$	$\sqrt{3}V_{dc}/3$
S_3, V_3	0	1	0	$-V_{dc}/3$	$\sqrt{3}V_{dc}/3$
S_4, V_4	0	1	1	$-2V_{dc}/3$	0
S_5, V_5	0	0	1	$-V_{dc}/3$	$-\sqrt{3}V_{dc}/3$
S_6, V_6	1	0	1	$V_{dc}/3$	$-\sqrt{3}V_{dc}/3$
S_7, V_7	1	1	1	0	0

simplicity in analysis, three-phase grid voltage (e_{Ga}, e_{Gb}, e_{Gc}), grid current (i_{Ga}, i_{Gb}, i_{Gc}), and rectifier voltages ($v_{RECa}, v_{RECb}, v_{RECc}$) are converted into space vector forms in the $\alpha\beta$ stationary coordinate as follows:

$$e_G = 2\left(e_a + e_b e^{j2\pi/3} + e_c e^{j4\pi/3}\right)/3, \tag{1}$$

$$i_G = 2\left(i_a + i_b e^{j2\pi/3} + i_c e^{j4\pi/3}\right)/3, \tag{2}$$

$$v_{REC} = 2\left(v_{RECa} + v_{RECb} e^{j2\pi/3} + v_{RECc} e^{j4\pi/3}\right)/3, \tag{3}$$

where e_G, v_{REC}, and i_G are the grid voltage, rectifier voltage and grid current vectors, respectively.

Due to the switching constraint of the PWM rectifier, only 8 switching states ($S_1 - S_6$) are allowed in the operation, and the voltage vectors (VV) of the PWM rectifier are limited to 6 active voltage vector ($V_1 - V_6$) and 2 zero-voltage vectors (V_0, V_7) as shown in Table 1 [3]. The switching function of each phase-leg is defined in (4):

$$S_j = \begin{cases} 1 & S_{ju} \text{ is on \& } S_{jl} \text{ is off} \\ 0 & S_{ju} \text{ is off \& } S_{jl} \text{ is on} \end{cases}, j \in \{a, b, c\}. \tag{4}$$

From Fig. 1, the dynamic equation of the grid current is derived as

$$L_f \frac{di_G}{dt} = e_G - R_f i_G - v_{REC}. \tag{5}$$

To predict the future grid current, (5) is discretized with a sampling time T_S by using the forward Euler method:

$$i^p_{G,k+1} = \left(1 - \frac{R_f T_S}{L_f}\right) i_{G,k} + \frac{T_S}{L_f}\left(e_{G,k} - v_{REC}|_{V_k}\right), \tag{6}$$

where $i^p_{G,k+1}$ is future grid current at instant $(k+1)$, and subscription "p" represents the predicted value. $v_{REC}|_{V_k}$ is the rectifier voltage with respect to the voltage vector V_k, whose value is listed in two last columns in Table 1.

To avoid performance degradation caused by the one-step delay in practice [16], $i^p_{G,k+2}$ is controlled rather than $i^p_{G,k+1}$. By shifting (6) one step ahead, the future grid current at instant $(k+2)$ can be predicted as follows:

$$i^p_{G,k+2} = \left(1 - \frac{R_f T_S}{L_f}\right) i^p_{G,k+1} + \frac{T_S}{L_f}\left(e_{G,k+1} - v_{REC}|_{V_{k+1}}\right). \tag{7}$$

In (7), $i^p_{G,k+1}$ is obtained from (6), and $e_{G,k+1}$ is replaced by $e_{G,k}$ since the sampling frequency used in FCS-MPC is much higher than the grid frequency. By using power theory in [17], the instantaneous active and reactive powers at instant $(k+2)$ are calculated as

$$\begin{cases} P_{k+2} \approx 1.5\left(e_{G\alpha,k} i^p_{G\alpha,k+2} + e_{G\beta,k} i^p_{G\beta,k+2}\right) \\ Q_{k+2} \approx 1.5\left(e_{G\beta,k} i^p_{G\alpha,k+2} - e_{G\alpha,k} i^p_{G\beta,k+2}\right) \end{cases}. \tag{8}$$

In order to control the active and reactive powers following their respective references, the cost function is designed in (9):

$$g = \left(P^{ref} - P_{k+2}\right)^2 + \left(Q^{ref} - Q_{k+2}\right)^2, \tag{9}$$

where P^{ref} and Q^{ref} are the references of active and reactive powers, respectively.

Figure 2 shows the control block diagram of the conventional FCS-MPC for the PWM rectifier in which the output voltage V_{DC} is regulated by using a proportional-integral (PI) controller to generate the active power reference P^{ref}, and the reactive power reference is set to be zero to control unity power factor at the input side. After measuring $i_{G,k}$ and $e_{G,k}$, $i^p_{G,k+1}$ is predicted from (6) to compensate the one-step delay, and then P_{k+2} and Q_{k+2} are obtained using (7) and (8) for each voltage vector in Table 1 [3]. To control the PWM rectifier, the cost function in (9) is evaluated for all 8 voltage vectors, and the voltage vector resulted in minimum cost function is applied in $(k+1)^{th}$ sampling period.

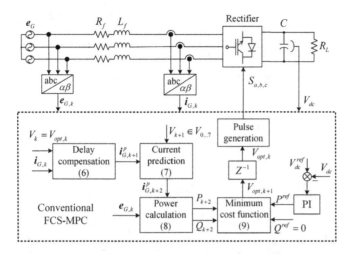

Fig. 2. Control block diagram of the conventional FCS-MPC for the PWM rectifier.

3 Proposed FCS-MPC for PWM Rectifier

Generally, due to the power-loss constraint of the filter inductor, the filter resistance is much smaller than the filter inductance impedance [18], and (6) can be simplified by neglecting the resistive term:

$$i^p_{G,k+1} = i_{G,k} + \frac{T_S}{L_f}\left(e_{G,k} - v_{REC}\big|_{V_k}\right). \tag{10}$$

In case that the filter inductance is mismatched, (10) becomes (11):

$$\tilde{i}^p_{G,k+1} = i_{G,k} + \frac{T_S}{\tilde{L}_f}\left(e_{G,k} - v_{REC}\big|_{V_k}\right), \tag{11}$$

where $\tilde{i}^p_{G,k+1}$ is the predicted grid current in the presence of the inductance mismatch, $\tilde{L}_f = L_f + \Delta L_f$ is the mismatched inductance, L_f is the accurate inductance value, and ΔL_f represents mismatch error. In case of the inductance mismatch, the predicted grid current becomes inaccurate, and the prediction error is obtained by subtracting (11) from (10):

$$error\big|_{V_k} = \tilde{i}^p_{G,k+1} - i^p_{G,k+1} = \frac{\Delta L_f T_S}{(L_f + \Delta L_f)L_f}\left(e_{G,k} - v_{REC}\big|_{V_k}\right). \tag{12}$$

In (12), $e_{G,k}$ changes with the grid frequency $(50 - 60\ \text{Hz})$ that is much lower compared to high sampling frequency used in FCS-MPC (typically $\geq 10\ \text{kHz}$), and $v_{REC}\big|_{V_k}$ is constant for a given voltage vector as shown in Table 1 [3]. Therefore,

LUT

Fig. 3. LUT in the proposed FCS-MPC.

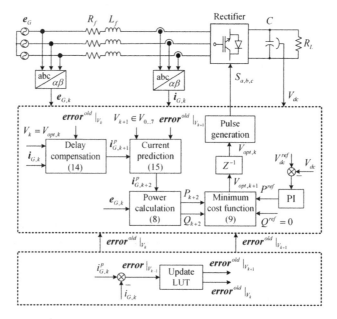

Fig. 4. Control block diagram of the proposed FCS-MPC for the PWM rectifier.

$error|_{V_k}$ changes slowly compared to the sampling frequency, and it can be approximated as follows:

$$error|_{V_k} \approx error^{old}|_{V_j = V_k},\tag{13}$$

where $error^{old}|_{V_k}$ is the old value of $error|_{V_k}$ which has been computed in the last time that the voltage vector V_k ($V_k = V_j$) was applied. Corresponding to 8 voltage vectors ($V_k \in \{V_0...V_7\}$), there are 8 prediction-error terms $error|_{V_k}$, respectively.

By substituting (13) into (12), the predicted grid current at instant $(k+1)$ can be corrected regardless of the parameter mismatch:

$$i^p_{G,k+1} \approx \tilde{i}^p_{G,k+1} - error^{old}|_{V_k}.\tag{14}$$

Similarly, the predicted grid current at instant $(k+2)$ is corrected as follows:

$$i^p_{G,k+2} \approx \tilde{i}^p_{G,k+2} - error^{old}|_{V_{k+1}}, \tag{15}$$

where $error^{old}|_{V_{k+1}}$ is stored in a look-up table (LUT) as shown in Fig. 3. To update the LUT, $error|_{V_{k-1}}$ is computed at every sampling period and used for replacing $error^{old}|_{V_{k-1}}$:

$$error|_{V_{k-1}} = \tilde{i}^p_{G,k} - i_{G,k}, \tag{16}$$

where $\tilde{i}^p_{G,k}$ has been predicted in $(k-1)^{th}$ sampling period, and $i_{G,k}$ is measured grid current at instant k.

It should be noted that when the accurate inductance is used in the predictive model (11), the prediction-error term $error^{old}|_{V_k}$ becomes zero, and (15) is fully consistent with (7). Therefore, there is no performance difference between proposed method and conventional method in case of accurate parameter.

Figure 4 shows the control block diagram of the proposed FCS-MPC for the PWM rectifier. After sampling $e_{G,k}$ and $i_{G,k}$, $error|_{V_{k-1}}$ is computed using (16) for replacing $error^{old}|_{V_{k-1}}$ in the LUT. Based on the LUT, $i^p_{G,k+2}$ can be predicted accurately from (15) irrespective of the inductance mismatch.

4   Experimental Results

In order to verify the effectiveness of the proposed FCS-MPC, the proposed FCS-MPC is experimentally compared with the conventional FCS-MPC. All methods were implemented based on a digital signal processor (DSP) TMS28F379D to control the PWM rectifier under the same operation conditions. The PWM rectifier is constructed from 6 IGBTs FGH40N60SFD from Fairchild. All experimental results are recorded.

Table 2. System parameters

Grid voltage	$e_G = 110\,\text{V}/60\,\text{Hz}$
Filter inductance and resistance	$L_f = 10\,\text{mH}, R_f = 0.2\,\Omega$
Output capacitance	$C = 660\,\mu\text{F}$
DC load	$R_L = 80\,\Omega$
Output-voltage reference	$V_{DC}^{ref} = 250\,\text{V}$
Sampling time	$T_S = 30\,\mu\text{s}$

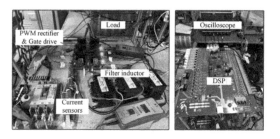

Fig. 5. Experimental setup for verifications.

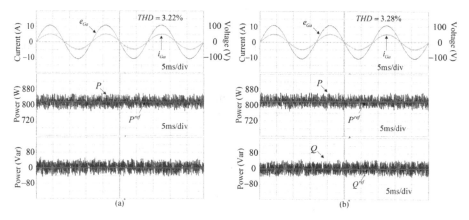

Fig. 6. Steady-state performance with accurate parameter (a) Conventional FCS-MPC, (b) Proposed FCS-MPC.

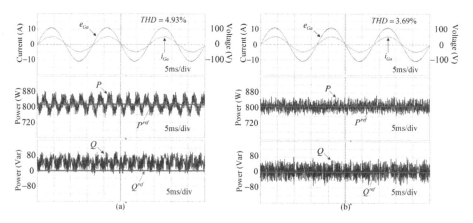

Fig. 7. Steady-state performance with 50% underestimated filter inductance (a) Conventional FCS-MPC, (b) Proposed FCS-MPC.

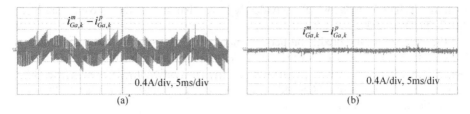

Fig. 8. Prediction performance with accurate parameter (a) Conventional FCS-MPC, (b) Proposed FCS-MPC.

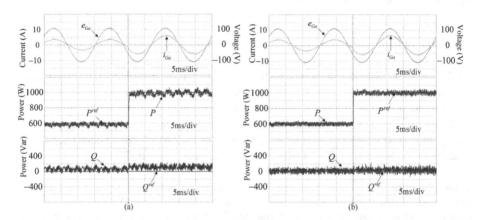

Fig. 9. Dynamic performance with 50% underestimated filter inductance when *Pref* is changed from 600W to 1kW (a) Conventional FCS-MPC, (b) Proposed FCS-MPC.

using oscilloscope Wave Runner 610Zi. The system parameters used for verifications are listed in Table 2, and a photo of the experimental setup is shown in Fig. 5.

Figure 6 shows the steady-state performance of the conventional FCS-MPC and proposed FCS-MPC with the accurate inductance, respectively. From Fig. 6, the proposed FCS-MPC achieves similar performance with the conventional FCS-MPC, and active and reactive powers track their respective references accurately with both control methods. The grid current with the proposed FCS-MPC is sinusoidal and smooth with low total harmonic distortion (THD) of 3.28%, which is almost the same as THD = 3.22% achieved from the conventional FCS-MPC.

Figure 7 shows the steady-state performance of the conventional FCS-MPC and proposed FCS-MPC with 50% underestimated inductance. As shown in Fig. 7, the performance of the conventional FCS-MPC is deteriorated significantly, and the grid current is highly distorted with THD = 4.93%. From Fig. 7(a), both active and reactive powers with the conventional FCS-MPC track their respective references with high tracking errors.

In contrast, the performance of the proposed FCS-MPC is guaranteed regardless of the inductance mismatch as shown in Fig. 7(b). Both active and reactive powers track their references accurately, and the grid current with the proposed method is sinusoidal with low THD of 3.69%, which is evidently lower than that in the conventional FCS-MPC (THD = 4.93%).

Figure 8 shows the prediction performance comparisons between the conventional FCS-MPC and proposed FCS-MPC. It is clearly that the prediction error in the proposed FCS-MPsC is much lower than that in the conventional FCS-MPC. That is the reason why the proposed FCS-MPC achieves higher control performance than the conventional FCS-MPC.

Figures 9(a) and (b) show the dynamic performance of the conventional FCS-MPC and proposed FCS-MPC, respectively. As we can see, beside achieving low grid current distortion and good power tracking, the proposed FCS-MPC shows similar dynamic response with the conventional FCS-MPC.

5   Conclusion

An improved FCS-MPC strategy for PWM rectifiers with filter inductance mismatch has been proposed in this paper by adding a correction term into the predictive model to improve the prediction accuracy. In the proposed method, high prediction accuracy is achieved regardless of the parameter mismatch, and its control performance is significantly improved compared to that of conventional FCS-MPC. The effectiveness and robustness of the proposed FCS-MPC are experimentally verified.

Acknowledgments. This work was supported in part by the National Research Foundation of Korea Grant funded by the Korean Government under Grant NRF-2018R1D1A1A09081779 and in part by the Korea Institute of Energy Technology Evaluation and Planning and the Ministry of Trade, Industry and Energy under Grant 20194030202310.

References

1. Zhang, Z., Fang, H., Gao, F., Rodriguez, J., Kennel, R.: Multiple-vector model predictive power control for grid-tied wind turbine system with enhanced steady-state control performance. IEEE Trans. Ind. Electron. **64**(8), 6287–6298 (2017)
2. Blaabjerg, F., Teodorescu, R., Liserre, M., Timbus, A.V.: Overview of control and grid synchronization for distributed power generation systems. IEEE Trans. Indus. Electron. **53**(5), 1398–1409 (2006)
3. Rodríguez, J.R., Dixon, J.W., Espinoza, J.R., Pontt, J., Lezana, P.: PWM regenerative rectifiers: State of the art. IEEE Trans. Ind. Electron. **52**(1), 5–22 (2005)
4. Blasko, V., Kaura, V.: A new mathematical model and control of a three-phase AC-DC voltage source converter. IEEE Trans. Power Electron. **12**(1), 116–123 (1997)
5. Guzman, R., De Vicuña, L.G., Morales, J., Castilla, M., Matas, J.: Sliding-mode control for a three-phase unity power factor rectifier operating at fixed switching frequency. IEEE Trans. Power Electron. **31**(1), 758–769 (2016)
6. Malesani, L.: Robust dead-beat current control for PWM rectifiers and active filters. IEEE Trans. Ind. Appl. **35**(3), 613–620 (1999)
7. Choi, D.K., Lee, K.B.: Dynamic performance improvement of AC/DC converter using model predictive direct power control with finite control set. IEEE Trans. Ind. Electron. **62**(2), 757–767 (2015)

8. Song, Z., Tian, Y., Chen, W., Zou, Z., Chen, Z.: Predictive duty cycle control of three-phase active-front-end rectifiers. IEEE Trans. Power Electron. **31**(1), 698–710 (2016)

9. Cortés, P., Rodríguez, J., Antoniewicz, P., Kazmierkowski, M.: Direct power control of an AFE using predictive control. IEEE Trans. Power Electron. **23**(5), 2516–2523 (2008)

10. Xue, C., Ding, L., Li, Y., Zargari, N.R.: Improved model predictive control for high-power current-source rectifiers under normal and distorted grid conditions. IEEE Trans. Power Electron. **35**(5), 4588–4601 (2020)

11. Young, H.A., Perez, M.A., Rodriguez, J.: Analysis of finite-control-set model predictive current control with model parameter mismatch in a three-phase inverter. IEEE Trans. Ind. Electron. **63**(5), 3100–3107 (2016)

12. Xia, C., Wang, M., Song, Z., Liu, T.: Robust model predictive current control of three-phase voltage source PWM rectifier with online disturbance observation. IEEE Trans. Ind. Inform. **8**(3), 459–471 (2012)

13. Lai, Y.S., Lin, C.K., Chuang, F.P., Te Yu, J.: Model-free predictive current control for three-phase AC/DC converters. IET Electr. Power Appl. **11**(5), 729–739 (2017)

14. Le, V.-T., Lee, H.-H.: An enhanced model-free predictive control to eliminate stagnant current variation update for PWM rectifiers. IEEE J. Emerg. Sel. Top. Power Electron. **99**, 1–1 (2021)

15. Kwak, S., Moon, U.C., Park, J.C.: Predictive-control-based direct power control with an adaptive parameter identification technique for improved AFE performance. IEEE Trans. Power Electron. **29**(11), 6178–6187 (2014)

16. Cortes, P., Rodriguez, J., Silva, C., Flores, A.: Delay compensation in model predictive current control of a three-phase inverter. IEEE Trans. Ind. Electron. **59**(2), 1323–1325 (2012)

17. Akagi, H., Kanazawa, Y., Nabae, A.: Instantaneous reactive power compensators comprising switching devices without energy storage components. IEEE Trans. Ind. App. **IA-20**(3), 625–630 (1984). https://doi.org/10.1109/TIA.1984.4504460

18. Razali, A.M., Rahman, M.A., George, G., Rahim, N.A.: Analysis and design of new switching lookup table for virtual flux direct power control of grid-connected three-phase PWM AC-DC converter. IEEE Trans. Ind. App. **51**(2), 1189–1200 (2015)

Deep Integration Navigation Technique Based on Strong Tracking UKF Algorithm

Cheng Xuwei$^{(\boxtimes)}$, Zhang Zaitian, Ren Haoyu, Qiu Fengqi, and Chen Jianzhou

Department 21, Unit 32272, PLA, Chengdu 610214, China

Abstract. In order to improve precision and reliability of deep integration system of Inertial Navigation System (INS) and Global Satellite Navigation System (GNSS), an adaptive strong tracking unscented Kalman filter algorithm was adopted, which utilize suboptimal multiple fading factors to achieve parameter adaptation task for various dynamic characteristics, and improve algorithm precision and real-time ability simultaneously. Simulation experiment are provided for illustrating the effective improvement in navigation estimation accuracy. The results show that STUKF algorithm is better than single UKF and STF, which can alternate traditional algorithm for INS/GNSS deep integration.

Keywords: Integration navigation · Deep integration · Strong tracking · UKF

1 Introduction

In the field of navigation, INS (Inertia Navigation System) and GNSS (Global Navigation Satellite System) has complementary advantages [1]. In order to improve system precision and reliability, utilizing data fusion method two system data can be integrated to make each subsystem learn from each other [2, 3]. While, the deep integrated navigation system based on INS/GNSS can overcome the defect that loose and tight combination relies heavily on GNSS to assist INS [4], and improve receiver's acquiring and tracking ability of weak signals [5, 6].

In fact, to improve accuracy and stability of the (deep) integrated system of INS/GNSS and reduce the influence of carrier dynamics on system performance effectively, many nonlinear filtering algorithms are adopted successively. For example, representative algorithm are Extended Kalman Filter (EKF) and Unscented Kalman Filter (UKF) based on UT proposed by Julier [7]. The estimation accuracy of state mean and variance of latter is significantly higher than the former, for there is no needed to solving Jacobian Matrix, so it avoids truncation error in linearization process. In addition, in order to avoid declining accuracy caused by system model uncertainty, even filter divergence, Strong Tracking Filtering (STF) strategy is widely introduced into nonlinear filtering by Zhou [8, 9]. However, for INS/GNSS integrated system, algorithm accuracy, complexity, cost and real-time performance all need to be taken into consideration for choosing proper filter algorithm.

This paper mainly studies the Strong Tracking Unscented Kalman Filter (STUKF) which combines UKF algorithm with STF strategy, based on which simulation

© Springer Nature Switzerland AG 2021
D.-S. Huang et al. (Eds.): ICIC 2021, LNCS 12837, pp. 172–182, 2021.
https://doi.org/10.1007/978-3-030-84529-2_15

experiments are carried out in deep integrated INS/GNSS. The experimental results show that this algorithm has excellent performance, and can deal with sudden change of carrier motion state. Although the suboptimal weakening factor of the algorithm is selected, the real-time performance can also meet system demand, and has good application value, which is of guiding significance for designing integrated system.

2 Deep Integrated Navigation System Model

In the deep integration structure, the In-phase, Quadrature branch signals and navigation information are directly input into navigation filter to processing, which are produced by GNSS receiver base-band and INS respectively. Estimating parameters of satellite signal tracking process and fusing navigation and positioning results output to user is accomplished by navigation filter. Figure 1 shows the structure of integrated SINS/BDS deep integrated navigation system based on I and Q signals.

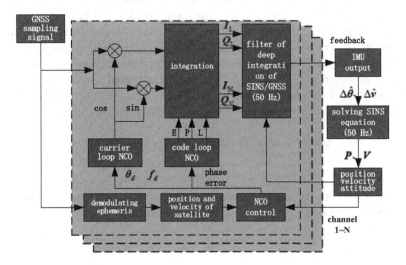

Fig. 1. Deep integrated structure of SINS/GNSS based on I, Q signals

The state and observation equations of system can be expressed as

$$X(k) = \Phi(k, k-1)X(k-1) + W(k-1) \tag{1}$$

$$Z(k) = H(k)X(k) + V(k) \tag{2}$$

Where Φ and H are system state transition matrix and observation matrix respectively, W and V are process noise and measurement noise respectively, both are zero mean Gaussian white noise and not correlated with each other.

For the deep integration structure shown in Fig. 1, state vector of the navigation filter is similar to tight integration filter, in which the clock deviation and random drift of receiver are increased, compared with loosely integrated navigation filter. Which are given in the form of equivalent pseudo-range and pseudo-range rate, that is, the pseudo-range δt caused by clock deviation between satellite and receiver and its change rate δt_f caused by clock drift. If navigation system is solved in the form of pseudo-range and pseudo-range rate, the INS error, GNSS clock error effect and its drift equivalent rate are selected as the system states. Meanwhile, the state variables can be expressed as [10]

$$
X = \left\{ \begin{matrix} \delta L & \delta \lambda & \delta h & \delta V_E & \delta V_N & \delta V_U & \varphi_E & \varphi_N & \varphi_U \\ \varepsilon_x & \varepsilon_y & \varepsilon_z & \Delta_x & \Delta_y & \Delta_z & \delta t_u & \delta t_f \end{matrix} \right\}^T \tag{3}
$$

In which $[\delta L \quad \delta \lambda \quad \delta h]$ is system position error, $[\delta V_E \quad \delta V_N \quad \delta V_U]$ is system velocity error, $[\varphi_E \quad \varphi_N \quad \varphi_U]$ is system attitude error; $[\varepsilon_x \quad \varepsilon_y \quad \varepsilon_z]$ and $[\Delta_x \quad \Delta_y \quad \Delta_z]$ are bias error of gyroscope and accelerometer respectively.

The foundation of engaging state matrix and observation matrix lies in the corresponding relationship between I and Q signals and position and velocity variables of system state. However, internal structure and completion process of navigation filter is determined directly by observation variables selection. For the deep integration based on I and Q signals, its filter directly uses the most original tracking loop information, that is, the observation differences of two signals are used as observation. In the integration system based on SINS/BDS, firstly I and Q signals are predicted through output information of SINS, then, for ith tracking channel, the measurement equation is

$$
\begin{aligned}
Z_k &= (I + dI_{INS}, Q + dQ_{INS})|_i - (I + \eta_1, Q + \eta_Q)|_i \\
&= (dI - \eta_1, dQ - \eta_Q)|_i
\end{aligned} \tag{4}
$$

Where dI, dQ are the signal variation of I and Q signal caused by sensor error predicted by INS respectively, accordingly, the measurement matrix is

$$
\mathbf{H}_k = \left[\begin{array}{cccccc|ccc|cc}
h_{Ix1} & h_{Iy1} & h_{Ih1} & h_{Ix1} & h_{Iy1} & h_{Ih1} & 0 & \cdots & 0 & 1 & 0 \\
\vdots & \vdots & \vdots & \vdots & \vdots & \vdots & 0 & \ddots & 0 & \vdots & \vdots \\
h_{Ixm} & h_{Iym} & h_{Ihm} & h_{Ixm} & h_{Iym} & h_{Ihm} & 0 & \cdots & 0 & 1 & 0 \\
\hline
h_{Qx1} & h_{Qy1} & h_{Qh1} & h_{Qx1} & h_{Qy1} & h_{Qh1} & 0 & \cdots & 0 & 0 & 1 \\
\vdots & \vdots & \vdots & \vdots & \vdots & \vdots & 0 & \ddots & 0 & \vdots & \vdots \\
h_{Qxm} & h_{Qym} & h_{Qhm} & h_{Qxm} & h_{Qym} & h_{Qhm} & 0 & \cdots & 0 & 0 & 1
\end{array} \right]_{2N \times 17} \tag{5}
$$

Because of the nonlinear relationship between measurement signal and state variables, in order to take accuracy and calculation cost into account, advanced filtering algorithm must be adopted to improve system performance.

3 UKF Algorithm

UKF is a kind of filter algorithm which uses sampling points to approximate distribution of nonlinear state parameters. Kalman filtering structure is still used in filtering framework, but the difference is that it is based on Unscented transfer (UT). When state variables are similar to Gaussian distribution, a group of sigma points are used to obtain mean and covariance of Gaussian density function.

For nonlinear measurement model with additive noise, basic process of discrete UKF filtering includes UT transform and filtering process.

3.1 UT Transformation

The first step of UKF is to obtain prior distribution of state variables by sampling. For n dimensional state variables $\mathbf{x}$ with mean $\hat{\mathbf{x}}$ and variance are $\mathbf{P}$, symmetric $2n+1$ sampling sigma points are obtained by UT transformation. Then, adopting nonlinear function, these sigma points can get vector $\mathbf{X}$ and weighted points W, namely

$$
\begin{aligned}
\boldsymbol{\xi}_{(0)} &= \overline{\mathbf{x}} \\
\boldsymbol{\xi}_{(i)} &= \overline{\mathbf{x}} + \left(\sqrt{(n+\lambda)\mathbf{P}}\right)_i^{\mathrm{T}}, \qquad i = 1, \cdots, n \\
\boldsymbol{\xi}_{(i+n)} &= \overline{\mathbf{x}} - \left(\sqrt{(n+\lambda)\mathbf{P}}\right)_i^{\mathrm{T}}, \quad i = 1, \cdots, n
\end{aligned}
\tag{6}
$$

Where $\left(\sqrt{(n+\lambda)\mathbf{P}}\right)_i^{\mathrm{T}}$ is the square root of matrix in row, $\sqrt{(n+\lambda)\mathbf{P}}$ is obtained by Cholesky decomposition of lower triangular matrix, $\lambda = \alpha^2(n+\kappa) - n$ is a normalization parameter, in which parameter α mainly adjusts the distance between sigma points and $\overline{\mathbf{x}}$, to keep its value moderate, which is usually set to a small positive number, such as $1 \times 10^{-4} \leq \alpha \leq 1$, κ is a secondary normalization factor, generally set to 0. When the state is multivariable, it can be set to $\kappa = 3 - n$. When the proportion correction algorithm is used, the weight coefficients corresponding to Eq. (1) are

$$
\begin{aligned}
(\boldsymbol{\xi}_{(i)})' &= \boldsymbol{\xi}_{(0)} + \alpha(\boldsymbol{\xi}_{(i)} - \boldsymbol{\xi}_{(0)}) \\
(W_0^{(m)})' &= \lambda/(n+\lambda) \\
(W_0^{(c)})' &= (W_0^{(m)})' + (1 - \alpha^2 + \gamma) \\
(W_i^{(m)})' &= (W_i^{(c)})' = 1/2(n+\lambda), \ i = 1, \cdots, 2n
\end{aligned}
\tag{7}
$$

In the formula, the prior distribution knowledge of $\overline{\mathbf{x}}$ mainly reflected by parameter γ, which can combine motion difference of high-order covariance term. For normal distribution $\mathbf{x}$ with Gaussian distribution, $\gamma = 2$ is optimal. $(W_i^{(m)})'$ and $(W_i^{(c)})'$ represent weights of mean and variance corresponding to ith point correlation respectively.

In this case, after sigma vector propagates with nonlinear function, transformed point set will be obtained

$$
(\boldsymbol{\zeta}_k^-)_i = f_k((\mathbf{X}_{k-1})_i), \ i = 0, \cdots, 2n
\tag{8}
$$

Where mean and variance $(\zeta_k^-)_i$ of are approximated by weighted mean and variance of transformed sigma point set, namely

$$
\begin{aligned}
\hat{\mathbf{x}}_k^- &= \sum_{i=0}^{2n} (W_0^{(c)})'(\zeta_k^-)_i' \\
\mathbf{P}_{zz} &= \sum_{i=0}^{2n} (W_i^{(c)})'[(\zeta_k^-)_i - \hat{\mathbf{x}}_k^-][(\zeta_k^-)' - \hat{\mathbf{x}}_k^-]^{\mathrm{T}} \\
\mathbf{P}_{xz} &= \sum_{i=0}^{2n} (W_i^{(c)})'[(\mathbf{X}_{k-1})_i - \overline{\mathbf{x}}_k)][(\zeta_k^-)' - \hat{\mathbf{x}}_k^-]^{\mathrm{T}}
\end{aligned}
\tag{9}
$$

3.2   UKF Filtering

Step 1. Adopting state equation, sigma point set after unscented transfer can obtain

$$
(\zeta_k^-)_i = f_k((\mathbf{X}_{k-1})_i), i = 0, \cdots, 2n
\tag{10}
$$

Step 2. Predicting mean and variance

$$
\hat{\mathbf{x}}_k^- = \sum_{i=0}^{2n} W_i^{(\mathrm{m})} (\zeta_k^-)_i
\tag{11}
$$

$$
\mathbf{P}_k^- = \sum_{i=0}^{2n} W_i^{(c)}[(\zeta_k^-)_i - \hat{\mathbf{x}}_k^-][(\zeta_k^-)_i - \hat{\mathbf{x}}_k^-]^{\mathrm{T}} + \mathbf{Q}_{k-1}
\tag{12}
$$

Step 3. Prediction point set is obtained by the observation model and observations

$$
(\mathbf{Z}_k^-)_i = h_k((\zeta_k^-)_i)
\tag{13}
$$

$$
(\hat{\mathbf{z}}_k^-)_i = \sum_{i=0}^{2n} W_i^{(\mathrm{m})} (\mathbf{Z}_k^-)_i
\tag{14}
$$

Step 4. Updating variance and covariance

$$
\begin{aligned}
\mathbf{P}_{zz} &= \sum_{i=0}^{2n} W_i^{(c)}[(\mathbf{Z}_k^- - \hat{\mathbf{z}}_k^-)(\mathbf{Z}_k^- - \hat{\mathbf{z}}_k^-)]^{\mathrm{T}} + \mathbf{R}_k \\
\mathbf{P}_{xz} &= \sum_{i=0}^{2n} W_i^{(c)}[(\zeta_k^-)_i - \hat{\mathbf{x}}_k^-][((\mathbf{Z}_k^-)_i - \hat{\mathbf{z}}_k^-)]^{\mathrm{T}}
\end{aligned}
\tag{15}
$$

Step 5. Complete update

$$
\begin{aligned}
\mathbf{K}_k &= \mathbf{P}_{xz}\mathbf{P}_{zz}^{-1} \\
\hat{\mathbf{x}}_k &= \hat{\mathbf{x}}_k^- + \mathbf{K}_k(\mathbf{Z}_k - \hat{\mathbf{z}}_k) \\
\mathbf{P}_k &= \mathbf{P}_k^- - \mathbf{K}_k\mathbf{P}_{zz}\mathbf{K}_k^{\mathrm{T}}
\end{aligned}
\tag{16}
$$

When nonlinear system model has higher accuracy, and initial condition $\hat{\mathbf{x}}_0^-, \mathbf{P}_0^-$ and filtering algorithm are appropriate, UKF can get more accurate state estimation. But general system has uncertainty of different degrees, such as unpredictability of structure or parameters of mathematical model, unknown disturbance of external environment and inaccuracy of measurement noise statistical result, which will cause filtering accuracy decline of EKF and UKF, even divergence, for both are derived based on traditional Kalman filtering structure.

4   UKF Algorithm with Strong Tracking

4.1   Strong Tracking Filtering Strategy

In order to make filter optimal and avoid system divergence, various adaptive Kalman filtering strategies have emerged. Among them, strong tracking filter with fading factor can make algorithm have strong tracking filtering characteristics, when system model is uncertain. The time-varying fading factor is used to weaken old observation measurement influence on current filtering results, and increase effect of current measurement. The fading factor $\mathbf{K}_k$ is adjusted in real time, and residual sequence of forced filtering keeps orthogonality at all times, and variance matrix is updated to

$$
\mathbf{P}_k^- = \lambda_k \mathbf{\Phi}_{k-1} \mathbf{P}_{k-1} \mathbf{\Phi}_k^{\mathrm{T}} + \mathbf{Q}_{k-1}
\tag{17}
$$

In which $\lambda_k \geq 1$ is the time-varying fading factor.

In fact, very complicated calculate are required to designate optimal fading factor λ_k, orthogonality principle and iterative calculation are both applied, and the real-time problem also need to take into account. While, suboptimal fading factor can be determined by

$$
\lambda_k = \begin{cases} \lambda_0, & \lambda_k \geq \\ 1, & \lambda_k < \end{cases} \qquad \lambda_0 = \frac{\mathrm{tr}(\mathbf{N}_k)}{\mathrm{tr}(\mathbf{M}_k)}
\tag{18}
$$

Where $\mathrm{tr}(\,\cdot\,)$ represents the operator of matrix trace, $\mathbf{M}_k = \mathbf{H}_k \mathbf{P}_k^l \mathbf{H}_k^{\mathrm{T}} + \mathbf{R}_k - \mathbf{V}_k + \mathbf{N}_k$, and $\mathbf{N}_k = \mathbf{V}_k - \mathbf{H}_k \mathbf{Q}_k \mathbf{H}_k^{\mathrm{T}} - \mathbf{R}_k$. $\mathbf{P}_k^l$ is state prediction variance without fading factor, obviously, $\mathbf{P}_k^l = \mathbf{\Phi}_k^- \mathbf{P}_k (\mathbf{\Phi}_k^-)^{\mathrm{T}} + \mathbf{Q}_k$, $\mathbf{V}_k$ is covariance matrix of actual output residual sequence. Because it is unknown, so estimator is used as

$$V_k = \begin{cases} \boldsymbol{\varepsilon}_0\boldsymbol{\varepsilon}_0^{\mathrm{T}}, & k = 1 \\ (\rho V_{k-1} + \boldsymbol{\varepsilon}_k\boldsymbol{\varepsilon}_k^{\mathrm{T}})/(1+\rho), & k \geq 2 \end{cases} \tag{19}$$

Where $0 < \rho \leq 1$ is forgetting factor, set to 0.95.

$\mathbf{P}_k^-$ is given in Eq. (17), in fact, after elimination factor is substituted, output covariance $\mathbf{P}_{zz}$ and cross covariance $\mathbf{P}_{xz}$ can be expressed as

$$\begin{aligned} \mathbf{P}_{zz} &= \mathbf{H}_k\mathbf{P}_k^l\mathbf{H}_k^{\mathrm{T}} + \mathbf{R}_k \\ \mathbf{P}_{xz} &= \mathbf{P}_k^l\mathbf{H}_k^{\mathrm{T}} \end{aligned} \tag{20}$$

Where superscript l represents result without fading factor. So far, Eq. (16) can be rewritten as

$$\begin{aligned} \mathbf{K}_k &= \mathbf{P}_{xz}\mathbf{P}_{zz}^{-1} \\ \hat{\mathbf{x}}_k &= \hat{\mathbf{x}}_k^- + \mathbf{K}_k(\mathbf{z}_k - \hat{\mathbf{z}}_k) \\ \mathbf{P}_k &= \mathbf{P}_k^- - \mathbf{K}_k\mathbf{P}_{zz}\mathbf{K}_k^{\mathrm{T}} \end{aligned} \tag{21}$$

Correspondingly, the equivalent expressions of $\mathbf{N}_k$ and $\mathbf{M}_k$ are $\mathbf{N}_k = V_k - (\mathbf{P}_{xz}^l)^{\mathrm{T}}(\mathbf{P}_k^l)^{-1}\mathbf{Q}_k(\mathbf{P}_k^l)^{-1}(\mathbf{P}_{xz}^l)^{\mathrm{T}} - \mathbf{R}_k$ and $\mathbf{M}_k = \mathbf{P}_{zz}^l - V_k + \mathbf{N}_k$, respectively.

4.2   UKF Algorithm Based on Strong Tracking (STUKF)

Combining UKF algorithm with strong tracking strategy, STUKF is obtained. The process can be described in detail as follows.

Step 1. Time update

$$(\boldsymbol{\zeta}_k^-)_i = \boldsymbol{f}_k((\mathbf{X}_{k-1})_i), \quad i = 0, \cdots, 2n \tag{22}$$

$$\hat{\mathbf{x}}_k^- = \sum_{i=0}^{2n} W_i^{(\mathrm{m})}(\boldsymbol{\zeta}_k^-)_i + \mathbf{q}_{k-1} \tag{23}$$

$$\mathbf{P}_k^- = \lambda_k \sum_{i=0}^{2n} W_i^{(\mathrm{c})}[(\boldsymbol{\zeta}_k^-)_i - \hat{\mathbf{x}}_k^-][(\boldsymbol{\zeta}_k^-)_i - \hat{\mathbf{x}}_k^-]^{\mathrm{T}} + \mathbf{Q}_{k-1} \tag{24}$$

Where $(\boldsymbol{\zeta}_k^-)_i$ is the sigma point set composed of $\hat{\mathbf{x}}_k$ and $\mathbf{P}_k$.

Step 2. Measurement update

$$(\mathbf{Z}_k^-)_i = h_k((\boldsymbol{\zeta}_k^-)_i) \tag{25}$$

$$(\hat{\mathbf{z}}_k^-)_i = \sum_{i=0}^{2n} W_i^{(\mathrm{m})}(\mathbf{Z}_k^-)_i + \mathbf{r}_k \tag{26}$$

$$\mathbf{P}_{zz} = \sum_{i=0}^{2n} W_i^{(c)} [(\mathbf{Z}_k^- - \hat{\mathbf{z}}_k^-)(\mathbf{Z}_k^- - \hat{\mathbf{z}}_k^-)]^{\mathrm{T}} + \mathbf{R}_k \tag{27}$$

$$\mathbf{P}_{xz} = \sum_{i=0}^{2n} W_i^{(c)} [(\boldsymbol{\zeta}_k^-)_i - \hat{\mathbf{x}}_k^-)][((\mathbf{Z}_k^-)_i - \hat{\mathbf{z}}_k^-)]^{\mathrm{T}} \tag{28}$$

Then filter update is obtained.

$$\begin{aligned} \hat{\mathbf{x}}_k &= \hat{\mathbf{x}}_k^- + \mathbf{K}_k(\mathbf{z}_k - \hat{\mathbf{z}}_k^-) \\ \mathbf{K}_k &= \mathbf{P}_{xz}\mathbf{P}_{zz}^{-1} \\ \mathbf{P}_k &= \mathbf{P}_k^- - \mathbf{K}_k\mathbf{P}_{zz}\mathbf{K}_k^{\mathrm{T}} \end{aligned} \tag{29}$$

In fact, if optimal fading factor consumes a lot of calculation cost in the process of determination, it is difficult to be applied in real time, and cannot guarantee system convergence at every time. Therefore, the suboptimal fading factor λ_k is usually used. For each channel, the fading factor can be expressed as

$$\begin{aligned} \lambda_{i,k} &= \frac{\mathrm{tr}(\eta \mathbf{V}_k - \beta \mathbf{R}_k)}{\mathrm{tr}(\mathbf{P}_{zz})} = \begin{cases} \lambda_{i,k}, & \lambda_{i,k} > 1 \\ 1, & \lambda_{i,k} \le 1 \end{cases} \\ \mathbf{V}_k &= \begin{cases} \boldsymbol{\varepsilon}_0 \boldsymbol{\varepsilon}_0^{\mathrm{T}}, & k = 1 \\ (\rho \mathbf{V}_{k-1} + \boldsymbol{\varepsilon}_k \boldsymbol{\varepsilon}_k^{\mathrm{T}})/(1+\rho) & k \ge 2 \end{cases} \end{aligned} \tag{30}$$

Where $\boldsymbol{\varepsilon}_k = \mathbf{z}_k - \hat{\mathbf{z}}_k^-$ is residual sequence. In order to avoid the over regulation and make state estimation result smoother, softening factor β introduced and usually given by $\beta = \min(\sum_{k=1}^{2n} \sum_{i=1}^{n} |x_k^i - \hat{x}_k^i|)$, which is determined by computer simulation. variance matrix in formula (12) is updated according to correction factor, and the results are obtained

$$\mathbf{P}_k^- = \lambda_k \sum_{i=0}^{2n} W_i^{(c)} [(\boldsymbol{\zeta}_k^-)_i - \hat{\mathbf{x}}_k^-)][(\boldsymbol{\zeta}_k^-)_i - \hat{\mathbf{x}}_k^-)]^{\mathrm{T}} + \mathbf{Q}_{k-1} \tag{31}$$

Similarly, the covariance matrix $\mathbf{P}_{zz}$ and $\mathbf{P}_{xz}$ can be expressed as

$$\begin{aligned} \mathbf{P}_{zz} &= \lambda_k \sum_{i=0}^{2n} W_i^{(c)} [(\mathbf{Z}_k^- - \hat{\mathbf{z}}_k^-)(\mathbf{Z}_k^- - \hat{\mathbf{z}}_k^-)]^{\mathrm{T}} + \mathbf{R}_k \\ \mathbf{P}_{xz} &= \lambda_k \sum_{i=0}^{2n} W_i^{(c)} [(\boldsymbol{\zeta}_k^-)_i - \hat{\mathbf{x}}_k^-)][((\mathbf{Z}_k^-)_i - \hat{\mathbf{z}}_k^-)]^{\mathrm{T}} \end{aligned} \tag{32}$$

Where $\mathbf{S}_k = \mathrm{diag}(s_1, s_2, \cdots, s_m)$, while $s_i \le 1, i = 1, \cdots, m$, the filter is in steady-state operation process, while $s_i > 1$, the filter is in unstable state. Under special condition $s_i = 1$, the filter degenerates into general UKF filtering process.

5   Simulation and Result Analysis

The experiments are based on SINS/BDS integration adopting nonlinear filtering algorithm. Simulation experiment is mainly to verify feasibility of STUKF algorithm, and provide theoretical guidance for developing actual system. Parameters in experiment are as follows: the output period of SINS is 0.01 s, the BDS output and the system filter period are 1 s, variance of BDS receiver in horizontal and altitude directions is 5 m and 10 m respectively, and velocity variance is 0.1 m/s. The motion trajectory is generated by simulation at the initial position (116.186°, 39.843° and 200 m), and then performances of classic UKF, fading memory UKF and STUKF are compared respectively. Filtering results are shown in Fig. 2 and Fig. 3 respectively.

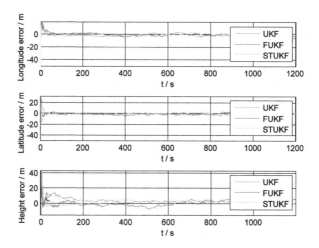

Fig. 2. Position error of three different algorithms

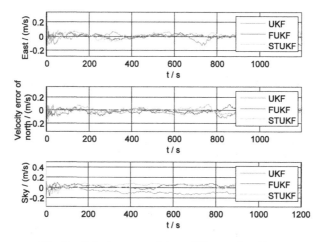

Fig. 3. Velocity error of three different algorithms

From Figs. 2 and 3, it can be seen that accuracy of STUKF algorithm is higher than UKF, because it uses the characteristics of strong tracking filter to approximate the mapping function of nonlinear model. Table 1 shows error (variance) of each parameter of STUKF algorithm, classical UKF and fading memory UKF algorithms in 1–1200 epoch.

Table 1. Performance comparison of three kinds of different algorithms

Algorithm	Position error		Speed error		Calculation time/s
	East/m	North/m	East/m $\cdot$ s^{-1}	North/m $\cdot$ s^{-1}	
UKF	7.4275	7.5172	0.0016	0.0011	9.0671
Fading memory UKF	7.4994	8.4625	0.0022	0.0015	9.2810
STUKF	5.5911	5.9177	0.0015	0.0013	8.9507

Table 1 shows operation time of three algorithm. It can be seen that position accuracy of STUKF algorithm is improved and its running time is reduced under same parameters. In fact, STUKF can be used more fully when system's running time longer, especially in terms of unpredictable structure and parameters of dynamic mathematical model, unknown disturbance of external environment and inaccurate statistical results of noise measurement, it can still hold good performance without big errors. The simulation results show that the STUKF adaptive filtering scheme can handle the integrated navigation problem very well, which provides theoretical and practical guidance for further experiments.

6 Conclusion

In order to improve accuracy and stability of INS/GNSS integrated navigation system, deep integrated system model is given. Based on traditional UKF filtering algorithm and strong tracking strategy, STUKF algorithm is deduced, which solves the problem of numerical stability and noise mutation of traditional algorithm. Experimental results show that the algorithm improves performance of INS/GNSS (deep) integrated navigation system.

The algorithm can be used for state estimation of moving carrier in high dynamic environment with proper parameter. It can provide guidance for INS/GNSS integrated navigation design, which is of great significance to improve navigation precision and guidance ability of weapon equipment.

References

1. Zaidi, A.S., Suddle, M.R.: Global navigation satellite systems: a survey. In: Advances in Space Technologies, pp. 84–87 (2006)
2. Tang, X.Q., Cheng, X.W., Gao, J.Q.: Hybrid theoretical variance analysis for random error properties of optic gyroscope. Acta Armamentarii **36**(9), 1688–1695 (2015)

3. Lashley, M., Bevly, D.M.: Performance comparison of deep integration and tight coupling. Navigation **60**(3), 159–178 (2013)
4. Kaplan, E.D., Hegarty, C.J.: Understanding GPS: Principle and Applications, 2nd edn. Artech House Inc., Boston (2006)
5. Borre, K., Akos, D.: A Software-defined GPS and Galileo receiver: single-frequency approach. In: ION 18th International Technical Meeting of the Satellite Division, Long Beach, CA, USA, pp. 1632–1637 (2005)
6. Soloviev, A.: Tight coupling of GPS and INS for urban navigation. IEEE Trans. Aerosp. Electron. Syst. **46**(4), 1731–1746 (2010)
7. Julier, S.J., Uhlmann, J.K., Durrant-Whyte, H.F.: A new approach for the nonlinear transformation of means and co-variances in linear filters. IEEE Trans. Autom. Control **45**(3), 477–482 (2000)
8. Zhou, D.H., Frank, P.M.: Strong tracking Kalman filtering of nonlinear time-varying stochastic systems with coloured noise: application to parameter estimation and empirical robustness analysis. Int. J. Control **65**, 295–307 (1996)
9. Zhou, D.H., Wang, Q.L.: Strong tracking filtering of nonlinear systems with colored noise. J. Beijing Inst. Technol. **17**(3), 321–326 (1997)
10. Zhao, S.H., Lu, M.Q., Feng, Z.M.: Application of EKF and UKF in tightly-coupled integrated navigation system. Syst. Eng. Electron. **31**(10), 2450–2454 (2009)

The Application of Theoretical Variance#1 Method and Lifting Wavelet for Optic Gyroscopes

Cheng Xuwei$^{(\boxtimes)}$, Li Yuan, Zhou Min, Yan Zitong, and Xie Can

Department 21, Unit 32272, PLA, Chengdu 610214, China

Abstract. In order to achieve a particular desired result for de-noising the random noises of optic gyroscopes in Inertial Measurement Unit under disturbing circumstance, the Theoretical variance#1 was applied to determine the characteristics of the various types of error terms in the gyroscope gyros output by performing certain operations on the entire length of data, which improves the confidence efficiently and evaluation precision than conventional variance methods. Meanwhile, the application of the lifting wavelet transform has been implemented in de-noising the stochastic noise generated for a gyroscope gyro. Experimental results have validated the feasibility of the proposed filtering method, which is more effective than the traditional de-noising methods.

Keywords: Fiber optic gyro · Random error · Lifting wavelet · Allan variance · Theoretical variance#1

1 Introduction

An inertial navigation system based on optic gyroscope has been widely applied to various fields for its high accuracy and reliability. While, the gyroscope outputs are usually corrupted by different types of error sources [1]. The regular error can be compensated based on drift testing with hardware or software, while the exact mathematical model of the random drift of gyroscope is hardly been built [2]. So precise modeling, estimating and de-noising of the gyroscope noise components with various advanced filter algorithms in needed [3, 4].

Allan variance (Avar) describes the root mean square random-drift errors of data as a function of averaging time which is applied by the IEEE standard [5]. It has two fatal weakness: (1) it cannot possibly characterize frequency stability over an interval greater than 50% of the data length, and (2) has lower confidence at long averaging time [6, 7]. In order to solve such problems, Theoretical variance#1 [8] is adopted, which is not only consistent with the power law noises in estimating long-term τ-values, but also improves the confidence efficiently. Furthermore, it has a highest evaluation accuracy than other variances. By optimizing the performance of Theo1, the Hybrid Variance method can be adopted to characterize various types of noise terms of gyroscope outputs, complementing with the frequency domain analysis.

© Springer Nature Switzerland AG 2021
D.-S. Huang et al. (Eds.): ICIC 2021, LNCS 12837, pp. 183–193, 2021.
https://doi.org/10.1007/978-3-030-84529-2_16

Wavelet analysis is especially suited for analyze and process non-stationary signals than FFT or STFT for its excellent features. Based on the so-called lifting scheme, Sweldens proposed a general, and simple second generation wavelets with particular properties. Consider the non-stationary and nonlinear behavior of gyroscope signal and random disturbances, lifting wavelet filtering algorithm and a kind of threshold function combining with soft and hard threshold was designed. Adopting data moving window realizes recursive computation solves filtering signal's boundary problem with symmetric periodic extension method.

A brief review of Theo1 and new variance method is firstly described, then, a kind of second generation wavelet de-noising strategy is proposed. At last, we present an application of eliminating stochastic noises of optic gyroscope under stationary environment. Experimental results are given with original and filtered gyroscope outputs separately, which validates performance of the proposed filter method.

2 Brief Review of Allan Variance and Theoretical Variance#1

In the following, the mathematical definition of the Theoretical variance #1 variance analysis method and new hybrid method are given, and the relationship between variance and the noise is established. Using the unique relationship, the behavior of the characteristic curve for a number of prominent noise terms can be determined.

2.1 Theoretical Variance#1

By using Avar, the uncertainty of the data is assumed to be generated by noise sources of specific character. The magnitude of each noise source covariance is then estimated from the data. Similar to Avar, Theo1 is a novel kind of variance technique that character reliable estimation of frequency stability for longer sample periods compared to the length of a data run, which has two advantages over other variance estimating methods of frequency stability: evaluate frequency stability at a sample period (τ) of 3/4 the length of a data run, and presently attains the highest edf of any estimator of frequency stability.

Consider a sequence $x(i)$ with a sampling period between adjacent observations given by τ_0. We can accomplish reforming the samples with given span or stride $\tau = m\tau_0$ as shown in Fig. 1. Thus, Theo1 is defined in terms of $x(i)$ data by

$$\text{Theo1}(m, \tau_0) = \frac{1}{2} \left\langle \frac{2}{0.75m} \sum_{\delta=0}^{m/2-1} \frac{1}{m(m/2-\delta)} \cdot \frac{1}{\tau_0^2} [(x_{i+m/2} - x_{i+\delta}) - (x_{i-\delta} - x_{i-m/2})] \right\rangle \tag{1}$$

Where, δ is the averaging factor, $\tau_s = 0.75m\tau_0$, τ_s denotes the same meaning as traditional sampling period τ in Avar estimator, which indicates the last τ_s equals $0.75(N_x - 1)\tau_0$. Thus, averaging times is approximately 3/4 of data length, while the Avar can estimate only up to 1/2 of total data run. For the purpose of act on as many points as possible for a given data, we want m to be large for a good evaluation of Theo1 at a particular τ_S. Taking the relationship between edfs and confidence interval into account, for given τ, we choose even m, generally $10 \leq m \leq N_x - 1$ [9].

Figure 1 shows the sampling process of Theo1 in terms of $\{y(i)\}$ defined as $y_n = (x_n - x_{n-1})/\tau$. It computes frequency differences in interval T at varying stride $\tau_{S_1,S_2,etc.}$, and corresponding averaging time $\tau_{1,2,etc.}$ given by the inner summation. While stride τ_{S_1} equals averaging time $\tau_1 = T/2$, the summation's first term is the classical Avar. For $1 < \delta < m/2$, $\tau_{S(\cdot)} > T/2$ intermediate sampling functions are illustrated. The last sampling function is $\tau_{S(N)} > T - \tau_0$. Therefore, the effective τ-value of individual frequency differences averaged in Theo1 is limited between $T/2$ and $T - \tau_0$.

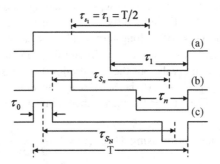

Fig. 1. Sampling process at varying stride using Theo1

In practice, we adopt the estimated value of the Theo1, it can be estimated as

$$\hat{\text{Theo}}1(m, \tau_0, N_x) = \frac{1}{0.75(N_x - m)(m\tau_0)^2} \sum_{i=1}^{N_x-m} \sum_{\delta=0}^{m/2-1} \frac{1}{m/2 - \delta}[(x_i - x_{i-\delta+m/2}) + (x_{i+m} - x_{i+\delta+m/2})] \quad (2)$$

An estimation error also be assigned to Theo1 by using of edf, which are substantially greater than max-overlap estimates of the Avar, particularly at long-term τ-values. A set of rule of thumb formulae calculating confidence intervals with 10% fitting accuracy of edfs, corresponding to Theo1 for the five noises. Their edf are

$$\underbrace{edf}_{\text{WHFM}} = (\frac{4.1N_x + 0.8}{\tau_s} - \frac{3.1N_x + 6.5}{N_x})(\frac{\tau_s^{3/2}}{\tau_s^{3/2} + 5.2})$$

$$\underbrace{edf}_{\text{FLFM}} = (\frac{2N_x^2 - 1.3N_x\tau_s - 3.5\tau_s}{N_x\tau_s})(\frac{\tau_s^3}{\tau_s^3 + 2.3})$$

$$\underbrace{edf}_{\text{RWFM}} = (\frac{4.4N_x - 2}{2.9\tau_s})(\frac{(4.4N_x - 1)^2 - 8.6\tau_s(4.4N_x - 1) + 11.4\tau_s^2}{(4.4N_x - 3)^2}) \qquad (3)$$

$$\underbrace{edf}_{\text{WHPM}} = (\frac{0.86(N_x + 1)(N_x - 4/3\tau_s)}{N_x - \tau_s})(\frac{\tau_s}{\tau_s + 1.14})$$

$$\underbrace{edf}_{\text{FLPM}} = (\frac{4.798N_x^2 - 6.374N_x\tau_s + 12.387\tau_s}{(\tau_s + 36.6)^{1/2}(N_x - \tau_s)})(\frac{\tau_s}{\tau_s + 0.3})$$

Clearly, the edfs depend on two factors, correlation time and total sampling length. The edfs of traditional Avar estimator are low, so evaluating value are always fluctuate; While Theo1 has more high estimating edfs, thus, it can replace Avar to evaluate the characteristics of optical gyros within long-term, and to evaluate modeling and filtering results of random process. In this case, the χ^2 distribution function is used for the Avar for calculating confidence intervals, but the distribution functions are narrowing using Theo1 at long averaging periods. At the same way, an empirical formula to obtain percentage upper-bound Theo1($\tau_s = 0.75m\tau_0$) uncertainty is [10]

$$\text{Percentage error using Thêo1}(\tau) = \frac{100}{\sqrt{2(edf + 6.6)}} \qquad (4)$$

Where $edf \geq 1$ is calculated from the Eq. (3). The uncertainty is meant to provide some information on the spread of possible values of Theo1 that are consistent with 90% confidence. Table 1 compares edf between Avar and Theo1 variance from simulation experience of random process in which $N_x = 20\,000$ using the three FM noises from the short- to long-term range $16 \leq m \leq 512$. In all cases, the increased edf using Thoe1 is significant. The initial sample size is $m \geq 10$.

Table 1. The edfs with Theo1 of noises and percentage error

m	WHFM		FLFM		RWFM	
	Edf (m)	Percentage error	Edf (m)	Percentage error	Edf (m)	Percentage error
16	240.09	6.37	126.75	8.66	94.29	9.96
32	124.66	8.73	62.75	12.01	45.68	13.836
64	61.94	12.08	30.73	16.37	21.40	18.90
128	29.63	16.61	14.71	21.66	9.30	25.08
256	13.30	22.42	6.70	27.42	3.33	31.73
512	5.10	29.232	2.70	32.79		

2.2 Hybrid Variance Method

Because data acquisition method of #1 Theoretical variance is different from Avar, there is a significant deviation between them, which is expressed with ratio as

$$\text{nbias}(m, \tau_0) = \frac{\sigma_y^2(\tau)}{\text{Theo1}(m, \tau_0)} - 1 \tag{5}$$

In reference [11], Monte Carlo method was used to fit the relationship between the deviation in equation and m, we can easily obtain

$$\text{bias}(\tau) = a + b/m^c - 1 \tag{6}$$

Where: a, b and c are constants. Theo1 and Avar deviation values corresponding to 5 kinds of conventional noise, namely, frequency modulated white noise (WHFM), frequency modulated flicker noise (FLFM), frequency modulated random walk (RWFM), phase modulated white noise (WHPM) and phase modulated flicker noise (FLPM), which are shown in Table 1. It should be noted that except for the unbiased estimation of WHFM, other noises are biased.

Table 2. Coefficient of the bias functions of Theo1

	WHPM	FLPM	WHFM	FLFM	RWFM
a	0.09	0.14	1	1.87	2.70
b	0.74	0.82	0	−1.05	−1.53
c	0.40	0.30	0	0.79	0.85

In order to eliminate the bias of theo1, an automatic unbiased Theo1 bias-removed method (TheoBR) constructed from the maximum value of Avar

$$\text{ThêoBR}(m, \tau_0, N_x) = \frac{1}{n+1} \cdot \text{Thêo1}(m, \tau_0, N_x) = \sum_{i=0}^{n} \frac{\text{Avar}(m = 9 + 3i, \tau_0, N_x)}{\text{Thêo1}(m = 12 + 4i, \tau_0, N_x)} \tag{7}$$

Where, $n = floor(0.1N_x/3 - 3)$, in order to prevent decrease of confidence of Avar in long correlation time, τ value should be selected within a certain proportion of the total length of data (such as $\tau = 10\% \sim 20\%n$).

Combining Avar (m, τ_0, N_x) and Theo1#(m, τ_0, N_x) to construct TheoH variance on a complete data segment, which is called mixed theoretical variance, whose estimated value is expressed as

$$\text{ThêoH}(m, \tau_0, N_x) = \begin{cases} \text{Avar}(m, \tau_0, N_x) & 1 \leq m \leq k/\tau_0 \\ \text{ThêoBR}(m, \tau_0, N_x) & k/0.75k\tau_0 \leq m \leq N_x - 1 \end{cases} \tag{8}$$

Where $k = 10\%T$ is maximum value within the reliable confidence range of Avar (m, τ_0, N_x); TheoBR (m, τ_0, N_x) is used after the data length of 10%. Because of longer correlation time, statistical characteristics of Avar will be worse and worse, so TheoBR is used instead. The results obtained in this way are fully compatible with Avar in the short correlation time, and in the total correlation time, the average value becomes $0.75m\tau_0$. At this time, the span of value will be 50% longer than that of Avar, so it can more completely describe noise characteristics of whole data segment.

2.3   Analyzing Random Error Characteristics of FOG

Thoe1 variance techniques can be used to investigate stochastic processes, because there is a direct relationship that exists between $\sigma_A^2(\tau)$ and the power spectral density (PSD) of the intrinsic random processes. Thus, for one-sided PSD, the unique relationship is described as

$$\sigma_A^2(\tau) = 4 \int_0^\infty S_\omega(f) \frac{\sin^4(\pi f \tau)}{(\pi f \tau)^2} df \qquad (9)$$

In which τ is the correlation time, $S_\omega(f)$ is the PSD. Equation (9) is the result to calculate Avar from noise PSD, which can also be used to Theo1# or TheoBR. It is seen that different types of random processes can be examined by adjusting τ. Applying such relationship, the types and magnitude of prominent noise terms can be obtained directly from variance curve [12]. Thus, the Avar and Theo1 both provide a complete means of identifying and quantifying various noise terms (angle random walk, rate random walk, bias instability, quantization noise, and rate ramp) that exist in output.

Consider the relationship between Avar and signal PSD [13], applying the relationship between the PSD and the correlation time τ, we can readily obtain

$$\sigma_A^2(\tau) = \frac{3Q^2}{\tau^2} + \frac{N^2}{\tau} + (0.6643B)^2 + \frac{K^2\tau}{3} + \frac{R^2\tau^2}{2} = \sum_{n=-2}^{n=2} A_n\tau^n \qquad (10)$$

Where, A_n of Eq. (10) is obtained adopting least square fitting, each noise coefficients can be calculated. In practice, noises source is statistically independent of all the random processes, that is, each noise term appears in different regions of τ. Avar or Theo1 variance always be expressed as the square sum of several types of noises, and plotted in a log-log plot looks like the one shown in Fig. 2.

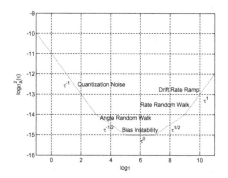

Fig. 2. Log-log plot of $\tau \sim \sigma^2$

Because of taking the different sampling process, there is certain biases between Avar and Theo1, which is described by the ratio their estimator values ratio. Equation (5) gives each standard variance coefficient and the biases between Avar and Theo1 corresponding to every noise term respectively. Consider the biases in Table 2, and assume noises source is statistically independent, the Theo1 is obtained as follows:

$$\text{Theo1}\,\sigma_A^2(\tau) = \frac{1}{0.4}A_{-2}\tau^{-2} + A_{-1}\tau^{-1} + \frac{1}{1.7}A_0\tau^0 + \frac{1}{2.24}A_1\tau^1 + \frac{1}{3.24}A_2\tau^2 \quad (11)$$

Where, a_n is fitting coefficients of Theo1 variance, while A_n is noise parameters using Avar. Calculating the Theo1 variance, we can readily obtain the complete curve of loglog $\tau \sim \sigma^2$ in the total correlation time, thus, each random noise be separated from varying slopes in the loglog plot; Lastly, fitting coefficients a_n are obtained (Table 3).

Table 3. Allan std. variance vs Theo1 bias of FOG random error

Noise pattern	Parameters (unit)	Allan std. variance	Variance bias[a]	Std. variance bias
Quantization noise	Q (μrad)	$\frac{10^6 \cdot 3600\pi\sqrt{3A_{-2}}}{180}$	0.4Theo1	0.63Theo1dev
Angle random walk	N ($^\circ/h^{\frac{1}{2}}$)	$\sqrt{A_{-1}/60}$	Theo1	Theo1dev
Bias instability	B ($^\circ/h$)	$\sqrt{A_0/0.6643}$	0.17Theo1	0.31Theo1dev
Rate random walk	K ($^\circ/h^{\frac{3}{2}}$)	$60\sqrt{3A_1}$	2.24Theo1	1.5Theo1dev
Drift rate ramp	R ($^\circ/h^2$)	$3600\sqrt{2A_2}$	3.24Theo1	1.8Theo1dev

3   Lifting Wavelet Transform

Wavelet analysis method has localization characteristics changed with time and frequency window, which is particularly suitable for analyzing and processing nonstationary signals [14, 15]. To deal with the situations where first-generation wavelet is not applicable, such as optic gyroscopes output, which is irregularly sampled data, second-generation wavelets is proposed based on the lifting scheme, which provides a flexible and efficient framework for building a new construction of wavelet.

3.1   Lifting Scheme

Lifting can be considered as building a new wavelet with improved properties, by gradually adding new basis functions or multi-resolution analysis. Which optimizes the computations to utilize limited memory efficiently and speed up calculation. It only involves integer coefficients instead of complex float point coefficients. In order to get signal decomposition without loss, we construct second generation wavelet transform with lifting scheme for gyroscopes output. A normative lifting scheme consists of four steps: split, predict, update, and scale (omitted), as an illustration, we give in Fig. 3 a block diagram of the forward and inverse wavelet lifting transform.

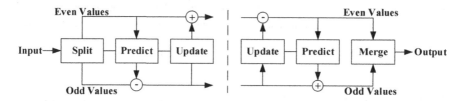

Fig. 3. Lifting scheme (one stage only), left part: forward lifting; right part: inverse lifting.

(1) Split: This step splits original signal $s_j = \{s_{j,k}\}$(of even length) into two disjoint sets of sequence, those with even and those with odd index, indicated by e_{j-1} and o_{j-1} respectively. Each subset length equals half of original signal's length. This is called the *Lazy Wavelet Transform.*

$$Spit(s_j) = (e_{j-1}, o_{j-1}), e_{j-1} = \{s_{j,2k}\}, o_{j-1} = \{s_{j,2k+1}\} \tag{12}$$

(2) Predict: As the even sequence and odd sequence o_{j-1} are correlated, we can predict e_{j-1} from o_{j-1}

$$d_{j-1} = o_{j-1} - P(e_{j-1}) \tag{13}$$

P is a prediction operator, which is applied to the even sequence and the result is subtracted from the odd sequence to get detail band d_{j-1}, we call such process as *dual lifting*. It can be expressed as function P_k, such as $e_{j,2k}$, $(e_{j-1,k} + e_{j-1,k+1})/2$ or other more complex functions.

(3) Update: Similarly, in order to keep some essential features of original data, an update operator U is applied to define s_{j-1}, which is called *primal lifting*. U can also be expressed as different functions, thus, the updating result can be expressed as:

$$s_{j-1} = e_{j-1} - U(d_{j-1}) \tag{14}$$

(4) Scale: The scale step is always omitted.

A significant feature of the lifting scheme is that the inverse transform can be constructed simply. This is accomplished by "inverse lifting" diagram, see Fig. 3 (right). The unique P and U describes different wavelet transforms, signal s_j can be decomposed as approximation band s_{j-1} and detail band d_{j-1} after lifting wavelet transform. While the process of spit, predict and update can repeatedly applied to the s_{j-1}, thus, we obtain s_{j-2} and d_{j-2}. For a multi-level transform, the process is repeatedly applied until a desired number of decomposition level is reached.

Denote the decomposition level n, the original signal s_j then be expressed as $\{s_{j-n}, d_{j-n}\}$. Reconstructing each band processed separately, signals with different time-frequency domain characteristics can be obtained. Considerate the useful signal is narrow band signal with low frequency, while the noises are wide band signal. Wavelet de-noise aimed to separate useful signal and disturbing signal in original output, that is, the wavelet transform of real signal and noise can be separated effectively in wavelet domain.

4   Simulation and Experiments

In order to verify algorithm effectiveness, static experiments are analyzed. 72000 static test sampling data was held of single z-axis optic gyroscope at room temperature. For the sake of contrast, both the Avar and Theo1 methods are applied to the whole data sets respectively, a log-log plot of each standard deviation versus the cluster time is shown in Figs. 4 and 5 (only the first 2000 sampling points are given).

By performing operation on entire length of data, the characteristic curves are obtained which are easily used to identify different types and magnitude of error terms. Figure 5 shows that the quantization noise is dominant noise for short cluster time, while, the angle random walk is dominant error source for long-term test. Thus, proper measurement needs to be taken in order to improve gyroscope characteristics. Because of quantization noise estimating is based on very short cluster times, independent cluster number is large enough, so estimation quality is relatively good. However, the Avar gives unstable results even fluctuate at large τ values, which is obviously noticed in Fig. 5. At short-term (such as 1 000) and medium-term (such as 30 000) τ averaging intervals, the estimator for the Avar has good enough confidence, so its results amount to Theo1. Particularly, the advantage of Thoe1 is useful for evaluating frequency noises stability at large τ values, namely, it is higher edfs that make results more accurate, especially for long data runs.

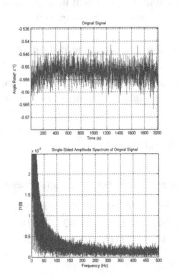

Fig. 4. Original gyro output signal of and its Frequency spectrum

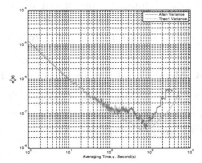

Fig. 5. Avar and Theo1 of gyro output (loglog)

Using TheoH, we can see that the stability up to 3/4 of data length in Fig. 6; In this case, the duration of data-run length is 72 000, longest averaging time is $\tau_s = 0.75(N_x - 1)\tau_0$ corresponding to 54 000. That is 50% longer than longest τ value of Avar. The goodness of fit is improved effectively for Theo1 with certain linearity. Thus, although there is still the bias between Theo1 and Avar, however, Theo1 describes frequency drift noises more valid than Avar. TheoH is compatible with Avar in short-term, and particularly suitable to estimate gyroscope random error in long-term.

In order to verify filter effect of the new filtering algorithm, adopt new wavelet transform to filter the noises. The filtered results are shown as Fig. 7 (only the first 2000 sampling points are given).

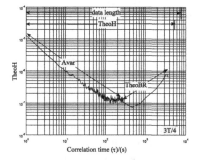

Fig. 6. TheoH of gyro output (loglog)

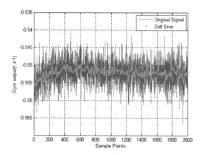

Fig. 7. Before and after filtering with lifting wavelet transform

Each noise coefficient is obtained and shown as Table 4. It is worth noting that the noise coefficients (except B) are halved, which demonstrates that the proposed method can effectively eliminate the effect of drift error.

The method presented in this paper allow a systematic characterization of the various random errors contained in optic gyroscope output data. The static test results prove that the filtering algorithm is valid, and for this gyroscope the quantization noise is the dominant error term in short cluster times, and the angular random walk and drift rate ramp are dominant errors in long cluster times. By using wavelet filtering, most of noise are removed, the proposed filtering method is effective to eliminate static errors of gyroscope, signals after filtering are more smooth and steady.

Table 4. Mean value, deviations and noise coefficients of gyroscope (before and after filtering)

Data pattern	Mean value	Deviation	$Q(\mu rad)$	$N(^{\circ}/h^{1/2})$	$B(^{\circ}/h)$	$K(^{\circ}/h^{3/2})$	$R(^{\circ}/h^2)$
Before filtering	0.5556618	0.0074	16.2301	0.1922	0.1250	3.3742	2.6247
After filtering	0.5556785	0.0022	1.1936	0.0048	0.2415	0.7239	1.9216

5 Conclusion

This work proposes a new variance evaluation method and second wavelet filter algorithm constructed by lifting scheme, which is employed to solve the drift error modeling problem of optic gyroscope in static test. The new second generation wavelet filter algorithm is applied to decompose and reconstruct gyroscope output signal. Experiments results are carried out to verify the theoretical results and clearly show that the parameters of each noise terms are decreased, and indicated that the proposed filter method is valid. This study also clearly reveals that Theo1 is a more powerful technique to investigate gyroscope error behavior on different timescales than Avar, which provides a new idea for noise parameters estimation of gyroscopes. Future work will involve testing the algorithm with time-varying condition and other dynamic applying environments.

References

1. Wang, X.L., Ma, S.: Applicability compensation method for random drift of fiber optic gyro scopes. J. Beijing Univ. Aeronaut. Astronaut. **34**(6), 681–685 (2008)
2. Liu, Y., Li, L.L., Liu, J., et al.: Study on gray neural network drift modeling for piezoelectric gyro. J. Syst. Simul. **19**(20), 4676–4679 (2007)
3. Liu, J.Y., Yu, Z.Y., Ma, X.W.: Modeling and compensation of static temperature error synthetically for fiber optic gyroscope. Acta Optica Sinica **32**(8), 823005 (2012)
4. Tang, X.Q., Cheng, X.W., Guo, L.B., et al.: Error modeling and compensating of fiber optic gyro based on wavelet analysis and grey neural network. Chin. Lasar **39**(10), 1008003 (2012)
5. IEEE STD: IEEE Standard Specification Format Guide and Test Procedure for Single-Axis Interferometric Fiber Optic Gyros (1998)
6. Wang, X.L., Du, Y., Ding, Y.B.: Investigation of random error model for fiber optic gyroscope. J. Beijing Univ. Aeronaut. Astronaut. **32**(7), 769–772 (2006)
7. Han, J.L., Ge, S.M., Shen, Y.: Research on the random error properties of fog based on total variance. J. Harbin Inst. Technol. **39**(5), 708–711 (2007)
8. Howe, D.A.: Method of improving the estimation of long-term frequency variance. In: Proceedings of 1997 European Frequency and Time Forum, pp. 91–99 (1997)
9. Tasset, T.N., Howe, D.A., Percival, D.B.: Theo1 confidence intervals. In: 2004 IEEE International Frequency Control Symposium (Canada), pp. 725–728 (2004)
10. Howe, D.A.: ThêoH: a hybrid, high-confidence statistic that improves on the Allan deviation. Metrologia **43**, S322–S331 (2006)
11. Howe, D.A., Tasset, T.N.: Theo1: characterization of very long-term frequency stability. In: 18th European Frequency and Time Forum (EFTF 2004), pp. 581–587 (2004)
12. Tang, X.Q., Cheng, X.W., Gao, J.Q.: Hybrid theoretical variance analysis for random error properties of optic gyroscope. Acta Armamentarii **36**(9), 1688–1695 (2015)
13. Naser, El.-S., Hou, H.Y., Niu, X.J.: Analysis and modeling of inertial sensors using Allan variance. IEEE Trans. Instrum. Meas. **57**(1), 140–149 (2008)
14. Liu, Y., Li, Y., Xu, J.J.: Application of a multi-algorithm fusion real time filter in FOGs. Acta Photonica Sinica **9**(6), 1116–1119 (2010)
15. Dang, S.W., Tian, W.F., Qian, F.: De-noising fractional noise in fiber optic gyroscopes based on lifting wavelet. Chin. J. Lasers **36**(3), 625–629 (2006)

Proposing a Novel Fixed-Time Non-singular Terminal Sliding Mode Surface for Motion Tracking Control of Robot Manipulators

Anh Tuan Vo[1] , Thanh Nguyen Truong[1] , Hee-Jun Kang[1(✉)] ,
and Tien Dung Le[2]

[1] Department of Electrical, Electronic and Computer Engineering,
University of Ulsan, Ulsan 44610, South Korea
hjkang@ulsan.ac.kr
[2] The University of Danang – University of Science and Technology,
54 Nguyen Luong Bang Street, Danang 550000, Vietnam

Abstract. This article proposes a new fixed-time non-singular terminal sliding mode (FxNTSM) surface for synthesizing the tracking control algorithm of robotic manipulators. The new sliding mode surface has no singularity, improved convergence speed. It contains dynamic coefficients that can be adapted to control errors that drive the system states quickly converge to the equilibrium point within bounded convergence time regardless of initial system states. The settling time of the sliding motion phase is fully calculated in both cases, including the initial state of the system is far/near the equilibrium stability point. A novel finite-time NTSMC (FnNTSMC) is formed from a combination of the proposed sliding surface and auxiliary reaching control law (ARCL). Due to the advantages of the proposed sliding surface and mentioned reaching control law, therefore, the novel finite-time control system provides superior characteristics such as high tracking accuracy, improved fast convergence, and robustness to cope with the lumped uncertainties. The fixed time/finite-time stability and convergence of the control method are validated by using the Lyapunov criteria. Computer simulations applied to a 3-DOF robotic manipulator are presented to verify the effectiveness and outstanding properties of the proposed control method.

Keywords: Robot manipulators · Non-singular terminal sliding mode control · Fixed-time control

1 Introduction

Due to the outstanding technical potential, the applications of robotic manipulators are getting popular day by day in the military, scientific, medical field, commercial, or industrial applications such as automotive industry, ocean exploration, surgery, agriculture, bomb detection, etc. [1]. With the increase in the number of applications, there are growing concerns in control research for robot operators that control the robot trajectory to tracks the desired trajectory with effective control methods. In general, robot control is not difficult. However, robot control operated with high precision and

© Springer Nature Switzerland AG 2021
D.-S. Huang et al. (Eds.): ICIC 2021, LNCS 12837, pp. 194–206, 2021.
https://doi.org/10.1007/978-3-030-84529-2_17

safety is also a big challenge. Because the dynamic model of robotic manipulators is very complicated with coupling components, high nonlinearities, the effect of external disturbances, the payload variations, frictions, sensor noises, etc. Therefore, it seems impossible to model the uncertain components using mathematical equations.

Sliding mode control (SMC) is one of the most powerful nonlinear controllers which can be used in uncertain nonlinear dynamic systems or estimation systems [2]. SMC can cope with presence of many uncertain terms existing inside/outside the system. Therefore, SMC has been broadly employed in real applications such as robotic manipulators, cable-driven manipulators, nonlinear systems, etc. Unfortunately, it only guarantees asymptotic stability while a lot of chattering exists in the control input. The SMC's asymptotic stability means that it does not achieve a finite-time convergence for the system's state trajectories. To secure the finite-time convergence of the system state trajectories, a lot of attempts have been paid. In studies [3, 4], or [17], the terminal SMC (TSMC) has been introduced to provide higher tracking accuracy of the system states with finite-time stability. Therefore, the convergence performance of TSMC is significantly improved. However, the singularity problem can appear with some classic TSMC. There are some modified TSMC for robot manipulators in which the singularity problem was fully solved such as the non-singular TSMC (NTSMC) [5]. However, the convergence speed of NFTSMC has not been significantly improved yet. Furthermore, finite-time TSMC (FnTSMC) can control the system state trajectories of the system to reach the equilibrium point within a bounded time related to the initial states of the system [6]. This problem leads to additional difficulty in case of unknown initial conditions that prevent its applications. To increase the stability speed and dynamic performance of TSMC and FnTSMC, the fast TSMC (FTSMC) and global FTSMC (GFTSMC) have been developed as in [7, 8] and [9, 10], respectively. Unfortunately, FTSMC or GFTSMC do not guarantee a fixed-time convergence and stability. Furthermore, serious chattering occurs in the control signals of FTSMC or GFTSMC. As an extension of finite-time stability theory, fixed-time stability theory has been proposed. The fixed-time controller not only provides the desirable convergence performance but also removes dependence on the initial states of the system. Due to its good properties, fixed-time TSMC (FxTSMC) has been extensively studied in recent years. FxTSMC has been broadly applied to many applications [11, 12].

Based on the stated motivation, this paper firstly proposes a new FxNTSM surface for improving convergence speed of NTSMC and synthesizing the tracking control algorithm of robotic manipulators. The new FxNTSM surface has no singularity, improved convergence speed. It contains dynamic coefficients that can be adapted to control errors that drive the system states quickly converge to the equilibrium point within bounded convergence time regardless of initial system states. The settling time of the sliding motion phase is fully calculated in both cases, including the initial state of the system is far/near the equilibrium stability point. A novel FnNTSMC is formed from a combination of the proposed FxNTSM surface and ARCL. Due to the advantages of the proposed sliding surface and mentioned ARCL, therefore, the novel finite-time control system provides superior characteristics such as high tracking accuracy, fixed-time fast convergence, and robustness to cope with the lumped uncertainties. The fixed-time/finite-time stability and convergence of the control system are validated by using the Lyapunov criteria.

The organization of our article is as follows. After the introduction section, the essential preliminaries, notations, and problem formulations are presented in Sect. 2. Next, a new FxNTSM surface is developed in Sect. 3. In Sect. 4, the new FnTSMC is synthesized. In Sect. 5, Computer simulations applied to a 3-DOF robotic manipulator are presented to verify the effectiveness and outstanding properties of the proposed control method. Finally, Sect. 6 is conclusions.

2 Notations, Problem Formulations, and Preliminaries

2.1 Notations

To explicitly format the critical issues and the main results, several definitions will be given as R^n means the real n-dimensional space, R_+ denotes the set of positive real numbers, and $R^{n \times m}$ stands for the set of m by n real matrices. For any given vector, matrix, and its transpose, they are defined as $\mathbf{x}$, $\mathbf{x}$, and $\mathbf{x}^T$, respectively. $|x|$ and $\|\mathbf{x}\|$ represent absolute value of x and Euclidean norm of a vector $\mathbf{x}$, respectively.

The sign function has a following expression:

$$\text{sgn}(x) = \begin{cases} 1 & \text{if } x > 0 \\ 0 & \text{if } x = 0 \\ -1 & \text{otherwise} \end{cases} \tag{1}$$

$$\text{sig}(x)^\varphi = |x|^\varphi \text{sgn}(x), \quad \text{where } \varphi > 0 \tag{2}$$

It can be clearly confirmed that as $\varphi \geq 0$

$$\frac{d}{dt}\text{sig}(x)^\varphi = \varphi |x|^{\varphi-1}\dot{x} \tag{3}$$

2.2 Manipulator Dynamics

We consider the n-DOF manipulator dynamics with the following expression in Lagrangian form as [6, 16]:

$$\mathbf{M}(\mathbf{q})\ddot{\mathbf{q}} + \mathbf{C}(\mathbf{q}, \dot{\mathbf{q}})\dot{\mathbf{q}} + \mathbf{G}(\mathbf{q}) + \mathbf{f}_r(\dot{\mathbf{q}}) + \tau_d = \tau \tag{4}$$

where $\mathbf{q}, \dot{\mathbf{q}}, \ddot{\mathbf{q}}$ are the $n \times 1$ vectors of joint angular position, velocity, and acceleration, respectively; $\mathbf{M}(\mathbf{q})$ is the $n \times n$ positive-definite and symmetric matrix of inertia parameters; $\mathbf{C}(\mathbf{q}, \dot{\mathbf{q}})$ is the $n \times 1$ vector of the Coriolis and centripetal forces; τ is the $n \times 1$ vector of the control input torque; $\mathbf{G}(\mathbf{q}), \mathbf{f}_r(\dot{\mathbf{q}})$, and τ_d are the $n \times 1$ vector of the gravitational forces, friction forces, and the unknown time-varying external disturbance, respectively.

In fact, it is not easy to achieve a precise dynamic model of the robot. Therefore, we assume   that: $\mathbf{M}(\mathbf{q}) = \hat{\mathbf{M}}(\mathbf{q}) + \Delta\mathbf{M}(\mathbf{q})$, $\mathbf{C}(\mathbf{q}, \dot{\mathbf{q}}) = \hat{\mathbf{C}}(\mathbf{q}, \dot{\mathbf{q}}) + \Delta\mathbf{C}(\mathbf{q}, \dot{\mathbf{q}})$,   and

$G(q) = \hat{G}(q) + \Delta G(q)$. $\hat{M}(q) \in R^{n \times n}$, $\hat{C}(q, \dot{q}) \in R^{n \times n}$, and $\hat{G}(q) \in R^{n \times 1}$ are corresponding to estimation values of the real values of $M(q)$, $C(q, \dot{q})$, and $G(q)$. $\Delta M(q) \in R^{n \times n}$, $\Delta C(q, \dot{q}) \in R^{n \times n}$, and $\Delta G(q) \in R^{n \times 1}$ are uncertain dynamics.

We set $x = \begin{bmatrix} x_1^T, x_2^T \end{bmatrix}^T = \begin{bmatrix} q^T, \dot{q}^T \end{bmatrix}^T$ and $u = \tau$; accordingly, the modelling of the robot dynamic (4) is depicted in state space by:

$$\begin{cases} \dot{x}_1 = x_2 \\ \dot{x}_2 = a(x)u + b(x) + \delta(x, \Delta, \tau_d) \end{cases}, \tag{5}$$

where $b(x) = -\hat{M}^{-1}(q)\left(\hat{C}(q, \dot{q})\dot{q} + \hat{G}(q)\right)$ represents the lumped known terms of the robot, $a(x) = \hat{M}^{-1}(q)$ stands for a smooth function, while $\delta(x, \Delta, \tau_d) = -\hat{M}^{-1}(q)\begin{pmatrix} f_r(\dot{q}) + \Delta M(q)\ddot{q} \\ + \Delta C(q, \dot{q})\dot{q} + \Delta G(q) + \tau_d \end{pmatrix}$ indicates the lumped unknown uncertainty.

Motivated by the observation, the philosophy of this article is to develop a new robust control algorithm for manipulators (4) in position tracking control. So far, the system states of manipulators will be tracked the desired trajectories q_d with high tracking performance, including rapid convergence, non-singular, strong to cope with the uncertainty components, and impressive small control errors.

The control errors between the system states and the reference trajectories are determined as follows:

$$e_1 \triangleq q - q_d \tag{6}$$

From the control errors (6), their first and second derivative with respect to time are:

$$e_2 \triangleq \dot{q} - \dot{q}_d \text{ and } \dot{e}_2 \triangleq \ddot{q} - \ddot{q}_d \tag{7}$$

The manipulator dynamics (4) can be transferred to state-space equation with the following form related to the control errors:

$$\begin{cases} \dot{e}_1 = e_2 \\ \dot{e}_2 = b(e) + a(x)u + \delta(e, \Delta, \tau_d) \end{cases} \tag{8}$$

Here, vector of control errors is $e = \begin{bmatrix} e_1^T & e_2^T \end{bmatrix}^T \in R^{2n}$, the smooth nonlinear function $b(e) = -\hat{M}^{-1}(q)\left(\hat{C}(q, \dot{q})\dot{q} + \hat{G}(q) - \ddot{q}_d\right)$.

2.3 Preliminaries

The following assumptions, lemmas, and definitions are required for control design as well as finite-time/fixed-time stability evidence of the proposed control methodology.

Assumption 1 [13]: The total unknown uncertainties are assumed to be bounded by:

$$|\delta(\mathbf{e}, \Delta, \tau_d)| < \Lambda \tag{9}$$

where the positive constant Δ is considered to be foreknowledge.

Definition 1 [14]: The following autonomous system is considered:

$$\dot{\mathbf{x}}(t) = \mathbf{f}(\mathbf{x}, t), \mathbf{x}(0) = \mathbf{x}_0, \mathbf{f}(\mathbf{x}_0, 0) = 0, \mathbf{x} \in R^n \tag{10}$$

with $\mathbf{f}: D \times R_+ \to R^n$ is on an open neighborhood $D \subseteq R^n$ of the origin $\mathbf{x} = 0$, the zero solution of the system (10) is finite-time convergent if there is an open neighborhood $U \subseteq D$ of the origin and a function $T: U \backslash \{0\} \to R_+$, such that $\forall \mathbf{x}_0 \in U$, the solution trajectory $\psi(t, \mathbf{x}_0)$ of the system (11) is defined and $\psi(t, \mathbf{x}_0) \in U \backslash \{0\}$ for $t \in [0, T(\mathbf{x}_0))$, and $lim_{t \to T(\mathbf{x}_0)} \psi(t, \mathbf{x}_0) = 0$. Then, $T(\mathbf{x}_0)$ is called the settling time.

Definition 2 [14]: The zero solution of the system (10) is called to be a globally finite-time stable if it is Lyapunov stable with bounded time function $T(\mathbf{x}_0)$, i.e., there exists $T_{max} > 0$ such that $T(\mathbf{x}_0)$ satisfies the term $T(\mathbf{x}_0) < T_{max}$.

Lemma 1 [14]: Assume that there exists a continuous positive definite function $V(x)$: $D \to R$ such that the following condition hold:

1. $V(x)$ is positive definite.
2. There exist real numbers $\kappa > 0, 0 < \alpha < 1$ and an open neighborhood $D_0 \subseteq D$ of the origin such that:

$$\dot{V}(x) + \kappa V^\alpha(x) \leq 0, \ x \in D_0 \backslash \{0\} \tag{11}$$

Then, the origin is a finite-time-stable equilibrium of the system (11) and the finite settling time T satisfies:

$$T \leq \frac{1}{\kappa(1-\alpha)} V^{1-\alpha}(x(0)) \tag{12}$$

If in addition $D_0 = D = R^n$, the origin is a globally finite-time-stable equilibrium of system.

3   Design of a FxNTSM Surface

For the system (8), the novel FxNTSM surface is launched as:

$$\mathbf{s} = \varsigma + \Gamma \mathbf{e}_2^\theta \tag{13}$$

where $\mathbf{s} = [s_1 \quad \cdots \quad s_n]^T \in R^n$ is FxNTSM surface, $\mathbf{e}_1 = [e_{11}, \quad \cdots, \quad e_{1n}]^T$, $\mathbf{e}_2 = [e_{21}, \quad \cdots, \quad e_{2n}]^T$, $\mathbf{e}_2^\theta = [e_{21}^\theta, \quad \cdots, \quad e_{2n}^\theta]^T$, $\Gamma = diag(\Gamma_1, \quad \cdots, \quad \Gamma_n)$, $\Gamma_i > 0$, and $\varsigma = \left[(1+e_{11}^2)^\theta \arctan(e_{11}), \quad \cdots \quad ((1+e_{1n}^2)^\theta \arctan(e_{1n})) \right]^T$. θ is chosen to be satisfied the following condition:

$$\theta = 0.5(\eta/\rho + \rho/\eta) + 0.5(-\eta/\rho + \rho/\eta)\mathrm{sgn}(|e_{1i}| - 1) \tag{14}$$

where η and ρ are defined as are positive odd integers and $1 < \eta/\rho < 2$.

$$\theta = \begin{cases} \rho/\eta & |e_{1i}| > 1 \\ \eta/\rho & |e_{1i}| \le 1 \end{cases} \tag{15}$$

Once the proposed FxNTSM surface $\mathbf{s} = [s_1 \quad \cdots \quad s_n]^T = \mathbf{0}$ is accomplished, a set of the following differential equations is achieved:

$$(1 + e_{1i}^2)^\theta \arctan(e_{1i}) + \Gamma_i e_{2i}^\theta = 0, \quad i = 1, \ldots, n \tag{16}$$

Conducting a simple computation, for $i = 1, \ldots, n$, Eq. (16) becomes:

$$(1 + e_{1i}^2) \arctan(e_{1i})^{1/\theta} = -\Gamma_i^{1/\theta} \dot{e}_{1i} \tag{17}$$

Letting $\upsilon = \arctan(e_{1i})$, thus, its derivative is calculated as:

$$\dot{\upsilon} = (1 + e_{1i}^2)^{-1} \dot{e}_{1i} \tag{18}$$

Substituting $\dot{e}_{1i}$ which is obtained from Eq. (18) into Eq. (17), it results:

$$\upsilon^{1/\theta} + \Gamma_i^{1/\theta} \dot{\upsilon} = 0 \tag{19}$$

We consider respectively the following two cases:

Case 1: The initial state of the system is far from the equilibrium stability point $|e_{1i}| > 1$, Eq. (19) becomes:

$$\dot{\upsilon} = -\Gamma_i^{-\eta/\rho} \upsilon^{\eta/\rho} \tag{20}$$

Selecting the Lyapunov function $V_1 = 0.5\upsilon^2$, hence, its time derivative of V_1 is calculated along with the result from Eq. (20) as:

$$\dot{V}_1 = \upsilon\dot{\upsilon}$$
$$= -\Gamma_i^{-\eta/\rho} \upsilon^{\eta/\rho + 1} \tag{21}$$

It can be seen that $V_1 > 0$ and $\dot{V}_1 < 0$. Therefore, v and $\dot{v}$ can be stabilized to the equilibrium point.

To drive $e_{1i}(0) \rightarrow |e_{1i}| = 1$ then, $v(0) \rightarrow |v| = \pi/4$. Therefore, the settling time of the sliding motion in this stage is calculated by:

$$\int_0^{t_{1s1}} dt = \Gamma_i^{\eta/\rho} \int_{\pi/4}^{v(0)} v^{-\eta/\rho} d(|v|)$$

$$t_{1s_i} = \Gamma_i^{\eta/\rho} \int_{\pi/4}^{v(0)} v^{-\eta/\rho} d(|v|) = \Gamma_i^{\eta/\rho} \frac{\rho}{\rho - \eta} \left(|v| - \frac{\pi}{4} \right)^{1-\eta/\rho} \tag{22}$$

$$\leq \Gamma_i^{\eta/\rho} \frac{\rho}{\rho - \eta} \left(\left(\frac{\pi}{2} \right)^{1-\eta/\rho} - \left| \frac{\pi}{4} \right|^{1-\eta/\rho} \right)$$

Case 2: The trajectory error between the initial state of the system and the desired trajectory is $|e_{1i}| \leq 1$, Eq. (19) is written as:

$$\dot{v} = -\Gamma_i^{-\rho/\eta} v^{\rho/\eta} \tag{23}$$

Selecting a positive semi-definite function $V_2 = 0.5v^2$, hence, its time derivative of V_2 is given by referring to Eq. (23):

$$\dot{V}_2 = v \left(-\Gamma_i^{-\rho/\eta} v^{\rho/\eta} \right)$$

$$= -\Gamma_i^{-\rho/\eta} v^{\rho/\eta+1} \tag{24}$$

$$= -\sqrt{2}^{\rho/\eta+1} \Gamma_i^{-\rho/\eta} V_2^{(\rho/\eta+1)/2}$$

Because the value of η/ρ is selected to be satisfied the condition $1 < \eta/\rho < 2$, consequently, $(\rho/\eta + 1)/2 \in (3/4, 1)$. Referring to Lemma 1; it is concluded that v can reaching to origin in finite-time. Otherwise stated, due to $v = \arctan(e_{1i})$, e_{1i} can converge to zero in finite-time with the settling time is given by:

$$t_{2s_i} \leq \frac{\Gamma_i^{\rho/\eta}}{(1 - \rho/\eta)} |\arctan(e_{1i})|^{1-\rho/\eta} \leq \frac{\Gamma_i^{\rho/\eta}}{(1 - \rho/\eta)} \left(\frac{\pi}{2} \right)^{1-\rho/\eta} \tag{25}$$

Synthesizing both cases, the total time convergence of the sliding motion is:

$$t_s = \max_{1 \leq i \leq n} \{t_{si}\} \leq \max_{1 \leq i \leq n} \{t_{1si} + t_{2si}\} \tag{26}$$

Based on Eqs. (22) and (25), it is easily observed that t_s only depends on the design constants. Therefore, it can conclude that e_{1i} can converge to zero in fixed-time.

4   Design of a FnNTSMC Method

Firstly, calculating the time derivative of FxNTSM surface, one has:

$$
\begin{aligned}
\dot{\mathbf{s}} &= \dot{\varsigma} + \theta \mathbf{\Gamma} diag\left(e_{21}^{\theta-1}, \quad \ldots, \quad e_{2n}^{\theta-1} \right) \dot{\mathbf{e}}_2 \\
&= \dot{\varsigma} + \theta \mathbf{\Gamma} diag\left(e_{21}^{\theta-1}, \quad \ldots, \quad e_{2n}^{\theta-1} \right)(\mathbf{b}(\mathbf{e}) + \mathbf{a}(\mathbf{x})\mathbf{u} + \delta(\mathbf{e}, \Delta, \tau_d)) \\
&= \dot{\varsigma} + \theta \mathbf{\Phi}(\mathbf{b}(\mathbf{e}) + \mathbf{a}(\mathbf{x})\mathbf{u} + \delta(\mathbf{e}, \Delta, \tau_d))
\end{aligned}
\tag{27}
$$

in which $\mathbf{\Phi} = \mathbf{\Gamma} diag\left(e_{21}^{\theta-1}, \quad \ldots, \quad e_{2n}^{\theta-1} \right)$.

Then, the control torques are formed from Eq. (27) and ARCL as follows:

$$
\mathbf{u} = \mathbf{u}_{eq} + \mathbf{u}_r
\tag{28}
$$

$$
\mathbf{u}_{eq} = -\mathbf{a}^{-1}(\mathbf{x})(\mathbf{b}(\mathbf{e}) + \Sigma)
\tag{29}
$$

where $\Sigma = \left[\frac{(1+e_{11}^2)^{\theta-1}}{\Gamma_1 \theta}(1 + 2\theta e_{11}\arctan(e_{11}))e_{21}^{2-\theta}, \quad \ldots, \quad \frac{(1+e_{1n}^2)^{\theta-1}}{\Gamma_n \theta}(1 + 2\theta e_{1n}\arctan(e_{1n}))e_{2n}^{2-\theta} \right]^T$.

The ARCL is designed to improve robustness in coping with the total unknown uncertainties.

$$
\mathbf{u}_r = -\mathbf{a}^{-1}(\mathbf{x})(K\mathbf{s} + \Lambda \mathrm{sgn}(\mathbf{\Phi}\mathbf{s}))
\tag{30}
$$

where K is the selected positive constant.

Choose the following Lyapunov functional candidate $V_3 = 0.5\mathbf{s}^T\mathbf{s}$. Then, differentiating Lyapunov functional candidate gives:

$$
\begin{aligned}
\dot{V}_3 &= \mathbf{s}^T\dot{\mathbf{s}} \\
&= \mathbf{s}^T(\dot{\varsigma} + \theta\mathbf{\Phi}(\mathbf{b}(\mathbf{e}) + \mathbf{a}(\mathbf{x})\mathbf{u} + \delta(\mathbf{e}, \Delta, \tau_d)))
\end{aligned}
\tag{31}
$$

Applying control torques (28)–(30) to Eq. (31) obtains:

$$
\begin{aligned}
\dot{V}_3 &= \mathbf{s}^T\mathbf{\Phi}(\mathbf{a}(\mathbf{x})\mathbf{u}_r + \delta(\mathbf{e}, \Delta, \tau_d)) \\
&= -\theta\mathbf{s}^T\mathbf{\Phi}(K\mathbf{s} + \Lambda\mathrm{sgn}(\mathbf{\Phi}\mathbf{s})) + \theta\mathbf{s}^T\mathbf{\Phi}\delta(\mathbf{e}, \Delta, \tau_d) \\
&= -\theta\mathbf{s}^T\mathbf{\Phi}K\mathbf{s} - \theta\Lambda|\mathbf{\Phi}\mathbf{s}| + \theta\mathbf{s}^T\mathbf{\Phi}\delta(\mathbf{e}, \Delta, \tau_d)
\end{aligned}
\tag{32}
$$

Because of $\theta\mathbf{s}^T\mathbf{\Phi}K\mathbf{s} = K\theta\sum\limits_{i=1}^{n}\Gamma_i e_{2i}^{\theta-1}s_i^2 \geq 0$, therefore,

$$
\begin{aligned}
\dot{V}_3 &\leq -\theta\Lambda|\mathbf{\Phi}\mathbf{s}| + \theta\mathbf{s}^T\mathbf{\Phi}\delta(\mathbf{e}, \Delta, \tau_d) \\
&\leq -\theta(\Lambda - |\delta(\mathbf{e}, \Delta, \tau_d)|)|\mathbf{\Phi}\mathbf{s}| \\
&\leq -\varepsilon|\mathbf{s}|
\end{aligned}
\tag{33}
$$

with $\varepsilon = \theta(\Lambda - |\delta(\mathbf{e}, \Delta, \tau_d)|) \min_{1 \leq i \leq n} \{\Gamma_i e_{2i}^{\theta-1}\}$. By Lemma 1, it can be concluded that the proposed FxNTSM surface will be converged to zero in finite-time t_r and the convergence time t_r is bounded by [15]:

$$t_r \leq |\mathbf{s}(0)|/\varepsilon \tag{34}$$

Total time of the attainment of the equilibrium point is:

$$t = t_r + t_s \tag{35}$$

5   Illustrative Example and Discussion

Computer simulation is implemented for a 3 DOF manipulator shown in Fig. 1 using the proposed algorithm. A detailed description of the robot and its specifications was presented in our newly published article [6]. Simulation performance is compared between the proposed controller and another controller [5] called NFTSMC1 to verify their effectiveness in motion tracking. In path tracking, the given trajectory is set as:

$$\begin{cases} x = 0.85 - 0.01t \\ y = 0.2 + 0.2\sin(0.5t) \quad (m) \\ z = 0.7 + 0.2\cos(0.5t) \end{cases} \tag{36}$$

Fig. 1. 3D sketch of a 3-DOF manipulator with SolidWorks.

To verify the fixed time convergence of the control errors in both cases ($|e_{1i}| > 1$ and $|e_{1i}| \leq 1$) as described in Sect. 3, the initial state values of the manipulator are configured as follows: $q_1 = -1.6$ (rad), $q_2 = -1$ (rad), and $q_3 = -0.5$ (rad).

The unknown dynamic models are assumed to be $\Delta \mathbf{M}(\mathbf{q}) = 0.2\mathbf{M}(\mathbf{q})$, $\Delta \mathbf{C}(\mathbf{q}, \dot{\mathbf{q}}) = 0.2\mathbf{C}(\mathbf{q}, \dot{\mathbf{q}})$, and $\Delta \mathbf{G}(\mathbf{q}) = 0.2\mathbf{G}(\mathbf{q})$ for all cases of the simulation.

The friction and external disturbance are

$$\mathbf{f}_r(\dot{\mathbf{q}}) = [\, 0.1\mathrm{sgn}(\dot{q}_1) + 2\dot{q}_1 \quad 0.1\mathrm{sgn}(\dot{q}_2) + 2\dot{q}_2 \quad 0.1\mathrm{sgn}(\dot{q}_3) + 2\dot{q}_3 \,]^T (N.m) \qquad (37)$$

and

$$\tau_d(t) = [\, 4\sin(t) \quad 5\sin(t) \quad 6\sin(t) \,]^T (N.m). \qquad (38)$$

The proposed system has the selected control parameters: $\eta = 5$, $\rho = 3$, $\Gamma = diag(0.4, \quad 0.4, \quad 0.4)$, $K = 0.2$, and $\Lambda = 13$. The paper in [5] has the chosen control parameters like the proposed system to guarantee a fair comparison.

The tracking performance of both controllers is shown in Figs. 2, 3 and 4. Look at Fig. 2, it is seen that the initial state values of the manipulator were configured far from the desired trajectory to confirm the fixed-time convergence regardless of initial states. The results from Fig. 3 show that the proposed controller has a convergence rate of the control errors (at the first joint) much faster than the convergence rate of those in the study [5]. Because it contains dynamic coefficients that can be adapted to control errors that drive the system states quickly converge to the equilibrium point. It should be noted that the study [5] only considers the case $|e_{1i}| \leq 1$ and ignores case $|e_{1i}| > 1$. Consequently, the proposed controller provides convergence speed of the control errors along x-direction and y-direction much faster than the convergence rate of those in the study [5] as exhibited in Fig. 4.

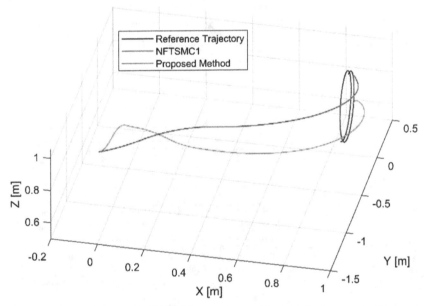

Fig. 2. The desired trajectory and real trajectory under NFTSMC1 and the proposed method.

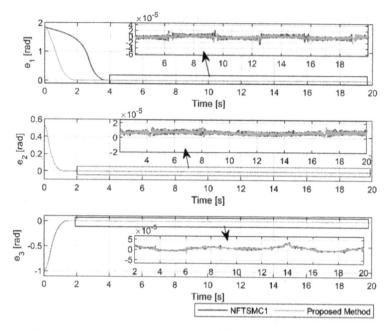

Fig. 3. The control errors in Joint space under NFTSMC1 and the proposed method.

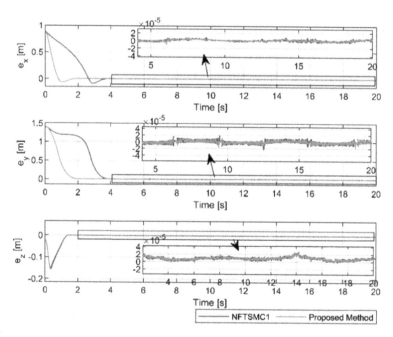

Fig. 4. The control errors in the Cartesian space under NFTSMC1 and the proposed method.

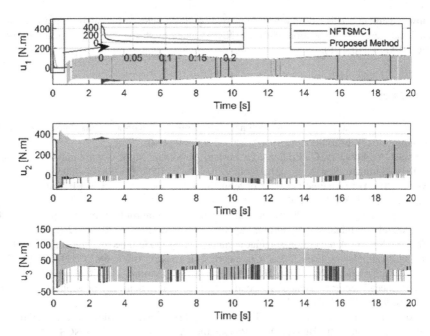

Fig. 5. The control torque from the controller [5] and the proposed controller.

Considering the ability to track trajectory, both controllers provide high tracking precision for manipulator (Fig. 5).

In order not to lose the generality of both controllers, both methods apply a discontinuous control law to cope with uncertain components. Therefore, both control systems appear the same chattering. The chattering phenomenon can be easily eliminated/reduced using a boundary technique as shown in the paper [5]. Although the oscillation amplitude of the chattering is the same, the proposed controller has a smaller initial torque value (at the first joint) than that of controller [5].

From simulation results, it is concluded that the proposed system has significantly improved the control performance compared to the study [5].

6   Conclusion

In this paper, a novel FnNTSMC was formed from a combination of the proposed FxNTSM surface and ARCL. Through theoretical analysis and computer simulations, the proposed FnNTSMC has proven its efficiency and outstanding features for robot manipulators such as high tracking accuracy, improved fast convergence in the bounded time regardless of initial system states, and robustness to cope with the lumped uncertainties. The fixed time/finite-time stability and convergence of the control method have been validated by using the Lyapunov criteria.

Acknowledgement. This research was supported by Basic Science Research Program through the National Research Foundation of Korea (NRF) funded by the Ministry of Education (NRF-2019R1D1A3A03103528).

References

1. Sciavicco, L., Siciliano, B.: Modelling and Control of Robot Manipulators. Springer, London (2012). https://doi.org/10.1007/978-1-4471-0449-0
2. Shtessel, Y., Edwards, C., Fridman, L., Levant, A.: Sliding Mode Control and Observation, vol. 10. Springer, New York (2014). https://doi.org/10.1007/978-0-8176-4893-0
3. Mu, C., He, H.: Dynamic behavior of terminal sliding mode control. IEEE Trans. Industr. Electron. **65**(4), 3480–3490 (2017)
4. Vo, A.T., Kang, H.J.: An adaptive terminal sliding mode control for robot manipulators with non-singular terminal sliding surface variables. IEEE Access. **7**, 8701–8712 (2018)
5. Zhai, J., Xu, G.: A novel non-singular terminal sliding mode trajectory tracking control for robotic manipulators. IEEE Trans. Circuits Syst. II Express Briefs **68**(1), 391–395 (2020)
6. Vo, A.T., Truong, T.N., Kang, H.J.: A novel tracking control algorithm with finite-time disturbance observer for a class of second-order nonlinear systems and its applications. IEEE Access. **9**, 31373–31389 (2021)
7. Truong, T.N., Kang, H.-J., Vo, A.T.: An active disturbance rejection control method for robot manipulators. In: Huang, D.-S., Premaratne, P. (eds.) ICIC 2020. LNCS (LNAI), vol. 12465, pp. 190–201. Springer, Cham (2020). https://doi.org/10.1007/978-3-030-60796-8_16
8. Mobayen, S.: An adaptive fast terminal sliding mode control combined with global sliding mode scheme for tracking control of uncertain nonlinear third-order systems. Nonlinear Dyn. **82**(1–2), 599–610 (2015). https://doi.org/10.1007/s11071-015-2180-4
9. Chen, Z., Yuan, X., Wu, X., Yuan, Y., Lei, X.: Global fast terminal sliding mode controller for hydraulic turbine regulating system with actuator dead zone. J. Franklin Inst. **356**(15), 8366–8387 (2019)
10. Truong, T.N., Vo, A.T., Kang, H.J.: A backstepping global fast terminal sliding mode control for trajectory tracking control of industrial robotic manipulators. IEEE Access. **9**, 31921–31931 (2021)
11. Zhang, J., Yu, S., Wu, D., Yan, Y.: Nonsingular fixed-time terminal sliding mode trajectory tracking control for marine surface vessels with anti-disturbances. Ocean Eng. **217**, 108158 (2020)
12. Vo, A.T., Truong, T.N., Kang, H.J.: A novel fixed-time control algorithm for trajectory tracking control of uncertain magnetic levitation systems. IEEE Access. **9**, 47698–47712 (2021)
13. Nguyen, V.C., Vo, A.T., Kang, H.J.: A non-singular fast terminal sliding mode control based on third-order sliding mode observer for a class of second-order uncertain nonlinear systems and its application to robot manipulators. IEEE Access. **8**, 78109–78120 (2020)
14. Tran, X.-T., Kang, H.-J.: A novel adaptive finite-time control method for a class of uncertain nonlinear systems. Int. J. Precis. Eng. Manuf. **16**(13), 2647–2654 (2015). https://doi.org/10.1007/s12541-015-0339-z
15. Chen, S.Y., Lin, F.J.: Robust nonsingular terminal sliding-mode control for nonlinear magnetic bearing system. IEEE Trans. Control Syst. Technol. **19**(3), 636–643 (2010)
16. Vo, A.T., Kang, H.J.: A new finite time control solution for robotic manipulators based on nonsingular fast terminal sliding variables and the adaptive super-twisting scheme. J. Comput. Nonlinear Dyn. **14**(3), (2019)
17. Vo, A.T., Kang, H.J.: A chattering-free, adaptive, robust tracking control scheme for nonlinear systems with uncertain dynamics. IEEE Access **7**, 10457–10466 (2019)

A Neural Terminal Sliding Mode Control for Tracking Control of Robotic Manipulators in Uncertain Dynamical Environments

Thanh Nguyen Truong[1] , Anh Tuan Vo[1] , Hee-Jun Kang[1(✉)] ,
and Tien Dung Le[2]

[1] Department of Electrical, Electronic and Computer Engineering,
University of Ulsan, Ulsan 44610, South Korea
hjkang@ulsan.ac.kr
[2] The University of Danang – University of Science and Technology,
54 Nguyen Luong Bang Street, Danang 550000, Vietnam

Abstract. This article investigates a control algorithm for trajectory tracking control of robot manipulators in uncertain dynamical environments. To deal with chattering behavior that always still exists in the conventional sliding mode control, to remove the requirement about a dynamic model of the uncertain robot system, and prior knowledge of the unknown function such as its upper boundary, the proposed solution is to develop a control method that combines the advantages of both terminal sliding mode control and radial basis function neural network. Furthermore, the proposed controller's fast stabilization and convergence are also significantly improved by using a novel adaptive fast reaching control law. Hence, the proposed controller's performance expectations are always guaranteed such as high tracking accuracy, fast stabilization, chattering reduction, fast convergence, and robustness to uncertain dynamical environments. Especially, the proposed controller can operate without the robot's dynamic model. Both theoretical investigations based on Lyapunov stability theory and computer simulation using MATLAB/Simulink are presented to confirm the effectiveness of the proposed control solution.

Keywords: Terminal sliding mode control · Radial basis function neural network · Robotic manipulators

1 Introduction

The role of robot manipulators is becoming increasingly important and essential in industrial applications, security, agriculture, health care, underwater exploration, military, food preparation, etc. Robots have been used for performing dangerous missions in place of humans. Nevertheless, the dangerous applications can lead to malfunction of the robot, which not only degrades the product quality but also harms users and others. Hence, to control effectively and safely the robots with a powerful controller has always been an interesting subject and difficult challenge in scientific communities. In real applications, robot manipulators are multi-input multi-output non-linear systems and in the working process, they are often affected by the nonlinear friction, payload

© Springer Nature Switzerland AG 2021
D.-S. Huang et al. (Eds.): ICIC 2021, LNCS 12837, pp. 207–221, 2021.
https://doi.org/10.1007/978-3-030-84529-2_18

change, external perturbation, etc. For that reason, designing a control method based on the robot's exact dynamic model was no easy challenge.

To address the mentioned problems, Several typical control methods have been developed and successfully applied to the robot systems, such as sliding mode control (SMC) [1–3], computed torque control [4], PID [5], adaptive algorithm [6], and neural network (NN) based control schemes [7, 8] etc.

In previous decades, the applications of SMCs for the approximation of nonlinear systems have been of interest. A core feature of the SMC method is its ability to provide robustness to control systems so that closed-loop systems are completely insensitive to nonlinearities, indeterminate dynamics, and input disturbances in sliding mode. Recently, convergence speed, tracking accuracy, and robustness of the motion tracking achieved from the SMC-based control methods for the nonlinear systems have been enhanced by proposing terminal SMC (TSMC) and its enhanced methods [9–18]. However, because TSMC is developed from SMC theory, thus it generates chattering phenomena in its control signals, it also requires the upper bound of the unknown function, an exact dynamic model of the robot system be known.

For unknown dynamics of the robot system, several control methods based on NN techniques have been proposed to estimate its dynamic model or unknown uncertainties. For example, in [19], this study suggested a controller using a feedforward neural network (FNN) compensator in compensation for the influence of unknown uncertainties and exterior perturbations. In [20], an adaptive NN controller has been proposed for tracking control of the robot with unknown dynamics. In comparison with several NNs such as FNN, recurrent NN, RBFNN appears to be more widely applied to nonlinear systems due to simplicity in network structure, fast convergence speed, and excellent approximation ability. Recently, many studies have been proposed, as shown in [17, 21, 22]. The article in [17] applied an RBFNN in efficiently improving the control performance against the great uncertain components of the magnetic levitation system. The updating laws of RBFNN and global stability are derived based on the Lyapunov algorithm. Consequently, the weight adaptation of NN is always guaranteed with convergence along with the global stability of the control method.

To solve all mentioned problems including chattering, prior knowledge of the upper bound of the unknown uncertain functions, and dynamic model of the robot system be known, a neural TSMC (NTSMC) that combines RBFNN and TSMC could be a good solution due to both methods' robustness and capability. Therefore, we propose a novel NTSMC for motion tracking of robot manipulators. This control scheme is an integration of the advantages from TSMC, RBFNN, and adaptive algorithm. Accordingly, the designed controller achieves the high precision position tracking under various environments. Some of the new and important contributions of the proposed controller can be mentioned as follows:

- The proposed control solution combines the advantages of TSMC and RBFNN for tracking control of robotic manipulators which provides high tracking performance in presence of uncertain dynamical environments.
- The proposed controller can operate without the robot's dynamic model.

- A novel auxiliary adaptive fast reaching control law has been applied to improve fast stabilization and convergence speed. In addition, the requirement of the upper boundary of total unknown uncertain terms is rejected.
- The adaptation laws of RBFNN and adaptive fast reaching control law are gained according to the Lyapunov algorithm.

The remainder of this paper is organized as follows. In Sect. 2, the robot dynamics are investigated, and problems are described. Control design procedures are presented in Sect. 3. Analysis and evaluation of the computer simulation results are presented in Sect. 4. Conclusions are summarized in Sect. 5.

2 Problem Statements

The dynamic model of n-degree-of-freedom (DOF) robotic manipulators are de-scribed as based on [23]:

$$M(q)\ddot{q} + C(q, \dot{q})\dot{q} + G(q) + F(\dot{q}) + d = \tau \tag{1}$$

where $q = [\, q_1 \quad q_2 \quad \cdots \quad q_n \,]^T \in \mathbb{R}^n$, $\dot{q} = [\, \dot{q}_1 \quad \dot{q}_2 \quad \cdots \quad \dot{q}_n \,]^T \in \mathbb{R}^n$, and $\ddot{q} = [\, \ddot{q}_1 \quad \ddot{q}_2 \quad \cdots \quad \ddot{q}_n \,]^T \in \mathbb{R}^n$ represents the joint angle position, the joint angle velocity, and the joint angle acceleration, respectively. $M(q) \in \mathbb{R}^{n \times n}$ is a inertia matrix, $C(q, \dot{q}) \in \mathbb{R}^{n \times n}$ consists of Coriolis and centrifugal force, and $G(q) \in \mathbb{R}^{n \times 1}$ is a gravity matrix. $F(\dot{q}) \in \mathbb{R}^{n \times 1}$ and $d \in \mathbb{R}^{n \times 1}$ are friction and external disturbance matrices, respectively. $\tau \in \mathbb{R}^{n \times 1}$ is a control input vector.

The dynamic model of n-DOF robotic manipulator can be represented as:

$$M(q)\ddot{q} + C(q, \dot{q})\dot{q} + G(q) + \Phi = \tau \tag{2}$$

where $\Phi = F(\dot{q}) + \tau_d \in \mathbb{R}^{n \times 1}$ is the whole system uncertainties and external disturbances.

Assumption 1: The whole system uncertainties and external disturbances is bounded by:

$$|\Phi| \leq \Delta \tag{3}$$

This paper proposes a non-model-based-control algorithm termed NTSMC for trajectory tracking control of industrial robotic manipulators to provide a high control performance in the presence of uncertainties and external disturbances.

3   Design of the Proposed Control Algorithm

3.1   Terminal Sliding Mode Control Method

We define the position error and velocity error as follows:

$$e_i = q_{di} - q_i \text{ and } \dot{e}_i = \dot{q}_{di} - \dot{q}_i, \quad i = 1, 2, ..., n. \tag{4}$$

where q_d, $\dot{q}_d \in \mathbb{R}^n$ are the desired position and the desired velocity, respectively.
   The terminal sliding mode surface is selected as [12]:

$$s_i = \dot{e}_i + \alpha_1 |e_i|^{h_i} \text{sgn}(e_i) + \beta_1 |e_i|^{m_i} \text{sgn}(e_i), \quad i = 1, 2, ..., n \tag{5}$$

where $h_i = 0.5(p_1 + 1) + 0.5(p_1 - 1)\text{sgn}(|e_i| - 1)$, $m_i = 0.5(p_1 + q_1) + 0.5(p_1 - q_1)$ $\text{sgn}(|e_i| - 1)$, $p_1 > 1$, $0 < q_1 < 1$, $\alpha_1 > 0$, $\beta_1 > 0$ $\beta_1 > 0$.
   From Eq. (5), we have:

$$\dot{q}_i = \dot{q}_{di} + \alpha_1 |e_i|^{h_i} \text{sgn}(e_i) + \beta_1 |e_i|^{m_i} \text{sgn}(e_i) - s_i, \quad i = 1, 2, ..., n \tag{6}$$

Let $U_i = \dot{q}_{di} + \alpha_1 |e_i|^{h_i} \text{sgn}(e_i) + \beta_1 |e_i|^{m_i} \text{sgn}(e_i)$, $U = [U_1 \quad U_2 \quad ... \quad U_n]^T$. The Eq. (6) in matrix form as follows:

$$\dot{q} = U - \dot{s} \tag{7}$$

Taking time derivative of the sliding surface (5), we can gain:

$$\dot{s}_i = \ddot{e}_i + \left(h_i \alpha_1 |e_i|^{h_i - 1} + m_i \beta_1 |e_i|^{m_i - 1} \right) \dot{e}_i, \quad i = 1, 2, ..., n \tag{8}$$

To simplify, let $Y_i = \left(h_i \alpha_1 |e_i|^{h_i - 1} + m_i \beta_1 |e_i|^{m_i - 1} \right) \dot{e}_i$, $Y = [Y_1 \quad Y_2 \quad ... \quad Y_n]^T$. Equation (8) in matrix form as follows:

$$\dot{s} = \ddot{e} + Y \tag{9}$$

Multiplying both sides of the Eq. (9) by $M(q)$ and using Eqs. (2) and (7), we can get as follows:

$$\begin{aligned} M(q)\dot{s} = M(q)(\ddot{e} + Y) &= M(q)(\ddot{q}_d + Y) - M(q)\ddot{q} \\ &= M(q)(\ddot{q}_d + Y) - M(q)\left(M^{-1}(q)(-C(q,\dot{q})\dot{q} - G(q) - \Phi + \tau)\right) \\ &= M(q)(\ddot{q}_d + Y) + C(q,\dot{q})U - C(q,\dot{q})s + G(q) + \Phi - \tau \end{aligned} \tag{10}$$

The Lyapunov function is selected for stability investigation as:

$$V_1 = 0.5 s^T M(q) s \tag{11}$$

Calculating the time derivative of the Lyapunov function (11)

$$\begin{aligned}
\dot{V}_1 &= s^T M(q)\dot{s} + 0.5 s^T \dot{M}(q)s \\
&= s^T(M(q)(\ddot{q}_d + Y) + C(q,\dot{q})U - C(q,\dot{q})s + G(q) + \Phi - \tau) \\
&\quad + 0.5 s^T \dot{M}(q)s \\
&= s^T(M(q)(\ddot{q}_d + Y) + C(q,\dot{q})U + G(q) + \Phi - \tau) \\
&\quad + 0.5 s^T(\dot{M}(q) - 2C(q,\dot{q}))s
\end{aligned} \tag{12}$$

It is known that the characteristic $s^T(\dot{M}(q) - 2C(q,\dot{q}))s = 0$ always exists in the robot. Therefore, Eq. (12) can be obtained as:

$$\dot{V}_1 = s^T(M(q)(\ddot{q}_d + Y) + C(q,\dot{q})U + G(q) + \Phi - \tau) \tag{13}$$

If the dynamic model of robotic manipulator, including $M(q)$, $C(q,\dot{q})$, and $G(q)$ are known, we can design the control input law based on Eq. (13) as follows:

$$\tau = u_{eq} + u_r \tag{14}$$

in which, the equivalent control law is:

$$u_{eq} = M(q)(\ddot{q}_d + Y) + C(q,\dot{q})U + G(q) \tag{15}$$

and the fast-reaching control law is:

$$u_{ri} = \Delta_i \text{sgn}(s_i) + \alpha_2 |s_i|^{x_i} \text{sgn}(s_i) + \beta_2 |s_i|^{y_i} \text{sgn}(s_i), \quad i = 1,2,3,...,n \tag{16}$$

where $u_r = [u_{r1} \quad u_{r2} \quad ... \quad u_{rm}]^T$, $x_i = 0.5(p_2 + 1) + 0.5(p_2 - 1)\text{sgn}(|s_i| - 1)$, $y_i = 0.5(p_2 + q_2) + 0.5(p_2 - q_2)\text{sgn}(|s_i| - 1)$, $p_2 > 1$, $0 < q_2 < 1$, $\alpha_2 > 0$, $\beta_2 > 0$.
Substituting the control rule from Eqs. (14)–(16) into (13), we can obtain:

$$\begin{aligned}
\dot{V}_1 &= s^T(\Phi - u_r) \\
&= \sum_{i=1}^{n} s_i(\Phi_i - \Delta_i \text{sgn}(s_i) - \alpha_2 |s_i|^{x_i} \text{sgn}(s_i) - \beta_2 |s_i|^{y_i} \text{sgn}(s_i)) \\
&\leq \sum_{i=1}^{n} (\Phi_i - \Delta_i)|s_i| - \alpha_2 \sum_{i=1}^{n} |s_i|^{x_i + 1} - \beta_2 \sum_{i=1}^{n} |s_i|^{y_i + 1} \leq 0
\end{aligned} \tag{17}$$

In practice, the calculation of the robot manipulator's dynamic model is very complicated due to the complicated mechanical structure of the robot. For example, the robot has many degrees of freedom or variation of payload. In case of the robot's dynamic model $f = M(q)(\ddot{q}_d + Y) + C(q,\dot{q})U + G(q)$ is unknown. Hence, an RBFNN is applied to estimate this term f.

3.2   RBFNN Approximation

An RBFNN is used to approximate the unknown term f. The structure of RFNN is shown in Fig. 1.

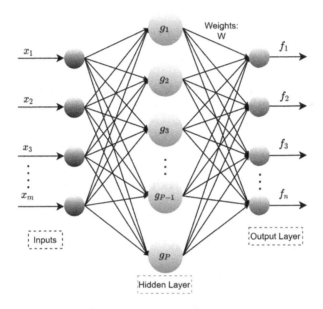

Fig. 1. The structure of RBFNN

The algorithm of RBFNN is described as:

$$g_j = \exp\left(\|x - c_j\|^2 / b_j^2\right), \quad j = 1, 2, \ldots, P$$
$$f = W^T g(x) + \varepsilon \tag{18}$$

where x is the input state of the NN, m is the input number of the NN, P is the number of hidden layer nodes in the NN, $g(x) = \begin{bmatrix} g_1 & g_2 & \cdots & g_P \end{bmatrix}^T$ is the output of Gaussian function, $W \in \mathbb{R}^{P \times n}$ is the NN weights, ε is approximation error of NN, and $|\varepsilon| \leq \varepsilon_N$. The input of RBFNN is selected as:

$$x = \begin{bmatrix} e^T & \dot{e}^T & q_d^T & \dot{q}_d^T & \ddot{q}_d^T \end{bmatrix}^T \in \mathbb{R}^{5n \times 1} \tag{19}$$

The output of RBFNN is presented as:

$$\hat{f}(x) = \hat{W}^T g(x) \in \mathbb{R}^{n \times 1} \tag{20}$$

where $g(x)$ is the Gaussian function of RBFNN.

Let $\tilde{W} = W - \hat{W}$ is estimation error of the weights, one has:

$$f - \hat{f}(x) = W^T h(x) + \varepsilon - \hat{W}^T h(x) = \tilde{W}^T g(x) + \varepsilon \qquad (21)$$

3.3 Design of the Proposed Non-model Terminal Sliding Mode Control Algorithm

$$\tau = \hat{f}(x) + u_r \qquad (22)$$

in which, $\hat{f}(x)$ is the output of RBFNN, which is used to approximate the equivalent term, and the new adaptive fast reaching law is designed as:

$$u_{ri} = \hat{\Delta}_i \mathrm{sgn}(s_i) + \alpha_2 |s|^{x_i} \mathrm{sgn}(s_i) + \beta_2 |s|^{y_i} \mathrm{sgn}(s_i), \quad i = 1, 2, ..., n \qquad (23)$$

where $u_r = [u_{r1} \quad r_{r2} \quad ... \quad u_{rn}]^T$, $x_i = 0.5(p_2 + 1) + 0.5(p_2 - 1)\mathrm{sgn}(|s_i| - 1)$, $y_i = 0.5(p_2 + q_2) + 0.5(p_2 - q_2)\mathrm{sgn}(|s_i| - 1)$, $p_2 > 1$, $0 < q_2 < 1$, $\alpha_2 > 0$, $\beta_2 > 0$. $\hat{\Delta}_i$ is adapted by the following law:

$$\dot{\hat{\Delta}}_i = \begin{cases} \psi_i |s_i| & |s_i| > \gamma_i \\ 0 & |s_i| \le \gamma_i \end{cases}, \quad i = 1, 2, ..., n \qquad (24)$$

in which, $\hat{\Delta} = [\hat{\Delta}_1 \quad \hat{\Delta}_2 \quad ... \quad \hat{\Delta}_n]^T$, $\psi = diag(\psi_1, \psi_2, ..., \psi_n)$, $\psi_i > 0$ and $\gamma_i > 0$ are the user defined constants. Let $\tilde{\Delta} = \Delta + \varepsilon_N - \hat{\Delta}$ is estimation error of boundary value of the whole approximation error of NN and uncertainties.

Assume that approximation error of NN and uncertainties to be bounded by:

$$|\Phi + \varepsilon| \le \Delta + \varepsilon_N \qquad (25)$$

From Eqs. (21), (22), (23), (24) and (10), we have

$$\begin{aligned} M(q)\dot{s} &= f - C(q, \dot{q})s + \Phi - \hat{f}(x) - u_r \\ &= \tilde{W}^T g(x) + \varepsilon - C(q, \dot{q})s + \Phi - u_r \end{aligned} \qquad (26)$$

Selecting the Lyapunov function for stability investigation as

$$V_2 = \frac{1}{2} s^T M(q)s + \frac{1}{2} tr\{\tilde{W}^T Q^{-1} \tilde{W}\} + \frac{1}{2} \tilde{\Delta}^T \psi^{-1} \tilde{\Delta} \qquad (27)$$

where Q is a positive matrix.

Taking the time derivative of the Lyapunov function Eq. (27) and using Eqs. (26), (24). We can obtain:

$$
\begin{aligned}
\dot{V}_2 &= s^T M(q)\dot{s} + \frac{1}{2} s^T \dot{M}(q)s + tr\left\{ \tilde{W}^T Q^{-1} \dot{\tilde{W}} \right\} + \tilde{\Delta}^T \psi^{-1} \dot{\tilde{\Delta}} \\
&= s^T \left(\tilde{W}^T g(x) + \varepsilon - C(q,\dot{q})s + \Phi - u_r \right) \\
&\quad + \frac{1}{2} s^T \dot{M}(q)s - tr\left\{ \tilde{W}^T Q^{-1} \dot{\hat{W}} \right\} - \left(\Delta + \varepsilon_N - \hat{\Delta} \right)^T |s| \\
&= s^T (\varepsilon + \Phi - u_r) + \frac{1}{2} s^T \left(\dot{M}(q) - 2C(q,\dot{q}) \right) s \\
&\quad - tr\left\{ \tilde{W}^T \left(Q^{-1} \dot{\hat{W}} + g(x)s^T \right) \right\} - \left(\Delta + \varepsilon_N - \hat{\Delta} \right)^T |s|
\end{aligned}
\tag{28}
$$

From Eq. (28), it is known that the characteristic $s^T (\dot{M}(q) - 2C(q,\dot{q}))s = 0$ always exists in the robot. In addition, the adaptive rule of NN is selected to guarantee the stability as follows:

$$
\dot{\hat{W}} = Qg(x)s^T
\tag{29}
$$

Therefore, Eq. (28) can be obtained as:

$$
\begin{aligned}
\dot{V}_2 &= s^T (\varepsilon + \Phi - u_r) - \left(\Delta + \varepsilon_N - \hat{\Delta} \right)^T |s| \\
&= \sum_{i=1}^{n} ((\varepsilon_i + \Phi_i) - (\varepsilon_{Ni} + \Delta_i))|s_i| - \sum_{i=1}^{n} \left(\alpha_2 |s_i|^{x_i+1} + \beta_2 |s_i|^{y_i+1} \right) \le 0
\end{aligned}
\tag{30}
$$

It is observed that $V_2 \ge 0$ and $\dot{V}_2 \le 0$. Consequently, the system runs stable.

To eliminate the chattering phenomenon in the control input signal, we use the saturation function to replace the $\text{sgn}(\cdot)$ function in the reaching control stated in Eq. (23). The saturation function is defined as:

$$
sat(s_i) = \begin{cases} 1 & s_i > \mu_i \\ s/\mu_i & |s_i| \le \mu_i \\ -1 & s_i < -\mu_i \end{cases}, \quad \text{with } \mu_i > 0, \quad i = 1, 2, ..., n
\tag{31}
$$

Diagram of the proposed control method is shown in Fig. 2.

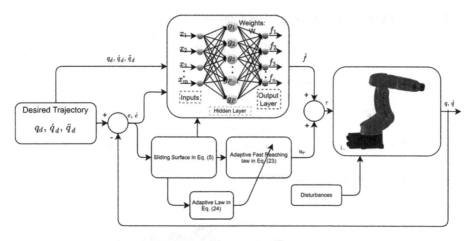

Fig. 2. Diagram of the proposed control system

4   Simulation and Discussion

Computer simulation using a 3-DOF FARA Robot Manipulator in MATLAB/Simulink environment is presented to confirm the effectiveness of the proposed control solution. The robot's kinematic description is exhibited in Fig. 3. This robot is a type of industrial robot using in the laboratory for research and development. The system parameters of the robot are given in Table 1. The simulation configuration is set with a fixed step 0.001 s (ODE5 dormand-prince). To verify the advanced capability and outstanding efficiency of the proposed system, the proposed system is compared with CTC, SMC.

Table 1. The designed parameters of the robotic system

		Link 1	Link 2	Link 3
Length (m)		0.15	0.255	0.3
Weight (kg)		37.985	21.876	16.965
Center of Mass (mm)	l_{cx}	68.067	95.045	71.496
	l_{cy}	−1.185	5.399	−72.007
	l_{cz}	64.931	−0.002	−1.004
Inertia $(kg.m^2)$	I_{xx}	0.252	0.359	0.306
	I_{yy}	0.395	0.623	0.853
	I_{zz}	0.356	0.319	0.306

The control torque of CTC method as follows:

$$\tau = M(q)\left(\ddot{q}_d + K_p e + K_v \dot{e}\right) + C(q,\dot{q})\dot{q} + G \tag{32}$$

where K_p and K_v are the positive diagonal matrices.
The control torque of SMC as follows:

$$\begin{cases} s = \dot{e} + ce \\ \tau = M(q)(\ddot{q}_d + c\dot{e}) + C(q,\dot{q})(q_d + ce) + G + diag(\Delta)\text{sgn}(s) \end{cases} \tag{33}$$

where c is a positive diagonal matrix.

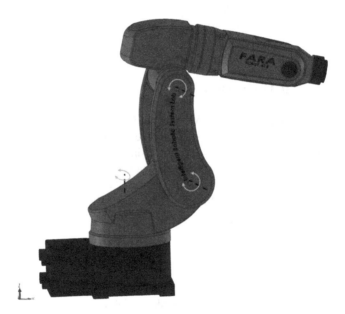

Fig. 3. 3D SOLIDWORK model of the 3-DOF FARA Robot Manipulator

The robot is required to perform the task of following the desired trajectory in the Cartesian space:

$$\begin{cases} x_d = 0.43 + 0.01\sin(0.75t) \\ y_d = 0.06\sin(0.75t) \\ z_d = 0.26 + 0.06\sin(0.375t) \end{cases} \tag{34}$$

Friction and disturbance terms are added to each joint to investigate the influence of nonlinear uncertainties, as follows:

$$F(\dot{q}) + d = \begin{bmatrix} 2\dot{q}_1 + 0.01\text{sgn}(\dot{q}_1) \\ 2\dot{q}_2 + 0.01\text{sgn}(\dot{q}_2) \\ 2\dot{q}_3 + 0.01\text{sgn}(\dot{q}_3) \end{bmatrix} + \begin{bmatrix} 3\sin(0.4t) + 4\sin(0.6t) \\ 2\sin(0.4t) + 3\sin(0.6t) \\ 1\sin(0.4t) + 2\sin(0.6t) \end{bmatrix} \qquad (35)$$

The control parameters of three controllers are shown in Table 2.

Table 2. Selected parameters of three controllers

Controllers	Control parameters	Value
CTC	K_P, K_v	$2000 \times I^{3\times3}$, $200 \times I^{3\times3}$
SMC	c, Δ	$10 \times I^{3\times3}$, $diag(7.5, 5.5, 3.5)$
Proposed method	m, P, n, b_j, Q, W_0	15, 7, 3, 10, $50 \times I^{7\times7}$, $0.1 \times ones(7,3)$
	c	$0.1 \begin{bmatrix} -1.5 & -1 & -0.5 & 0 & 0.5 & 1 & 1.5 \\ \vdots & \vdots & \vdots & \vdots & \vdots & \vdots & \vdots \\ -1.5 & -1 & -0.5 & 0 & 0.5 & 1 & 1.5 \end{bmatrix} \in \mathbb{R}^{15\times7}$
	α_1, β_2, p_1, q_1, α_2, β_2,	5, 5, 1.6, 0.6, 80, 80,
	p_2, q_2, γ_i, ψ_i, μ	1.6, 0.8, 0.05, 600, 0.5

The simulation results are shown in Figs. 4, 5, 6 and 7. From Figs. 4 and 5, the proposed control solution presents the best tracking control performance including superior tracking precision, fastest convergence speed, and strongest property to uncertain terms among the three control methods.

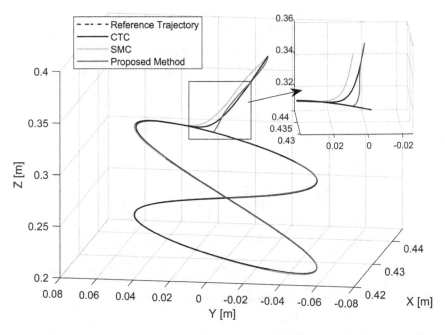

Fig. 4. Desired trajectory and actual trajectory of the end-effector under three controllers

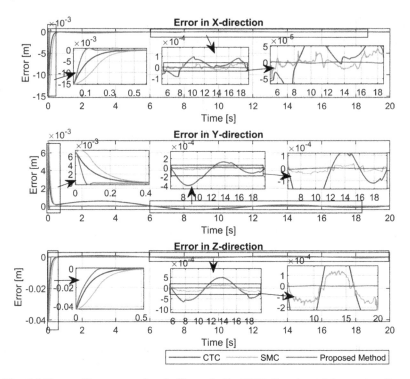

Fig. 5. The tracking errors of end-effector in the XYZ direction of three controllers

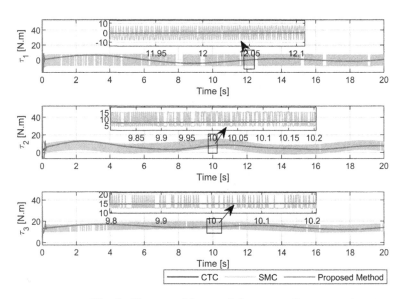

Fig. 6. The control inputs of three controllers

It is seen from Fig. 6; CTC offers a chattering-free control signals due to the non-existence of sgn($\cdot$) function; SMC with the presence of sgn($\cdot$) function and a large enough value of the sliding gain generates oscillation control signals; While the proposed control solution provides a smooth control signal by applying the boundary technique (31).

Obviously, the adaptive values of the upper bound of uncertainties and disturbances at each joint were adapted according to variations of those uncertain terms and quickly stabilized to constants as shown in Fig. 7.

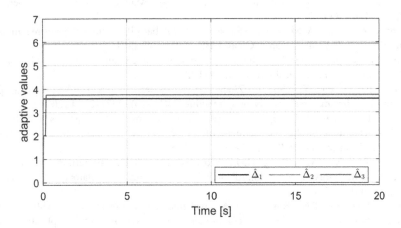

Fig. 7. The adaptation values

5 Conclusions

This article proposed a control algorithm for trajectory tracking control of manipulators without using a dynamic model. The proposed controller's performance expectations have been guaranteed such as high tracking accuracy, fast stabilization, chattering reduction, and fast convergence, robustness to uncertain dynamical environments. Especially, the proposed controller operated with no dependence on the robot's dynamic model and the upper bound of uncertain terms. Both theoretical investigations based on Lyapunov stability theory and computer simulation using MATLAB/ Simulink have been fully presented to confirm the effectiveness of the proposed solution. From simulation results, the proposed solution presented the best tracking control performance including superior tracking precision, fastest convergence speed, and strongest property to uncertain terms among the three control methods.

Acknowledgements. This research was supported by Basic Science Research Program through the National Research Foundation of Korea (NRF) funded by the Ministry of Education (NRF-2019R1D1A3A03103528).

References

1. Shtessel, Y., Edwards, C., Fridman, L., Levant, A.: Sliding Mode Control and Observation. Springer, New York (2014). https://doi.org/10.1007/978-0-8176-4893-0
2. Islam, S., Liu, X.P.: Robust sliding mode control for robot manipulators. IEEE Trans. Industr. Electron. **58**(6), 2444–2453 (2010)
3. Baek, J., Jin, M., Han, S.: A new adaptive sliding-mode control scheme for application to robot manipulators. IEEE Trans. Industr. Electron. **63**(6), 3628–3637 (2016)
4. Llama, M.A., Kelly, R., Santibañez, V.: Stable computed-torque control of robot manipulators via fuzzy self-tuning. IEEE Trans. Syst. Man Cybern. Part B (Cybern.) **30**(1), 143–150 (2000)
5. Hernández-Guzmán, V.M., Orrante-Sakanassi, J.: Global PID control of robot manipulators equipped with PMSMs. Asian J. Control **20**(1), 236–249 (2018)
6. Wang, H.: Adaptive control of robot manipulators with uncertain kinematics and dynamics. IEEE Trans. Autom. Control **62**(2), 948–954 (2016)
7. Sun, T., Pei, H., Pan, Y., Zhou, H., Zhang, C.: Neural network-based sliding mode adaptive control for robot manipulators. Neurocomputing **74**(14–15), 2377–2384 (2011)
8. Truong, T.N., Kang, H.-J., Le, T.D.: Adaptive neural sliding mode control for 3-DOF planar parallel manipulators. In: Proceedings of the 2019 3rd International Symposium on Computer Science and Intelligent Control, pp. 1–6 (2019)
9. Truong, T.N., Vo, A.T., Kang, H.-J.: A backstepping global fast terminal sliding mode control for trajectory tracking control of industrial robotic manipulators. IEEE Access **9**, 31921–31931 (2021)
10. Vo, A.T., Truong, T.N., Kang, H.J.: A novel tracking control algorithm with finite-time disturbance observer for a class of second-order nonlinear systems and its applications. IEEE Access **9**, 31373–31389 (2021)
11. Van, M., Ceglarek, D.: Robust fault tolerant control of robot manipulators with global fixed-time convergence. J. Franklin Inst. **358**(1), 699–722 (2021)
12. Vo, A.T., Truong, T.N., Kang, H.-J.: A novel fixed-time control algorithm for trajectory tracking control of uncertain magnetic levitation systems. IEEE Access **9**, 47698–47712 (2021)
13. Vo, A.T., Kang, H.-J., Truong, T.N.: A fast terminal sliding mode control strategy for trajectory tracking control of robotic manipulators. In: International Conference on Intelligent Computing, pp. 177–189 (2020)
14. Wang, Y., Zhu, K., Chen, B., Jin, M.: Model-free continuous nonsingular fast terminal sliding mode control for cable-driven manipulators. ISA Trans. **98**, 483–495 (2020)
15. Van, M., Mavrovouniotis, M., Ge, S.S.: An adaptive backstepping nonsingular fast terminal sliding mode control for robust fault tolerant control of robot manipulators. IEEE Trans. Syst. Man Cybern. Syst. **49**(7), 1448–1458 (2018)
16. Truong, T.N., Kang, H.-J., Vo, A.T.: An active disturbance rejection control method for robot manipulators. In: International Conference on Intelligent Computing, pp. 190–201 (2020)
17. Truong, T.N., Vo, A.T., Kang, H.-J.: Implementation of an adaptive neural terminal sliding mode for tracking control of magnetic levitation systems. IEEE Access **8**, 206931–206941 (2020)
18. Ma, Z., Sun, G.: Dual terminal sliding mode control design for rigid robotic manipulator. J. Franklin Inst. **355**(18), 9127–9149 (2018)
19. Vo, A.T., Kang, H.: Neural integral non-singular fast terminal synchronous sliding mode control for uncertain 3-DOF parallel robotic manipulators. IEEE Access **1** (2020)

20. Vo, A.T., Kang, H.-J.: An adaptive neural non-singular fast-terminal sliding-mode control for industrial robotic manipulators. Appl. Sci. **8**(12), 2562 (2018)
21. Rani, K., Kumar, N.: Intelligent controller for hybrid force and position control of robot manipulators using RBF neural network. Int. J. Dyn. Control **7**(2), 767–775 (2019)
22. Kumar, N., et al.: Finite time control scheme for robot manipulators using fast terminal sliding mode control and RBFNN. Int. J. Dyn. Control **7**(2), 758–766 (2019)
23. Craig, J.J.: Introduction to Robotics: Mechanics and Control, 3rd edn. Pearson Education India, Noida (2009)

Fuzzy PID Controller for Accurate Power Sharing and Voltage Restoration in DC Microgrids

Duy-Long Nguyen[1] and Hong-Hee Lee[2(✉)]

[1] Graduate School of Electrical Engineering, University of Ulsan,
Ulsan, South Korea
[2] School of Electrical Engineering, University of Ulsan, Ulsan, South Korea
hhlee@mail.ulsan.ac.kr

Abstract. In DC microgrids, conventional droop control is widely used to perform power sharing of distributed energy resources. Although, this method is communication less and reliable, it cannot achieve accurate power sharing and voltage restoration due to missing global information such as average output voltage and average power per unit. To solve this problem, this paper proposes a distributed secondary control level based on a Fuzzy logic controller. Thanks to the proposed method, accurate power sharing and average voltage restoration are achieved simultaneously. Based on voltage shifting method, secondary control level is simplified with the only one Fuzzy logic controller. Furthermore, with outstanding feature of the Fuzzy controller, both voltage and power allocation performances are enhanced in comparison with a conventional PID controller or a simple integrator. The proposed control scheme is verified by simulation in Matlab & Simulink.

Keywords: DC Microgrid · Power sharing · Droop control · Fuzzy logic control

1 Introduction

In these days, to integrate multiple distributed generators (DGs) and clusters of loads, microgrids have been emerged as a promising solution with enhanced efficiency and high system reliability [1]. MGs can be categorized as alternative current (AC) and direct current (DC) MGs [2, 3]. In comparison with AC MGs, DC MGs has been attracting more interests due to its outstanding feature such as direct interface with many types of renewable energy sources, no problems with reactive power flow, harmonic compensation [4, 5]. A DC MGs can operate flexibly in either grid connected mode [6] or islanded mode [7]. In islanded DC MGs, droop control scheme is widely used to obtain power allocation in a decentralized manner. The droop control scheme is simple since it is carried out without communication links and only local measurements

© Springer Nature Switzerland AG 2021
D.-S. Huang et al. (Eds.): ICIC 2021, LNCS 12837, pp. 222–232, 2021.
https://doi.org/10.1007/978-3-030-84529-2_19

are needed [8]. However, this method suffers from inherent problems: load dependent voltage deviation and poor power sharing performance due to mismatched line resistances [9, 10].

To handle the drawbacks of the droop control, a secondary controller based on communication network is adopted [11]. Particularly, the secondary controller can be implemented by centralized manner or distributed manner [12, 13]. In centralized control scheme, a central controller is developed based on communication with all local controllers. Even though it is convenient to facilitate optimization and supervisory operation, it experiences serious problems such as single point of failure which decreases system reliability [14, 15]. To solve the shortcomings of centralized control, distributed control has been proposed in which local controllers are connected with each other to cooperate harmoniously [16, 17]. This is more effective solution with higher reliability and flexibility since there is no central controller and plug-play performance is achieved.

In order to achieve voltage restoration and accurate power sharing, voltage shifting method is popularly used. In this method, output voltages of converters are compensated by two controllers. The first controller is responsible for voltage restoration, while the second controller ensures accurate power sharing. In addition, information shared among local controllers includes at least two qualities: output voltage and output current or power of converter. Recently, G. Silva et al. [18] has given a simple integrator to achieve both voltage restoration and accurate power sharing. This control scheme depends on only one transmitted variable which increases efficiency of communication network. However, dynamic performance of the integrator is slow and worse in comparison with PI controller or PID controller.

In this paper, based on the distributed control scheme, we proposed a Fuzzy PID controller for secondary control level to achieve both voltage restoration and accurate power sharing in DC MGs. According to the proposed control scheme, only one variable is transmitted in communication network, which helps to decrease the communication burden. Besides, with outstanding feature of the Fuzzy controller, both voltage and power allocation performances are enhanced compared with a conventional PID controller and a simple integrator. The effectiveness of the proposed Fuzzy PID controller is verified by simulation in Matlab and Simulink.

2 Distributed Secondary Control Scheme with an Integrator

A typical DC MG is illustrated in Fig. 1.

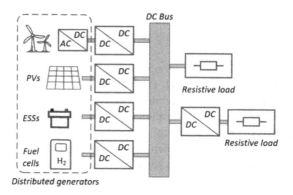

Fig. 1. A typical structure of DC MG.

In Fig. 1, all DGs are cooperated to balance total power consumed by loads and losses of DC MG. To avoid overuse of a certain DG, power sharing among DGs are regulated according to their own power ratings. In addition, due to droop effect, output voltages of converters are decreased. Therefore, average output voltage of converters should be restored to the nominal voltage.

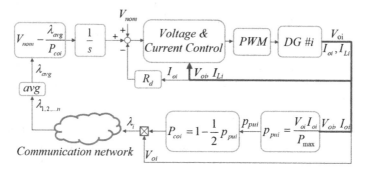

Fig. 2. Simple integrator for achieving voltage restoration and accurate power sharing.

To achieve voltage restoration and accurate power sharing, G. Silva et al. [18] has proposed a control strategy with a simple integrator as shown in Fig. 2. This control scheme is based on a factor λ which is transmitted information in communication network. The factor λ is calculated as follow:

$$\lambda_i = P_{coi} V_{oi}, \tag{1}$$

$$P_{coi} = 1 - \frac{1}{2} P_{pui}, \tag{2}$$

where P_{coi}, and V_{oi} are power quality and output voltage of converter i, and p_{pui} is power per unit of converter i:

$$p_{pui} = \frac{V_{oi}I_{oi}}{P_{max}} \tag{3}$$

where I_{oi} is output current of converter i, P_{max} is maximum power of converter i.

In steady state, due to the effect of the integrator, we have:

$$V_{nom} = \frac{\lambda_{avg}}{P_{coi}} \tag{4}$$

where V_{nom} is nominal voltage and λ_{avg} is average of factor λ. From (4), we have:

$$\frac{\lambda_{avg}}{P_{co1}} = \frac{\lambda_{avg}}{P_{co2}} = \dots = \frac{\lambda_{avg}}{P_{con}} \tag{5}$$

where P_{co1}, P_{co2}, ..., P_{con} are power quality of converter 1, 2, ..., n respectively. It means that accurate power sharing is achieved:

$$P_{co1} = P_{co2} = \dots = P_{con} \tag{6}$$

or $p_{pu1} = p_{pu2} = \dots = p_{pun}$.

From (1) and (6), we have:

$$\lambda_{avg} = P_{coi}V_{avg} \tag{7}$$

Substituting to (4), (8) is derived as

$$V_{nom} = V_{avg} \tag{8}$$

In conclusion, with a simple integrator, the control scheme in Fig. 2 can guarantee both voltage restoration and accurate power sharing, and it becomes so simple since only one integrator is adopted as secondary control level. In addition, only one variable is transmitted, which helps to decrease communication burden. However, dynamic performance of the integrator is quite slower and worse than that with PI controller or PID controller. To overcome this problem, we propose Fuzzy PID controller to enhance the performance of the integrator as well as the secondary control level.

3   Proposed Secondary Control Scheme with Fuzzy PID Controller

In order to improve dynamic performance of voltage and power of converters in DC MGs, the secondary control level with Fuzzy PID controller is proposed as shown in Fig. 3.

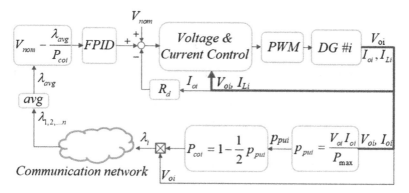

Fig. 3.   Secondary control level used fuzzy PID controller.

The operation of Fuzzy PID controller is illustrated in Fig. 4, where r, y, e, and u are reference value, output value, error, and control signal, respectively.

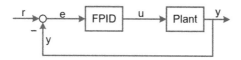

Fig. 4.   Feedback control system with fuzzy PID controller.

The Structure of Fuzzy PID controller is shown in Fig. 5 [19, 20]. In Fig. 5, G_{CE}, G_{CU}, G_E, and G_U are gain factors of the controller. These gains and fuzzy inference system are considered when designing Fuzzy PID controller.

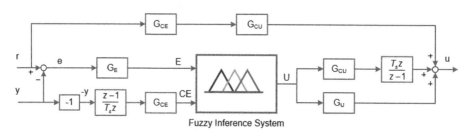

Fig. 5.   The structure of fuzzy PID controller.

We assume output of Fuzzy Inference System U which depends on E and CE linearly:

$$U = E + CE \tag{9}$$

From (9) and Fig. 5, the control signal u can be expressed as:

$$u = (G_{CE}G_{CU} + G_E G_U)e + G_E G_{CU}e\frac{T_S z}{z-1} - G_{CE}G_U y\frac{z-1}{T_S z} \tag{10}$$

Since:

$$-\frac{y(k) - y(k-1)}{T_S} = \frac{(r - y(k)) - (r - y(k-1))}{T_S}$$
$$= \frac{e(k) - e(k-1)}{T_S} \tag{11}$$

We have

$$-y\frac{z-1}{T_S z} = e\frac{z-1}{T_S z} \tag{12}$$

Substituting (12), (10) become:

$$u = (G_{CE}G_{CU} + G_E G_U)e + G_E G_{CU}e\frac{T_S z}{z-1} + G_{CE}G_U e\frac{z-1}{T_S z} \tag{13}$$

(13) can be rewritten as the form of PID controller:

$$u = K_P e + K_I e\frac{T_S z}{z-1} + K_D e\frac{z-1}{T_S z} \tag{14}$$

where:

$$K_P = G_{CE}G_{CU} + G_E G_U$$
$$K_I = G_E G_{CU} \tag{15}$$
$$K_D = G_{CE}G_U$$

Therefore, if the condition (9) and (15) are satisfied, the performance of Fuzzy PID controller becomes exactly the same as the performance of PID controller. From that, the gain factors of Fuzzy PID controller are designed as followed: firstly parameters of PID controller K_P, K_I, K_D are tuned with desired performance, and the gain factors of Fuzzy PID controller G_{CE}, G_{CU}, G_E, G_U are chosen according to Eqs. (15). To improve the performance of Fuzzy PID controller, linear relation in (9) is changed to nonlinear relation by modifying membership function and Fuzzy control rule [20].

4 Design Example

To verify the effectiveness of the proposed Fuzzy PID controller, a typical DC MG in Fig. 6 is simulated by Matlab and Simulink. The DG unit is boost converter with parameters are shown in Table 1.

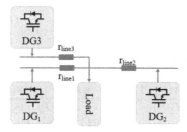

Fig. 6. The structure of DC MGs used for simulation.

Table 1. System Parameters

DC MG parameters		
Parameters	Symbol	Value
$P_{DG1,2,3}$	Rated power of $DG_{1,2,3}$	5 kW
V_{nom}	Nominal voltage	200 V
$r_{lline,2,3}$	Line resistance 1,2, 3	0.1 Ω, 0.2 Ω, 0.3 Ω
Load	Load resistance	4 Ω
DC-DC converter parameters		
Parameters	Symbol	Value
V_{in}	Input voltage	100 V
C	Output capacitor	2200 μF
L	Inductor	0.5 mH
f_{sw}	Switching frequency	20 kHz

The output power is normalized to per unit, the error e is within the range $[-1,1]$. The input range of Fuzzy Inference System is chosen as $[-10, 10]$, therefore the gain G_E is 10. The parameters of PID controller is tuning with desired performance: $K_P = 20$, Ki = 150, $K_D = 0$. From (15), gains factor of Fuzzy PID controller are calculated: $G_U = 2$, $G_{CU} = 15$, and $G_{CE} = 0$.

To improve the performance of Fuzzy PID controller, the relationship between input and output is modified to nonlinear characteristic by modifying input membership function and Fuzzy control rule [20]. Membership function is chosen as Gaussian function as shown in Fig. 7.

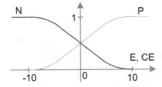

Fig. 7. Gaussian membership input function

The output membership functions are chosen simply: Negative (N) is −20, Zero (Z) is 0, and Positive (P) is 20. In addition, the Fuzzy control rules are chosen as in Table 2.

Table 2. Fuzzy control rule for nonlinear input output mapping

E	CE	
	N	P
N	N	Z
P	Z	P

From these membership functions and Fuzzy control rules, nonlinear input output relation is plotted as shown in Fig. 8.

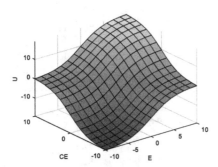

Fig. 8. Nonlinear input output mapping

In Fig. 8, when error or change of error is large, gains become higher than in case of linear input output mapping, and then nonlinear Fuzzy PID controller achieve faster performance comparing with linear Fuzzy PID controller.

5   Simulation Result

Figure 9 shows average output voltage performance with three controllers including: integrator, PID controller and Fuzzy PID controller. At the beginning, average output voltage is smaller than nominal voltage $V_{nom} = 200$ V due to the effect of the droop control. After secondary control level is activated at 0.5 s, average output voltage increase. At 1.5 s when R_{load2} is connected to the system, a sag in average output voltage appears. Particularly, Fuzzy PID controller guarantee shallowest sag in comparison with other controllers. In addition, Fuzzy PID controller achieve fastest voltage restoration among three controllers.

Figure 10 illustrates power per unit performance with three kinds of controllers. In the first time, there are differences in power per units among three DGs due to mismatched line resistances. At 0.5 s, secondary control level is active and power sharing is realized. However, there exist oscillations in power per unit performance with integrator. On the other hand, Fuzzy PID controller provides the best performance with the fastest power allocation among three controllers due to nonlinear characteristic.

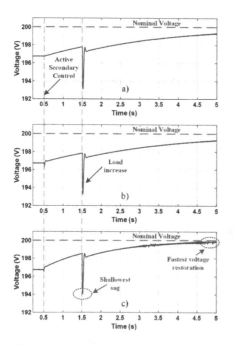

Fig. 9. Average output voltage performance with a) Integrator, b) PID controller and c) Fuzzy PID controller.

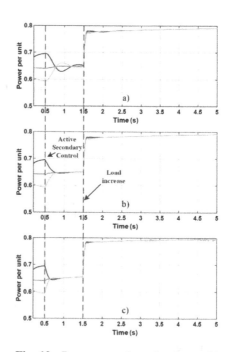

Fig. 10. Power per unit performance with a) Integrator, b) PID controller and c) Fuzzy PID controller.

From simulation, we can see Fuzzy PID provides outstanding voltage restoration and accurate power sharing with faster and smaller overshoot in comparison with PID controller or integrator.

6   Conclusion

This paper presents a secondary control scheme to achieve both accurate power sharing and voltage restoration in DC MGs. In the secondary control level, a Fuzzy PID controller is proposed to enhance performance of voltage restoration and power allocation.

Thanks to the proposed Fuzzy controller, the performance of both the voltage restoration and the power sharing are improved without any oscillation in comparison with the integrator and the conventional PID controller. The effectiveness of the proposed controller is verified by Matlab and Simulink simulation.

Acknowledgments. This work was supported in part by the NRF of Korea Grant under Grant NRF-2018R1D1A1A09081779 and in part by the KETEP and the MOTIE under Grant 20194030202310.

References

1. Palma-Behnke, R., et al.: A microgrid energy management system based on the rolling horizon strategy. IEEE Trans. Smart Grid **4**(2), 996–1006 (2013)
2. Nguyen, D. L., Lee, H. H.: Cooperative control strategy for voltage restoration and power allocation in DC microgrids. J. Power Electron. **20**, 4 (2020).
3. Lin, P., Zhang, C., Wang, P., Xiao, J.: A decentralized composite controller for unified voltage control with global system large-signal stability in DC microgrids. IEEE Trans. Smart Grid **10**(5), 5075–5091 (2019)
4. Dragicevic, T., Vasquez, J.C., Guerrero, J.M., Skrlec, D.: Advanced LVDC electrical power architectures and microgrids: a step toward a new generation of power distribution networks. IEEE Electrif. Mag. **2**(1), 54–65 (2014)
5. Dam, D., Lee, H.: A power distributed control method for proportional load power sharing and bus voltage restoration in a DC microgrid. IEEE Trans. Ind. Appl. **54**(4), 3616–3625 (2018)
6. Nguyen, T. L., Guerrero, J. M., Griepentrog, G.: A self-sustained and flexible control strategy for islanded DC nanogrids without communication links. IEEE J. Emerg. Sel. Top. Power Electron, 1 (2019).
7. Tucci, M., Riverso, S., Vasquez, J.C., Guerrero, J.M., Ferrari-Trecate, G.: A decentralized scalable approach to voltage control of DC islanded microgrids. IEEE Trans. Control Syst. Technol. **24**(6), 1965–1979 (2016)
8. Gao, F., Bozhko, S., Asher, G., Wheeler, P., Patel, C.: An improved voltage compensation approach in a droop-controlled DC power system for the more electric aircraft. IEEE Trans. Power Electron. **31**(10), 7369–7383 (2016)
9. Chen, F., Burgos, R., Boroyevich, D., Vasquez, J.C., Guerrero, J.M.: Investigation of nonlinear droop control in dc power distribution systems: Load sharing, voltage regulation, efficiency, and stability. IEEE Trans. Power Electron. **34**(10), 9404–9421 (2019)

10. Nguyen, D.-L., Lee, H.-H.: Fuzzy PID controller for adaptive current sharing of energy storage system in DC microgrid. In: Huang, D.-S., Premaratne, P. (eds.) ICIC 2020. LNCS (LNAI), vol. 12465, pp. 213–223. Springer, Cham (2020). https://doi.org/10.1007/978-3-030-60796-8_18
11. Dragičević, T., Lu, X., Vasquez, J.C., Guerrero, J.M.: DC microgrids—Part I: a review of control strategies and stabilization techniques. IEEE Trans. Power Electron. 31(7), 4876–4891 (2016)
12. Guerrero, J.M., Vasquez, J.C., Matas, J., de Vicuna, L.G., Castilla, M.: Hierarchical control of droop-controlled AC and DC microgrids—a general approach toward standardization. IEEE Trans. Ind. Electron. 58(1), 158–172 (2011)
13. Nasirian, V., Moayedi, S., Davoudi, A., Lewis, F.L.: Distributed cooperative control of DC microgrids. IEEE Trans. Power Electron. 30(4), 2288–2303 (2015)
14. Hoang, T. V., Lee, H. H.: Distributed control scheme for accurate reactive power sharing with enhanced voltage quality for islanded microgrids. J. Power Electron. 20(2) 2020.
15. Wang, C., Duan, J., Fan, B., Yang, Q., Liu, W.: Decentralized high-performance control of DC microgrids. IEEE Trans. Smart Grid 10(3), 3355–3363 (2019)
16. Fan, B., Peng, J., Yang, Q., Liu, W.: Distributed periodic event-triggered algorithm for current sharing and voltage regulation in DC microgrids. IEEE Trans. Smart Grid (2020).
17. Zhou, J., Shi, M., Chen, X., Chen, Y., Wen, J., He, H.: A cascaded distributed control framework in DC microgrids. IEEE Trans. Smart Grid. 1 (2020).
18. Silva, W.W.A.G., Oliveira, T.R., Donoso-Garcia, P.F.: An improved voltage-shifting strategy to attain concomitant accurate power sharing and voltage restoration in droop-controlled DC microgrids. IEEE Trans. Power Electron. 36(2), 2396–2406 (2021)
19. Xu, J.-X., Hang, C.-C., Liu, C.: Parallel structure and tuning of a fuzzy PID controller. Automatica 36(5), 673–684 (2000)
20. Jantzen, J.: Tuning of fuzzy PID controllers. (1998).

Sensor-Less Contact Force Estimation in Physical Human-Robot Interaction

Quang Dan Le and Hee-Jun Kang[✉]

Department of Electrical, Electronic and Computer Engineering,
University of Ulsan, Ulsan, South Korea
hjkang@ulsan.ac.kr

Abstract. In this paper, the contact force estimation in physical human-robot interaction without an explicit force sensing device based on the adaptive third order sliding mode observer is proposed. Firstly, the identification parameter of a robot manipulator is required. Then an adaptive third order sliding mode observer is designed to estimate contact force. Finally, the experimental results with the immediate stop control and admittance control using the proposed estimation contact force for 3-DOF AT2-FARA robot manipulator are shown to illustrate the capability of the proposed in the real application.

Keywords: Physical human-robot interaction · Sliding mode observer · Third order sliding mode observer · Admittance control sensor-less

1 Introduction

In collaboration solution, human and robots share the workspace and their skills. The advantage of collaborative robots is the combination of the advantages of automation with the flexibility and soft skills of human workers. The traditional industrial robots cannot adapt when the environment has a small change. Robots can collision avoidance, collision detection and reaction are one of the challenges in physical human-robot interaction (pHRI) subject. This research topic has been attended attracted by numbers of research nowadays.

In industrial, to prove the contact force in the interaction robot with the environment, the force/torque sensor is mounted on the end-effector of the robot. However, by using an external force/torque sensor, the price of the robot is increased, and the payload capacity reduces effective. Addition, the force/torque sensor is sensitive to different environments such as temperatures. Therefore, this solution carefully considers before applying to balance between economy and benefit. With purpose effectiveness economy, a large variety of controllers have been proposed without the force/torque sensor. For example, in [1], the direct motor current and velocity of robot is used to detect the contact and allows an immediate stop. Moreover, the authors in [2] not only detect the collisions but also introduce reaction schemes for collaboration human- robot by using the motor current and joint position. A dithering feed forward torque [3] is one method to estimate contact force when robot not in motion without a sensor.

© Springer Nature Switzerland AG 2021
D.-S. Huang et al. (Eds.): ICIC 2021, LNCS 12837, pp. 233–244, 2021.
https://doi.org/10.1007/978-3-030-84529-2_20

During human-robot collaboration, the collision can occur at the whole-body of the robot manipulator. Therefore, in [4, 5] the tactile sensitive skin in a different location is proposed. Another approach called virtual force sensing [6, 7] use the stereo camera to detect the collision and estimate contact force. Both methods above combining with mode-base have high effectiveness to estimate contact force. However, as mentioned above, the additional external sensor in the robot manipulator system will increase the price. Therefore, the solution estimate contact force without a sensor is more attention in industrial application.

Disturbance observer-based contact force estimation has been widely applied in robot manipulator control. It treats the uncertainties and disturbance like an external force [8–10]. In a conventional disturbance observer, the observer requires to accelerate measurement [8]. However, in robot control, it is not popular and difficult to get correct accelerate signal. Therefore, [9–12] proposed the disturbance observer based contact force estimation without accelerating measurement. Moreover, an extended state observer can estimate the disturbance and external force [13, 14]. In these days, sliding mode observer has been attracted in research with finite time convergence feature [15, 16]. In the position control or low-level control areas, sliding mode observer was combined with low-level control got high effectiveness in control [17, 18]. By using extend state observer or high order sliding mode observer, the controller only requires position measurement. It can control the system with high quality. The most disadvantage of high order sliding mode observer came from the stability proof. Fortunately, in [19, 20], the authors proposed the method to prove the stability of the system when it uses high order sliding mode observer. In addition, the conventional sliding mode observer requires the knowledge of the upper bound of uncertainties/disturbances and external force. This information is difficult to know in practice. Therefore, an adaptive method is proposed to handle this issue [21, 22].

In this paper, the adaptive third order sliding mode observer is proposed to estimate contact force in pHRI. Then, the online contact force estimation will use for stopping the robot, and admittance control for the robot in pHRI. Firstly, the complete dynamic model of robot is required. The rescuer least square method is used to identify the parameter of the robot. Next, the adaptive third order sliding mode observer is proposed to estimate the contact force in pHRI with threshold technique. From contact force information, the pHRI controller is introduced such as the robot stop and admittance control for robot manipulator. Finally, experimental results of pHRI based on the contact force estimation proposed on 3-DOF manipulator are shown to illustrate the capability in real application.

The main contributions of this paper are briefly listed as:

1) Physical human-robot interaction without a force sensor.
2) The estimation contact force in pHRI only requires position measurement.
3) Experimental results are shown to illustrate the capability of proposed in the real application.

The rest of this paper is presented following as. Section 2, the dynamic and identification of robot manipulator is shown. Section 3, an adaptive third order sliding mode observer to estimate contact force in pHRI is proposed. In Sect. 4, experimental results of proposed on 3-DOF robot manipulator with the robot stop and admittance control for pHRI is presented. Finally, the conclusion and future work are given in Sect. 5.

2 Dynamics Model of Robot Manipulator

The dynamics model of robot manipulator for n-degree in contact with human is given as

$$M(q)\ddot{q} + C(q,\dot{q})\dot{q} + G(q) + F_f(\dot{q}) = \tau + \tau_{\text{int}} \tag{1}$$

where $\ddot{q}, \dot{q}, q \in \Re^n$ are the vector of joint accelerations, velocity and position, respectively. $M(q) \in \Re^{n \times n}$ is inertia matrix, $C(q,\dot{q}) \in \Re^n$ represents the centripetal and Coriolis matrix, $G(q) \in \Re^n$ represents the gravitation torques, F_f is friction term. τ is torque provided at joint. τ_{int} is interaction force.

Property 1: The matrix $\dot{M}(q) - 2C(q.\dot{q})$ is skew-symmetric.

Property 2: $\|M^{-1}(q)\| < \alpha$ where α is a positive constant.

3 Contact Force Estimation

In this section, the contact force estimation based on adaptive third order sliding mode observer is proposed.

The dynamic model (1) can be rewritten in state space as

$$\begin{cases} \dot{x}_1 = x_2 \\ \dot{x}_2 = f(x_1, x_2, u) + \phi(x_1, x_2, t) \end{cases} \tag{2}$$

where $u = \tau, f(x_1, x_2, u) = M^{-1}(q)(\tau - C(q,\dot{q}) + G(q) + F_f)$ and $\phi(x_1, x_2, t) = M^{-1}(q)(-\tau_{\text{int}})$.

The adaptive third order sliding mode observer is designed as

$$\begin{cases} \dot{\hat{x}}_1 = k_1|e_1|^{\frac{2}{3}}sgn(e_1) + \hat{x}_2 \\ \dot{\hat{x}}_2 = k_2|e_1|^{\frac{1}{3}}sgn(e_1) + f(x_1, \hat{x}_2, u) + \hat{x}_3 \\ \dot{\hat{x}}_3 = k_3 sgn(e_1) \end{cases} \tag{3}$$

where $e_1 = x_1 - \hat{x}_1$ and $\hat{x}_3$ is the estimation of $\phi(x_1, x_2, t)$. Let $L = k_3, k_2 = 5.3L^{\frac{2}{3}}$ and $k_1 = 3.34L^{\frac{1}{3}}$ and the update law

$$\dot{L} = \begin{cases} \bar{L}|e_1|sign(|e_1| - \varepsilon) & \text{if } L > \lambda \\ 0 & \text{if } L \le \lambda \end{cases} \tag{4}$$

Lemma 1: Let $\eta : \Re^n \to \Re$ and $\gamma : \Re^n \to \Re_+$, that is $\gamma(x) \ge 0 \ \forall x$, be two continuous homogeneous functions, with weights $r = (r_1, ..., r_2)$ and degrees m, such that $\{x \in \Re^n \backslash \{0\} : \gamma(x) = 0\} \subseteq \{x \in \Re^n \backslash \{0\} : \eta(x) < 0\}$. Then, there exists a real number

λ^* such that, for all $\lambda \geq \lambda^*$ for all $x \in \Re^n \backslash \{0\}$, and some $c > 0$, $\eta(x) - \lambda\gamma(x) < -c\|x\|_{r,p}^m$.

Lemma 2 (Young's inequality): For any positive real numbers $a > 0$, $b > 0$, $c > 0$, $p > 1$ and $q > 1$ with $\frac{1}{q} + \frac{1}{p} = 1$,

$$ab \leq \frac{c^p}{p}a^p + \frac{c^{-q}}{q}b^q,$$

and equality holds if and only if $a^p = b^q$.

Lemma 3: For any real numbers $a_1, ..., a_n$, $if\, 0 < p < q$,

$$\left(\sum_{i=1}^n |a_i|^p\right)^{\frac{1}{p}} \leq \left(\sum_{i=1}^n |a_i|^q\right)^{\frac{1}{q}}$$

Theorem: Considering the system (2) with observer (3) and adaptive law (4) then $\hat{x}_1 \to x_1, \hat{x}_2 \to x_2, \hat{x}_3 \to \phi$.

Proof: From (2) and (3) we have the error dynamics

$$\begin{cases} \dot{e}_1 = -k_1|e_1|^{\frac{2}{3}}\mathrm{sgn}(e_1) + e_2 \\ \dot{e}_2 = -k_2|e_1|^{\frac{1}{3}}\mathrm{sgn}(e_1) + e_3 \\ \dot{e}_3 = -k_3\mathrm{sgn}(e_1) \end{cases} \tag{5}$$

where $e_3 = \phi(x_1, x_2, t) - \hat{x}_3$. Let $z_1 = \frac{e_1}{1}$; $z_2 = \frac{e_2}{k_1}$; $z_3 = \frac{e_3}{k_2}$ then (5) become as

$$\begin{cases} \dot{z}_1 = -k_1\left(|z_1|^{\frac{2}{3}}sgn(z_1) - z_2\right) \\ \dot{z}_2 = -k_2\left(|z_1|^{\frac{1}{3}}sgn(z_1) - z_3\right) \\ \dot{z}_3 = -k_3 sgn(z_1) \end{cases} \tag{6}$$

(6) can be rewritten as

$$\begin{cases} \dot{z}_1 = -k_1\left([z_1]^{\frac{2}{3}} - z_2\right) \\ \dot{z}_2 = -k_2\left([z_1]^{\frac{1}{3}} - z_3\right) \\ \dot{z}_3 = -k_3[z_1]^0 \end{cases} \tag{7}$$

where $[z_k]^n = |z_k|^n sgn(z_1)$, $z_k = [z_k]^1 = |z_k|sgn(z_k)$ and $[z_k]^0 = sgn(z_k)(n, k \in \Re)$. Let define the Lyapunov function as

$$V = V_1 + V_2 \tag{8}$$

where

$$V_1(z) = Z_1(z_1, z_2) + \beta_1 Z_2(z_2, z_3) + \beta_2 \frac{1}{5} |z_3|^5 \tag{9}$$

which β_1, and β_2 are positive.

$$\begin{cases} Z_1(z_1, z_2) = \frac{3}{5}|z_1|^{\frac{5}{3}} - z_1 z_2 + \frac{2}{5}|z_2|^{\frac{5}{2}} \\ Z_2(z_2, z_3) = \frac{2}{5}|z_2|^{\frac{5}{2}} - z_2[z_3]^3 + \frac{3}{5}|z_3|^5 \end{cases} \tag{10}$$

Using Lemma 2 for $z_1 z_2$ and $z_2[z_3]^3$ term we have

$$\begin{cases} z_1 z_2 \leq \frac{3}{5}|z_1|^{\frac{5}{3}} + \frac{2}{5}|z_2|^{\frac{5}{2}} \\ z_2[z_3]^3 \leq \frac{2}{5}|z_2|^{\frac{5}{2}} + \frac{3}{5}|z_3|^{\frac{5}{3}} \end{cases} \tag{11}$$

Therefore,

$$\begin{cases} Z_1(z_1, z_2) \geq 0 \\ Z_2(z_2, z_3) \geq 0 \end{cases} \tag{12}$$

From (9) and (12), we have $V_1 \geq 0$ and

$$V_2 = \frac{1}{2\Lambda}(L - L^*)^2 \geq 0 \tag{13}$$

where L^* is positive constants.
Derivative (8) we have

$$\dot{V} = \dot{V}_1 + \dot{V}_2 \tag{14}$$

Let consider $\dot{V}_1$

$$\dot{V}_1 = \sigma_1 \dot{z}_1 - |z_2|^0 \lambda_1 \dot{z}_2 + \beta_1 \sigma_2 \dot{z}_2 - 3\beta_1 |z_3|^2 \lambda_2 \dot{z}_3 + \beta_2 [z_3]^4 \dot{z}_3 \tag{15}$$

where $\sigma_1 = [z_1]^{\frac{2}{3}} - z_2$, $\sigma_2 = [z_2]^{\frac{3}{2}} - [z_3]^3$, $\lambda_1 = z_1 - [z_2]^{\frac{3}{2}}$ and $\lambda_2 = z_2 - [z_3]^2$. We can rewrite (15) as

$$\dot{V}_1 = -W_1(z) \tag{16}$$

where

$$W_1(z) = -\sigma_1 \dot{z}_1 + |z_2|^0 \lambda_1 \dot{z}_2 - \beta_1 \sigma_2 \dot{z}_2 + 3\beta_1 |z_3|^2 \lambda_2 \dot{z}_3 - \beta_2 [z_3]^4 \dot{z}_3 \tag{17}$$

Substituting (7) into (17) we have

$$W_1(z) = k_1 \underbrace{\left([z_1]^{\frac{2}{3}} - z_2\right)\left([z_1]^{\frac{2}{3}} - z_2\right)}_{\eta_1} - |z_2|^0 k_2 \left([z_1]^{\frac{1}{3}} - z_3\right)\left(z_1 - [z_2]^{\frac{3}{2}}\right)$$

$$+ k_2\beta_1\left([z_2]^{\frac{3}{2}} - [z_3]^3\right)\left([z_1]^{\frac{1}{3}} - z_3\right) - 3\beta_1 k_3 [z_1]^0 |z_3|^2 \left(z_2 - [z_3]^2\right) \tag{18}$$

$$+ \beta_2 k_3 [z_1]^0 [z_3]^4$$

$$= k_1 \eta_1(z_1, z_2) + \mu_1(z)$$

In (18) η_1 is non negative and it vanishes when $[z_1]^{\frac{2}{3}} = z_2$. According to Lemma 1 there exists a sufficiently large positive value of k_1 such as $W_1(z) > 0$ if the value of μ_1 restricted to $[z_1]^{\frac{2}{3}} = z_2$ is positive. From (18) it is seen that $W_2(z)$ can be written is

$$W_2(z) = k_2\beta_1\left([z_2]^{\frac{3}{2}} - [z_3]^3\right)\left([z_1]^{\frac{1}{3}} - z_3\right) - 3\beta_1 k_3 [z_1]^0 |z_3|^2 \left(z_2 - [z_3]^2\right)$$

$$+ \beta_2 k_3 [z_1]^0 [z_3]^4$$

$$= k_2\beta_1\left([z_2]^{\frac{3}{2}} - [z_3]^3\right)\left([z_2]^{\frac{1}{2}} - z_3\right) - 3\beta_1 k_3 [z_1]^0 |z_3|^2 \left(z_2 - [z_3]^2\right) \tag{19}$$

$$+ \beta_2 k_3 [z_1]^0 [z_3]^4$$

$$= k_2 \eta_2 + \mu_2(z)$$

In (19) η_2 is positive on and it vanishes when $[z_2]^{\frac{1}{2}} = z_3$. Lemma 1 there exists a sufficiently large positive value of k_2 such that $W_2(z)$ is positive definite on $[z_2]^{\frac{1}{2}} = z_3$ if $\mu_2(z)$ is positive. From (21) we see that $\mu_2 = k_3\beta_2[z_2]^0[z_3]^4 = k_3\beta_2|z_3|^4$ is positive on $[z_2]^{\frac{1}{2}} = z_3$. We conclude that $W(z) \geq 0$. Therefore, $\dot{V}_1 \leq 0$.

Let consider $\dot{V}_2$

$$\dot{V}_2 = \frac{1}{\Lambda}(L - L^*)\dot{L} \tag{20}$$

Substituting (4) into (20) we have

$$\dot{V}_2 = \frac{1}{\Lambda}(L - L^*)\overline{L}|e_1|sign(|e_1| - \varepsilon) \tag{21}$$

Substituting (21) and (16) into (14) we have

$$\dot{V} = -W_1(z) + \frac{1}{\Lambda}(L - L^*)\overline{L}|e_1|sign(|e_1| - \varepsilon)$$

$$\leq -W_1(z) + \underbrace{\frac{1}{\Lambda}|L - L^*|\overline{L}|e_1|sign(|e_1| - \varepsilon)}_{\zeta} \tag{22}$$

Case 1: $|e_1| < \varepsilon$. In this case $\zeta \leq 0$ so that $\dot{V} \leq -W_1(z)$.

Case 2: $|e_1| \geq \varepsilon$. In this case $\zeta \geq 0$ so that (20) can be positive that mean it can not conclude system stability. Therefore, $|e_1|$ can be decrease. As soon as $|e_1|$ less than ε and the system return the previous case. Theorem is proven.

4   Experimental Results

4.1   Hardware Setup

The experimental setup is shown in Fig. 1 with a 3-DOF FARA-AT2 robot manipulator. This robot manipulator has 6-DOF, but for these experiments, joints 4-5-6 are blocked. The 3-DOF FARA-AT2 robot has a CSMP series motor at each joint, and the CSMP-02BB driver is used for joints 1 and 2 while the CSMP-01BB driver is used for joint 3. The gear box at each joint is 120:1,120:1, 100:1 at joints 1, 2 and 3, respectively. The encoder at each joint is a 2048 line count incremental encoder. The controller runs on Labview-FPGA NI-PXI-8110 and NI-PXI-7842R PXI cards with the frequency control set at 500 Hz. NI-PXI-8110 is run on a Windows operating system.

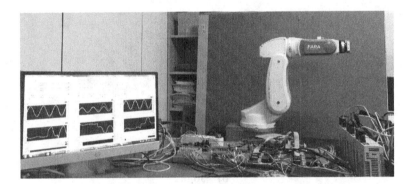

Fig. 1. Real-system 3-DOF robot manipulator

4.2   Contact Detection

In this section, the results of contact detection between human and robot are presented. The safety is a fundamental of pHRI. Therefore, the detection in whole-body of the robot is necessary and pays high attention by numbers of research in pHRI with variable methods. In this experiment, each time a contact is detected, the robot stops and resumes its trajectory tracking after 3 s. To detect the contact force and false alarm, the threshold technique is used. For simplest, the threshold at each joint is fixed in this experiment.

Desired tracking trajectory of the end-effector:

$$\begin{cases} x_d = 0 \\ y_d = 0.15\sin\left(\frac{\pi t}{1600}\right) \\ z_d = 0.15\sin\left(\frac{\pi t}{1600}\right) - 0.15 \end{cases} \tag{23}$$

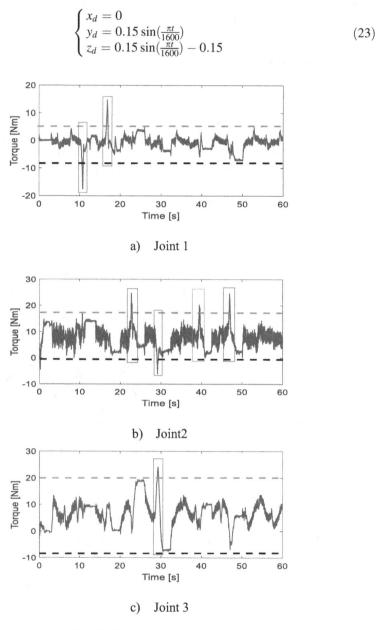

a)   Joint 1

b)   Joint2

c)   Joint 3

Fig. 2. Detection contact force

In Fig. 2, the collision detection is shown. The collision occurs at t = 11,17,23, 29,39.5,48 s. It can be detected in whole-body robot. In Fig. 3, the joint position in case collision. The snapshot of touching robot with position of touching were shown in Fig. 4.

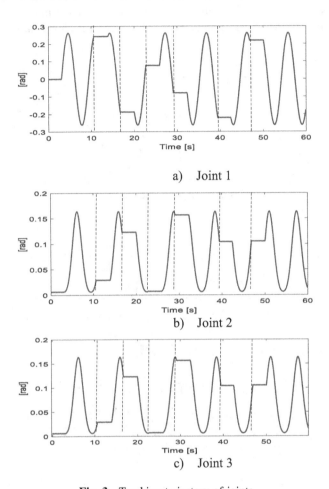

a)   Joint 1

b)   Joint 2

c)   Joint 3

Fig. 3. Tracking trajectory of joints

Fig. 4. Snapshot of the touching robot experiment

4.3   Human Push/Pull

In this section, the interaction human-robot is based on the admittance control. The human pushes/pulls robot in the different points and on different links. The estimation contact force during the physical human-robot interaction is shown in Fig. 6 (Fig. 5).

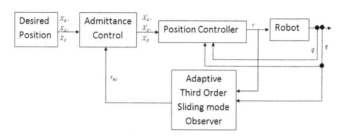

Fig. 5. Admittance control

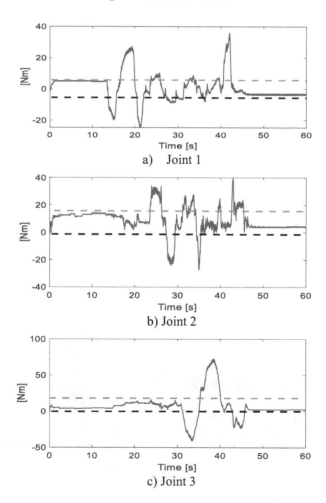

a)   Joint 1

b) Joint 2

c) Joint 3

Fig. 6. Contact force estimation

The results in Fig. 6 show that the contact force estimation can be used for admittance control in physical human-robot interaction. The estimation accuracy of the interaction force depends on the accuracy of the dynamic model robot manipulator. Therefore, the identification process is one of importance in contact force estimation based on model. The snapshot of push and pull were shown in Fig. 7.

Fig. 7. Snapshot of push and pull robot experiment

5   Conclusion

In this paper, the contact force estimation was proposed by using the adaptive third order sliding mode observer to estimate the contact force interaction. The experimental results in two scenarios detection contact and push/pull were shown to illustrate the effectiveness of the approach method in physical human-robot interaction. However, the accuracy of this method highly depends on the accuracy of the identification step in model dynamic. To overcome this problem the threshold is use to detect contact and estimate contact force to get this information to apply in admittance control.

Acknowledgment. This research was supported by Basic Science Research Program through the National Research Foundation of Korea (NRF) funded by the Ministry of Education (NRF-2016R1D1A3B03930496).

References

1. Suita, K., Yamada, Y., Tsuchida, N., Imai, K., Ikeda, H., Sugimoto, N.: A failure-to-safety "Kyozon" system with simple contact detection and stop capabilities for safe human-autonomous robot coexistence. In: Proceedings of the 1995 IEEE International Conference on Robotics and Automation, pp. 3089–3096 (1995)
2. Geravand, M., Flacco, F., De Luca, A.: Human-robot physical interaction and collaboration using an industrial robot with a closed control architecture. In: 2013 IEEE International Conference on Robotics and Automation, pp. 4000–4007 (2013). https://doi.org/10.1109/ICRA.2013.6631141
3. Stolt, A., Robertsson, A., Johansson, R.: Robotic force estimation using dithering to decrease the low velocity friction uncertainties. In: 2015 IEEE International Conference on Robotics and Automation (ICRA), pp. 3896–3902 (2015)

4. Duchaine, V., Lauzier, N., Baril, M., Lacasse, M.-A., Gosselin, C.: A flexible robot skin for safe physical human robot interaction. In: IEEE International Conference on Robotics and Automation, ICRA 2009, pp. 3676–3681 (2009)

5. Cirillo, A., Ficuciello, F., Natale, C., Pirozzi, S., Villani, L.: A conformable force/tactile skin for physical human–robot interaction. IEEE Robot. Autom. Lett. **1**, 41–48 (2016)

6. Magrini, E., Flacco, F., De Luca, A.: Estimation of contact forces using a virtual force sensor. In: 2014 IEEE/RSJ International Conference on Intelligent Robots and Systems (IROS 2014), pp. 2126–2133 (2014)

7. Ebert, D., Henrich, D.: Safe human-robot-cooperation: Image-based collision detection for industrial robots (2002)

8. Eom, K.S., Suh, I.H., Chung, W.K., Oh, S.-R.: Disturbance observer based force control of robot manipulator without force sensor. In: Proceedings of the 1998 IEEE International Conference on Robotics and Automation, pp. 3012–3017 (1998)

9. Gaz, C., Magrini, E., De Luca, A.: A model-based residual approach for human-robot collaboration during manual polishing operations. Mechatronics (2018)

10. Magrini, E., Flacco, F., De Luca, A.: Control of generalized contact motion and force in physical human-robot interaction. In: 2015 IEEE International Conference on Robotics and Automation (ICRA), pp. 2298–2304 (2015)

11. De Luca, A., Mattone, R.: Sensorless robot collision detection and hybrid force/motion control. In: Proceedings of the 2005 IEEE International Conference on Robotics and Automation, ICRA 2005, pp. 999–1004 (2005)

12. Chen, S., Luo, M., He, F.: A universal algorithm for sensorless collision detection of robot actuator faults. Adv. Mech. Eng. **10**, 1687814017740710 (2018)

13. Hu, J., Xiong, R.: Contact force estimation for robot manipulator using semiparametric model and disturbance Kalman filter. IEEE Trans. Ind. Electron. **65**, 3365–3375 (2018)

14. Khalil, H.K.: Cascade high-gain observers in output feedback control. Automatica **80**, 110–118 (2017)

15. Kommuri, S.K., Rath, J.J., Veluvolu, K.C.: Sliding-mode-based observer-controller structure for fault-resilient control in DC servomotors. IEEE Trans. Ind. Electron. **65**, 918–929 (2018)

16. Fridman, L., Shtessel, Y., Edwards, C., Yan, X.-G.: Higher-order sliding-mode observer for state estimation and input reconstruction in nonlinear systems. Int. J. Robust Nonlinear Control IFAC Affiliated J. **18**, 399–412 (2008)

17. Jung, S., Hsia, T.C., Bonitz, R.G.: Force tracking impedance control of robot manipulators under unknown environment. IEEE Trans. Control Syst. Technol. **12**, 474–483 (2004). https://doi.org/10.1109/TCST.2004.824320

18. Van, M., Ge, S.S., Ren, H.: Robust fault-tolerant control for a class of second-order nonlinear systems using an adaptive third-order sliding mode control. IEEE Trans. Syst. Man Cybern. Syst. **47**, 221–228 (2017)

19. Ortiz-Ricardez, F.A., Sánchez, T., Moreno, J.A.: Smooth Lyapunov function and gain design for a second order differentiator. In: 2015 IEEE 54th Annual Conference on Decision and Control (CDC), pp. 5402–5407 (2015)

20. Cruz-Zavala, E., Moreno, J.A.: Lyapunov functions for continuous and discontinuous differentiators. IFAC-PapersOnLine **49**, 660–665 (2016)

21. Bahrami, M., Naraghi, M., Zareinejad, M.: Adaptive super-twisting observer for fault reconstruction in electro-hydraulic systems. ISA Trans. **76**, 235–245 (2018)

22. Luo, D., Xiong, X., Jin, S., Kamal, S.: Adaptive gains of dual level to super-twisting algorithm for sliding mode design. IET Control Theory Appl. **12**, 2347–2356 (2018)

Model-Free Continuous Fuzzy Terminal Sliding Mode Control for Second-Order Nonlinear Systems

Van-Cuong Nguyen, Phu-Nguyen Le, and Hee-Jun Kang$^{(\boxtimes)}$

Department of Electrical, Electronic and Computer Engineering,
University of Ulsan, Ulsan 44610, South Korea
hjkang@ulsan.ac.kr

Abstract. This paper proposes a continuous fuzzy terminal sliding mode control (C-F-TSMC) for model-free second-order nonlinear systems. This controller inherited the advantages of three techniques: fuzzy logic system, terminal sliding mode and super-twisting algorithm (STA). First, a T-S fuzzy logic system is used to model the unknown second-order nonlinear systems. Then, a terminal sliding mode controller based on STA is implemented to achieve the highest tracking performance. The proposed controller provides the superior control properties such as robustness, high tracking accuracy, finite-time convergence, and less chattering phenomenon. Moreover, it is designed without the system model. The theory analysis is accomplished to prove the stability and finite-time convergence of the system. The proposed C-F-TSMC is applied to a serial rigid link robot manipulator. Simulation results are presented to validate the analysis.

Keywords: Terminal sliding mode control · T-S fuzzy · Super-twisting algorithm

1 Introduction

In recent years, improving tracking performance of nonlinear systems have been attracted the interest from a lot of researchers [1, 2]. The target systems can be magnetic levitation, helicopters, aircraft, motors, robot manipulator systems, etc. Their common characteristics are that they have a highly nonlinear dynamic and complex mathematical models. In literature, fuzzy logic has been recognized as one of the most powerful tools for system design due to their characteristics such as reduced mathematical complexity, good approximation, and easy implementation for describing complex nonlinear systems [3–5]. Takagi-Sugeno (T-S) fuzzy approach has attracted considerable attention because of its particular advantages, which include: simplicity of concept, ease of construction, allowance of offline computing and being justified as a universal approximation [5]. The basic idea of the T-S fuzzy model is to first decompose the model of a nonlinear system into a set of linear subsystems, and then smoothly connect them by a fuzzy membership function. Although the T-S fuzzy model can approximate nonlinear system well by the suitable selection of the

© Springer Nature Switzerland AG 2021
D.-S. Huang et al. (Eds.): ICIC 2021, LNCS 12837, pp. 245–258, 2021.
https://doi.org/10.1007/978-3-030-84529-2_21

membership function, it creates additional uncertainties due to the mismatch between the T-S fuzzy model and the original nonlinear system.

To deal with uncertainty problem, various control approaches have been developed such as PID control [6], adaptive control [7], neural network control [8], and sliding mode control (SMC) [9–11], etc. Among them, the SMC is the most attractive and effective method because of its robustness, fast transient response, and simple design procedure [12–14]. However, the conventional SMC adopts a linear sliding surface, which only provides asymptotic stability of the system in the sliding phase. To over-come this drawbacks, terminal sliding mode control (TSMC) was developed, which used a nonlinear sliding surface instead of the linear one [15, 16]. Compared with conventional SMC, TSMC offers some superior properties such as finite-time con-vergence and obtaining higher accuracy with carefully designing parameters. However, the same as the SMC, the TSMC still suffers from chattering phenomenon due to the high frequency switching of the control signal.

To decrease the chattering phenomenon, it is necessary to reduce or eliminate the discontinuous control element in the switching control law. The natural idea is to replace the discontinuous switching law by a continuous one; for example: saturating approximation [17] or boundary layer technique [18]. By using these methods, the chattering phenomenon can be decreased; however, as a trade-off, the tracking accuracy of the system is reduced. Another approach to reduce chattering phenomenon is using high-order sliding mode (HSMC) [19, 20], which moves the discontinuous to the higher-order of derivative that leads to the reduction of chattering. Furthermore, HOSM can bring better accuracy than conventional SMC while the robustness of the control system is similar to SMC. Many high-order sliding algorithms require the derivative of the sliding variable, however, super-twisting algorithm (STA) method [21, 22] just depends on the sliding variable, so it quite easy to implemented.

Motivated by all the above concerns, in this paper, a T-S fuzzy system is applied to model the nonlinear second-order system. In this step, the system parameters are computed offline. Then, a controller based on terminal sliding surface and STA will be employed to deal with the effects of modelling uncertainty and to reduce the chattering phenomenon. The proposed controller provides the superior control properties such as robustness, high tracking accuracy, finite-time convergence, and less chattering phe-nomenon. Moreover, it is designed without the system model.

The remainder of this paper is designed as follows. Section 2 describes the for-mulation of problem. Then, Sect. 3 presents the design of proposed C-F-TSMC. And Sect. 4 gives the study of simulation. Finally, some conclusion is shown in Sect. 5.

2   Problem Formulation

Consider the nonlinear second-order systems with dynamic equation as following

$$
\begin{aligned}
\dot{x}_1 &= x_2 \\
\dot{x}_2 &= f(x) + g(x)u + \Delta(x,t) \\
y &= x_1
\end{aligned}
\tag{1}
$$

where $f(x)$, $g(x)$ are smooth and nonlinear functions; $g(x)$ is invertible; $x = \left[x_1^T, x_2^T\right]^T$ is system state; u is the control input; $\Delta(x, t)$ is the unknown but bounded uncertainty and/or disturbance; and y is the measurement output of system.

Assumption 1: The desired state vector $x_d(t)$ is a twice continuously differentiable function respect to time t.

The nonlinear system in Eq. (1) can be modeled by a T-S fuzzy system as follows [23]:

Plant rule i: If $z_1(t)$ is M_{i1} and ... and $z_p(t)$ is M_{ip}, then

$$\dot{x}_1 = x_2$$
$$\dot{x}_2 = A_i x + B_i u \quad i = 1, 2, ..., r \tag{2}$$
$$y = x_1$$

where M_{ij} is the fuzzy set; r is the number of model rules; $z_1(t), ..., z_p(t)$ are known premise variables that may be functions of state variables; A_i, B_i are system matrices.

The fuzzy system then is inferred as:

$$\dot{x}_1 = x_2$$
$$\dot{x}_2 = \frac{\sum_{i=1}^{r} w_i(z)\{A_i x + B_i u\}}{\sum_{i=1}^{r} w_i(z)} = \sum_{i=1}^{r} h_i(z)\{A_i x + B_i u\} \tag{3}$$
$$y = x_1$$

where $\quad z(t) = \left[z_1(t) \quad z_2(t) \quad \cdots \quad z_p(t)\right], \quad w_i(z) = \prod_{j=1}^{p} M_{ij}\left(z_j(t)\right), \quad$ and

$h_i(z(t)) = \frac{w_i(z(t))}{\sum_{i=1}^{r} w_i(z(t))}$.

Finally, the nonlinear system in Eq. (1) can be expressed by a T-S fuzzy system as follows

$$\dot{x}_1 = x_2$$
$$\dot{x}_2 = \sum_{i=1}^{r} h_i(z)\{A_i x + B_i u\} + d(x, t) \tag{4}$$
$$y = x_1$$

where $d(x, t)$ denotes the uncertainty, which consists of the approximation error of the fuzzy model and the uncertainty and/or disturbance of the system in Eq. (1).

Assumption 2: The system uncertainty in Eq. (4) is bounded and there exists a known constant D such that $\|d(x, t)\| < D$.

3 Design of Controller

3.1 Continuous Fuzzy Terminal Sliding Mode Controller

A terminal sliding surface is designed as follows [16]:

$$s = \dot{e} + \beta e^{\frac{q}{p}} \tag{5}$$

where β is a positive define matrix; p and q are positive odd integers which satisfy $p > q$; and e, $\dot{e}$ are position and velocity error, respectively, which are defined as follows

$$e = x_1 - x_d \tag{6}$$

$$\dot{e} = \dot{x}_1 - \dot{x}_d \tag{7}$$

Take derivative both sides of Eq. (5)

$$\dot{s} = \ddot{e} + \beta \frac{q}{p} e^{\frac{q}{p}-1}\dot{e} = (\dot{x}_2 - \ddot{x}_d) + \beta \frac{q}{p} e^{\frac{q}{p}-1}\dot{e} \tag{8}$$

Substitute Eq. (4) into Eq. (8)

$$\dot{s} = \left(\left(\sum_{i=1}^{r} h_i(z)\{A_i x + B_i u\} + d(t) \right) - \ddot{x}_d \right) + \beta \frac{q}{p} e^{\frac{q}{p}-1}\dot{e} \tag{9}$$

The controller is proposed as follow

$$u = u_{eq} + u_{st} \tag{10}$$

The equivalent control term u_{eq} is obtained when $\dot{s} = 0$ and $d(x,t) = 0$

$$u_{eq} = \left(\sum_{i=1}^{r} h_i(z)B_i \right)^{-1} \left[\ddot{x}_d - \beta \frac{q}{p} e^{\frac{q}{p}-1}\dot{e} - \sum_{i=1}^{r} h_i(z)A_i x \right] \tag{11}$$

The switching control term is designed using STA as

$$u_{st} = \left(\sum_{i=1}^{r} h_i(z)B_i \right)^{-1} \left(-k_1 |s|^{1/2} sign(s) + z \right)$$

$$\dot{z} = -k_2 sign(s) \tag{12}$$

Substituting Eq. (10), (11) and (12) into Eq. (9) leads to

$$\dot{s} = -k_1|s|^{1/2}sign(s) + z + d(x,t) \tag{13}$$

3.2 Stability Analysis

Theorem 1: For system described by Eq. (4) satisfies Assumption 2, the sliding surface s will converge to zero in finite-time by controller designed as Eq. (12) with coefficients are chosen as

$$k_1 > 2D \text{ and } k_2 > k_1 \frac{5k_1 + 4D}{2k_1 - 4D} D \tag{14}$$

Proof: Follow the references [22, 24], consider a Lyapunov function as.

$$V = \zeta^T P \zeta \tag{15}$$

where $\zeta = \left[|s|^{1/2}sign(s), z\right]^T$ and $P = \frac{1}{2}\begin{bmatrix} k_1^2 + 4k_2 & -k_1 \\ -k_1 & 2 \end{bmatrix}$.

V is positive definite and can be rewritten in the following form

$$V = 2k_2|s| + \frac{1}{2}\left(k_1|s|^{1/2}sign(s) - z\right)^2 + \frac{z^2}{2} \tag{16}$$

Take derivative of V

$$\dot{V} = -\frac{1}{|s|^{1/2}}\left(\zeta^T Q_1 \zeta - d(x,t)Q_2^T \zeta\right) \tag{17}$$

Using condition in Assumption 2, Eq. (17) can be inferred to

$$\dot{V} \leq -\frac{1}{|s|^{1/2}}\zeta^T Q \zeta \leq -\frac{1}{|s|^{1/2}}\lambda_{min(Q)}\|\zeta\|^2 \tag{18}$$

where $\lambda_{min(Q)}$ is minimum Eigen value of Q and

$$Q = \frac{k_1}{2}\begin{bmatrix} 2k_2 + k_1^2 - \left(\frac{4k_2}{k_1} + k_1\right)D & -k_1 - 2D \\ -k_1 - 2D & 1 \end{bmatrix} \tag{19}$$

If the condition in Eq. (14) is satisfied, then $Q > 0$; so $\dot{V}$ is negative definition. On the other hand, the following condition is satisfied

$$\lambda_{min(P)}\|\zeta\|^2 \leq V \leq \lambda_{max(P)}\|\zeta\|^2 \tag{20}$$

where $\lambda_{min(P)}$ and $\lambda_{max(P)}$ are respectively minimum and maximum Eigen value of P.

Using Eq. (20) and the fact that

$$|s|^{1/2} \le \|\zeta\| \le \frac{V^{1/2}}{\lambda_{min}^{1/2}} \tag{21}$$

$$\|\zeta\| \ge \frac{V^{1/2}}{\lambda_{max}^{1/2}} \tag{22}$$

Then, substituting Eq. (21) and (22) into (18), we obtain

$$\dot{V} \le -\gamma V^{1/2} \tag{23}$$

where $\gamma = \frac{\lambda_{min(P)}^{1/2} \lambda_{min(Q)}}{\lambda_{max(P)}^{1/2}}$.

Since the solution of the differential equation

$$\dot{v} = -\gamma v^{1/2}, \quad v(0) = v_0 > 0 \tag{24}$$

is given as

$$v(t) = \left(v_0^{1/2} - \frac{\gamma}{2}t\right)^2, \tag{25}$$

therefore, $v(t)$ converges to zero in finite-time and reaches zero after $t_r = 2v_0^{1/2}/\gamma$. It follows from the comparison principle [18] that $V(t) \le v(t)$ when $V(x_0) \le v_0$. From Eq. (25), we can conclude that $V(t)$ and therefore s converge to zero in finite-time and reaches that value at most after t_r units of time

$$t_r = \frac{2V^{1/2}(x_0)}{\gamma} \tag{26}$$

3.3   Analysis of System Finite-Time Convergence

Suppose that after reaching time t_r, sliding surface reaches value zero ($s = 0$), the attaining time (time for $e \rightarrow 0$) is t_s. The Eq. (8) will become

$$\dot{e} + \beta e^{\frac{q}{p}} = 0 \quad \Rightarrow \quad e^{-\frac{q}{p}}de = -\beta dt \tag{27}$$

Integrating the above differential equation in Eq. (27)

$$\int_{e(t_r)}^{e(t_r + t_s)} e^{-\frac{q}{p}}de = \int_{t_r}^{t_r + t_s} -\beta dt \tag{28}$$

Then, Eq. (28) leads to

$$t_s = \frac{p}{\beta(p-q)} \left| e(t_r) \right|^{\frac{p-q}{p}} \tag{29}$$

That means after sliding surface reaches to value zero, the tracking error converges to zero after finite-time. Or the whole close-loop system will converge after finite-time t_c as $t_c = t_r + t_s$.

4 Simulation Results

In this part, the proposed C-F-TSMC is applied on a two link with the dynamic equation as following

$$M(\theta)\ddot{\theta} + C\left(\theta, \dot{\theta}\right)\dot{\theta} + G\left(\dot{\theta}\right) + F_r\left(\theta, \dot{\theta}\right) + \tau_d = \tau \tag{30}$$

where

$$M(\theta) = \begin{bmatrix} (m_1 + m_2)l_1^2 & m_2 l_1 l_2 (s_1 s_2 + c_1 c_2) \\ m_2 l_1 l_2 (s_1 s_2 + c_1 c_2) & m_2 l_2^2 \end{bmatrix}$$

$$C\left(\theta, \dot{\theta}\right) = \begin{bmatrix} -m_2 l_1 l_2 (c_1 s_2 - s_1 c_2)\dot{\theta}_1 \\ -m_2 l_1 l_2 (c_1 s_2 - s_1 c_2)\dot{\theta}_2 \end{bmatrix}$$

$$G(\theta) = \begin{bmatrix} -(m_1 + m_2)l_1 g s_1 \\ -m_2 l_2 g s_2 \end{bmatrix}$$

The dynamic parameters of the robot are assigned as: $m_1 = m_2 = 1(kg)$, $l_1 = l_2 = 1(m)$. The disturbance is assumed as $\tau_d = [0.5sin(3\theta_1 + \pi/2) \quad -0.3sin (1.5\theta_2)]^T$ and the friction is assumed as $F_r = \left[5sin\left(\dot{\theta}_1\right) + 0.2sign(\theta_1) \quad 5sin\left(\dot{\theta}_2\right) + 0.2sign(\theta_2)\right]^T$. The initial states of the robot are assigned as $\theta = [\theta_1 \quad \theta_2]^T = [0.15 \quad -0.15]^T$, $\dot{\theta} = [\dot{\theta}_1 \quad \dot{\theta}_2]^T = [0 \quad 0]^T$.

The dynamic equation of the robot in Eq. (30) can be rewritten in state space form as

$$\dot{x}_1 = x_2$$
$$\dot{x}_2 = f_1(x) + g_{11}(x)u_1 + g_{12}u_2 + \eta_1$$
$$\dot{x}_3 = x_4$$
$$\dot{x}_4 = f_2(x) + g_{21}(x)u_1 + g_{22}u_2 + \eta_2$$

where $x_1 = \theta_1$, $x_2 = \dot{\theta}_1$, $x_3 = \theta_2$, $x_4 = \dot{\theta}_2$ and

$$f_1(x) = \frac{(s_1c_2 - c_1s_2)(m_2l_1l_2(s_1s_2 + c_1c_2)x_2^2 - m_2l_2^2x_4^2)}{l_1l_2((m_1 + m_2) - m_2(s_1s_2 + c_1c_2))}$$
$$+ \frac{((m_1 + m_2)l_2gs_1 - m_2l_2gs_2(s_1s_2 + c_1c_2))}{l_1l_2((m_1 + m_2) - m_2(s_1s_2 + c_1c_2))}$$

$$f_2(x) = \frac{(s_1c_1 - c_1s_2)(-(m_1 + m_2)l_1^2x_2^2 + m_2l_1l_2(s_1s_2 + c_1c_2)x_4^2)}{l_1l_2((m_1 + m_2) - m_2(s_1s_2 + c_1c_2))}$$
$$+ \frac{(-(m_1 + m_2)l_1gs_1(s_1s_2 + c_1c_2) + (m_1 + m_2)l_1gs_2)}{l_1l_2((m_1 + m_2) - m_2(s_1s_2 + c_1c_2))}$$

$$g_{11}(x) = \frac{m_2l_2^2}{m_2l_1^2l_2^2\left((m_1 + m_2) - m_2(s_1s_2 + c_1c_2)^2\right)}$$

$$g_{12}(x) = g_{21}(x) = \frac{-m_2l_1l_2(s_1s_2 + c_1c_2)}{m_2l_1^2l_2^2\left((m_1 + m_2) - m_2(s_1s_2 + c_1c_2)^2\right)}$$

$$g_{22}(x) = \frac{(m_1 + m_2)l_1^2}{m_2l_1^2l_2^2\left((m_1 + m_2) - m_2(s_1s_2 + c_1c_2)^2\right)}$$

By defining $X_1 = [x_1 \ x_3]^T, X_2 = [x_2 \ x_4]^T, X = [X_1 \ X_2]^T$, the T-S fuzzy model for the robot is described by nine rules as in [3].

Rule 1: If x_1 is about $-\pi/2$ and x_2 is about $-\pi/2$ then

$$\begin{cases} \dot{X}_1 = X_2 \\ \dot{X}_2 = A_1X + B_1u \end{cases}$$

Rule 2: If x_1 is about $-\pi/2$ and x_2 is about 0 then

$$\begin{cases} \dot{X}_1 = X_2 \\ \dot{X}_2 = A_2X + B_2u \end{cases}$$

Rule 3: If x_1 is about $-\pi/2$ and x_2 is about $\pi/2$ then

$$\begin{cases} \dot{X}_1 = X_2 \\ \dot{X}_2 = A_3X + B_3u \end{cases}$$

Rule 4: If x_1 is about 0 and x_2 is about $-\pi/2$ then

$$\begin{cases} \dot{X}_1 = X_2 \\ \dot{X}_2 = A_4X + B_4u \end{cases}$$

Rule 5: If x_1 is about 0 and x_2 is about 0 then

$$\begin{cases} \dot{X}_1 = X_2 \\ \dot{X}_2 = A_5 X + B_5 u \end{cases}$$

Rule 6: If x_1 is about 0 and x_2 is about $\pi/2$ then

$$\begin{cases} \dot{X}_1 = X_2 \\ \dot{X}_2 = A_6 X + B_6 u \end{cases}$$

Rule 7: If x_1 is about $\pi/2$ and x_2 is about $-\pi/2$ then

$$\begin{cases} \dot{X}_1 = X_2 \\ \dot{X}_2 = A_7 X + B_7 u \end{cases}$$

Rule 8: If x_1 is about $\pi/2$ and x_2 is about 0 then

$$\begin{cases} \dot{X}_1 = X_2 \\ \dot{X}_2 = A_8 X + B_8 u \end{cases}$$

Rule 9: If x_1 is about $\pi/2$ and x_2 is about $-\pi/2$ then

$$\begin{cases} \dot{X}_1 = X_2 \\ \dot{X}_2 = A_9 X + B_9 u \end{cases}$$

The matrixes $A_i, B_i (i = 1 : 9)$ are as follow

$$A_1 = \begin{bmatrix} 5.927 & -0.001 & -0.315 & -8.4 \times 10^{-6} \\ -6.859 & 0.002 & 3.155 & 6.2 \times 10^{-6} \end{bmatrix};$$

$$A_2 = \begin{bmatrix} 3.0428 & -0.0011 & 0.1791 & -0.0002 \\ 3.5436 & 0.0313 & 2.5611 & 1.14 \times 10^{-5} \end{bmatrix};$$

$$A_3 = \begin{bmatrix} 6.2728 & 0.003 & 0.4339 & -0.0001 \\ 9.1041 & 0.0158 & -1.0574 & -3.2 \times 10^{-5} \end{bmatrix};$$

$$A_4 = \begin{bmatrix} 6.4535 & 0.0017 & 1.2427 & 0.0002 \\ -3.1873 & -0.0306 & 5.1911 & -1.8 \times 10^{-5} \end{bmatrix};$$

$$A_5 = \begin{bmatrix} 11.1336 & 0 & -1.8145 & 0 \\ -9.0918 & 0.0158 & 9.1638 & 0 \end{bmatrix};$$

$$A_6 = \begin{bmatrix} 6.1702 & -0.001 & 1.687 & -0.0002 \\ -2.3559 & 0.0314 & 4.5298 & 1.1 \times 10^{-5} \end{bmatrix};$$

$$A_7 = \begin{bmatrix} 6.1206 & -0.0041 & 0.6205 & 0.0001 \\ 8.8794 & -0.0193 & -1.0119 & 4.4 \times 10^{-5} \end{bmatrix};$$

$$A_8 = \begin{bmatrix} 3.6421 & 0.0018 & 0.0721 & 0.0002 \\ 2.4290 & -0.0305 & 2.9832 & -1.9 \times 10^{-5} \end{bmatrix};$$

$$A_9 = \begin{bmatrix} 6.2933 & -0.0009 & -0.2188 & -1.2 \times 10^{-5} \\ -7.4649 & 0.0024 & 3.2693 & 9.2 \times 10^{-5} \end{bmatrix}.$$

$$B_1 = B_5 = B_9 = \begin{bmatrix} 1 & -1 \\ -1 & 2 \end{bmatrix}, \qquad\qquad B_2 = B_4 = B_6 = B_8 = \begin{bmatrix} 0.5 & 0 \\ 0 & 1 \end{bmatrix},$$

$$B_3 = B_7 = \begin{bmatrix} 1 & 1 \\ 1 & 2 \end{bmatrix}.$$

The parameters of the proposed C-F-TSMC are chosen as: $\beta = diag(1,1)$, $p = 5, q = 3, k_1 = 9, k_2 = 11$.

To verify the effectiveness, the proposed C-F-TSMC is compared to the two controllers. The first controller is fuzzy SMC (F-SMC), which is designed based on T-S fuzzy model, conventional sliding surface, and the discontinuous switching law (sign (s)). The second controller is fuzzy TSMC (F-TSMC), which uses the terminal sliding surface in Eq. (5) instead of the conventional one. The simulation results are shown in below Figs. 1, 2 and 3. The output tracking and tracking error of joint 1 and joint 2 are shown in Fig. 1 and Fig. 2, respectively. As shown in the result, by using the terminal sliding surface, the F-TSMC provides better tracking performance than that of F-SMC. In term of convergence rate, the F-TSMC has higher convergence speed when system states are near equilibrium point. However, the F-TSMC converges slower when the system states are far from the equilibrium point. The proposed C-F-TSMC not only maintains the convergence speed of F-TSMC but also increases the tracking performance by using the STA. The control input at each joint is shown in Fig. 3. Thanks to the ability to provide a continuous control signal of the STA, the chattering phenomenon of proposed controller is reduced.

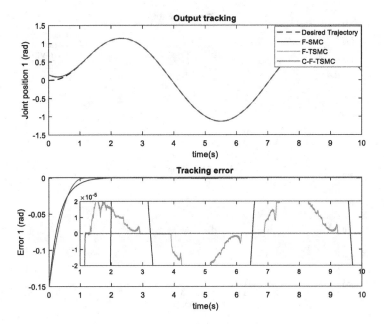

Fig. 1. Tracking performance at joint 1.

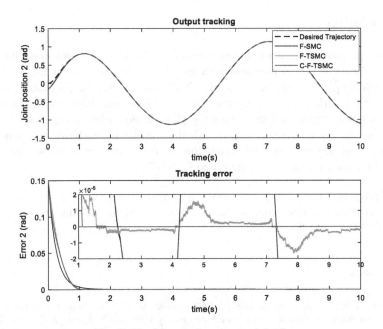

Fig. 2. Tracking performance at joint 2.

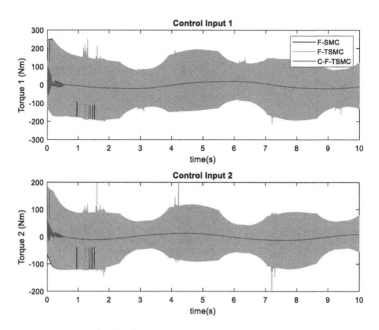

Fig. 3. Control input signal at each joint.

5 Conclusion

This paper proposed a model-free continuous fuzzy terminal sliding mode control (C-F-TSMC) for second-order nonlinear systems. This controller inherited the advantages of three techniques: fuzzy logic system, TSMC and STA. The proposed controller provides the superior control properties such as robustness, high tracking accuracy, finite-time convergence, and less chattering phenomenon without the need of system model. The theory analysis is accomplished to prove the stability and finite-time convergence of the system. The simulation on the 2-link serial robot manipulator validated the effectiveness of the proposed C-F-TSMC.

Acknowledgment. This research was supported by Basic Science Research Program through the National Research Foundation of Korea (NRF) funded by the Ministry of Education (2019R1D1A3A03103528).

References

1. Truong, T.N., Vo, A.T., Kang, H.-J.: Implementation of an adaptive neural terminal sliding mode for tracking control of magnetic levitation systems. IEEE Access **8**, 206931–206941 (2020)
2. Nguyen, V.-C., Vo, A.-T., Kang, H.-J.: A finite-time fault-tolerant control using non-singular fast terminal sliding mode control and third-order sliding mode observer for robotic manipulators. IEEE Access **9**, 31225–31235 (2021)

3. Tseng, C.-S., Chen, B.-S., Uang, H.-J.: Fuzzy tracking control design for nonlinear dynamic systems via TS fuzzy model. IEEE Trans. Fuzzy Syst. **9**(3), 381–392 (2001)
4. Liang, Y.-W., Xu, S.-D., Liaw, D.-C., Chen, C.-C.: A study of T-S model-based SMC scheme with application to robot control. IEEE Trans. Industr. Electron. **55**(11), 3964–3971 (2008)
5. Liang, Y.-W., Xu, S.-D., Ting, L.-W., et al.: TS model-based SMC reliable design for a class of nonlinear control systems. IEEE Trans. Ind. Electron. **56** (9), 3286–3295 (2009)
6. Zhang, J., Guo, L.: Theory and design of PID controller for nonlinear uncertain systems. IEEE Control Syst. Lett. **3**(3), 643–648 (2019)
7. Deng, W., Yao, J., Ma, D.: Adaptive control of input delayed uncertain nonlinear systems with time-varying output constraints. IEEE Access **5**, 15271–15282 (2017)
8. Vo, A.T., Kang, H.-J.: Adaptive neural integral full-order terminal sliding mode control for an uncertain nonlinear system. IEEE Access **7**, 42238–42246 (2019)
9. Tran, X.-T., Kang, H.-J.: Adaptive hybrid high-order terminal sliding mode control of MIMO uncertain nonlinear systems and its application to robot manipulators. Int. J. Precis. Eng. Manuf. **16**(2), 255–266 (2015)
10. Nguyen, V.-C., Kang, H.-J.: A fault tolerant control for robotic manipulators using adaptive non-singular fast terminal sliding mode control based on neural third order sliding mode observer. In: Huang, D.-S., Premaratne, P. (eds.) ICIC 2020. LNCS (LNAI), vol. 12465, pp. 202–212. Springer, Cham (2020). https://doi.org/10.1007/978-3-030-60796-8_17
11. Vo, A.T., Kang, H.-J.: Neural integral non-singular fast terminal synchronous sliding mode control for uncertain 3-DOF parallel robotic manipulators. IEEE Access **8**, 65383–65394 (2020)
12. Utkin, V.I.: Sliding Modes in Control and Optimization. Springer, Heidelberg (2013)
13. Nguyen, V.-C., Vo, A.-T., Kang, H.-J.: Continuous PID sliding mode control based on neural third order sliding mode observer for robotic manipulators. In: Huang, D.-S., Huang, Z.-K., Hussain, A. (eds.) ICIC 2019. LNCS (LNAI), vol. 11645, pp. 167–178. Springer, Cham (2019). https://doi.org/10.1007/978-3-030-26766-7_16
14. Nguyen, V.-C., Vo, A.-T., Kang, H.-J.: A non-singular fast terminal sliding mode control based on third-order sliding mode observer for a class of second-order uncertain nonlinear systems and its application to robot manipulators. IEEE Access (2020)
15. Zhihong, M., Yu, X.H.: Terminal sliding mode control of MIMO linear systems. IEEE Trans. Circuits Syst. I Fundam. Theory Appl. **44**(11), 1065–1070 (1997)
16. Zhihong, M., Paplinski, A.P., Wu, H.R.: A robust MIMO terminal sliding mode control scheme for rigid robotic manipulators. IEEE Trans. Autom. Control **39**(12), 2464–2469 (1994)
17. Li, T.-H.S., Huang, Y.-C.: MIMO adaptive fuzzy terminal sliding-mode controller for robotic manipulators. Inf. Sci. **180**(23), 4641–4660 (2010)
18. Slotine, J.-J.E., Li, W., et al.: Applied nonlinear control. Prentice Hall, Englewood Cliffs (1991)
19. Levant, A.: Higher-order sliding modes, differentiation and output-feedback control. Int. J. Control **76**(9–10), 924–941 (2003)
20. Ding, S., Wang, J., Zheng, W.X.: Second-order sliding mode control for nonlinear uncertain systems bounded by positive functions. IEEE Trans. Ind. Electron. **62**(9), 5899–5909 (2015)
21. Xu, S.S.-D.: Super-twisting-algorithm-based terminal sliding mode control for a bioreactor system. In: Abstract and Applied Analysis, vol. 2014 (2014)
22. Tran, M.-D., Kang, H.-J.: Nonsingular terminal sliding mode control of uncertain second-order nonlinear systems. Math. Probl. Eng. **2015** (2015)

23. Tanaka, K., Wang, H.O.: Fuzzy Control Systems Design and Analysis: A Linear Matrix Inequality Approach. Wiley, Hoboken (2004)
24. Moreno, J.A., Osorio, M.: A Lyapunov approach to second-order sliding mode controllers and observers. In: 2008 47th IEEE Conference on Decision and Control, pp. 2856–2861 (2008)

Deep Q-learning with Explainable and Transferable Domain Rules

Yichuan Zhang⑩, Junkai Ren⑩, Junxiang Li⑩, Qiang Fang⑩,
and Xin Xu(✉)⑩

National University of Defence Technology, Changsha 410000, China
{zhangyichuan15, xinxu}@nudt.edu.cn

Abstract. Recent research on deep reinforcement learning (RL) has shown its capability in automating sequential decision-making and control tasks. However, RL agents require a large number of interactions with the environment and are lacking in transferability. These problems severely restrict the adoption of deep RL in real-world tasks. Taking the learning process of human beings as inspiration, a promising way to solve these problems is to integrate transferable domain knowledge into deep RL. In this work, we propose a method called Deep Q-learning with transferable Domain Rules (DQDR) that combines transferable domain knowledge to enhance the sample efficiency and transferability of RL algorithms. We extract domain knowledge from human beings and express it into a set of rules, then couple this knowledge with the deep Q-network (DQN). The experiments are conducted by comparing this DQDR with other proposed knowledge-based methods and applying this approach to a series of CartPole and FlappyBird tasks with different system dynamics. The empirical results show that our approach can accelerate the learning process and improve the transfer capability of RL algorithms.

Keywords: Reinforcement learning · Domain knowledge · Transfer reinforcement learning · Explainability

1 Introduction

Deep Q-network [1] has been studied as an important framework in deep reinforcement learning (RL) algorithms. Till now, deep RL has been applied in a wide range of challenging domains, including robotic control [2], game playing [3], and autonomous driving [4]. However, deep RL algorithms suffer from the challenge of sample complexity issue, which dramatically limits its adoption in real-world domains. In contrast, human beings can leverage prior domain knowledge and rarely learn to master a new task from scratch. Therefore, the learning process of human beings is much quicker than that of an RL agent.

Researchers have been working on the sample inefficient problem in RL. Till now, solutions that have been proposed to address this problem can be broadly classified into two streams: (i) Learning the environment dynamic model that predicts the future states; (ii) Integrating the learning process with prior domain knowledge. The first stream is referred to as model-based RL, which learns the system dynamic model

© Springer Nature Switzerland AG 2021
D.-S. Huang et al. (Eds.): ICIC 2021, LNCS 12837, pp. 259–273, 2021.
https://doi.org/10.1007/978-3-030-84529-2_22

during iteration. Once a well-performed system dynamic model is learned, the RL agent would not need to iterate with the real-world environment frequently. Our work falls into the second stream, which integrates domain knowledge into the RL process.

There are some previous researches focusing on combining domain knowledge in the learning process. And one of the most important lines of methods that leverage human knowledge in solving sequential decision-making and control problem is imitation learning (IL) [5], which leverages the demonstration trajectories to learn the corresponding policy in the same task (and then improve the policy performance in RL process [6]). Recent works in autonomous driving [7] and robotic manipulation tasks [8] have shown that the IL can significantly improve the sample efficiency of the policy learning process. But since the demonstration can be only illustrated based on a concrete task, demonstration data can only be seen as the instance of human knowledge and is hard to transfer to different tasks.

As mentioned above, another problem that RL agents are confronted with is the lack of transferability [5]. Imitation learning can only learn knowledge with low-level representation from demonstration data by supervised learning and this always causes negative transfer. Because of the negative transfer, policies learned from source Markov Decision Process (MDP) are not capable to perform well in target MDPs. When we want to solve a series of similar tasks, RL agents still need to be trained from scratch separately. Therefore, one promising way to solve this problem is to decouple the general domain knowledge (e.g., common sense) and the task-specific knowledge in the learning process and only share some general knowledge to the target MDPs to avoid negative transfer and help the learning process without demonstration.

To address these problems, we propose a novel RL algorithm called DQDR in this paper, which empowers original DQN with transferable domain knowledge. Specifically, we take domain knowledge from human beings and express it in terms of transparent general rules and leverage these rules to guide the learning process of RL by initializing the Q-network. When the RL agent encounters situations covered by these general rules, it chooses an action without learning by enormous exploration. Thus, the agent only learns in the task-specific state-action space. We evaluate the sample efficiency of DQDR by comparing it with other proposed methods, and the performance of DQDR is also evaluated in the sparse reward setting. Then a series of CartPole [9] and FlappyBird [10] tasks with different system dynamics are constructed to verify the transferability of DQDR. Besides, with the help of rules, DQDR shows a better explainability than ordinary Deep RL.

The rest of this paper is organized as follows. In Sect. 2, we present an overview of the related work. In Sect. 3, we introduce the architecture and details of our DQDR method. The performance and comparison of the proposed method are shown in Sect. 4 and conclusions are given in Sect. 5.

2   Related Work

2.1   Value-Based Reinforcement Learning

Reinforcement learning provides an ideal method for sequential decision-making tasks. To solve problems in the RL framework, the problem should be described as an MDP, which consists of (S, A, T, R, γ) where S represents the state space, A stands for the action space, T represents the system dynamics, R is the reward function and γ represents the discount factor [11].

An RL agent interacts with the MDP following its policy $\pi(a_t|s_t)$, which maps the state space to the action space. The goal of RL is to find an optimal policy that maximizes the expected cumulative reward. And $V_\pi(s)$ is usually used to estimate the expected cumulative future reward which can be expressed as:

$$V_\pi(s_t) = E_\pi\left[\sum_{k=0}^{\infty} \gamma^k r_{t+k+1}\right] \tag{1}$$

where $E_\pi[\cdot]$ denotes the expectation with respect to policy π.

For value-based RL, the optimal policy is derived by the optimal Q function learned by the agent. Temporal Difference learning (TD) is commonly used to evaluate the value function. The traditional Q-Learning process updates the Q function by bootstrapping in an off-policy way (α denotes the learning rate):

$$Q(s_t, a_t) \leftarrow (1 - \alpha) Q(s_t, a_t) + \alpha[r_t + \gamma \max_a Q(s_{t+1}, a)] \tag{2}$$

With the learned optimal Q function, the policy can be expressed as:

$$\pi(s) = \operatorname*{argmax}_a Q(s, a) \tag{3}$$

With the help of deep neural network, DQN becomes one of the most commonly used value-based RL algorithms. To alleviate the instability and non-convergence caused by the neural network approximator, experience replay, and fixed Q-targets mechanisms are proposed. The experience replay mechanism allows the DQN agent to store state, action, reward, and next state tuples in the replay buffer D. In the training process, the agent updates the Q-network by sampling experience batches from the replay buffer, which improves the training speed and stability. By introducing the fixed Q-targets mechanism, the DQN agent has the prediction Q-network parameterized as θ to predict the Q-value of each action corresponding to the current state and the target Q-network parameterized as θ^- to predict the Q-value of each action in the next states. The fixed Q-targets mechanism aims to reduce the correlation between the current Q value and the target Q value and thus improves the algorithm stability.

The loss function of the Q-network can be expressed as:

$$\mathcal{L}(\theta) = \mathbb{E}_{s,a,r,s' \sim \mathcal{D}} \left[\left(r + \gamma \max_{a'} Q(s',a';\theta^-) - Q(s,a;\theta) \right)^2 \right] \tag{4}$$

The gradient of the prediction Q-network is:

$$\Delta\theta = \alpha \left[r + \gamma \max_{a'} Q(s',a';\theta^-) - Q(s,a;\theta) \right] \nabla Q(s,a;\theta) \tag{5}$$

and then the parameter θ is updated following:

$$\theta = \theta + \Delta\theta \tag{6}$$

For every fixed time steps C, the target Q-network is updated by:

$$\theta^- = \theta \tag{7}$$

Recently, some further value-based methods like Double-DQN [12], Dueling-DQN [13], and Rainbow [14] have been proposed to enhance the original DQN.

2.2   Learning from Demonstration

Learning from Demonstration (LfD) originated in the field of robotics, where demonstrations can be seen as a form of assistance for humans to provide.

Imitation Learning can be seen as a mainstream of LfD. Imitation learning is an effective method of extracting the knowledge of experts from demonstration trajectories and it can be broadly classified into Behavioural Cloning (BC) and Inverse Reinforcement Learning (IRL). In the RL framework, the knowledge extracted by BC and IRL corresponds to the policy function and reward function of the expert. BC is the simplest paradigm of imitation learning as it extracts the optimal policy from demonstration data through Supervised Learning (SL) directly. As a high-capacity function approximator, the deep neural network is used in BC to mimic the expert policy recently. However, because of the compounding error caused by covariate shift [15], the agent can drift away from the demonstrated states. Thus, the supervised policy training method demands large amounts of data in solving sequential decision-making tasks. A method to address this problem is imitative RL. In imitative RL [6], the policy network is trained in the BC approach by demonstration data and then is fine-tuned in the RL process to acquire the ability to solve sequential decision-making tasks in high data efficiency. Another IL framework is IRL, which tries to recover the reward function of the expert and then learns a policy within the RL framework, thus this method avoids compounding error. Recently, [16] proposes a model-free IRL method called Generative Adversarial Imitation Learning (GAIL) by leveraging Generative Adversarial Networks (GAN) [17], and some variants of GAIL like InfoGAIL [18] and [19] are proposed to avoid the mode collapse of GAIL.

Another stream of LfD is to put expert demonstrations into the replay buffer of off-policy RL for value estimation. The representative work in this area is DQN from Demonstration (DQfD) [20] and DDPG from Demonstration (DDPGfD) [21]. For DQfD, at the pre-training phase of DQfD, the agent samples mini-batches from the demonstration data and updates the network by applying four loss functions. Once the pre-training phase is complete, the agent starts collecting self-generated data and updating its Q-network once the replay buffer is full. Similar to DQfD, DDPGfD also leverages the demonstration data by putting them into the replay memory while DDPGfD is based on an Actor-Critic RL framework.

However, a general limitation of the existing LfD methods is that demonstrations can only be seen as the low-level representation of knowledge for a fixed task, thus different demonstrations are required for different tasks. Therefore, the LfD method is more desirable to learn the policy to solve a particular task.

3   The Proposed DQDR Approach

In this section, we present the novel RL method DQDR that leverages prior domain knowledge in the value-based RL paradigm. To demonstrate our proposed DQDR, we also compare DQDR with the rule-based RL and transfer RL later.

3.1   Algorithm Framework

The proposed DQDR method consists of the incomplete and transferable domain knowledge rules and the Q-network of the agent. The overall architecture is shown in Fig. 1.

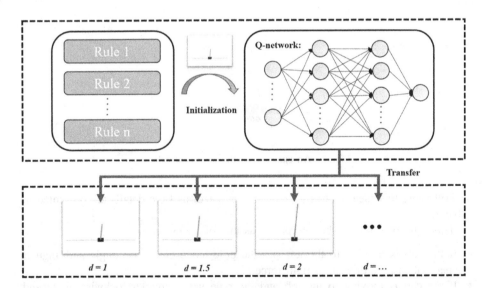

Fig. 1. The overall architecture of DQDR. It consists of a set of rules for human knowledge representation and a Q-network initialized by the human knowledge.

The traditional DQN algorithm has no background knowledge of the task to be addressed, and therefore initializes the Q-network randomly, which results in massive unnecessary exploration at the beginning of the learning process. However, in many control and decision-making tasks, human beings have empirical knowledge to help to speed up the learning process even if the knowledge is incomplete. Inspired by this, our method which incorporates incomplete domain knowledge with DQN is proposed.

To illustrate the algorithm clearly, an easily understood task (i.e., the OpenAI Gym CartPole task) is used here as an example. In the CartPole task, a pole is attached by a joint to a cart and the aim is to balance the pole by applying left or right force to the cart. Each episode has a maximum of 200 steps and each episode ends when the cart moves more than 2.4 units away from the center or the angle of the pole is more than 15° from vertical. For each step, the agent receives +1 as the reward, and thus the total reward upper bound of an episode is 200. This task has the state input as $\left[x, \dot{x}, \theta, \dot{\theta}\right]$ and discrete action output as $\{0, 1\}$. For the state input, x and $\dot{x}$ are the position and speed of the cart; θ and $\dot{\theta}$ are the angle and angular speed of the pole. For the action output, 0 means giving the cart a force to the left, and 1 means giving the cart a force to the right. For this task, the policy needs to keep the pole balanced in the vertical direction. We aim to learn a policy that could control the CartPole system efficiently and has the transfer capability to apply this policy to other CartPole tasks with different system dynamics.

In the CartPole task, we assume that the positive direction of the cart is right and the positive direction of the pole is clockwise. For human beings, it is easy to determine some general control guidelines according to our experience. For example, if both the angle and angular velocity are positive and the cart moves in the negative direction, the desired action should be chosen to give the cart a positive force, which is shown in Fig. 2.

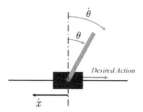

Fig. 2. An example of the proposed rules.

Following this idea, more rules can be extracted, and the details are presented in Table 1.

This form of rules can be interpreted as the following terms:

- If the cart is moving to the left and the pole has a clockwise angle and angular speed, then give this cart a right force.
- If the cart is moving to the left and the pole has a counterclockwise angle and angular speed, then give this cart a left force.

Table 1. Rules table for CartPole task, where P means positive direction; N means negative direction; $[-]$ means we can not confirm which action to choose in that corresponding state.

	Cart speed	Pole angle	Pole angular speed	Action
1	N	P	P	P
2	N	P	N	$[-]$
3	N	N	P	$[-]$
4	N	N	N	N
5	P	P	P	P
6	P	P	N	$[-]$
7	P	N	P	$[-]$
8	P	N	N	N

- If the cart is moving to the right and the pole has a clockwise angle and angular speed, then give this cart a right force.
- If the cart is moving to the right and the pole has a counterclockwise angle and angular speed, then give this cart a right force.

The whole learning processes are as follows:

I. Rules Extraction
According to the human domain knowledge and the characteristics of the task, some proper control rules are formulated as shown in Table 1.

II. Obtain a transferable prior Q-network
The agent starts interacting with the environment following the instruction of some model-agnostic rules and saves transition data into the replay buffer D_{init} to update the Q-network. By leveraging these rules as the instructor, the Q-network can be well initialized and saved for the following different tasks.

III. Cancel rules and interact independently
The rules are terminated after the initialization stage, to achieve better performance in target tasks, DQN is executed to interact with the environment and update the Q-network.

The pseudo-code is shown in Algorithm 1.

We argue that in some environmental states, with the help of this general knowledge initialized Q-network, the RL agent can avoid massive exploration in the learning process. In other complicated environmental states, it is hard for human beings to define complete rules for making decisions. However, this can be fine-tuned by RL, so we can leave corner cases to RL algorithms to explore and learn. Therefore, the RL agents only need to learn the control policy in a part of the state space, which enables the policy to converge fast. More importantly, these rules based on domain knowledge will not change with the system dynamics (e.g., the length of the pole or the weight of the cart), so it is suitable to adopt this knowledge initialized Q-network to different tasks.

3.2   Analysis and Discussions

To evaluate the proposed DQDR, we discuss the differences and similarities between DQDR and previous work in rule-based RL and transfer RL.

Algorithm 1 Training process of DQDR

Input: Initialized D_{init} and D with the capacity N respectively
        Initialized Q-network with random weights θ
        Initialized target Q-network with random weights θ^-
Output: Optimal Q-network.
 1: **for** each episode in knowledge initialization stage **do**
 2:    initialize state s
 3:    **while** s is not the terminal state **do**
 4:       choose an action a based on the **Table 1**
 5:       execute action a and observe reward r and the next state s'
 6:       store (s, a, r, s_0) in D_{init}
 7:       sample random batch of experience from D_{init}
 8:       update the Q functions by **Equation 4-7**
 9:       $s = s'$
10: **end for**
11: save the initialized Q-network
12: **for** each episode of DQN **do**
13:    initialize s
14:    **while** s is not the terminal state **do**
15:       choose an action a based on the the Q-network through ε-greedy
16:       execute action a and observe reward r and the next state s'
17:       store (s, a, r, s_0) in D
18:       sample random batch of experience from D
19:       update the Q functions by **Equation 4-7**
20:       $s = s'$
21: **end for**

The rule-based reinforcement learning framework tries to incorporate human knowledge into RL. Rule-based RL has been applied to a wide range of engineer problems because of its high interpretability and transparency.

The mainstream of rule-based RL is to adopt the rules as the main architecture of the RL algorithm. The representative work is [22], which creates fuzzy IF-THEN rules to design the Actor-Critic controller architecture and this framework allows the engineer to manually design the structure and adjustable parameters in the sense of engineering. And the other stream is using RL to fine-tune the rules by adjusting the weighting factor. For example, [23] uses RL to learn the parameters of rules to achieve transparent decision-making. [24] uses a series of fuzzy rules as the control policy and optimize these rules based on Q value.

The work [25] combines the above two streams by incorporating fuzzy rules as part of its policy architecture and fine-tune the weight factor of rules. In more detail, they propose a policy framework by using a set of fixed fuzzy rules as the core knowledge controller and fine-tune these fuzzy rules by leveraging the hyper-network refine module. The whole architecture proposed in their work is considered as a general policy module. Integrating this policy module with Proximal Policy Optimization (PPO) [26] framework, the experimental results show that this method achieves better performance than baseline PPO.

However, comparing with DQDR, this rule-based architecture focuses on usability and transparency but does not evaluate the transferability of the rule-based learning method, and also this rule-based method is mostly confined to policy-based RL.

Another related research area is transfer reinforcement learning, which aims to leverage knowledge learned from original MDPs to improve the RL performance in target MDPs. Transfer reinforcement learning can be classified into inter-task transfer learning and inter-domain transfer learning. We focus on the first category, which corresponds to the setting that the source MDPs and target MDPs have the same state-action space but different system dynamics. In inter-task transfer reinforcement learning, the methods can be classified into three categories: instance-based transfer reinforcement learning, feature-based transfer reinforcement learning, and model-based transfer reinforcement learning [27], where feature-based transfer reinforcement learning and model-based transfer reinforcement learning are closely related to our method.

Feature-based transfer reinforcement learning leverages the high-level knowledge from the source MDPs by adapting the feature representation of a target MDP. A representative method is the option-based transfer RL [28], which assumes that the options (can be seen as abstract actions) can be scaled to the target MDP. However, this feature-based method focuses more on the hierarchical representation of tasks while DQDR tries to reuse the generic knowledge for different MDPs.

Model-based transfer reinforcement learning learns the shared parameters from the source MDPs and initializes the target MDP with the shared parameters. Policy distillation is a classical method to accomplish the model transferring from the teacher policy to the student policy. Previous work [29] transfers the fully connected layers across MDPs since the task-specific convolutional feature functions are crucial for the performance. The main difference between this method and DQDR is that model-based transfer reinforcement learning can not provide the real meaning that the shared parameter stands for and it also needs tons of experimental work to determine what parameter to transfer.

Besides, a general drawback of the above methods is that those transfer RL methods require more prior computational processes to achieve good performance in the target task, which results in the slow learning process. But with the transferable domain rules, the learning process is significantly accelerated without extra resource-intensive training process.

4 Experimental Results

In this section, we investigate the data efficiency and transferability of our proposed DQDR on CartPole task and FlappyBird task respectively. Experiments in Sect. 4.1 and Sect. 4.2 are performed to evaluate the data efficiency in normal reward setting and sparse reward setting. Experiments in Sect. 4.3 are performed to evaluate the transferability. All the experiments are conducted on Ubuntu 16.04 and the DQDR learning framework is built with Pytorch.

The experimental setup is as follows: for all the experiments, we use the full-connected Q-network that has 3 layers with 2 hidden layers that have 128 nodes and the activation function is *tanh*. We choose Adam as our optimizer [30] with a learning rate of 5×10^{-4} and the discount factor γ is set to 0.98. In every training process, the Q-network is trained 10 times. And the target Q-network update frequency C is equal to 10. In the CartPole task, we make the rules to interact with the environment 200 episodes in the Q-network initialization stage and update the Q-network simultaneously. Considering the difficulty of the FlappyBird task, we make the rules to interact with the environment 2000 episodes to initialize the Q-network.

4.1 Experimental Results on the Validity of Knowledge

In this section, we compare our DQDR method with vanilla PPO, vanilla DQN, and Heuristic DQN on CartPole task and FlappyBird task. Both tasks here have the default setting.

The Heuristic exploration method in RL was proposed in [31] on the Grid-World navigation task. They introduce some heuristic strategies to accelerate the learning process. And the way to do this is to make sure that the random action selected in the exploration stage satisfies the proposed behavioral rules. We adopt this method in these tasks to compare with our method in data efficiency.

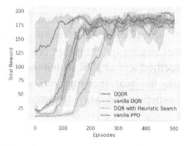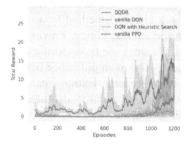

Fig. 3. Data efficiency experimental results. We compare DQDR with other methods in the CartPole task (left figure) and FlappyBird task (right figure). The shaded region denotes the standard deviation of average evaluation over 5 trials.

From Fig. 3, the DQN with heuristic search can help with the data efficiency compared with the original DQN since the heuristic exploration in ε-greedy is guided by the rules. And the DQDR method has better data efficiency performance than the original PPO, original DQN, and DQN with heuristic search. From the reward curve, DQDR also shows a warm start at the beginning of the learning stage.

Therefore, with the help of domain knowledge, it is easier for DQDR to explore in the smaller state space and gain better policy.

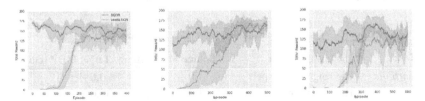

Fig. 4. Experimental results in the setting of sparse reward in the CartPole task. We set sparse interval $T = 25$ (left figure), 50 (middle figure), and 100 (right figure). The shaded region denotes the standard deviation of average evaluation over 5 trials.

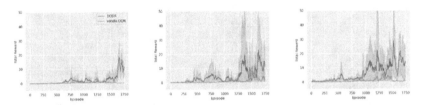

Fig. 5. Experimental results in the setting of sparse reward in FlappyBird task. We set sparse interval $T = 2$ (left figure), 3 (middle figure), and 4 (right figure). The shaded region denotes the standard deviation of average evaluation over 5 trials.

4.2 Sparse Reward Setting

In this section, we demonstrate the effectiveness of our DQDR in the sparse reward setting by comparing it with the baseline DQN in the CartPole task and FlappyBird task.

In the CartPole sparse reward setting, we consider that the agent only receives the multi-step accumulated rewards at sparse time steps T. Similarly, in the FlappyBird sparse reward setting, the agent receives the accumulated rewards after passing every T pipes. We carry out the experiment under different sparse intervals.

Figure 4 and Fig. 5 illustrate the experimental results in two tasks. From the result, DQDR shows its great performance in sparse reward settings. For vanilla DQN, it is hard to learn a policy effectively when the agent does not receive the reward from the environment frequently. And it is shown clearly from Fig. 4 that as the sparse interval increases in the CartPole task, the learning process of the DQN agent becomes more difficult. However, with human knowledge, DQDR can avoid some unnecessary exploration and receive sparse reward signals easily to learn effective policies.

4.3   Experimental Results on the Transferability

In this experiment, we evaluate the transferability of the proposed DQDR method in both CartPole and FlappyBird tasks, and the transfer settings are shown in Table 2. The source MDP and the target MDP have the same state and action space but different system dynamics. We choose DQN-Finetune and DQN-Retrain as comparison. DQN-Finetune method denotes that the Q-network is trained in the original MDP and then fine-tuned in the target MDP. DQN-Retrain denotes that the performance of vanilla DQN in the target MDP, which acts as the lower bound.

Table 2. The transfer settings in the CartPole and FlappyBird tasks.

Task	The changed item	Source MDP setting	Target MDP setting
CartPole	The length of the pole	1.0	3.0
	The weight of the cart	1.0	10.0
FlappyBird	The gravity of the bird	9.0	11.0
	The gap between two pipes	100.0	80.0

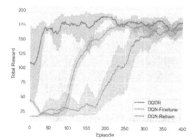 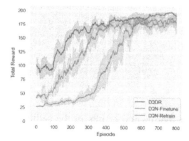

Fig. 6. Experimental results of the pole length transfer (left figure) and cart weight transfer (right figure) on **CartPole task**. The shaded region denotes the standard deviation of average evaluation over 5 trials.

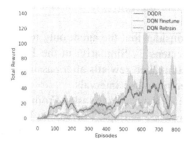 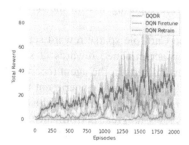

Fig. 7. Experimental results for bird gravity transfer (left figure) and pipe gap transfer (right figure) on **FlappyBird task**. The shaded region denotes the standard deviation of average evaluation over 5 trials.

In the CartPole task, we conduct two transfer experiments over the pole length and cart weight respectively. For the pole length transfer experiment, the DQDR method initializes the Q-network by the rules in the source MDP setting where the pole length is equal to 1 and then applies this Q-network to the target MDP setting where the pole length is equal to 3. DQN-Finetune curve shows the result that the Q-network trained in the source MDP setting and deploy this Q-network to the target MDP setting. DQN-Retrain curve shows the result that the vanilla DQN is trained directly in the target MDP setting. For the cart weight transfer experiment, the basic idea is similar to the pole length transfer where the source cart weight is equal to 1 and the target cart weight is equal to 10.

In the FlappyBird task, we conduct transfer experiments over the gravity of the bird and the distance between two pipes. For the bird gravity transfer experiment, the DQDR method initializes the Q-network by the general rules in the source MDP setting with the gravity is equal to 9 and then applies this Q-network to the target MDP setting with the gravity is equal to 11. DQN-Finetune curve shows the result that the Q-network trained in the source MDP and deploy this Q-network to the target MDP. DQN-Retrain curve shows the result that the vanilla DQN is trained directly in the target MDP setting. For the pipe gap transfer experiment, the settings are similar to the gravity transfer transfer experiment where the original pipe gap is equal to 100 and the target pipe gap is equal to 80.

From the results shown in Fig. 6 and Fig. 7, the DQN-Finetune method can achieve better transfer capability than the original DQN which trains from scratch. The proposed DQDR can avoid the negative transfer by only transferring the general knowledge. Thus, DQDR achieves better performance than the DQN-Finetune method which transfers all the knowledge learned from previous tasks.

5 Conclusions

In this paper, Deep Q-learning with transferable Domain Rules (DQDR) method is proposed. Firstly, we extract a set of transferable rules from human knowledge to act as the model-agnostic policy to initialize the Q-network and then update this Q-network in the DQN process to improve the decision ability in some corner cases. We evaluate our DQDR method in the CartPole and FlappyBird task. The experimental results show that our approach demonstrates significant improvement in the learning efficiency and transfer capability of the DQN algorithm.

In the future, we will focus on introducing automated knowledge extracting methods to replace the manual knowledge expressing process to reduce the human intervention in the whole framework. And this DQDR framework can also be extended to high dimensional state input tasks such as pixel input tasks by integrating dimension reduction module.

Acknowledgement. This work is supported by the National Natural Science Foundation of China under Grant 61825305, and the National Key R&D Program of China 2018YFB1305105.

References

1. Mnih, V., et al.: Human-level control through deep reinforcement learning. Nature **518**, 529–533 (2015). https://doi.org/10.1038/nature14236
2. Mirowski, P., et al.: Learning to navigate in cities without a map. In: NeurIPS (2018)
3. Du, Y., Narasimhan, K.: Task-agnostic dynamics priors for deep reinforcement learning. In: International Conference on Machine Learning, pp. 1696–1705. PMLR (2019)
4. Li, J., Yao, L., Xu, X., Cheng, B., Ren, J.: Deep reinforcement learning for pedestrian collision avoidance and human-machine cooperative driving. Inf. Sci. **532**, 110–124 (2020). https://doi.org/10.1016/j.ins.2020.03.105
5. Hussein, A., Gaber, M.M., Elyan, E., Jayne, C.: Imitation learning: a survey of learning methods. ACM Comput. Surv. (CSUR). **50**, 1–35 (2017). https://doi.org/10.1145/3054912
6. Liang, X., Wang, T., Yang, L., Xing, E.: CIRL: controllable imitative reinforcement learning for vision-based self-driving. In: Ferrari, V., Hebert, M., Sminchisescu, C., Weiss, Y. (eds.) ECCV 2018. LNCS, vol. 11211, pp. 604–620. Springer, Cham (2018). https://doi.org/10.1007/978-3-030-01234-2_36
7. Bojarski, M., et al.: Explaining how a deep neural network trained with end-to-end learning steers a car. arXiv preprint arXiv:1704.07911 (2017)
8. Xie, X., Li, C., Zhang, C., Zhu, Y., Zhu, S.-C.: Learning virtual grasp with failed demonstrations via Bayesian inverse reinforcement learning. In: 2019 IEEE/RSJ International Conference on Intelligent Robots and Systems (IROS), pp. 1812–1817. IEEE (2019). https://doi.org/10.1109/IROS40897.2019.8968063
9. Brockman, G., Cheung, V., Pettersson, L., Schneider, J., Schulman, J., Tang, J., Zaremba, W.: OpenAI Gym. arXiv:1606.01540 [cs] (2016)
10. Tasfi, N.: PyGame Learning Environment. https://github.com/ntasfi/PyGame-Learning-Environment
11. Sutton, R.S., Barto, A.G.: Reinforcement Learning: An Introduction. MIT Press, Cambridge (2018)
12. Van Hasselt, H., Guez, A., Silver, D.: Deep reinforcement learning with double q-learning. In: Proceedings of the AAAI Conference on Artificial Intelligence (2016)
13. Wang, Z., Schaul, T., Hessel, M., Hasselt, H., Lanctot, M., Freitas, N.: Dueling network architectures for deep reinforcement learning. In: International Conference on Machine Learning, pp. 1995–2003. PMLR (2016)
14. Hessel, M., et al.: Rainbow: combining improvements in deep reinforcement learning. In: Proceedings of the AAAI Conference on Artificial Intelligence (2018)
15. Ross, S., Bagnell, D.: Efficient reductions for imitation learning. In: Proceedings of the thirteenth international conference on artificial intelligence and statistics. In: JMLR Workshop and Conference Proceedings, pp. 661–668 (2010)
16. Ho, J., Ermon, S.: Generative adversarial imitation learning. In: NIPS (2016)
17. Goodfellow, I.J., et al.: Generative adversarial nets. In: NIPS (2014)
18. Li, Y., Song, J., Ermon, S.: InfoGAIL: interpretable imitation learning from visual demonstrations. In: Proceedings of the 31st International Conference on Neural Information Processing Systems, pp. 3815–3825 (2017). https://doi.org/10.1007/978-3-319-70139-4
19. Hausman, K., Chebotar, Y., Schaal, S., Sukhatme, G., Lim, J.J.: Multi-modal imitation learning from unstructured demonstrations using generative adversarial nets. In: Proceedings of the 31st International Conference on Neural Information Processing Systems, pp. 1235–1245 (2017)
20. Hester, T., et al.: Deep q-learning from demonstrations. In: Proceedings of the AAAI Conference on Artificial Intelligence (2018)

21. Vecerik, M., et al.: Leveraging demonstrations for deep reinforcement learning on robotics problems with sparse rewards. arXiv preprint arXiv:1707.08817 (2017)
22. Treesatayapun, C.: Knowledge-based reinforcement learning controller with fuzzy-rule network: experimental validation. Neural Comput. Appl. **32**(13), 9761–9775 (2019). https://doi.org/10.1007/s00521-019-04509-x
23. Likmeta, A., Metelli, A.M., Tirinzoni, A., Giol, R., Restelli, M., Romano, D.: Combining reinforcement learning with rule-based controllers for transparent and general decision-making in autonomous driving. Robot. Auton. Syst. **131**, 103568 (2020). https://doi.org/10.1016/j.robot.2020.103568
24. Vincze, D., Tóth, A., Niitsuma, M.: Antecedent redundancy exploitation in fuzzy rule interpolation-based reinforcement learning. In: 2020 IEEE/ASME International Conference on Advanced Intelligent Mechatronics (AIM), pp. 1316–1321. IEEE (2020). https://doi.org/10.1109/AIM43001.2020.9158875
25. Zhang, P., et al.: KoGuN: accelerating deep reinforcement learning via integrating human suboptimal knowledge. arXiv preprint arXiv:2002.07418 (2020)
26. Schulman, J., Wolski, F., Dhariwal, P., Radford, A., Klimov, O.: Proximal Policy Optimization Algorithms. arXiv:1707.06347 [cs] (2017)
27. Yang, Q., Zhang, Y., Dai, W., Pan, S.J.: Transfer Learning. Cambridge University Press, Cambridge (2020)
28. Sutton, R.S., Precup, D., Singh, S.: Between MDPs and semi-MDPs: a framework for temporal abstraction in reinforcement learning. Artif. Intell. **112**, 181–211 (1999). https://doi.org/10.1016/S0004-3702(99)00052-1
29. Yin, H., Pan, S.: Knowledge transfer for deep reinforcement learning with hierarchical experience replay. In: Proceedings of the AAAI Conference on Artificial Intelligence (2017)
30. Kingma, D.P., Ba, J.: Adam: a method for stochastic optimization. arXiv preprint arXiv:1412.6980 (2014)
31. Li, S., Xu, X., Zuo, L.: Dynamic path planning of a mobile robot with improved Q-learning algorithm. In: 2015 IEEE International Conference on Information and Automation, pp. 409–414. IEEE (2015). https://doi.org/10.1109/ICInfA.2015.7279322

Influence of Interference and Noise on Indoor Localization Systems

Huy Q. Tran[1] [iD], Chuong Nguyen Thien[2], and Cheolkeun Ha[3]([✉])

[1] Faculty of Automotive, Mechanical, Electrical and Electronic Engineering, Nguyen Tat Thanh University, Ho Chi Minh City, Vietnam
[2] Nha Trang University, Nha Trang, Vietnam
[3] Robotics and Mechatronics Lab, University of Ulsan, Ulsan 44610, Republic of Korea

Abstract. Recently, visible light based positioning systems have been a promising solution for indoor environment. The outstanding advantages of this positioning method are the very high accuracy and low installation cost. However, one of the most serious limitations of an indoor positioning system using visible light is to maintain the stability of positioning accuracy despite the interference from unavoidable noise and interference sources such as Gaussian noise, multipath reflection. In this paper, we are the first to analyze and demonstrate the effects of all the above-mentioned noises in a typical simulation environment using Matlab/Simulink. Differences in experimental results with and without separate types of noise, as well as the influence of multiple noises at the same time, are analyzed in detail in the paper.

Keywords: Visible light · Indoor localization · Noise

1 Introduction

Nowadays, almost human activities take place indoors, therefore, the need to estimate the location of people and related objects has become more urgent. The lack of precise positioning devices for indoor environment is the main cause attracting much attention from researchers. Indoor positioning methods using wireless signals are favorite solutions, in which WiFi, Zigbee, Bluetooth [1, 2], and visible light are the most used. A common difficulty in establishing indoor positioning systems is the diversity of building structures. Therefore, a positioning system that ensures robust and stability in all housing structures is a serious challenge. Besides the traditional positioning systems, these days, LED-based positioning system is also a promising solution. High positioning accuracy and low cost are one of the significant advantages of the visible light positioning (VLP) systems [3, 4].

In all positioning applications from outdoor to indoor, the influence of noises on the positioning quality of the system is one of the inevitable factors. In any case, the researchers try to reduce and gradually eliminate them by developing more advanced technologies in near future [5, 6]. In the field of indoor positioning using visible light, the positioning accuracy is highly dependent on the type of interference and the intensity of the interference it receives. For the LED-based positioning system, ambient

© Springer Nature Switzerland AG 2021
D.-S. Huang et al. (Eds.): ICIC 2021, LNCS 12837, pp. 274–283, 2021.
https://doi.org/10.1007/978-3-030-84529-2_23

light such as sunlight, artificial light from lamps, directly affects the quality of the desired signals. Moreover, the LEDs' manufacturing technology, the microchips used for sensors and signal amplification circuits are also one of the main causes to increase the system disturbance.

Recently, several studies have focused on minimizing the impact of noise on the quality of LED positioning systems [5, 7–10]. For the visible light positioning system, the influence of Gaussian noise arising from the optical transmitter, the optical receiver, the ambient light and the effects of multipath reflections arising from the reflective objects such as walls, ceilings, floors, furniture, and people are the main issues. Moreover, in many fields, the low frequency noise of LED light is also particularly noticeable, especially in the field related to biomedical. In this case, optimizing the certain current level and applying a large number of external series resistance are affective solutions to reduce the noise level [7]. To reduce the Gaussian noise from above mentioned light sources, Yong-Juk Won et al. [8] proposed an adaptive differential equalization technique that uses to increase the signal-to-noise ratio (SNR) by attenuating the magnitude of the noise. Furthermore, the Poisson functions are used as a weight function to increase the efficiency of reducing shot noise. Finally, the authors measured the SNR, bit error rate, and eye diagram of the processed signals to evaluate the level of noise attenuation. In another approach, Dehao Wu et al. [9] applied multiple periods averaging technique to remove noise. By using this method, the sensing distance was extended, and the estimation error was also considerably reduced. To evaluate the noise effect on visible light communication, Alin-Mihai et al. [10] focused on the pulse width distortions when using Manchester and Miller coding. This work helped to recognize the modifications of the data pulse under the effect of noise.

It is worth noting that the above publications only focus on mitigating the impact of various types of interference independently, without specifically analyzing their impact on the positioning quality of the system. Especially, the effects of multiple disturbances at the same time. In this paper, we, therefore, focus on analyzing the impact of many types of noise on the localization accuracy. From these results, we recognize the influence level of each type of noise on the positioning process. These positioning results are the prerequisite for setting up navigation systems in the real environment.

2   Interference and Noise Analysis for VLP Systems

In this section, the common types of noise affecting the positioning accuracy of visible light based indoor positioning systems, including Gaussian noise, reflection are presented in details.

2.1   Gaussian Noise

Among the data collection methods to perform the positioning process, the positioning method based on the received signal strength (RSS) is adversely affected by noise, in which Gaussian noises as shown in Eq. (1), including shot noise, thermal noise and inter-symbol interference have a detrimental effect. Data transmission for positioning purposes takes place in a short time, hence the negative influences of inter-symbol interference can be ignored. Meanwhile, shot noise and thermal noise are types of noise that always exist in real VLP applications. In most cases, shot noise comes from the transmission side including the LED lights and ambient light (e.g. sunlight and artificial light sources), whereas thermal noise appears on the receiver side.

$$N = N_s + N_T + N_I \tag{1}$$

where N_S, N_T, and N_I are the shot noise, thermal noise, and inter-symbol interference, respectively. These noises are given as follows [11]:

$$N_s = 2q\gamma P_T B + 2qI_B I_2 B \tag{2}$$

$$N_T = \frac{8\pi\kappa T_k}{G}\eta A I_2 B^2 + \frac{16\pi^2\kappa T_k\Gamma}{g}\eta^2 A^2 I_3 B^3 \tag{3}$$

$$N_I = \gamma^2 P_I^2 \tag{4}$$

where q is the electronic charge; γ is the PD responsibility; I_B is the photocurrent due to background radiation; I_2 is the noise bandwidth factor; and B is the equivalent noise-bandwidth of the PD; κ is the Boltzmann's constant; T_k is the absolute temperature; G is the open-loop voltage gain; η_{PD} is the fixed capacitance of PD per unit area; Γ is the FET channel noise factor, g is the trans-conductance; and I_3 is the noise-bandwidth factor; and the received optical power P_I due to inter-symbol interference is given by [11]:

$$P_I = \int_T^\infty \left(\sum_{i=1}^N h_i(t) \otimes X(t) \right) dt \tag{5}$$

The negative effect of the above mentioned is expressed based on the signal-to-noise ratio (SNR), which shows the ratio of the desired information to the undesired information or noise and is illustrated in Fig. 1 [12]:

$$SNR = \frac{R^2 R_T^2}{N} \tag{6}$$

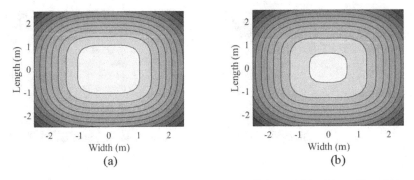

Fig. 1. Signal-to-noise ratio: (a) without ambient light and (b) with ambient light

2.2 Multipath Reflection

In most real indoor spaces, the existence of multipath reflection is an inevitable problem because objects that cause reflection such as walls, ceilings, floors, objects or people exist everywhere. Therefore, the effect of this type of noise is an integral part of real positioning systems. In many cases, researchers ignore the effects of the None-line-of-sight (NLOS) signals and consider only the optimal environments in the presence of a line-of-sight (LOS) signals. Such studies do not demonstrate the practicality of localization systems using light-emitting-diode (LED) lights. In this paper, the influence of multipath reflection with different types of reflective materials is considered. To illustrate that effect, the received optical powers are showed in Fig. 2 and are given as follows [12]:

$$P_{LOS} = P\frac{(m+1)A}{2\pi l^2}\cos^m(\phi)T(\psi)g(\psi)\cos(\psi), \quad 0 \leq \psi \leq FOV \tag{7}$$

$$P_{NLOS} = P\frac{A(m+1)}{2\pi l_1^2 l_2^2}\rho\cos^m(\phi)dA_w\cos(\alpha)\cos(\beta)\cos(\psi_r)T(\psi_r)g(\psi_r) \tag{8}$$

where P is the transmitted optical power; m is the Lambertian order; A is the photo-detector (PD) active area, l is the transmission distance from the LED to the PD; ϕ is the angle between the transmitter and the receiver; ψ is the incidence angle; $T(\psi)$ is the gain of the optical filter; $g(\psi)$ is the gain of the optical concentrator; FOV is the receiver field of view; ρ is the reflectance factor; l_1 and α are the distance and the angle of irradiance from an LED to the reflective point; l_2 and β are the distance and the angle between a reflective point and the receiver; and ψ_r is the incidence angle of the light from the wall.

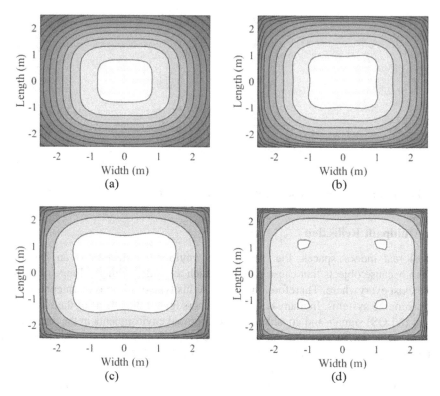

Fig. 2. Optical power: (a) without reflective object, (b) low reflective rate, (c) medium reflective rate, and (d) high reflective rate

3 Experimental Results and Discussion

To analyze the negative impact of Gaussian noise and other interferences on the accuracy of the LED-based indoor positioning systems, in this paper, we apply trilateration algorithm to measure the distances from the target to the reference points as shown in Fig. 3. Although the positioning accuracy is not outstanding, this algorithm is quite simple and popular. The data used to implement the suggested algorithm are RSS values. One of the highlights of RSS-based methods is that collected information is very vulnerable to noise. Thus, the combination of RSS and trilateration method is an uncomplicated way to evaluate the effect of different types of noise on the quality of the positioning system.

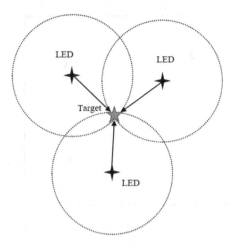

Fig. 3. Trilateration algorithm

The proposed positioning environment is an empty room with dimensions of $5 \times 5 \times 2.5$ m. The data transmission system consists of 4 LEDs, each with a transmitted optical power of 10 W. These lights are suspended from the ceiling in fixed positions as given in Fig. 4. In addition, we leverage a PD as an optical receiver. Because we analyze not only the effect of Gaussian noise but also the multipath reflection, the VLP channel is depicted as shown in Fig. 5. Other important specifications are shown in Table 1.

Table 1. VLP parameters

Parameter	Value
Active area of PD	1 cm^2
FOV	$70°$
Filter gain	1
LED power	10 W
LED data rate	2 Mbps
LED bandwidth	3 MHz
LED semi angle	$60°$
LED position	LED 1 (1.2×1.2)
	LED 2 (1.2×3.8)
	LED 3 (3.8×3.8)
	LED 4 (3.8×1.2)

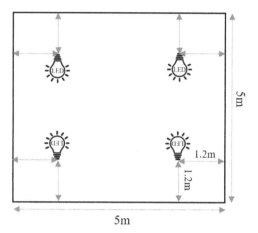

Fig. 4. LED arrangement

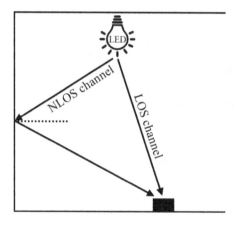

Fig. 5. VLP channel

In this section, we conduct experiments to evaluate the error level of the positioning system under the negative influence of different types of noise. The analysis is carried out under four specific cases as follows:

- Case 1st: without ambient light and without reflection
- Case 2nd: with ambient light and without reflection
- Case 3rd: without ambient light and with reflection
- Case 4th: with ambient light and with reflection

As shown in Fig. 6 and Fig. 7, The error intensity distribution increases with each case. The error intensity varies considerably from 0.03 m in the absence of interference to approximately 0.6 m when the system is affected by all kinds of disturbances. This result clearly shows that with different types of noise, the positioning quality of the system will be completely different.

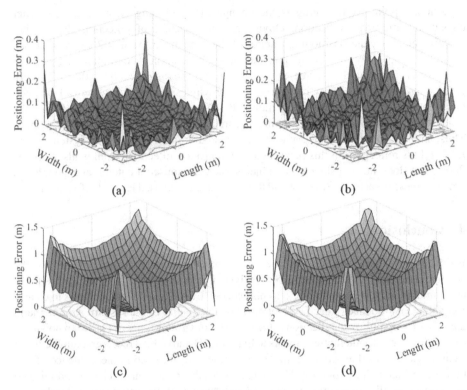

Fig. 6. Positioning errors obtained from trilateration method: (a) without reflective object and without ambient light, (b) without reflective object and with ambient light, (c) with reflective object and without ambient light, (d) with reflective object and with ambient light

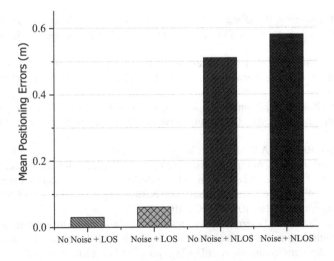

Fig. 7. Positioning errors comparison

Another point worth noting is the change of positioning error before and after considering multipath reflection. Obviously, the presence of reflections has a very negative effect on the positioning quality of the system. In the first 2 cases (first two columns in Fig. 7) when we only consider noise from ambient light including sunlight and other artificial lighting source, the error level is in the range of 0.03–0.06 m. This is a pretty impressed level of error compared to other wireless navigation systems. However, this trend changes markedly when multipath noise is taken into account. Under the influence of reflection, the positioning accuracy decreased approximately 17 times. Thus, an important factor that cannot be eliminated in indoor navigation systems using LED light is reflection. In this paper, we only consider reflections from walls, ceilings, floors, but does not consider other objects such as household items and people. In practice, these objects always exist and that leads to an increased level of reflection noise.

4   Conclusion

In this paper, we analyzed the influence of Gaussian noise and multipath reflection on the positioning quality of the system. The analysis is performed not only for each type of noise individually, but also for their combinations. The obtained results help us to clearly recognize the influence of each type of noise on the quality of the system. Among them, the influence of multipath reflection is the largest. For various types of Gaussian noise, we can gradually reduce their effect thanks to higher LED and optical receiver manufacturing technologies, as well as we can remove these noise by eliminating the influence of sun light or artificial light sources. The obtained results create an important premise to improve the quality of positioning in real localization systems based on LED light.

References

1. Obeidat, H., Shuaieb, W., Obeidat, O., et al.: A review of indoor localization techniques and wireless technologies. Wireless Pers. Commun. (2021). https://doi.org/10.1007/s11277-021-08209-5
2. Farid, Z., Nordin, R., Ismail, M.: Recent advances in wireless indoor localization techniques and system. J. Comput. Netw. Commun. **2013**, 1–12 (2013). https://doi.org/10.1155/2013/185138
3. Hassan, N., Naeem, A., Pasha, M.A., Jadoon, T., Yuen, C.: Indoor positioning using visible LED lights: a survey. ACM Comput. Surv. 48 (2015). https://doi.org/10.1145/0000000.000000
4. Mohaimenur Rahman, A.B.M., Li, T., Wang, Y.: Recent advances in indoor localization via visible lights: a survey. Sensors **20**(5), 1382 (2020). https://doi.org/10.3390/s20051382
5. Costanzo, A., Loscri, V.: Visible light indoor positioning in a noise-aware environment. In: 2019 IEEE Wireless Communications and Networking Conference (WCNC), pp. 1–6 (2019). https://doi.org/10.1109/WCNC.2019.8885607
6. Kazama, S., Ikeda, H.: Determination of noise sources affecting wireless communications: a noise identification method using coherence. In: 2013 IEEE Electrical Design of Advanced Packaging Systems Symposium (EDAPS), pp. 209-212 (2013). https://doi.org/10.1109/EDAPS.2013.6724426

7. Rumyantsev, S.L., Sawyer, S., Pala, N., Shur, M.S., Bilenko, Y., Zhang, J.P., Hu, X., Lunev, A., Deng, J., Gaska, R.: Low frequency noise of light emitting diodes. In: Proc. SPIE 5844, Noise in Devices and Circuits III (23 May 2005). https://doi.org/10.1117/12.608559

8. Won, Y.-Y., Yoon, S.M., Seo, D.: Ambient LED light noise reduction using adaptive differential equalization in Li-Fi wireless link. Sensors 21(4), 1060 (2021). https://doi.org/10.3390/s21041060

9. Dehao, W., Wende, Z., Chen, C.: A noise cancellation approach for an illuminating LEDs based short-range visible light sensing system. Procedia Eng. 140, 195–202 (2016). https://doi.org/10.1016/j.proeng.2016.07.345

10. Cailean, A., Cagneau, B., Chassagne, L., Popa, V., Dimian, M.: Evaluation of the noise effects on visible light communications using Manchester and Miller coding. In: International Conference on Development and Application Systems (DAS), pp. 85–89. Suceava, Romania, May 2014. hal-01207160

11. Komine, T., Nakagawa, M.: Fundamental analysis for visible-light communication system using LED lights. IEEE Trans. Consum. Electron. 50, 100–107 (2004)

12. Ghassemlooy, Z., Popoola, W., Rajbhandari, S.: Optical Wireless Communications, System and Channel Modeling with MATLAB. CRC Press, Boca Raton, FL, USA (2012). ISBN 9781439851883

Exploration of Smart Medical Technology Based on Intelligent Computing Methods

Sijia Wang[✉] and Yizhang Jiang

School of Artificial Intelligence and Computer Science, Jiangnan University,
1800 Lihu Avenue, Wuxi 214122, Jiangsu, People's Republic of China
6201613008@stu.jiangnan.edu.cn

Abstract. As intelligent computing has been an important application in the medical field, there are more and more examples of smart medical technology improving healthcare services. Smart medical technology is a kind of artificial intelligence diagnostic technology that can simulate the diagnosis experience of medical experts and combine the thinking process of medical knowledge through the study and analysis of data. With the help of machine learning algorithms, doctors can diagnose different kinds of diseases more conveniently and accurately, which conforms to the needs of modern society's developments. Smart medical technology can effectively solve shortage problems of medical resources and inexperience of young doctors, and reduce the labor intensity of medical staffs. It can make a more scientific and objective diagnosis and can expand the database by keeping collecting data and truly keeping up with the times. At the same time, it also provides scientific experts with better auxiliary functions and better solutions and generate positive guidance and influence for people's future lifestyles through the analysis of attribute indicators. In this paper, several intelligent computing methods are selected to model and analyze the epileptic EEG signal dataset and thyroid dataset, establish their respective classification models by parameter optimizations, find out the most suitable intelligent computing classification algorithms by comparing a series of parameter indicators, analyze the advantages and disadvantages of these algorithms, and provide selection suggestions.

Keywords: Intelligent Computing · Smart Medical Technology · Classification Algorithms

1 Introduction

In the era of big data, data analysis and machine learning classification can be better used and applied. The rapid development of artificial intelligence technology has demonstrated irreplaceable advantages in disease detection and medical imaging. Smart medical technology has been used in a large range and also in the spotlights. As a research hotspot, it also has important applications and good development prospects. With the continuous development of artificial intelligence technology, more and more researchers have discovered that more accurate diagnosis conclusions can be obtained if we could use some existing intelligent calculation methods, when modeling and analyzing existing medical data.

© Springer Nature Switzerland AG 2021
D.-S. Huang et al. (Eds.): ICIC 2021, LNCS 12837, pp. 284–293, 2021.
https://doi.org/10.1007/978-3-030-84529-2_24

For medical data, it is very important to find a suitable data analysis method. Data mining [1] technology can manipulate data with complex structures and large amounts. Data mining is to capture the potential connections contained in the data. Using data mining technology means that it is easy to get valuable information in the data. The method of intelligent computing provides research methods and effective tools for data mining technology. Its main learning method is to learn the known information so as to be able to judge the unknown information. Intelligent computing methods can learn from the data simultaneously, capture the information contained in the data and the distribution structures between the data, discover a certain pattern hidden in the data and the relationships between the data. It can research and analyze existing data simultaneously, improve its learning ability, and make intelligent judgments with high accuracy from the validation dataset. When we use intelligent computing methods to analyze medical data, we can make decisions and analysis on unknown disease cases.

Smart medical technology is a kind of artificial intelligence diagnostic technology that can simulate the diagnosis experience of medical experts and combine the thinking process of medical knowledge by studying and analyzing the data. With the continuous development and wild application of machine learning, as well as the increase amount of data, massive amounts of medical data provide a prerequisite by using existing intelligent computing methods to design decision-making systems to assist diagnosis. Moreover, there is an important problem of insufficient medical resources in today's society, and the problem of difficulty in seeing a doctor is widespread. The uneven entry time and level of doctors have also led to the lack of medical diagnosis experience for most newly recruited doctors. The applicability of smart medical technology in artificial intelligence technology is very high. One technology can be applied to the diagnosis of multiple diseases and it is very easy to promote. Therefore, the exploration of smart medical technology based on intelligent computing methods can reduce the labor intensity of medical staffs. It also plays a very important role in improving the accuracy and instantaneity of disease diagnosis. With the development of computer science and artificial intelligence technology, the use of computer intelligent computing methods which help doctors diagnose various diseases can make diagnosis more conveniently and accurately, thereby achieving the development and needs of smart cities [2].

2 Related Work

2.1 Introduction to the Data Set

The data used in this paper are divided into two major categories, an epileptic EEG signal dataset and a thyroid dataset.

The EEG dataset of epileptic EEG signals was obtained from the database of the University Medical Center Bonn, Germany, and the EEG signal measurement method and its complete description are detailed in reference [3]. The EEG dataset includes five groups of data, named A ~ E, and the detailed descriptions of the five groups are shown in Table 1.

Table 1. EEG dataset description.

Testers	Group	Data set description
Health	A	EEG signal received from a healthy person with eyes open
	B	EEG signal received from a healthy person with eyes closed
Illness	C	EEG signal received in the area of the patient's interictal epileptic lesion
	D	EEG signal received outside the area of the patient's interictal epileptic lesion
	E	EEG signal measured during the patient's seizure

For the EEG dataset, BD two of the five groups are selected after feature extraction by NoD technique. The data contains 200 sets of data samples, 160 sets are extracted as training dataset to train the classification model, and the remaining 40 sets are used for validation.

We selected 970 patients' thyroid-related data from a hospital as the thyroid dataset, and the dataset was divided into two types, a thyroidx32 dataset containing 31 indicators and a thyroid19 dataset obtained by merging several related binary features from the previous dataset. The indicators in the dataset contain date, age, gender, and some physiological indicators, which are divided into two classes: data class and attribute class, for example, gender and other indicators are numerically classified into two classes of 0 and 1, while some other indicators are just numerical. Lastly, the thyroid19 dataset containing 18 indicators as features of the model was obtained, and the indicators are: age, gender, upper and lower, anterior, and posterior, left and right, site, morphological irregularity, unclear boundary, aspect ratio <1, echogenicity, structure, strong echogenicity, abnormal dark band, breakthrough envelope, satellite foci, blood flow grade $3 \sim 4$, lymph node abnormality, and pathological malignancy. The current data set is divided into two groups, 900 groups of data as the training set for building the model and 70 groups as the validation set for testing the accuracy of the model.

2.2   Main Methods

The K-NN algorithm [4] calculates the distance between data in training and validation datasets. Three important factors in the K-NN algorithm are distance metric, K-value selection, and decision making. In the K-NN algorithm, the distance metric we use is the Minkowski Distance.

There are three points to be noted in the selection of k values. In general, the value of k starts from a small value, such as k = 3. It is better to choose an odd number of k values because when the value is even, it is very likely that the distribution of the nearest points in the two categories is the same, which is not easy to make a classification. Moreover, it is recommended that the maximum k-value should not exceed 20. However, it is still possible to exceed it because we will adjust the value of k in the actual K-NN model. Generally, it will gradually increase from the minimum value. At this time, we will see the accuracy of the model, which will rise at first but will fall slowly after. And we will choose the most suitable k-value during the tuning process.

The final procedure is decision-making. The majority decision principle is adopted. The class with a larger amount decides the final result. Regarding the weight of the samples, in general, we take the same weight, that the distance is the same regardless of the impact on the judgment results. There is another method: The closer the distance, the greater the weight. The closer sample in the final result have a greater impact.

K-NN algorithm is easy to implement and especially suitable for multi-classification problems. However, it is very computationally intensive for the validation data sets, with high memory space demand and high time complexity.

Logistic regression [5] is a combination of linear regression and Sigmoid function. Linear regression is the application of a straight line which fits the relationship between the independent and dependent variables. The Sigmoid function is used when we want to change the linear regression to a classification.

We want a good outcome of the parameters, which are the parts of linear regression previously described, and at this point we transform the linear regression problem into a least squared error problem. The least squares error is finding the smallest sum of squared error to determine the parameters of a function in order to solve for the most appropriate function. Using least squared error, we can easily obtain the unknown data and use them to predict so that the sum of the squared errors between the predicted and actual data can be minimized.

When studying the relationship between two variables, we need to learn a set of data. We correspond this set of data to the Cartesian coordinate system. If we find that this set of data is near a straight line, we define the straight line as:

$$\hat{Y}i = a_0 + a_1 x_i \tag{1}$$

To get the optimal function, we need to minimize the difference between the predicted value of the regression equation and the theoretical actual value for each independent variable x. For the whole training process, it is to minimize the sum of squared errors. The reason for using the squared errors instead of the simple difference is that we want to get rid of different signs of values and we want a scalar, so we use the squared errors. So, we get the following formula:

$$\sum_{i=1}^{n} (Y_i - \hat{Y}_i)^2 = Q(a_0, a_1) \tag{2}$$

Apply Eq. (1) into Eq. (2):

$$\sum_{i=1}^{n} (Yi - (a_0 + a_1 x_i))^2 = Q(a_0, a_1) \tag{3}$$

The subsequent solution process is to minimize the function, that is, the parameters obtained are for the target function to obtain the minimum value. To minimize this function, we must find the first-order partial derivatives of this function on each of the two parameters and make the derivative value equals to 0. Meanwhile, the value of this function can be minimized, the error between the predicted value and the theoretical value is a minimum.

Logistic regression algorithm is easy to understand and simple to implement. However, the logistic regression algorithm is very sensitive to abnormal data, and if there exists an outlier, it would have a great impact on the algorithm.

There are two types of decision trees [6], one is classification tree and the other is regression tree. The output of classification tree is qualitative, and the output of regression tree is quantitative. It is an algorithm used to solve classification problems efficiently. Three steps are required to construct a decision tree, the first is the feature selection of the decision tree, the second is the generation of the decision tree, and the last is the pruning of the decision tree. Since the decision tree itself is the process of classification and decision making, those things that influence our decision are the features of the decision tree. The technical term in decision tree algorithm is called information gain or information gain ratio. After we finish feature selecting, then we could construct a decision tree, that is, find out which feature is the internal node of the tree and which feature is the leaf node of the tree. After we calculate the feature values, we use the decision tree algorithm to construct a tree. The information gain corresponding to decision tree algorithm is the ID3 algorithm, and the information gain ratio is C4.5. In the calculation of the information gain, the most important feature has the largest information gain value (as the leaf node).

To calculate the feature value, the first thing is the information gain which is the most basic task of the decision tree. The input of information gain are the training set D and the feature set A, then calculate the relationship between the training set and the feature set, and the output is the information gain g(D, A) of the feature A to the training data set D. The entropy describes the degree of uncertainty, the greater the uncertainty, the greater the entropy. The first step is to calculate the empirical entropy of the dataset, and the formula for calculating the empirical entropy is as follows:

$$H(D) = -\sum_{i=1}^{n} p_i \log p_i \qquad (4)$$

The second step is to calculate the empirical conditional entropy of the feature to the data set, which is calculated by probability. The formula for calculating the empirical conditional entropy is as follows:

$$H(D|A) = \sum_{i=1}^{n} p_i H(D|A = a_i) \qquad (5)$$

The third step is to calculate the information gain, which is the value obtained by taking a difference between the empirical entropy and the empirical conditional entropy.

Decision trees can be visualized to understand the operation process more intuitively. However, when we did not apply pruning process, the test results may not be accurate.

The neural network algorithm [7] is based on the principle of biological neural network, which is a mathematical model that simulates the nervous system of the human brain. The structure of a neural network is composed of an input layer, a hidden layer, and an output layer. Many nodes are added between the input and output, and each node calculates the data input to it from the previous node according to its own weight, which means that each node is a specific output function, which is also known as an activation function, and the connection between every two nodes contains a weighted value, which is a weight.

Neural networks mainly use three functions. First is the activation function, which is the nonlinear transformation, after each layer of output, there is an activation function to do the nonlinear transformation. Then there is the loss function, which is the objective function of the neural network for optimization. In order to minimize the loss function, the neural network must be trained and optimized. The smaller the value of the loss function, the smaller the deviation of the predicted result from the actual value. Finally, there is the optimization function. Through the optimization function, we can make the value of the loss function smaller.

The neural network algorithm performs better with larger data sets, but it is not easy to understand, and the algorithm is more complex, and it takes longer time.

Support vector machine [8] is a binary classification algorithm. Its basic model is a linear classifier whose maximum interval depends on the entropy of the feature space. It can be used as a linear classifier or a nonlinear classifier. The learning strategy of a support vector machine is to maximize the interval, which can also be understood as a problem of seeking convex quadratic programming. For example, there are many points from a training set of binary classification, we want to find a hyperplane to classify these points into two classes. There may be many hyperplanes, but we must find an optimal hyperplane by support vector machines.

Support vector machines [9] have three models from simple to complex. First is the linearly separable support vector machine. When the data are linearly separable, the classifier is learned by hard interval maximization; then on the linear support vector machine, when the data are approximately linearly separable, the support vector machine can be solved by adding slack variables to the solution process and learning the classifier by soft interval maximization; finally, when the data are linearly inseparable, a nonlinear support vector machine is used, which means, in the low-dimensional space, the data is nonlinearly separable and the data are mapped from the low-dimensional space to the high-dimensional space by kernel functions. In this case, the data can be linearly separable in the high-dimensional space.

SVM algorithm simplifies the difficulty of solving high-dimensional space problems and has good generalization ability. However, it is difficult to solve multi-classification problems and the training time is long when the data volume is large.

TSK [10] defines the rules of a fuzzy system in terms of "IF-then" rules, which can be expressed as follows:

$$
\begin{aligned}
&\text{IF } x1 \text{ is } A_1^k \wedge x2 \text{ is } A_2^k \wedge \cdots \wedge xd \text{ is } A_d^k \\
&\text{THEN } f^k(x) = p_0^k + p_1^k x1 + \cdots + p_d^k xd \text{ for } k = 1, \cdots, K
\end{aligned}
\tag{6}
$$

In the formula above, k represents the number of fuzzy rules in the fuzzy system, d represents the number of inputs, A_i^k is a fuzzy set in the i-th input for the k-th rule, and

$\wedge$ is a fuzzy connection operator. The output of the TSK fuzzy system can then be expressed as:

$$y^0 = \sum_{k=1}^{K} \frac{\mu^k(x)f^k(x)}{\sum_{k'=1}^{K} \mu^{k'}(x)} = \sum_{k=1}^{K} \tilde{\mu}^k(x)f^k(x) \tag{7}$$

There are many affiliation functions that can be used, but the commonly used one is the Gaussian affiliation function which can be expressed as:

$$\mu A_i^k(x_i) = exp(\frac{-(x_i - c_i^k)^2}{2\delta_i^k}) \tag{8}$$

In the formula above, c_i^k is the cluster center of the i-th dimension under the k-th rule, which is the center parameter and δ_i^k is the width parameter. Ultimately, the output of the TSK fuzzy system can be written as the following linear equation:

$$y^o = p_g^T x g \tag{9}$$

TSK is rule-based and it does not require a mathematical model.

3   Experimental Details

3.1   Evaluation Index

This article uses the accuracy rate, the prediction accuracy of benign cases and the prediction accuracy of malignant cases, the recall rate and the f1 score, the training time, and the runtime to evaluate· the effect of the models. First, we calculate the predictive classification accuracy of each classification model as the most important indicator to judge the suitability of the model. Secondly, we use the prediction accuracy of benign cases and the prediction accuracy of malignant cases as well as the recall rate and f1 score as auxiliary indicators. The recall rate is the prediction of all positive cases divided by all the positive cases; the f1 score is the harmonic mean of the recall rate and the accuracy of malignant cases (also known as the precision rate). Because some algorithms may have a high accuracy rate of case prediction in one category, but a very low rate of prediction in others. We use these indicators as evaluation criteria. If the number of samples in one category is much higher than that it in another category, and all samples are classified into the category that contains more samples, the accuracy rate is also very high, and the recall rate and precision rate scores may also be very high, but the model is inapplicable. In this case, the category which is counted as heterogeneous is often the positive sample we want, as the accuracy is high, but the recall and precision scores calculated in this way are very low, so that the model can be said to be completely useless. Lastly, we should also compare the runtime and the training time of each algorithm.

3.2   Experimental Comparison

To compare the feasibility of different intelligent computational methods, we selected the six above methods that can be used for classification to conduct comparative experiments and observe the performance of different methods on the dataset. These six algorithms are: K-NN algorithm, logistic regression algorithm (LR), decision tree algorithm (DT), neural network algorithm (NNs), SVM algorithm and TSK fuzzy system.

Using different algorithms or using different parameters for the same algorithm, their results are different, so continuous learning and training are needed to discover the best algorithm and the optimal parameters. When building a K-NN classification model, we find the best k-value by comparing odd k-values with even k-values; when building a decision tree model, we find the best model by comparing different pruning thresholds; when building a neural network model, we compare the performance obtained by modifying the number of layers, the number of units in one layer, the number of iterations and the learning rate to find the optimal parameter values; when building a support vector machine classification model, the RBF kernel function is used and the optimal values of c and g are found by cross-validation methods. The parameters of each algorithm are adjusted to the best, and then we conduct a comparison experiment.

3.3   Analysis of Results

We compare the accuracy, benign accuracy, malignant accuracy, recall, f1 score, training time and runtime between different models. Table 2 and Table 3 compare the indicators of the six algorithms on the EGG dataset and the thyroid dataset.

The indicators on the EGG dataset show that the K-NN algorithm, the logistic regression algorithm, the SVM algorithm and the TSK fuzzy system are all impressive, except for the SVM algorithm which takes longer time because it requires cross-validation to find c and g.

Table 2. Experimental comparison of EGG dataset.

	KNN	LR	SVM	TSK
Accuracy	100.00%	97.50%	100.00%	95.00%
Benign accuracy	100.00%	95.00%	100.00%	90.00%
Malignant accuracy	100.00%	100.00%	100.00%	100.00%
Recall	100.00%	95.24%	100.00%	100.00%
f1 score	100.00%	97.56%	100.00%	100.00%
Training time		0.0799	27.9480	0.0087
Runtime	0.1017	0.1023	28.5448	0.0094

Table 3. Experimental comparison of thyroid dataset.

	KNN	LR	DT	NNs	SVM	TSK
Accuracy	85.68%	84.86%	81.08%	80.81%	86.76%	81.08%
Benign accuracy	90.38%	85.58%	91.83%	85.58%	91.35%	79.63%
Malignant accuracy	79.63%	83.95%	67.28%	74.69%	80.86%	82.21%
Recall	85.07%	87.25%	78.28%	81.28%	85.97%	77.71%
f1 score	87.65%	86.41%	84.51%	83.37%	88.58%	79.90%
Training time		0.1441	11.3435	9.9379	652.8370	0.0428
Runtime	0.3174	0.1970	17.7348	10.3007	653.4430	0.0438

Looking at the indicators on the thyroid dataset, by calculating the two indicators of recall and f1 score, we find that all algorithms perform well in all metrics except for the decision tree algorithm and TSK fuzzy system, which have a somewhat low recall, and the best performer is the support vector machine algorithm. The low recall of the decision tree indicates that even though the accuracy of the model is not low, the accuracy gap may be too large for the two types respectively, so let's look at the accuracy, malignant accuracy, and benign accuracy of the decision tree algorithm. The decision tree algorithm has the highest malignant accuracy rate, while it has the lowest benign accuracy rate. During the experiments, we found that when the threshold value is set to be some certain values, there is even a malignant accuracy rate of 1 and a benign accuracy rate of less than 30%. Although the decision tree algorithm has high accuracy and malignant accuracy, the benign accuracy is still too low, and it is easy to misdiagnose benign patients as malignant, and its training time and runtime are the longest except for the support vector machine algorithm which requires cross-validation, so we exclude the decision tree algorithm first.

Although the accuracy of neural network algorithm is acceptable, it has the lowest accuracy compared to other algorithms except decision tree algorithm, and its training time and runtime are longer than the other two algorithms except support vector machine algorithm, so it is better not to use neural network algorithm until further optimization. The accuracy of the support vector machine algorithm is the highest, the malignant accuracy is the highest except for the decision tree algorithm, and the benign accuracy is the highest except for the logistic regression algorithm and the TSK fuzzy system, so that the accuracy of the support vector machine algorithm is optimal, but its cross-validation process is too time-consuming, and if the size of data is too large, it may take several days to find out the best c and g, and the time cost is too high.

4   Conclusion

In this article, we use several intelligent computational methods for disease diagnosis prediction. In consideration of accuracy and time cost, K-NN algorithm and logistic regression algorithm perform the best, both have good data indicators. Relatively speaking, K-NN algorithm has higher malignant accuracy and logistic regression algorithm has good performance in both types of accuracy. And K-NN algorithm and logistic regression algorithm are also the most user-friendly because they are the easiest to understand. To sum up, if there is enough time for model training, support vector machine algorithm is the best choice. But if you do not want high time cost, then K-NN algorithm and logistic regression algorithm are the best choices. In this research, we only present a few basic algorithms with small amounts of data sets, so there is still room for a further research. Specifically, we can optimize the used model further with more data to test the generalization performance of the method better. We will carry out more in-depth research on this topic in the future.

References

1. Wagholikar, K.B., Sundararajan, V., Deshpande, A.W.: Modeling paradigms for medical diagnostic decision support: a survey and future directions. J. Med. Syst. **36**(5), 3029–3049 (2012)
2. Qureshi, K.N., Din, S., Jeon, G., et al.: An accurate and dynamic predictive model for a smart M-health system using machine learning. Inf. Sci. **538**, 486–502 (2020)
3. Andrzejak, R.G., Lehnertz, K., Mormann, F., et al.: Indications of nonlinear deterministic and finite-dimensional structures in time series of brain electrical activity: dependence on recording region and brain state. Phys. Rev. E Stat. Nonlinear Soft Matter Phys. **64**(6), 061907 (2001)
4. Losing, V., Hammer, B., Wersing, H.: KNN classifier with self adjusting memory for heterogeneous concept drift. In: IEEE International Conference on Data Mining. IEEE (2017)
5. Zhang, Z.: Model building strategy for logistic regression: purposeful selection. Ann. Transl. Med. **4**(6), 111 (2016)
6. Shao, M.H., University, S.M.: A review of typical decision tree algorithms. Comput. Knowl. Technol. **14**, 175–177 (2018)
7. Ishioka, J., Matsuoka, Y., Uehara, S., et al.: Computer-aided diagnosis of prostate cancer on magnetic resonance imaging using a convolutional neural network algorithm. BJU Int. **122** (3), 411–417 (2018)
8. Zhang, Y.D., Wang, J., Wu, C.J., et al.: An imaging-based approach predicts clinical outcomes in prostate cancer through a novel support vector machine classification. Oncotarget **7**(47), 78140 (2016)
9. Surangsrirat, D., Thanawattano, C., Pongthornseri, R., et al.: Support vector machine classification of Parkinson's disease and essential tremor subjects based on temporal fluctuation. In: Engineering in Medicine and Biology Society, pp. 6389–6392. IEEE (2016)
10. Zhang, Y., Dong, J., Zhu, J., et al.: Common and special knowledge-driven TSK fuzzy system and its modeling and application for epileptic EEG signals recognition. IEEE Access **7**, 127600–127614 (2019)

Blockchain Based Trusted Identity Authentication in Ubiquitous Power Internet of Things

Yiming Guo[1], Xi Chen[1], Shuang Tian[1], Le Yang[3], Xiao Liang[4],
Jie Lian[2], Dianwei Jin[1(✉)], Aleksei Balabontsev[2],
and Zhihong Zhang[2]

[1] State Grid Shaanxi Electric Power Company Information and Communication
Co., LTD., Xi'an 710065, China
[2] School of Informatics, Xiamen University, Xiamen 361005, China
[3] Information and Communication Branch, Shaanxi Daqin Electric Power Group
Co., Ltd., Xi'an 710000, China
[4] Shaanxi Electric Power Company, Xi'an 710000, China

Abstract. To solve the problem of data loss and malicious tampering caused by
IoT devices without security verification during the implementation of smart
grid projects, this paper utilizes the distributed ledger and non-tampering
characteristics of blockchain technology to ensure the security of the stored data
and to solve the problem of data reliability before winding. We design a
hardware module based on the Ed25519 signature protocol combined with
identity authentication technology to provide the identity key for IoT devices.
By combining the two solutions, we can effectively solve the data security
problems in the implementation of smart grid projects.

Keywords: Blockchain · Identity authentication · Smart power grids ·
Distributed ledger

1 Introduction

In recent years, the State Grid Corporation of China has actively increased its emphasis
on smart grid projects and development and construction to promote the continuous
integration of power flow, information flow[1], and business flow. Its advantage is that it
can integrate grid infrastructure equipment with information technology and sensor
technology. Moreover, the combination of automatic control and other technologies,
and with the help of intelligent systems, allows obtaining panoramic information in the
power grid in real-time to avoid failures.

In 2019, State Grid Corporation proposed a power Internet of Things system
architecture [1] consisting of "cloud, pipe, and edge". This series of significant reforms
undoubtedly declared the advent of the era of Energy Internet [2] in the power industry.
It means an advanced sensor controls and software applications, billions of devices,
machines, and systems are connected at the energy-producing, energy-transmitting, and
energy-consuming ends, to construct the foundation of the Internet of Things. Through

© Springer Nature Switzerland AG 2021
D.-S. Huang et al. (Eds.): ICIC 2021, LNCS 12837, pp. 294–302, 2021.
https://doi.org/10.1007/978-3-030-84529-2_25

the integration of operating data, weather data, grid data, power market data with Big Data analysis, load forecasting, power generation forecasting, machine learning to open up and optimize the operational efficiency of energy production and energy consumption, which can be based on demand and supply. The corresponding optimization plan is dynamically adjusted at any time and used as "technical support". Therefore, it is necessary to carry out a series of upgrades of traditional power grid equipment such as introducing intelligent multi-functional electricity meter, environmental comprehensive monitoring device, intelligent monitoring AC substation box, and other intelligent Internet of Things devices. In this context, traditional identity authentication methods can no longer guarantee the credibility of the device's data, which can easily lead to data fraud and malicious tampering, which will bring significant social risks and economic losses.

With the improvement of blockchain technology, many traditional central database applications are now moving towards it to ensure data security, such as Shopping platform application [3], Financial markets [4], and IoT [5–7]. However, most of the applications in the current industry carry on the chain processing of the data in the previous central database without considering the identity reliability of the terminal device and the reliability of the transmitted data. If the terminal equipment is interconnected, it requires many resources to build an extensive network, so it cannot be regarded as a feasible scheme. To solve the "last mile" problem of blockchain data on the chain, the sensor side is modified. The sensor-side hardware is added to generate its public and private keys for the sensor, and the corresponding edge-side hardware is also added on the edge side. The edge-side hardware has an allowlist verification function. The allowlist verification function is used to verify whether the sent data device is at the edge within the acceptance range of the device and verify whether the sent data device is a counterfeit device through the signature verification data encrypted by the public and private keys, to increase the reliability and security of the data before the data is chained. It's done to ensure the security of equipment and equipment data during the development of smart grid. The contributions of this paper are as follows:

- Mainly design the decryption hardware module based on encrypting of Ed25519 agreement to solve the data blockchain chain before the "last mile", further promote the important industry with the depth of the Internet of things;
- The terminal side hardware and the fusion terminal side hardware are used to replace the intelligent terminal device as the carrier of identity information transmission, and the upper computer sends the allowlist to the fusion terminal side hardware to verify the identity of the terminal device.

2   Related Work

2.1   Blockchain Technology

In 2008, Satoshi Nakamoto elaborated the definition of block and chain in the Bitcoin white paper [8], in which the concept of block [9] had been proposed by Professor Stuart Haber and Professor W. Scott Stornetta as early as 1991. Moreover, with the development of the currency, people gradually understood the blockchain [10]: one of its underlying technologies. In a narrow sense, it means to use each data block to create a time sequence by linking blocks together into an encrypted chain-like data structure; essentially, there should not be any modifications of distributed database [11]. Through blockchain technology, network members can copy and share, and thus the problem of double payment [12] can be solved. Since every transmission for data is regarded as a transaction in the blockchain [13], the Bitcoin blockchain contains an authoritative transaction ledger that determines the transaction process and the owned funds. The chain structure is shown in the Fig. 1, each block in the chain contains a transaction list and the hash key value of the previous block. Each block shown in Fig. 1 represents a node in the blockchain, and users operate on the same blockchain through the ledger held by each node.

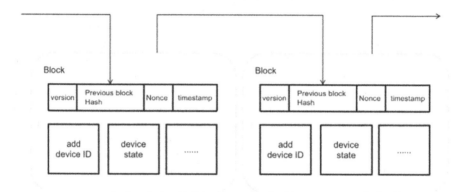

Fig. 1. Blockchain structure

2.2   Ed25519 Signature Authentication Algorithm and Curve Asymmetric Cryptographic Algorithm

EdDSA [14] is a variant of the signature mechanism for Schnorr signatures defined on (warped) Edwards curves. Ed25519 [15] is an EdDSA signature mechanism constructed by Bernstein et al., in 2011, based on the twisted Edwards25519 curve with bidirectional rational equivalence to Montgomery curve Curve25519 [16].

Only two levels of headings should be numbered. Lower level headings remain unnumbered; they are formatted as run-in headings. Ed25519, as a digital signature algorithm, has high signature and verification performance, the hardware requirements are low, it has robust adaptability (e.g., a 4-core 2.4 GHz Westmere CPU can verify 71000 signatures per second), and high-security level equivalent to about 3000-bit RSA. Curve25519 has the highest level of the Diffie-Hellman [17] function and suitable for a wide range of scenarios. Chiefly applicable to the public key and private key generated by the sixty-four length hexadecimal string. Moreover, as an asymmetric encryption algorithm, it belongs to the same module as the Ed25519 protocol.

2.3   Research Status of Trusted Identity Authentication in the Internet of Things

In the trusted identity authentication system of the Internet of Things, identity authentication between various entities is a crucial link. Many researchers have studied this. The main content of this paper is to use traditional cryptography for authentication. In 1981, Lamport L first proposed a remote authentication protocol scheme based on user password authentication [18], which laid a foundation for future remote authentication schemes. Still, it is not suitable for resource-limited Internet of Things devices. Fu-tai Zhang et al. researched certificateless public key cryptography and has obtained certain results, for the first time proposing two certificateless remote anonymous authentication protocols [19] to protect privacy in the Internet of Things field. However, the need to be stored on the server authentication protocol table for identity authentication leads to data being susceptible to tampering and forgery. Simultaneously, due to bilinear peer cryptography [20] operation, the calculation burden on the sensor is significant and is not suitable for clients with limited resources. In 2014, Hu Xiong proposed a remote anonymous authentication protocol based on certificateless encryption [21]. This protocol uses many dot product operations on the client side and does not consider the retraction of the key, so it has a significant security problem and computing burden. Considering that, to improve the security of authentication in the Internet of Things and reduce the cost of computing, Wang Lizhen et al. in 2015 proposed a comprehensive authentication protocol for the Internet of Things, which includes an efficient multi-layer authentication protocol and a secure session key generation scheme for the Internet of Things. With the help of PDA and certificateless authentication technology, the protocol can effectively reduce the energy consumption and computing burden of the sensor and has a remarkable practical significance.

To sum up, in the Internet of things, the identity authentication of the paper in the world has achieved definite results, which can accomplish anonymity, safety, and efficient operations. It has been applied in practice using the traditional identity authentication form of cryptography. At the same time, it is possible to avoid the defect of biological information authentication. However, it also has a high computational cost and Transmission speed is slow. It is vulnerable to collusion attacks and man-in-the-middle attacks, and data is vulnerable to tampering, which is still a big challenge for resource-limited sensor nodes.

3   Key Technology of Power Terminal Trusted Identity Authentication

3.1   Identity Information of Power System Devices on Blockchain

The most basic and core part of the blockchain-based power terminal trusted security system platform is to transmit the identity information of the device to the blockchain. Figure 2 is a transaction flow chart describing data on-chain or on-chain applications from the side of the blockchain platform. This process is encapsulated by the blockchain platform and provides business-related interfaces to external programs.

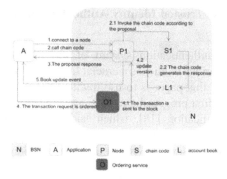

Fig. 2. Data link flow chart

The detailed steps are described as follows:

a) Application A connects to ledger P1 and calls chain code S1 to query or update the grid device identity information on L1. At this time, the chain code cache service information will be generated.

b) P1 calls S1 to generate a proposal response that contains the query result or the ledger to be updated.

c) Application A receives the proposal response, and for the query of the identity information of the grid equipment, the process has been completed here.

d) For grid equipment identity information update, A will generate a transaction based on all responses and send it to sorting service O1 for sorting. O1 collects all transactions on the network into a block and distributes them to all nodes, including P1. P1 validates the transaction and then posts it to ledger L1.

e) When L1 is updated, P1 generates an event, and A receives the event, marking the end of the process.

3.2 Hardware Module Design Based on ED25519 Protocol

The traditional data security of a hardware module is mainly based on the encryption and decryption function provided by the TPM security chip. Still, the the following problems will be encountered when it is applied in power grid projects: 1) The cost of this solution is relatively high, and the third-party hardware manufacturer cannot develop the corresponding TPM driver for the system integrated with its hardware and reconstruct the interface program in the data acquisition process; 2) Due to the need to ensure the security of the data environment of power grid equipment, its network environment is generally a local area network; And blocks in the chain process has many external package dependencies; If the TPM chip is used as a hardware design solution for security assurance, it is required locally download and link the dependent packages required by the blockchain program, and repeat the process when the dependent packages are updated, the convenience will be significantly reduced, and the local dependent packages need to be updated regularly; 3) The fusion terminal and sensor devices need to implant blockchain-related programs, but the compilation of these programs requires a third-party compilation environment. Therefore, a cross-compilation environment needs to be established, but in the process of cross-compilation, the external network also needs to be connected to download relevant dependent packages;

Considering the above problems, we specially designed the related hardware module based on ED25519 to solve the problem of public and private key allocation and encryption/decryption for terminal devices under the condition of ensuring security and high efficiency. The related design of the hardware module mainly includes the upper computer, the fusion terminal side blockchain hardware (from now on referred to as hardware A), and the sensor side blockchain hardware (from now on referred to as hardware B), of which the ED25519 protocol is adopted as the signature algorithm for hardware A and hardware B. Since most of the current fusion terminal-side hardware are 32-bit machines, at the same time, the application background of the related signature algorithm is made for 64-bit machines. In this background, we use the combination of software and hardware to make the algorithm that can be used on a 32-bit machine.

As a signature algorithm used in Stellar networks, ED25519 has very high performance in both signature and verification. The signature process does not depend on the random number generator, does not depend on the anti-collision of a hash function, and does not have the problem of time channel attack. The design of this module is shown in Fig. 3. Dual-channel RS-485 interfaces are used between hardware A and B. The uplink is used to pass between the fusion terminal and the acquisition terminal. The downlink is used for the mutual communication between modules A and B, and the Wi-Fi interface is used to communicate with the upper computer.

In Fig. 3, the upper computer can set the public and private keys for the end-to-side hardware and the fusion end-to-side hardware and query them. The end-to-side hardware can send data to the fusion end-to-side hardware, but the reception of the data must first pass the allowlist verification. The upper computer sets the allowlist of the hardware on the fusion terminal side, and the query interface is provided.

This hardware module mainly solves the problem of traditional TPM security chips as the security guarantee of terminal equipment. It puts forward a scheme of using middleware to encrypt and decrypt data to ensure data reliability. It has practical reference value in the actual engineering problems.

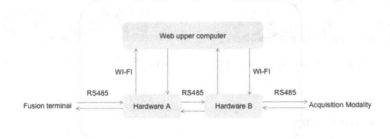

Fig. 3. Hardware module design based on ED25519 protocol experiment

4   Experiment

To ensure the overall functionality and stability of hardware modules, our tests mainly include single module tests and integrated stability tests during development. At the same time as the integration test, we also carried out related tests on the delay of the whole hardware module.

4.1   Single Module Testing

To ensure the overall functionality and stability of the designed hardware module, it is necessary to carry out development testing. he purpose of development testing is to ensure that each hardware module can run normally. The development test is shown in Fig. 4, and the test steps are as follows. Firstly, we developed and tested the upper computer (see Figure 6). The main functions of the test include: 1) setting and querying the allowlist; 2) Setting and query of public and private keys; 3) Allowlist verification results monitoring for trusted authentication. After that, we tested the fusion terminal side hardware, and the main tests included: 1) The message receiving function of the hardware on the fusion terminal is verified by setting the allowlist message; 2) Finally we carried out the development and test of the terminal side hardware, including the test steps: 1) Passing the public and private key through the interface of the hardware module to query the message and view the message outgoing by the hardware module; 2) Passing the validation failed message through the interface of the hardware module, view the outgoing message of the hardware module (Fig. 5).

   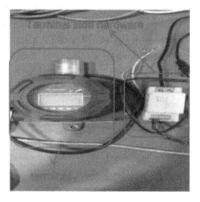

Fig. 4. Fusion terminal side hardware testing   **Fig. 5.** End-side hardware testing

4.2   Integration Testing

To verify the function of the designed hardware module, we carried out a system stability test and functional stability test in the real environment in Shaanxi Communications Company of China. A setup for a live test is shown in Figure 6. The funtional stability tests are as follows: during the system test process, the host computer can assign an allowlist and receive sensor information from the sensor-side hardware, and the sensor terminal-side hardware can perform allowlist verification on the sensor. The entire verification process does not affect the final result being transmitted to the IoT platform. It can reliably verify the identity of the device. System stability is mainly used to detect whether the function of the system is stable under different environments. The test steps are as follows: Every 1 s, 2 s, 3 s, or more (at most 10 s), a custom interface transmits verification information and observes the delay variation. According to the results of multiple tests, the delay was 1.5 s. After modifying the fusion terminal delay, the desired effect was achieved.

4.3   Time Delay Analysis

After the functionality and stability of the hardware module are verified, we analyze the delay of the module. This module's basic unit of data includes 1 byte plus two header fields and 1 bit CRC. Therefore, the delay of each data transmission (see Eq. 1) is multiplied by 8 because the terminal device and the fusion terminal device are dual-channel RS485 interfaces, and each verification needs to go through the dual-channel interface four times. The 9600 represents the number of bits per second sent through the port.

$$T_{delay} = \frac{n * 11}{9600} * 8 \tag{1}$$

After performing field testing of hardware module system stability and functional stability of the integration several times, we averaged the test results and found that the delay was 0.15 s. When the module is not added, the terminal delay is 0.1 s, and the test result is still within the allowable range.

5   Conclusion

This paper designed a hardware module based on the Ed25519 signature algorithm to guarantee the reliability of the terminal device. The experimental evidence proves that the combination of blockchain technology and identity authentication-related technologies can improve the security of related IoT devices and their data in smart grid projects and meet the security requirements of smart grid projects.

Acknowledgment. This work was supported by Research Funds of State Grid Shaanxi Electric Power Company and State Grid Shaanxi Information and Telecommunication Company (contract no. SGSNXT00GCJS2000104).

References

1. Wang, Q., Wang, Y.G.: Research on power Internet of Things architecture for smart grid demand. In: 2018 2nd IEEE Conference on Energy Internet and Energy System Integration (EI2). IEEE (2018)
2. Sun, Q.: Energy Internet and We-Energy. Springer, Singapore (2019). https://doi.org/10.1007/978-981-13-0523-8
3. Lim, Y.H., Hashim, H., Poo, N., Poo, D.C.C., Nguyen, H.D.: Blockchain technologies in E-commerce: social shopping and loyalty program applications. In: Meiselwitz, G. (ed.) HCII 2019. LNCS, vol. 11579, pp. 403–416. Springer, Cham (2019). https://doi.org/10.1007/978-3-030-21905-5_31
4. Wu, B., Tingting, D.: The application of blockchain technology in financial markets. J. Phys. Conf. Ser. **1176**(4), 042094 (2019). IOP Publishing
5. Zhang, Y., Wen, J.: The IoT electric business model: using blockchain technology for the internet of things. Peer-to-Peer Netw. Appl. **10**(4), 983–994 (2016). https://doi.org/10.1007/s12083-016-0456-1
6. Li, D., et al.: A blockchain-based authentication and security mechanism for IoT. In: 2018 27th International Conference on Computer Communication and Networks (ICCCN). IEEE (2018)
7. Huang, J.C., Shu, M.H., Hsu, B.M., et al.: Service architecture of IoT terminal connection based on blockchain identity authentication system. Comput. Commun. **160**, 411–422 (2020)
8. Nakamoto, S.: Bitcoin: A Peer-to-Peer Electronic Cash System. Manubot (2019)
9. Haber, S., Scott , W.: How to time-stamp a digital document. J. Cryptol. **3**, 99–111 (1991)
10. Swan, M.: Blockchain: Blueprint for a New Economy. O'Reilly Media Inc., Newton (2015)
11. Zyskind, G, Nathan, O.: Decentralizing privacy: using blockchain to protect personal data. In: 2015 IEEE Security and Privacy Workshops, pp. 180–184. IEEE (2015)
12. Stoft, S.: 1. The Double-Payment Conundrum of Order 745 (2011)
13. Watanabe, H., Fujimura, S., Nakadaira, A., et al.: Blockchain contract: a complete consensus using blockchain. In: 2015 IEEE 4th global conference on consumer electronics (GCCE), pp. 577–578. IEEE (2015)
14. Bernstein, D.J., et al.: EdDSA for more curves. Cryptology ePrint Archive 2015 (2015)
15. Bernstein, D.J., et al.: High-speed high-security signatures. J. Crypt. Eng. **2**(2), 77–89 (2012)
16. Bernstein, D.J.: Curve25519: new Diffie-Hellman speed records. In: Yung, M., Dodis, Y., Kiayias, A., Malkin, T. (eds.) PKC 2006. LNCS, vol. 3958, pp. 207–228. Springer, Heidelberg (2006). https://doi.org/10.1007/11745853_14
17. Joux, A.: A one round protocol for tripartite Diffie–Hellman. In: Bosma, W. (ed.) ANTS 2000. LNCS, vol. 1838, pp. 385–393. Springer, Heidelberg (2000). https://doi.org/10.1007/10722028_23
18. Lamport, L.: Password authentication with insecure communication. Commun. ACM **24**, 770–772 (1981)
19. Tao, F., Liang, Y.X.: Provably secure certificate less blind proxy re-signatures. J. Commun. **33**(Z1), 58 (2012)
20. Zhang, Y., Jiguo, L., Yuan, H.: Certificateless proxy signature scheme. Nanjing Xinxi Gongcheng Daxue Xuebao **9**(5), 490–496
21. Xiong, H.: Cost-effective scalable and anonymous certificateless remote authentication protocol. IEEE Trans. Inf. Forensics Secur. **9**(12), 2327–2339 (2014)

Intelligent Modeling Technologies
for Smart Cities

A YOLOv3-Based Learning Strategy for Vehicle-Thrown-Waste Identification

Zhichao Dai[(⊠)] and Zhaoliang Zheng

School of Artificial Intelligence and Computer Science,
Jiangnan University, Wuxi, Jiangsu, China

Abstract. At present, throwing objects from car windows is increasingly becoming a major illegal conduct that pollutes the urban environment and affects the city's image. Apart that, the rubbish casually thrown from car windows is left in the motorway, which seriously threatens the lives of sanitation workers. Moreover, it is time-consuming and laborious to manage through manual monitoring. In response to this phenomenon, we propose an improved yolov3-based real-time target detection for throwing objects from car windows. In order to solve the problem of deep hierarchy in the YOLOv3 network and improve the accuracy and speed of small target detection, the following improvements are made: Firstly, random feature sampling and interpolation are added to the residual blocks of the YOLOv3 backbone network to reduce the computational effort of the CNN network while maintaining high performance. Secondly, the Soft NMS is changed to Matrix NMS to speed up the identification of the best candidate boxes. Finally, a residual layer was removed to improve the accuracy of small target detection. The experimental study evaluates and demonstrates a number of improvements in accuracy and speed compared to current mainstream target detection methods.

Keywords: YOLOv3 · Matrix NMS · Stochastic Feature Sampling and Interpolation

1 Introduction

"Throwing objects out of windows" may seem like a small thing, but it is a big thing that destroys environmental cleanliness and threatens the safety of driving and life safety. With the rapid growth in the number of vehicles, "throwing objects out of windows" has become a "major hazard" on the road. It not only affects the appearance of the city, but also makes it hard for sanitation workers to complete their tasks [1]. On the other hand, if the workers don't have time to dodge, the consequences are unthinkable. The problem has therefore become a difficult issue for the authorities [2, 3]. At present, the current treatment methods are similar from place to place. There are two main measures to control. Firstly, increasing the cost of violations of throwing objects out of windows. That is to say, high fines will be issued to those who violate the law, so that people will be supposed to obey the official laws and regulations. Secondly, relevant video surveillance equipment should be laid on the roads with high violation rate, and then photos or real-time videos should be taken to save the video, and finally

© Springer Nature Switzerland AG 2021
D.-S. Huang et al. (Eds.): ICIC 2021, LNCS 12837, pp. 305–315, 2021.
https://doi.org/10.1007/978-3-030-84529-2_26

the supervision personnel should be responsible for monitoring. However, the method of monitoring and observing through personnel consumes a lot of manpower and material resources, and we have not seen many favorable results. Therefore, it is imperative to adopt an intelligent and efficient real-time target detection method on the problem of "Throwing objects out of windows".

At present, the target detection algorithm models based on deep learning mainly include SPP Net [6], Fast R-CNN [7], SSD [8], YOLO [4, 9] and other models. Among them, YOLOv3 [4] real-time target detection and recognition framework was proposed by Redmon, Joseph and Farhadi in 2018. This target detection method has high detection accuracy and can simultaneously detect targets with multi-scale resolutions, which is suitable for the field of multi-target recognition with the same lens. At present, there are few video detection methods for garbage identification, so we summarize relevant studies. Siddhant Bansal1, Seema Patel uses CNN [22–24] network and machine to design a robotic system to recognize and capture garbage. In the process of target detection, once garbage is detected, the system only calculates the location of garbage by using the camera. The system can distinguish between valuables and garbage. Ying Wang1 and Xu Zhang [10] train a Faster R-CNN framework by using region proposal network and ResNet network algorithms to quickly and accurately detect garbage from city images. X Zhang and Y Gao [11] use the YOLOv3-tiny open source framework to realize real-time monitoring and data statistics of garbage trucks. These methods have high accuracy and recall rate for the process, but too many neural network layers lead to a huge amount of calculation, and the time needs to be improved. In order to solve this problem, this paper draws on the idea of random feature sampling and interpolation [12] proposed by Z Xie, Z Zhang and others. Based on the YOLOv3 network structure, an improved YOLOv3 method is proposed, which combines the volume in the residual network. The layers are randomly sampled and then reconstructed by interpolation. This can greatly reduce the amount of calculation in the network layer.

In short, in this article, our contribution lies mainly in the following two points:

In the backbone network of YOLOv3, there are a total of 1×1 convolution and a 3×3 convolution used in the residual module [14, 21]. We will use two random convolutions in the process of these two convolution operations. The interpolation sampling method reduces the computation and improves the network performance.

We use Matrix NMS to replace Soft NMS [13, 25, 26], Matrix NMS calculates the IOU [30] between any two boxes through a matrix parallel operation. Each bounding box has confidence. Soft NMS is to lower the threshold of the bounding box with lower confidence for two bounding boxes of the same category. Liner and Gauss are the methods to lower the threshold. The method is serial. Matrix NMS calculates the probability of each box being suppressed through the matrix, and updates the confidence. It can improve the detection accuracy, and avoid the decrease of the reasoning speed.

In the following content, the second section reviews YOLOv3, Matrix NMS and Stochastic Feature Sampling and Interpolation. The third section proposes an improved YOLOv3 algorithm. Section 4 describes the experimental results. Finally, conclusions are given in the fifth section.

2   Related Work

2.1   YOLOv3

The backbone of YOLOv3 network adopts a 53-layer convolutional network [15]. The YOLOv3 neural network is also known as Darknet-53, whose residual module consists of 1×1 and 3×3 convolutional layers. YOLOv3 contains 23 residual modules and the full connection layer of detection channelsYolov3 also uses a multi-scale prediction method. YOLOv3 has three scales: 52×52, 26×26 and 13×13, of which 52×52 is used to detect small targets. Corresponding 26×26 is used to detect medium targets and 13×13 is used to detect large targets. The feature size obtained by each detection task is $N * N * [B * (4 + 1 + C)]$, where N is the size of the grid. B is the number of detected boundaries of each grid C is the number of detected categories, 4 is the number of coordinates of the bounding box, and 1 is the target prediction. Finally, NMS is used to delete redundant bounding boxes for multiple identified bounding boxes. Figure 1 below shows the network architecture of Yolov3.

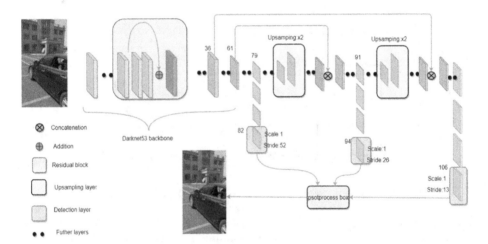

Fig. 1. YOLOv3 network architecture

2.2   Matrix NMS

Matrix NMS is improved on the basis of Soft NMS. Soft NMS recursively updates the confidence of the bounding box every time, and finally finds out the bounding box within the range of threshold value. On the other hand, the Matrix NMS looks at the problem from a different perspective. It considers is how the predicted mask is suppressed. It is influenced by two aspects. One is the penalty coefficient of every two bounding boxes, and the other is the probability of being suppressed. Through these two aspects of parallel computation, one-time calculation of attenuation parameters.

2.3 Stochastic Feature Sampling and Interpolation

When the neural network carries out the convolution operation, the feature map has redundancy in space. By predicting the coordinates with a median value of 0 in the feature map, only the value of the feature map with a value other than 0 can be calculated to reduce the computational workload. For the convolution layer in a CNN network, the input feature map is $X \in R^{C_{in} \times H \times W}$ And the output is $Y \in R^{C_{out} \times H \times W}$, the following formula represents the convolution operation:

$$Y(p) = \sum_{p\prime \in R_k} W_c(p\prime)X(p+p\prime), p \in \Omega \tag{1}$$

In the above formula, H and W respectively represent the height and width of the feature map, and C represents the number of channels. W represents weight, and represents the coordinate set of elements in the feature map in space. R is used to build the window for the convolution operation.

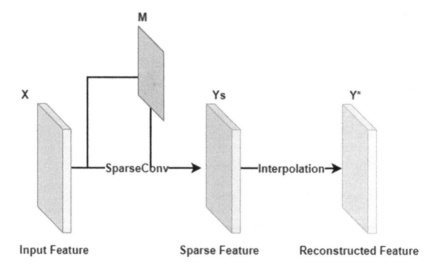

Fig. 2. Stochastic feature sampling and interpolation module.

This is shown in Fig. 2, the sampling module generates a binary mask matrix, expressed as $M \in R^{H \times W}$, Convolution operation is performed on this matrix to obtain the corresponding sparse matrix, and the resulting feature map is denoted as $Y_s(p)$. It is expressed as follows:

$$Y_s(p) = \begin{cases} 0 \ M(p) = 0 \\ Y(p) \ M(p) = 1 \end{cases} \tag{2}$$

In the matrix mask, if the value is 0, it means that the value at this position can be obtained through the interpolation module. The formula output by the interpolation module is as follows:

$$Y^*(p) = \begin{cases} C(Y_s)(p)M(p) = 0 \\ Y_s(p)\,M(p) = 1 \end{cases} \tag{3}$$

If the mask matrix is sparse enough, most of the elements are obtained by interpolation methods with a small amount of calculation. Using the module in the above figure to replace the traditional convolution in CNN can reduce the amount of calculation.

3 The Proposed Method

In YOLOv3, there are a total of 53 convolution layers, which contain a large number of convolution operations, and the process of convolution operation is time-consuming. We thought of optimizing the time from the perspective of simplifying the amount of computation, and improved YOLOv3 from two aspects: The first one, inspired by the paper of random sampling interpolation, the method is used to reduce the calculation amount of convolution layer and is suitable for optimizing the residual network. There are 53 convolution layers in the backbone network of YOLOv3, where the number of residual modules is $1 + 2 + 8 + 8 + 4 = 23$, and there are 46 convolution layers. The use of random sampling interpolation method can greatly simplify the calculation. What's more, a layer of residual blocks is also reduced, The Second improvement. In YOLOv3, a multi-scale feature network structure was adopted to reduce the information loss layer by layer in order to reuse multi-layer features. The 13×13, 26×26 and 52×52 size resolutions of the network were predicted respectively, and all the results were fused by maximum suppression. Matrix NMS method is used for fusion in the improved YOLOv3.

If a deep structure is just enough to solve the problem, it is not possible to solve it with a shallower, equally compact structure. To solve a complex problem, the method is to increase the depth or increase the width. However, in general, neural network is not as deep as possible, nor as wide as possible, and due to the limitation of computation or the demand for speed, how to obtain better accuracy with fewer parameters is undoubtedly an eternal pursuit.

In the application of traffic target recognition, the recognition and tracking of small targets is one of the key research issues. For conventional targets, classical detection methods can be used to obtain better detection results. However, the detection of small targets is not ideal. YOLOv3 has a poor detection effect on small targets such as birds and ships, yet it has a good detection effect on larger but more complex targets. On the one hand, the feature network lacks the ability to learn specific features and express features, on the other hand, the small target features extracted by the network can provide too little information for the model. Because the volume of garbage discarded from the car window is relatively small, when the shooting equipment is in video surveillance, a frame of picture contains many small and medium targets but few large targets, so in the improved YOLOv3, the residual of the first layer is removed. The module uses down sampling twice at the beginning.

The picture on the left in Fig. 3 is the optimized network structure, including a total of 22 residual module. In the shallow layer, the number of convolutional layers and residual connection structures is reduced, which can enhance the characterization ability of low-level features. Then, in the deeper network, three different feature information is extracted and input into the detection channel. The detection channel of the deep network is up-sampled to the shallow layer to enrich the feature information of the shallow layer to improve the detection accuracy of small targets, then from the deeper in the network, to extract the characteristic information input respectively the three different test channels.

The picture on the right in Fig. 3 shows the newly designed residual module, which has two convolution layers of 1×1 and 3×3 respectively, and two randomly sampled matrices. The 1×1 and 3×3 convolutions use the ReLU [20] function and Batch normalization [16] internally. Two randomly generated matrices m1 m2 are used for convolution in the sampling module to generate sparse feature matrices, and then features are reconstructed through interpolation module.

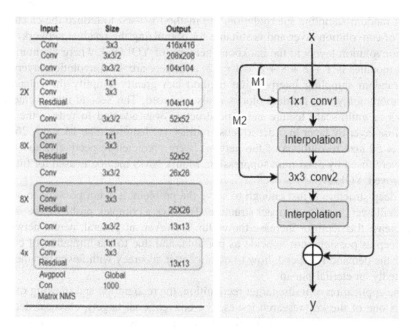

Fig. 3. The figure on the left is our improved YOLOv3 network result graph, and the figure on the right is the new residual module constructed.

4 The Experimental Studies

4.1 Dataset and Experimental Environment

In the video resource database currently developed, no relevant video of throwing objects from car windows has been found. In order to better test the experiment and get the experimental results, we obtained the pictures by two ways: taking pictures manually and recording video by ourselves. Then using the annotation tool LabellMG to manually mark the data and generate XML files. In the prepared data, there are 2424 pictures in the training set and 337 pictures in the test set. These samples include the three most common types of traffic garbage, namely: cans, bottles and paper boxes. In addition, three traffic videos were prepared as test videos. All of these traffic images were pre-processed into the same size: 1080×1920, and were labeled as from 0 to 2. The following Table 1 is the picture of the sample and the corresponding label information and Fig. 4 is a video we prepared for target detection. In this video, people recognize abandoned items from the car, and we randomly intercepted four photos as detection results.

Table 1. Data set sample information

Sample Type	bottole	box	can
Sample picture			
Sample picture	0	1	2

Fig. 4. A screenshot of the garbage disposal video from the car window

Our experimental studies were carried out on a workstation with an Intel i7-6850K 3.60 GHz CPU, 128 GB of RAM, NVIDIA TITAN XP (12 GB) GPU, Ubuntu16.04 (64 bit), Python 2.7, and Tensorflow 1.12.0 (GPU).

4.2   Training

After the data set is annotated and sorted out, a file about the coordinate frame will be generated with the software. We need to conduct unsupervised clustering on the frames of the data set, and K-means algorithm [19, 29] is used to obtain the frames obtained by our 9 clusters. Then the training is carried out according to the generated prediction box. The data will be put into our improved network for training. The specific training operations are shown in the following paragraphs. The network parameters generated by the training will be saved. In the second module, we need to make use of the video we recorded, get the frames of the video, and then input them into our network for training, and use the network parameters in the first module. In the second module, the network will output the prediction boxes in each frame of the picture, and there are three possible prediction boxes for each possible object. At this time, our third module is used to screen the boxes through Matrix NMS and mark the boxes in the frame picture. Finally, each frame is reassembled into a video. Figure 5 below shows the training process.

Training method: 6 pictures were selected each time. The initial learning rate was $1e-4$ and gradually decreasing, but not less than $1e-6$, and IOU was set as 0.5. Back propagation was adopted to fine-tune network parameters. The first 20 epochs first optimized the network parameters of the last layer, and the last 100 epochs adjusted the network parameters of the whole network.

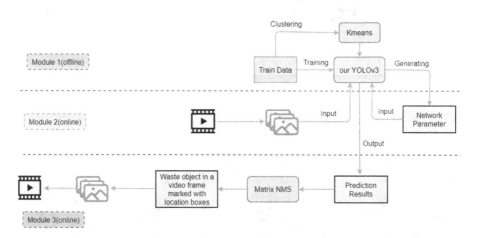

Fig. 5. Training operation step process

4.3   Performance

Through experiments, the detection effect of our improved YOLOV3 algorithm is shown in Fig. 6. As can be seen from Fig. 6, for detected targets, the position of prediction box marks is good, and the confidence of each detected target is high. It shows that the network model designed in this paper can detect the target effectively.

Fig. 6. The result of frame image training

In order to verify the effectiveness of the algorithm in this paper, this paper compares the accuracy and training time for each picture calculated by the YOLOv2 algorithm, YOLOv3, YOLOv3 + Matrix NMS, and the improved algorithm of this paper on the self-made data set. The experimental results are shown in Table 2 below.

Table 2. Compare it to other algorithms.

Model	Average accuracy	Time
YOLOV2	0.887	16.9 s per frame
YOLOv3	0.964	18.3 s per frame
YOLOv3 + Matrix Nms	0.964	17.6 s per frame
Improved YOLOv3	0.982	15.6 s per frame

As can be seen from Table 2, the improved YOLOv3 improves by 1.8% on accuracy with YOLOv3, and after the model is trained, the module 2 and module 3 in Fig. 5 are used for target detection. It can be found that the running time of YOLOv3 to detect a frame of image is 18.3, and our improved YOLO time is 15.6. Our improved YOLOv3 has improved its speed to a certain degree. In addition, both YOLOv2 and YOLOv3 adopt multi-scale and cross-scale feature fusion, but simple multi-scale is not the key to improving target detection performance. In YOLOv2, anchor was used to make regression prediction box. The receptive field of larger subsampling factor was larger, which was beneficial to the classification task, but detrimental to target detection and location. To solve the above problems, the YOLOv3 network used three different scales for box prediction, and used the idea of FPN for target prediction in the early stage of subsampling to improve the detection and positioning of small targets. In addition, the input size of YOLOv2 is an image, and only the feature image is used in the end, resulting in poor detection performance. However, the detection speed of the relatively simple network structure is faster. This paper mainly changes the residual network based on the YOLOv3 network structure, reduces the amount of calculation, improves the detection reading, and removes a residual layer, which can improve the detection accuracy of small targets, and enhances the generalization ability.

5 Conclusion

This paper proposes to use the target detection network YOLOv3 to detect the abandoned objects in the car window. Our experimental research verifies the effectiveness and superiority of our proposed method. The improved YOLOv3 network enhances the detection speed and increases the detection accuracy. Compared with other network structures in the experiment, it has a good detection effect on the detection of objects abandoned from the car window. In the course of the experiment, some pictures were blurred and the detection failed. However, the target detection of throwing objects from car windows is mainly aimed at the situation of discarding garbage under high-speed driving. The video export frame pictures in this situation are blurry, which is not conducive to detection. Therefore, how to improve the accuracy of target detection under blurred images is the focus of our next research.

References

1. Carvalho, V.F., Silva, M.D., Silva, L.M.S., Borges, C.J., Silva, L.A., Robazzi, M.L.C.C.: Occupational risks and work accidents: perceptions of garbage collectors. J. Nurs. UFPE Online **10**(4), 1185–1193 (2016)
2. Kim, C., Hwang, J.-N.: Object-based video abstraction for video surveillance systems. IEEE Trans. Circuits Syst. Video Technol. **12**(12), 1128–1138 (2002)
3. Meng, Y., Wu, H.: Highway visibility detection method based on surveillance video. In: Proceedings IEEE 4th International Conference on Image, Vision and Computing (ICIVC), pp. 197–202, July 2019
4. Redmon, J., Farhadi, A.: YOLOv3: An Incremental Improvement. arXiv e-prints (2018)
5. Wang, X., Zhangm R,, Kongm T,, et al.: SOLOv2: Dynamic, Faster and Stronger (2020)
6. Purkait, P., Zhaom C., Zach, C.: SPP-Net: deep absolute pose regression with synthetic views. In: British Machine Vision Conference (BMVC 2018) (2017)
7. Girshick, R.: Fast R-CNN. Computer Science (2015)
8.. Liu, W., et al.: SSD: single shot multibox detector. In: Leibe, B., Matas, J., Sebe, N., Welling, M. (eds.) ECCV 2016. LNCS, vol. 9905, pp. 21–37. Springer, Cham (2016). https://doi.org/10.1007/978-3-319-46448-0_2
9. Redmon, J., Farhadi, A.: YOLO9000: better, faster, stronger. In: IEEE 2017, pp. 6517–6525. IEEE

10. Bansal, S., Patel, S., Shah, I., et al.: AGDC: Automatic Garbage Detection and Collection (2019)
11. Zhang, X., Gao, Y., Xiao, G., et al.: A real-time garbage truck supervision and data statistics method based on object detection. Wirel. Commun. Mob. Comput. **2020**, 1–9 (2020)
12. Xie, Z., Zhang, Z., Zhu, X., et al.: Spatially Adaptive Inference with Stochastic Feature Sampling and Interpolation arXiv e-prints. arXiv (2020)
13. Bodla, N., Singh, B., Chellappa, R., et al.: Soft-NMS – Improving Object Detection with One Line of Code (2017)
14. He, K., Zhang, X., Ren, S., Sun, J.: Deep residual learning for image recognition. In: Proceedings of the IEEE Conference on Computer Vision Pattern Recognition (CVPR), June 2016, pp. 770–778
15. Chua, L.O., Roska, T.: CNN paradigm. IEEE Trans. Circuits Syst. I Fundam. Theory Appl. **40**(3), 147–156 (1993)
16. Ba, J.L., Kiros, J.R., Hinton, G.E.: Layer Normalization (2016)
17. Hopkins, D.L., Rutherford, J.C.: Animal Identification Marking, USA (2008)
18. Wang, Y., Zhu, S., Hong, W.: Characteristic analysis of pedestrian violation crossing behavior based on logistics model. In: International Conference on Intelligent Computation Technology and Automation. IEEE Computer Society (2010)
19. Wong, J.A.H.A.: Algorithm AS 136: A K-Means clustering algorithm. J. Roy. Stat. Soc. **28**(1), 100–108 (1979)
20. Xu, B., Wang, N., Chen, T., Li, M.: Empirical evaluation of rectified activations in convolutional network (2015), arXiv:1505.00853.
21. Zia, T., Razzaq, S.: Residual recurrent highway networks for learning deep sequence prediction models. J. Grid Comput. **18**(1), 169–176 (2018). https://doi.org/10.1007/s10723-018-9444-4
22. Gu, J., et al.: Recent advances in convolutional neural networks (2015), arXiv:1512.07108.
23. Passalis, N., Tefas, A.: Training lightweight deep convolutional neural networks using Bag-of-features pooling. IEEE Trans. Neural Netw. Learn. Syst. **30**(6), 1705–1715 (2019)
24. LeCun, Y., Kavukcuoglu, K., Farabet, C.: Convolutional networks and applications in vision. In: Proceedings of the IEEE International Symposium on Circuits Systems, June 2010, pp. 253–256
25. Qiu, S., Wen, G., Deng, Z., Liu, J., Fan, Y.: Accurate non-maximum suppression for object detection in high-resolution remote sensing images. Remote Sens. Lett. **9**(3), 237–246 (2018)
26. Neubeck, A., Van Gool, L.: Efficient non-maximum suppression. In: Proceedings of the 18th International Conference on Pattern Recognition (ICPR), August 2006, pp. 850–855
27. Wang, N., Zhu, X., Zhang, J.: License plate segmentation and recognition of Chinese vehicle based on BPNN. In: Proceedings of the 12th International Conference on Computational Intelligence and Security (CIS), December 2016, pp. 403–406
28. Ashtari, A.H., Nordin, M.J., Fathy, M.: An Iranian license plate recognition system based on color features. IEEE Trans. Intell. Transp. Syst. **15**(4), 1690–1705 (2014)
29. Jain, A.K.: Data clustering: 50 years beyond K-means . Pattern Recogn. Lett. **31**(8), 651–666 (2010)
30. Girshick, R., Donahue, J., Darrell, T., Malik, J.: Rich feature hierarchies for accurate object detection and semantic segmentation. In: Proceedings of the IEEE Conference on Computer Vision and Pattern Recognition, June 2014, pp. 580–587

Research on Chinese Word Segmentation Based on Conditional Random Fields

Chao Fan[1,2](✉) ⓘ and Yu Li[1,2]

[1] The School of Artificial Intelligence and Computer Science,
Jiangnan University, Wuxi 214122, China
fanchao@jiangnan.edu.cn
[2] Jiangsu Key Laboratory of Media Design and Software Technology,
Jiangnan University, Wuxi 214122, China

Abstract. Word segmentation is the first step in Chinese natural language processing. The accuracy of segmentation has substantial impacts on subsequent tasks such as part-of-speech tagging, semantic analysis, etc. This research explores the Chinese word segmentation based on Conditional Random Fields (CRFs). First of all, we apply different character templates to conduct feature selection. Baseline and eight templates are combined to construct nine groups of features. Also, Tsai's feature set and our proposed feature set are raised to refine the experimental results. Moreover, several parameters of model are adjusted to make the CRF model more effective and efficient. Finally, different tag set of characters are verified according to different dataset. The 2-Tag set is chosen for character tagging in this research. Experimental results demonstrate that the CRF-based Chinese word segmentation with proposed feature set achieves the best F score.

Keywords: Chinese word segmentation · Conditional random fields · Feature template

1 Introduction

Since Chinese does not have an apparent "space" separator, Chinese word segmentation is a challenging issue compared with English tokenization. Also, word segmentation is the basis of many tasks in natural language processing, information retrieval, search engine, etc. For instance, an inverted index is the most important data structure in search engine. When building the inverted index, the sentences need to be segmented into words. Hence, Chinese word segmentation is an indispensable part to create the complicated information processing applications.

In this paper, Conditional Random Fields (CRFs) are utilized to study the Chinese word segmentation. Initially, we take full advantage of baseline character template and eight other templates to construct nine groups of features. Some n-gram features (Unigram, Bigram) are designed by Tsai et al. in order to eliminate the noise information in the feature set. Then we improve Tsai's feature set by adding extra feature templates 6, 7, 8. Additionally, some parameters of the CRF model (f and c) are tuned in this research, so that we can acquire a better segmentation result. Finally, three tag

© Springer Nature Switzerland AG 2021
D.-S. Huang et al. (Eds.): ICIC 2021, LNCS 12837, pp. 316–326, 2021.
https://doi.org/10.1007/978-3-030-84529-2_27

sets (2-Tag, 4-Tag, 6-Tag) of character are compared in four groups of datasets. Thus, we can obtain the best tag set which is suitable for our experiments.

The remainder of this article is discussed as follows. Related work of Chinese word segmentation is reviewed in Sect. 2. CRF model, feature selection, and tag set of Chinese word are given in Sect. 3. Dataset, preprocessing, and method of evaluation are illustrated in Sect. 4. Experimental results and some discussion are presented in Sect. 5. Section 6 draws the conclusion and provides some challenging work in future.

2   Related Work

Chinese word segmentation is a prevalent research direction in natural language processing. Chen et al. [1] developed a gated recursive neural network to process the Chinese word segmentation task. Their method can reset and update gates to include the complicated combinations of the context characters. Yao et al. [2] made use of bi-directional RNN with long short-term memory units for this task. The approach specialized in retaining the contextual information in both directions. Wang et al. [3] brought forward a dilated convolution neural networks to segment Chinese word. Radical information of Chinese was added to enrich the input. Also, they leveraged the convolution neural network to perform feature extraction. The proposed method can comprehend semantic information and enhance the efficiency of training.

As for the Chinese word segmentation in CRF, there exists a number of studies considering different aspects of CRF model. In the field of feature selection, Xue [4] proposed the earliest features for character tagging: characters and POS tags. Low [5] introduced the feature set of punctuations, numbers, letters, and dates. They are offered to the maximum entropy approach in order to segment Chinese sentences. According to some literature [6, 7], these features can be applied to conditional random fields (CRFs). In order to reduce the noise information in the features, Tsai et al. [8] employed n-gram features which contain three types: Unigram, Bigram, and Jump Bigram. Recently, Huang et al. [9] proposed a multi-criteria learning method to complete the Chinese word segmentation. They exploited a BERT-based model to incorporate open-domain knowledge and utilized a CRF tag inference layer to gain the best tag sequence. Deng et al. [10] researches on domain adaption of Chinese word segmentation based on semi-supervised CRFs. They did the supervised learning on the labeled text in common domain. Unsupervised learning was carried out on the unlabeled text in the professional domain.

3   Methodology

3.1   Segmentation Based on CRFs

Conditional random fields (CRFs), proposed by Lafferty [11], are widely used in various tasks in natural language processing. The graph structure of a linear chain CRF is presented in Fig. 1. It utilizes a single exponential model to calculate the joint probability of the entire sequence of labels **s** given the observation sequence **o**.

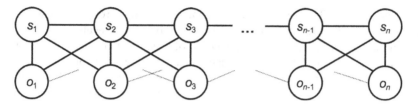

Fig. 1. A graph structure of a linear chain CRF

Taking into account the context information, CRF avoids the strong independence assumption of Hidden Markov Model (HMM). Therefore, it has a good effect on the recognition of ambiguous words and unregistered words. As a discriminative model, CRF has the advantage of flexible feature design as Maximum Entropy Model (MEM). Furthermore, it is an undirected graph model, which does not normalize each node but calculates the conditional probability of the global optimal output node. It can obtain the global optimal value, overcoming the label bias problem of Maximum Entropy Markov Model (MEMM). Nevertheless, the training process of a CRF is expensive and complex, so it is not suitable for the task with excessive labels.

When applied in the Chinese word segmentation task, CRFs regard the segmentation problem as a process of adding tags to each word in a sentence. Tags "B" and "I" are adopted to mark whether two adjacent Chinese characters in the sentence are divided or not, deciding the word boundary. The best sequence of labels can be found according to the given Chinese character sequence.

3.2   Feature Selection

In this research, five types of features are extracted to the CRFs. They are introduced in detail in the following parts.

Single Character Features. The single character feature is a character in a certain position, which is a Unigram Token. For instance, C_{-1} indicates the previous character, and C_0 represents the current character.

Single String Features. The single string feature is a string which is composed of a character at a certain position and its adjacent characters. For example, $C_{-1}C_0$ is a Bigram Token with previous character and current character. $C_{-1}C_0C_1$ is a Trigram Token which involves previous character, current character, and next character.

Combination Features. The combination feature is a combination of characters or strings in different positions. $C_{-1}C_1$, a combination of two characters before and after the current character, is often used. It is also called Jump Bigram. $C_{-2}C_0$ is another case of combination feature.

Punctuation Features. A character can be tagged by PUCN and NOPU with a meaning of punctuation and non-punctuation. The $Pu(C_0)$ indicates whether a character at a certain position is a punctuation or not.

Character Type Features. The character at a certain position has the following four types: NUMR (number), DATE (date), ALPH (English letter), and OTHR (others). The $Ty(C_{-1})Ty(C_0)Ty(C_1)$ gives the types of three characters: the previous character, the current character, and the next character.

3.3   Tag Set of Chinese Word

There are three types of tag sets used for character tagging in Chinese word segmentation based on CRFs. They are displayed in Table 1.

Table 1. Tag set names in Chinese word segmentation based on CRFs

Tag set name	Content
2-Tag	B/I
4-Tag	B/M/E/S
6-Tag	B/B2/B3/M/E

Peng et al. [12] first raised the 2-Tag on CRF segmentation. Tag "B" is utilized to denote the beginning of a word. Tag "I" represents a non-head part of a word, e. g. the middle or the end of a word. Xue [4] developed a 4-Tag set in Maximum Entropy Model, which are the beginning character "B", end character "E", middle character "M", and a single character "S". Zhao [6] further added "B2" and "B3" (6-Tag) to conduct more accurate learning on segmentation based on CRFs.

2-Tag character tagging is the simplest tag set for Chinese word segmentation. An example of words handled by 2-Tag character tagging is described in Fig. 2. The result of word segmentation is illustrated with Fig. 3.

脸/B 上/I 没/B 有/I 什/B 么/I 表/B 情/I ，/B

Fig. 2. An example of 2-Tag character tagging

脸上(on the face)/I 没有(there is not)/I 什么(any)/I 表情(expression)/I ，
(comma)/B

Fig. 3. The result of Chinese word segmentation based on CRFs

Through the character tagging, we can learn the knowledge of word position from the training data, including probability and context of the character in a specific position. Take "没(not)" in Fig. 2 as an example, the tagging result is equivalent to obtaining the following six rules in Fig. 4.

1. If the previous character is "上(on)", the current character is marked as B;

2. If the current character is "没(not)", the current character is marked as B;

3. If the next character is "有(there is)", the current character is marked as B;

4. If the previous character is "上(on)" and the current character is "没(not)", the current character is marked as B;

5. If the current character is "没(not)" and the next character is "有(there is)", the current character is marked as B;

6. If the previous character is "上(on)" and the next character is "有(there is)", the current character is marked as B.

Fig. 4. Six rules for "没(not)" with character tagging

4   Preparation

4.1   Dataset

The dataset is from the Institute of Computational Language, Peking University. It incorporates 32,099 segmented sentences in Chinese character. The character encoding is utf-16 with little endian. We divide the dataset (Train_utf16.seg) into two parts: training data (70%) and test data (30%).

4.2   Preprocessing of Dataset

A toolkit of conditional random fields (CRFs) named CRF++ is exploited in this research. This toolkit provides a unified template to build conditional random field models on standardized data. The trained model can be used to deal with various problems, such as word segmentation, part-of-speech tagging, recognition of named entities, and so forth.

We process the dataset into the format required by CRF++. Both training data and test data are composed of multiple columns. An example of feature template for training data is presented in Fig. 5. The contents in the columns from left to right are as follows: the Chinese character, whether it is punctuation, the type of the character, and the tag label. Each row represents the information of a character. The columns are splitted by the space punctuation.

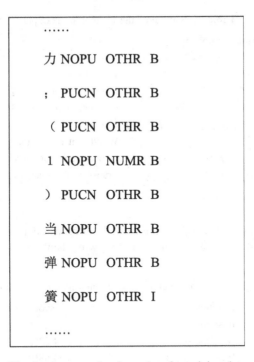

......

力 NOPU OTHR B

； PUCN OTHR B

（ PUCN OTHR B

1 NOPU NUMR B

） PUCN OTHR B

当 NOPU OTHR B

弹 NOPU OTHR B

簧 NOPU OTHR I

......

Fig. 5. An example of template for training data

4.3   Evaluation

An evaluation tool of CoNLL-2000 is used to measure the experimental results of segmentation. There measurements of precision, recall, and F score are calculated for training data and test data. F score is the harmonic mean of precision and recall.

5   Experiment and Analysis

5.1   Feature Optimization

We make a comparison of several groups and try to find the best features to perform the segmentation based on CRFs. The feature template $C_{-1}C_0C_1$ (the character window is 1) is selected as the baseline. The three characters contain previous character, current character, and next character. Besides the baseline, the following features in Table 2 are used for training.

Table 2. Sample of the features in example

No.	Feature templates	Contents
1	$C_nC_{n+1}, n = -2, -1, 0, 1$	Bigram
2	$C_{-1}C_1$	Jump Bigram
3	$C_{-2}C_2$	Character window is 2
4	$C_{-3}C_3$	Character window is 3
5	$C_nC_{n+1}C_{n+2}, n = -2, -1, 0$	Trigram
6	$Pu(C_0)$	Whether it is punctuation
7	$Ty(C_{-1})Ty(C_0)Ty(C_1)$	Cooccurrence of character type
8	$Ty(C_{-1})Ty(C_0)$	Transfer of character type

The experiments of segmentation are carried out on test data. Experimental results of evaluating the test data are shown in the following Table 3.

Table 3. Result of different feature combinations on test data

Features	F score
Baseline	87.53%
Baseline + 1	94.53%
Baseline + 1 + 2	94.58%
Baseline + 1 + 2 + 3	94.68%
Baseline + 1 + 2 + 3 + 4	94.60%
Baseline + 1 + 2 + 3 + 5	94.63%
Baseline + 1 + 2 + 3 + 6	94.72%
Baseline + 1 + 2 + 3 + 6 + 7	95.07%
Baseline + 1 + 2 + 3 + 6 + 7 + 8	95.16%

The F score increases as the character window size becomes larger. However, the F score declines when we expand the character window to 3 or introduce the Trigram features. This is because too much noise information is introduced in the feature set. Hence, we consider how to reduce the noise information in character and Bigram features. Tsai et al. [8] designed groups of n-gram features to remove the noise information and chose the best combination of n-gram features. Tsai's feature set are written in Table 4.

Table 4. Features designed by Tsai et al.

Feature templates	Contents
C_{-2}, C_{-1}, C_0, C_1	Unigram
$C_{-2}C_{-1}, C_{-1}C_0, C_0C_1$	Bigram
$C_{-3}C_{-1}, C_{-2}C_0, C_{-1}C_1$	Jump Bigram

Tsai's feature set is tested in this research. Moreover, we devise an improved feature set by employing Tsai's feature set, $Pu(C_0)$, and $Ty(C_n)$. We are able to obtain the following evaluation results of two feature sets in Table 5. As can be seen from the table, the results of Tsai's features and our proposed features are better than other features because the noise information is removed.

Table 5. Result of Tsai's feature set and an improved feature set on test data

Features	F score
Tsai	94.84%
Tsai + 6 + 7 + 8	95.21%

5.2 Parameter Optimization

The two most important parameters of CRF++ are the f and c. The f is the threshold value "cut-off". The number of iterations in training is not less than f value, which is particularly useful when applied to large-scale data. The c controls the extent of over-fitting. It inclines to be over-fitted if the c is larger. However, the training data and test data are selected from the same corpus, so a larger c may lead to a better result. The experimental results of parameter f and c are given in Tables 6 and 7.

Table 6. F score of parameter f when c is fixed

Parameter f	F score
f = 1	95.15%
f = 2	95.19%
f = 3	95.07%

Table 7. F score of parameter c when f is fixed

Parameter c	F score
c = 1	95.15%
c = 4	95.36%
c = 14	95.42%

5.3 Tag Set Optimization

The same CRF model is employed to evaluate the feature templates of Tasi (2-Tag), Xue (4-Tag), and Zhao (6-Tag) on four datasets. Results of three feature templates are listed in the following Table 8.

Table 8. Comparison of three feature templates

Kernel function	AS	CityU	CTB	MSRA
2-Tag	94.6%	96.2%	92.4%	94.6%
4-Tag	95.2%	96.7%	93%	95.5%
6-Tag	95.4%	96.9%	93.2%	96.1%

According to Table 8, 6-Tag feature template is better than others. However, the training time increases dramatically. Zhao et al. [6] spent $2 \sim 3$ times longer based on 6-Tag than 2-Tag in a small dataset. The time consumed in 6-Tag training on our dataset is more than 10 times of 2-Tag, thus further experiments did not continue.

5.4 Analysis

Experimental results show CRF-based segmentation is capable of recognizing the unregistered words. For instance, "李家梅 (Li Jiamei)" in the first line of the test data is a Chinese personal name. It can be effectively identified by the CRF model in segmentation. Huang et al. [13] discussed that the accuracy drop caused by out-of-vocabulary words is at least five times greater than that of segmentation ambiguities. Therefore, CRF based on character tagging is better than word-based approaches in Chinese segmentation.

As for the feature selection, the window size of sub-features and features of characters is generally around 2. Too large window sizes are likely to bring noise information. Also, the window size does not necessarily have to be symmetrical. Experiments in 5.1 demonstrate this conclusion.

In addition, the Jump Bigram features also affect the word segmentation accuracy to a certain extent. Consider the example of a Chinese word "(1)", "1" is supposed to be current character. The feature template $C_{-1}C_1$ represents two Chinese punctuation marks before and after "1", so the tag of C_0 must be "S" or "B".

For the multi-tags such as 4-Tag and 6-Tag, they can improve the results, but the complexity of calculation is high and it spends too much time to train the model.

6   Conclusions

This paper attempted to study the Chinese word segmentation based on conditional random fields. We explored the feature selection, model's parameter optimization, and different tag sets verification. At first, eight feature templates plus baseline feature were mixed to produce the best feature for CRF character tagging. Nine groups of features were tested by combining the different feature templates. We discovered that the feature "Baseline + 1 + 2 + 3 + 6 + 7 + 8" outperformed other groups of features. Furthermore, a feature set proposed by Tsai was exploited to reduce the noise information. We improved the feature selection by combining Tsai's feature set with three features (feature 6, 7, 8). Experimental results verified our proposed feature set. Besides the feature selection, parameter and tag set optimization are performed to refine the results.

Two parameters f and c are tested for the CRF++ tool. 4-Tag and 6-Tag were taken into account in order to improve the accuracy of segmentation. Results show that the method using multi-tags consumes too much running time and it is not applicable in this research.

In future, resolving the time limitation of 6-Tag feature is a research direction. High-level CRF (e.g. second order markov CRF) or FCRF can also be adopted for Chinese segmentation, even though it may increase the training time. Finally, additional dictionaries will also be added to enhance the effect of word segmentation.

Acknowledgement. This work was supported by the High-level Innovation and Entrepreneurship Talents Introduction Program of Jiangsu Province of China, 2019.

References

1. Chen, X., Qiu, X., Zhu, C., Huang, X.: Gated recursive neural network for Chinese word segmentation. In: Proceedings of the 53rd Annual Meeting of the Association for Computational Linguistics (ACL), pp. 1744–1753 (2015)
2. Yao, Y., Huang, Z.: Bi-directional LSTM recurrent neural network for Chinese word segmentation. In: Hirose, A., Ozawa, S., Doya, K., Ikeda, K., Lee, M., Liu, D. (eds.) ICONIP 2016. LNCS, vol. 9950, pp. 345–353. Springer, Cham (2016). https://doi.org/10.1007/978-3-319-46681-1_42
3. Wang, X., Li, C., Chen, J.: Chinese word segmentation method based on expanded convolutional neural network model. J. Chin. Inform. **33**(9), 154–161 (2019)
4. Xue, N.: Chinese word segmentation as character tagging. Comput. Linguist. Chin. Lang. **8**(1), 29–47 (2003)
5. Low, J., Ng, H., Guo, W.: A maximum entropy approach to Chinese word segmentation. In: Proceedings of the 4th SIGHAN Workshop on Chinese Language Processing, pp. 161–164 (2005)
6. Zhao, H., Huang, C., Li, M.: An improved Chinese word segmentation system with conditional random field. In: Proceedings of the 5th SIGHAN Workshop on Chinese Language Processing, pp. 162–165 (2006)
7. Mao, X., et al.: Chinese word segmentation and named entity recognition based on conditional random fields. In: Proceedings of the 6th SIGHAN Workshop on Chinese Language Processing (2008)
8. Tsai, R., et al.: On closed task of Chinese word segmentation: an improved CRF model coupled with character clustering and automatically generated template matching. In: Proceedings of the 5th SIGHAN Workshop on Chinese Language Processing, pp. 108–117 (2006)
9. Huang, W., et al.: Toward fast and accurate neural Chinese word segmentation with multi-criteria learning. In: Proceedings of the 28th International Conference on Computational Linguistics, pp. 2062–2072 (2020)
10. Deng, L., Luo, Z.: Domain adaptation of Chinese word segmentation on semi-supervised conditional random fields. J. Chin. Inform. **31**(4), 9–19 (2017)
11. Lafferty, J., Mccallum, A., Pereira, F.: Conditional random fields: probabilistic models for segmenting and labeling sequence data. In: Proceedings of the 18th International Conference on Machine Learning (ICML) (2002)

12. Peng, F., Feng, F., McCallum, A.: Chinese segmentation and new word detection using conditional random fields. In: Proceedings of the 20th International Conference on Computational Linguistics, COLING, pp. 562–568 (2004)
13. Huang, C., Zhao, H.: Chinese word segmentation: a decade review. J. Chin. Inform. Process.. **21**(3), 8–19 (2007)

Knowledge Discovery and Data Mining

Financial Distress Detection and Interpretation with Semi-supervised System

Xiaoqing Zhu[1], Fangfang Liu[1(✉)], and Zhihua Niu[1,2]

[1] School of Computer Engineering and Science, Shanghai University,
Shanghai 200444, China
{lilyzhu, ffliu, zhniu}@shu.edu.cn
[2] Key Laboratory of Applied Mathematics (Putian University),
Fujian Province University, Fujian Putian 351100, China

Abstract. Financial distress prediction is of great importance for companies to take timely measures. It is also useful for banks and investors to avoid potential risk. Mining potential distressful companies in this fast-changing business environment is crucial for timely intervention, but it is also very challenging. Because in real scenario, only few companies have been assessed to be financial distressful, while most companies remain unlabeled. Traditional supervised learning and unsupervised learning are no longer suitable under such circumstances. Since this problem can be viewed as an anomaly detection problem with partially observed anomalies, this paper proposes a semi-supervised learning framework adopting PU-learning method and unsupervised method combined with improved feature selection procedure. The proposed system makes full use of limited observed data and is robust to unknown novel anomalies at the same time. This system outperforms traditional supervised and unsupervised methods as well as some other semi-supervised methods. Meanwhile, the framework provides explanation for the detected anomalous companies which is useful for further analysis.

Keywords: PU learning · SHAP model interpretation · Feature selection · Anomaly detection

1 Introduction

Nowadays, companies are under unprecedented complicated market environment with continuous emerging risks. The 'black swan' incident like COVID-19 and financial crisis can have huge impact on many small and medium-sized enterprises [1]. It is crucial to detect financial distressful company to avoid great loss. The problem is, even with a lot of metric describing the companies' operating situation, it is still hard to distinguish the distressful company from the normal ones. Traditional machine learning methods make prediction based on historic bankruptcy companies [2]. This is reasonable while this method will fail when the unprecedented crisis occurs because the new anomalous pattern can be different from the past ones. How to deal with the

Sponsored by Key Laboratory of Applied Mathematics of Fujian Province University (Putian University) (NO.SX202102)

sudden events and predict accurately is important. The difficulty lies in two aspects: firstly, the label is not complete. Only a small part of the financial distressful companies has been observed or assessed when the crisis just arrives. Secondly, the data is high-dimensional, the data involves lots of financial metrics and non-financial metrics. On one hand, it helps to analyze the problem from various aspects, but on the other hand, it is hard to detect anomalous companies in such high-dimensional space.

Financial distress detection can be viewed as the anomaly detection problem where the financial distressful companies are the anomaly points to be detected. Anomaly detection is widely used to solve different types of problems including credit card fraud detection [3], web intrusion detection [4], mechanical fault detection, etc. Machine learning methods have been widely adopted in anomaly detection, the methods traditionally fall into two categories: supervised learning and unsupervised learning. Typically, unsupervised methods are considered when all the data are without labels. Isolation-based method [5], proximity-based method [6] are commonly used approaches. When labeled data is sufficient, various supervised classification models can be trained to solve the problem, such as SVM [7] and ensemble techniques [8].

However, in our scenario, the problem is either suitable for supervised or unsupervised learning, because anomalies have only been partially discovered in this dataset while we want to detect other anomalies hidden in the large amount of unlabeled data. When solving the company distress anomaly detection problem, it is impossible to detect all the distressful companies given limited prior information and fast changing market environment. We only have the ground truth of a small fraction of the positive (financial distressful) companies, while most data are not labeled. Among those unlabeled companies, some may run well, others may also face severe financial distress or even bankrupt soon. Due to the lack of labels, supervised methods cannot be directly employed. Meanwhile, unsupervised methods cannot fully utilize the labeled data and will perform badly in solving this high dimensional financial data.

Given the discussed financial distress detection problem scenario and the drawback of prevalent supervised and unsupervised learning methods, this paper proposes a semi-supervised system which combines a semi-supervised learning method called PU learning and an unsupervised learning method improved by SHAP feature selection to detect emerging financial distressful companies.

PU learning (positive unable learning) is an important branch of semi-supervised learning methods [9]. It is often used to detect anomalies when only positive labels are accessible which is consistent with our scenario. There are mainly two types of PU learning methods, one stage and two stage strategy respectively. For one-stage methods, all unlabeled data are regarded as negative instances in training and biased classifier is used. Two-stage strategy is more often applied, it consists of two steps: Firstly, the reliable normal samples are selected, and then a classifier is trained with the pseudo labels generated in the first step [10]. PU learning is suitable for this financial distress detection scenario because it makes full use of the observed anomalies to detect other anomalies with similar patterns.

When novel anomaly patterns are quite different from the observed ones, the distribution of the observed positive samples is shifted from the real distribution, so bias is inevitable. Under such circumstance, PU learning methods may fail [10]. Unsupervised methods are useful when we have no prior knowledge of the novel anomalies' data distribution and can be applied to solve this problem. However, if the financial data has too many features, samples become sparse in high dimensional space and this will cause the ineffectiveness of unsupervised methods [11]. This paper proposes an improved unsupervised method which can alleviate this problem by adopting effective feature selection with the help of model interpretation algorithm SHAP [12]. This procedure fully utilizes the knowledge learned from the semi-supervised model and can greatly improve the performance of the unsupervised learning model. With unsupervised method, the framework is more robust for those unprecedented emerging anomalies.

Combining PU learning and unsupervised learning methods together, the proposed system achieves better result than supervised methods and unsupervised methods alone. It also outperforms some other PU learning methods like ADOA [13] and PU-bagging [14]. The following parts will be organized as follow. In Sect. 2, the proposed financial distress detection framework is discussed in detail. In Sect. 3, the result of this framework is analyzed. Finally, Sect. 4 reaches the conclusions.

2 The System Architecture

In this section, the architecture of the developed framework is presented. As showed in Fig. 1, the system consists of four parts. 1, The semi-supervised module applies PU learning method to train classifiers and uses bagging method to make the prediction more robust. 2, The model explanation module calculates the contribution of all the features. This module not only explains the anomalies detected by the PU method but also provides the reliable feature selection for the unsupervised module which can greatly improve the unsupervised model's prediction. 3, The unsupervised module uses the improved isolation forest with feature selection. 4, The semi-supervised module and unsupervised model are blended to get the final prediction.

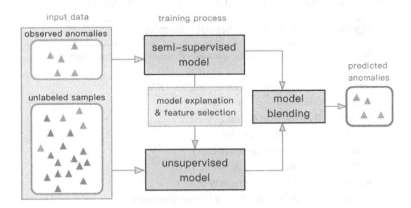

Fig. 1. Overview of the framework

2.1 The Semi-supervised PU Module

Since no negative labels is accessible in our scenario, supervised learning methods cannot be directly deployed to this problem. Thus, the PU learning method is adopted to solve the problem of lacking reliable negative samples. Two-stage strategy is one of the widely used PU learning method, it selects some reliable negative samples from the unlabeled data, and then the problem is transformed into the traditional supervised classification problem.

Different from the traditional classification problem, the data of anomaly detection problem is often imbalanced because positive samples only accounts for a small part of the whole dataset, and this will cause the weak performance of the classifier in detecting the positive samples in the second stage. Some methods have been proposed to alleviate this problem, like oversampling of the minority class. However, the oversampling methods doesn't work well when the observed positive sample is too little in quantity, the observed positive samples cannot truly represent the real positive samples' distribution, so the oversampling methods will involve bias to the distribution of the positive class. Another commonly used method is cost-sensitive learning. In anomaly detection scenario, lowering the cost of FN(False Negative) is more important than lowering the cost of FP(False Positive). Under this circumstance, the cost sensitive objective function is applied to train the classifier with different misclassification cost. Minimizing balanced cross entropy is a common method for addressing such class imbalanced problem. In 2018, focal loss was proposed to further prevent the easily classified negative samples dominate the gradient [15]. With the help of addressing positive/negative, hard/easy sample with corresponding loss, the classifier will lay more emphasis on the hard-negative samples during the training process to achieve better result in imbalanced problem. Therefore, in the second stage of PU learning module, lightGBM classifier with focal loss objective function was adopted.

First Stage: Select reliable negative samples from unlabeled instances using 'Spy' technique. Part of the positive samples are randomly chosen as spies and are added to the unlabeled set and were pretended to be 'negative samples. The reliable negative samples are selected using the threshold score of 'spy' samples [16].

Algorithm1: select reliable negative samples

Input: observed positive samples set- P, unlabeled samples set - U, sample ratio – s%
Output: a set of reliable negative samples - RN
1: $RN \leftarrow \emptyset$
2: sample s% from P to get the spy set S
3: set $P \leftarrow P - S$, label as positive (1)
4: set $N \leftarrow U \cup S$, label as negative(0)
5: Train a classifier with P, N.
6: Get the class predict probability of N.
7: Select anomaly threshold $t = \min(\text{probability}(S))$.
8: $RN = RN \cup (\text{probability}(N) < t)$
9: return RN

Second Stage: A lightGBM model is trained using the observed positive samples as well as the reliable negative samples selected in the first stage. α-balanced Focal loss was implemented as the model's objective function to solve class imbalance. The objective function is as follow.

$$FL(p, y) = -\alpha y(1-p)^{\gamma} log(p) - (1-\alpha)(1-y)p^{\gamma} log(1-p) \tag{1}$$

α and $1 - \alpha$ denote the misclassification penalty of real Positive samples and negative samples respectively. α is set larger than 0.5 which means that misclassifying positive samples will cause more loss than misclassifying negative samples. γ is the parameter to penalize the hard samples. When γ is set to 0, the objective function degenerates to normal α-balanced cross entropy loss. The larger γ is, the hard samples' misclassification account for more loss. In our case, since we regard the selected reliable sample as true negative. If the selection in step one doesn't perform well and involves some positive samples, the wrongly selected samples can turn into 'hard' samples in the second stage. Thus, we can adjust γ to a small number in this case. If the first stage selection is quite reliable, we can set γ to a relative larger number to put more emphasis on hard samples which can lead to a clear classification boundary.

The semi-supervised PU methods involves some randomness in the first stage because the 'spy' samples are randomly selected from the positive samples. To get a stable prediction from the model, average bagging is used to ensemble the classifiers which can lower the variance of the prediction.

2.2 Model Interpretation and Feature Selection

Anomaly detection problems require model interpretability to help users understand the black box model and gain more confidence towards the model's prediction. At the same time, understanding the model helps us get more insights about the data and makes feature selection easier. Recent years, many researches have been made to realize the model explanation. Many methods like information gain, split count and permutation can solve the global feature attribution and individual feature attribution problem for tree-based models. However, their reliability has not been strictly proved. SHAP [12] method has attracted much attention since it was proposed because It utilizes the Shapley value estimation which has been proved theoretically to be the only possible consistent and locally accurate approach for feature attribution. As a strict additive feature attribution method, it satisfies three desirable properties: 1, local accuracy property: It means that the sum of attribute value of a sample equals to the model's prediction. 2, missingness property: The missing feature won't be attributed any importance. 3, consistency property: feature importance assigned by the method is consistent. When the feature has more impact on the model's prediction, the importance attribution will never decrease. Even these are very basic requirement for feature importance attribution, many widely used methods fail to satisfy these three properties [12].

SHAP method combines the conditional expectation with traditional Shapley value comes from game theory to evaluate feature importance when making prediction using the black box model. As shown in the following Eq. 2, the specific feature i's contribution to the prediction is calculated by computing the weighted average of the marginal contribution of every single feature under all feature combinations. In Eq. 2,

M denotes the full feature space. S represents the feature subset selected from M and the specific feature i is removed from set S. The effect of feature i can be evaluate through this permutation $f_x(S \cup i) - f_x(S)$.

Besides, to overcome the computation complexity of this algorithm, tree SHAP optimize the calculation within $O(TL2^M)$ time [12].

$$\phi_i = \sum_{S \subseteq M - \{i\}} \frac{|S|!(|M| - |S| - 1)!}{|M|!} [f_x(S \cup i) - f_x(S)] \tag{2}$$

Since the prediction of semi-supervised model is after a bagging ensemble, SHAP values need to be calculated for each base classifier, and then the averaged SHAP value is used as feature importance score.

By computing the SHAP value of every feature, we can learn more about the model prediction and find some important aspects for further analysis. Meanwhile, the ranking of feature importance also serves as a feature selection method well.

The financial distress problem always involves a lot of metrics to reflect the company's operation. Among such many features (83 features in our case), it is hard to analyze them directly. Traditional feature extraction method including dimension reduction techniques like PCA can transform the high dimensional dataset into a smaller subspace, however after such transformation, the meaning of features also change, and this may lead to the difficulty of explaining the outcome. In financial distress problem, the metrics have specific meaning, so it is better to use feature selection instead of dimension reduction. The lightGBM classifier automatically select useful features during the training process which saves us the labor for feature selection, but as for the unsupervised model, it is crucial to choose the top-K features given by the SHAP value. According to the experiment given in Sect. 3, this feature selection method can largely improve the result of isolation forest which is the most used unsupervised anomaly detection algorithm.

2.3 The Unsupervised Module

Many unsupervised anomaly detection methods are distance-based, while these methods will be unreliable in high dimensional space due to the curse of dimensionality. The noise of different dimensions will interfere the detection of real anomalies [17]. Selecting the proper subspace is important for the outlier detection task. Isolation forest is also closely related to subspace outlier detection because the data is recursively partitioned with axis-parallel cuts at random partition points in randomly selected attributes. The different branches correspond to different local subspace regions of the data. The smaller paths refer to lower dimensionality of the subspace for the sample to be isolated. The random isolation tree construction approach is repeated many times and the tree path length is averaged as the final anomaly score of certain samples. Outliers can be isolated in much lower-dimensional subspaces than normal samples. The algorithm of isolation forest is briefly displayed in Algorithm 2.

However, when there are too many features, some useless feature combination will interfere the performance of isolation forest. To solve this problem, some methods have been proposed to pre-select the features for isolation forest including multidimensional kurtosis feature selection and Forward Selection Independent Variables (FSIV) algorithm [11]. The previous methods are based on unsupervised scenario, we can indeed achieve better result with more prior knowledge in our semi-supervised scenario by simply using the top features selected by SHAP values. This procedure serves as the global subspace selection, at the same time Isolation forest is still able to explore different local subspaces with its random split approach. The result after feature selection is much better which is shown in Sect. 3.

Algorithm2: Isolation Forest

Input: data - X, number of iTrees - t, ψ – subsampling size
Output: a set of iTrees - *Forest*
1: *Forest* ← ∅
2: **for** 1 to t **do:**
3:     X' ← sample(X, ψ)
4:     **construct *iTree(X')* iteratively**
        until leaf-node or reaches max height:
5:         Randomly select a feature x_i
6:         Randomly select a split $p \in (\min(x_i), \max(x_i))$
7:         X_l ← $filter(X', xi < p)$
8:         X_r ← $filter(X', xi > p)$
9:     *Forest* ← *Forest* ∪ *iTree(X')*
10: **end for**
11: **return** *Forest*

---

3   Experiment and Evaluation

3.1   Data and Setup

The real-world data used in the experiment is company distress evaluation data provided by Ebrahimi [18]. This data consists 83 features including financial and non-financial characteristics of the sampled companies. To adjust the data to our scenario, the data was divided into two parts: positive samples are randomly selected from the full set of recorded bankrupt companies, unlabeled samples are the rest companies in the dataset. A validation set is used for the selection of hyperparameters. The labeled data is rare, and the class is imbalanced. In our experiment setting, the observed positive samples (bankrupt companies) account for 30% of total anomalies. The validation set with ground truth knowledge of both positive samples and negative samples is very small which is in accordance with real world scenario where we have little knowledge of the reliable negative label. Thus, 0.05% data is selected as validation set. With the growing size of validation set in real scenario, the proposed can achieve better result.

3.2   Result and Analysis

As is described in Sect. 2, final prediction is the blending result of a semi-supervised classifier and an unsupervised isolation forest classifier improved with SHAP feature selection. The result will be analyzed separately in our experiment. As for the semi-supervised part, the experiment was conducted to compare the performance of the proposed method with supervised method. As for the unsupervised part, the performance is compared between the isolation forest with the proposed SHAP feature selection and the original isolation forest.

The distribution of classes is imbalanced where most samples belong to the negative class (136 firm-year observations are financially distressed while 3546 are not distressful in our case). To better reflect the effectiveness of classifier, both AUC and f1-score are adopted to evaluate the performance. The AUC value is equivalent to the probability that a randomly chosen positive example is ranked higher than a randomly chosen negative example [19]. Instead of averaged by both normal and positive class, the f1-score applied here is only calculated for the positive class to clearly investigate model performance in detecting potential financial distressful companies. Higher AUC and f1-score indicates better model performance.

Table 1 shows the comparison between proposed semi-supervised classifier and supervised method: the semi-supervised classifier was introduced in Sect. 2. For fair comparison, the supervised method is a LightGBM classifier which uses the same validation set and regards all the unsupervised samples as negative; oversampling was also applied in the supervised method to avoid the imbalance problem.

Table 1. Comparison between proposed semi-supervised classifier and supervised method.

Experiment result	Proposed semi-supervised	Supervised method
AUC	**0.833357**	0.517241
F1-score	**0.370417**	0.066667

As we can see from the result, the proposed method performs significantly better than supervised method. The result indicates that it is not suitable to simply regards all the unlabeled samples as negative because the true positive samples will contaminate the real negative class and the noise information will seriously deteriorate the performance. The experiment demonstrates the effectiveness of the semi-supervised module in this anomaly detection system.

To help further analysis, most informative features given by SHAP explanation are shown in Fig. 2. This is the summary plot given by SHAP method, it interprets the learned model by assigning feature importance. In the ensembled model, feature 35 takes the most important role in the model's prediction. Using the small evaluation set, top 6 features are selected for the unsupervised module.

Table 2 shows the comparison between the isolation forest combined with the proposed SHAP feature selection method and the traditional isolation forest without feature selection. The isolation forest method is greatly improved by the feature selection process. AUC was improved about 13% and F1 score was improved 15%.

The result shows that it is useful to combine the prior knowledge extracted from the semi-supervised training with the unsupervised model by applying informative feature selection, especially when the data is high dimensional.

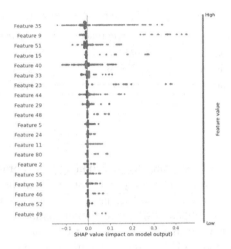

Fig. 2. SHAP model interpretation

Table 2. Comparison between improved isolation forest and traditional isolation forest

Experiment result	Improved isolation forest	Traditional isolation forest
AUC	**0.631516**	0.507400
F1-score	**0.192043**	0.041026

The final model output is result after blending the semi-supervised and unsupervised classification model together. Table 3 compares the blending outcome of the proposed system with two PU learning based methods ADOA [13] and PU bagging [14]. The same validation set is also utilized to select best hyperparameters for each method.

Table 3. Comparison between proposed method and other PU learning methods

Experiment result	Blending outcome	ADOA	PU bagging
AUC	0.836597	0.512868	**0.861213**
F1-score	**0.403846**	0.080477	0.267327

AUC of the blending result reaches 0.836597 and the F1-score is 0.403846. The result after blending ensemble is better than using the semi-supervised module and unsupervised module alone. In addition, the proposed method is much better than ADOA. Even though AUC is less than that of PU bagging, the F1-score is much higher. The above experiments validate the effectiveness of the proposed structure.

To further analyze the difference between these two methods, precision and recall are studied to explain the f1-score. As displayed in Table 4, the proposed method achieves higher F1 score by achieving a better balance between precision and recall.

Table 4. Outcome analysis proposed method and PU bagging

Experiment result	Blending outcome	PU bagging
F1-score	**0.403846**	0.267327
Precision	**0.280000**	0.158824
Recall	0.724138	**0.843750**

4 Conclusion

In this work, a financial distress company detection system is proposed based on both semi-supervised and unsupervised anomaly detection method. The paper focuses on the problem we often encounter in real world scenario that we have already identified some of the financial distressful companies while we are not confident about whether the rest companies are not financial distressful in terms of operation situation. The problem setting can be formulated as a PU learning problem where we are only accessible with positive samples. Therefore, this paper adopts the 2-stage PU learning method combined with cost-sensitive loss function. Meanwhile, an enhanced unsupervised module is also adopted to detect unprecedented novel anomalies with feature selection procedure by using SHAP model explanation.

The proposed system consists of three parts. Firstly, the semi-supervised module is a 2-step PU learning method including reliable normal sample selection and a lightGBM classifier with the alpha balanced focal loss objective function. Secondly, the model explanation and feature selection module adopt the novel black-box model interpretation method SHAP to better understand the prediction of the semi-supervised model. At the same time, the calculated SHAP values also serve as the useful metric in selecting informative feature for the unsupervised module. Lastly, improved isolation forest with SHAP feature selection is used as the unsupervised module which greatly alleviate the dimensional disaster.

Experiments result proves the effectiveness of the proposed system. The system outperforms both supervised and unsupervised anomaly detection methods. It also has higher F1 score compared with other PU learning methods. Further improvement can be made by implementing more semi-supervised and unsupervised methods into the structure to get more robust results. In addition to finance distressful company detection, the system can be employed to various anomaly detection tasks including network intrusion and fraud detection.

References

1. Tatiana, D., Federico, H., Mauricio, L., Sergio, L.: Financing firms in hibernation during the COVID-19 pandemic. J. Finan. Stabil. **53** (2021)
2. Geng, R., Bose, I., Chen, X.: Prediction of financial distress: an empirical study of listed Chinese companies using data mining. In: European Journal of Operational Research, pp. 236–247. Elsevier, Netherlands (2015)
3. Kou, Y., Lu, C.T., Sirwongwattana, S., Huang, Y.P.: Survey of fraud detection techniques. In: IEEE International Conference on Networking, Sensing and Control, 2004, vol. 2, pp. 749–754. IEEE (2004)
4. Eskin, E., Arnold, A., Prerau, M., Portnoy, L., Stolfo, S.: A geometric framework for unsupervised anomaly detection. In: Barbará, D., Jajodia, S. (eds.) Applications of Data Mining in Computer Security. Advances in Information Security, vol. 6, pp. 77–101. Springer, Boston, MA (2002). https://doi.org/10.1007/978-1-4615-0953-0_4
5. Liu, F.T., Ting, K.M., Zhou, Z.H.: Isolation forest. In: Proceeding of the 8th IEEE International Conference on Data Mining, ICDM, pp. 413–422. Institute of Electrical and Electronics Engineers Inc., Pisa, Italy (2008)
6. Markus, M.B., Kriegel, H.P., Raymond, T.N., Sander, J.: LOF: identifying density-based local outliers. ACM. Sigmod. Record. **29**, 93–104 (2000)
7. Juszczak, P., Duin, R.P.: Uncertainty sampling methods for one-class classifiers. In: Proceedings of ICML 2003, Workshop on Learning with Imbalanced Data Sets II, pp. 81–88. AAAI, Washington (2003)
8. Liu, X., Wu, J., Zhou, Z.H.: Exploratory undersampling for class-imbalance learning. IEEE. Trans. Syst. Man. Cybern. Part B **39**(2), 539–550 (2009)
9. Elkan, C., Noto, K.: Learning classifiers from only positive and unlabeled data. In: Proceedings of the 14th ACM SIGKDD International Conference on Knowledge Discovery and Data Mining, pp. 213–220. Association for Computing Machinery, Nevada USA (2008)
10. Aggarwal, C.: An Introduction to Outlier Analysis. Presented at the (2017). https://doi.org/10.1007/978-3-319-47578-3_1
11. Puggini, L., McLoone, S.: An enhanced variable selection and Isolation Forest based methodology for anomaly detection with OES data. Eng. App. Artif. Intell. **67**, 126–135 (2018)
12. Lundberg, S., Lee, S.I.: A unified approach to interpreting model predictions. In: 31st Annual Conference on Neural Information Processing Systems, NIPS 2017, pp. 4766–4775. Neural information processing systems foundation, CA, USA (2017)
13. Zhang, Y.L., Li, L., Zhou, J., Li, X., Zhou, Z.H.: Anomaly detection with partially observed anomalies. In: 27th International World Wide Web, pp. 639–646. Association for Computing Machinery, Lyon (2018)
14. Mordelet, F., Vert, J. P.: A bagging SVM to learn from positive and unlabeled examples. Pattern Recognition Letters, pp.201–209. Elsevier, Netherlands (2014)
15. Lin, T.Y., Goyal, P., Girshick, R., He, K., Dollár, P.: Focal loss for dense object detection. In: Proceedings of the IEEE International Conference on Computer Vision, pp. 2980–2988. Elsevier, England (2017)
16. Liu, B., Lee, W. S., Yu, P.S., Li, X.: Partially supervised classification of text documents. In: ICML, pp. 387–394. Morgan Kaufmann, Sydney (2002)
17. Aggarwal, C.: High-Dimensional Outlier Detection: The Subspace Method. Presented at the (2017). https://doi.org/10.1007/978-3-319-47578-3_5
18. Financial distress detection dataset. https://www.kaggle.com/shebrahimi/financial-distress. Accessed 26 Apr 2021
19. Fawcett, T.: An introduction to ROC analysis. In: Pattern Recognition Letter, pp. 861–874. Elsevier, Netherlands (2006)

Solving Online Food Delivery Problem via an Effective Hybrid Algorithm with Intelligent Batching Strategy

Xing Wang[1(✉)], Ling Wang[1], Shengyao Wang[2], Yang Yu[2],
Jing-fang Chen[1], and Jie Zheng[1]

[1] Department of Automation, Tsinghua University, Beijing, China
{wang-x17, j-zheng18}@mails.tsinghua.edu.cn,
wangling@mail.tsinghua.edu.cn
[2] Meituan, Beijing, China
{wangshengyao, yuyang47}@meituan.com

Abstract. Online food ordering and delivery has been gaining increasing attention owing to the rapid development of internet and O2O (online to offline) business. In this paper, we address the emerging online food delivery problem (OFDP) where n orders are considered to be served by m drivers. We decompose the OFDP into order assignment and route planning and design an effective hybrid algorithm to solve the problem. For order assignment, a modified Kuhn-Munkres algorithm is developed to find the best matching between orders and drivers. Besides, in order to avoid inappropriate matching, we build a machine learning classification model with eXtreme Gradient Boosting (XGBoost) to describe the similarity between different orders and predict the order batching results. In addition, a fast rule-based route planning method is adopted to generate feasible routes for drivers. Through conducting experiments on real datasets from Meituan, we validate the satisfying performance of the classification model and demonstrate the effectiveness of the proposed algorithm for solving the OFDP.

Keywords: Online food delivery · Hybrid algorithm · Order assignment · XGBoost · Order batching

1 Introduction

Online food ordering and delivery, as an essential component of O2O (online to offline) business, is becoming more and more prevalent in modern daily life. A statistical report from Meituan, which is the biggest food delivery company in China, reveals that the maximum amount of orders in a single day has reached over 40 million [1]. Worldwide companies such as Grubhub, Uber Eats, Yelp and Just Eat are also developing a tremendous scale of transactions with the valuation of 94 billion dollars [2]. With such a broad market and bright prospect, online food ordering and delivery is destined to attract the investors and enterprises, as well as the researchers from fields of scheduling and optimization.

© Springer Nature Switzerland AG 2021
D.-S. Huang et al. (Eds.): ICIC 2021, LNCS 12837, pp. 340–354, 2021.
https://doi.org/10.1007/978-3-030-84529-2_29

Although with great potential, it is not easy to successfully run an online food delivery platform. Several stakeholders need to be considered, i.e., the customer, the restaurant, and the driver. Customers want their food to be fresh and delivered punctually. Restaurants need to guarantee the quality of food to avoid the negative comments from customers, which will cause bad influence to their reputation. Drivers hope to finish more food delivery tasks with shorter route length. Figure 1 illustrates the relationship between them in a typical food ordering and delivery process. The platform collects orders from customers and pushes them to the restaurants and drivers. Then the restaurants start to prepare the food and drivers shuttle between the restaurants and the customers to finish the pickup and delivery tasks.

To well organize the stakeholders and make reasonable arrangements, food delivery platforms need to efficiently handle the online food delivery problem (OFDP), where two optimization problems are involved. One is the order assignment problem, which assigns proper orders to drivers. The other is the route planning problem, which decides feasible routes for the drivers. Strictly speaking, these two problems are coupled intensely since the evaluation of the assignment scheme involves the route planning results while the route planning requires the information of the assigned orders. Nevertheless, solving the order assignment and route planning problem simultaneously is difficult and time-consuming, which is not acceptable for an online food delivery platform, since the platform receives hundreds of orders and matches them with thousands of drivers under limited computational resources. Therefore, in this paper, we decompose the OFDP and develop an effective hybrid algorithm (HA) to solve the problem. Specifically, we present a modified Kuhn-Munkres (KM) algorithm with intelligent order batching strategy to match order batches with drivers and utilize a rule-based route planning method to generate high-quality routes for drivers. The contributions of the paper are summarized as follows.

- The OFDP is decomposed into order assignment problem and the route planning problem. Efficient algorithms are designed to construct solutions for each problem.
- The order assignment is represented as a weighted bipartite graph and the classic KM algorithm is modified to quickly find matchings between orders and drivers.
- An order batching strategy is developed to determine whether orders should be assigned to the same driver, based on the classification model that is established through supervised learning.
- An effective rule-based route planning method is adopted to quickly generate feasible routes for drivers.
- Extensive experiments are conducted on real datasets to validate the effectiveness of the proposed algorithm.

Accordingly, the rest of the paper is organized as follows. Section 2 reviews the related work and Sect. 3 describes the OFDP in detail. The proposed algorithm is elaborated in Sect. 4 and the corresponding computational results are presented and discussed in Sect. 5. Finally, Sect. 6 completes the paper with conclusions and some future work ideas.

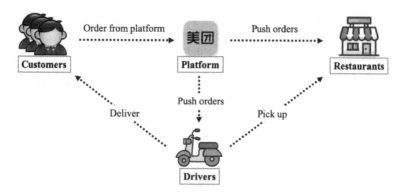

Fig. 1. A typical food ordering and delivery process

2 Literature Review

The OFDP is thought to be very similar to the well-known pickup and delivery problem with time windows (PDPTW), where a set of m identical vehicles are available to serve the transportation requests from n customers. Algorithms for PDPTW can be generally divided into three categories, i.e., exact algorithms, heuristics and meta-heuristics. Exact algorithms such as branch-and-bound [3], branch-and-cut-and-price [4] and column generation [5], often require large amounts of time to find the optimal solution, due to which it is only suitable for small-scale problem instances. Heuristics run faster than exact algorithms but cannot guarantee the optimality of the generated solutions. Lim et al. [6] presented an "squeaky wheel" optimization algorithm. Lu and Dessouky [7] developed an insertion-based heuristic for the PDPTW. Meta-heuristics can usually find satisfying solutions for large-scale instances but they also need plenty of time to converge. Typical meta-heuristics for PDPTW include tabu search [8], genetic algorithm [9], simulated annealing [10], adaptive large neighborhood search algorithm [11] and so on.

Though sharing many similarities with PDPTW, OFDP has some exclusive characteristics. First, there is no depot in OFDP, which means that drivers do not gather together at a fixed location but scatter over the delivery region. This indicates that the route of each driver does not have to be close looped—they stay where they are after finishing the last delivery task. Second, the drivers may have unaccomplished delivery tasks when new assignments and routes are considered. Last but the most important, the number of orders and drivers is large but the available computational time is extremely limited. Recently, there are some researches discussing the food delivery business [12–14], but few of them consider the aforementioned properties simultaneously. Therefore, effective algorithms need to be developed to better facilitate the online food delivery sector.

3 Problem Description

In OFDP, there are n orders to be served by m drivers. Let $N = \{1, 2, \ldots, n\}$ be the set of orders and $M = \{1, 2, \ldots, m\}$ be the set of drivers. Each order $i \in N$ specifies a pickup location l_i and a corresponding delivery location l_{i+n}. The demand of order $i \in N$ is denoted as q_i, which needs to be transferred from the pickup location to the delivery location. Besides, a time window (EPT_i, ETA_i) is specified for each order i, where EPT_i functions as the earliest visiting time of the pickup location l_i while ETA_i functions as the latest visiting time of the corresponding delivery location l_{i+n}. This time window is soft because the pickup location can be visited before EPT_i but extra waiting time will occur if doing so, which will impede the improvement of efficiency. Similarly, if the delivery location is visited after ETA_i, there will be a positive time delay which will frustrate our customers. For every driver $j \in M$, there are possibly some uncompleted orders previously assigned to them, which is denoted as the uncompleted order set U_j, where U_j is not necessarily to be non-empty. Since the drivers may be in the middle of executing previous delivery tasks when assigning new orders, starting points of drivers are generally not the same, either. The trunk capacity of each driver is identically set as Q, which limits the maximum demands the driver can load.

To solve the OFDP problem, we need to find the best assignment between orders and drivers and plan a satisfactory route for each driver, such that the following objective function, i.e., total cost (TC), is minimized,

$$\min TC = \sum_{i \in N} \max(T_i - ETA_i, 0) + \sum_{j \in M} d_j \tag{1}$$

where T_i is the estimated delivery time of order i and $\max(T_i - ETA_i, 0)$ is the time delay. d_j is the route length of driver j. The objective function measures the satisfaction of customers (time delay) and the workload of drivers (route length) simultaneously.

In order to construct a feasible solution for OFDP, there are several constraints that need to be respected. First, each order has to be served by exactly one driver and each location is allowed to be visited exactly once. Second, the pickup location has to be visited before the corresponding delivery location. Third, it is better that each location is visited within the soft time window so that severe time delay is avoided. Finally, the total load of each driver cannot exceed Q at any time.

4 Methodology

The OFDP cannot be solved optimally in polynomial time due to its NP-hard property. Besides, the scale of problem is usually large but very little computational time is available, as mentioned in Sect. 1. Under this situation, generating a satisfactory solution quickly while consuming as little computational resource as possible is of great significance in terms of application. Therefore, we decompose the OFDP and handle the order assignment and route planning in a hierarchical way.

Specifically, order assignment is represented as a weighted bipartite graph and a modified KM algorithm is adopted to find the maximum weighted perfect matching of the graph. What's more, since the assignment problem studied in this paper is special that different orders are allowed to be assigned to the same driver, it is necessary to batch these orders in advance so that they will not be separated and assigned to different drivers, which may cause the damage to the quality of solutions. Figure 2 gives an example where 3 orders are considered and order 1 and 2 are geographically close. It is reasonable and also efficient to assign both of them to driver 1. Without the procedure of order batching, they will be assigned to different drivers by the KM algorithm, which can possibly increase the total route length of all drivers. To precisely batch orders together, a machine learning classification model is established to evaluate the similarity of orders and help make the final order batching decision due to the. As for route planning, an effective rule-based route planning method is adopted to efficiently generate a high-quality route for each driver. The overall framework of the proposed HA is presented in Fig. 3.

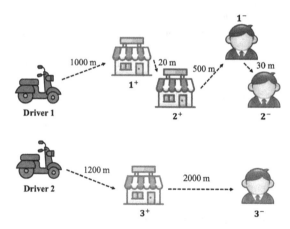

Fig. 2. An example of order batching

4.1   Order Assignment with Modified KM Algorithm

The KM algorithm (also known as the Hungarian algorithm) [15, 16] is famous for solving a $z \times z$ linear assignment problem in polynomial time ($O(z^3)$). The key idea of the KM Algorithm is to start from the best objective function and gradually compromise in subsequent steps so as to find a feasible solution. Due to its simplicity and effectiveness for bipartite graph matching, we adopt the KM algorithm to efficiently

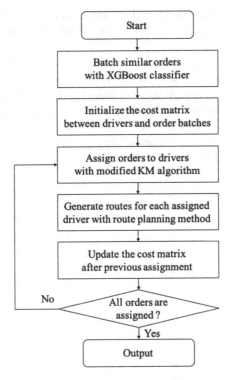

Fig. 3. The overall framework of HA

generate the matching between orders and drivers. The main steps of the classic KM algorithm [16] are briefly described as follows.

Step 1: Given the cost function matrix of the linear assignment problem, find the smallest element of each row and subtract it from each element in that row.

Step 2: Similarly, for each column, find the smallest element and subtract it from each element in that column.

Step 3: Cover all zeros in the resulting matrix using a minimum number of horizontal and vertical lines. If a total of z lines is required, the zeros in the resulting matrix describe a complete set of unique assignments. The algorithm stops in this case. Otherwise, continue with Step 4.

Step 4: Find the smallest element, denoted as λ, that is not covered by any line in Step 3. Subtract λ from all the uncovered elements, and add λ to all the elements that are covered twice. Return to Step 3.

Step 5: Each zero in the matrix indicates a possible component of the whole assignment. Find the most suitable zeros that do not violate the problem constraints and calculate the final objective function, which is the sum of the costs whose locations are the same as the locations of selected zeros.

Although the order assignment problem in OFDP is similar to the linear assignment problem, there are still some differences between them.

- First, the number of orders is generally unequal to the number of drivers in OFDP while in the linear assignment problem the number of agents and tasks holds the same.
- Second, the bipartite graph of the order assignment problem in OFDP is not complete, which means that not every driver can serve every order. This is because considering the convenience of drivers, some faraway and inappropriate orders are filtered by the platform when constructing possible matchings.
- Lastly, in OFDP, conflicts may occur between orders when the number of orders is larger than the number of candidate drivers, i.e., many orders "scramble" for few drivers, as shown in Fig. 4.

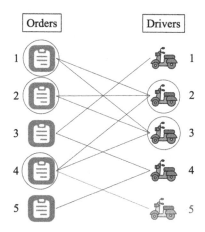

Fig. 4. Conflicts between orders

Under such circumstance, the matching process of KM algorithm will be interrupted, resulting an infeasible solution. To cope with the differences mentioned above, the KM algorithm needs to be modified accordingly. For the first difference, considering $k = \min(n, m), l = \max(n, m), (l - k)$ virtual lines with the same element φ are added to the initial $n \times m$ cost function matrix, resulting a $l \times l$ square matrix. It is obvious that φ can be set as an extremely large number so that no zero will appear in the virtual lines before a feasible solution is generated. As a matter of fact, it is not necessary to generate the $l \times l$ square matrix [17]. Secondly, to deal with an incomplete bipartite graph, the matrix is modified so that the edges between filtered drivers and orders are assigned with an infinite cost. In this way, the selection of these edges is prevented. Finally, to settle the conflicts between orders, dummy drivers are added when the conflicts occur. As shown in Fig. 4, a dummy driver 5 is added so that order 1, 2 and 4 are correctly assigned. With this modification, successful matching in each iteration is guaranteed. To acquire a feasible solution, another matching iteration is required to select real drivers for those orders who select dummy drivers.

4.2   Order Batching with Binary Classification Model

To successfully batch similar orders, a classification problem is solved by establishing the classification model in the way of supervised learning. Generally speaking, data used in supervised learning usually consist of features (also known as attributes) and label (also known as the ground truth), which can be expressed as (x, y), where $x = [x_1, x_2, \ldots, x_p]$ represents the vector of p attributes and $y \in \{0, 1\}$ is the corresponding binary label. For data in the training set, the ground truth is available while for the test dataset, it remains unknown. The training process of the classification model is to expose the model to data in the training set and adjust the parameters of the model accordingly so that the output of the model is consistent with the ground truth. Through repetitive exposure, a mapping function is formed to describe the correlation between the input and output variables. Consequently, once trained, the classification model can classify the data whose label is not known. Due to the effectiveness of the supervised learning in prediction and classification, in this study, we use large amounts of real data and form a binary classification model to predict whether two different orders should be batched together based on different types of features.

Features

The total number of features used in constructing the classification model equals 34, and they can be broadly divided into three categories. The first type is time-based features, mainly including the difference and average between different types of time of orders, such as the creation time, the latest delivery time, the preparation time, etc. Second, distance-based features are considered by calculating the distances between different locations of orders. This type of features affects the similarity of two orders largely since it determines the length of possible routes of the driver. In addition to aforementioned features, other 15 features related to the basic information of orders are also involved to depict the portrait of data more comprehensively, such as whether two orders come from the same restaurant, the workload of the recipient area, the sum of the order price, and so on.

Labelling

To acquire high-quality labels for the datasets, we use the following greedy dispatch algorithm (GDA) to generate a satisfactory assignment scheme, from which we check whether orders are assigned to the same driver and generate the corresponding positive and negative labels. The pseudocode of labelling using GDA is shown in algorithm 1. The GDA dispatches orders in an iterative way so that orders are assigned to the most suitable drivers. It is obvious that GDA can assess the matching of order-driver pairs more accurately than KM algorithm since it updates the cost matrix each time an order is dispatched. But it requires much computational time due to frequent callbacks of the evaluation function.

Algorithm 1 Label generation with GDA

Input: the set of orders N, the set of drivers M
Output: the set of labels Φ
1:      Initialize the cost matrix $S_{|N| \times |M|}$ with route planning method.
2:      Set $R = \emptyset$, $k = 0$.
3:      **While** $k < |N|$ **do**
4:          **For** $i = 1$ to $|N|$ **do**
5:              Find the best driver i^* with the smallest cost for order i;
6:              **If** $i^* \in R$ **do**
7:                  Sequentially compare the time delay and route length between order i and the
8:                  order that is previously assigned to driver i^*;
9:                  Assign the better one to driver i^*;
10:             **Else**
11:                 Assign order i to driver i^*;
12:                 $R = R \cup \{i^*\}$;
13:             **End if**
14:          Update the cost matrix accordingly;
15:         **End for**
16:         Eliminate the assigned orders from the cost matrix.
17:         $k = k + |R|$;
18:     **End while**
19:     Denote the former assignment scheme as π.
20:     **For** $u = 1$ to $|N|$ **do**
21:         **For** $v = u + 1$ to $|N|$ **do**
22:             **If** order u and v are assigned to the same driver **do**
23:                 $label_{u,v} = 1$;
24:             **Else**
25:                 $label_{u,v} = 0$;
26:             **End if**
27:             $\Phi = \Phi \cup \{label_{u,v}\}$;
28:         **End for**
29:     **End for**

Model Architecture

The classification model introduced before is implemented by using an ensemble learning technique that consists of a sequence of decision trees. An eXtreme Gradient Boosting (XGBoost) algorithm is used to train and optimize the parameters of the model. XGBoost does not train all the trees simultaneously but tries to optimize the loss function one tree at each step. A best tree that minimizes the loss function based on current trees is added. Compared with other tree boosting methods, XGBoost presents huge advantages in precision, flexibility, model complexity control and parallel computing [18]. It is also effective in reducing the computing time and providing optimal use of memory resources, which is the reason that many researchers utilize it after first proposed by Chen and Guestrin [19].

4.3   Rule-Based Route Planning Method

In OFDP, route planning serves not only as a module that generates the real route for drivers, but also as an evaluation component that offers information of objective function for the order assignment part to decide the quality of an assignment scheme. Under such circumstance, fast generating feasible routes for current assignment is of great significance. Therefore, we adopt a rule-based route planning method [20], whose pseudocode is shown in Algorithm 2.

Algorithm 2 Rule-based route planning method

Input: the driver j and the set of carrying orders O_j
Output: the planned route

1:     Sort the orders in an increasing order of ETA (O_j^{eta}).
2:     Sort the orders in a decreasing order of urgency (O_j^{urg}).
3:     **For** $k = 1$ **to 2 do**
4:         **If** $k = 1$ **do**
5:             Sequentially take the orders out from O_j^{eta} and greedily insert each order into the route;
6:         **Else**
7:             Sequentially take the orders out from O_j^{urg} and greedily insert each order into the route;
8:         **End if**
9:     **End for**
10:    Select the better one between two routes as the current route.
11:    Find delivery locations with the largest time delay and move them backward to an optimal posi-
12:    tion.
13:    Find delivery locations with the most sufficient time and move them forward to an optimal posi-
       tion.
       Output the final route.

---

5   Computational Experiment

5.1   Datasets

To generate the training and test data for the classification model, we collect the historical data from Meituan delivery platform from November 9, 2020 to November 16, 2020. The areas involved in the dataset belong to two different cities (Beijing and Kunming), resulting two different datasets with different characteristics, denoted as Set_b and Set_k, respectively. The ground truth for every piece of data is calculated by GDA mentioned earlier. In this way we obtain two original datasets, containing 2845442 and 420744 samples after eliminating the samples with missing values. Through date-based splitting, the training set and test set are generated without overlapping. The details of the datasets can be found in Table 1.

Besides the machine learning classification model, the performance of the HA for solving OFDP also needs to be tested. To keep consistent with the settings of the aforementioned test sets, we select the same experimental areas and dates.

Table 1. Information of datasets

Dataset	Type	Number of samples	Date
Set_b	Train	2486195	2020.11.09–2020.11.15
	Test	359247	2020.11.16
Set_k	Train	391467	2020.11.09–2020.11.15
	Test	29277	2020.11.16

5.2 Performance Metrics

On the one hand, to effectively measure the performance of the classification model, multiple indicators are introduced that are widely used in the field of machine learning. In our study, 6 important scalar metrics are adopted to evaluate the XGBoost classifier, which are listed in Table 2. To calculate these metrics, a confusion matrix is defined with True Positive (TP), True Negative (TN), False Positive (FP) and False Negative (FN), where TP and TN are correct predictions while FP and FN are incorrect predictions.

Table 2. Performance metrics for XGBoost classifier

Metric	Calculation	Range
Accuracy	(TP + TN)/(TP + FP + FN + TN)	[0,1]
Precision	TP/(TP + FP)	[0,1]
Recall	TP/(TP + FN)	[0,1]
F1 score	(2·Precision·Recall) / (Precision + Recall)	[0,1]
AUPR	Area under PRC	[0,1]
AUC	Area under ROC	[0,1]

Here, Accuracy generally reflects the correctness of the prediction results. Precision reflects the capability of model for recognizing negative samples while Recall reflects the capability of recognizing positive samples. F1 Score is the integration of Precision and Recall, which can evaluate the model comprehensively. Moreover, with Precision and Recall, we can plot a Precision-Recall Curve (PRC) under different decision threshold. Consequently, the Area Under PRC, denoted as AUPR, reflects the performance of the model where the larger the AUPR, the better the model. Similarly, the Receiver Operating Characteristic (ROC) is plotted by plotting the true positive rate (TPR) against the false positive rate (FPR), which are calculated by Eq. (2) and (3). Correspondingly, the area under ROC, denoted as AUC, can also represent the capability of the model. The difference between AUC and AUPR is that AUPR still works when there is a significant quantitative imbalance between positive and negative samples.

$$TPR = TP/(TP + FN) \tag{2}$$

$$FPR = FP/(FP + TN) \tag{3}$$

On the other hand, to evaluate the performance of the proposed algorithm for solving OFDP, we compare the HA with the GDA and the KM algorithm without order batching procedure. The results are presented in relative percentage deviation (RPD), which is calculated as follows,

$$RPD = (value_{alg} - value_{best})/(value_{best}) \times 100 \tag{4}$$

where $value_{alg}$ is the objective function value of a certain algorithm alg, and $value_{best}$ is the smallest objective function value among all compared algorithms. The smaller the RPD, the better the solution. Moreover, as an online algorithm, the computational performance should also be considered. In our experiments, we use the number of callbacks for evaluation (NCE) to approximately measure the computational burden of an algorithm since evaluating a solution is generally the most time-consuming part.

5.3   Computational Results and Discussions

In this section, the statistical results of the designed experiments in this study are presented. Table 3 shows the performance of the XGBoost classifier in terms of the metrics mentioned in the previous section. Figure 5 gives the graphic results of the ROC and PRC, from which AUC and AUPR can be visually perceived. From the results we can see that the classification model performs well with the AUC of 0.848 and the AUPR of 0.822 on Set_b, which indicates that the classifier is able to catch the latent information between different orders and gives correct predictions under most cases. In addition, the results on Set_k also show a satisfactory AUC and AUPR, which demonstrates the robustness and scalability of the classification model and the effectiveness of the designed features.

Table 3. Experimental performance of XGBoost classifier

Dataset	Metric					
	Accuracy	Precision	Recall	F1 score	AUC	AUPR
Set_b	0.782	0.799	0.679	0.734	0.848	0.822
Set_k	0.813	0.791	0.848	0.819	0.880	0.866

As for the comparison between algorithms for solving OFDP, we present the comparative results in Table 4. It can be observed that the GDA achieves the smallest RPD among all compared algorithm on both datasets. This is expected because GDA updates the cost matrix immediately after current matching procedure. Therefore, the calculation of costs between drivers and orders is more accurate and it is possible to assign different orders to the same driver. However, frequently updating the cost matrix also results in a large number of callback times, which increases the computational burden of the

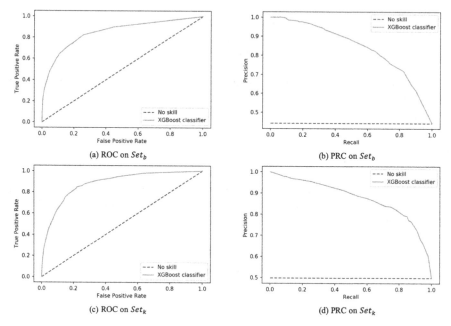

(a) ROC on Set_b (b) PRC on Set_b

(c) ROC on Set_k (d) PRC on Set_k

Fig. 5. ROC and PRC on Set_b and Set_k. "No skill" represents a random classifier and "XGBoost classifier" represents our trained model.

Table 4. Comparative results of different algorithms

Dataset	Algorithm	Time delay	Route length	TC	NCE
Set_b	GDA	0.000	0.000	0.000	8662
	KM	2.626	9.267	3.597	276
	HA	0.605	1.509	1.190	275
Set_k	GDA	0.000	0.000	0.000	4193
	KM	1.594	8.513	3.012	131
	HA	0.583	1.367	0.762	129

dispatching system. KM algorithm without order batching has fewer callback times, but it also suffers from inferior objective function value, resulting from possible inappropriate matching, where two orders are separately assigned to different drivers but it is actually better to assign them to the same driver. Compared with KM algorithm without order batching, HA has similar callback times but has a better objective function value, as the contribution of the classification model and the order batching method since they can make the correct prediction and batch those similar orders together so that they are not separately dispatched. It mitigates the damage of the solution quality while keeping the advantage of very few callback times. Hence, it can be concluded that HA can effectively and efficiently generate satisfactory solutions for the OFDP, which is of great practical meaning for online food delivery platforms and companies.

6   Conclusions and Future Works

In this paper, we proposed an effective hybrid algorithm to solve the online food delivery problem, which was decomposed into order assignment and route planning. Specifically, the order assignment was formulated as a bipartite graph and the Kuhn-Munkres algorithm was modified to generate feasible matching between drivers and orders. Moreover, to avoid inappropriate matching, we built a machine learning classification model with XGBoost to describe the similarity between different orders and predict the order batching results. As for the route planning, a rule-based fast route planning method was designed, which serves as an essential character of evaluating candidate solutions. Through experiments on real datasets from Meituan, we found that the designed algorithm could generate satisfactory solutions for the OFDP while drastically reducing the callback times of the evaluation function. This result suggested the great value of the designed algorithm in practical application. Besides, the successful utilization of XGBoost classifier also showed the great potential of combining machine learning technology with traditional optimization methods.

In the future work, there are mainly two aspects to extend our research. On the one hand, the order batching procedure can be implemented with more machine learning models, which may possibly result in better performance, such as using unsupervised learning to automatically form the order batching results, which will also save the process of labelling. On the other hand, considering more realistic constraints for the online food delivery problem, such as the uncertainty of travel time and dynamic arrival of orders, is worth studying, both practically and academically.

Acknowledgment. This study is supported by the National Natural Science Foundation of China (No. 61873328), and Meituan.

References

1. Meituan Delivery Homepage. https://peisong.meituan.com/about. Accessed 20 Mar 2021
2. The changing market for food delivery. https://www.mckinsey.com/industries/high-tech/our-insights/the-changing-market-for-food-delivery. Accessed 3 Dec 2020
3. Kalantari, B., Hill, A.V., Arora, S.R.: An algorithm for the traveling salesman problem with pickup and delivery customers. Eur. J. Oper. Res. 22(3), 377–386 (1985)
4. Baldacci, R., Bartolini, E., Mingozzi, A.: An exact algorithm for the pickup and delivery problem with time windows. Oper. Res. 59(2), 414–426 (2011)
5. Venkateshan, P., Mathur, K.: An efficient column-generation-based algorithm for solving a pickup-and-delivery problem. Comput. Oper. Res. 38(12), 1647–1655 (2011)
6. Lim, H., Lim, A., Rodrigues, B.: Solving the pickup and delivery problem with time windows using "squeaky wheel" optimization with local search. In: AIS (2002)
7. Lu, Q., Dessouky, M.: A new insertion-based construction heuristic for solving the pickup and delivery problem with time windows. Eur. J. Oper. Res. 175(2), 672–687 (2006)
8. Lau, H.C., Liang, Z.: Pickup and delivery with time windows: algorithms and test case generation. Int. J. Artif. Intell. Tools 11(03), 455–472 (2002)

9. Lu, Y., Wu, Y., Zhou, Y.: Order assignment and routing for online food delivery: two meta-heuristic methods. In: Proceedings of the 2017 International Conference on Intelligent Systems, Metaheuristics & Swarm Intelligence, pp. 125–129 (2017)

10. Li, H., Lim, A.: A metaheuristic for the pickup and delivery problem with time windows. Int. J. Artif. Intell. Tools 12(02), 173–186 (2003)

11. Ropke, S., Pisinger, D.: An adaptive large neighborhood search heuristic for the pickup and delivery problem with time windows. Transp. Sci. 40(4), 455–472 (2006)

12. Ulmer, M.W., Thomas, B.W., Campbell, A.M., Woyak, N.: The restaurant meal delivery problem: dynamic pickup and delivery with deadlines and random ready times. Transp. Sci. 55(1), 75–100 (2021)

13. Reyes, D., Erera, A., Savelsbergh, M., Sahasrabudhe, S., O'Neil, R.: The meal delivery routing problem. Optimization Online (2018)

14. Zhou, Q., et al.: Two fast heuristics for online order dispatching. In: 2020 IEEE Congress on Evolutionary Computation, pp. 1–8. IEEE (2020)

15. Kuhn, H.W.: The Hungarian method for the assignment problem. Naval Res. Logist. Q. 2, 83–97 (1955)

16. Munkres, J.: Algorithms for the assignment and transportation problems. J. Soc. Indus. Appl. Math. 5(1), 32–38 (1957)

17. Bourgeois, F., Lassalle, J.C.: An extension of the Munkres algorithm for the assignment problem to rectangular matrices. Commun. ACM 14(12), 802–804 (1971)

18. Dhaliwal, S.S., Nahid, A.A., Abbas, R.: Effective intrusion detection system using XGBoost. Information 9(7), 149 (2018)

19. Chen, T., Guestrin, C.: Xgboost: a scalable tree boosting system. In: Proceedings of the 22nd ACM SIGKDD International Conference on Knowledge Discovery and Data Mining, pp. 785–794 (2016)

20. Wang, X., et al.: An effective iterated greedy algorithm for online route planning problem. In: 2020 IEEE Congress on Evolutionary Computation, pp. 1–8. IEEE (2020)

Graph Semantics Based Neighboring Attentional Entity Alignment for Knowledge Graphs

Hanchen Wang$^{(\boxtimes)}$ ⓘ, Jianfeng Li, and Tao Luo ⓘ

Beijing Laboratory of Advanced Information Networks, and Beijing Key
Laboratory of Network System Architecture and Convergence,
School of Information and Communication Engineering,
Beijing University of Posts and Telecommunications, Beijing 100876, China
{hanchenwang, lijf, tluo}@bupt.edu.cn

Abstract. Entity alignment is the task of matching entities in different knowledge graphs if they refer to the same real-world identity. A promising method for entity alignment is to use embedding methods to learn knowledge graph representations and align entities by measuring their embedding distance. However, when dealing with the challenge of structural heterogeneity between knowledge graphs, most existing entity alignment methods ignored the potential evidence provided by entity and relation semantics. In this paper, an entity alignment framework that incorporates graph semantic information with neighboring attention is proposed. The framework leverages both entity and relation semantic information by introducing the attention mechanism into a graph convolutional network module. In particular, an attention mechanism about neighboring relation semantic information is developed in the proposed framework to learn entity representations as well as to ignore unimportant neighborhoods. The experimental results on the real-world dataset WK3L demonstrates that the proposed framework consistently outperforms other state-of-the-art models.

Keywords: Knowledge graph · Entity alignment · Knowledge graph embedding · Graph convolutional network · Attention mechanism

1 Introduction

Knowledge graphs (KGs) like DBpedia [1], YAGO [2], and XLore [3], support various knowledge-driven applications, e.g., semantic search [4], question-answering [5], and recommender systems [6]. KGs organize structured knowledge in the form of a triple like (e_h, r, e_t) indicating that head entity e_h and tail entity e_r are connected by relation r. Different KGs contain different but complementary content in the same domain. Thus, it is essential to integrate KGs into a unified and compact KG to benefit the knowledge representation and application.

Entity alignment, the key technology to merge heterogeneous KGs, seeks to link entities in different KGs that refer to the same real-world identity. The classical entity

D.-S. Huang et al. (Eds.): ICIC 2021, LNCS 12837, pp. 355–367, 2021.
https://doi.org/10.1007/978-3-030-84529-2_30

alignment methods mainly rely on feature design [7] and much human involvement [8] to find equivalent entity pairs.

Recently, more research attention on entity alignment is attracted by embedding-based methods due to their greater universality and efficiency. The embedding-based methods aim to encode KGs into a unified vector space and align entity pairs according to their embedding distance. In general, these methods can be divided into two categories: translation-based approaches and GNNs-based approaches.

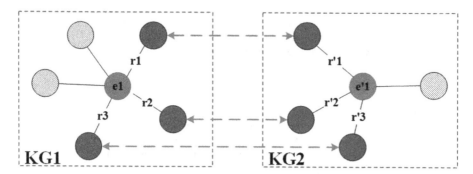

Fig. 1. Illustration of structural heterogeneity between KGs. The common entities of $e1$ and $e'1$ are linked by the blue dotted lines, ri and $r'i$ are their relations. Light colored vertexes are unique neighboring entities in each KGs that are not conducive to alignment (Color figure online).

Translation-based approaches embed KGs via translational models like TransE [9]. These methods use relation triples to learn entity and relation representations jointly but are defective in preserving KG structural information and utilizing relation information to enhance entity alignment. GNNs-based approaches aggregate KG neighboring topological features by using various graph neural networks (GNNs), e.g., Graph Convolutional Networks (GCN) [10] and Graph Attention Networks (GAT) [11]. However, the vanilla GCN models depict KG as an unlabeled single-relational graph and produce entity embeddings by aggregating structural features. Hence, the GCN-based methods are usually sensitive to structural heterogeneity between KGs, which is a major challenge for entity alignment. In particular, due to the independence and incompleteness of KGs, equivalent entities may have different neighbors in their KG. As shown in Fig. 1, $e1$ and $e'1$ are different entities from different KGs that refer to the same real-world object. Although the above two entities share some common neighbor entities and relations, there also exist some different neighbors that may cause incorrect alignment during feature aggregation. Therefore, every neighbor of entity should have different importance and attention for entity alignment.

Recent studies introduced the attention mechanism to address the heterogeneity problem and enhance entity alignment [12, 13]. While promising, these prior works fail to take advantage of neighboring relational semantic information. Although equivalent entities and relations between KGs have different surface forms, they are equivalent in semantics. Hence, we consider the graph semantic information can provide useful evidence in entity alignment for heterogeneous KGs.

The main contribution of this work is that we propose a graph semantics based neighboring attentional entity alignment framework (GSNA) that jointly considers the relation semantic information and the attention mechanism in a GCN-based framework. In particular, we first define the entity semantics as the input feature of GSNA, then use the relation semantics as the basis for the calculation of the neighboring attention coefficient. Our key contributions include:

- We propose a GSNA framework for entity alignment to alleviate the impact of structural heterogeneity by jointly considering relation semantics and neighboring attention.
- The GSNA model extracts semantic information of the relations as the evidence for the attention mechanism to learn the embedding that integrates neighboring features with a weighted combination.
- We conduct extensive experiments on the WK3L dataset, and the GSNA achieves better performance in terms of *Hits@k* and MRR. The experimental results and further analysis validate that GSNA achieves significant improvements over the state-of-the-art methods.

The remainder of the paper is organized as follows. Section 2 discusses the related work. The problem definition of entity alignment is introduced in Sect. 3. Section 4 describes the proposed GSNA framework. The experiment and evaluation results are present in Sect. 5. Finally, Sect. 6 is the conclusion and future work.

2 Related Work

Entity alignment for KGs is a long-standing task. The traditional entity alignment approaches mainly require hand-crafted features or crowdsourcing. These methods require a lot of time and labor costs, so they are inevitably limited by efficiency and flexibility. Recently, a large number of embedding-based approaches have been proposed and proved to be feasible means of entity alignment. The embedding-based methods aim to represent entities and relations in a unified low-dimensional vector space through iterative training on pre-aligned entity pairs and they can roughly be classified into two categories: translation-based and GNNs-based methods.

2.1 Translation-Based Entity Alignment

Translation-based methods utilize translational distance model such as TransE to represent both entities and relations as vectors into the same embedding space and characterize a triple as a translation from head entity through relation to tail entity.

JE [14] is one of the first attempts in translation-based entity alignment. It jointly encodes different KGs into a unified vector space by adding a margin-based loss with alignment seeds. Following the main principle of JE, JAPE [15] introduces attribute triples as additional information for entity alignment and embeds attribute triples with Skip-gram model. Meanwhile, MTransE [16] utilizes TransE to embed the triples of each KG into an independent space and designs three different transitions to map each embedding space with its counterpart. To address the lack of alignment seeds, BootEA

[17] applies the bootstrapping strategy to generate new aligned entities as training data. Moreover, it converts entity alignment into a classification task and achieves significant performance improvements. SEA [18] leverages TransE to embed KGs and tries to address the impact of degree difference by adapting adversarial training process. It maps KGs in both directions and adds a cycle-consistency loss to generate the final embedding space.

Though effective, all the aforementioned methods focus too much on modeling of translation relation between entities and inevitably ignore useful structural information. Therefore, these methods are not sufficient to capture the complex KG structures.

2.2 Graph Neural Networks Based Entity Alignment

Graph neural networks learn vertex representations and preserve structural information by recursively aggregating the neighboring nodes features. Recently, several variants of GNN have been used in entity alignment and have achieved significant results, including Graph Convolutional Networks (GCN) and Graph Attention Networks (GAT), etc. Considering the powerful graph structure modeling capabilities, we also encode the structural information of KGs by using GNNs.

GCN-Align [19] is the representative work that first introduces GCNs to the task of entity alignment. It uses GCN to embed the entities in two KGs into a unified vector space and combined attributes of entities in the framework by embedding attributes with another GCN to improve the alignment results. In order to get better performance, the approach requires pre-aligned entities and attributes as training data. KECG [12] combines translation-based and GNNs-based methods to enhance entity alignment performance. To alleviate the negative impact of heterogeneity, it utilizes an extended GAT to embed the KGs while paying different attention to their neighbors. Furthermore, KECG also utilizes TransE to model relational constraints among inner-graph entities. More recently, RDGCN [13] constructs dual relation graph and utilizes the attention mechanism to update representations by encouraging interactions between the dual relation graph and the primal graph. It extends the idea of Dual-Primal Graph CNN (DPGCNN) to incorporate relation information into the entity representations and achieves great performance on entity alignment.

The above methods all utilize GNNs to embed KGs and tackle the challenge of structural heterogeneity with different efforts. However, the relation and entity semantic information contained in KGs are not considered. Therefore, inspired by GAT that integrates neighboring features with different attention, our approach leverages both entity and relation semantic information by introducing a graph semantics based attention mechanism.

3 Problem Definition

A knowledge graph can be represented as $G = (E, R, T)$, where E is denoted as the set of entities, R is the set of relations, and T is the set of triples. Specifically, $T = \{(e_h, r, e_t), e_h \in E, r \in R, e_t \in E\}$, where e_h, r, and e_t are head entity, relation, and tail entity, respectively.

Let $G_1 = (E_1, R_1, T_1)$ and $G_2 = (E_2, R_2, T_2)$ be two heterogeneous KGs that need to be aligned. $L = \{(e_{i1}, e_{i2}) | e_{i1} \in E_1, e_{i2} \in E_2\}$ is the set of entity pairs between two heterogeneous KGs that are equivalent in semantics. Without loss of generality, we randomly select some alignment seeds $S \in L$ as the training data. According to the above definition, the goal of entity alignment can be summarized as discovering the remaining equivalent entity pairs in KGs as possible.

4   Methodology

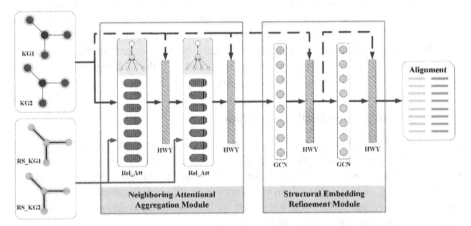

Fig. 2. The overall framework of GSNA for entity alignment. The RS_KG1 and RS_KG2 denote the relation semantics of KG1 and KG2 respectively, the Rel_Att indicates relation attention layer, and the HWY is the customized highway gate.

As described in Sect. 1, structural heterogeneity poses a great challenge towards entity alignment. To tackle this problem, we propose GSNA to combine GCNs and attention mechanism with consideration of graph semantic information. The framework of our proposed approach is shown in Fig. 2, the GSNA mainly consists of a neighboring attentional aggregation module and a structural embedding refinement module. Given the input KGs, we first extract the semantic information of entity and relation by word embedding. Next, GSNA employs graph attention mechanism to carry out neighboring attentional aggregation of semantic features. Finally, the entity embedding is fed to highway GCN to refine their representations by KG structural information.

4.1   Graph Semantics Extraction

Graph semantics extraction seeks to leverage the semantic information of entity and relation to help the model better capture the neighboring features of entities. Different from the classic GCN-based models which randomly initialize the structural features, GSNA uses the entity semantic information as the initial entity features of the framework. As illustrated in Fig. 2, relation semantic features also play a vital role in the process of neighboring attention interaction. Therefore, it is necessary to embed entity and relation names to extract the graph semantic information.

In the field of NLP technology, many word embedding algorithms have successfully captured fine-grained semantic and grammatical rules. Word embedding models represent each word with a real-valued vector to benefit multiple downstream tasks as semantic features. Among all the word embedding algorithms, GloVe [20] is one of the most widely used methods. The GloVe is an unsupervised global log-bilinear regression model for word representations that outperforms other models on word similarity and named entity recognition tasks.

For graph semantics extraction process, we first translate entity and relation names of different KGs into English by using Google Translate API. Thus, we unified the language of each dataset. Then we use the pre-trained word vectors model glove.840B.300d to extract both relation and entity semantics. Specifically, for the name of each entity or relation that consists of several words, we extract the embedding representation of each word in the pre-training model, and then average all embedding vectors to obtain the final semantic embedding representation.

4.2 Neighboring Attentional Aggregation

The purpose of neighboring attentional aggregation is to conduct feature aggregation with relation semantics based weighted combination. Therefore, we apply an attention mechanism to produce different attention coefficients during iteration. Note that this module consists of two types of layers, multi-head relation attention layer, and customized highway gate. And we stack two multi-head relation attention layers with customized highway gates in the framework.

Multi-head Relation Attention Layer. In this layer, we perform self-attention on the nodes to compute a linear combination of the neighborhood features. To stabilize the learning process of self-attention, we also extend self-attention to multi-head attention. And the output features for every node is express as:

$$h_{i'} = \sigma\left(\frac{1}{M}\sum_{m}^{M}\sum_{j\in N_i} a_{ij}^{(m)} h_j\right), \tag{1}$$

where $h_{i'}$ is the output feature of node i, M is the number of attention head, N_i denotes the neighborhood entities set of node i, $a_{ij}^{(m)}$ is the normalized attention coefficient between node i and j in the m - th head, and h_j is the input feature of neighboring node j.

For the normalized attention coefficient in each head, let c_{ij} denote the attention coefficient between node e_i and its neighbor e_j in relation attention layer, we normalize the coefficient by using the softmax function as below:

$$a_{ij} = soft\max(c_{ij}) = \frac{\exp(LeakyReLU(q(s_{ij}^r))}{\sum_{k\in N_i} \exp(LeakyReLU(q(s_{ik}^r))}, \tag{2}$$

where c_{ij} is the attention coefficient between entity e_i and its neighbor e_j, q is a single-layer fully connected neural network, s_{ij}^r is the semantic vector of relation r between e_i and e_j which is obtained by graph semantics extraction process.

Customized Highway Gate. To Control the Noise Propagation and Error Accumulation, We Introduce a Customized Highway Gate Between GCN Layers. Inspired by the Principle Proposed in Work [21] and Extend Its Spirit to a General Situation, Our Customized Highway Gate Layer is Formally Described Below:

$$T(X^{(i)}) = \sigma(X^{(i)} W_T^{(i)} + b_T^{(i)}), \tag{3}$$

$$X^{(j)} = T(X^{(i)}) \cdot X^{(j)} + (1 - T(X^{(i)})) \cdot X^{(i)}, \tag{4}$$

where $X^{(i)}$ is the output of the i - th layer, $i < j$, σ is a sigmoid activation function, $W_T^{(i)}$ is the weight matrix of transform gate $T(X^{(i)})$, and $b_T^{(i)}$ is a bias vector. In particular, when $j = i+1$, the customized highway gate can be regarded as a general highway layer [21]. In our framework, the customized highway gate is more conducive to retaining useful structural and semantic information.

4.3   Structural Embedding Refinement

In this stage, the model employs a two-layer GCN with customized highway gates to refine the vertex representation and produce the final entity embedding results that incorporate structural feature.

Given a set of vertex feature representation $X^{(l)} = \{x_1^{(l)}, x_n^{(l)}, \ldots, x_n^{(l)} | x_i^{(l)} \in \mathbb{R}^{d^{(l)}}\}$ as the input to the l - th GCN layer, where n is the number of entities, and $d^{(l)}$ is the number of features in the l - th layer, the output $X^{(l+1)}$ of the forward propagation can be calculated as:

$$X^{(l+1)} = \varphi(\tilde{D}^{-\frac{1}{2}} \tilde{A} \tilde{D}^{-\frac{1}{2}} X^{(l)} W^{(l)}), \tag{5}$$

where φ is a ReLU activation function, $\tilde{A} = A + I$ is the adjacency matrix with self-connections, A is the adjacency matrix of KGs and I is the identity matrix, $\tilde{D}$ is the diagonal degree matrix of KGs and $\tilde{D}_{ii} = \sum_{k=1}^{n} \tilde{A}_{ik}$, and $W^{(l)} \in \mathbb{R}^{d(l) \times d(l+1)}$ is the layer-specific trainable weight matrix in the l - th GCN layer.

4.4   Model Training

The training process is implemented successively by neighboring attentional aggregation and structural embedding refinement. For entity alignment, equivalent entity pairs are expected to have a closer embedding distance than their negative counterparts. Therefore, in each step, we introduce a margin-based scoring function as the training objective. Then we minimize the objective function to enforce the embeddings of equivalent entity pairs to become as close as possible. The above loss function is defined as:

$$L_S = \sum_{(e,v)\in L} \sum_{(e',v')\in L'} [d(h(e), h(v)) - d(h(e'), h(v')) + \gamma]_+ , \qquad (6)$$

where L is the set of alignment seeds, L' is the negative entity alignment set that constructed by replacing $(e, v) \in L$ by e' or v' in G_1 or G_2, $d(x, y) = \|x - y\|_{L_1}$, $h(e)$ denotes the embedding represent of entity e, $\gamma > 0$ is a margin hyper-parameter, $[x]_+ = \max\{0, x\}$. As for negative sampling, we select the top K nearest entities of e or v to replace e or v as the negative samples.

5 Experiments

In this section, we conduct comprehensive experiments on WK3L-15k to evaluate our GSNA framework for entity alignment task. Further, the alignment performance and sensitivity to the training data size of different approaches are assessed.

5.1 Dataset and Baseline

To comprehensively evaluate our GSNA model, we conduct experiments on the real-world heterogeneous knowledge graphs dataset WK3l-15k. WK3l dataset [16] is selected from DBpedia and consists of English (En), French (Fr), and German (De) knowledge graphs with known alignments as ground truth. In each KG, WK3l-15k matches about 15,000 nodes with FB15k, which is the largest monolingual graph widely used by many previous works. For WK3l, the alignments are provided as directed mapped entities and triples from one KG to its counterpart. To better adapt to the requirements of this task, we extract the bidirectional aligned entities from the above alignments with a common assumption that an alignment entity pair should be symmetric. Statistics of the dataset is given in Table 1, and the number of equivalent entity pair we collected is shown in Table 2. Note that the number of symmetric equivalent entity pairs is the same as the results in work [22].

Table 1. Statistics of the WK3l-15k dataset

Dataset	Triples	Entities	Relations
En-De	En:203,502	En:15,127	En:1,841
	De:145,616	De:14,603	De:596
En-Fr	En:203,502	En:15,170	En:2,228
	Fr:170,605	Fr:15,393	Fr:2,422

Table 2. Number of aligned entities in WK3l-15k dataset

Dataset	Left to Right	Right to Left	Symmetric
En-De	11,594	11,445	10,383
En-Fr	10,108	10,164	8,024

In the experiments, we comprehensively verify the effectiveness of the proposed approach, and the following methods are included as the baseline model:

- **MTransE** [16] embeds triples with TransE and explores different transformations of entities and relations between KGs.
- **ITransE** [23] encodes entities and relations jointly and iteratively into a unified low-dimensional space.
- **JAPE** [15] learns embeddings for triples with TransE and further incorporates attribute embedding to refine entity embeddings.
- **BootEA** [17] utilizes bootstrapping strategy to iteratively enlarge the training data for learning alignment-oriented embeddings based on TransE.
- **GCN-Align** [19] first introduces GCN into the task of entity alignment and generates neighborhood-aware embeddings of entities based on a two-layer GCN.
- **RDGCN** [13] constructs dual relation graphs and learns both entity and relation representations by a GCN-based framework.
- **NA** is our proposed framework without structural embedding refinement stage. It only embeds entities with neighboring attentional aggregation process.

5.2   Experimental Settings

Evaluation Metrics. In our work, we adopt two popular metrics, *Hits@k* score and mean reciprocal rank (MRR), as the evaluation measures to assess the entity alignment performance of all the approaches. *Hits@k* score measures the proportion of correctly aligned entities in top-k ranked candidates, and MRR is the average of the reciprocal ranks of correct alignment results. Moreover, we report *Hits@1*, *Hits@5*, *Hits@10*, and MRR in the overall alignment experiment. In the process of sensitivity to training size evaluation, *Hits@1* and *Hits@10* are shown in the experimental line charts.

Parameter Settings. For the parameters of our GSNA, We set the embedding dimension of KGs $d = 300$, learning rate $lr = 0.1$, and conduct a grid search with margin γ and attention head number M. The range of margin γ among $\{0.5, 1, 1.5, 2.0, 2.5, 3.0\}$, and attention head number M among $\{1, 2, 4, 6, 8\}$, the best configuration is $\gamma=1.5$ and $M=4$. following the previous works [13, 18, 19], we randomly select 30% of equivalent entity pairs as alignment seeds for training and the rest 70% of them for testing. And different proportions of alignment seeds are also further analyzed in Sect. 5.3. During the training process, each positive alignment seed has 125 negative alignment pairs generated by the nearest sample every 10 epochs.

Table 3. Overall alignment performance on the WK3L-15k

Model	En → De				De → En			
	hits@1	hits@5	hits@10	MRR	hits@1	hits@5	hits@10	MRR
ITransE	15.98	28.63	32.71	0.218	13.42	25.63	31.17	0.205
MTransE	6.17	8.48	10.39	0.078	4.69	6.61	7.74	0.059
JAPE	16.85	27.32	34.74	0.226	13.92	22.15	29.68	0.189
BootEA	33.13	54.13	61.70	0.435	30.47	45.33	53.52	0.381
GCN-Align	18.25	31.30	37.26	0.248	15.70	27.53	33.31	0.217
RDGCN	57.42	71.08	74.51	0.635	58.80	72.17	75.18	0.647
NA	62.25	74.10	77.08	0.675	63.36	74.76	77.31	0.684
GSNA	**65.61**	**76.42**	**79.12**	**0.706**	66.10	76.97	79.41	**0.711**
Model	En → Fr				Fr → En			
	hits@1	hits@5	hits@10	MRR	hits@1	hits@5	hits@10	MRR
ITransE	18.21	24.34	27.41	0.214	18.61	33.64	36.28	0.248
MTransE	16.77	21.64	25.35	0.198	19.85	31.27	38.21	0.261
JAPE	15.68	23.45	28.69	0.208	16.22	28.93	34.71	0.219
BootEA	29.72	52.92	61.19	0.395	30.77	55.44	63.67	0.428
GCN-Align	17.24	27.29	31.16	0.220	17.58	30.82	36.21	0.237
RDGCN	58.74	68.57	71.56	0.633	58.53	68.55	71.00	0.632
NA	61.30	70.27	73.02	0.655	61.57	70.58	73.05	0.658
GSNA	**62.12**	**71.68**	**75.03**	**0.668**	63.17	72.07	75.04	0.675

5.3 Results and Discussion

Overall Alignment Performance. Table 3 summarizes the experimental results of all compared approaches on the WK3L-15k dataset. The results of first five baselines are taken from work [18] and the remaining results are generated by grid search on parameters. The best scores are denoted in bold in the table. Noteworthily, our GSNA model consistently outperforms all the baseline methods under different evaluation metrics. For example, in De → En, our approach achieves the best performance, GSNA exceeds at least 12.4%, 6.7%, 5.6%, and 9.9% than RDGCN and BootEA on the *Hits*@1, *Hits*@5, *Hits*@10, and MRR respectively. Meanwhile, NA also achieves more promising results than other baselines. The above two observations indicate that the neighboring attentional aggregation process effectively utilizes the graph semantic information extracted by GloVe.

Except for the NA model, RDGCN achieves better performance than the other four baselines under all evaluation metrics due to its powerful attention interaction strategy between primal and dual graphs. Among all the translation-based methods, BootEA attains a better result than ITransE, MTransE, and JAPE by employing bootstrapping strategy to overcome the lack of training data. Due to the powerful awareness of neighborhood information, the GCN-based methods generally outperform translation-based models by capturing rich structural information in most cases of the above experiment.

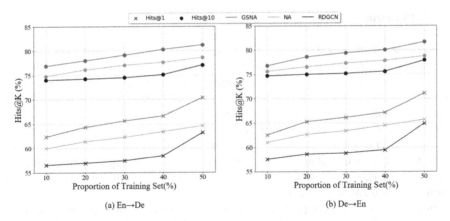

Fig. 3. *Hits*@1 and *Hits*@10 of GSNA, NA, and RDGCN using different proportions of aligned entity pairs for training in En-De dataset. The x-axes are the proportions of alignment seeds for training, and the y-axes are *Hits*@*K* scores.

Impact of Structural Embedding Refinement. Regarding the experimental results of NA and GSNA, both methods can achieve satisfactory entity alignment performance. For instance, the *Hits*@1 values of GSNA and NA are all above 60% in all datasets, but the other methods are far below this value. Compared with the GSNA model, NA always results in a significant drop on all evaluation metrics by removing the highway GCN module, E.G., the *Hits*@1 of GSNA exceeds NA by 3.36% in En → De. As the key component, multi-head relation attention layers can sufficiently integrate the entity and relation semantic information to conduct feature aggregation. Also, the Highway GCN layers can effectively capture the structural features and are complementary to the key component.

Sensitivity to the Proportion of Alignment Seeds. To investigate the model sensitivity to the size of alignment seeds, we further test our GSNA, NA, and RDGCN on the En-De dataset. Concretely, we randomly select aligned entity pairs with the proportion from 10% to 50% with a step of 10%, and the rest equivalent pairs are test data. Figure 3 shows the *Hits*@1 and *Hits*@10 of GSNA, NA, and RDGCN when varying the training data size. As excepted, all approaches get better performance with an increased proportion of the alignment seeds. In general, GSNA consistently outperforms other baselines, and NA seems to be more insensitive than GSNA and RDGCN. Particularly, given a small proportion of training set like 10%, GSNA and NA can still achieve satisfactory results because sufficient semi-supervised graph semantic information is available. For example, *Hits*@1 on De → En of NA is 61.05%, and GSNA is 62.55% When the Proportion is 10%. Therefore, the results confirm the model robustness of GSNA for entity alignment with a limited number of alignment seeds.

6 Conclusion

In this paper, we propose a novel GSNA framework for entity alignment. The GSNA is designed to leverage relation semantic information to mitigate the impact of heterogeneity between KGs. It joint relation semantics and neighboring attention mechanism to learn the entity representations that aggregate neighboring features with different attention. Besides, the entity semantics is also extracted as the input vertex features to improve the alignment performance. To better integrate graph semantics with the structural information of KGs, GCN layers with customized highway gates are applied to the GSNA model, leading to a further improvement. The experimental results on real-world datasets show that our model consistently outperforms the state-of-art baselines on the entity alignment task.

In future work, we will explore more advanced powerful GCN models and introduce more useful semi-supervised information in KG to enhance the model. Meanwhile, the semantic ambiguity and fine-grained graph information will also be considered to deal with the impact of structural heterogeneity.

Acknowledgement. This work is supported by the National Key Research and Development Program of China No.2019YFC1709202 and 2019YFC1709200.

References

1. Lehmann, J., et al.: Dbpedia–a large-scale, multilingual knowledge base extracted from Wikipedia. Semant. Web **6**(2), 167–195 (2015)
2. Suchanek, F.M., Kasneci, G., Weikum, G.: Yago: a core of semantic knowledge. In: Proceedings of the 16th international conference on World Wide Web, pp. 697–706 (2007)
3. Wang, Z., et al.: Xlore: a large-scale English-Chinese bilingual knowledge graph. Int. Semant. Web Conf. (Posters & Demos). **1035**, 121–124 (2013)
4. Guha, R., McCool, R., Miller, E.: Semantic search. In: Proceedings of the 12th International Conference on World Wide Web, pp. 700–709 (2003)
5. Zhang, Y., Dai, H., Kozareva, Z., Smola, A., Song, L.: Variational reasoning for question answering with knowledge graph. In: Proceedings of the AAAI Conference on Artificial Intelligence, vol. 32 (2018)
6. Zhang, F., Yuan, N.J., Lian, D., Xie, X., Ma, W.Y.: Collaborative knowledge base embedding for recommender systems. In: Proceedings of the 22nd ACM SIGKDD International Conference on Knowledge Discovery and Data Mining, pp. 353–362 (2016)
7. Nguyen, T., Moreira, V., Nguyen, H., Nguyen, H., Freire, J.: Multilingual schema matching for Wikipedia infoboxes. Proc. VLDB Endow. **5**(2), 133–144 (2011)
8. Mahdisoltani, F., Biega, J., Suchanek, F.M.: Yago3: a knowledge base from multilingual Wikipedias (2013)
9. Bordes, A., Usunier, N., Garcia-Duran, A., Weston, J., Yakhnenko, O.: Translating embeddings for modeling multi-relational data. In: Advances in Neural Information Processing Systems, pp. 2787–2795 (2013)
10. Kipf, T.N., Welling, M.: Semi-supervised classification with graph convolutional networks. In: International Conference on Learning Representations (ICLR) (2017)
11. Velickovi´c, P., Cucurull, G., Casanova, A., Romero, A., Liò, P., Bengio, Y.: Graph attention networks. Int. Conf. Learn. Represent. (2018)

12. Li, C., Cao, Y., Hou, L., Shi, J., Li, J., Chua, T.S.: Semi-supervised entity alignment via joint knowledge embedding model and cross-graph model. In: Proceedings of the 2019 Conference on Empirical Methods in Natural Language Processing and the 9th International Joint Conference on Natural Language Processing (EMNLP-IJCNLP), pp. 2723–2732. Association for Computational Linguistics, Hong Kong, China (2019)

13. Wu, Y., Liu, X., Feng, Y., Wang, Z., Yan, R., Zhao, D.: Relation-aware entity alignment for heterogeneous knowledge graphs. In: Proceedings of the Twenty-Eighth International Joint Conference on Artificial Intelligence, IJCAI 2019, pp. 5278–5284 (2019)

14. Hao, Y., Zhang, Y., He, S., Liu, K., Zhao, J.: A joint embedding method for entity alignment of knowledge bases. In: Chen, H., Ji, H., Sun, L., Wang, H., Qian, T., Ruan, T. (eds.) CCKS 2016. CCIS, vol. 650, pp. 3–14. Springer, Singapore (2016). https://doi.org/10.1007/978-981-10-3168-7_1

15. Sun, Z., Hu, W., Li, C.: Cross-lingual entity alignment via joint attribute-preserving embedding. In: d'Amato, C., et al. (eds.) ISWC 2017. LNCS, vol. 10587, pp. 628–644. Springer, Cham (2017). https://doi.org/10.1007/978-3-319-68288-4_37

16. Chen, M., Tian, Y., Yang, M., Zaniolo, C.: Multilingual knowledge graph embeddings for cross-lingual knowledge alignment. In: Proceedings of the 26th International Joint Conference on Artificial Intelligence (IJCAI) (2017)

17. Sun, Z., Hu, W., Zhang, Q., Qu, Y.: Bootstrapping entity alignment with knowledge graph embedding. IJCAI 18, 4396–4402 (2018)

18. Pei, S., Yu, L., Hoehndorf, R., Zhang, X.: Semi-supervised entity alignment via knowledge graph embedding with awareness of degree difference. In: The World Wide Web Conference (2019)

19. Wang, Z., Lv, Q., Lan, X., Zhang, Y.: Cross-lingual knowledge graph alignment via graph convolutional networks. In: Proceedings of the 2018 Conference on Empirical Methods in Natural Language Processing, pp. 349–357 (2018)

20. Pennington, J., Socher, R., Manning, C.D.: Glove: global vectors for word representation. In: Empirical Methods in Natural Language Processing (EMNLP). pp. 1532–1543 (2014)

21. Srivastava, R.K., Greff, K., Schmidhuber, J.: Highway Networks. Computer Science (2015)

22. Berrendorf, M., Faerman, E., Melnychuk, V., Tresp, V., Seidl, T.: Knowledge graph entity alignment with graph convolutional networks: Lessons learned. In: Jose, J.M. (ed.) ECIR 2020. LNCS, vol. 12036, pp. 3–11. Springer, Cham (2020). https://doi.org/10.1007/978-3-030-45442-5_1

23. Zhu, H., Xie, R., Liu, Z., Sun, M.: Iterative entity alignment via joint knowledge embeddings. IJCAI. 17, 4258–4264 (2017)

An Improved CF Tree Clustering Based on Tissue-Like P System

Qian Liu[1,2] and Xiyu Liu[1,2(✉)]

[1] Business School, Shandong Normal University, Jinan, China
xyliu@sdnu.edu.cn
[2] Academy of Management Science, Shandong Normal University, Jinan, China

Abstract. With the increase of data sets, clustering in data analysis is also facing huge challenges. Clustering based on the CF tree is a sequential clustering method, that is, the data cannot be clustered at the same time. In addition, it is more suitable for clustering of circular clusters from the distribution of the data set. Therefore, this study proposes an improved CF tree clustering method (IP-ICF) based on tissue-like P system to solve these problems. The algorithm integrates the structure of the CF tree into the improved class organization P system, and takes advantage of the parallelism of the P system. Moreover, the shared nearest neighbor (SNN) similarity is introduced to merge clusters after the construction of CF tree, so that clusters of arbitrary shapes can be identified. Experiments on three UCI datasets and four artificial datasets verify the effectiveness of the proposed algorithm.

Keywords: CF tree · Tissue-like P system · Membrane computing

1 Introduction

Membrane computing, as a new type of biological computing model, simulates the function and structure of cells and organs and abstracts their biochemical reactions and material exchanges to realize computation. Păun [1] firstly proposed the mem-brane computing model, which has the largest parallelism, and has since attracted attention from people at home and abroad. Membrane system (also called P system) [2] is a distributed parallel computing model inspired by living cells. The P system is roughly divided into three types: cell-like P system, tissue-like P system and neuron-like P system. Each P system corresponds to a membrane structure. A group of objects are placed in the membrane structure, and these objects are calculated and evolved through defined rules [3, 4]. Many variations of membrane computing have been proposed to solve different problems, and they have proven to have powerful computing power and high efficiency. The tissue-like P system is proposed on the basis of the cell-like P system. The cell-like P system has only one cell, while the tissue-like P system is composed of multiple cells [5]. Literature [6] proposed a tissue-like P system with fission rules. In the calculation process, the number of cells doubled, and verified that the tissue-like P system has powerful computing capabilities. In the [7], a tissue-like P system with a promoter was proposed, and the promoter in the system regulates the evolutionary rules. A similarly organized P system with the same direction/reverse rule

© Springer Nature Switzerland AG 2021
D.-S. Huang et al. (Eds.): ICIC 2021, LNCS 12837, pp. 368–381, 2021.
https://doi.org/10.1007/978-3-030-84529-2_31

was proposed by [8]. Cells communicate with each other to change the position of the object in the system, but not the object itself.

Clustering analysis is to group data objects according to the information of describing the objects and their relationships in the data, so that the objects are similar in the group, but the objects are different in different groups. Clustering algorithms are mainly divided into the following categories: partitioning methods, hierarchical methods, density-based methods, grids-based methods and model methods etc. In the hierarchical method, Zhang proposed the BIRCH algorithm [9]. It was a comprehensive hierarchical clustering algorithm that scans the database to build a clustering feature tree initially stored in memory, and then clustering the leaf nodes of the clustering feature tree. In 2011, literature [10] proposed a time series clustering algorithm based on Birch algorithm. It used multiresolution transformation as feature extraction technology to cluster time series data, and used the clustering feature tree to solve the problems related to the initial center selection, which significantly reduced the execution time and improves the quality of clustering. A parallel R tree indexing method based on BIRCH algorithm and Hilbert filling curve was proposed by Yang et al. [11]. This method can not only maintain the spatial relationships and attributes within the geographic datasets, but also has good performance in the division and retrieval of spatial data. In [12], an automatic threshold estimation method of birch tree clustering algorithm based on GAP statistics was proposed. The algorithm set the initial threshold heuristically according to the distribution law of seismic impedance data, dynamically increases the threshold, and used the agglomerative algorithm for global clustering. Hc Ryu et al. [13] proposed an effective multirange query based clustering method (ERC) as a global clustering method, and proposed a CF+ tree for the clustering of very large datasets.

Due to the characteristics of the P system, people combine it with other algorithms. In recent years, membrane computing has begun to be applied to clustering problems. For example, Zhao et al. [14] proposed a special P system, whose rules are used to express the improved k-medoids algorithm. Literature [15] proposed the Chameleon algorithm based on the tissue P system with promoters and inhibitors, which performed well in cluster analysis, and the time complexity of the algorithm was also reduced. Liu [16] proposed a chain-like structure of membrane system based on the idea of multilayer framework of cells to improve the performance of the clustering algorithm. The SCBK-CP method was proposed by Liu et al. [17], which combined with the tissue-like P system to improve the spectrum clustering to reduce the complexity of the entire algorithm and improve the efficiency.

In this paper, we have improved the CF tree and constructed a tissue-like P system as the framework of the algorithm to improve the efficiency of the algorithm, which is called IP-ICF. The main contributions of this paper are as follows: (1) A new tissue-like P system is proposed, which integrates SNN similarity and CF tree clustering into the tissue-like membrane system to perform clustering tasks. (2) Four synthetic datasets and four UCI datasets are used to simulate and verify the clustering performances of IP-ICF.

2 Related Work

2.1 The Tissue-Like P System

The tissue-like P system regards cells as the vertices of the graph in the system, and cells can communicate and transmit information. A tissue-like P system has the following forms:

$$\Pi = (O, \sigma_1, ..., \sigma_m, ch, i_0)$$

Where:

(1) O is an alphabet, which contains all objects in the system;
(2) ch represents the connection between cells, and information can be exchanged between connected cells;
(3) i_0 indicates the label of the output cell of the tissue-like P system;
(4) $\sigma_1, ..., \sigma_m$ represents the m cells in the tissue-like P system, and its form is as follows: $\sigma_i = (Q_i, s_{i,0}^t, \omega_{i,0}, R_i), 1 \leq i \leq m$, Q_i is a finite set of states, $s_{i,0}^t \in Q_i$ represents a state at time t, $\omega_{i,o} \in O^*$ represents a multiple set of initial objects, R_i represents a finite set of rules of the form $(s, x/y, s')$ with $s, s' \in B, x, y \in O^*$.

2.2 CF Tree

The CF tree uses a tree structure to perform hierarchical division, and then optimizes the clustering results. In the process of the algorithm, two concepts are introduced: clustering feature (CF) and clustering feature tree (CF tree) [18]. The information of each cluster is described using cluster characteristics, which is defined as follows:

Definition 1: CF suppose there are N D-dimensional data points in a cluster, $\{\vec{x_n}\}(n = 1, 2, ..., N)$, then the clustering $\overrightarrow{CF}$ feature define is a triple: $\overrightarrow{CF} = \left\langle N, \overrightarrow{\Sigma_l}, \Sigma_s \right\rangle$, where N is the number of data points in the cluster, and the vector $\overrightarrow{\Sigma_l}$ is linear summation of each data point, as shown in formula (1):

$$\overrightarrow{\Sigma_l} = \sum_{n=1}^{N} \overrightarrow{xn} = \left(\sum_{n=1}^{N} x_{n1}, \sum_{n=1}^{N} x_{n2}, ..., \sum_{n=1}^{N} x_{nD} \right)^{\mathrm{T}} \tag{1}$$

Scalar Σ_s is the sum of squares of each data point, as shown in formula (2):

$$\Sigma_s = \sum_{n=1}^{N} \overrightarrow{x_n^2} = \sum_{n=1}^{N} \overrightarrow{x_n^T} \overrightarrow{x_n} = \sum_{n=1}^{N} \sum_{i=1}^{D} x_{n_i}^2 \tag{2}$$

According to the above definition, there are the following theorems:

Theorem 1: Let $\overrightarrow{CF}_1 = \langle N_1, \overrightarrow{\Sigma_{l1}}, \Sigma_{s1} \rangle$ and $\overrightarrow{CF}_2 = \langle N_2, \overrightarrow{\Sigma_{l2}}, \Sigma_{s2} \rangle$ respectively denote the clustering characteristics of two disjoint clusters. If these two clusters are merged into one cluster, then the clustering characteristics of the large cluster are

$$\overrightarrow{CF}_1 + \overrightarrow{CF}_2 = \left\langle N_1 + N_2, \overrightarrow{\Sigma_{l1}} + \overrightarrow{\Sigma_{l2}}, \Sigma_{s1} + \Sigma_{s2} \right\rangle \tag{3}$$

The clustering feature is essentially a statistical summary of a given cluster, which can effectively compress the data. Based on the clustering feature, many statistics and distance measures of the cluster can be easily derived. Assuming that there are N D-dimensional data points in a given cluster, $\{\overrightarrow{x_n} : n = 1, 2, ..., N\}$, the following statistics can be derived:

$$\overrightarrow{x} = \frac{1}{N} \sum_{n=1}^{N} \overrightarrow{x_n} = \frac{\overrightarrow{\Sigma_l}}{N} \tag{4}$$

$$\rho = \sqrt{\frac{\sum_{n=1}^{N} (\overrightarrow{x_n} - \overrightarrow{x})^2}{N}} = \sqrt{\frac{N \Sigma_s - \overrightarrow{\Sigma_l^2}}{N_2}} \tag{5}$$

$$\delta = \sqrt{\frac{\sum_{m=1}^{N} \sum_{n=1}^{N} (\overrightarrow{x_m} - \overrightarrow{x_n})^2}{N(N-1)}} = \sqrt{\frac{2N \Sigma_s - 2\Sigma_l^2}{N(N-1)}} \tag{6}$$

Where $\overrightarrow{X}$ is the centroid. Radius ρ is the average distance from the data point in the cluster to the cluster centroid. Diameter δ is the average distance between two data points in the cluster. Both of these statistics reflect the compactness within the cluster.

Theorem 2: The CF tree is a highly balanced tree. Each node of the tree is composed of several clustering features. The CF of the internal node has pointers to child nodes, and all leaf nodes are linked by a doubly linked list.

The CF tree has three parameters: branch balance factor B, leaf balance factor L and spatial threshold T. The CF tree is composed of root nodes, branch nodes and leaf nodes, as shown in Fig. 1. The non-leaf node contains no more than B elements, where $\overrightarrow{CF_i}$ represents the clustering feature information of the i-th sub-cluster on the node, and the pointer $child_i$ points to the i-th child node of the node. The leaf node contains no more than L elements of the form $\overline{[CF_i]}$, and each leaf node contains a pointer $prev$ to the previous leaf node and a pointer $next$ to the next leaf node. Each node represents a cluster formed by merging the corresponding sub-clusters of clustering features in each element. The spatial threshold T is used to limit the size of the sub-clusters of leaf nodes, that is to say, the diameter or radius of the corresponding sub-clusters of each element of all leaf nodes shall not be greater than T.

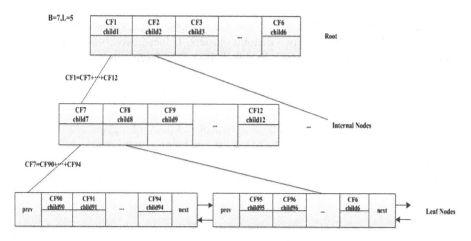

Fig. 1. CF tree

2.3   Shared Nearest Neighbor (SNN) Similarity

Definition 2: (SNN similarity) The SNN similarity between objects a and b is the number of k-nearest neighbors shared by a and b, $Similarity(a, b) = size$ $(nn[a] \cap nn[b])$, Among them: $nn[a]$ and $nn[b]$ are the k-nearest neighbor sets of a and b respectively, and $size(A)$ is the size of set A [19]. If the number of points in a given neighborhood between two points (SNN similarity) exceeds the threshold *MinPts*, the two points are merged.

As shown in Fig. 2, each red point (a and b) has 8 nearest neighbors containing each other, of which 4 points are shared, represented by blue points, so these two points (a and b) the shared nearest neighbor similarity is 4.

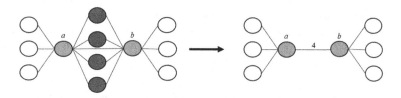

Fig. 2. SNN similarity

3   An Improved CF Tree Based on Tissue-like P System

The membrane structure formed by the tissue-like P system is a network structure, which is also based on the three angles of the membrane object, the membrane structure and the membrane ruler. There are mainly two types of rules in the similar tissue-like P system: communication rules and evolution rules. Communication rules complete the exchange of objects from one membrane to another. The purpose of the evolution rule is to evolve the object in the cell and realize the change of the object state. The proposed algorithm in this paper is to apply the tissue-like P system to the clustering process, apply evolution rules to realize the calculation in the clustering process, and realize the transfer of objects between adjacent membranes through communication rules. The change and movement of the object are controlled by specific rules, which not only change the state of the object in the membrane, but also determine the subsequent application rules.

3.1 The Structure of the Membrane System

We construct a tissue-like P system with 6 cells as shown in Fig. 3, and its form is as follows:

$$\Pi = (O, \sigma_1, ..., \sigma_6, ch, i_0)$$

where:

(1) $O = \{x_1, x_2, ...x_n\}$ is a set of objects, where $x_j (j = 1, 2, ..., n)$ is the j-th data point;

(2) $ch = \{(1, 2), (2, 3), (1, 3), (3, 4), (4, 3), (3, 5), (5, 6)\}$ represents the connection channel between different cells. For example, cell 2 can only receive information from cell 1, but cells 3 and 4 can communicate with each other;

(3) $i_0 = 6$ indicates that cell 6 is the output cell of the entire system;

(4) $\sigma_i (i = 1, 2, 3, 4, 5, 6)$ represents the cell, its form is as follows: $\sigma_i = (Q_i, s_{i,0}^t, \omega_{i,0}, R_i), 1 \leq i \leq 6$, $Q_i = (s_{i,1}^t, s_{i,2}^t, ..., s_{i,6}^t)$, where $t = (0, 1, ...t_{max})$, $s_{i,0}^t (i = 1, 2, ..., 6)$ represents the state of the i-th object in membrane i at time t, $\omega_{i,0} \in O^*$ represents the multiset of initial objects, and R_i represents the limited set of rules, including communication rules and evolution rules. Communication rules realize the transfer of objects between cells, and evolution realizes the change of objects in cells.

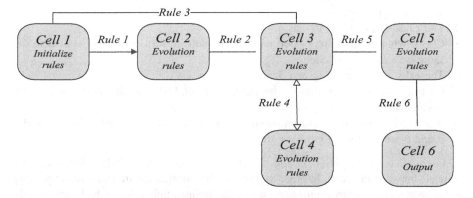

Fig. 3. The framework structure of the tissue-like P system

3.2   Description of the Rules

The CF tree uses diameter or radius to control the boundaries of clusters, so it can only find clusters that are close in distance and are spherical. However, the distance between the data exceeds the set threshold, and the clustering method based on the CF tree cannot reflect very good results. In order to solve this problem, we introduce SNN similarity and *MinPts* threshold parameters in the merging process. It divides closely related and highly similar data into one category to obtain good clustering results.

The algorithm is initialized by inputting parameters at first. Under the condition of the initially formed cluster, the SNN similarity between clusters is calculated. If the SNN similarity is not less than the set threshold, then the two clusters are merged into one cluster, otherwise the original cluster division is still used.

The data clustering process is placed in the P system, and each data is regarded as an object in the cell. Objects in the cell evolve through evolutionary rules, and then change the original state to achieve clustering. Therefore, the evolution rules in each cell are different. The evolution rules in each cell will be described in detail below.

(1) The rules of evolution in cell 2:
- Using the parallelism of membrane system calculation, and the distance between data points is calculated.
- Get a sample point from the datasets as the root node.
- Find the leaf node that is closest to the root node and the CF node in the leaf node from the datasets, and add new samples.

(2) The rules of evolution in cell 3:
- If the radius corresponding to this CF node is less than T, all CF triples on the path are updated. Otherwise, transfer to the next step.
- If the number of CF nodes in the current leaf node is less than L, create a new CF node and place it in the leaf node. All CF triples on the path are updated. Otherwise, transfer to the next step.
- The current leaf node is divided into two new leaf nodes. From all CF triples in the old leaf node, select the two most distant triples, and assign them as the first CF node in the two new leaf nodes. Put all triples into the corresponding leaf nodes according to the distance.

(3) The rules of evolution in cell 4:
- Check whether the parent node needs to be split in turn, and if so, split it in the same way as the leaf node.

(4) The rules of evolution in cell 5:
- CF tree is formed according to the above rules, and SNN similarity of each cluster is calculated.
- If the SNN similarity of cluster a and b is greater than the threshold *MinPts*, cluster a and b are combined into one cluster.

Inter-cell communication in the system can be completed only when there are channels between the cells. In order to improve the information transmission capability of the system, this paper proposes seven communication rules, which are mainly divided into two types, namely, one-way and two-way transmission rules.

(1) *Rule* $= (i, u/j, \lambda)$, one-way transmission rule. This rule indicates that the string u can be transmitted from the cell i to the cell j, and the cell j cannot transmit the cell i. Because λ is an empty string. In this system, Rules 1, 2, 3, 5, and 6 are all one-way transmission rules.

(2) *Rule* $= (i, u/j, v)$, two-way transmission rule. It means that the string u can be transmitted from cell i to cell j, and cell j can also transfer the string v to cell i. *Rule* $= (3, u/4, v)$, it is a two-way transmission rule. It can transfer the object u from cell 3 to cell 4, and cell 4 can also transfer the object v to cell 3.

3.3   The Realization Process of the Membrane System

The implementation process of the IP-ICF clustering method is shown in Fig. 3. In cell 2, the distance between the data points is calculated. In cell 3 and cell 4, leaf nodes and parent nodes can be checked at the same time, which effectively improves calculation efficiency. The clustering process in the membrane system is as follows:

Step1: Initially, we input the datasets Y and the parameter branch balance factor B, leaf balance factor L and spatial threshold T as initial objects into cell 1. Cell 2, cell 3, cell 4, cell 5 and cell 6 are all empty, and cell 6 is the output cell.

Step 2: The data object in cell 1 is transported to cell 2 according to Rule 1. Evolutional rule is applied in cell 2 to calculate and save the distance between data points. The object with the minimum distance is transported to cell 3 via Rule 2.

Step 3: While Rule 2 is running, Rule 3 transports the parameter B, L and T in cell 1 to cell 3 to construct a tree structure.

Step 4: In cell 3, the child node is checked to see if it needs to divide node according to the evolution rule. At the same time, the tree structure constructed in cell 3 is transferred to cell 4 via Rule 4. The parent node is checked to see if it needs to divide, and the result is transferred back to cell 3.

Step 5: Steps 2 to 4 are repeated until each node no longer divides. Finally, the constructed CF tree is transferred to cell 5 through Rule 6 to merge the clusters.

Step 6: When the combined structure is no longer changed in cell 5, the system stops operating and the results are delivered to the output cell 6 via Rule 7.

3.4   Termination Conditions

When the merged structure no longer changes in the cell 5, the system stops running and the final result is delivered to the output cell 6.

4   Experimental Results and Analysis

4.1   Evaluation Method

In order to evaluate whether the algorithm is effective for the clustering effect, this paper uses the two indicators – Silhouette Coefficient and Davies-Bouldin Index.

(1) Silhouette Coefficient (SI)

SI is for a single sample, let a be the average distance of other samples in the same category with it, and b be the average distance of samples in the different categories closest to it. The SI is defined as:

$$SI = \frac{b - a}{\max(a, b)} \tag{7}$$

For a sample set, its SI is the average of the SI of all samples. The range of the SI is [−1, 1]. When the SI is larger, it suggests the result is better.

(2) Davies-Bouldin Index (DBI)

DBI measures the mean value of the maximum similarity of each cluster. Suppose there is a bunch of data points, divide them into n clusters. The calculation formula is as follows:

$$DBI = \frac{1}{N} \sum_{i=1}^{N} \max_{j \neq i} \left(\frac{\overline{S_i} + \overline{S_j}}{||w_i - w_j||_2} \right) \tag{8}$$

where $\overline{S_i}$ and $\overline{S_j}$ calculate the average distance between the data and the centroid of the cluster. $||w_i - w_j||_2$ is the distance between the center of clusters i and the j. In fact, DBI is to calculate the ratio between the sum of distance within the class and distance between the classes. When the value of the DBI is smaller, it indicates that the effect is better.

4.2   Dataset

In order to evaluate the performance of the algorithm, the experiment uses three UCI datasets and four artificial datasets. The specific information of the datasets is shown in Table 1. Each dataset includes four parts: object, attribute, clusters and source. Figure 4 is the original data of the artificial datasets.

Table 1. Datasets information

Datasets	Objects	Attributes	Clusters	Source
Iris	150	4	3	UCI
Digits	1797	64	10	UCI
Wine	178	13	3	UCI
D1	1000	2	4	Artificial
D2	2000	2	3	Artificial
D3	3000	2	3	Artificial
D4	5000	2	3	Artificial

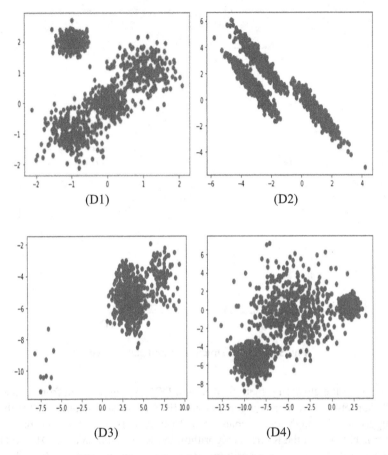

Fig. 4. The original data on artificial datasets

4.3   Experimental Result

As we all know, the implementation of CF tree clustering algorithm requires input parameters B, L and T. Different datasets have different requirements for parameters, and the same dataset will also produce different clustering effects under different parameters. This paper conducted 100 experiments on each dataset, and selected the best value of the dataset division through continuous adjustment of parameters.

Hierarchical structure can better identify the level of clustering, so that the results are more accurate. Through this algorithm, the artificial dataset can be hierarchically divided, and the tree structure obtained is shown in Fig. 5. It can be clearly seen from the tree structure that the D3 dataset can be divided into 3 classes. The D1 dataset can be classified as 3 categories at first, but as the hierarchical structure deepens, it can be divided more accurately. The D2 and D4 datasets can be divided into 2 classes at the beginning, and further classified as 3 categories according to the tree structure.

Birch algorithm is a clustering algorithm based on CF tree, and the IP-ICF algorithm proposed in this paper is also based on CF tree. Therefore, we compared the three

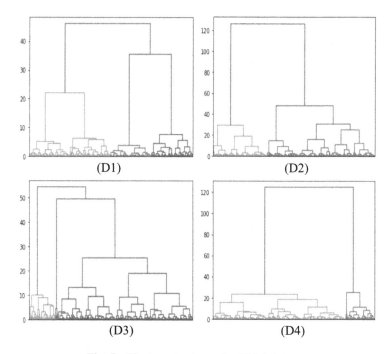

Fig. 5. The tree structure of artificial datasets

algorithms with the clustering metrics of SI and DBI. As can be seen from Table 2, for the UCI datasets, the K-means algorithm and IP-ICF are significantly better than the Birch algorithm. Through the comparison of SI indicators, the algorithm proposed in this paper is better than the K-means algorithm. In the artificial datasets D1 and D3, the clustering results of the three algorithms are relatively good. But for the D2 and D4 datasets, the clustering effect of K-means and Birch algorithm is relatively confusing, and our IP-ICF algorithm is more accurate.

Table 2. Experimental results

Dataset	K-means		Birch		IP-ICF	
	SI	DBI	SI	DBI	SI	DBI
Iris	0.649	0.503	0.453	0.821	0.659	0.483
Digits	0.445	1.127	0.142	0.395	0.542	0.339
Wine	0.459	0.833	0.453	0.821	0.539	0.784
D1	0.800	0.302	0.534	0.564	0.834	0.326
D2	0.621	0.528	0.595	0.629	0.675	0.401
D3	0.738	0.479	0.775	0.388	0.800	0.379
D4	0.538	0.421	0.347	0.659	0.747	0.345

The results of the three algorithms on four artificial datasets are given in Fig. 6, 7, 8 and Fig. 9. As shown in Fig. 6, it is the clustering result of the D1 data set on the three algorithms. It can be seen from the figure that the three algorithms on this dataset show good performance. Figure 7 shows the clustering results of the D2 dataset. Obviously, the K-means and Birch algorithms cannot divide the data set well, while IP-ICF shows a good clustering effect. The clustering results of D3 dataset are shown in Fig. 8. The clustering effect of Birch and IP-ICF algorithm are slightly better than K-means algorithm. Figure 9 shows the clustering results of the D4 dataset. The Birch algorithm has the worst effect among the three algorithms, followed by the K-means algorithm, and the result of the IP-ICF we proposed is the best. In summary, it can be seen that the IP-ICF algorithm has better clustering performance in the comparison of the three algorithms.

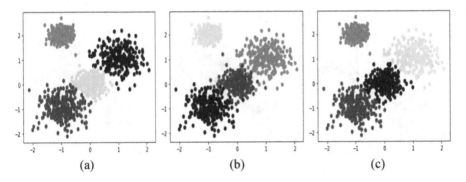

    (a)   (b)   (c)

Fig. 6. Results of D1 (**a** result of K-means, **a** result of Birch, **c** result of IP-ICF)

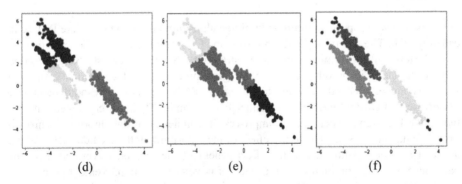

    (d)   (e)   (f)

Fig. 7. Results of D2 (**d** result of K-means, **e** result of Birch, **f** result of IP-ICF)

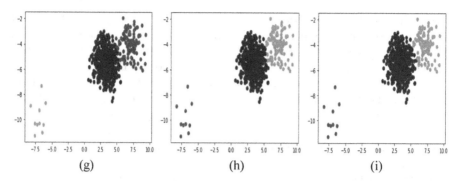

Fig. 8. Results of D3 (**g** result of K-means, **h** result of Birch, **i** result of IP-ICF)

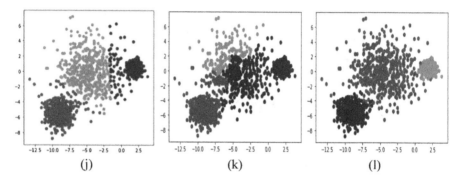

Fig. 9. Results of D4 (**j** result of K-means, **k** result of Birch, **l** result of IP-ICF)

5 Conclusion

In this paper, we propose an improved CF tree algorithm based on tissue-like P system, called IP-ICF. The algorithm firstly improves the CF tree clustering method and merges the clusters after forming the tree structure. Shared Nearest Neighborhood (SNN) similarity is introduced in the merging process, and the regions with high similarity are divided into clusters by using the SNN similarity threshold parameter *Minpts*, so that clustering of arbitrary shape can be found. Then it is combined with the tissue-like P system, which greatly improves the efficiency of the algorithm. Through experimental analysis, the clustering effect of the algorithm on the UCI datasets and the artificial datasets is shown, and the effectiveness of the method is verified. But this paper also has some problems, such as lack of novelty in the improvement points, and the experimental results are not very convincing. In the future, we will conduct deeper research on clustering methods based on the CF tree structure, such as improving the structure of the CF tree and automatically determining the number of clusters.

References

1. Păun, G.: Computing with membranes. J. Comput. Syst. Sci. **61**(1), 108–143 (2000)
2. Pan, L., Perez-Jimenez, M.: Computational complexity of tissue-like P systems. Complexity **26**(3), 296–315 (2010)
3. Frisco, P.: Computing with cells: advances in membrane computing. Comput. Rev. **51**(9), 527–528 (2010)
4. Díaz-Pernil, D., Pérez-Jiménez, M., Romero-Jiménez, A.: Efficient simulation of tissue-like P systems by transition cell-like P systems. Nat. Comput. **8**(4), 797–806 (2009)
5. Ye, L., Zheng, J., Guo, P.: Solving the 0–1 knapsack problem by using tissue P system with cell division. IEEE Access **7**, 66055–66067 (2019)
6. Leporati, A., Manzoni, L., Mauri, G.: Characterising the complexity of tissue P systems with fission rules. Comput. Syst. Sci. **90**, 115–128 (2017)
7. Song, B., Zhang, C., Pan, L.: Tissue-like P systems with evolutional symport/antiport rules. Inf. Sci. **378**, 177–193 (2017)
8. Song, B., Pan, L.: The computational power of tissue-like P systems with promoters. Theor. Comput. Sci. **641**, 43–52 (2016)
9. Zhang, T., Ramakrishnan, R., Livny, M.: BIRCH: an efficient data clustering method for very large databases. ACM SIGMOD Rec. **25**(2), 103–114 (1999)
10. Le Quy Nhon, V., Anh, D.T.: A BIRCH-based clustering method for large time series databases. In: Cao, L., Huang, J.Z., Bailey, J., Koh, Y.S., Luo, J. (eds.) PAKDD 2011. LNCS (LNAI), vol. 7104, pp. 148–159. Springer, Heidelberg (2012). https://doi.org/10.1007/978-3-642-28320-8_13
11. Yang, Y., Wu, L., Guo, J.: Research on distributed Hilbert R tree spatial index based on BIRCH clustering. In: 20th International Conference on Geoinformatics, Hong Kong, pp. 1–5. IEEE (2012)
12. Lorbeer, B., Kosareva, A., Deva, B., Softić, D., Ruppel, P., Küpper, A.: A-BIRCH: automatic threshold estimation for the BIRCH clustering algorithm. In: Angelov, P., Manolopoulos, Y., Iliadis, L., Roy, A., Vellasco, M. (eds.) INNS 2016. AISC, vol. 529, pp. 169–178. Springer, Cham (2017). https://doi.org/10.1007/978-3-319-47898-2_18
13. Ryu, H.C., Jung, S., Pramanik, S.: An effective clustering method over CF+ tree using multiple range queries. IEEE Trans. Knowl. Data Eng. **32**(9), 1694–1706 (2020)
14. Zhao, Y., Liu, X., Qu, J.: The K-medoids clustering algorithm by a class of P system. Inf. Comput. Sci. **9**(18), 5777–5790 (2012)
15. Zhao, Y., Liu, X., Yan, X.: A grid-based chameleon algorithm based on the tissue-like P system with promoters and inhibitors. Comput. Theor. Nanosci. **13**(6), 3652–3658 (2016)
16. Liu, X., Jiang, Z., Zhao, Y.: Chain membrane system and the research progress of direct (indirect) membrane algorithm in clustering analysis. Anhui Univ. (Nat. Sci. Ed.) **42**(03), 11–17 (2018)
17. Zhang, Z., Liu, X.: An improved spectral clustering algorithm based on cell-like P system. In: Milošević, D., Tang, Y., Zu, Q. (eds.) HCC 2019. LNCS, vol. 11956, pp. 626–636. Springer, Cham (2019). https://doi.org/10.1007/978-3-030-37429-7_64
18. Huang, D., Yang, C., Song, L.: CF-Tree: a structure to fast index and retrieve multi-dimension data. In: 2010 2nd International Conference on Advanced Computer Control, pp. 500–504 (2010)
19. Jarvis, R.A., Patrick, E.A.: Clustering using a similarity measure based on shared near neighbors. In: IEEE Transactions on Computers, vol. C-22, no. 11, pp. 1025–1034 (1973)

Classification Method of Power Consumption Periods Based on Typical Daily Load Curve

Yuhang Qiu[1] , Dexin Li[2], Xin Liu[3], Chang Liu[2], Shang Wang[1],
and Tao Peng[1(✉)]

[1] College of Computer Science and Technology, Jilin University,
Changchun 130012, China
qiuyh18@mails.jlu.edu.cn,
{wangshang, tpeng}@jlu.edu.cn
[2] State Grid Jilin Electric Power Research Institute, Changchun, China
[3] State Grid Changchun Supply Company, Changchun, China

Abstract. With the rapid development of new energy, new energy has the characteristics of high volatility and anti-peak regulation, and high penetration of new energy has brought new challenges to the power system. The range of peak and valley period of new energy such as wind and solar energy and overall load in a region can be known. The power system guides users to use electricity in low valley through sectional electricity price, so as to help the power grid cut peak and fill valley and make efficient use of resources. In this paper, a load classification method for period of power consumption based on typical daily load curve is proposed. So it can effectively distinguish the low and peak period of power consumption of users, so as to better understand users load characteristics, and lay the foundation for the source-load interaction. In this paper, the k-means algorithm is used to obtain the typical daily load curve of users, and the curve symbol aggregation approximation based on the typical daily load curve is used to classify the load periods. This method is very fast and needs less historical data. In this paper, the historical data of electric boiler with heat reservoir in Jilin Province are used to test the effectiveness of this method, which verifies that this method can effectively analyze the load operation characteristics, classify load users in different power consumption periods.

Keywords: Load classification · Typical daily load curve · Electric boiler · k-means method

1 Introduction

In recent years, due to the shortage of fossil energy and environmental pollution seriously threaten human survival, renewable new energy sources have been vigorously developed around the world, such as wind power, photovoltaic, hydropower and bioelectricity. However, wind power and photovoltaic have the characteristics of generation in special period, which more wind power generates in windy seasons. Photovoltaic generates more in sunny days and does not generate electricity at night, which often leads to peak period of power consumption is in low period of new energy generation, low period of power consumption is in high period of new energy

D.-S. Huang et al. (Eds.): ICIC 2021, LNCS 12837, pp. 382–394, 2021.
https://doi.org/10.1007/978-3-030-84529-2_32

generation, resulting in the curtailment of new energy power [5]. Because of mismatching source and load, after the new energy is connected with the grid, the power utilization efficiency of the power system has decreased greatly, and the supply and demand in the power period is unbalanced. Therefore, we not only should consider multiple new energy sources for complementary [6], such as solar and wind power, but also should analyze the load operation characteristics, tap the adjustable potential of load, and carry out the load time-sharing price and load response management based on the load characteristics of users, encourage peak period of load users to use electricity in low-level and peaceful sections, so as to ensure the safe and stable operation of power system and more consumption of new energy. When the power supply and demand is unbalanced, the power system usually uses the switch to limit the power according to the user level, and transfers the peak load to the valley, thus cutting the peak energy generation and filling the valley, smoothing the load curve, so as to balance the power supply and demand. But obviously, the power demand of users is limited and the utilization rate of power energy is reduced by the methods of closing and limiting power and reducing the new energy station. From the perspective of power system optimization, to solve the imbalance between supply and demand of power system, it is necessary to study the change characteristics of source and load. Among them, the research focus is to classify the load and formulate the relevant load and energy management.

2 Related Work

In order to manage the load, it is necessary to classify the load, so as to better respond to the load and manage the time-sharing price of the load. The paper focuses on two main problems: the drawing method of typical daily load curve of users and the classification method of load users.

Drawing method of typical daily load curve: the main purpose of drawing typical daily load curve is to reduce the computational complexity caused by a large number of historical data while retaining the effective characteristics of historical load. There are a lot of methods to draw typical daily load curve, such as manual selection method [6]: select typical days from historical data through subjective selection rules; clustering method [8]: take cluster center as typical daily load curve; heuristic scenario reduction method [9]: reduce computational complexity of big data and improve algorithm efficiency; sampling method [10]: conduct probability statistical sampling of load to obtain cluster center.

Load classification method: In order to learn the rules of load users, there are many traditional studies to classify load, mainly in two ways, one is unsupervised learning method, that is, the category label in the sample is unknown, and it is determined according to the clustering results, and the results of classification are changing with data. For example, fuzzy c-means clustering [11], stacked auto-encoder [12]-a deep learning method, the k-nearest neighbor algorithm [13]-classifying load by the distance between samples in the clusters, A semi-supervised feature selection algorithm [14], clustering framework for the automatic classification of electricity customers' loads [15], K-means clustering algorithm [16]. The other is to construct different classifiers according to the characteristics of electricity power of users and category tags, get the final results by supervised learning. Such as decision tree [17, 18], SVM [19], local characterization-based method [20]. Supervised learning method and unsupervised learning method have their own characteristics: supervised learning method requires higher requirements for training data, and the category of each sample must be known; while unsupervised learning method does not need the label of the known category, its clustering results often depend on the number of categories designated by human.

3 Problem Definition

3.1 Time Period Definition

In Jilin power grid, a day is divided into three periods that are described below:

- 8 h in peak period: 7:30–11:30, 17:00–21:00.
- 10 h in valley period: 21:00–7:00.
- 6 h in flat period: 7:00–7:30, 11:30–17:00.

The heating stage is divided into three period that are described below:

- Initial stage of heating: October 15 to November 15.
- Heating medium term: November 16 to March 15.
- End of heating period: March 16 to April 15.

3.2 Daily Load Characteristic Index

- Daily load curve: the load curve is drawn with all counting time points in a day as the abscissa axis and the load value at each counting time point as the ordinate axis, which can directly reflect the load characteristics and can be used to find the change law of load characteristics.
- Daily load rate: the ratio of the average value and the maximum value of the load at all counting time points in a day. This index is used to reflect the stability of the load change in a day. The smaller the load change is, the higher the daily load rate is. A higher load rate is conducive to the economic operation of the power system.

4 Model

4.1 Drawing Method of Typical Daily Load Based on k-means

There are three traditional methods for drawing typical daily load curve: under a certain time scale, (1) select the most similar and non-abnormal daily load curve as the typical daily load curve; (2) select the daily load curve with the largest daily electricity consumption as the typical daily load curve; (3) select a fixed working day in the time scale as typical daily load curve. These methods have great limitations. It is possible that the selection principle of typical daily load can be made manually through the analysis of load characteristics, but the universality is not strong, and it is biased to take people's subjective will as the evaluation standard.

Aiming the defects of traditional typical daily load curve drawing, many scholars study it from different angles. There are three main methods: selecting typical daily load curve considering daily load rate, daily minimum load rate and daily peak valley difference, combining particle swarm optimization algorithm with classification algorithm, using short distance clustering method and semantic clustering method, The typical daily load curve is selected by fuzzy clustering algorithm, K-means clustering algorithm and other clustering methods.

Most of the existing typical daily load selection methods only consider the load characteristics such as daily load rate and average load. The typical curve of total load is reasonable for the power grid with stable power consumption period, but it cannot well represent the user load with large difference in power consumption period. This method combines user daily load rate and load clustering to draw typical user daily load curve.

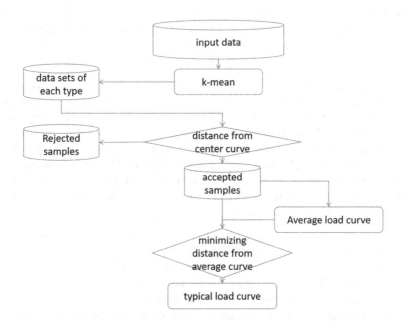

Fig. 1. Flow chart of typical daily load drawing method based on k-means

The flow chart of the method is shown in Fig. 1. The typical daily load drawing method is as follows:

- Step 1: preprocess the data.
- Step 2: eliminate all 0 value loads
- Step 3: remove the edge value. The k-means [21] clustering method is used to cluster the load curve, and 5% of the edge value of each class is eliminated. From the general error is positive probability distribution, 95% probability samples can

represent the whole sample, so it is reasonable to eliminate 5% of the edge value of each class.

- Step 4: select the load closest to the average load as the typical daily load curve in the remaining sample set.

4.2 Load Classification Method Based on Approximate Representation of Curve Symbol Aggregation

Compared with the regional aggregate load, the single user load has greater uncertainty and volatility, so the analysis of the curve needs to be "refined". The curve symbol aggregation approximation can roughly represent the power consumption of users for a period of time with simple symbols, eliminating the impact of analysis complexity caused by the fluctuation of single user load curve. In this method, the curve symbol aggregation is used to approximate the daily load curve, and the power consumption status of each period is counted. The typical daily load curve of each user is divided into five types: pure valley power consumption type, valley power consumption and horizontal power supplement type, all day power consumption type, daytime power consumption type, valley power consumption and peak power supplement type, idle type. The flow chart of the method is shown in Fig. 2. The basic operation flow of load classification method based on curve symbol aggregation approximation is as follows:

- Step 1: aggregate curve symbols to approximate the representation. According to the power consumption, the user power is divided into two states: idle and power consumption. The two states of the electricity consumption curve and the idle state are expressed by A and B symbol aggregation.
- Step 2: daily load curve classification. The sign aggregation of electricity consumption curve of daily load curve is approximate. The daily load curves are divided into six type: 1.pure valley period type, 2.valley-flat period type, 3.peak-valley-flat period type, 4.peak-flat type, 5.peak-valley type, 6. idle type. The categories are explained as follows:

  (1) Pure valley period type: only when user consumes electricity in the valley period, that is, only the valley period has A.
  (2) Valley-flat period type: when user consumes electricity in the valley and flat periods, that is, only the valley-flat period has A.
  (3) Peak-valley-flat period type: all periods have A.
  (4) Peak-flat type: when user consumes electricity in the peak and flat periods, that is, only the peak-flat period has A.
  (5) Peak-valley type: when user consumes electricity in the peak and valley periods, that is, only the peak-valley period has A.
  (6) Idle type: all periods are B.

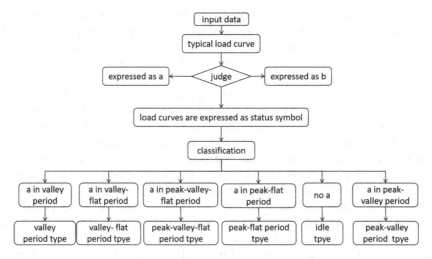

Fig. 2. Flow chart of Load classification method based on approximate representation of curve symbol aggregation.

5 Experimental Results

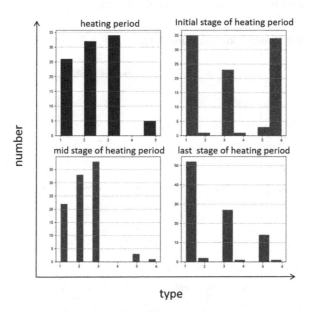

Fig. 3. Type statistics of different periods.

This part is the data of 97 electric boilers with heat reservoir in Jilin Province from 2019 to 2020. The typical daily load curve of each time scale is drawn by using the K-means based typical daily load drawing method in Sect. 4.2, and the load classification method based on the approximate representation of curve symbol aggregation is shown in Sect. 4.3.

There are six types of electricity consumption, which are identified as: 1. pure valley period type, 2. valley-flat period type, 3. peak-valley-flat period type, 4. peak-flat type, 5. peak-valley type, 6. idle type. The statistics of users in each heating period is shown in Fig. 3. Through the overall analysis, more users are pure valley period type, valley-flat period type and peak-valley-flat period type, the more uses consume in valley, the less electricity price is, so most of users are price driven. In the initial stage of heating, the most of users are valley-flat period type, peak-valley-flat period type and idle type. In the middle stage of heating, the most of users are pure valley period type, valley-flat period type and peak-valley-flat period type, which is same as the overall heating period. At the end of the heating period, the most of users are pure valley period type, peak-valley period type and peak-valley-flat period type. Through statistical comparison, it can be seen that there is more electricity consumption in the middle stage of heating, and the electricity consumption is relatively regular. At the beginning and end of heating, there is less electricity consumption, and the electricity consumption is irregular.

The following analyzes the characteristics of typical daily load curve of five types of useful electricity in each heating period except idle type.

5.1   Pure-Valley Power Consumption Type

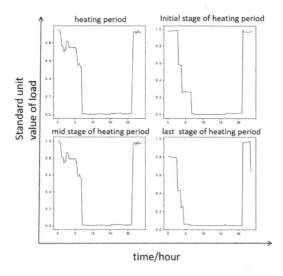

Fig. 4. Typical daily load curve of pure-valley power consumption type in each period.

Table 1. Characteristics of pure valley power consumption type

Heading level	Heating period	Initial stage of Heating Period	Mid stage of heating period	Last stage of heating period
Daily load rate	0.34	0.29	0.34	0.26

The daily load rate of typical daily load curve of pure valley power consumption type is shown in Table 1, and the daily load rate of each period is relatively low. According to the typical daily load curve of pure valley power consumption type in each period in Fig. 4, it can be seen that the power consumption at the beginning and end of heating is relatively low, and the power consumption is relatively consistent.

5.2 Power Consumption in Valley Section and Power Supplement in Horizontal Section

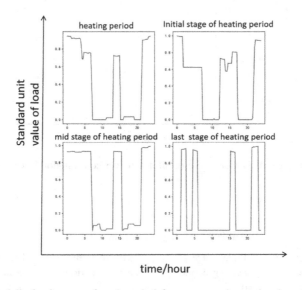

Fig. 5. Typical daily load curve of each period for power consumption in valley section and power supplement in horizontal section.

Table 2. Load characteristics of power consumption in valley section and power supplement in horizontal section.

Heading level	Heating period	Initial stage of heating period	Mid stage of heating period	Last stage of heating period
Daily load rate	0.44	0.45	0.51	0.26

Table 2 shows the daily load rate of typical daily load curve of valley section power consumption and horizontal section power supplement type, while the daily load rate at the end of heating period is relatively low, as shown in Fig. 5, The trend of supplementary electricity consumption is relatively more in the middle and middle periods, and the supplementary electricity consumption is less in the period from 2:00 to 12:00.

5.3 All-Day Electricity Type

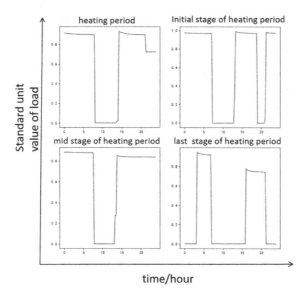

time/hour

Fig. 6. Typical daily load curve of all-day power consumption in each period.

Table 3. All-day load characteristics.

Heading level	Heating period	Initial stage of heating period	Mid stage of heating period	Last stage of heating period
Daily load rate	0.7	0.65	0.73	0.33

The daily load rate of typical daily load curve of all-day electricity consumption type is shown in Table 3. The daily load rate is lower at the end of heating period, and higher at the beginning and middle of heating period. As is shown in Fig. 6,which is typical daily load curve of all day power consumption type in each period, it can be seen that in heating period, power consumption is from 15:00 to 7:30 which is peak and valley section; in initial heating period, power consumption is from 21:00 to 7:00 which is in valley section, supplementary power consumption is from 13:00 to 19:30 which is in peak and valley section; in mid heating period, power consumption is from 15:00 to 7:30 which is in peak and valley section. At the end of the heating period, the power consumption is from 15:30 to 21:00 which is in the peak-valley section, and is from 3:00 to 7:00 which is in the valley section.

5.4 Daytime Power Consumption

The daily load rate of typical daily load curve of daytime power consumption type is shown in Table 4. The daily load rate of each period is relatively low, and there is no daytime power consumption type load in heating period and middle heating period. In the initial stage of heating, the peak and average power consumption is 7:30–17:00. At the end of the heating period, the power consumption is from 9:30 to 17:30 (Fig. 7).

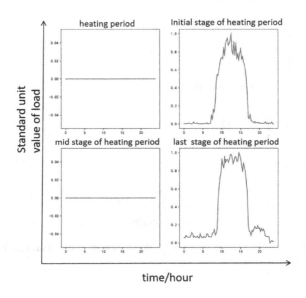

Fig. 7. Typical daily load curve of daytime power consumption in each period.

Table 4. Load characteristics of daytime power consumption type.

Heading level	Heating period	Initial stage of heating period	Mid stage of heating period	Last stage of heating period
Daily load rate	0	0.28	0	0.35

5.5   Power Consumption in Valley Section and Power Supplement in Peak Section

The daily load rate of typical daily load curve of valley section power consumption and peak section power supplement type is shown in Table 5, and the daily load rate of each period is low. As is shown in Fig. 8,which is typical daily load curve of valley power consumption and peak power supplement, it can be seen that this type of power consumption is roughly same with that of pure valley power consumption, but the power consumption is in 20:30–7:00 which is peak-valley period, only half an hour is the peak period.

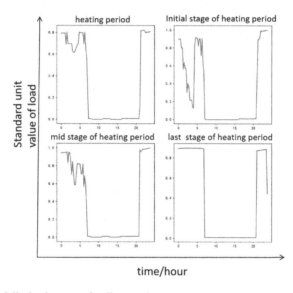

Fig. 8. Typical daily load curve of valley section power consumption and peak section power supplement type in each period.

Table 5. Load characteristics of valley power consumption and peak power supplement.

Heading level	Heating period	Initial stage of heating period	Mid stage of heating period	Last stage of heating period
Daily load rate	0	0.38	0.3	0.36

6   Summary

In this paper, the load classification method based on the approximate representation of curve symbol aggregation is used to classify the massive power consumption patterns. Through the overall analysis for the data of 97 electric boilers with heat reservoir in

Jilin Province from 2019 to 2020,The situation of overall heating period is similar to that of middle heating period, which most of users are pure valley section power consumption type, valley time flat section supplementary power consumption type and all day power consumption type. At the beginning and end of the heating period, the pure valley section uses more electricity and the whole day type uses more electricity. The medium-term heating load rate is high. In the whole heating period, there are four types of users: pure-valley period type, valley-flat period type, peak-valley-flat period type, peak-valley type. The numbers of user type are 24, 32, 34 and 5, respectively.

Acknowledgement. This work was financially supported by State Grid Technology Project (522300190009): Research and application of the key technologies of source-demand partici-pation in market transactions based on data value discovery.

References

1. Mehrpooya, M., Ghorbani, B., Hosseini, S.S.: Thermodynamic and economic evaluation of a novel concentrated solar power system integrated with absorption refrigeration and desalination cycles. Energy Convers. Manage. **175**, 337–56 (2018)
2. Bird, L., Lew, D., Michael, M., Carlini, E.M.: Wind and solar energy curtailment: a review of international experience. Renew. Sustain. Energy Rev. **65**, 577–86 (2016)
3. Luo, G., Li, Y., Tang, W., Wei, X.: Wind curtailment of China's wind power operation: evolution, causes and solutions. Renew. Sustain. Energy Rev. **53**, 1190–201 (2016)
4. Jacobson Mark, Z., Delucchi Mark, A.: Providing all global energy with wind, water, and solar power, Part I: technologies, energy resources, quantities and areas of infrastructure, and materials. Energy Policy **39**, 1154–69 (2011)
5. Lund Peter, D., Lindgren, J., Mikkola, J., et al.: Review of energy system flexibility measures to enable high levels of variable renewable electricity. Renew. Sustain. Energy Rev. **45**, 785–807 (2015)
6. Guo, L., Yang, S., Liu, Y., et al.: Typical day selection method for capacity planning of microgrid with wind turbine-photovoltaic and energy storage. Proc. CSEE **40**(643(08)), 94–105 (2020). (in Chinese)
7. Liu, J., Guo, L., Yang, S., et al.: Optimal sizing for multi PV-ESS microgrids in distribution network. Power Syst. Technol. **42**(9), 2806–2815 (2018). (in Chinese)
8. Piao, M., Shon, H.S., Lee, J.Y., et al.: Subspace projection method based clustering analysis in load profiling. IEEE Trans. Power Syst. **29**(6), 2628–2635 (2014)
9. Wang, Y.S., Liu, Y.Z., Kirschen, D.S.: Scenario reduction with submodular optimization. IEEE Trans. Power Syst. **32**(3), 2479–2480 (2017)
10. Gao, Y., Hu, X.Y., Liang, W., et al.: Multi-objective bilevel coordinated planning of distributed generation and distribution network frame based on multiscenario technique. IEEE Trans. Sustain. Energy **8**(4), 1415–1429 (2017)
11. Shi, L., et al.: Load Classification method using deep learning and multi-dimensional fuzzy C-means clustering. Proc. CSU-EPSA **31**(7), 43–50 (2019). (in Chinese)
12. Huang, X., Hu, T., Ye, C., Xu, G., Chen, L.: Electric load data compression and classification based on deep stacked auto-encoders. Energies **12**(4), 653 (2019)
13. Yang, C.C., Soh, C.S., Yap, V.V: A systematic approach in load disaggregation utilizing a multi-stage classification algorithm for consumer electrical appliances classification. Front. Energy. **13**, (002), 386–398 (2019)

14. Yang, L., Yang, H., Yang, H., Liu, H.: GMDH-based semi-supervised feature selection for electricity load classification forecasting. Sustainability **10**(1), 1–16 (2018)

15. Biscarri, F., Monedero, I., García, A., Guerrero, J.I., León, C.: Electricity clustering framework for automatic classification of customer loads. Expert Syst. Appl. (2017)

16. Song, J., Cui, Y., et al.: Load curve clustering method combining improved piecewise linear representation and dynamic time warping. Autom. Electric Power Syst. **45**(2), 89–96 (2021). (in Chinese)

17. Piscitelli, M.S., Brandi, S., Capozzoli, A.: Recognition and classification of typical load profiles in buildings with non-intrusive learning approach. Appl. Energy **255**(1), 113727.1–113727.17 (2019)

18. Dehghan-Dehnavi, S., Fotuhi-Firuzabad, M., Moeini-Aghtaie, M., Dehghanian, P., Wang, F.: Decision-making tree analysis for industrial load classification in demand response programs. IEEE Trans. Ind. Appl. **PP** (99), 1 (2020)

19. Lin, S., Gu, X.: Power load profile classification method based on neural network of sparse automatic encode. Power Syst. Technol. **44**(9), 3508–3515 (2020)

20. Piao, M., Ryu, K.H.: Local characterization-based load shape factor definition for electricity customer classification. IEEJ Trans. Electr. Electron. Eng. **12**, S110–S116 (2017)

21. Macqueen, J.: Some methods for classification and analysis of multivariate observations. In: Proceedings of Berkeley Symposium on Mathematical Statistics & Probability (1965)

A Data Processing Method for Load Data of Electric Boiler with Heat Reservoir

Feng Xiao[1] , Zhenyuan Li[2], Baoju Li[2], Chang Liu[3],
Yuhang Qiu[1] , Shang Wang[1], and Tao Peng[1(✉)]

[1] College of Computer Science and Technology,
Jilin University, Changchun 130012, China
{xiaofeng19,qiuyhl8}@mails.jlu.edu.cn,
{wangshang,tpeng}@jlu.edu.cn
[2] State Grid Jilin Electric Power Company Limited, Changchun, China
[3] State Grid Jilin Electric Power Research Institute, Changchun, China

Abstract. Data preprocessing is the first and critical step in various tasks due to existed quite a few dirty records. Although there are some corresponding methods for data from electric power system, those methods are less systematic and hard to handle various situations. In this paper, we integrate a data cleaning procedure that can handle various dirty records types present in the heat-storage electric boiler load data, including duplicated values, missing values, and abnormal values. The procedure is clear and consists of three modules connected in the pipeline, including data filtering, data cleaning, and data normalization in turn. The data cleaning is the core module in this procedure and includes missing points complementation and outliers processing. In order to demonstrate the effectiveness of the procedure, two real load power datasets from the heat-storage electric boiler are used in our experiments and the results demonstrate the effectiveness and availability of this work.

Keywords: Data cleaning · Missing value complementation · Outlier processing

1 Introduction

The construction of the Ubiquitous Electric Internet of Things (UEIOT) is a crucial goal for the power grid revolution. In the architecture of UEIOT, the sensorial layer is the bottom of the overall system and is used to collect data, and these data are the basis of supporting the overall system. With the same experience of all other industrial collections, the sensorial layer always acquires some dirty records during data collecting, transmission, and use due to equipment faults, unsteady signal channel, and environment interference. The load power data is the most important collecting item for the grid power information collection system. Since the load power data collected have large amounts of missing and outlier values, the subsequent application and analyses based on them will lack of accuracy and even not work. Facing the situation, if we can

© Springer Nature Switzerland AG 2021
D.-S. Huang et al. (Eds.): ICIC 2021, LNCS 12837, pp. 395–405, 2021.
https://doi.org/10.1007/978-3-030-84529-2_33

effectively restore missing points and amend outlier points by the inner structure of data, not only the integrity of data can be promised but also the subsequent analyses are also more highly reliability. Although a few methods are used to identify and process dirty records, a comprehensive and systematic cleaning procedure is still dearth for those problems. Therefore, this work integrates a data preprocessing procedure relying on existed theories and methods, including data filtering, data cleaning, and data normalization. More specifically, in this paper, the research object is the raw collecting samples from heat-storage electric boilers, and the process mainly includes duplicate filtering, missing points complementation, outlier points identification and amendment, and data normalization.

The remainder of this paper is organized as follows. In Sect. 2, we briefly review related work. The overall data cleaning procedure is illustrated in Sect. 3. To validate the effectiveness and availability of the procedure, in Sect. 4, we display two experimental cases using two different load datasets. Finally, we give a concluding remark in Sect. 5.

2   Related Work

There are a lot of research results for data cleaning. As for missing value complementation, common methods include Mean Interpolation (MI) [1, 2], Similar Mean Interpolation (SMI) [3], multiple interpolation (MUI) [4, 5]. MI, based on distance metrics, and if the distance of sample properties is measurable, employs valid average from corresponding property to complete missing value, otherwise, use the mode to finish the complementation. SMI, firstly, classifies all the samples into a few certain types, and then complete missing value using the average of all samples from a single type. As for MUI, it is a method based on probability random. MUL assumes the interpolation is random. In practice, MUI firstly estimates a set of interpolations and adds noise for them, and then selects the fittest interpolation according to some principles. In addition, there are some other available methods to handle the problem, such as modeling prediction (MP), high-dimensional mapping method (HDM), compression-aware, and matrix complementation [6]. Modeling prediction employs missing attributes as prediction targets and divides the dataset into two categories according to whether it contains missing values for a particular attribute, uses existing machine learning algorithms to predict the missing values. The fundamental drawback of MP is that if the other attributes are not relevant to the missing attribute, the filling method by forecasting is meaningless; and if the evaluation index of MP is quite accurate, the missing attribute is not necessary to exist in the dataset; the general situation is somewhere in between. High-dimensional mapping uses the one-hot encoding to map attributes to high-dimensional space. Specifically, the attribute value including K discrete value ranges is expanded to K + 1. If the attribute value is missing, the K + 1 attribute value after the expansion is set to 1. HDM is the most accurate method because the method retains all the information and does not add any additional information in the whole processing. If all variables are changed in this way during preprocessing, it will greatly increase the dimensionality of the data. In practice, manual interpolation is also a good choice, and it directly uses operators' subjective estimate value to supplement. The manual interpolation may not completely conform to the objective facts, in many cases, however, it has a promising performance based on operators' understanding of related data field. There are also a few methods to solve outlier identification, including Neighbor-based, Model-based, and Density-based [7, 8]. The Neighbor-based [9] method has the advantage to handle multi-dimension data, while the method needs large amounts of resources to compute. The Model-based [10] method is suitable for low-dimension data and the distribution of the sample is a prerequisite.

With advances in artificial intelligence, some researchers try rebuilding missing value by using deep learning [11]. For example, an improved generative adversarial network [12, 13], based on conventional single interpolation complementation, employs adaptive neural fuzzy inference model [14] to complete missing load power values. However, such methods are resource-intensive and require large amounts of historical data for model training.

3   Data Pre-process Procedure

Due to sampling faults, some dirty records exist in the historical load dataset. As for electrical power information collection system, dirty records mainly include duplicated values, missing values, and abnormal values. The existence of bad load data not only increases the difficulty of subsequent data analysis and method modeling but also seriously affects the accuracy and credibility of the experimental results. Therefore, in order to ensure the smoothness of the subsequent analysis and modeling as well as the accuracy and reliability of the experimental results, a comprehensive data cleaning procedure is integrated for raw data from heat-storage electric boilers.

The procedure consists of three modules, including data filtering, data cleaning, and data normalization. The *raw data* denotes original collecting samples in Fig. 1. The whole procedure connects three modules in a pipeline. Firstly, filtering module removes duplicate parts the existed in the raw data, and then the cleaning module works to complete missing values and handle outliers. Finally, the *normal data* is obtained by the normalization module. The process of the data pre-process procedure is shown in Fig. 1. The overall procedure is general electrical load data. Note that the first module and the third module can be easily changed for real situations.

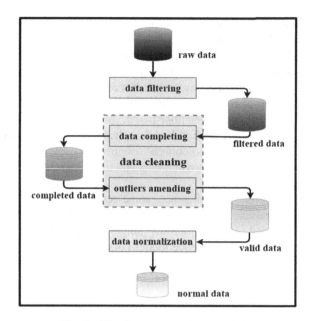

Fig. 1. The data pre-process procedure.

3.1   Filtering Duplicate

Although the module is not the core part of overall procedure, it is necessary and indispensable for facing situation. The duplicates not only waste the time and space resources but also influence subsequent analysis of electric power system. In experiments, the ID of customers is the unique identification for each sample point. Therefore, the feature can be employed when removing the duplicate parts. Whereas the method is used in our experiments needs to be changed according to different datasets. The specific implementation steps of data duplication module are as follows:

1) Set user ID and time as unique ID,

2) The load data matrix $M \in R^{d*f}$ is established, where d represents the time range of load data, f represents the sampling frequency of load data of regenerative electric boiler, initializes matrix m with −1. For example, if the time line is from 2019/11/10 to 2020/5/28, the initialization unique identification matrix value $M \in R^{183*96}$ with a total of −1 is established.

3) Input user data, if the matrix $M_n[t] > P_n[t]$, jump to 3, otherwise continue to cycle input, where $M_n[t]$ is the unique identification matrix value of user n on time t, where $P_n[t]$ is the original power value of user n on time t, which may be repeated;

4) No replacement, jump 3.

5) The end of the replacement cycle obtains the data m of the user load power curve without repeated data.

3.2 Outliers Identification and Mark

The feature vectors decide the classification of load customers so that it is necessary to select fit feature vectors before implementing data clustering. Due to existed quite a few outliers in original collecting samples, we need to handle outliers before selecting feature vectors. There are a few common processes such as the Empirical Correction method (EC), Threshold Discrimination method (TD), the Transversal Comparison method (TC), Vertical Comparison method (VC), and so on. The EC method identifies and amends outliers mainly relying on experts' knowledge. The TD method, based on probability statistics, set confidence intervals to handle abnormal data. The TC and VC method both employ load similarity of neighboring data, whereas the difference is that the transversal comparison is based on the electrical inertia of customer in a relative short period, while the vertical comparison is based on the electrical habits of customer. Generally, we need to select an appropriate process according to the different situations.

In this work, the method based on probability statistics is used to identify outliers. In the experiment, a threshold ε is set by data statistics. It can be considered as outlier, if the answer of Eq. (1) is True. More specifically, the ε is set to 0.1, i.e., if the differences between a certain point and its neighbor points are both greater than 10% of the average value of its neighbor points, it is an outlier. And then marking those outliers use a distinctive value that is set to -1 in experiment. The amendment of outliers is the same as the missing value complementation (Sect. 3.1), i.e., the outliers identified are regarded as missing points. Equation (1) is used to identify outliers, and I denotes discriminative answer for a pending outlier v_o, abs denotes the Absolute Value function. The v_l, v_r denote values of the left and right of v_o respectively.

$$I = \begin{cases} True, \ if \ abs(v_l - v_o) > K \ and \ abs(v_r - v_o) > K, K = \varepsilon * (v_l + v_r)/2 \\ False, \quad otherwise \end{cases} \quad (1)$$

3.3 Missing Values Complementation

The data complementation requires four consecutive steps in this module. The first step uses customers' ID concatenating frequency timestamp as unique identification. The second step constructs an initial matrix of the timeline for each customer and all elements are set -1. For example, there is a timeline between 2019/11/10 to 2020/05/28 for load dataset, generating a matrix $M \in R \times R$ that denotes load data from a certain customer. The row of matrix M denotes range of timeline and the column denotes frequency of sampling. The goal of the second step is to promise the timeline integrity of every customer. The third step replaces the negative values in initial M using real load data according to Eq. (2). And the left negative values in matrix M are considered as missing values. The fourth step is the key component and using a weighted average of neighbor to fill the missing values. Specifically, in the fourth step, three situations can be met by experimental statistics. With less than or equal to α consecutive missing values, the weighted average a_{w_r} of the normal neighbors from the row is used for filling. When more than α and less than γ, the a_{w_c} is the weighted average of normal neighbors from the column. As for more than γ consecutive missing values, the weighted average a_{w_h} can be generated by the same period load data if there are such historical data, otherwise using the a_{w_c} to fill the missing value. The steps 3 and 4 are implemented recurrently until there is no missing value in dataset.

$$M_i[t] = \begin{cases} P_i[t], & if\ P_i[t] > M_i[t] \\ M_i[t], & otherwise \end{cases} \tag{2}$$

$$f_i = \begin{cases} a_{w\_r}, & ifN_{missing} \le \alpha \\ a_{w\_c}, & if\ \alpha < N_{missing} < \gamma \\ a_{w\_h}, & if\ N_{missing} > \gamma \end{cases} \tag{3}$$

In Eq. (2), i denotes customer ID, and t denotes sampling time. $M_i[t]$ is the matrix generated in second step of this module, the $P_i[t]$ represents the actual load data point of the customer i at time t. In Eq. (3), the f_i denotes filling value, and $a_{w\_r}, a_{w\_c}, a_{w\_h}$ correspond weighted average from the row, the column, and from the same period historical data respectively. The $N_{missing}$ denotes the number of consecutive missing values, and α, γ denote two thresholds used to control operation.

$$a_w = w_1 v_1 + w_2 v_2 + \cdots + w_n v_n / n \tag{4}$$

$$w_i = d(a, v_i) \Big/ \sum_i d \tag{5}$$

As for Eqs. (4) and (5), a_w denotes weighted average, and w_i is the weight of neighbor data v_i, the token d is a distance function.

3.4 Data Normalization

To unify the data dimension, standardize the data interface of the subsequent modules of the electrical power system, data normalization is used in last procedure. Note that Min-Max Normalization (see Eq. (6)) is chosen in our experiment and other formula can be used relying on real requirements. In Eq. (6), x_{max}, x_{min} denote the minimum and maximum value of normalizing sequence respectively.

$$x' = x - x_{min}/x_{max} - x_{min} \tag{6}$$

4 Cases Analysis

4.1 Setups

The datasets are used in this work from State Grid Jilin Electric Power Company. Our experiments use two data sources, Source 1 and Source 2, that mainly include users' power, time, and a few other load information. The electricity consumption category includes 9 categories, covering 40 users, with a total data volume of 3085 in Source 1, and there are 97 users and 11939 records in Source 2. The statistical details of Source 1 and Source 2 are shown in Table 1. Note that the sampling frequency of the two data sources is the same that is one time per 15 min, and each record has 96 points.

Table 1. The statistical information of original data from Source 1 and Source 2.

Data sources	Users	Sampling frequency	Dimension	Size
Source 1	40	per 15 min	96	3085
Source 2	97	per 15 min	96	16939

5 Cases Analysis

The original data are continuously missing more and more domain-specific, which is not effective with the traditional interpolation method. By observing and analyzing the load data, two types of missing values can be statistically represented: (1) showing as null values in the records, (2) showing as negative values in the records. In addition to missing table content, i.e., showing missing values in the records, there is also statistical timeline data missing, e.g., the statistical timeline is from 2019/11/10 to 2020/5/28, but the data exists only from 2019/11/10 to 2020/3/21, or load data for intermediate dates are missing. If the load users with missing timelines are used for statistical analysis, there may be a gap with the real situation, so the timeline needs to be completed. Besides, there are also several days of continuous missing data that need to be filled in. Therefore, it is difficult to use a single model for missing values, for example, it is difficult for the predictive completion model to complete long continuous variables that are missing and for the matrix filling theory to complete data with large differences in electricity consumption behavior among customers. Figure 2 shows an example of duplicate parts existed in raw data.

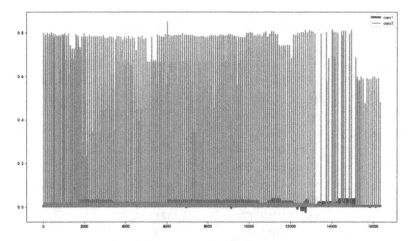

Fig. 2. The duplicate data from a customer. (Color figure online)

For Source 1, we found that the load records of the user with the ID 0065 have obvious duplicate data (see Fig. 2) in the data filtering phase. In Fig. 2, two different data distribution can be found, the orange line and the blue line. So, we process data of Source 1 that mainly includes (1) filtering 174 duplicate daily load data, (2) from 2017/10/15 to 2018/4/16 as a period to fill the missing points using −1 that indicates empty values, (3) using the data cleaning method mentioned in Sect. 3 to complete and modify abnormal data. The change of data in Source 1 by data cleaning procedure is shown in Table 2.

Table 2. The change of data in Source 1 by data cleaning procedure.

Users	Load records	Filtering records	Missing points	Outliers
40	3085	0	0	0
19	3477	174	0	0
19	3477	0	89726	0
19	3477	0	0	21757

For Source 2, the period of the dataset is 201 days from 2019/11/10 to 2020/5/28. The statistical daily load records less than 100 only have 8 users, and the number of users with duplicate records is 12. The dataset information for filtering the duplicate daily load records is shown in Table 3. We handle data of Source 2 that mainly includes (1) filtering 1820 duplicate daily load records, (2) from 2019/11/10 to 2020/5/28 as the period of the complementary data to fill missing points using −1 that denotes the empty value, (3) data complementation method is used to fill the negative values using the calculated values, (4) outliers are processed by identification and amendment method. The change of data in Source 2 by data cleaning procedure is shown in Table 4.

Table 3. The data filtering for Source 2.

Index	User	Before filtering	After filtering
1	0001	402	201
2	0002	344	145
3	0003	201	3
4	0004	224	103
5	0005	74	8
6	0005	367	166
7	0006	314	136
8	0007	378	177
9	0008	271	133
10	0009	402	201
11	0010	576	375
12	0011	341	170

Table 4. The change of data in Source 2 by data cleaning procedure.

Users	Load records	Filtering records	Missing points	Outliers
97	16939	0	0	0
97	19497	1820	0	0
97	19497	0	25705	0
97	19497	0	0	75677

Take the customer with ID 0059 as an example, an energy technology company, the missing value is set to −1, and the period is from November 10, 2020, to May 28, 2020. Figure 3 shows the comparison of the load power of the customer with ID 0059 before and after data complementation. The horizontal axis is 15 min as the time unit, the vertical axis is the power of the load (KW), without calculating the first and last missing values assigned to 0, there are 154 missing values.

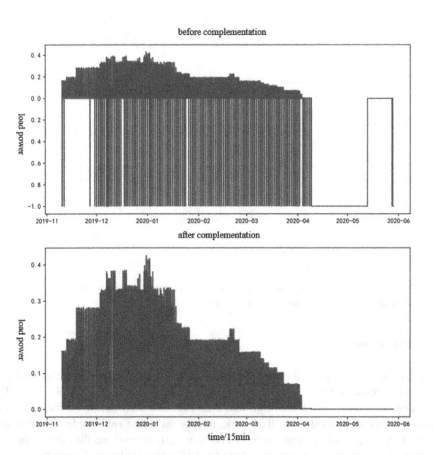

Fig. 3. The comparison of the load power of the customer with ID 0059 before and after data complementation.

Figure 4 shows the load curve before and after the amendment of the customer with ID 0074 for a certain period. Red rectangle circles the spikes that need to be repaired in Fig. 4 (left), if the difference with the neighbor points is greater than 10% of the average value of the neighbor points. It is believed that repairing small changes in the points will only smooth the curve and will not cause bad consequences. A total of 3 points were repaired on that day in Fig. 4 (right). The values of outliers and their nearest neighbors are shown in Table 5. Taking for instance point 12297, 10% of the mean value of the neighbors is 0.05, the difference between the left neighbor and the anomaly is 0.5, and the difference between the right neighbor and the anomaly is 0.5, both of which are much larger than 0.05, so it is an outlier. And remainder points also have the same character.

Table 5. Examples of outliers.

Point index	Left neighbor	Right neighbor	Outlier
12297	0.5094	0.5208	0.057
12299	0.5208	0.5128	0.057
12301	0.5128	0.5158	0.0571

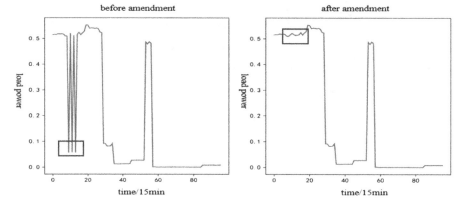

Fig. 4. Comparison before and after repair, the red rectangle indicates the change points by outliers' identification and amendment. (Color figure online)

6 Conclusion

In this paper, we integrate a data pre-process procedure relying on some existed theories and methods for heat-storage type electric boiler load data s. [4, 7–9, 11].` . The procedure consists of data filtering, data cleaning, and data normalization. The data cleaning is the core module in this work, mainly including outlier identification and amendment, and missing values complementation. To demonstrate the effectiveness of the procedure, two data sources from State Grid Jilin Electric Power Company are used in experiments. The two original datasets have quite a few missing values and outliers by analysis of statistics, and the time axis of some customers is not consistent. To make all load user data time axis aligned, the load is completed according to the longest heating period in dataset. Besides, load data have some missing values every 15 min, whereas the other are missing values for several hours. So, different situations should be completed by different methods, and long-missing load complementation and short-missing load complementation are used. For spiky points in the dataset, the method based on probability statistics is used in experiment. The experimental results (see Sect. 4) indicate the effectiveness and availability of this work.

Acknowledgement. This work was financially supported by State Grid Technology Project (522300190009): Research and application of the key technologies of source-demand partici-pation in market transactions based on data value discovery.

References

1. Jiahe, Y., Cui, J., Wang, F.: Research on interpolation method of wind measurement data of wind farm and analysis of interpolation results deviation. Sol. Energy **322**(02), 26–35 (2021)
2. Ang, L.: Research on missing data interpolation method based on electricity load. Distrib. Energy **5**(04), 74–80 (2020)
3. Shi, H., Yang, X., Wang, P.: An improved mean imputation clustering algorithm for incomplete data. J. Jiangsu Univ. Sci. Technol. (Nat. Sci. Ed.) **34**(04), 51–56 (2020)
4. Curnow, E., Hughes, R.A., Birnie K., Crowther, M.J., May, M.T., Tilling K.: Multiple imputation strategies for a bounded outcome variable in a competing risks analysis. Stat. Med. **40**(8), 1917–1929 (2021)
5. Jiang, C., Wang, H.: Research on AP point compensation and selection method based on multiple interpolation. Chain Comput. Commun. **32**(23), 47–49 (2020)
6. Chen, L., Chen, S.-C.: Survey on matrix completion models and algorithms. J. Softw. **28**(6), 1547–1564 (2017)
7. Li, S., Zhang, M., Chen, Y.: Abnormal traffic data recognition based on outlier algorithm. J. Shangdong Jiaotong Univ. **27**(02), 17–23, 37 (2019)
8. Zhao, T., Zhang, Y., Wang, J.: Load outlier identification method based on density-based spatial clustering and outlier boundaries. Autom. Electr. Power Syst. 1–12 (2021).
9. Li, M., Li, M., Ren, Q., Shen, Y.: Density-based clustering outlier detection in long-term monitoring data. J. Hydroelectr. Eng. 1–11 (2020)
10. Nie, B., Xue, H., Wang, X.: Outlier detection of air quality spatio-temporal data based on nearest neighbor analysis. Stat. Res. **34**(08), 61–70 (2017)
11. Xu, J.: Research on anomaly data recognition algorithm based on clustering and neural network. North China Electric Power University (Bei Jing) (2019)
12. Bao, P., Liu, Y.: Research on fault identification based on improved deep model in combination of generative adversarial networks under unbalanced data sets. J. Electron. Meas. Instrum. **33**(03), 176–183 (2019)
13. Zhang, Y., Lin, H., Guan, Y., Liu, C.: GAN image super-resolution reconstruction model with improved residual block and adversarial loss. J. Harbin Inst. Technol. **51**(11), 128–137 (2019)
14. Fenglin, P., Luo Minzhou, X., Xiaobin, T.Z.: Path tracking control of an omni-directional service robot based on model predictive control of adaptive neural-fuzzy inference system. Appl. Sci. **11**(02), 838 (2021)

Aggregate Model for Power Load Forecasting Based on Conditional Autoencoder

Yuhang Qiu[1] , Yong Sun[2], Chang Liu[3], Baoju Li[2], Shang Wang[1],
and Tao Peng[1(✉)]

[1] College of Computer Science and Technology,
Jilin University, Changchun 130012, China
qiuyh18@mails.jlu.edu.cn,
{wangshang, tpeng}@jlu.edu.cn
[2] State Grid Jilin Electric Power Company Limited, Changchun, China
[3] State Grid Jilin Electric Power Research Institute, Changchun, China

Abstract. Load forecasting is an important machine learning problem in the field of power system, which is of great significance to power system source load balance, power supply planning and system maintenance. With the development of power technology, fine-grained load data become more easily available, which puts forward new requirements for load forecasting accuracy. In this paper, we use a deep learning framework, namely conditional autoencoder, to forecast day-ahead load. And the conditional autoencoder extracts the characteristics of load time series through deep neural network and finds the sequence pattern of historical data. In addition, we use the load data of different time scales for training, so as to improve the accuracy of the model. In this paper, the real data of regenerative electric boiler in Jilin Province are used for load forecasting, and the effectiveness of the model is verified.

Keywords: Multivariate time series · Neural network · Conditional autoencoder · Electric boiler · Load forecasting

1 Introduction

The forecasting model [1] which based on machine learning has been developed rapidly, and the prediction effect is getting better and better. Time series prediction has been applied in many fields, such as traffic flow prediction [2] in the field of transportation, weather prediction [3] in the field of weather, human behavior prediction [4] using brain waves, intelligent financial forecasting [5], and machine learning can be used to predict the power consumption of consumers [6] and solar power generation [7] by using time series information. In the field of electric power, forecasting is particularly important [8]. Power planning needs to forecast the power consumption of user, so as to know how much installed capacity of the power plant in a certain area. New energy and load are traded in the market. Forecasting the power generation and consumption of new energy in the trading day is necessary, and the accuracy is important. The accuracy of forecasting is not only related to the power of buyers and sellers, but also related to safety of electricity system, because less power generation will affect the

© Springer Nature Switzerland AG 2021
D.-S. Huang et al. (Eds.): ICIC 2021, LNCS 12837, pp. 406–416, 2021.
https://doi.org/10.1007/978-3-030-84529-2_34

overall balance of the power system, leading to the reduction of new energy and power customer interruption.

With the fine-grained power load data getting better, the research on power user forecasting becomes more and more hot. Through user load forecasting, the accuracy of load forecasting of the whole area and power grid system is increased. In order to increase the consumption of new energy, Jilin province develops and uses regenerative electric boilers. In this paper, according to the real data of different user categories, different regions and multiple time scales of regenerative electric boilers in Jilin Province, a time aggregation prediction model based on deep neural network conditional autoencoder (CVAE) [9] is constructed. By training the data of different time scales, all models are aggregated to improve the prediction effect of neural network. The main contributions of this paper are as follows:

- 1) This paper proposes an aggregation prediction model ACVAE based on CVAE, which can aggregate different time scale data to improve the prediction effect of the model.
- 2) Experiments are carried out on the real data set of electric boiler users in Jilin Province to verify the effectiveness of ACVAE.

2 Related Work

There are many ways of load forecasting, which can be roughly divided into three types: (1) typical day type [10]; (2) regression forecasting type [11]; (3) deep learning type [12–14]. In the case of completed data, the regression model and deep learning model based on machine learning have better prediction effect. In the case of missing data or long-term prediction, the typical day prediction method is better. Typical daily forecasting methods mainly extract the representative actual daily load in a certain period of time by clustering or manual method as the forecast value. Regression method makes the load data of the past period as the features of input data, brings it into the linear regression equation to forecast the next day's data. Deep learning method uses deep network to mine the data feature for forecasting.

Conditional autoencoder (CVAE) is a deep generation network. By learning the probability distribution characteristics of data, it can sample and generate prediction data. CVAE has been applied in many fields, such as augmentation of medical data [15], classification [16], etc. Its original model is the Variational autoencoder (VAE). Variational autoencoder (VAE) [17] samples the original samples and obtains the latent variable space Z after training. Although it can fit the original data well and achieve the function of dimensionality reduction and feature extraction, Z uses Gaussian noise. Although VAE can produce high-quality data, the general recovered data cannot be randomly sampled in the latent variable space to achieve this effect, the model needs to traverse all the initial distributions. A large part of this problem is due to the VAE compression method, which makes the Z distribution unknown. In order to generate higher quality data, artificially generate data by condition, CVAE is constructed.

This paper uses conditional generation characteristics of CVAE, takes historical load curve as conditional characteristics, and forecasts load as output and input characteristics, so as to build forecasting model.

3   Framework

3.1   Problem Definition

In this paper, we focus on the task of load forecasting of electric boiler with heat reservoir. Let $z_t \in R^N$ denote the value of the user load power at time step t, N is the number of the user. Input a sequence of historical K time steps of the user load power, $X = \{z_{t1}, z_{t2}, \cdots, z_{tK}\}$, our goal is to predict a Q-step-away sequence of future load power $Y = \{z_{tK+1}, z_{tK+2}, \cdots, z_{tK+Q}\}$. Otherwise, other auxiliary features can input into the model, such as day of the week, day of the weather. The auxiliary features are concatenated with the input X as a condition C, and the input and output of the autoencoder are Y. By establishing the conditional joint probability distribution of Y, we can obtain the predicted value of Y under condition X. Finally, in the decoder, input condition x and median of latent variable, and output predicted values.

Two kinds of prediction at different time scales are carried out. They are day ahead forecasting and intra-day forecasting. They are defined as follows:

Day-ahead prediction: input the 96-point load power data of the previous day and output the 96-point load power data of the forecast day, where K and Q are 96.

Intra-day prediction: input 16-point power data in the first 4 h and output 16-point power data in the next 4 h, where K and Q are 16.

The mean square error MSE was used as the evaluation index, as is shown in formula 1.

$$\text{MSE}(x, y) = \frac{1}{m} \sum_{t=1}^{m} (x_t - y_t)^2 \tag{1}$$

3.2   Model Architecture

In this paper, the time aggregation prediction models of different user categories, different regions and different time scales are established. According to the aggregation of prediction results of different users, the multi-user aggregation prediction results are obtained. According to the power prediction results, the power prediction results of different time periods are obtained.

The load collection frequency is aggregated at different levels to construct different time series, and each time series is predicted separately. Finally, the sequence frequency is restored to the initial frequency, and the prediction results are integrated to obtain the final results.

The prediction model used is to fit the joint probability distribution of load in different time scales with the neural network CVAE model, which can get the joint probability distribution of load under artificial characteristics. The time aggregation is carried out in different time scales, and the aggregation prediction model is established, so as to carry out the prediction of load.

Our model framework is shown in Fig. 1. The coding layer of CVAE model is composed of one full connection layer, and the decoding layer is composed of a full connection layer. We divide the historical series into three different time scales for training, which are 15 min, 30 min and 60 min respectively. Combined with feature sequence, we input three different CVAE models in turn for training, and obtain three different prediction results, and then carry out weighted summation to obtain the final prediction results.

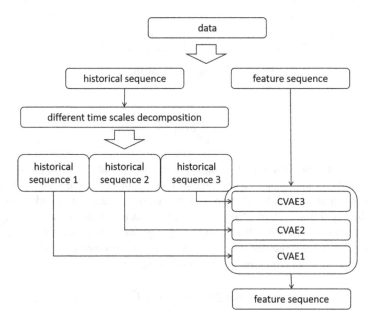

Fig. 1. The framework of the model.

3.3 Encode

The encode is used to reduce the dimension of historical load series data and future load series modeling data, and extract features from them. The structure of the encode is shown in Fig. 2.

We transform the historical load sequence data C and the future load sequence data Y into vectors, and then concatenate them, and input them into the encoder network in the form of vectors. The latent layer size of neural network is set to 80, the activation function is tanh, and the connection is made by full connection layer, which is in the day ahead forecasting. The final generation is the conditional joint probability distribution of 20-dimensional daily load series, which is hidden vector Z.

It is very difficult to calculate the distribution of Y based on condition C by using probability statistics, the distribution of Y based on condition x can be obtained simply and efficiently by combining probability distribution simulation with neural network encoder. This distribution can be expressed as P(Z|Y,C). The density function of Gaussian probability distribution is formula 2.

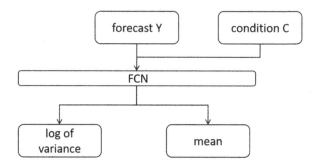

Fig. 2. The framework of encode.

$$f(x) = \frac{1}{\sqrt{2\Pi}\sigma}\exp(-\frac{(x-\mu)^2}{2\sigma^2})$$

(2)

It can be seen from formula 2 that the Gaussian probability distribution of Z can be obtained by solving the mean and variance of Z distribution, so the mean and variance of conditional probability distribution can be fitted by neural network encoder. In addition, the input and output of each layer of the neural network are floating-point numbers, which cannot control the positive and negative values of the network output, but the distribution variance needs to be positive, so the variance is set as the logarithm of the variance. The encoder can be expressed as formula 3.

$$(\mu_Z, \log \sigma_Z^2) = E(Y \oplus C)$$

(3)

3.4   Decode

Combining condition distribution Z and condition C, the decode samples from Z to generate data Y. its structure is shown in Fig. 3.

The decoder randomly samples from the joint probability distribution of the encoder output, adds the logarithm of the mean and variance to obtain Z, then inputs it to the fully connected neural network, and finally outputs y. The structure of the fully connected layer is consistent with that of the encoder. The decode is shown in formula 4.

$$Y = D(Sample(P(\mu_Z, \sigma_Z^2)))$$

(4)

Where d represents the encoder of the model proposed in this paper, from which the decoder learns the conditional joint probability distribution of the data. In the deep neural network model, gradient descent method is generally used to solve the optimization function, but in the VAE series model, because the hidden variable Z is used to simulate the joint probability distribution of data, the chain derivation is broken in Z, and the gradient descent algorithm needs to propagate forward continuously, so it needs to be improved, and the multiple parameter technique is used in CVAE.

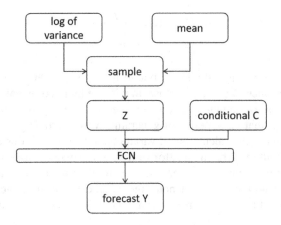

Fig. 3. The framework of decode.

3.5   Variational Derivation

With reference to, we derive the sequence prediction model of CVAE. The formal description of this problem is as follows: we obtain an i.i.d. sequence (Y,C), where Y are forecasting sequences, C are conditional sequences, and (y, c) denotes a sample in (Y,C). In this paper, we use c to forecast y. We assume $p(z|c)$ is distribution of Y and resample y from $p(y|z, c)$. In addition, we assume that the conditional distributions of the potential representations of Y and C are parameterized. In view of the description of the above model, our goal is to find the joint log likelihood $\log p(y|c)$ (called conditional log likelihood) of y and c for a given sequence of data points.

We can compute $p(y|c)$ in formula 5.

$$
\begin{aligned}
p(y|c) &= \int p(y, z|c)dz \\
&= \int p(y|z)p(z|c)dz
\end{aligned}
\tag{5}
$$

$\log p(y|c)$ can be rewritten as a formula 6.

The log of $p(y|z)$ and $p(z|c)$ is difficult to compute. Therefore, it can denote $p(z|y, c)$ by $q(z|y, c)$, which is easy to compute. The KL distance of $p(z|y, c)$ and $q(z|y, c)$ is regarded as fitting error, and its lower bound is required to compute. The KL distance is $D_{KL}(p||q)$, and denotes their distance. And the lower bound of variation can be rewritten as a formula 7.

$$L = \log p(y|c)$$
$$= L^V + D_{KL}(q(z|y,c)||p(z|y,c)) \qquad (6)$$
$$\geq L^V$$

$$L^V = -D_{KL}(q(z|y^{(i)}, c^{(i)})||p(z|c^{(i)}))$$
$$+ E_{q(z|y^{(i)}, c^{(i)})}(\log(p(y^{(i)}|z, c^{(i)}))) \qquad (7)$$

Finally, it is deduced that the lower bound is the sum of the two terms. The first ensures that the distance of q and p is close, that is, the input corresponding conditional c can output the corresponding y. The second term is the negative number of reconstruction error of y, that is, the error of generating y and sample y.

In order to simplify the calculation of variational lower bound. Ensuring $p(z|c)$, p, q are Gaussian distribution. And the feedforward neural network is used to fit them.σ and μ denote the value of variance and mean, which are fitted by neurons, the i denotes the sample label in batch processing, j denotes the parameter for fitting, generally set as 1.

The first term is fitted by compression model, as is shown in formula 8.

$$-D_{KL}(q(z|x^{(i)}, y)||p(z|y))$$
$$= \frac{1}{2} \sum_{j=1}^{J} (1 + \log(\sigma_{z_j}^{(i)^2}) - \mu_{z_j}^{(i)^2} - \sigma_{z_j}^{(i)^2}) \qquad (8)$$

The second term is fitted by generation model, as is shown in formula 9.

$$-\log(p(x^{(i)}|z^{(i)}, y)) = \sum_{j=1}^{D} \frac{1}{2} \log(\sigma_{x_j}^2) + \frac{(x_j^{(i)} - \mu_{x_j})^2}{2\sigma_{x_j}^2} \qquad (9)$$

4 Experimental Results

4.1 Data

The user load data is obtained from the voltage and power of Jilin power supply company and its subordinate counties. At present, the sampling frequency of electric power data in China is 15 min, which is provided by Jilin Power Grid. The data record is from 2018-10 to 2019-5, raw data has 3085 daily load records, and we fill reasonable data in missing value, and all data are measured by electric boiler with heat reservoir. The types of users are general industrial power, commercial power, large industrial power, rural residential power, urban residential power, non-industrial power, non-residential lighting power and primary and secondary school teaching power, etc. The users are from Baicheng city, Changchun city, Baishan city, Jilin city of Jilin Province. Cleaned data is shown in Table 1.

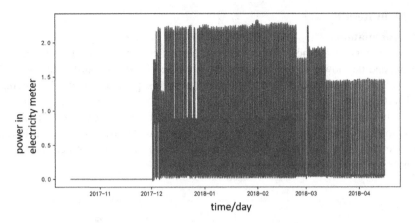

Fig. 4. Power curve of a user.

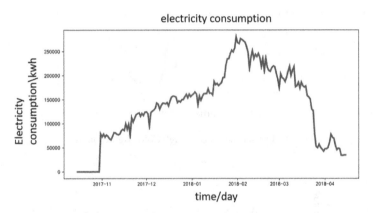

Fig. 5. Electricity consumption curve.

Table 1. Cleaned data

	Users	Daily load records	The number of user type
Cleaned data	19	3477	8

In order to better understand the data set, the power curve of user is drawn, as shown in Fig. 4, and the overall electricity consumption curve of user is shown in Fig. 5. It can be seen that the daily power of the data are similar, so it is reasonable to use the load power of previous day as the input data.

4.2   Main Result

4.2.1   Simulation

96 power points of the next day were modeled by conditional joint probability distribution, and the power of the next day was predicted by P50 quantile. Red is the true value, yellow is the predicted value, and blue line is predicted value by aggregate model in this paper, as is shown in Fig. 6.

16 power points in the next time were modeled by conditional joint probability distribution, and the power of 16 power points was predicted by P50 quantile. Red line is the true value, yellow line is the predicted value and blue line is predicted value by aggregate model in this paper, as is shown in Fig. 7.

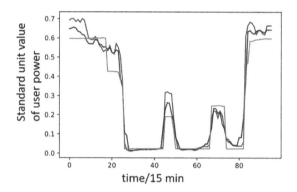

Fig. 6. Day-ahead forecasting (Color figure online)

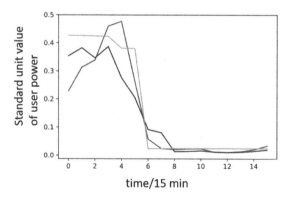

Fig. 7. Intra-day forecasting (Color figure online)

4.2.2   Compared Results

Table 2. MSE of forecasting models.

Model	MSE
ACVAE for intra-day forecasting (in this paper)	**0.007**
CVAE for intra-day forecasting	0.008
ACVAE for day-ahead forecasting (in this paper)	**0.009**
CVAE for day-ahead forecasting	0.01
The latest day load forecasting method for day-ahead forecasting	0.01

It can be seen from Table 2 that the effect of aggregation forecasting model is better than forecasting model without aggregation, no matter for intra-day forecasting or for day-ahead forecasting.

5   Summary

In this paper, we proposed an aggregate model for Power load forecasting based on Conditional autoencoder (ACVAE), and we set two kinds of forecasting scenarios, which are day-ahead and intra-day forecasting, and forecast 96 points and 16 points respectively. Experiments are carried out on the real data of electric boiler with heat reservoir in Jilin Province, which verify the effectiveness of the proposed model, and show that the time aggregation model can effectively improve the accuracy of forecasting. In the next step, we will combine the weather information, improve the data cleaning method, and mine more load time series features to improve the prediction accuracy in the further.

Acknowledgement. This work was financially supported by State Grid Technology Project (522300190009): Research and application of the key technologies of source-demand participation in market transactions based on data value discovery.

References

1. Deb, C., Zhang, F., Yang, J., Lee, S.E., Shah, K.W.: A review on time series forecasting techniques for building energy consumption. Renew. Sustain. Energy Rev. **74**(February), 902–924 (2017)
2. Xia, D., Wang, B., Li, H., Li, Y., Zhang, Z.: A distributed spatial-temporal weighted model on MapReduce for short-term traffic flow forecasting. Neurocomputing **179**, 246–263 (2016)
3. Rezaeian-Zadeh, M., Zand-Parsa, S., Abghari, H., Zolghadr, M., Singh, V.P.: Hourly air temperature driven using multi-layer perceptron and radial basis function networks in arid and semi-arid regions. Theor. Appl. Climatol. **109**(3–4), 519–528 (2012)
4. Shen, F., Liu, J., Wu, K.: Multivariate time series forecasting based on elastic net and high-order fuzzy cognitive maps: a case study on human action prediction through EEG signals. IEEE Trans. Fuzzy Syst. **6706**(c), 1–13 (2020)

5. Niu, T., Wang, J., Lu, H., Yang, W., Du, P.: Developing a deep learning framework with two-stage feature selection for multivariate financial time series forecasting. Expert Syst. Appl. 148,113237 (2020).

6. Javeed Nizami, S., Al-Garni, A.Z.: Forecasting electric energy consumption using neural networks. Energy Policy 23(12), 1097–1104 (1995)

7. Cheng, J., Huang, K., Zheng, Z.: Towards better forecasting by fusing near and distant future visions. In: The Thirty-Fourth AAAI Conference on Artificial Intelligence. pp. 3593–3600 (2020)

8. Suganthi, L., Samuel, A.A.: Energy models for demand forecasting - A review. Renew. Sustain. Energy Rev. 16(2), 1223–1240 (2012)

9. Sohn, K., Yan, X., Lee, H.: Learning structured output representation using deep conditional generative models. In: Advances in Neural Information Processing Systems. pp. 3483–3491 (2015)

10. Rhodes, J.D., Cole, W.J., Upshaw, C.R., Edgar, T.F., Webber, M.E.: Clustering analysis of residential electricity demand profiles. Appl. Energy. 135(2014), 461–471 (2014)

11. Jain, R.K., Smith, K.M., Culligan, P.J., Taylor, J.E.: Forecasting energy consumption of multi-family residential buildings using support vector regression: Investigating the impact of temporal and spatial monitoring granularity on performance accuracy. Appl. Energy. 123, 168–178 (2014)

12. Chen, P., Liu, S., Shi, C., Hooi, B., Wang, B., Cheng, X.: NeUCAST: seasonal neural forecast of power grid time series. IJCAI International Joint Conference Artificial Intelligence, July 2018, pp. 3315–3321 (2018)

13. Wu, L., Kong, C., Hao, X., Chen, W.: A short-term load forecasting method based on GRU-CNN hybrid neural network model. Math. Probl. Eng. 2020, 1428104 (2020)

14. Tan, M., Yuan, S., Li, S., Su, Y., Li, H., He, F.H.: Ultra-short-term industrial power demand forecasting using LSTM based hybrid ensemble learning. IEEE Trans. Power Syst. 35(4), 2937–2948 (2020)

15. Pesteie, M., Abolmaesumi, P., Rohling, R.: Adaptive augmentation of medical data using independently conditional variational auto-encoders. IEEE Trans. Med. Imaging 38(12), 2807–2820 (2019)

16. Zhao, Y., Hao, K., Tang, X.S., Chen, L., Wei, B.: A conditional variational autoencoder based self-transferred algorithm for imbalanced classification. Knowl. Based Syst. 128(9), 106756 (2021)

17. Kingma, D.P., Welling, M.: Auto-encoding variational Bayes. In: 2nd International Conference on Learning Representations, ICLR 2014 - Conference Track Proceedings, pp. 1–14 (2014)

Geographical Entity Community Discovery Based on Semantic Similarity

Miao Yu, Zhanquan Wang[(⊠)], Yajie Pang, and Yesheng Xu

East China University of Science and Technology, Shanghai 200237, China
zhqwang@ecust.edu.cn

Abstract. Geographical entity community discovery aims at discovering geographical entities that are closely related to each other. It has shown great practical value in many social applications such as economic development, administrative management, and resource utilization. Traditional research methods use network topology to represent the relationship between geographical entities, but cannot accurately define semantic relationships. To deal with this problem, this paper uses the network news that provides massive semantic information, and integrates the semantic information and spatial information into the geographic entity community discovery algorithm. An edge weight calculation method based on semantic association strength, geographic entity influence, and boundary connection distance is proposed. Experimental analyses show that compared with existing methods, the new algorithm improves the accuracy of spatial community division. A case study on real-world datasets shows that the experimental results of the new method are better than the existing methods and are more satisfied with common sense.

Keywords: Geographical entity · Community division · Semantic similarity · Edge weight

1 Introduction

Community discovery is an important research work in complex network analysis. A collection of tightly connected nodes is called a community. With the development of geographic information system (GIS) [1], cartography [2] and remote sensing technology, the importance of geographic entity similarity measurement in geographic information and other fields has become increasingly prominent. The methods of measuring the correlation of geographical entities are divided into spatial correlation and semantic correlation. In general, these two measures are modeled separately to check the similarity between entities from different perspectives, or to make a linear combination of the similarity results [3]. However, the semantic and spatial attributes of geographic entities interact. The above analysis method ignores the internal correlation between the two attributes to a certain extent, and the single research method does not conform to people's decision-making habits. At the same time, semantic information is incomplete and immeasurable, which makes it difficult to explore the deep relationship between geographical entities. The actual needs of geographical entity correlation can

© Springer Nature Switzerland AG 2021
D.-S. Huang et al. (Eds.): ICIC 2021, LNCS 12837, pp. 417–429, 2021.
https://doi.org/10.1007/978-3-030-84529-2_35

no longer be satisfied only through the spatial relevance of entities or the relevance only through semantic computation.

With the development of the mobile internet, network news provides people with a large amount of information. As a key element of geographical entities, place names contain rich semantic information, which is often ignored by researchers. In this paper, news semantic information and spatial geographic information are fused, and a geographical entity community discovery method based on semantic correlation strength, geographical entity influence and boundary connection distance is proposed.

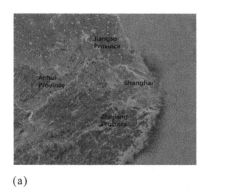
(a)

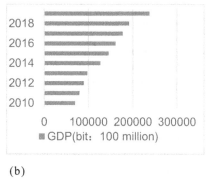
(b)

Fig. 1. (a) Geographical satellite map of the Yangtze River Delta. (b) GDP trend chart of the Yangtze River Delta from 2010 to 2019.

Application Domains. Regional economic policy formulation is a key strategy for development. For example, Fig. 1(a) shows a geographical satellite map of the Yangtze River Delta region. After the scope of the Yangtze River Delta was determined in 2010, the GDP curve from 2010 to 2019 is shown in Fig. 1(b). Due to reasonable regional division, the economy of the Yangtze River Delta has developed rapidly. This paper provides an efficient method of community division to help define reasonable areas for policy making in economic development. In terms of administration, it helps countries and regions to conduct better administration. In addition, the algorithm also helps more reasonable division of resources, improve the utilization of resources.

Contribution. This paper formulates a spatial community discovery algorithm that integrates semantic association strength, geographical entity influence, and boundary connection distance. A case study on real-world datasets shows that the proposed method has a higher accuracy of community division, and finds geographical entities that are easily missed before. Experimental analyses show that the proposed method is better than the existing methods and more practical.

Challenges. Challenges to the method proposed in this paper mainly come from the following aspects:

1) The selection of news dataset, the preprocessing of dataset
2) Geographic boundary irregularities, length of data acquisition
3) Complete node relationship requires a huge node network

Research Scope and Structure of Paper. This paper focuses on community discovery based on geographical entities as nodes in the graph network and combined with semantic relationships. Modeling of non-overlapping communities [4] and dynamic communities [5] is not considered. Furthermore, a thorough study of the division of communities into potential geographical entities can only be done by domain experts and is outside the scope of this article. The boundary number based geographical community discovery algorithm has been widely used [6], and the comparative analysis with these methods is beyond the research scope of this paper. The rest of this paper is structured as follows: Sect. 2 describes the related research. Section 3 states the problems of the study. Section 4 puts forward a new geographical entity partition model. Section 5 describes an experiment based on a real dataset. Section 6 discusses the evaluation of experimental results for the proposed method. Section 7 concludes the paper and previews future work.

2   Related Research

Community discovery stems from graphic segmentation in computer science and hierarchical clustering in sociology. In the past decade, community discovery has received extensive attention from researchers at home and abroad, and many pioneering research results have been achieved. Traditional community discovery algorithms include the method based on graph segmentation, such as Kernighan-Lin algorithm [7], and geographic entity clustering algorithm, the typical algorithms are hierarchical clustering, partition clustering, spectral clustering, etc. GN algorithm [8] is a classic community discovery algorithm, which belongs to split hierarchical clustering algorithm and was originally proposed by Newman M E J and Girvan M. Vincent D Blondel et al. proposed a community discovery algorithm based on modularity [9], which is one of the best community discovery algorithms at present.

The division of geographic entity community is an important theoretical problem of geographic information system (GIS). At present, there are many algorithms for community division of geographical entities, generally from the perspective of geospatial similarity and semantic similarity of geographical entities.

Research based on geospatial correlation uses geospatial relationship (distance relationship, direction relationship and network topological relationship) model to calculate the community division of geographic entities. The Spirit [10] system presents spatially dependent models of spatial inclusion, proximity, and orientation, and uses overlapping regions, euclidean distance, and azimuth models for quantitative calculations.

The research based on semantic similarity mainly focuses on the similarity of geographical entities in terms of categories and topics. However, the semantic computation of geographical entity is subjective and too simple. Simply name classification and feature description of geographical entities cannot accurately cover the complex semantic information of geographical entities, and it lacks practicability to meet the practical application needs. The relationship between a large number of things in social life can be described by various network structures. Because of its flexibility and powerful revealing ability, it is widely used in user-product relationships, social networks, co-citation of articles, etc.

Although the current network mining algorithms have been very mature, the correlation of geographical entities studied in this paper is deeply affected by spatial factors. Therefore, we integrate the spatial characteristics of geographical entities into the relevant algorithms of network mining, hoping to more accurately show the internal correlation of geographical entities. Community division of geographical entities can better classify, merge and study target geographical entities.

3   Problem Description

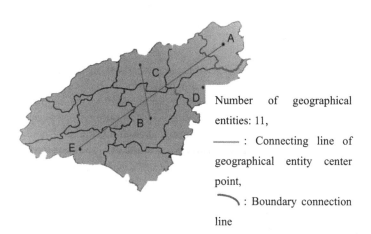

Number of geographical entities: 11,

———— : Connecting line of geographical entity center point,

⌐ : Boundary connection line

Fig. 2. Geographic entity symbiosis network diagram

3.1   Basic Concepts

In order to explain the basic concepts, the geographical entity network in Fig. 2 is used as an example.

Definition 1. A **geographic entity symbiosis network** G = (N, E) is composed of a node set N and an edge set E, where each node in the set N represents a geographic entity. Each edge E represents the co-occurrence relationship of two geographical entities. In Fig. 2, the geographical entity network is composed of geographical entities A, B, etc.

Definition 2. **Semantic association strength** Sij consists of the co-occurrence times of two geographical entity names in the news data set. For example, if the names of two geographical entities are co-occurrence times in n news texts in the dataset, the Sij value is n.

Definition 3. **Geographical entity influence** Pi is obtained from the influence factors of geographical entity i, such as economic strength, scientific and technological level, etc.

Definition 4. **Boundary connection distance** Cij is determined by the boundary connection distance of two geographic entities i and j. If there is no boundary connection distance, set Cij as 1. In Fig. 2, the red curve represents the boundary connection distance of two geographical entities. For example, the boundary connection distance between geographical entity B and geographical entity D is C_{bd}, whereas the boundary connection distance between geographical entity A and geographical entity D is longer.

Definition 5. **Geographic center distance** Dij, consisting of the distance between geographical entity i and j at the geographical center of the satellite map. In Fig. 2, the blue line represents the distance from the center of the geographical entity. For example, the distance between the center of the geographical entity A and E is D_{ae}, while the distance between the center of the geographical entity B and the center point of the geographical entity C is shorter.

3.2   Problem Statement

Given:

1) A geographical entity symbiosis network G = (N, E), which consists of a group of geographical entities,
2) The semantic association strength Sij between any two geographical entities i and j in the network,
3) The influence of geographical entity i, Pi,
4) Boundary connection distance between geographical entity i and geographical entity j in the network Cij,
5) The distance of geographical center point between any two geographical entities i and j in the network Dij.

Find: Geographical entity community division results.

Objective: To improve the accuracy of geographical entity community division.

Constraints:

1) The semantic correlation strength Sij between each two geographical entities is greater than or equal to a minimum threshold θS
2) The results are correct and complete.

4   Spatial Community Division

4.1   New Weight Calculation Formula

Inspired by Newton's gravitation formula, Stewart John Q proposed a gravitation model [11] to study the influence of distance on the correlation of things. Based on this feature, Mengyu Yan et al. [12] proposed the calculation method of edge weight in Louvain algorithm, which is defined as follows:

$$A_{ij} = \frac{S_{ij}}{D_{ij}^2} \tag{1}$$

Where, A_{ij} represents the weight of the connection edge between node i and node j, Sij represents the semantic association strength of node i and j, and Dij is the geographical center distance between node i and node j.

The edge of the weight calculation formula of the effective ways to incorporate semantics and geographical information, but because of the influence of the gravity model, part of the geographic shape special geographical entity due to the unsuitable selection of center distance makes the weight value, the deflection distance on the other hand some but semantic correlation strong geographical entity will be due to the large D^2ij value of formula to calculate the weight value of smaller than the actual. In view of the above two points, this paper proposes a new calculation formula for edge weight, whose definition is as follows:

$$A_{ij} = P + \frac{S_{ij}C_{ij}}{D_{ij}^2} \tag{2}$$

Where the new weight calculation formula adds the influence P of geographical entities and the boundary connection distance Cij of geographical entities. If there is no boundary connection between two geographical entities, Cij will be set to 1 in order not to affect the calculation.

Compared to the original weight calculation formula, a new weighted formula considering the influence geographical entities, semantic strength is closely related to the influence of the geographical entity, after adding the boundary connection distance, making connections between geographical entity with boundary more accord with the actual weights, also reduces the distance is far, but has a certain semantic relation intensity of geographical entity between the edge of the weight is more accurate. After considering more accurate geographical factors, the new formula can effectively improve the accuracy of edge weight, and then improve the accuracy of community division.

4.2   Strength of Semantic Association

Semantic similarity has always been the focus of natural language processing (NLP) [13]. For example, semantic similarity of search terms and search results in the search scenario, semantic similarity of recommended content and interest content in the recommendation system scenario, semantic similarity of statement A and statement B in the machine translation scenario, etc. At present, advanced algorithms include DSSM, CNN-DSSM, LSTM-DSSM and other deep learning models [14]. With the development of the Internet, various news media and hundreds of millions of web pages every day have produced a huge amount of text information that can not be efficiently processed by people. Therefore, from the perspective of news text, this paper defines the number of co-occurrence of geographical entity names as semantic correlation strength, and integrates semantic information into the geographic entity discovery algorithm.

4.3   Influence of Geographical Entities

In the early 20th century, the study of influence has attracted the attention of researchers. In the 1950s, Kate [15] et al. found that influence plays a crucial role in both daily life and voting activities. With the development of economy and Internet, the connection between regions is getting closer, thus forming spatial communities with certain characteristics. Regional influence refers to the power of a region to change the development of other geographical entities due to its own economy, science and technology, culture and geographical position. For example, Shanghai, Nanjing and Hangzhou rely on strong geographical entity influence to promote the formation of the Yangtze River Delta region. Therefore, the study of measuring, analyzing and modeling the influence of geographical entities has important theoretical value and practical role.

In this paper, random forest algorithm [16] was used to sort the importance of a large number of features of geographical entities and extract the most important N important factors. For the assignment of feature weight, the analytic hierarchy process is adopted. Analytic Hierarchy Process (AHP) [17] refers to a decision-making method that divides decision-related elements into goals, criteria, programs and other levels, and then conducts qualitative and quantitative analysis.

Geographical entity influence P is defined as follows:

$$P = N \cdot M \tag{3}$$

$$N = \overrightarrow{facroe\ 1\ score, factor\ 2\ score, \ldots, factor\ n\ score} \tag{4}$$

$$M = \begin{pmatrix} factor\ 1\ weight \\ factor\ 2\ weight \\ \cdot \\ \cdot \\ \cdot \\ factor\ n\ weight \end{pmatrix} \tag{5}$$

Where influence P is the dot product of vector N and vector M, N represents the score of N feature factors screened out by geographical entities, and M represents the feature weight of feature factors after decision by analytic hierarchy process. When the edge weight between two geographical entities is calculated, the influence P of geographical entities is calculated in the way of accumulation.

4.4 Boundary Connection Distance

Boundaries are the boundaries defined between geographical entities and territorial units, such as national boundaries and provincial boundaries. Borders have the following two most fundamental characteristics: geographical characteristics and political characteristics. Based on the above, the boundary plays a very important role in defining regional relations, dividing regional boundaries and coordinating regional management. As the most intuitive boundary feature, boundary connection distance also plays an important role in it. This paper attempts to explore the correlation of geographical entities from the perspective of boundary connection distance. The current advanced geographic information system (GIS) platform ArcGIS is used to obtain the accurate boundary connection distance between geographical entities.

5 Experiment

5.1 Data and Processing

In this paper, news text is used for semantic similarity analysis. Due to the different directions of news events transmitted by different websites and the overlapping of news events, and the low quality of news text of many websites, the blind selection of news text data sets may contain a lot of noise data. Therefore, the experiment selects high-quality news text data from Jiangsu news websites from March to May 2012. For the acquisition of semantic association strength, the co-occurrence number of place names was selected as the semantic association strength of geographical entities.

The preprocessed data was experimented in accordance with the experimental process, and the experimental flow chart is shown in Fig. 3.

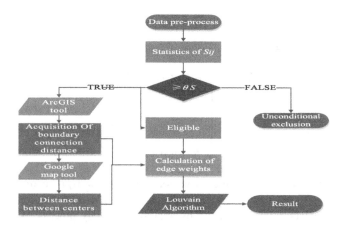

Fig. 3. Experimental flow chart

5.2   Semantic Association Strength and Center Distance

Selection of open source in Chinese word segmentation tools NLPIR to achieve geographical entity name recognition, after pretreatment, in a total of 37562 news text, the number of three types of co-occurrence modes are 78, 1308 and 5457, respectively, for the final result by merger way, add the statistical result of type two and the statistical result of type three to the statistical result of type one, such as Luhe district, Nanjing and other parts of the co-occurrence frequency of unity to the number of co-occurrence between Nanjing and other cities. The results of co-occurrence intensity of experimental statistics are regional division with attribution characteristics. The co-occurrence of place names used in this experiment refers to entities at important geographical locations in news texts, and its co-occurrence statistical results are more representative.

For the geographic center distance, this paper uses the electronic map service provided by Google Map to obtain the geographical center distance between geographical entities through the ranging tool.

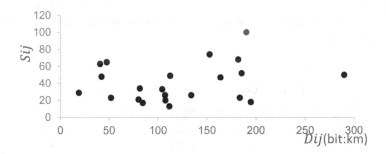

Fig. 4. Distribution of semantic intensity with distance

The distribution relationship between the semantic association intensity and the distance from the center of the geographical entity is shown in Fig. 4. It can be found from the figure that some points do not meet the requirement of the farther the distance is, the smaller the semantic association intensity is, such as the red dot in the figure, which is closely related to the influence of the geographical entity.

5.3   Calculation of Geographical Entity Influence

The experiment selects the GDP index, science and technology score and permanent resident population of geographical entities as the research objects through the sorting of characteristics. The feature weights were assigned through the analytic hierarchy process, and the feature weights of the above three features were 0.6/0.3/0.1 respectively (keeping one decimal). As for the characteristics of GDP, the GDP of each city in Jiangsu Province from 2012 to 2016 was queried by Jiangsu Provincial Bureau of Statistics. After the above processing, we can calculate the geographical entity influence of each city in Jiangsu Province.

5.4   Obtain Boundary Connection Distance

In this experiment, the current advanced geographic information system (GIS) platform ArcGIS is used to obtain the accurate boundary connection distance between geographical entities.

5.5   Geographical Entity Partitioning Based on Louvain Algorithm

In the experiment, the geographic entity symbiosis network was synthesized in each city of Jiangsu Province. According to the above experimental data, the edge weights calculated only by semantic correlation strength, the edge weights calculated by the existing gravity model weight formula and the results calculated by the new weight calculation formula were respectively divided into geographical entities using Louvain algorithm.

6   Experimental Results Evalution

The results of the geographical entity division of Jiangsu Province in this experiment are shown in Table 1. In Figs. 5, 6 and 7, the experimental division results are visualized respectively.

Table 1. Geographical entity division results

Result	Only semantic relevance strength	The gravity model	The new model
1	NJ,SZ,WX,CZ,NT,YZ,ZJ,TZ	NJ,CZ	NJ,SZ,CZ,WX,SZ,NT
2	HA,YC,SQ,LYG,XZ	SZ,WX	ZJ,YZ,TZ
3		ZJ,TZ,YZ,NT	HA,YC,SQ,LYG,XZ
4		HA,YC,SQ,LYG,XZ	

Fig. 5. Partition results of semantic correlation strength only considered

Fig. 6. Partition results of gravity model

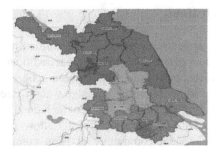

Fig. 7. Partition results of the new model

The results of the experiment were compared from the perspectives of Purity, Jaccard Coefficient, NMI and ARI, as shown in Table 2.

Table 2. Comparison of the results based on Purity, Jaccard coefficient, NMI and ARI

Methods	Purity	Jaccard coefficient	NMI	ARI
Only Sij	0.769	0.769	0.793	0.689
Gravity model	0.689	0.689	0.689	0.500
The new method	**0.923**	**0.846**	**0.829**	**0.792**

It can be seen from the experimental results that the geographical entity division results are different through different edge weight calculation methods. The main results are as follows:

1) In addition to the edge weight calculation method which only considers the strength of semantic correlation, the cities in Jiangsu Province are roughly divided into two communities, and the result is not accurate.
2) From the perspectives of Purity, Jaccard Coefficient, NMI, and ARI, the classification results of the new method all have the highest accuracy.
3) The original gravity model does not divide Suzhou, Nanjing and Wuxi together, but the new edge weight calculation method is more accurate. This reason is the gravity model is affected by the value D^2ij in the formula is too large, resulting in the calculated weight value is smaller than the actual value, thus leading to the incorrect division.
4) The special area Nantong. Due to its geographical location close to Shanghai and the port, Nantong develops rapidly and has a close connection with southern Jiangsu.

In general, the new edge weight algorithm improves the accuracy of geographical entity partitioning. Through boundary connection distance and geographical entity influence, it refines the results of geographical entity partitioning considering only semantic association strength algorithm, corrects the limitations of the original gravity model, and is more in line with the reality.

7 Summary and Outlook

This paper studies the correlation of geographical entities based on news text and geographical factors. In order to solve the problem that traditional research methods use network topology to represent the relationship between geographic entities, but cannot accurately define the semantic relationship, an edge weight calculation method based on semantic association strength, influence of geographic entities and boundary connection distance is proposed. Experimental results show that the new algorithm improves the accuracy of spatial community division compared with the existing methods. The method in this paper combines semantic information with geographic entity information to explore potential geographic patterns to a certain extent. In the future, we plan to make the selection of geographic information more detailed. Can we add other geographic information, such as geology and geomorphology? Semantic information mined from news texts has a strong timeliness, so we also plan to study geographic physical communities that change over time, which may reveal the life cycle and mobile trends of social life. The research on these issues has good prospects, but much more work needs to be done.

References

1. Goodchild, M.F.: Geographic information systems. Prog. Hum. Geogr. **15**(2), 194–200 (1991)
2. Daosheng, D.: The science of cartography in China. Cartogr. J. **21**(2), 145–147 (1984)
3. Larson, R.R.: Ranking approaches for GIR. Sigspatial Spec. **3**(2), 37–41 (2011)
4. Ni, L., Luo, W., Zhu, W., Hua, B.: Local overlapping community detection. ACM Trans. Knowl. Discov. Data (TKDD) **14**(1), 1–25 (2019)
5. Zhang, R., Jin, Z., Xu, P., Liu, X.: A dynamic clustering based on community detection. Cluster Comput. **22**(3), 5703–5711 (2019)
6. Qian, K.: Research on hierarchical clustering community discovery algorithm based on diverging number. Shanxi University (2015)
7. Meknassi, M., Aboulhamid, E.M., Cerny, E.: Algorithm for the graph-partitioning problem using a problem transformation method. Comput. Aided Des. **24**(7), 397–398 (1992)
8. Newman, M.E.J., Girvan, M.: Finding and evaluating community structure in networks. Phys. Rev. E **69**(2), 026113 (2004)
9. Blondel, V.D., Guillaume, J.L., Lambiotte, R., et al.: Fast unfolding of communities in large networks. J. Stat. Mech. Theory Exp. **2008**(10), P10008 (2008)
10. Purves, R.S., et al.: The design and implementation of SPIRIT: a spatially aware search engine for information retrieval on the Internet. Int. J. Geogr. Inf. Sci. **21**(7) 717–745 (2007)
11. Stewart, J.Q.: Demographic gravitation: evidence and applications. Sociology **11**(1/2) 31–58 (1948)
12. Yan, M., Jing, N., Zhong, Z., et al.: Geographical entity community mining based on spatial and semantic association. In: Proceedings of the 3rd International Conference on Computer Science and Application Engineering, pp. 1–6 (2019)
13. Socher, R., Bengio, Y., Manning, C.D.: Deep Learning for NLP (without magic) Tutorial Abstracts of ACL 2012, p. 5 (2012)
14. Shrestha, A., Mahmood, A.: Review of deep learning algorithms and architectures. IEEE Access **7**, 53040–53065 (2019)

15. Katz, E., Lazarsfeld, P.F.: Personal Influence, The Part Played by People in the Flow of Mass Communications, pp. 1–12. Free Press, New York (1955)
16. Verikas, A., Gelzinis, A., Bacauskiene, M.: Mining data with random forests: a survey and results of new tests. Pattern Recogn. **44**(2), 330–349 (2011)
17. Wind, Y., Saaty, T.L.: Marketing applications of the analytic hierarchy process. Manage. Sci. **26**(7), 641–658 (1980)

Many-To-Many Chinese ICD-9 Terminology Standardization Based on Neural Networks

Yijia Liu[(✉)], Shasha Li, Jie Yu, Yusong Tan, Jun Ma,
and Qingbo Wu

College of Computer, National University of Defense Technology,
Changsha, Hunan, China

Abstract. The ICD-9 terminology standardization aims to standardize collo-quial medical terminologies into ICD-9 standard terminologies. Due to the arbitrariness of natural language, these colloquial medical terminologies may be original terminologies or original terminologies combinations and correspond to standard terminologies or standard terminology combinations. Therefore, the ICD-9 terminology standardization task needs to solve the mapping of multiple original terminologies to multiple standard terminologies (namely the problem of many-to-many standardization). In this problem, when the top N (i.e. Top-N) ICD-9 terminologies with the highest probability are output as standard termi-nologies by the ICD-9 terminology standardization method based on BERT sorting, the output will be affected by the proportion of original terminologies in the original terminology combination. Due to the influence of different pro-portions, it is possible that multiple ICD-9 terminologies contained in Top-N are derived from a certain original terminology in the original terminology com-bination, resulting in a significant decline in the prediction effect. Therefore, this paper proposes a method for the standardization of many-to-many Chinese ICD-9 terminology based on neural networks: 1) The original terminology combi-nation split method based on named entity recognition and part-of-speech tag-ging; 2) The candidate terminology set construction and terminology standardization method based on N-gram and BERT. In order to better evaluate the effect of the method, based on the CHIP public dataset and a real-world electronic medical record, this paper constructs two many-to-many datasets, named as CHIP-MTM and SCD. The experimental results show that the method in this paper has achieved an accuracy improvement of 7.7% on both CHIP-MTM and SCD.

Keywords: ICD-9 · Similarity calculation · BERT classification

1 Introduction

ICD-9 [1] is the ninth edition of the International Classification of Diseases developed by the WHO, which classifies diseases according to surgical operations.

At present, standards of disease terminology used by hospitals at all levels across the country are different, and different doctors have different writing habits. The mapping of these different terminologies to unified standard terminologies is helpful for the research and communication of the medical community, and also helps the

© Springer Nature Switzerland AG 2021
D.-S. Huang et al. (Eds.): ICIC 2021, LNCS 12837, pp. 430–441, 2021.
https://doi.org/10.1007/978-3-030-84529-2_36

definition of medical insurance claims. Therefore, automatic standardization of disease terminology has become an urgent need [2].

The main content of the ICD-9 terminology standardization task is to find the corresponding standard terminology or standard terminology combination given in the ICD-9 standard terminology set for the original terminology or the original terminology combination. In order to achieve this task, our team proposed an ICD-9 terminology standardization method based on N-gram and BERT, named as ABTSBM [3]: Construct a Candidate Terminology Set (namely CTS) composed of top N (i.e. Top-N) ICD-9 terminologies with the highest similarity for each original terminology through the N-gram algorithm, and input the original terminology and its candidate terminology into BERT for 0–1 classification, and finally sort the classification results of BERT, and select the highest probability as the standard terminology output. However, the quality of CTS in this method directly restricts the performance of the method. The quality assessment of CTS includes Candidate Terminology Set Accuracy (CTSA) which is the probability that the correct standard terminology is included in the CTS; and Candidate Terminology Set Scale (CTSS) which is the value of N in Top-N. Obviously, CTSA directly determines the upper limit of the model's performance, while CTSS determines CTSA. However, expanding CTSS in order to obtain a higher CTSA will have limited improvement in model performance, and may even have a negative impact.

At the same time, in actual application scenarios, each standard terminology is indivisible, but the doctor usually uses abbreviations and other operations to merge multiple terminologies into one when recording; and the doctor may use some punctuation marks to connect two words which may construct a standard terminology together or correspond to a standard terminology separately. As shown in Table 1, for the standard terminologies "经尿道输尿管/肾盂取石术" ("transureteral ureter/renal pelvic lithotripsy") and "经尿道输尿管/肾盂激光碎石术" ("transurethral ureter/renal pelvic laser lithotripsy"), because they are only different in the final several words "取石术"和"激光碎石术" ("lithotripsy" and "laser lithotripsy"), the original terminology written by the doctor is often abbreviated as "经尿道输尿管/肾盂碎石取石术" ("Transurethral ureter/renal pelvic lithotripsy"),while for the original terminology "宫腔镜检查 + 诊刮术" ("hysteroscopy + curettage") and "经尿道前列腺电切术 + 尿道扩张术" ("transurethral resection of the prostate + urethral dilatation"), there are punctuation marks for linking, and the number of standard terminologies they correspond to is 1 and 2 respectively. In fact, most of the original terminologies and standard terminologies correspond to one-to-one, but one-to-many, many-to-one, and many-to-many (hereinafter collectively referred to as many-to-many) are still not to be ignored.

Table 1. Instances of the correspondence between the original and the standard terminology combination.

Original terminology	Standard terminology
经尿道输尿管/肾盂碎石取石术	经尿道输尿管/肾盂取石术 + 经尿道输尿管/肾盂激光碎石术
宫腔镜检查 + 诊刮术	宫腔镜诊断性刮宫术
经尿道前列腺电切术 + 尿道扩张术	经尿道前列腺切除术(TURP) + 尿道扩张

When ABTSBM is applied to many-to-many data, although the number of standard terminologies corresponding to the original terminology combination (which is the N) can be predicted by BERT, and then the Top-N of the BERT ranking result is taken to construct the standard terminology combination output. But the difficulty of many-to-many tasks is that the Top-N is directly sorted by BERT, and it is very likely that the ICD-9 terminologies obtained are similar terminologies to a certain original terminology in the original terminology combination. Such as the case of "经尿道前列腺电切术 + 尿道扩张术" ("transurethral resection of the prostate + urethral dilation") in Table 1, the length of "transurethral resection of the prostate" is much longer than "urethral dilation", and the two terminologies are not closely related. The results obtained by Top-N are likely to be irrelevant for "urethral dilation". If the predicted N is split based on punctuation and then predicted, it is necessary to ensure that the word information obtained after split is complete.

At the same time, in ABTSBM, the N-gram algorithm can greatly screen out literally similar terminologies from nearly 10 thousand ICD-9 standard terminologies. But in this task, there are still a lot of data that the original terminology and its corresponding standard terminology are not literally similar. These data require certain medical expertise to judge that the body structure and other information involved can be matched, so as to judge that they are similar terminologies. But some types of data have obvious characteristics and appear multiple times in the dataset. This paper believes that through semantic learning, we can learn that they are semantically similar. So they can be further included in the CTS. For the three examples shown in Table 2, the string based similarity between these original terminologies and their standard terminologies is low. For example, the standard terminologies of "右额钻孔硬脑膜下血肿引流术" ("right frontal drilling subdural hematoma drainage") rank only 81st in terminologies of similarity. Some ICD-9 terminologies that are far more similar than standard terminologies have become candidate standard terminologies. At the same time, take the data with the standard terminology "脑膜切开术" ("meningotomy") as an example. It appears many times in the dataset. And its original terminologies all have the words "血肿引流术" ("hematoma drainage") and have obvious characteristics. Training through BERT and reordering the CTS can effectively advance the ordering of these standard terminologies with insufficient literal similarity and make them enter the CTS.

Table 2. Instances of low similarity between the original and the standard terminology.

Original terminology	Standard terminology (Ranking)	Candidate standard terminologies selected based on N-gram (Ranking)
右额钻孔硬脑膜下血肿引流术	脑膜切开术 (81)	硬脑膜下钻孔引流术 (1)
右肾动脉栓塞术	经导管肾血管栓塞术 (38)	肾动脉取栓术 (1)
右额钻孔侧脑室穿刺外引流术	脑池穿刺 (63)	脑室钻孔引流术 (1)

In summary, the main content of this paper is: 1) Through the original terminology combination split method based on named entity recognition and part-of-speech tagging, solving the prediction bias caused by the different proportions of the original terminologies in the original terminology combination; 2) Through CTS construction and terminology standardization methods based on N-gram and BERT, constructing a higher CTSA, increasing the performance upper limit of the model, and finally achieving a higher terminology standardization accuracy. In the end, the method in this paper achieved 7.7% accuracy improvement on both the CHIP-MTM and SCD datasets.

2   Related Work

This paper believes that the ICD-9 terminology standardization task can be completed by natural language processing. Text similarity calculation, named entity recognition, and text classification can be applied to this task. There are some summaries of the methods of text similarity calculation based on string [4–6], such as N-gram [7], Longest Common Subsequence [8] and Levenshtein Distance [9]. The calculation of similarity based on semantics can be done by using pre-trained word embedding. Kenter et al. [10] used word embedding of different dimensions to train the classifier to predict the similarity score between short texts. Mikolov et al. [11] and Pennington et al. [12] proposed Word2Vec and GloVe to generate word embedding, and Duan et al. [13] used word embedding generated by Word2Vec to calculate similarity through cosine distance. Devlin et al. [14] proposed a pre-training model BERT, which can predict the similarity between sentence pairs through text classification.

3   Method

When the ABTSBM method is used to deal with many-to-many problems, since the proportions of the original terminologies in the original terminology combination are different, if the original terminology combination is directly input into BERT, there may be multiple standard terminologies in the final standard terminology combination, all of which are derived from the one original terminology which has the largest proportion in the combination, and the original terminology with the smallest proportion may not correspond to the standard terminology combination. At the same time, in the ABTSBM method, as a key factor affecting model performance, CTSA limits the upper limit of the model's performance, while CTSS determines CTSA.

In order to solve the above problems, the method in this paper is mainly divided into two parts: 1) split the original terminology combination based on named entity recognition and part-of-speech tagging; 2) CTS construction and terminology standardization based on N-gram and BERT. The framework as shown in Fig. 1.

The first part is used to solve the problem that the difference in the proportion of original terminologies in the original terminology combination leads to deviations in the prediction results. Named entity recognition is used to supplement the information of body structure for the split terminologies. And the part-of-speech is used to supplement some non-body structure information for the split terminology, such as surgical procedure, approach, endoscopic, and the use of surgical procedure to further split the original terminology combination. The second part advances the ranking of candidate standard terminologies that could not be selected in Top-N in order to improve CTSA while maintaining CTSS. Finally use the ABTSBM method for terminology standardization.

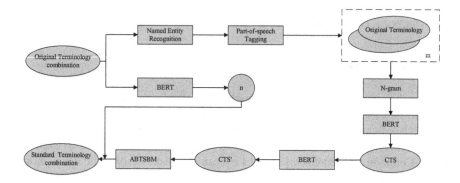

Fig. 1. The framework of many-to-many Chinese ICD-9 terminology standardization

The detailed diagram of the original terminology combination split method based on named entity recognition and part-of-speech tagging is shown in Fig. 2. This method first uses the "+" or "," two punctuation marks to initially segment the original terminology combination, and set the number of original terminologies after each data segmentation is m_1. Set the number of corresponding standard terminologies as N, and the N is predicted by BERT. Then the data with $N > 1$ is further divided into the original terminology combination based on named entity recognition and part-of-speech tagging. Finally, if the number of original terminologies after split is equal to N, the split of the original terminology combination is deemed complete, and each original terminology is input into the terminology standardization model as a piece of data. The rest of the original terminology combination will be regarded as a whole and input into the terminology standardization model as a split original terminology.

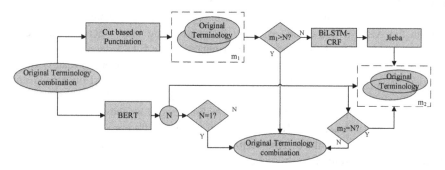

Fig. 2. The original terminology combination split framework based on named entity recognition and part-of-speech tagging

Through the named entity recognition based on BiLSTM-CRF [15], the method can identify the body structure that appear in the original terminology combination, such as "Lung", "Thyroid", etc. If the body structure information is missing in an original terminology, the body structure information in the previous original terminology is added to the original terminology.

This paper chooses Jieba as a part-of-speech tagging tool, and first determines that the verb as the surgical procedure, such as "切除" ("cut"), "取出" ("take out") and so on. Then take out the tokens containing "经" ("through") and "镜" ("mirror") and the tokens marked as nouns before and after them. Set these token as the supplementary information for the approach and the endoscopic respectively.

Generally speaking, the appearance position of body structure in the original terminology is not fixed, nor is the appearance position of surgical procedures, approaches and endoscopes absolutely fixed. However, this method is to deal with the original terminology, and the original terminology is an inaccurate terminology. Therefore, when inserting supplementary information, there is no need to find the most accurate position or the right position to make the original terminology have a smooth word order. Only need to ensure that the information contained in the original terminology is enough to find its corresponding standard terminology. Eventually, the body structure, approach and endoscope will be inserted into the head of the original terminology, and the surgical procedure will be inserted into the tail of the original terminology.

Table 3. Instances of split the original terminology combination.

Original terminology combination	Original terminology combination after split	Standard terminology combination
鼻内镜下腺样体切除术,双下甲部分切除术	鼻内镜下腺样体切除术 鼻双下甲部分切除术	鼻甲部分切除术 + 内镜下腺样体切除术
鼻内镜下腺样体切除术,双下甲部分切除术	鼻内镜下腺样体切除术 内镜鼻双下甲部分切除术	鼻甲部分切除术 + 内镜下腺样体切除术
腹腔镜下全子宫 + 双侧输卵管切除术	腹腔镜下全子宫切除 腹腔镜双侧输卵管切除术	腹腔镜双侧输卵管切除术 + 腹腔镜经腹全子宫切除术

Table 3 is an example of split the original terminology combination. In the original terminology combination "鼻内镜下腺样体切除术,双下甲部分切除术" ("endoscopic adenoidectomy of nose, partial double inferior thyroidectomy"), "鼻" ('nose') is a body structure. So add it to the head of '双下甲部分切除术" so that this original terminology combination is split into "鼻内镜下腺样体切除术" and "鼻双下甲部分切除术" ("nose partial double inferior thyroidectomy"),then further add the endoscopic method to "内镜" ("endoscopic") at the head of "鼻双下甲部分切除术', a new original terminology '内镜鼻双下甲部分切除术" ("endoscopic nose partial double inferior thyroidectomy") was constructed, and the split was finally completed. At the same time, in the case of the original terminology combination '腹腔镜下全子宫 + 双侧输卵管切除术" ("laparoscopic total uterus + bilateral salpingectomy"), the name of the former original terminology was omitted by the doctor because it was the same as the name of the latter terminology, but the name of the operation was directly affect the mapping of the original terminology to the standard terminology, so we will add "切除" ("cut off") to the tail of the previous original terminology.

In the CTS construction and terminology standardization method, the first step is to obtain two CTS A and B of similarity Top-N_1 and Top-N_2 based on the N-gram algorithm, where A $\subseteq$ B, and the similarity calculation formula is shown in formula 1, where T_i and T_j represent terminology i and terminology j, respectively. The second step is to mark the candidate standard terminology belonging to CTS A as label "1", and the candidate standard terminology belonging to CTS B-A as label "0", and input the original terminology and the candidate standard terminology one by one into BERT for 0–1 classification. When performing 0–1 classification, BERT will predict "0" or "1" for each set of data, but in this task, the reconstructed CTS is not composed of candidate standard terminologies that are predicted to be "1", but it sorts the probability that each candidate standard terminology output by BERT is predicted to be "1" and Top-N is re-taken to construct a new CTS. In other words, even if the probability of a candidate standard terminology being predicted as "1" is less than 0.5 which is being predicted as "0", as long as it is within Top-N, it will be added to the new CTS. The third step is to match the original terminology with the new candidate standard terminology one by one and enter BERT based on Focal Loss to perform 0–1 classification again, where the Focal Loss [16] formula is shown in formula 2, at this time, only the correct CTS The standard terminology is marked as "1", and the remaining candidate standard terminologies are marked as "0". Finally, as before, sort the probability of each candidate standard terminology being predicted to be "1", and take the first one as the standard terminology output.

$$\text{sim}(T_i, T_j) = \frac{2 \times Ngram(T_i, T_j)}{|T_i| + |T_j|} \tag{1}$$

$$L_{FL} = \begin{cases} -\alpha(1 - y')^{\gamma} \log y', & y = 1 \\ -(1 - \alpha)y'^{\gamma} \log(1 - y'), & y = 0 \end{cases} \tag{2}$$

4   Experimental and Analysis

4.1   Experiments Data

The CTS construction and terminology standardization model proposed in this paper uses the dataset CHIP One to One (namely CHIP-OTO), which is constructed from one-to-one data in the V2 dataset of CHIP 2019 evaluation task one. At the same time, the ICD-9 standard terminology collection contains a total of 9866 terminologies.

Since more than 90% of the data in CHIP 2019 are one-to-one data, in order to better evaluate the effect of this method on many-to-many problems, this paper first integrates the two versions of the dataset released by CHIP 2019 task one and named as CHIP Many To Many (namely CHIP-MTM), and then further integrated data from a real-world electronic medical record, and built a batch of data named as Self Constructed Dataset (namely SCD) from the above dataset. The data volume of the above three datasets is shown in the Table 4.

Table 4. The size of the datasets.

Dataset	CHIP-OTO	CHIP-MTM	SCD
Training dataset	3642	1144	1408
Testing dataset	1850	629	825

CHIP-MTM. This dataset comes from the two versions (namely V1, V2) of the dataset released during the evaluation period of CHIP 2019 evaluation task one. According to the players, the official has replaced the many-to-many data in the V1 dataset and the data containing "其他" ("other") in the standard terminology without changing the total amount of data. Therefore, in order to study the many-to-many ICD-9 terminology standardization task, this paper first extracts a total of 879 many-to-many data in the V1 dataset that does not contain "其他" ("other"), and then extracts a total of 494 many-to-many data in V2. Finally, 400 one-to-one data in the V2 dataset were randomly extracted. In the end, there is a total of 1773 pieces of data.

SCD. This dataset has a total of 2233 pieces of data, consisting of 1773 pieces of the above CHIP-MTM dataset, 189 pieces from a real-world electronic medical record, and 271 pieces of self-constructed data.

Observing the CHIP 2019 dataset, there are some data with the same standard terminology in either one-to-one data or many-to-many data. Therefore, according to the corresponding relationship between the original terminology and the standard terminology in the one-to-one data, the original terminology corresponding to a standard terminology in the many-to-many data can be replaced. The many-to-many data constructed in this way not only conforms to the mapping relationship between the original terminologies and the standard terminologies, and the combination of these original terminologies also conforms to the medical law, which are diseases that will occur at the same time. The self-constructed data base of this paper is the CHIP 2019 V2 dataset and a real-world electronic medical record data.

4.2 Evaluation Metrics

This paper uses accuracy as the evaluation standard, supplemented by CTSA to evaluate the quality of the candidate terminology set, as shown in formulas 3 and 4, where A represents the standard terminology, TP represents the number of standard terminologies with correct prediction, and FN represents the number of standard terminologies with error prediction. Because the predictive ability of negative samples is not considered in the terminology standardization experiment, FP and TN are 0.

$$Accuracy = \frac{TP + TN}{TP + FP + TN + FN} \tag{3}$$

$$CTSA = \frac{|\{CTS|A \in CTS\}|}{|CTS|} \tag{4}$$

4.3   Experimental Results and Analysis

CTS Construction and Terminology Standardization. The CTS construction and terminology standardization model based on N-gram and BERT is trained using CHIP-OTO dataset. Table 5 shows the two CTS construction method CTS based on N-gram (namely, CTS-N) and the CTS based on N-gram and BERT (namely CTS-NB) under the different Top-N's CTSA. Among them, in the CTS-NB method, N_2 is set to 100, and training is performed for different N_1. As shown in Table 6, when $N_1 = 20$, a higher CTSA can be achieved under a smaller CTSS, and achieve the best under Top-40. Continuing to expand the value of N, the improved of CTSA is limited and introduces new interference items, so this paper chooses $N_1 = 20$, N = 40.

Table 5. CTSA of CTS-N and CTS-NB methods under different N.

Method	Top-30	Top-40	Top-50	Top-100
CTS-N	89.5	90.4	91.4	93.7
CTS-NB	92.5	93.1	93.4	–

Table 6. CTSA of CTS-NB under different N_1 and N.

N1	Top-20	Top-30	Top-40	Top-50
20	89.0	**92.5**	**93.1**	**93.4**
30	–	91.0	92.8	93.1
40	–	–	91.4	92.9

This paper compares the ICD-9 terminology standardization accuracy of the CTS-N method and CTS-NB method under the same CTSS. At the same time, since the CTS-NB method is generated based on the CTS-N method of Top-100, it also compares with CTS-N with Top-100. The results are shown in Table 7.

Table 7. The accuracy of ICD-9 terminology standardization.

Training Dataset	Testing Dataset				
	CTS-N			CTS-NB	
	Top-30	Top-40	Top-100	Top-30	Top-40
Top-30	83.4	83.6	83.1	86.3	–
Top-40	83.4	83.9	84.5	–	**86.5**
Top-100	–	–	85.8	–	–

From the results of ICD-9 terminology standardization based on the CTS-N method, it can be seen that CTSA directly affects the accuracy of standardization, but the expansion of CTSS in order to improve CTSA has limited performance improvement:

1) When the training dataset and test dataset are the same CTSS, a better result can be obtained, but as the CTSS increases, the improvement tends to be flat. The CTSS from Top-30 to Top-40 increases by 10, and the accuracy increases by 0.5%., But the CTSS has increased by 60 from Top-40 to Top-100, but the accuracy has only increased by 1.9%;

2) When the size of the training dataset and the test dataset are different, the CTSS of the expanded test dataset will introduce interference items with low similarity. At this time, the ability of BERT to deal with these interference items is limited, and there will be no significant increase due to the significant increase in CTSA, or even accuracy may not rise but fall. When referring to the CTSA under different N in Table 5, and horizontally analyzing the experimental results under the CTS-N method in Table 7, under the training dataset Top-30, when the test dataset changes from Top-30 to Top-40, CTSA increased by 0.9%, but accuracy only increased by 0.2%, and when the test dataset became Top-100, CTSA increased by 4.2% over Top-30, but accuracy decreased by 0.3%.

Through the CTS-NB method, the Top-40 CTS improved 2% accuracy than CTS-N method under Top-40. The CTS of Top-40 from CTS-NB method was constructed by reordering the CTS of Top-100 from CTS-N method, and it was 0.6% CTSA lower than the CTS of Top-100 from CTS-N method. However, there was 0.7% accuracy improvement. It actually shows that the method in this paper is effective in removing interference items through reordering to maintain a small CTSS and shifting the correct standard terminology order forward.

Split Original Terminology Combination Method Based on Named Entity Recognition and Part-Of-Speech Tagging. In the many-to-many problem, first it needs to predict the number of standard terminologies (namely n) corresponding to the original terminology combination, and its prediction accuracy is shown in Table 8.

Table 8. Accuracy for predicting the value of n.

Dataset	Accuracy (%)
CHIP-MTM	99.5
SCD	99.6

It can be seen from Table 8 that the accuracy of using BERT to predict the number of standard terminologies on the existing dataset is extremely high, and there are almost no errors. Therefore, the decision result of many-to-many ICD-9 terminology standardization is mainly the result of BERT sorting. This result depends on the sorting model itself on the one hand, and on the input of the sorting model on the other hand. The experiments in this paper use the one-to-one ICD-9 terminology standardization model under Top-40 described above, and its accuracy is 86.5%. We compared the method of ABTSBM based on Original Terminology Union (namely ABTSBM-OTU) and the method of ABTSBM based on split Original Terminology Union (namely ABTSBM-SOTU) on the datasets CHIP-MTM and SCD. And the accuracy as shown in Table 9.

Table 9. Results of ABTSBM-OTU and ABTSBM-SOTU on the datasets CHIP-MTM and SCD.

Dataset	Accuracy (%)	
	CHIP-MTM	SCD
ABTSBM-OTU	61.3	56.1
ABTSBM-SOTU	**69.0**	**63.8**

It can be seen from Table 9 that even though the model's prediction accuracy for one-to-one data reaches 86.5%, after inputting many-to-many data, the accuracy drops by more than 30%. The ABTSBM-SOTU method splits the original terminology combination and supplements the original terminology. It achieves an accuracy improvement of 7.7% on the datasets CHIP-MTM and SCD. It fully shows the effectiveness of the method proposed in this paper, and significantly improves the accuracy of many-to-many ICD-9 terminology standardization.

5 Conclusion

Aiming at the many-to-many problem in the ICD-9 terminology standardization task, this paper proposes a many-to-many ICD-9 terminology standardization method: 1) split the original terminology combination based on named entity recognition and part-of-speech tagging; 2) CTS construction and terminology standardization based on N-gram and BERT. The method effectively solves the prediction bias caused by the different proportions of the original terminologies in the original terminology combination, effectively improves the CTSA without expanding the CTSS, and directly increases the upper limit of model performance. At the same time, this paper constructs two many-to-many datasets, named as CHIP-MTM and SCD. The experimental results show that the method in this paper has achieved an accuracy improvement of 7.7% on both CHIP-MTM and SCD.

References

1 Slee, V.N.: The international classification of diseases: ninth revision (ICD-9). Ann. Intern. Med. **88**(3), 424–426 (1978)

2 Qiu, J., Ma, H.: Analysis of the causes of coding errors in the operation of ICD-9-CM-3 in a hospital. Chin. Med. Rec. **21**(11), 28–30 (2020)

3 Liu, Y., Ji, B., Yu, J., et al.: An advanced ICD9 terminology standardization method based on BERT and text similarity. In: ICNC-FSKD (2020). https://doi.org/10.1007/978-3-030-70665-4_202

4 Nitesh, P., Manasi, G., Rajesh, W.: A Review on Text Similarity Technique used in IR and its application. Int. J. Comput. Appl. **120** (2015)

5 Wang, C., Yang, Y., et al.: A review of text similarity approaches. Inf. Sci. **37**(03), 158–168 (2019)

6 Erjing, C., Enbo, J.: A survey of research on text similarity calculation methods. Data Anal. Knowl. Disc. **1**(6), 1–11 (2017)

7 Brown, P.F., Pietra, V.J.D., Souza, P.V.D., et al.: Class-based n-gram models of natural language. Comput. Lingus **18**(4), 467–479 (1992)

8 Irving, R., Fraser, C.: Two algorithms for the longest common subsequence of three (or more) strings. In: DBLP (1992). https://doi.org/10.1007/3-540-56024-6_18

9 Navarro, G.: A guided tour approximate string matching. ACM Comput. Surv. (CSUR) (2001)

10 Kenter, T., Rijke, M.D.: Short text similarity with word embeddings. In: Acm International on Conference on Information and Knowledge Management. ACM (2015)

11 Mikolov, T.: Distributed representations of words and phrases and their compositionality. Adv. Neural. Inf. Process. Syst. **26**, 3111–3119 (2013)

12 Pennington. J., Socher. R., Manning C.D.: GloVe: global vectors for word representation. In: Proceedings of the 2014 Conference on Empirical Methods in Natural Language Processing, pp. 1532-1543 (2014)

13 Duan, X., Zhang, Y., Sun, Y.: Research on sentence vector representation and similarity calculation method about microblog texts. Comput. Eng. **043**(005), 143–148 (2017)

14 Devlin, J., Chang, M.W., Lee, K., et al.: BERT: pre-training of deep bidirectional transformers for language understanding (2018)

15 Chalapathy, R., Borzeshi, E.Z., Piccardi, M.: Bidirectional LSTM-CRF for clinical concept extraction (2016)

16 Lin, T.Y., Goyal, P., Girshick, R., et al.: Focal loss for dense object detection. IEEE Trans. Pattern Anal. Mach. Intell. **99**, 2999–3007 (2017)

Chinese Word Sense Disambiguation Based on Classification

Chao Fan[1,2]([⊠]) [iD] and Yu Li[1,2]

[1] The School of Artificial Intelligence and Computer Science,
Jiangnan University, Wuxi 214122, China
fanchao@jiangnan.edu.cn
[2] Jiangsu Key Laboratory of Media Design and Software Technology,
Jiangnan University, Wuxi 214122, China

Abstract. Word sense disambiguation (WSD) is a well-known task in the field of natural language processing. It attempts to determine a meaning of a word that has a couple of senses. This paper studies the Chinese word sense disambiguation by employing supervised classification method. Initially, feature selection is performed based on feature windows. Three types of features are extracted in this research: part-of-speech (POS), words, and 2-gram collocation. Further, we make a comparison of different classification algorithms, including sequential minimal optimization (SMO), naïve bayes, and multilayer perceptron (MLP). Different parameters are tested in order to obtain best precision of dataset. Additionally, a punctuation optimization approach is proposed to refine the final classification precision. Experimental results show that our method can achieve a good effect. The proposed approach contributes a lot to the WSD task by exploring feature selection as well as punctuation optimization.

Keywords: Chinese word sense disambiguation · Classification · Feature selection

1 Introduction

Word sense disambiguation is a challenging issue in natural language processing. A great number of application systems highly rely on the results of word sense disambiguation, such as speech recognition, machine translation, information retrieval, text classification, etc. Take translating an ambiguous Chinese word "苹果" as an example. It represents two senses in English: "Apple Inc." and "an edible apple". It is necessary to distinguish between two different word senses. Hence, the motivation of this research is trying to discover the specific sense of an ambiguous word according to its context. This article pays attention to the word sense disambiguation in Chinese characters. We attempt to handle this task by adopting supervised classification. The novelty of this work is to consider rich feature selection and introduce punctuation optimization, so that we can achieve a good precision and make a contribution to the WSD task.

Firstly, feature selection is completed by adjusting the POS and TOKEN windows around the ambiguous words. Secondly, many classification algorithms are introduced to construct classifiers to perform on training data based on extracted features. Thirdly,

D.-S. Huang et al. (Eds.): ICIC 2021, LNCS 12837, pp. 442–453, 2021.
https://doi.org/10.1007/978-3-030-84529-2_37

both parameter and punctuation optimization are enforced for advancing the results of classification. Finally, we compare different algorithms by choosing many groups of features so as to discover the best effect of word sense disambiguation.

The article can be organized as follows. Section 2 discusses the related work of word sense disambiguation topics. Section 3 studies the extraction of important features. Section 4 and Sect. 5 introduce the classifiers and the preprocessing. The core parts of experiments and discussions are given in Sect. 6. Conclusions are drawn and future work are brought forward in Sect. 7.

2   Related Work

As the number of word sense with an ambiguous word is already known for the WSD task, it is usually accomplished by a supervised classification approach. Wang et al. [1] designed an interactive learning method using expert labelling instances and features. The most informative instances are chosen by algorithm for future labels. Edilson et al. [2] utilized an approach of complex network to deal with word sense disambiguation. They built a bipartite network to represent feature words and ambiguous words. Tripodi et al. [3] studied the word sense disambiguation from a perspective of evolutionary game theory. Regarded as a constraint satisfaction problem, this task can be solved by tools from game theory. Abualhaija et al. [4] developed a D-Bees algorithm based on the bee colony optimization meta-heuristic, which is compared to the simulated annealing and ant colony algorithm.

As for Chinese word sense disambiguation, Wu et al. [5] researched on this issue based on the ensemble methods of classifiers. They selected nine types of combining schemes to run on two different datasets. Results show that three kinds of proposed methods outperform other six methods. Zhang et al. [6] borrowed the idea of ensembled classifier in the pattern recognition and constructed a classifier ensembled by dynamic weight adaptation. Meng et al. [7] presented an approach based on textual similarity with parts-of-speech tagging. They devised a context2vec model with POS features to distinguish between different meanings. Aiming at solve the data sparseness problem in the WSD task, Yang [8] advanced a technique based on context translation, which combined both authentic and pseudo training data to train a Bayesian model. The Bayesian model can be used for disambiguating the word sense.

3   Feature Selection

Feature selection has a significant impact on the performance of word sense disambiguation algorithms. In this paper, all classifiers employ features in the same feature space. Three types of features commonly utilized in the word sense disambiguation are selected, including "TOKEN", "POS" and "2-gram collocation". The meanings of the three feature types can be shown as follows:

(1)   TOKEN: word with location information;
(2)   POS: part of speech with location information;
(3)   2-gram collocation: co-occurrence feature of bigrams.

今天(today)/t ，/w 现任(current)/v 中国(China)/ns 中医(Chinese Medical Sciences)/n 研究院(Academy)/n 长城(Great Wall)/nz 医院(hospital)/n 院长 (Dean)/n 的(of)/u 周文志(Zhou Wenzhi)/nr 教授(professor)/n 来到(come)/v 了(-ed)/u 这里(here)/r 。(.)/w

Fig. 1. An example of a segmented sentence with parts of speech ("中医" is a word to be disambiguated)

Take a Chinese sentence as an example (see Fig. 1). There are two word senses for target noun "中医": "traditional Chinese medical science" and "practitioner of Chinese medicine". A sample of selected features can be presented in Table 1. The window opened involves 3 words before and after the target word in context.

According to the Table 1, it displays three types of features for ambiguous word "中医" with a window size of 3. "TOKEN" and "POS" denote words and parts of speech. 2-gram of words are also utilized within the window around the word to be disambiguated.

Table 1. Sample of the features in example

Feature types	Features
TOKEN	W_{-3} =,; W_{-2} = 现任; W_{-1} = 中国; W_1 = 研究院; W_2 = 长城; W_3 = 医院
POS	P_{-3} = w; P_{-2} = v; P_{-1} = ns; P_1 = n; P_2 = nz; P_3 = n
2-gram collocation	W_{-3}–W_{-2} =, -现任; W_{-2}–W_{-1} = 现任-中国; W_{-1}–W_0 = 中国-中医; W_0–W_1 = 中医-研究院; W_1–W_2 = 研究院-长城; W_2–W_3 = 长城-医院

TOKEN and POS at the same location are not combined together as a feature. As for the order of features, TOKEN and POS are arranged in the order of appearance in the sentence with POS in the front and TOKEN in the back. Besides, the category label is placed at the end of features. An example of selected features and the corresponding word category label is listed in Fig. 2.

ns,n,现任,中国,研究院,长城,医院,traditional_Chinese_medical_science

Fig. 2. An example of selected features and the corresponding label

It takes the first and last word around the ambiguous word "中医" as POS features. At the same time, the TOKEN features are extracted by selecting the first two and last three words around the target word. The selected features can be simplified by [−1, 1, −2, 3], where the figure represents the start or end position of a window for TOKEN and POS. In this way, different groups of features can be chosen to disambiguate the target word sense. The size of a TOKEN or POS window varies according to the specific algorithm, which will be discussed in the experiment part.

4 Classifiers

A number of classification algorithms can be employed to distinguish between different word senses. Three algorithms are outlined in the following parts.

4.1 Sequential Minimal Optimization (SMO)

Vapnik et al. designed another best criterion for linear classifiers after researching on statistical learning theory for many years. It starts from the linear separability, and then extends to the linear inseparable case or even the use of non-linear functions. This classifier is called Support Vector Machine (SVM). The Sequential Minimal Optimization (SMO) algorithm [9], an optimized implementation of SVM, is an effective approach to handle the challenge of support vector learning in large data sets.

4.2 Naïve Bayes

Naïve Bayes [10] is a simple probabilistic classification algorithm based on applying Bayes' theorem with strong independence assumptions between the features. The Bayes method is a pattern classification method provided that the prior probability and conditional probability are known. The classification result depends on the entire sample in various domains. It determines the best hypothesis in the hypothesis space H when the training data D is given. Best hypothesis is defined as the most probable hypothesis given the prior probabilities of different hypotheses in data D and H. Bayesian theory provides a method for calculating the probability of a hypothesis, based on the prior probability of the hypothesis, the probability of different data observed under a given hypothesis, and the observed data.

4.3 Multilayer Perceptron (MLP)

Multilayer perceptron (MLP) [11] is a mature structure in artificial neural networks. This structure is generally composed of an input layer, a hidden layer and an output layer. As an extension of perceptron, MLP can solve the non-linear separability problem that a single-layer perceptron cannot do. The most important thing in MLP is the back propagation (BP) algorithm, whose principle can be roughly described as follows. The working signal propagates forward, whereas the error signal propagates back and modifies the weight. Finally, the error gradually converges.

5   Preprocessing

When the size of feature selection window is larger than the number of POS or TOKEN before or after the ambiguous word, we define a tag "*" for padding. For instance, the word "中医" to be disambiguated is at the beginning of a sentence in Fig. 3. If the POS window of "中医" is [−2, 2] and the TOKEN window is [−2, 3], the extracted features can be described by Fig. 4.

中医/n 不/d 是/v "/w 慢/a 郎中/n "/w 对/v 症/Ng 急诊/n 有/v 绝招/n 中医院/n 遍/v 设/v 急诊科/n

Fig. 3. Another example of ambiguous word "中医"

,,d,v,*,*, 不, 是, "

Fig. 4. Extracted features with the POS window [−2, 2] and TOKEN window [−2, 3]

Punctuation marks are regarded as sentence breaks. We are enabled to decide whether to perform the following optimization processing. For the punctuation before the ambiguous word in the POS window and the TOKEN window, the POS and TOKEN before punctuation are considered to be "*", but the POS and TOKEN of the punctuation are retained. For the punctuation after the target word, the POS and TOKEN of the following words are considered to be "*", but the POS and TOKEN of the punctuation itself are also retained. For example, considering the POS window in Fig. 5, the parts-of-speech of the words after the comma is tagged with "*", but the part-of-speech "w" of the comma is retained.

钻研(study)中医理论(theory)，试图(try)从(from)前人(forerunner)
　　　v　　　n　　n　　　　w　　*　　*　　　*

Fig. 5. An optimization of POS window considering punctuation marks ("中医" is a word to be disambiguated)

6   Experiment and Discussion

6.1   Dataset

The dataset[1] incorporates 40 Chinese words to be disambiguated, including 19 nouns and 21 verbs. The training data (Chinese_train_pos.xml) contain a total of 2686 training samples. The test data (Chinese_test_pos.xml) contains a total of 935 test samples.

---

[1] http://www.icl.pku.edu.cn/icl_news/postinfo_detail.asp?id=385.

6.2   Evaluation

Two indicators, "macro-average precision" and "micro-average precision", are employed to evaluate the results of multi-class classification in the background of word sense disambiguation.

Macro-average precision score is calculated as arithmetic mean of different ambiguous words' precision scores. n is the number of words to be disambiguated. $Precision_i$ denotes the precision of disambiguation for each ambiguous word i. The calculation method of the macro average can be written as formula (1).

$$Pmar = PrecisionMacroAvg = \frac{1}{n}\sum_{i=1}^{n} Precision_i \tag{1}$$

Micro-average precision score is defined as sum of true positives (*TP*) for all ambiguous words divided by all positive predictions. The positive prediction is sum of all true positives (*TP*) and false positives (*FP*). Thus, the micro-average method of precision can be calculated as formula (2). This paper utilizes the micro-average score for evaluating the results of classification.

$$Pmir = PrecisionMicroAvg = \frac{\sum_{i=1}^{n} TP_i}{\sum_{i=1}^{n}(TP_i + FP_i)} \tag{2}$$

6.3   Feature Optimization

It is crucial to set an appropriate size of POS window or TOKEN window. A small window cannot capture enough information to describe the sense of an ambiguous word. However, as the window size expands, the problem of data sparseness surfaces. The over-fitting of training samples may lead to unsatisfactory results. For the POS feature, it is better to take a window of $[-1, 1]$ or $[-1, 2]$, which means the POS of latter words are more important than the previous ones in some cases. On the other hand, a window of $[-1, 1]$ or $[-2, 2]$ can achieve better results for the TOKEN feature in many experiments. The experimental results of testing different feature selection are listed in Table 2. From the example, both previous and latter words in the context have a balanced influence on classification, which can be confirmed by the result. Figure 6 presents the micro-average precision of the three classification algorithms under different feature windows.

Table 2. Comparison of classification algorithms with different features

Feature window	SMO	NaiveBayes	MLP
[−1, 1, 0, 0]	67.17%	65.67%	67.06%
[−1, 2, 0, 0]	68.77%	65.35%	67.17%
[−2, 2, 0, 0]	67.59%	65.67%	65.88%
[−1, 1,−1,1]	73.26%	70.80%	71.98%
[−1, 2,−1,1]	73.26%	**71.44%**	**72.62%**
[−1, 1,−2,2]	73.80%	**71.44%**	68.24%
[−1, 2,−2,2]	**74.65%**	70.91%	67.49%
[−2, 2,−2,2]	72.09%	70.05%	69.09%

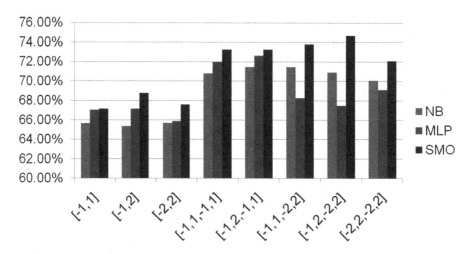

Fig. 6. Comparison of Pmir for three classifiers

The Pmir of MLP is better than that of Naïve Bayes method when the feature dimension is low. The performance of Naïve Bayes is better than MLP when the dimension increases. Naïve Bayes method has a better Pmir and a less running time than MLP in high dimension. SMO method outperforms the other two classifiers in micro-average precision, displaying a good generalization capability.

In addition, bi-gram feature is taken into account for the Naïve Bayes model. Specifically, parts-of-speech of ambiguous words and nearby words are combined as bi-gram. The bi-gram feature can partially improve the precision. Figure 7 shows the micro-average precision scores of naïve bayes model considering bi-gram features. Results of feature window [−1, 2, 0, 0] and [−1, 2, −2, 2] perform better than the baseline.

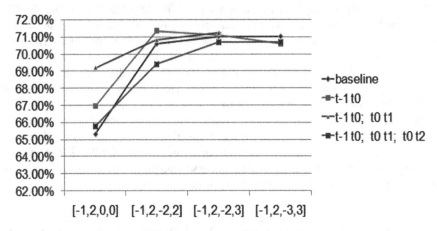

Fig. 7. Pmir of Naïve Bayes method using bi-gram features

6.4 Punctuation Optimization

When analyzing and selecting features, the punctuation optimization can be done by replacing the POS and TOKEN of punctuation marks with "*". The following Table 3 shows the statistical data before and after punctuation optimization of the three algorithms SMO, NaiveBayes and MLP for the feature window $[-1, 2, -2, 2]$.

Table 3. Result of punctuation optimization with a window $[-1, 2, -2, 2]$

	SMO	NaiveBayes	MLP
Pmir	72.83%	70.90%	65.88%
Pmir with punctuation optimization	74.01%	70.58%	67.49%
Improvement	1.18%	−0.32%	1.61%

According to the Table 3, punctuation optimization is capable of enhancing the effect of classification except for Naïve Bayes method. Experiments on other window features exhibited similar results. We give some examples to explain why punctuation optimization can improve the final precision.

他(He)/r 特别(particularly)/d 强调(emphasize)/v ，(punctuation mark)/w 出

(publish)/v 书(book)/n 要(need)/v 考虑(consider)/v 人民(people)/n 群众

(masses)/n 的('s)/u 实际(practical)/a 需要(necessity)/n

Fig. 8. An example of punctuation optimization for ambiguous word "出"

According to the example in Fig. 8, the word "强调(emphasize)" before the Chinese comma has no connection with the ambiguous word "出(publish)" in meaning. It has no contribution to disambiguate the word sense. Nonetheless, the word "书(book)" after the ambiguous word is a good indicator to recognize the word sense of "出 (publish)". The Chinese comma is just a conjunction between the two sentences before and after the punctuation mark. Therefore, the POS and TOKEN before the punctuation should not be considered in order to improve the effect of classification.

There are some exceptions exist for punctuation optimization. For example, Chinese punctuation " 、 " is a conjunction that joins elements of equal syntactic importance. In Fig. 9, a strong parallel relationship of meaning exists for "车主(car owner)", "单位(organization)" and "住址(address)". These exceptions do not affect the effectiveness of proposed punctuation optimization method.

车主(car owner)/n 、 (punctuation mark)/w 单位(organization)/n 、 (punctuation

mark)/w 住址(address)/n 及(and)/c 联系(contact)/vn 方法(information)/n 迅速

(quickly)/ad 显示(show)/v 出来(out)/v

Fig. 9. An exception of punctuation optimization for ambiguous word "单位"

6.5   Parameter Optimization

Kernel Functions. In SMO algorithm, polynomial kernel function shows better average performance and higher accuracy. Further, the training process applying PolyKernel is faster than other kernel functions. A comparison of different kernel functions using POS windows is given in Table 4. The micro-average precisions of PolyKernel are better than others on average, so PolyKernel is the best kernel function in the SMO algorithm.

Table 4. Comparison of kernel functions in SMO

Kernel function	[−1, 1]	[−1, 2]	[−2, 1]	[−2, 2]
PolyKernel	67.17%	68.77%	65.67%	67.59%
Puk	66.20%	63.64%	63.10%	62.14%
RBFKernel	47.81%	49.95%	45.78%	50.59%

C Parameter. According to the results of the experiment, it is better to choose a smaller value for the penalty factor C in the SMO, such as 0.43, which means errors to a certain extent are allowed in the algorithm.

The results of C parameter selection are depicted in Fig. 10. Setting a too high penalty factor will lead to an overfitting to the training data, even though the accuracy can be improved. "punc" in Fig. 8 indicate the use of punctuation optimization. The result of window [−1, 2, −2, 2] with "punc" outperforms other feature windows.

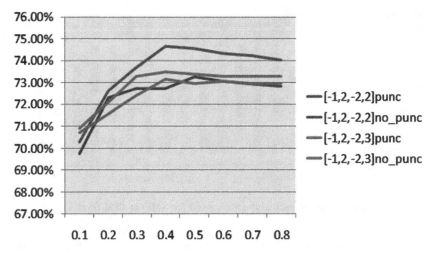

Fig. 10. Comparison of Pmir using different C parameter in the SMO method

6.6   Precision of Each Word Sense

The Pmir reaches a best value 74.43% (corresponding Pmar is 78.02%) when the parameters of SMO method are set as follows: PolyKernel, a window [−1, 2, −2, 2], C = 0.43, punc. The precision for each ambiguous word can be illustrated in Fig. 11 and Fig. 12.

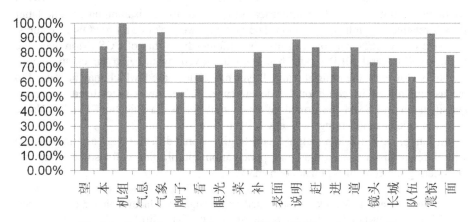

Fig. 11. Pmir of each ambiguous word for SMO method (Part 1)

As can be seen from Fig. 11 and Fig. 12, the precision scores of many ambiguous words are close to or equal to 100%. The majority of them are 2-class words, such as "中医" with two senses of "traditional Chinese medical science" and "practitioner of Chinese medicine", "儿女" with "young man and woman" and "children", "动摇" with

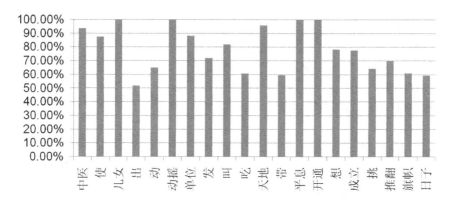

Fig. 12. Pmir of each ambiguous word for SMO method (Part 2)

"shake" and "vacillate". As for the multi-class word sense disambiguation, the effect is not as good as 2-class disambiguation, such as "出" (9-class), "带" (8-class), "吃" (4-class), etc.

7 Conclusions

In this paper, we explored the Chinese word sense disambiguation based on classification approaches. Words, parts-of-speech and bi-gram features in the context of target ambiguous word were extracted as features. Feature window was set to control the size of feature, involving POS window and TOKEN window. The POS and TOKEN windows are not necessarily symmetrical to the word to be disambiguated. The latter POS is relatively significant for nouns and verbs, so we took one more POS as features. The size of TOKEN window can also be different from that of POS. Moreover, the Pmir for SMO and MLP model has been improved by utilizing punctuation optimization. However, the Pmir of Naïve Bayes model decreased, which may be related to the assumption of its conditional independence. Naïve Bayes specialized in 2-class classification and SVM performed better on multi-class classification. Finally, Naïve Bayes is efficient in training, whereas MLP is time-consuming.

In future, we will add a weight factor to give words different degree of importance with different distances. Furthermore, punctuation optimization will also be refined by considering the type of a punctuation mark.

Acknowledgement. This work was supported by the High-level Innovation and Entrepreneurship Talents Introduction Program of Jiangsu Province of China, 2019.

References

1. Wang, Y., et al.: Interactive medical word sense disambiguation through informed learning. J. Am. Med. Inform. Assoc. **25**(7), 800–808 (2018)
2. Edilson, A., et al.: Word sense disambiguation: A complex network approach. Inf. Sci. **442–443**, 103–133 (2018)
3. Tripodi, R., Pelillo, M.: A game-theoretic approach to word sense disambiguation. Comput. Linguist. **43**(1), 31–70 (2017)
4. Abualhaija, S., Zimmermann, K.: D-Bees: A novel method inspired by bee colony optimization for solving word sense disambiguation. Swarm Evol. Comput. **27**, 188–195 (2016)
5. Wu, Y., et al.: Ensembles of classifier for chinese word sense disambiguation. J. Comput. Res. Dev. **45**(8), 1354–1361 (2008)
6. Zhang, Y., Guo, J.: Word sense disambiguation based on ensembled classifier with dynamic weight adaptation. J. Chin. Inf. Process. **026**(001), 3–8+36 (2012)
7. Meng, Y., et al.: Word sense disambiguation based on context similarity with POS tagging. J. Chin. Inf. Process. **32**(8), 9–18 (2018)
8. Yang, Z.: Supervised WSD method based on context translation. Comput. Sci. **44**(004), 252–255 (2017)
9. Berueco, J., et al.: Age estimation using support vector machine-sequential minimal optimization. Int. J. Image Graph. **2**(2), 145–150 (2015)
10. Vernekar, K., Kumar, H., Gangadharan, K.V.: Engine gearbox fault diagnosis using empirical mode decomposition method and Nave Bayes algorithm. Sadhana **42**(7), 1–11 (2017)
11. Bishop, C.: Exact calculation of the hessian matrix for the multilayer perceptron. Neural Comput. **4**(4), 494–501 (2014)

Research on the Authorship of Dream of the Red Chamber Based on Link Prediction

Chao Fan[1,2(✉)] 🆔 and Yu Li[1,2]

[1] The School of Artificial Intelligence and Computer Science,
Jiangnan University, Wuxi 214122, China
fanchao@jiangnan.edu.cn
[2] Jiangsu Key Laboratory of Media Design and Software Technology,
Jiangnan University, Wuxi 214122, China

Abstract. Dream of the Red Chamber (DRC), written in Qing dynasty, is a prestigious classical novel in Chinese literature. There exists a disputation over the authorship of DRC for the last 40 chapters. This research makes an effort to explore the DRC's authorship from the perspective of link prediction. At first, segmentation, part-of-speech tagging and named entity recognition are performed on the Chinese text of the DRC novel. A social network representing the relationship of characters is constructed based on the co-word analysis. Link is weighted according to the co-occurrence of two characters in a sentence or a paragraph. Furthermore, link prediction is completed on two groups of datasets: first 80 chapters and the whole 120 chapters. Two link prediction approaches are utilized in this research, including the similarity-based method and the classification-based method. Finally, the experiments lead to a conclusion that the author of the last 40 chapters is different from the first 80 chapters with a high probability.

Keywords: Link Prediction · Authorship · Dream of the Red Chamber

1 Introduction

The authorship of Dream of the Red Chamber (DRC) has always been in disputation since its birth in the middle of the 18th century. A universally acknowledged story about the authorship is that Cao Xueqin completed the first 80 chapters and Gao E finished the last 40 chapters. Many scholars believe the author of the last 40 chapters is different to the first 80 chapters while others disagree. There is no conclusion about who is the real author of the DRC, even though a significant number of researchers have studied this issue with enthusiasm.

This study tries to research on the authorship from the perspective of link prediction in social network. Firstly, we process the DRC corpus with the technique of natural language processing. Character name list is secured from the Internet and extended with the named-entity recognition. Secondly, character relationship networks of DRC are constructed based on co-word analysis. We design experiments on two groups of datasets: "the first 80 and last 40 chapters" and "the first 60 and 61–80 chapters". Four DRC networks are created with two training data and two test data. Thirdly, link

D.-S. Huang et al. (Eds.): ICIC 2021, LNCS 12837, pp. 454–464, 2021.
https://doi.org/10.1007/978-3-030-84529-2_38

prediction experiments are done on two groups of datasets. We take both the similarity-based method and the classification-based method into consideration. For the similarity-based method, a couple of similarities are calculated for prediction. For the classification-based method, extracted features are sent to three classifiers to complete the learning and prediction. Finally, the experimental results are displayed on two groups of datasets and an interesting conclusion of DRC's authorship is given based on the results of link prediction.

We organize this paper according to the following steps. Related work about the link prediction and the authorship of DRC is introduced in Sect. 2. The research framework and creation of the DRC network are discussed in Sect. 3. Section 4 presents the detailed approaches of link prediction in this paper. Section 5 gives the experimental results and some discussions. Conclusions and future work are given in Sect. 6.

2 Related Work

A lot of previous work of the novel concentrated on both qualitative and quantitative analysis. Recent studies explored the authorship of DRC from the perspective of statistic data. Wei [1] took five entities as scenario indicators where the frequency can be used as a writing style of a specific author. They drew a conclusion that the last 40 chapters should not be finished by Cao Xueqin. Shi [2] selected 42 Chinese function words as feature vectors and performed classification by utilizing the SVM. Chapter 20–29 and 110–119 were chosen as training data with class label 1 and 2, whereas the rest chapters were taken as test data. His research concluded that the first 80 chapters and the last 40 chapters were written by two people. Li et al. [3] divided the novel into three parts: 1–40 chapters, 41–80 chapters, 81–120 chapters. Language models such as N-gram were analyzed and Random forest was exploited for classification. Their experimental results showed the first 40 chapters and middle 40 chapters cannot be classified as the same group of the last 40 chapters. Xiao et al. [4] performed both hierarchical clustering and K-means clustering for DRC dataset. After finishing the lexical analysis, they chose Chinese function words, all content words, N-gram of words, N-gram of POS, and word length for clustering. Ye [5] discussed the authorship by building vector space model of stylistic features and clustering with the features. Jiang [6] analyzed the DRC with various methods in machine learning, including SVM, K-means, logistic regression, etc.

Contrary to the analysis of writing style, important words and scenario indicators, this work studies the DRC from a perspective of social network. Recently, many applications of link prediction have been found in many researches [7–9]. In this paper, we pay more attention to the storyline and organization style based on link prediction approach in the social network. The link prediction technique is adopted to examine the authorship of the classical novel.

3 Overview

3.1 Research Framework

In order to research the authorship of DRC, we focus on a different viewpoint from social network analysis. Different authors arrange the storyline and the appearance order of the multiple characters in different ways. Employing the evolution network of character relationship in DRC, we can infer which links will appear in the network in the next scene based on the current network. In other words, we can estimate the cooccurrence of two characters who have not been connected in previous chapters. If the whole novel is written by one author, the estimation will be accomplished with a higher accuracy. Otherwise, the link prediction will be less accurate if the author of first 80 chapters is different from that of last 40 chapters.

DRC corpus downloaded from Internet originates from a version published by People's Literature Publishing House in 2000 [10]. After securing the corpus, we conduct the lexical analysis, such as segmentation, part-of-speech tagging, named entity recognition, etc. All character names of DRC are obtained by using a main character name list and a name expansion technique [11].

Further, a weighted undirected network of DRC is established based on co-word analysis. Character names are reckoned as nodes and the cooccurrences of two characters in a paragraph are taken as links. Weighting schemes are adjusted according to the distance of character names.

Building the DRC network of character relationship, link prediction is leveraged to analyze the authorship of the novel. The first step is to predict the new character relationships of last 40 chapters using the data of first 80 chapters and the last 40 chapters. The second step involves the prediction of 61–80 chapters utilizing the data of first 60 chapters and 61–80 chapters. Two kinds of prediction methods are considered in this research: the similarity-based method and the classification-based method. Finally, we compare the prediction accuracy of two steps and draw a conclusion on authorship of DRC.

3.2 Construction of DRC Network

The text of each chapter is organized as sentences. Sentences are separated by Chinese punctuation marks such as "?", "。", "!", "…", etc. The weighting scheme for cooccurrence is presented in the following parts.

(1) weight = 1 if the cooccurrence of two characters appears in the same sentence;
(2) weight = 0.6 if two characters appear in the adjacent sentences;
(3) weight = 0.4 if two characters appear in the same paragraph except for the case of (1) and (2).

The weight of two characters' relationship in DRC can be defined by formula (1).

$$E_{xy} = \frac{C_{xy}}{C_x} \tag{1}$$

C_{xy} is the weight of cooccurrence for character name x and y in the whole novel. C_x is the total number of appearances for character name x. Hence, E_{xy} represents the normalized weight of cooccurrence for character name x and y.

As a preprocessing, some processing rules of character names are taken into account. For example, Lin Daiyu (the full name) and Daiyu (the given name) appearing in the novel are reckoned as the same node in the DRC network. Also, Shui Rong and Bei Jingwang represent the same character because Bei Jingwang is the title of Shui Rong.

We regard character names and their cooccurrences as nodes and links in the social network. Then a weighted undirected network of character relationship in DRC can be created. The network contains 724 nodes and 10,514 links. The weight of link can be defined by E_{xy}.

4   Link Prediction

Link prediction tries to predict missed or future relationships according to the existing links [12]. In this research, we predict new relationships of DRC characters in last 40 chapters based on observed connections in first 80 chapters. The result of prediction can help in identifying the authorship of DRC. Link prediction in social network incorporates two important methods: the similarity-based method and the classification-based method.

4.1   Similarity-Based Method

Methods Based on Node Neighborhoods. For a network G, $\Gamma(x)$ represents the set of neighbors of node x in G [13]. If node x and y do not have a connection but have many neighbors in common, they incline to be related to each other in the future. The similarity between two nodes can be calculated by following methods.

(1) Common neighbors. The simplest way to define the similarity between node x and y is counting the number of their neighbors in common, which is given by formula (2). It can be used for predicting possible links in future.

$$score(x, y) = |\Gamma(x) \cap \Gamma(y)| \tag{2}$$

(2) Jaccard's coefficient. This similarity metric is a normalized score divided by the number of the union of $\Gamma(x)$ and $\Gamma(y)$.

$$score(x, y) = \frac{|\Gamma(x) \cap \Gamma(y)|}{|\Gamma(x) \cup \Gamma(y)|} \tag{3}$$

(3) Adamic/Adar. The metric changes the weighting scheme by attaching more importance to rarer features.

$$score(x, y) = \sum_{z \in \Gamma(x) \cap \Gamma(y)} \frac{1}{log|\Gamma(z)|} \tag{4}$$

(4) Preferential attachment. The probability that a new link attaches node x is proportional to the number of $\Gamma(x)$. Therefore, the probability of a link appearing between node x and y is correlated with the product of $|\Gamma(x)|$ and $|\Gamma(y)|$.

$$score(x, y) = |\Gamma(x)| \cdot |\Gamma(y)| \tag{5}$$

Methods Based on the Ensemble of All Paths. The shortest-path distance can also be used to measure the similarity. The approach in this group considers all paths between two nodes.

(5) Katz. The measure attempts to weight short paths more heavily. It sums over all paths between node x and y, which are exponentially damped by length l. β is a very small value. In this research, β is set with 0.0085 to acquire a best result.

$$score(x, y) = \sum_{l=1}^{\infty} \beta^l \cdot \left| paths_{x,y}^{(l)} \right| \tag{6}$$

(6) Hitting time and variants. If we start a random walk from node x to y in network G, the hitting time $H_{x,y}$ is the expected number of steps. A normalized version of the hitting time in symmetric form can be described in formula (7).

$$score(x, y) = -\left(H_{x,y} \cdot \pi_y + H_{y,x} \cdot \pi_x \right) \tag{7}$$

(7) SimRank. It is assumed that the similarity between two nodes obeys a recursive definition, which can be shown in formula (8).

$$similarity(x, y) = \gamma \cdot \frac{\sum_{a \in \Gamma(x)} \sum_{b \in \Gamma(y)} similarity(a, b)}{|\Gamma(x)| \cdot |\Gamma(y)|} \tag{8}$$

where $\gamma \in [0, 1]$ and $similarity(x, x) = 1$. Hence the $score(x, y)$ can be calculated by solving the $similarity(x, y)$. In our studies, 0.8 is chosen for γ so as to obtain a better score.

$$score(x, y) = similarity(x, y) \tag{9}$$

The similarity-based method calculates the similarity between two nodes without a connection. The scores of missing links can be ranked by the similarity values. By comparing the predicted links to the gold standard, we are able to estimate the proportion of the top-N missing links that are correctly predicted.

4.2   Classification-Based Method

The similarity above can be chosen as features. We select degree of node x and y (Degree), score of common neighbors (ComNeigh) and Jaccard's coefficient (Jacc-Coeff) as features. Different combinations of features are provided to three classification algorithms. We carry out two groups of experiments. The first group of experiment takes features of first 80 chapters and last 40 chapters as training data and test data respectively. The second group of experiment uses features of first 60 chapters and 61–80 chapters as training data and test data. By comparing two groups of experiments, we can find which group can achieve a better accuracy than the other one. It can be used for estimating the authorship of DRC. Three classification algorithms are introduced in detail in the following parts.

Naïve Bayes. The naïve bayes algorithm simplifies the assumption that features are independent of each other. It supposes the features obey a certain distribution. Then, the parameters of the distribution are estimated. Even though many limitations exist in naïve bayes, it is still a simple and effective algorithm to make a prediction.

Support Vector Machine. Support Vector Machine (SVM), proposed by Vapnik, is a supervised learning model based on optimization theory. It utilizes a hypothesis space of linear functions in a high dimensional feature space, to which the kernel function is employed to map. SVM is a robust and efficient method for classification, regression, density estimation, etc. In this paper, an implement of LibSVM is applied to classify the data.

Decision Tree. Decision tree is a way to represent rules hidden in the training data. A decision tree can be used to represent a statistical probability. Each branch denotes a possible decision. Hence, we are able to comprehend the rules by utilizing the decision tree. This paper adopts C4.5 to generate a decision tree for classification.

5   Experiment and Discussion

5.1   Results of Similarity-Based Method

Link prediction are performed by both similarity-based method and classification-based methods. Two different indicators are devised to evaluate the results of two different methods separately.

For link prediction using similarity-based method, "Precision@N" is borrowed for evaluation. This method calculates the similarity between two nodes that have no connection. Then the scores of possible links are arranged in descending order. At this time, "Precision@N" is adopted to estimate the proportion of the top-N missing links that are correctly predicted. The correct label can be acquired by comparing the predicted links to the real links in dataset.

Precision@*N*

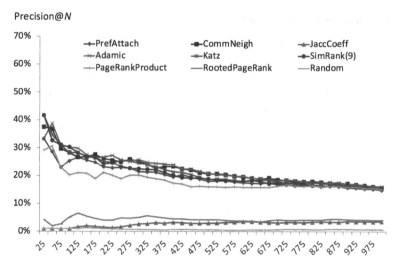

Fig. 1. Result of similarity-based method (first 80 and last 40 chapters)

120 chapters of DRC are divided into two parts. The first 80 chapters are used for creating the currently observed network. The links of last 40 chapters are reckoned as gold standard in link prediction. 1475 links exist in the last 40 chapters. 962 new links appear between nodes that have not been connected in the first 80 chapters. According to Fig. 1, the highest average precision is 15.30% when predicting 962 links.

Precision@*N*

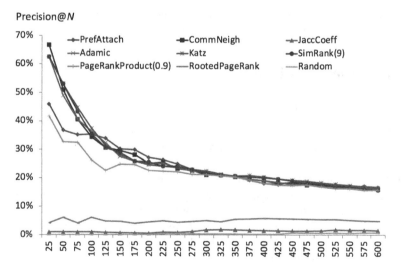

Fig. 2. Result of similarity-based method (first 60 and 61–80 chapters)

Further, the first 60 chapters are exploited to construct the currently observed network, while the 61–80 chapters are regarded as gold standard. There are 1424 links happening in the middle 61–80 chapters. 581 new links occur between nodes without connection in the first 60 chapters. As can be seen from Fig. 2, the highest average precision at 581 is 17.07%.

According to Fig. 1 and Fig. 2, we can discover that the method using first 60 and 61–80 chapters outperforms the other. This is to say, link prediction in first 80 chapters performs better than the whole 120 chapters.

5.2   Results of Classification-Based Method

Precision, recall and F-score are employed for evaluation. The precision is the number of true positive (*TP*) results divided by the number of all positive results, which are the sum of true positives (*TP*) and false positives (*FP*). The recall is the number of true positive (*TP*) results divided by the number of all samples that should have been identified as positive, which are the sum of true positives (*TP*) and false negatives (*FN*). The F-score is defined as the harmonic mean of precision and recall. They are presented in formula (10), (11) and (12).

$$Precision = \frac{TP}{TP + FP} \tag{10}$$

$$Recall = \frac{TP}{TP + FN} \tag{11}$$

$$Fscore = \frac{2 \cdot Precision \cdot Recall}{Precision + Recall} \tag{12}$$

Degree of node x and y (Degree), score of common neighbors (ComNeigh) and Jaccard's coefficient (JaccCoeff) are computed and combined as features. We then choose three features for classification: Degree, Degree + ComNeigh, and Degree + ComNeigh + JaccCoeff. Two groups of experiments using three classification algorithms are designed for comparison. The first group selects features of first 80 chapters and last 40 chapters as training data and test data respectively. The second group utilizes features of first 60 chapters and 61–80 chapters as training data and test data. Experimental results are summarized in tables (see Table 1 to Table 6).

Table 1 and Table 2 show the classification results of naïve bayes method. As the number of features increase, the corresponding F scores become greater. Besides, the classifier achieves a better F score in the second group (first 60 and 61–80 chapters) for three features.

Table 1. Result of Naïve Bayes (first 80 and last 40 chapters)

Features	Precision	Recall	F-score
Degree	81.81%	81.12%	81.02%
Degree + ComNeigh	87.10%	87.05%	87.05%
Degree + ComNeigh + JaccCoeff	89.13%	89.12%	89.12%

Table 2. Result of Naïve Bayes (first 60 and 61–80 chapters)

Features	Precision	Recall	F-score
Degree	83.61%	82.94%	82.85%
Degree + ComNeigh	92.26%	92.13%	92.13%
Degree + ComNeigh + JaccCoeff	93.16%	92.83%	92.82%

According to the Table 3 and Table 4, the classifier of SVM displays a similar result for three features. The second group of experiment performs better than the first group in the classification result. However, SVM is less effective than naïve bayes in this experiment.

Table 3. Result of SVM (first 80 and last 40 chapters)

Features	Precision	Recall	F-score
Degree	83.78%	83.42%	83.38%
Degree + ComNeigh	89.06%	87.97%	87.88%
Degree + ComNeigh + JaccCoeff	89.31%	88.51%	88.45%

Table 4. Result of SVM (first 60 and 61–80 chapters)

Features	Precision	Recall	F-score
Degree	87.80%	86.52%	86.40%
Degree + ComNeigh	91.45%	89.92%	89.83%
Degree + ComNeigh + JaccCoeff	91.95%	90.66%	90.59%

As can be seen from Table 5 and Table 6, the result of second group is better than the first group in the same way in the decision tree learning. Additionally, decision tree outperforms the other two algorithms from the effect of classification in our experiment.

Table 5. Result of Decision Tree (first 80 and last 40 chapters)

Features	Precision	Recall	F-score
Degree	85.60%	85.56%	85.56%
Degree + ComNeigh	91.72%	91.69%	91.69%
Degree + ComNeigh + JaccCoeff	91.36%	91.36%	91.36%

Table 6. Result of Decision Tree (first 60 and 61–80 chapters)

Features	Precision	Recall	F-score
Degree	88.32%	87.85%	87.81%
Degree + ComNeigh	94.73%	94.31%	94.30%
Degree + ComNeigh + JaccCoeff	94.51%	94.03%	94.01%

5.3   Discussion

Considering the results of similarity-based method, link prediction in first 80 chapters performs better than the whole 120 chapters. On the other hand, in classification-based method, it can be demonstrated that the link prediction in the second group (first 60 and 61–80 chapters) achieves a better accuracy than in the first group (first 80 and last 40 chapters) according to all experimental results of three classifiers.

Consequently, we can infer that the consistency in appearance order of the characters for the first 80 chapters is better than the first 80 chapters plus last 40 chapters. As different authors arrange the appearance order of the characters in different manners, the authorship of last 40 chapters may be different from the first 80 chapters from the perspective of storyline arrangement. In other words, there is a high probability that the traditional viewpoint of the DRC's authorship is true. Cao Xueqin and Gao E should be the authors of the first 80 chapters and the last 40 chapters respectively.

6   Conclusions

This research attempted to study the authorship of Dream of the Red Chamber with the technique of link prediction. At first, the DRC corpus was preprocessed by the lexical analysis tool such as segmentation, named-entity recognition, etc. A complete character name list of the novel was acquired by using the downloaded list and recognizing the named entities. Then a character relationship network was constructed from the corpus based on co-word analysis. The cooccurrence of two names was believed to be connected by a link. Different weighting schemes of links were decided according to the position of two characters appearing in the same paragraph. Moreover, link prediction was carried out on two groups of experiments: "the first 80 and last 40 chapters" and "the first 60 and 61–80 chapters". Both similarity-based method and classification-based method were exploited in this research. For classification method, experiments on three features were conducted by employing three classifiers. Final results demonstrated that the second group performed better than the first group in link prediction. As a consequence, the real author of last 40 chapters of DRC may differ from the first 80 chapters with a high possibility.

Future work will consider incorporating other groups of combination features and different classification approaches. In addition, combining other text analysis methods with the link prediction technique is another research direction in the future.

Acknowledgement. This work was supported by the High-level Innovation and Entrepreneurship Talents Introduction Program of Jiangsu Province of China, 2019.

References

1. Wei, B.: Statistical analysis on the differences of writing style between first 80 chapters and last 40 chapters in "dream of red mansions": an application of equivalent test on two independent binominal populations. Chinese J. Appl. Probab. Statist. **25**(4), 441–448 (2009)
2. Shi, J.: The authorship research on a dream of red mansions based on support vector machine. Stud. "A Dream Red Mansions" (5), 35–52 (2011)
3. Li, H., Liu, Y.: Language models and classification analysis for dream of the red chamber. In: Proceedings of the 2nd Conference on Cloud Computing & Intelligence Systems, Hangzhou, China, pp. 1459–1464 (2012)
4. Xiao, T., Liu, Y.: Words and n-gram models analysis for "A Dream of Red Mansions." New Technol. Libr. Inf. Serv. **4**, 50–57 (2015)
5. Ye, L.: The authorship research on A Dream of Red Mansions based on clustering of statistical stylistic features. Stud. "A Dream of Red Mansions" **5**, 312–324 (2016)
6. Jiang, N.: A study of the author of a dream of red mansions based on machine learning. Master thesis, Zhejiang University, Hangzhou (2018)
7. Hou, X., Liu, Y., Li, Z.: Convolutional adaptive network for link prediction in knowledge bases. Appl. Sci. **11**(9), 4270 (2021)
8. Wang, G., Wang, Y., Li, J., Liu, K.: A multidimensional network link prediction algorithm and its application for predicting social relationships. J. Comput. Sci. **53**, 101358 (2021)
9. Ajay, K., et al.: Link prediction techniques, applications, and performance: a survey. Physica A **553**, 124289 (2020)
10. Cao, X., Gao, E.: A Dream of Red Mansions. People's Literature Publishing House, Beijing (2000)
11. Fan, C.: Research on relationships of characters in the dream of the red chamber based on co-word analysis. ICIC Express Lett. Part B Appl. **11**(5), 1–8 (2020)
12. Martínez, V., Berzal, F., Cubero, J.C.: A survey of link prediction in complex networks. ACM Comput. Surv. **49**(4), 69.1–69.33 (2016)
13. Liben-Nowell, D., Kleinberg, J.: The link-prediction problem for social networks. J. Am. Soc. Inform. Sci. Technol. **58**(7), 1019–1031 (2007)

Span Representation Generation Method in Entity-Relation Joint Extraction

Yongtao Tang[(⊠)], Jie Yu, Shasha Li, Bin ji, Yusong Tan, and Qingbo Wu

College of Computer, National University of Defense Technology, Changsha, China

Abstract. Relation extraction (RE) is an important part of knowledge graph construction. The span-based entity-relation joint extraction model is an emerging model for Relation extraction. In the span-based entity-relation joint extraction model, the method of generating span representation vectors is usually relatively simple, and the semantic representation ability is insufficient. This paper studies the impact of four different span vector representation methods on the performance of the entity-relation joint extraction model, and enriches the features of span representation vectors by combining multiple span semantic representation methods. Compared with the baseline model, the combined span representation method can effectively improve the performance of the model on the CoNLL04 data set. Named entity recognition has achieved an F1 score of 89.37%, and relation extraction has achieved an F1 score of 72.64%. Compared with the baseline model, it has increased by 0.43% and 1.17% respectively.

Keywords: Relation extraction · Span-based model · Joint model

1 Introduction

In recent years, more and more researchers in relation extraction tasks have focused attention on the span-based entity-relation joint extraction model. This model generates vector representations of all possible spans, uses them in the named entity recognition step, and then classifies all span pairs in the relation classification step. The representation vector of the span pairs is generated based on the two span representation vectors. This method applies the same span representation to the named entity recognition and relation classification steps at the same time, and solves the relation extraction problem end-to-end.

The span representation vector is an important basis for named entity recognition and relation classification in the span-based entity-relation joint extraction model. The span vector representation method used in the existing span-based entity-relation joint extraction model is often simple. For example, the span representation vector is generated using max-pooling in the SPERT model proposed by Eberts et al. [4].

The generation of a span representation vector is essentially a process of obtaining the overall semantic features of the token sequence according to the embedding vector representation of tokens. This is similar to the method of generating the overall semantics of the sentence. Commonly used methods to generate overall semantic

© Springer Nature Switzerland AG 2021
D.-S. Huang et al. (Eds.): ICIC 2021, LNCS 12837, pp. 465–476, 2021.
https://doi.org/10.1007/978-3-030-84529-2_39

representation include convolutional neural network (CNN), long short-term memory network (LSTM), Attention mechanism, and max-pooling methods. It can be used as a span vector representation method by modifying these methods.

In addition, we believe that different span vector representation methods are complementary. These methods focus on different features. For example, a convolutional neural network mainly focuses on the local structural features of sentences, while a long short-term memory network focuses more on the overall semantics of sentences. different focus causes the model to show performance differences on named entity recognition and relation extraction tasks. By combining multiple span vector representation methods, different types of features can be merged, and the span semantic representation of the model can be enriched, which can improve the performance of the span-based joint extraction model.

Based on the above points, comparative experiments are conducted on the four span vector representation methods and the combined vector representation method on the CoNLL04 data set to compare the performance of different span representation vector generation methods under the same model framework. We achieved F1 scores of 89.37% in the named entity recognition task and 72.64% in the relation extraction task on the CoNLL04 data set by combining different span representation vector generation methods. Compared with the baseline model, the performance of the model using the combined span representation on named entity recognition and relation extraction tasks increased by 0.43% and 1.17%, respectively.

In this paper, our main contributions include:

—A method using Bi-LSTM hidden state difference to represent span semantics is proposed.

—The effects of several mainstream overall semantic generation methods on the performance of the span-based entity-relation joint extraction model are compared.

—A combined span vector representation method is proposed to enrich the semantic features of span vector representation.

—The performance of the baseline model both on named entity recognition and relation extraction tasks is improved.

2   Related Work

The main goal of relation extraction (RE) is to extract relational triples composed of two entities and a relation type from unstructured text. The mainstream relation extraction model can be divided into two-stage model and joint extraction model.

2.1   Two-Stage Model

Zeng et al. [21] first proposed the convolutional neural network for relation ex- traction tasks in 2014. And then, they introduced a piece-wise convolutional neural network (PCNN) on this basis to improve the performance of extraction in 2015 [20]. Shen and Huang [15] proposed a combination of convolutional neural networks and attention mechanism. Wang et al. [18] replaced the attention mechanism with multi-level attention to improve the performance in this task. Li et al. [7] proposed a model that uses entity perception embedding and self-attention [17] to enhancement the PCNN model for relation extraction.

Besides, Zhang et al. [23] proposed a position-aware attention mechanism on LSTM sequences for this task. Nayak and Ng [13] use a dependency-based distance separate multi-focus attention model to accomplish this task. Bowen et al. [19] use a paragraph-level attention mechanism in their model. Zhang et al. proposed an attention-based capsule network for relation extraction.

2.2 Joint Extraction Model

Miwa et al. [12] solved the entity-relation joint extraction task as a table filling problem. In 2016, Miwa et al. [11] used stacked RNNs to implement entity-relation joint extraction again. Li et al. [6] apply BERT [2] and Q&A models to the entity-relation joint extraction. Luan et al. [8] proposed the DyGIE model based on span realizes the entity-relation joint extraction. Wadden et al. [9] pro-posed DyGIE++ based on DyGIE. Eberts et al. [4] proposed a span-based entity relation joint extraction model SpERT, and achieved the SOTA results in the span-based model.

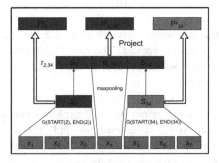

Fig. 1. The overall structure of the baseline model, Where $G()$ represents any neural network structure that can use token represent vectors to generate span represent vectors.

3 Model

In this paper, The model proposed by Eberts et al. is adopted as the model framework, which is the SOTA model in span-based entity-relation joint extraction models. Figure 1 shows The overall structure of the baseline model. And the formal representation of this model framework will be described in the following subsections.

3.1 Basic Model

For a given document D with T tokens, we used x_t to represent the task-independent embedding representation of the t-th token, where $1 < = t < = T$. The token representation is generated by concatenating the original context-independent word embedding and the context-sensitive token semantic representation using BERT.

$$x_t = \left\{ x_t^r, x_t^c \right\}$$

Where x_t^r is the original context-independent word embedding, which is generated by the word embedding table pre-trained in the BERT model [2]. x_t^c is the context-sensitive token semantic representation using the final output representation vector of the BERT model.

And then, the stacked Bi-LSTM is used to generate the task-related token embedding representation x_t'.

$$
\begin{aligned}
h_t^f &= LSTM \left(h_{t-1}^f, x_t \right) \\
h_t^b &= LSTM \left(h_{t+1}^b, x_t \right) \\
x_t' &= \left\{ h_t^f, h_t^b \right\}
\end{aligned}
$$

Where h_t^f and h_t^b are the hidden states of the last layer in the stacked Bi-LSTM. x_t' is the concatenation of h_t^f and h_t^b. Task-related token embedding representation is used to generate the span vector representation s_i.

$$s_i = G(START(i), END(i))$$

Where $G()$ represents any neural network structure that can use token represent vectors to generate span represent vectors, The formal description of different $G()$ will be explained in the following subsections. s_i is the span vector representation of the span pair $(START(i), END(i))$.

In the baseline system SPERT, only the maximum pooling is used to generate the span vector representation. In this paper, we will explore more methods for generating span vector representations. At the same time, we apply the same extra features added to the span vector representation, following the baseline model, such as span length embedding, span head and tail token representation, etc., to more directly compare the impact of span vector semantic representation on the performance of the span-based entity-relation joint extraction model.

We use the softmax function to calculate the distribution of entity types for each span.

$$
\begin{aligned}
score_i^s &= MLP_s(s_i) \\
p_i^s &= softmax\left(score_i^s \right)
\end{aligned}
$$

The output dimension of MLP_s and the dimension of p_i^s are equal to the number of entity types. Each $span_i$ predicted entity type corresponds to the entity type with the highest entity type score, $argmax\left(score_i^s \right)$. The span whose predicted entity type is not NONE is selected for the relation extraction task.

For each ordered span pair $\left(span_i, span_j \right)$, we use the same method as Eberts et al. to generate the ordered pair embedding representation $r_{i,j}$:

$$r_{i,j} = \{s_i, s_j, \theta(i,j)\}$$

Where s_i and s_j are the span embedding representations of the first and second entities respectively. $\theta(i,j)$ is the semantic representation of the context between the first and second entities, obtained according to the following formula.

$$\theta(i,j) = maxpooling\left(\left[x'_{END(i)+1}, \ldots, x'_{START(j)-1}\right]\right)$$

Where *maxpooling* is the max-pooling operator. The distribution of relation types for each ordered span pair is calculated by the softmax function, according to the ordered pair embedding representation.

$$score^r_{i,j} = MLP_r\left(r_{i,j}\right)$$
$$p^r_{i,j} = softmax\left(score^r_{i,j}\right)$$

The output dimension of MLP_r and the dimension of $p^r_{i,j}$ are equal to the number of relation types. The relation type predicted for each span pair $(span_i, span_j)$ corresponds to the relation type with the highest relation type score, $argmax\left(score^r_{i,j}\right)$.

3.2   Max-Pooling Based Method

The span representation vector generation model based on max-pooling is a span representation vector generation model adopted by the baseline system SPERT. It generates the representation vector of each span by using the max-pooling operator for all token representation vectors in the span.

$$s_i = maxpooling\left(\left[x'_{START(i)}, \ldots, x'_{END(i)}\right]\right)$$

3.3   CNN-Based Method

Convolutional neural networks generate feature vectors by capturing local features. In this paper, a recursive convolutional neural network is adopted. Figure 2 shows the overall neural network structure of the CNN-based method. This method uses shorter span vectors as the features of longer spans instead of calculating The low-level feature vector for each span separately. This method is first applied to the span-based entity-relation joint extraction model.

$$s_i = \begin{cases} x'_{START(i)} & if \quad START(i) = END(i) \\ CNN\left(s_p, s_q\right) & if \quad START(i) \neq END(i) \end{cases}$$

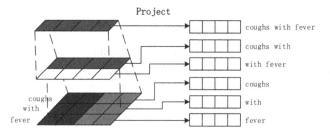

Fig. 2. Span vector representation method based on convolutional neural network (CNN)

Where $span_p$ and $span_q$ are sub-sequences of $span_i$ whose length is one token less than $span_i$.

$$
\begin{aligned}
START(p) &= START(i) + 1 \\
END(p) &= END(i) \\
START(q) &= START(i) \\
END(q) &= END(i) - 1
\end{aligned}
$$

3.4 LSTM-Based Method

A commonly used model for generating overall semantics is the bidirectional long short-term memory network(Bi-LSTM). However, it is unacceptable to use the Bi-LSTM for each span individually, because the number of all spans is too large.

Therefore, we propose a model that can obtain the representation vectors of all spans by using Bi-LSTM only once. Figure 3 shows the overall neural network structure of the Bi-LSTM-based span vector representation method. We noticed that after using a Bi-LSTM for the entire sentence, the forward hidden states represent the overall semantics of all tokens before the current token, while the backward hidden states represent the overall semantics of all tokens after the current token. In order to distinguish the hidden state generated by the Bi-LSTM mentioned above, we denote it as $\widehat{h_i^f}$ and $\widehat{h_i^b}$.

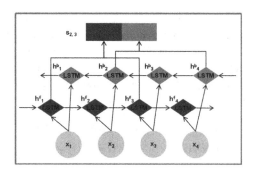

Fig. 3. Span vector representation method based on bidirectional long short-term memory network (LSTM)

The two tokens at the beginning and the end of each span have different hidden states. We believe that the difference between the two hidden states can represent the overall semantic state change caused by all tokens in the span. We use this state change as the feature vector of each span. This process can be formalized as:

$$
\widehat{h_i^f} = LSTM\left(\widehat{h_{i-1}^f}, x_i'\right)
$$
$$
\widehat{h_i^b} = LSTM\left(\widehat{h_{i+1}^b}, x_i'\right)
$$
$$
s_i = \left\{\widehat{h_{END(i)}^f} - \widehat{h_{START(i)-1}^f}, \widehat{h_{START(i)}^b} - \widehat{h_{END(i)+1}^b}\right\}
$$

3.5   Attention-Based Method

The Attention mechanism obtains the core semantics of each span by calculating the attention score for each token and strengthening the semantic representation of important tokens. In this paper, we use the scalar attention method proposed by Dixit et al. [3] to generate the overall semantic features of span.

$$
\alpha_t = MLP_\alpha\left(x_t'\right)
$$
$$
\beta_{i,t} = \frac{\exp(\alpha_t)}{\sum_{k=START(i)}^{END(i)} \exp(\alpha_k)}
$$
$$
s_i = \sum_{k=START(i)}^{END(i)} \beta_{i,t} x_t
$$

3.6   Combined Method

Combining different span representations can enrich the semantic features of span. In this paper, the span vector representation is generated by the combination of the three methods, Max-pooling based method, CNN-based method, and Bi-LSTM-based method, used as the final span vector representation.

$$
s_i = \left\{s_i^{MXP}, s_i^{CNN}, s_i^{BiLSTM}\right\}
$$

where s_i^{MXP}, s_i^{CNN}, s_i^{BiLSTM} correspond to the three span vector representations of Max-pooling based method, CNN-based method, and Bi-LSTM-based method respectively. The task-related token representations used in each span representation are calculated independently.

4   Experiment

4.1   Training

We use the $BERT_{base}$ model trained by Google as a pre-training language model, and use random initialization for Bi-LSTM, entity classifiers and relational classifiers. The training process uses the training data to learn the classifier parameters of entities and relations, and fine-tunes the BERT in the process. in this paper, the supervised learning method is used for training. Given a sentence with labeled entities and relations, we define a joint loss function for entity classification and relation classification:

$$L = L^s + L^r$$

Where L^s represents the loss of the span classifier using the cross-entropy of the entity types(including the NONE entity), and L^r represents the cross-entropy of the relations types. Both losses are calculated based on the average value of samples in each batch. No application type weight.

For the construction of training samples, we followed the Eberts et al.

- For the span classifier, we use all labeled entities as positive samples, denoted as s^{gt}, plus a fixed number of random non-entity spans of Ne as negative samples.
- In order to train the relations classifier, we use all the labeled relations as positive samples, and extract Nr negative samples from those entity pairs $s^{gt} * s^{gt}$ that have not labeled any relations.

The purpose is to control the percent of positive and negative samples, solve the problem of imbalance between positive and negative samples caused by using all samples, and improve the performance of the classifier.

4.2   Dataset

In this paper, we use the CoNLL04 dataset for experiments [13]. The CoNLL04 dataset contains annotated sentences with named entities and relations from news articles. The dataset includes four entity types (Location, Organization, People, Other) and five relation types (Work-For, Kill, Organization Based-In, Live-In, Located-In). We divide the data set into a training set (1153 sentences) and a test set (288 sentences), which follow Gupta et al. [5] 20% of the training set is reserved as the validation set to adjust the hyper-parameter.

4.3   Evaluation

In this paper, we use strict standards to evaluate the performance of the model in both entity recognition and relation extraction. If the predicted span and entity type of the span is correct, the entity classification is considered correct. If the relation type of a span pair is correct, and the two related entities are correct (considering span and type at the same time), then the relation is considered correct. The evaluation indicators are the accuracy, recall and F1 score of each entity and relation type. Finally, we report the micro-average F1 score of the model on the CoNLL04 dataset as the final performance indicator.

4.4 Hyper-parameter

In order to reduce experimental variables, unless otherwise specified, in this paper, we use SPERT's hyper-parameter settings [4] during the model training process. For the pre-trained language model, we use $BERT_{base}$ as the sentence encoder.

We use normally distributed random numbers ($\mu = 0$, $\sigma = 0.02$) to initialize the weights of the classifier. We use an optimizer with linear warm-up and linear decay learning rate, the dropout rate before the entity and relation classifier is 0.1, the batch size is 2. No further optimization of these parameters. Based on the CoNLL04 development set, we choose the relation filtering threshold is $\alpha = 0.4$, and the number of negative entities and relation samples for each sentence ($Ne = Nr = 100$).

In order to make the model performance reach the optimal value, we carefully adjusted the learning rate and the number of training epochs for each span vector representation method. The scope of the learning rate experiment is $5e - 6$, $1e - 5$, $5e - 5$, $1e - 4$, and $5e - 4$. And the scope of the number of training epochs is 15, 20, 25, 30, and 50. We conduct 5 experiments for any combination of two parameters, and take the average of all experimental results as the final result. The best-performing result among all parameter combinations is reported in this paper.

4.5 Experiment Results

We conducted two experiments on the CoNLL04 dataset. In the first experiments, the performance of the model is compared when using different span representation vector generation methods. And In the second experiment, the performance of the model using the combined span representation vector generation method is compared with the baseline model and other models to show the effect of the combined span representation vector.

In the first experiment, while keeping the framework unchanged, we only replaced the span representation vector generation method. The generation method of s_i mentioned in the previous section was used for experimentation. We compared four different span representation vector generation methods: the max-pooling based method, the CNN-based method, the Bi-LSTM-based method, and the attention-based method. These four methods are represented by MXP, CNN, Bi-LSTM, and Attention. The result of the max-pooling based method adopts the result reported by the baseline, SPERT.

Table 1 shows the performance of the model when different span representation is used. The performances of MXP, CNN and Bi-LSTM are basically similar, and both have advantages and disadvantages. CNN has achieved the best performance on named entity recognition tasks, but Bi-LSTM has the best performance on relation extraction tasks.

Table 1. The performance of the entity-relation joint extraction model using different spans representation in the CoNLL04 data set. The performance is evaluated by the micro F1 scores (%).

Method	Entity recognition	Relation extraction
MXP	88.94	71.47
CNN	**90.03**	70.29
Bi-LSTM	86.58	**73.12**
Attention	83.64	69.52

It can be noticed that Attention is significantly worse than the other three models. Through the analysis of the attention distribution of the Attention method in the calculation process, we found that most of the tokens get similar attention scores, which cannot effectively extract the semantic information of important tokens in the span.

In the second experiment, We combined three span vector representation methods, MXP, CNN and Bi-LSTM, as the final span vector representation to enrich the feature of span vector representation. Limited by the experimental conditions, we reduced the batch size of the model to 1.

Table 2. The performance of the entity-relation combined extraction model using combined spans representation in the CoNLL04 data set. The performance is evaluated by the micro F1 scores (%)

Method	Entity recognition	Relation extraction
Global Optimization [22]	85.60	67.80
Multi-turn QA [6]	87.80	67.90
Relation-Metric [16]	84.57	62.68
Bi-affine Attention [14]	86.20	64.40
Table-filling [10]	80.70	61.00
Hierarchical Attention [1]	86.51	62.32
SPERT [4]	88.94	71.47
Our Model	**89.37**	**72.64**

Table 2 shows the performance of the combined span representation method on the CoNLL04 data set. Compared with the baseline model, SPERT, the combined span representation method can effectively improve the performance of the model. Named entity recognition has achieved an F1 score of 89.37%, and relation extraction has achieved an F1 score of 72.64%. Compared with the baseline model, it has increased by 0.43% and 1.17% respectively. At the same time, it also has better performance than other relation extraction models.

5 Conclusion

In this paper, we study the impact of different span vector representation methods on the performance of entity-relation joint extraction, and find that the performance of different span vector representation methods is affected by specific tasks. For example, CNN performs better on entity recognition tasks. And Bi-LSTM performs well on relation extraction tasks. In addition, by combining different span vector representations to enrich the features, the performance of the span-based entity-relation joint extraction model on the two tasks of entity recognition and relation extraction. Through the above experiment, we found.

- The performance of different span representations is related to the task.
- The combined span vector representation can integrate the advantages of different vectors and improve the quality of the span representation vector.

References

1. Chi, R., Wu, B., Hu, L., Zhang, Y.: Enhancing joint entity and relation extraction with language modeling and hierarchical attention. In: Shao, J., Yiu, M.L., Toyoda, M., Zhang, D., Wang, W., Cui, B. (eds.) APWeb-WAIM 2019. LNCS, vol. 11641, pp. 314–328. Springer, Cham (2019). https://doi.org/10.1007/978-3-030-26072-9_24
2. Devlin, J., Chang, M.W., Lee, K., Toutanova, K.N.: BERT: pre-training of deep bidirectional transformers for language understanding. In: Proceedings of the 2019 Conference of the North American Chapter of the Association for Computational Linguistics: Human Language Technologies, Volume 1 (Long and Short Papers), pp. 4171–4186 (2018)
3. Dixit, K., Al-Onaizan, Y.: Span-level model for relation extraction. In: Proceedings of the 57th Annual Meeting of the Association for Computational Linguistics, pp. 5308–5314 (2019)
4. Eberts, M., Ulges, A.: Span-based joint entity and relation extraction with trans- former pre-training. In: ECAI, pp. 2006–2013 (2020)
5. Gupta, P., Schütze, H., Andrassy, B.: Table filling multi-task recurrent neural network for joint entity and relation extraction. In: Proceedings of COLING 2016, the 26th International Conference on Computational Linguistics: Technical Papers, pp. 2537–2547 (2016)
6. Li, X., et al.: Entity-relation extraction as multi-turn question answering. In: Proceedings of the 57th Annual Meeting of the Association for Computational Linguistics, pp. 1340–1350 (2019)
7. Li, Y., et al.: Self-attention enhanced selective gate with entity-aware embedding for distantly supervised relation extraction. Proc. AAAI Conf. Artificial Intell. **34**, 8269–8276 (2020)
8. Luan, Y., He, L., Ostendorf, M., Hajishirzi, H.: Multi-task identification of entities, relations, and co-reference for scientific knowledge graph construction. In: Proceedings of the 2018 Conference on Empirical Methods in Natural Language Processing, pp. 3219–3232 (2018)
9. Luan, Y., Wadden, D., He, L., Shah, A., Ostendorf, M., Hajishirzi, H.: A general framework for information extraction using dynamic span graphs. In: Proceedings of the 2019 Conference of the North American Chapter of the Association for Computational Linguistics: Human Language Technologies, Volume 1 (Long and Short Papers), pp. 3036–3046 (2019)

10. 1Miwa, M., Sasaki, Y.: Modeling joint entity and relation extraction with table representation. In: Conference on Empirical Methods in Natural Language Processing (2014)

11. Miwa, M., Bansal, M.: End-to-end relation extraction using lstms on sequences and tree structures. In: Proceedings of the 54th Annual Meeting of the Association for Computational Linguistics (Volume 1: Long Papers), vol. 1, pp. 1105–1116 (2016)

12. Miwa, M., Sasaki, Y.: Modeling joint entity and relation extraction with table representation. In: Proceedings of the 2014 Conference on Empirical Methods in Natural Language Processing (EMNLP), pp. 1858–1869 (2014)

13. Nayak, T., Ng, H.T.: Effective attention modeling for neural relation extraction. In: Proceedings of the 23rd Conference on Computational Natural Language Learning (CoNLL), pp. 603–612 (2019)

14. Nguyen, D.Q., Verspoor, K.: End-to-end neural relation extraction using deep biaffine attention. In: Azzopardi, L., Stein, B., Fuhr, N., Mayr, P., Hauff, C., Hiemstra, D. (eds.) ECIR 2019. LNCS, vol. 11437, pp. 729–738. Springer, Cham (2019). https://doi.org/10.1007/978-3-030-15712-8_47

15. Shen, Y., Huang, X.: Attention-based convolutional neural network for semantic relation extraction. In: Proceedings of COLING 2016, the 26th International Conference on Computational Linguistics: Technical Papers, pp. 2526–2536 (2016)

16. Tran, T., Kavuluru, R.: Neural metric learning for fast end-to-end relation extraction (2019)

17. Vaswani, A., et al.: Attention is all you need. In: Proceedings of the 31st International Conference on Neural Information Processing Systems, vol. 30, pp. 5998–6008 (2017)

18. Wang, L., Cao, Z., de Melo, G., Liu, Z.: Relation classification via multi-level attention cnns. In: Proceedings of the 54th Annual Meeting of the Association for Computational Linguistics (Volume 1: Long Papers), vol. 1, pp. 1298–1307 (2016)

19. Yu, B., Zhang, Z., Liu, T., Wang, B., Li, S., Li, Q.: Beyond word attention: Using segment attention in neural relation extraction. In: Proceedings of the Twenty- Eighth International Joint Conference on Artificial Intelligence, pp. 5401–5407 (2019)

20. Zeng, D., Liu, K., Chen, Y., Zhao, J.: Distant supervision for relation extraction via piece-wise convolutional neural networks. In: Proceedings of the 2015 Conference on Empirical Methods in Natural Language Processing, pp. 1753–1762 (2015)

21. Zeng, D., Liu, K., Lai, S., Zhou, G., Zhao, J.: Relation classification via convolutional deep neural network. In: Proceedings of COLING 2014, the 25th International Conference on Computational Linguistics: Technical Papers, pp. 2335–2344 (2014)

22. Zhang, M., Yue, Z., Fu, G.: End-to-end neural relation extraction with global optimization. In: Proceedings of the 2017 Conference on Empirical Methods in Natural Language Processing (2017)

23. Zhang, Y., Zhong, V., Chen, D., Angeli, G., Manning, C.D.: Position-aware attention and supervised data improve slot filling. In: Proceedings of the 2017 Conference on Empirical Methods in Natural Language Processing, pp. 35–45 (2017)

Machine Learning

Prediction of Railway Freight Customer Churn Based on Deep Forest

Danni Liu[1][(⊠)] [iD], Xinfeng Zhang[1], Yongle Shi[2], and Hui Li[1]

[1] Beijing University of Technology, Beijing, China
zxf@bjut.edu.cn
[2] China Railway Guangzhou Group Co. Ltd., Guangzhou, China

Abstract. With increasingly fierce competition in other transportation markets, the customer churn in railway freight becomes a very serious problem. Facing the high similarity and indistinguishability of railway freight data, the customer churn prediction (CCP) becomes one of the challenging tasks in this industry. In this paper, a deep forest-based model is developed which can achieve better accuracy of churn predicting in railway freight customer and effectively separate the churners from the non-churners. Inspired by the layer-by-layer processing of deep neural network, the cascade structure is adopted to the deep forest, in which the input of each layer includes the original features and the feature information processed by the previous layer. The deep forest in this paper achieves the best performance in railway freight data with the accuracy of 0.78, the precision of 0.75, the recall of 0.66, the F1-score of 0.7 and the AUC value of 0.86, which are higher than those of decision tree, AdaBoost and XGBoost. At the same time, this model has smaller standard error and makes more stable performance.

Keywords: Customer churn prediction · Churn in railway freight · Deep forest · Classification

1 Introduction

As an important customer identification technique, customer churn prediction (CCP) is widely used for customer retention management in various fields and industries [1]. Churn management is used by enterprises to retain customers who have the risk of abandoning this company [2].

With the promotion of supply side structural reforming and the implementation of reducing excessive steel and coal production capacity in China, the railway transportation industry has been affected to a certain extent. Furthermore, with the increasingly fierce competition in other transportation markets, the railway freight transport industry is facing a growing challenge. Under all these pressures, the churn of railway customer has become a very serious problem and the priority of railway transportation industry is to seize their customers to enhance the competition ability of the railway freight transportation enterprises [3].

Customer churn will decrease sales performance and attracting new customers will greatly increase the cost. Studies have shown that the cost of acquiring a new customer

© Springer Nature Switzerland AG 2021
D.-S. Huang et al. (Eds.): ICIC 2021, LNCS 12837, pp. 479–489, 2021.
https://doi.org/10.1007/978-3-030-84529-2_40

is 5 to 6 times than that of maintaining an existing customer [4, 5]. Therefore, the way to ensure the success of the enterprises is to accurately predict the churners, which are also urgent problems to be solved in railway freight transportation [6]. The accurately prediction will give timely warning for the management department and help them to formulate the magnetic marketing strategy for retaining.

Nowadays, many researches have been carried out on customer churn prediction and prediction models have been established. These models have achieved good application effects in many industries such as telecom [7, 8], finance [9], E-commerce [10] and so on. However, few scholars have conducted research on railway freight data.

In the field of machine learning, many effective models are used to tackle this problem, which include decision trees (DT) [11], logistic regression (LR) [12], naïve Bayes, support vector machine (SVM) [13], etc. With achieving great success in many fields, especially in image processing and behavior prediction, deep neural network has also been used to solve the these prediction problem and has made good performance in CCP [14].

As a new machine learning paradigm, ensemble learning [15] gains the high-accuracy and powerful generalization performance by training multiple base learners (also called weak learners) and then combining them for use. As we all know, an ensemble learner is usually much more accurate than a single learner. With Boosting and Bagging as representatives, ensemble methods are a kind of state-of-the-art learning approach and have make great achievements in many real-world tasks. They have been widely used in data mining competition such as KDD-Cup, Kaggle and Datacastle, due to the multi parameter and low speed of deep neural network. Idris [16] proposed an algorithm based on AdaBoost to solve the churn problem in telecom-munications. Huang et al. [17] leveraged the big data platform and Random Forest algorithm to solve the prediction of customer churn. [18] proposed a model for churn prediction using XGBoost (XGB).

The high similarity of railway freight data makes it is very difficult for separating the churners from the non-churners. The ensemble methods have only made a slight improvement in the face of indistinguishable railway freight data. The deep neural network can mine deep information through space mapping of vector based on the layer-by-layer processing, but the parameters of the neural network are complicated and difficult to confirm. The neural network used in the railway freight data without spatial continuity also has not achieved satisfactory results.

The work carried out in this paper is to develop a deep forest model that can achieve better accuracy of churn predicting in railway freight customer and effectively separate the churners from the non-churners. It is a novel decision tree ensemble learning method which ensembles forest to achieve the characterization learning. This method has a cascade structure inspired by that representation learning in the deep neural networks relies on the layer-by-layer processing of raw features. The deep forest model we used adopts the cascade structure, in which the input of each layer includes the original features and the feature information processed by the previous layer. This method provides a satisfactory scheme for the classification of railway customers.

The deep forest algorithm belongs to another deep learning model, which can automatically train model parameters in each layer without back propagation. Mean-while, the basic unit of the deep forest model is the decision tree in the random forest, this algorithm needs shorter time when training and has fewer parameters to manually adjust than the neural network. Above all, it is suitable for the establishment of prediction models with large amount of data, such as complaint prediction, load prediction and churn prediction. In addition, it can avoid overfitting effectively and show better generalization ability even with the small training sample.

Several algorithms were used to compare with our method in this experiments such as decision tree, AdaBoost and XGBoost. We found that the deep forest we used has the best evaluation measurements of the four models. Especially, the model shows a distinct improvement in recall, F1-score and the AUC value while making slight improvement in accuracy and precision. At the same time, this model has smaller standard error and makes more stable performance.

The paper is organized as follows. The characteristics of railway freight data and the model framework of deep forest are described in Sect. 2. Section 3 explains the results of our experiment and analysis will also be discussed in this Section. The conclusion is described in Sect. 4.

2 Research Methodology

The characteristics of railway freight data and the model framework of deep forest are detailed in this Section.

2.1 The Characteristics of Railway Freight Data

Different from telecom customer data, the customer data of railway freight transportation is usually composed of delivery records, including not only customer information and cargo information, but a lot of unavailable information, such as delivery registration personnel, registration personnel number, etc. So it is necessary to manually delete some irrelevant information and collate the data to obtain a good prediction dataset.

Railway freight data on annual cycle makes the prediction results non-real time, thus it is too late for retaining customers and a shorter time cycle is required. If the data is sorted out by day, the workload of railway operators will be greatly improved. So the monthly prediction is used to ensure the churners can be found in time which also effectively improves the practicability of our model.

The characteristics of railway freight data have been discovered that there are multiple delivery data in the same company and the company may ship and receive at different locations each time. Therefore, while sorting out the original forecast variables, it is better to make additional statistics on the number of shipping locations, receiving locations and railway freight routes and the times and days of customer monthly deliver goods.

2.2 Deep Forest

Deep forest [19] is a novel decision tree ensemble learning method which ensembles forest to achieve the purpose of letting the classifier perform characterization learning. This method has a cascade structure which enables representation learning by forests. Multi granularity scanning can be used to further improve the representation learning ability of the model, and make the deep forest have the ability of context or structure awareness. Inspired by that representation learning in the deep neural networks relies on the layer-by-layer processing of raw features, the deep forest we used adopts the cascade structure as shown in Fig. 1, in which the input of each layer includes the original features and the feature information processed by the previous layer. In Fig. 1, each layer of the cascade consists of four forests [20], including two random forests (black) and two completely-random tree forests (blue). Suppose each forest all contains 500 trees, the completely-random tree forest contains 500 completely-random trees [21], which are generated by randomly choosing a feature for splitting on each node of the tree, and keeping the tree growing until each leaf node contains only the same class of instances. While, the feature with the minimum Gini value is selected for segmentation among randomly selecting $\sqrt{D}$ features (D is the dimension of input features) in the trees of random forest.

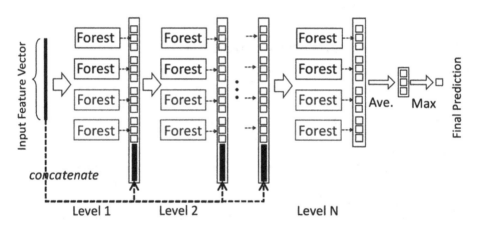

Fig. 1. Illustration of the cascade forest structure [19]. (Color figure online)

Given an instance, each forest will count the percentage of different training instance classes on the leaf node where the relevant instance is located, and then average across all trees in the same forest to get the estimated value of class distribution, as demonstrated in Fig. 2. In Fig. 2, the path from the instance traversing to the leaf node is highlighted in red and the different marks in leaf nodes imply different classes.

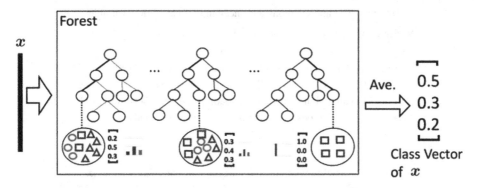

Fig. 2. Illustration of class vector generation [19]. (Color figure online)

The estimated class distribution will form a class vector. Then the class vector and the raw feature vector are concatenated together as the new input vector for the next layer. Take three classes for example, each forest will produce a three-dimensional class vector and each cascade layer will produce four three-dimensional class vectors. Therefore, the next cascade layer will receive 12(=3 × 4) enhanced features, as shown in Fig. 1.

In order to prevent overfitting, the k-fold cross validation is used to generate the class vector produced by each forest. Specifically, each instance will be used as training data for k − 1 times to obtain k − 1 class vectors, and then be averaged to produce the final class vector as the enhanced feature for the next cascade layer. The 3,5,10-fold cross validation are usually used to obtain the appropriate parameter. After extending a new layer, the validation set is used to estimate the performance of the entire cascade. If there is no significant performance gaining, the training process will be terminated; thus, the number of cascade layers will be automatically determined.

3   Experiment and Result

The proposed method has been implemented in Python 3.7. All the experiments are implemented on windows 10 operating system with AMD Ryzen 5 4600H having RAM of 16 GB memory and NVDIA GeForce GTX 1650 with 12GB of GPU memory.

Our dataset was sorted out by the original data of a Railway Bureau. While sorting out the original forecast variables, we also make additional statistics on the number of shipping locations, receiving locations and railway freight routes and the times and days of customer monthly deliver goods.

Feature selection can reduce the feature dimension, which plays an important role in improving classification accuracy and is useful for customer churn prediction. Correlation coefficient method [22] in feature selection was used in the paper to remove some irrelevant features and repetitive features and the final prediction features are shown in Table 1 and Table 2. Table 1 describes the abbreviations and meanings of categorical variables for prediction, while Table 2 describes the information about the continuous variable.

Table 1. The abbreviations and meanings of categorical variables

Variable	Meaning	Categorical
I-Sign	Inspection sign	0-Sign0; 1-Sign1
T-Sign	Transport sign	1-FCL; 2-LTL; 3-Container; 4-Others
S-Sign	Seal sign	0-Railway seal; 1-Owner seal; 2-No sealing
L-Sign	Loading sign	0-Railway loading; 1-Owner loading
Type	Type of goods	Numbering of 27 types of goods

Table 2. The abbreviations and meanings of continuous variables

Variable	Meaning
No.sl	The number of shipping locations
No.rfr	The number of railway freight routes
No.rl	The number of receiving locations
Tot.dm	The total delivery mileage
Spcl.dm	The delivery mileage of special railway line
Lim.pred	The longest railway freight transit period acceptable to customers
No.veh	The number of vehicles loaded of goods
Tot.wt	The total weight of the goods the customer needs to transport
No.tarp	The number of tarpaulins used for packaging the goods
S.I	The insured value selected by customer
Tot.cost	The total amount of freight
tickets	The number of times customer monthly deliver goods
days	The days of times customer monthly deliver goods

Company	time_mouth	I-Sign	T-Sign	No.sl	No.rfr	No.rl	Spcl.dm	Tot.dm	Lim.pred	No.veh	Tot.wt	No.tarp	S-Sign	L-Sign	Type	S.I.	Tot.cost	tickets	days	Label
C001	201711	0	4	1	1	1	0	1565	11	2	1180	0	0	0	22	100000	39100	1	1	1
C002	201809	1	1	1	1	2	10000	2195	3	5	0	0	2	1	17	0	657700	5	1	0
C002	201810	1	1	1	1	2	10000	1755	4	4	0	0	2	1	17	0	635700	4	1	1
C003	201903	1	1	2	2	3	0	1049	6	1	60000	2	0	0	11	100000	853650	1	1	0
...	...	...	...	...	...	...	...	...	...	...	...	...	...	...	...	...	...	...	...	...

Fig. 3. The sample data

The complete database is composed of label and feature set which usually consists of variables to predict customer churn. The Fig. 3 shows the part of data in the dataset used for this experiment.

In the experiments, our deep forest model uses the cascade structure, where each layer is composed of 2 random forests and each forest contains 100 trees. The three-fold cross validation is used to generate class vector. The number of cascade levels is automatically determined. Three algorithms including decision tree, AdaBoost and XGBoost were used as the classifier to compare with our model. The grid search and ten-fold cross validation were applied to select the suitable parameters for these algorithms.

The dataset was randomly divided into two subsets, including 70% for training and 30% for testing. The splitting process was done before the dataset standardization and the training dataset was used in the classification process. The dataset details are shown in Table 3.

Table 3. Dataset details

Type	Churners	Non-churners	All
Training dataset	14412	22620	37032
Test dataset	6176	9694	15870
Dataset	20588	32314	52902

The evaluation measurements of the experiment are the mean and standard error obtained by training 10 times, in which each time the training data and testing data are randomly with the fixed splitting ratio as described above.

Because our dataset was calculated according to the customer's delivery year and month, we considered that year and month can be used as features to predict according to the seasonal characteristics of railway freight. Table 4 demonstrated the experiment results of classification using deep forest model in datasets with and without time features. It can be clearly seen that the time features make the overall accuracy increased from 0.75 to 0.78.

Table 4. Time features results

Dataset	Accuracy ($\pm$std.)
without time features	0.75 ± 0.03
with time features	0.78 ± 0.01

Finally, the dataset consists of 52 features including delivery year and delivery month.

In this paper, several evaluation measurements were used to compare the performance of models, including accuracy, precision, recall, F1-score and the value of AUC. F1-score is obtained by calculating harmonic average between precision and recall, which is more often used to measure models than accuracy. The area under the ROC curve is the value of AUC.

Table 5 shows the experiment results of customer classification using decision tree, AdaBoost, XGBoost and deep forest. It can be seen that the deep forest we used achieves the best performance in railway freight data with the accuracy of 0.78, the precision of 0.75, the recall of 0.66, the F1-score of 0.7 and the AUC value of 0.86, which are higher than those of other model. Especially, the model shows a distinct improvement in the recall, the F1-score and the AUC value while making slight improvement in accuracy and precision. At the same time, this model has smaller standard error which make more stable performance.

Table 5. Experiment results

Model	DT	AdaBoost	XGB	Deep forest
Accuracy(std.)	0.74 ± 0.03	0.75 ± 0.02	0.76 ± 0.02	0.78 ± 0.01
Precision(std.)	0.70 ± 0.05	0.70 ± 0.04	0.73 ± 0.05	0.75 ± 0.04
Recall(std.)	0.58 ± 0.09	0.62 ± 0.09	0.62 ± 0.05	0.66 ± 0.05
F1-score(std.)	0.63 ± 0.05	0.65 ± 0.05	0.67 ± 0.05	0.70 ± 0.01
AUC(std.)	0.77 ± 0.05	0.80 ± 0.03	0.84 ± 0.02	0.86 ± 0.01

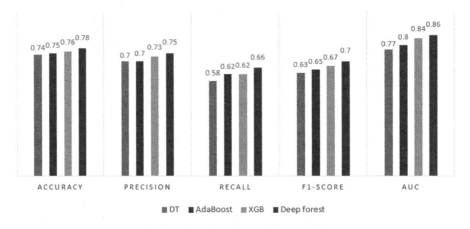

Fig. 4. Line chart of experiment results.

Among the five evaluation measurements, F1-score and AUC value are the two most important evaluation measurements to evaluate the overall performance of the model.

The best F1-score when using deep forest as the classification algorithm achieved of 0.70, which is 7% higher than that of decision tree of 0.63, 5% higher than that of AdaBoost of 0.65 and is 3% higher than that of XGBoost of 0.67.

Figure 5 shows the ROC curves of four models. The area under the ROC curve is the value of AUC. The AUC value of this deep forest reached the highest of 0.86, which was 9% points higher than decision tree, 6% points higher than AdaBoost and 2% points higher than XGBoost.

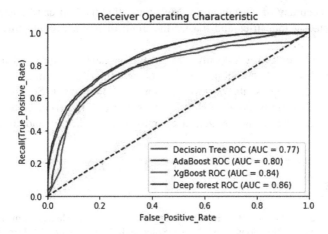

Fig. 5. ROC curve of four models.

As demonstrated above, the deep forest model shows its superiority in one dimensional data (sequence) classification through multiple ensemble. Through training layer-by layer, the best features can be found for classification. It also shows the great advantage in time due to the few parameters. So this model is undoubtedly the best choice compared with other ensemble models.

4   Conclusions

Combined with the seasonal characteristics of railway freight, this paper expands the time features of the dataset such as deliver year and month, and combines them with the deep forest model to predict customer churn.

The performance of ensemble learning will be corresponding improved when the differences of learning samples are fully reflected. Based on this ideal, deep forest we used adopts diverse structures to improve accuracy of classification.

The deep forest in this paper achieves the best performance in railway freight data with the accuracy of 0.78, the precision of 0.75, the recall of 0.66, the F1-score of 0.7 and the AUC value of 0.86, which are higher than those of decision tree, Ada-Boost and XGBoost. At the same time, this model has smaller standard error and makes more stable performance.

The best F1-score when using deep forest as the classification algorithm is 7% higher than that of decision tree of 0.63, 5% higher than that of AdaBoost of 0.65 and 3% higher than that of XGBoost of 0.67. The AUC value of this deep forest reaches the highest of 0.86, which was 9% points higher than that of other models.

Generally, the data is simply sorted according to each customer, but in this paper, we adopted a monthly recording pattern to subdivide customer delivery information in case the prediction results lag behind. The practicability of our model is improved effectively since the churners can be found in time due to the monthly prediction. At the same time, we build a model based on the deep forest framework, which greatly improves the accuracy of classification. Meantime, this method can be applied to the other transportation industries and master the trend of major account in e-commerce and financial.

In this paper, the deep forest only adapted the cascade structure, but multi granularity scanning had not been applied well. Multi granularity scanning can be used to further improve the representation learning ability of the model. So, there are some important work that can be conducted in the near future. It is also of great necessity to improve the data processing for getting a more perfect dataset and to get a more accurate model.

References

1. Ascarza, E., et al.: In Pursuit of Enhanced Customer Retention Management: Review, Key Issues, and Future Directions. Cust. Needs Solut. 5(1–2), 65–81 (2017). https://doi.org/10.1007/s40547-017-0080-0
2. Aggarwal, C.C.: Data streams: models and algorithms. J. Springer Science & Business Media 31, 169–207 (2007). https://doi.org/10.1007/978-0-387-47534-9
3. Wang, Q., Zeng, W.: Thought and discussion on customer relationship management of railway freight transport system construction. J. Railway Transp. Econ. 01, 35–38 (2011). https://doi.org/10.3969/j.issn.1003-1421.2011.01.008
4. Athanassopoulos, A.D.: Customer satisfaction cues to support market segmentation and explain switching behavior. J. Bus. Res. 47(3), 191–207 (2000). https://doi.org/10.1016/S0148-2963(98)00060-5
5. Bhattacharya, C.B.: When customers are members: customer retention in paid membership contexts. J. Acad. Mark. Sci. 26(1), 31–44 (1998). https://doi.org/10.1177/0092070398261004
6. Jones, T.O., Sasser, W.E.: Why satisfied customers defect. J. Harvard Business Review 73(6), 88–99 (1995). https://doi.org/10.1061/(ASCE)0742-597X(1996)12:6(11.2)
7. Huang, B., Kechadi, M.T., Buckley, B.: Customer churn prediction in telecommunications. J. Expert Systems with Applications 39(1), 1414–1425 (2012). https://doi.org/10.1016/j.eswa.2011.08.024
8. Verbeke, W., Dejaeger, K., Martens, D., Hur, J., Baesens, B.: New insights into churn prediction in the telecommunication sector: A profit driven data mining approach[J]. Eur. J. Oper. Res. 218(1), 211–229 (2012). https://doi.org/10.1016/j.ejor.2011.09.031
9. Bilal Zoric, A.: Predicting customer churn in banking industry using neural networks. Interdisc. Description Complex Syst. Sci. J. 14(2), 116–124 (2016). https://doi.org/10.7906/indecs.14.2.1
10. Yu, X.B., Cao, J., Gong, Z.W.: Review on customer churn issue. J. Comput. Integr. Manufacturing Syst. 18(10), 2253−2263 (2012). CNKI:SUN:JSJJ.0.2012-10-020
11. Caigny, A.D., Coussement, K., Bock, K.W.D.: A new hybrid classification algorithm for customer churn prediction based on logistic regression and decision trees. J. Eur. J. Oper. Res. 269, 760–772 (2018). https://doi.org/10.1016/j.ejor.2018.02.009
12. Martens, D., Vanthienen, J., Verbeke, W., Baesens, B.: Performance of classification models from a user perspective. J. Decision Support Syst. 51(4), 782–793 (2011). https://doi.org/10.1016/j.dss.2011.01.013
13. Hemalatha, P., Keshav, H.: A hybrid classification approach for customer churn prediction using supervised learning methods: banking sector. Banking Sector (2019)

14. Gong, J., Ju, J., Sun, Z., Ying, C., Tan, S., Sun, Z.: Research on customer churn prediction method based on variable precision rough set and BP neural network. In: Proceedings of the 2018 International Conference on Transportation & Logistics, Information & Communication, Smart City(TLICSC), vol. 161, pp. 298–304 (2018). https://doi.org/10.2991/tlicsc-18. 2018.46

15. Zhou, Z.H.: Ensemble Methods: Foundations and Algorithms. CRC, Boca Raton (2012). https://doi.org/10.1201/b12207

16. Idris, A., Khan, A., Lee, Y.S.: Genetic programming and adaboosting based churn prediction for telecom. In: IEEE International Conference on Systems, pp.1328–1332 (2012). https:// doi.org/10.1109/ICSMC.2012.6377917

17. Huang, Y., et al.: Telco churn prediction with big data. In: The 2015 ACM SIGMOD International Conference, pp. 607–618 (2015). https://doi.org/10.1145/2723372.2742794

18. Ahmad, A.K., Jafar, A., Aljoumaa, K.: Customer churn prediction in telecom using machine learning in big data platform. J. Big Data 6(1), 1–24 (2019). https://doi.org/10.1186/s40537-019-0191-6

19. Zhou, Z.H., Feng, J.: Deep forest. J. National Sci. Rev. 6(1), 74–86 (2019). https://doi.org/10.1093/nsr/nwy108

20. Breiman, L.: Random forests. Mach. Learn. 45(1), 5–32 (2001). https://doi.org/10.1023/A:1010933404324

21. Liu, F.T., Ting, K.M., Yu, Y., Zhou, Z.-H.: Spectrum of variable-random trees. J. Artif. Intell. Res. 32, 355–384 (2008). https://doi.org/10.1613/jair.2470

22. Chen, P., Li, F., Wu, C.: Research on intrusion detection method based on Pearson correlation coefficient feature selection algorithm. J. Phys. Conf. Ser. 1757(1), 012054 (2021). (10pp). https://doi.org/10.1088/1742-6596/1757/1/012054

Multi-view of Data for Auto Judge Model in Online Dispute Resolution

Qinhua Huang[(✉)] and Weimin Ouyang

School of AI and Law, Shanghai University of Political Science and Law,
Shanghai 201701, China
{hqh, oywm}@shupl.edu.cn

Abstract. Online dispute resolution (ODR) is a mechanism of solving legal or business by the online way. It was set to solve problems with less cost of attendant parities, or bring some convenience. ODR involved several information techniques to replace the traditional negotiations. Some problems for ODR as the trust problem on the judgment and the transparence of the whole process could lead negative affect on the application. Efficiency is also a key factor of the ODR system in some scenarios, like online internet courts in China and electronic commerce dispute, where there will be large piles of case generated every minute. The core factor to affect the efficiency and transparency is developing the reliable, trusty automatic judge system. In this paper, we proposed a method to build auto legal judge model from the view of ODR. This research combined information such as past information of case as well as textual evidence of the current transaction in model constructing. Experiment showed our method is effective and can have a promising scalability within the certain extent.

Keywords: Machine learning · Crime classification · Online dispute resolution

1 Introduction

With the development of online business, especially e-commerce, people have taken great advantage from the living model of home working. In 2020, the globally breakout of coronavirus could have damaged billions of peoples normal living. But with the help of AI, there were many kinds human activities move to be online in many area, which partly avoided more damages. Moving everything online has two-fold affects. One is it spared many people's time cost. On the other hand, it may have some impacts on people opinions. People do business and make various interactions online is easy. But how would they do if they have disputes under this online-working model? Back to normal face-to-face dispute resolution mechanism may cancel out all the advantage we had taken for all parties. In some scenarios, this is even impossible. Online dispute resolution (ODR) is to deploy applications and computer networks for resolving disputes. Since 1996 the term ODR was formally reported been seen [1], it was mainly thought to be in relation to resolving online disputes, especially in the area of e-commerce. Now various kinds of ODR tools have been intensively developed to facilitate settlement in limited range of legal issues and to determine some small claims.

© Springer Nature Switzerland AG 2021
D.-S. Huang et al. (Eds.): ICIC 2021, LNCS 12837, pp. 490–498, 2021.
https://doi.org/10.1007/978-3-030-84529-2_41

Big e-commerce companies as Alibaba, eBay have deployed ODR to deal with its customers' dispute. The core of these sorts of tools are the applications of auto judge model. AI algorithms make decisions using the attending parties' data as input based on the history records data. For an ODR tool, to make the decision model transparent and predictable is of the most important.

Ideas of using AI model as the judge can be traced back to 1958, when Lucien Mehi proposed the automatic judge problem of AI world [2]. In 1980's, the joint disciplines researches of AI and Law had come to its prosperity. In the early stages, people focused on developing AI with similar logic used in law, which emphasize the mechanism and tool of reasoning, representing, known a rule-based reasoning(RBR). This has the same idea with the NLP field. In 1980's researchers realized the complex attributes of natural language, and the law information retrieval method was put out. Experienced the basic stage of using keyword searching in AI, Carole Hafner proposed her novel method of semantic net representations. Then many law expert system were build, but there were one flaw that matters, which mainly due to the openness nature of law suit predicate. While these reasoning systems based on rules continued to be developed, the other way using case to do reasoning appeared which is called cased-based reasoning (CBR). Rissland developed the HYPO system, which was taken to be the first real CBR system in AI and Law. This research stream attracted many interests. To combine the benefits of RBR and CBR, Rissland proposed the CABARET system, a reasoned using hybrid of CBR-RBR. CABARET can dynamically apply RBR and CBR, rather than serially call it in a fixed order. With all these effort, it is broad consensus that automatically representation of law entity and concept searching are the core target of AI in law. Due to the limitation of data sparsity, most AI law models were far from success.

With the deep learning method developed in recent years, many researches are working on building legal AI models using pre-trained NLP models based on deep learning, trained and generated from big dataset of real legal documents. With the great progress made in the field of natural language processing, especially in the language model, such as ELMo, BERT, GPT-2, etc., the concept of knowledge graph is developed [3, 4]. The legal artificial intelligence based on knowledge graph can be used as a trial assistant in the ODR scene, providing accurate reference. In addition, the study is important through joint learning, ODR can provide important legal data, is seen as an effective solution to the sparseness of legal data. The technology has been piloted in the Hangzhou Court [5]. Liu Zonglin [6] and others noticed that the crime prediction and legal recommendation are important sub-tasks of legal judgment pre-diction, and proposed to use multi-task learning model to model the two tasks jointly, and to integrate the crime keyword information into the model, which can improve the accuracy of the model for these two tasks.

In the field of legal intelligence question-and-answer, especially legal advice, there is sometimes a need to answer questions that do not depend on a certain fact. McElvain has studied and established the West Plus system to provide question-and-answer capabilities by using IR and NLP technologies rather than relying on a structured knowledge base. This is a continuation of the idea of legal concept search. Due to the large degree of data openness, Chinese's study of automatic trial prediction is relatively active. Meanwhile for the data policy and other reasons, auto judge system in other languages are relatively few. Ilias Chalkidis combined the data of the European Court of Human Rights with other data, set up a deep learning model based on the BERT language model. From the perspective of human, the traditional feature model in machine learning is more comprehensible, while deep learning model may have better accuracy and performance. The bias of the model by deep learning is also hard to explain and may be unpredictable. This is very critical in legal area. Haoxi Zhong [7] discussed the application of topological learning in trial prediction. In the actual legal trial, the trial is generally composed of several sub-tasks, such as application of laws, charges, fines, sentences, etc., based on the dependence between sub-tasks, the establishment of topological mapping, and thus the construction of trial models. In recent years, natural language processing has made great progress in the field of general reading comprehension, such as BERT model application, attention mechanism, Shangbang Long [8] studied the problem of legal reading comprehension, according to the description of facts, complaints and laws, according to the judge's working mechanism, to give predictions, and thus to achieve automatic prediction. Wenmin Yang and other researchers based on the topological structure of sub-tasks, combined with the attention mechanism of word matching, proposed a deep learning framework obtained a greater performance improvement.

2 Related Works

In this section, we briefly listed the related models in knowledge graph embeddings [9]. Due to the space limitation, we mainly discuss two kinds of method. One the translational method. The other is tensor factorization method.

TransE firstly purposed projecting the entities into the same space, where the relation can be taken as a vector from head entity to tail entities [10]. Formally, we have a triple (h,r,t), where h, r, t $\in \mathbb{R}^k$, h is the head entity vector, r is the relation vector and t is the tail entity vector. The TransE model represents the a relationship by a translation from head entity to tail entity, thus it holds $h + r \approx t$. By minimizing the score function $f(h, r, t) = \| h + r - t \|_2^2$, which means h + r is the closest to t in distance. This representation has very clear geometric meaning as it showed in Fig. 1.

TransH was proposed to address the issue of N-to-1, 1-to-N and N-to-N relations [11]. It projected (h, r, t) onto a hyperplane of w_r, where w_r is the hyperplane normal vector of r. TransR noticed both TransE and TransH took the assumption that embeddings of entities and relations are represented in the same space $\mathbb{R}^k$. And relations and entities might have different semantic meaning. So TransE suggest project entities and relations onto different spaces in representation, respectively. The score function will be minimized by translating entity space into relation space.

There are some other models like Unstructured Model [9], which are some variant of simplified TransE. It suppose that all r = 0; Structured Embedding, it adopted L_1 as its distance measure since it has two relation-specific matrices for head and tail entities; Neural Tensor Network (NTN), which has some complexity that makes it only suit for small knowledge graphs. For the convenience of comparison, we listed the embeddings and score functions of some models in Table 1.

Table 1. Entity and relation embedding models: embeddings and score functions

Model name	Embeddings	Score function s(h,r,t)
Neural Tensor Network (NTN)	$M_{r,1}, M_{r,2} \in \mathbb{R}^{k \times d}$, $b_r \in \mathbb{R}^k$	$u_r^\top g\left(h^\top M_r t + M_{r,1}h + M_{r,2}t + b_r\right)$
Latent Factor Model (LFM) [12]		$h^\top M_r t$
Semantic Matching Energy (SME)	M_1, M_2, M_3, M_4 are weight matrices, $\otimes$ is the Hadamard product, b_1, b_2 are bias vectors	$((M_{1h}) \otimes (M_2 r) + b_1)\top((M_3 t) \otimes (M_4 r) + b_2)$
TranE	$h, r, t \in \mathbb{R}^k$	$\| h + r - t \|$
TransH	$h, t \in \mathbb{R}^k, w_r, d_r \in \mathbb{R}^k$	$\| (h - w_r^\top h w_r) + d_r - (t - w_r^\top t w_r) \|$
TransD	$\{h, h_p \in \mathbb{R}^k\}$ for entity h, $\{t, t_p \in \mathbb{R}^k\}$, for entity t, $\{r, r_p \in \mathbb{R}^d\}$ for relation r	$\| (h + h_p^\top h r_p) + r - (t + t_p^\top t r_p)) \|$
TransR [13]	$h, t \in \mathbb{R}^k, r \in \mathbb{R}^d, M_r \in \mathbb{R}^{k \times d}$, M_r is a projection matrix	$\| M_r h + r - M_r t \|$

In Table 2 the constraints of each models are presented. As we should point out that with the models developed, the embeddings and constraints actually become more complicated. One thing is sure that if the model is more complicated, the computation cost goes higher. This problem should be carefully considered in related algorithm design.

Table 2. Entity and relation embedding models: constraints

Model name	Constraints
TranE	$h, r, t \in \mathbb{R}^k$
TransH	$h, t \in \mathbb{R}^k, w_r, d_r \in \mathbb{R}^k$
TransD	$\{h, h_p \in \mathbb{R}^k\}$ for entity h, $\{t, t_p \in \mathbb{R}^k\}$, for entity t, $\{r, r_p \in \mathbb{R}^d\}$ for relation r
TransR	$h, t \in \mathbb{R}^k, r \in \mathbb{R}^d, M_r \in \mathbb{R}^{k \times d}$, M_r is a projection matrix

There are ways of freeing limitation on the entity embeddings. The main idea is to let head and tail embedding representation independent on each other. We give a possible implementation method here, as showed in Fig. 1.

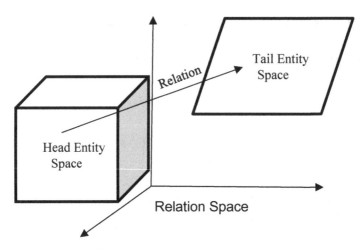

Fig. 1 A simple illustration of entity and relation spaces in embeddings model, where the space dimensions k, l and d might not be the same between any two spaces

For each triple (h, r, t), $h \in \mathbb{R}^k$, $t \in \mathbb{R}^l$, while $r \in \mathbb{R}^d$. Here k, l and d can be different. For the sake of calculation, like in TransR and TransE, we project head entities and tail entities into relation space. The projected vectors of head entities and tail entities are defined as

$$h_r = hM_{hr}, t_r = tM_{tr}$$

where $M_{hr} \in \mathbb{R}^{k \times d}$, $M_{tr} \in \mathbb{R}^{l \times d}$ are transition matrix.
Routinely the score function is defined as

$$f_r(h, t) = \| h_r + r - t_r \|_2^2$$

And there are constraints on the norms of embeddings h, r, t and the transition matrix. As it showed below

$$\forall h, \ r, t, \ \| h \|_2 \leq 1, \ \| r \|_2 \leq 1, \ \| t \|_2 \leq 1, \ \| hM_{hr} \|_2 \leq 1, \ \| tM_{hr} \|_2 \leq 1$$

The Canonical Polyadic method in link prediction take the head and tail entities by learning independently. After nalyzing the negative side of independency, the SimplE model simplified the freedom limitation of Canonical Polyadic decomposition. It let the two kind entities learn dependently, while both of them have the same idea that using two embedding representation for each one entity while it could take head or tail position.

3   Training Method

The transitional methods generally adopt margin-based score function, as showed in follow equation.

$$L = \sum_{(h,r,t)\in S} \sum_{(h',r,t')\in S'} \max(0, f_r(h,t) + \gamma - f_r(h',t'))$$

where max(x, y) aims to get the maximum between x and y, γ is the margin, S is the set of correct triples and S' is the set of incorrect triples.

The optimization method is using stochastic gradient descent (SGD) [14], while other method may have a little bit training performance difference [15]. Here the h and t may come from different weight list for one entity.

The SimplE defined a function f(h, r, t) = $\langle h, rt \rangle$. The $\langle h, rt \rangle$ is defined to be $\langle h,\ rt \rangle \underline{\text{def}} \sum_{j=1}^{d} h[j] * r[j] * t[j]$, for the brevity, $\langle h,\ r,\ t \rangle \underline{\text{def}} (v \odot w) \cdot x$, where the $\odot$ is Hadamrd multiplication (element-wise), the $\cdot$ stands for dot product. The optimization object is to minimize each batch with $\sum \text{softplus}(-1 \cdot \langle h, rt \rangle + \lambda\theta)$, here λ is the model hyper-parameter and θ is the regularization factor, softplus = $\log(1 + e^x)$ is a softened relu function. In addition, the l is in $\{-1, 1\}$, which is the labeled result.

For the translational method, take TransE for example, to make the predication model work, the parameters mainly are $\{h^{k\times d}, r^{n\times d}\}$. If we free the ties between head entity and tail entity for an entity, a $t^{k\times d}$ also is needed to represent the tail entity, for simplicity, assuming the 3 types has same embedding dimension. Let us consider the size of weight parameters in SimplE. We can rewrite parameter spaces here, parameter = $\{h^{k\times d}, r^{n\times d}, r^{n\times d}, t^{k\times d}\}$. As a comparison, the TransE method has a smaller weight spaces in size. Roughly, the SimplE's weight size is about twice as the TransE's if we specify the two embedding size for the same dimension. It could be an important factor that weakened TransE's expressiveness. If we simply try to increase TransE's embedding size to improve the expressiveness, it might cause problem of overfitting, which is not wanted. In general, there is no perfect theoretical result in deciding how the dimension d should be in this sort of problem. The parameter n, k are characters of the specified training data. Consider if the Knowledge Graph can grow, n, l, k can be very different. It will generate some impact on the performance of a trained model.

4   Training Models

We trained our model on top of BERT. To fully apply the information in the history case, we deploy Legal Reading Comprehension model (LRC) to make the model. A law may have many articles. To decide a judgement of a fact violation of law, we can do the supervised learning from labeled data. Formally, let $F = \{f_1, f_2, \ldots f_{|F|}\}$ denote all the facts, $W_{f1} = \{w_1, w_2, \ldots, w_k\}$ denote all words in a fact f_1, $L = \{l_1, l_2, \ldots, l_{|L|}\}$ denote all the law articles, and $P = \{p_1, p_2, \ldots, p_{|A|}\}$ denote all the pleas. The objective function would be:

$$r = \text{argmax}_{\{r \in \{\text{support,reject}\}\}} P(r|f, p, l) \, [16]$$

We deployed multi-view tasks in prediction. The first is binary violation, which will classify a fact into violated or not class. The multi-label classification task is to label a fact into multi articles.

During the train process, after our embedding representation, we apply self-attention reader to read fact embedding. Like in biGRU-Att, the fact embedding will pass to the output layer using a sigmoid layer to get the binary violation task label. The multi-label task is accomplished by using a ReLU layer after reading the fact embeddings.

5   Experiment and Result

Firstly we evaluate our embedding methods with two typical knowledge graphs, built with Freebase [17] and WordNet [18]. These two datasets are chosen in the same way with SimplE and other embedding models. The statics of the datasets is showed in Table 3 This step is to fix the optimization of our embedding settings Table 4 To get the optimized dimension size of the embedding, we tested size from 50 to 200, and set 150 as the embedding parameter.

Table 3. Datasets statics

Dataset	#Rel	#Ent	#Train	#Valid	#Test
FB15k	1,345	14,951	483,142	50,000	59,071
WN18	18	40,943	141,442	5,000	5,000

Table 4. Experiment Filtered results on WN18 of SimplE

Embedding dimension size	fil_mrr	fil_hit@1	fil_hit@3	fil_hit@10
50	0.8154	0.7281	0.9019	0.9372
100	0.9326	0.926	0.9391	0.9392
150	0.93832	0.937	0.9395	0.9398
200	0.938574	0.9373	0.9395	0.9405

Secondly, we conduct empirical experiments for crime classification task. Here is the experimental settings in Table 5 Our empirical experiments runs on the real-world legal datasets, i.e., Fraud and Civil Action dataset, which has 17,160 samples. The Statistics of the dataset is showed in Table 5 The learning rate, is set to be 0.0005.

We compared the performance of our embedding method with the original BoW in HMN.

Table 5. Statistics of experiment data set [19]

Dataset	#fact	#Laws	#Articles	AVG fact description size	AVG article definition size	AVG law set size perfact	AVG article set size perfact
Fraud and Civil Action	17,160	8	70	1,455	136	2.6	4.3

Table 6 showed that our embedding strategy have some positive effects on the performance. On performance of macro-P and Jaccard, our method can achieve higher value while using the original simple BoW strategy can have better performance in terms of macro-R and macro-F.

Table 6. Result and performance comparison

Dataset	macro-P	macro-R	macro-F	Jaccard	Model
Fraud and Civil Action	65.2	30.6	43.3	67.5	HMN using BoW
	67.4	28.1	40.2	69.3	HMN using our embedding

6 Conclusion

In this paper, we researched the automatic legal judgment problem in online dispute resolution. We carefully investigated the typical embedding algorithms, TransE and SimplE, and then we optimized our embedding strategy by unlink the head and tail entity. We adopt our embedding method to apply a multi-view crime classification model. By testing the performance on real dataset, we can find our embedding method has some privileges in result. There still some problem remained to solve. In the future, we will test on more dataset in terms of scalability and variety, and keep on optimizing some strategy to get better performance. We plan to test more models using our embedding strategy.

References

1. Katsh, E.: Online dispute resolution: building institutions in cyberspace. Univ. Connecticut Law Rev. **28**, 953–980 (1996)
2. Mehl, L.: Automation in the legal world, conference on the mechanisation of thought processes held at teddington. England (1958)
3. Ji, G.: Knowledge graph embedding via dynamic mapping matrix. In: ACL, pp. 687–696 (2015)
4. Singhal, A.: Introducing the knowledge graph: things, not strings. Google Official Blog16 May 2012. Accessed 6 September 2014
5. Zhejiang to comprehensively promote the construction of digital courts 2021. https://www.chinacourt.org/index.php/article/detail/2021/04/id/5949110.shtml

6. Liu, Z., et al.: Multi-task learning model for legal judgment predictions with charge keywords. J. Tsinghua Univ. Sci. Technol. **59**(7), 497–504 (2019)

7. Zhong, H.: Legal judgment prediction via topological learning. In: Proceedings of the 2018 Conference on Empirical Methods in Natural Language Processing, pp. 3540–3549

8. Shangbang, L.: Automatic judgment prediction via legal reading comprehension. In: Sun, M., Huang, X., Ji, H., Liu, Z., Liu, Y. (eds.) CCL 2019, LNAI 11856, pp. 558–572. Springer, Cham (2019). https://doi.org/10.1007/978-3-030-32381-3_45

9. Xiao, H., Huang, M., Yu, H., Zhu, X.: From one point to a manifold: knowledge graph embedding for precise link prediction. In: IJCAI, pp. 1315–1321 (2016)

10. Bordes, A.: Translating embeddings for modeling multi-relational data. In: NIPS, pp. 2787–2795 (2013)

11. Wang, Z., Zhang, J., Feng, J., Chen, Z.: Knowledge graph embedding by translating on hyperplanes. In: AAAI, pp. 1112–1119 (2014)

12. Jenatton, R., Roux, N.L., Bordes, A., Obozinski, G.R.: A latent factor model for highly multi-relational data. In: Proceedings of NIPS, pp. 3167–3175 (2012)

13. Yankai, L., Zhiyuan, L., Maosong, S., Yang, L., Xuan, Z.: Learning entity and relation embeddings for knowledge graph completion. In: AAAI, pp. 2181–2187 (2015)

14. Recht, B., Re, C., Wright, S., Niu, F.: HOGWILD: a lock-free approach to parallelizing stochastic gradient descent. In: NIPS, pp. 693–701 (2011)

15. Zhao, S.-Y., Zhang, G.-D., Li, W.-J.: Lock-free optimization for nonconvex problems. In: AAAI, pp. 2935–2941 (2017)

16. Kazemi, S.M., Poole, D.: Simple embedding for link prediction in knowledge graphs. In: Advances in Neural Information Processing Systems 2018

17. Bollacker, K., Evans, C., Paritosh, P., Sturge, T., Taylor, J.: Freebase: a collaboratively created graph database for structuring human knowledge. In: Proceedings of KDD, pp. 1247–1250 (2008)

18. Miller, G.A.: Wordnet: a lexical database for english. Commun. ACM **38**(11), 39–41 (1995)

19. Wang, P., Fan, Y., Niu, S., Yang, Z., Zhang, Y., Guo, J.: Hierarchical matching network for crime classification. In: SIGIR 2019, 21–25 July 2019, Paris, France (2019)

Multi-task Learning with Riemannian Optimization

Tian Cai, Liang Song[⊠], Guilin Li, and Minghong Liao

School of Informatics, Xiamen University, Xiamen 361005, China
songliang@xmu.edu.cn

Abstract. Multi-task learning (MTL) is a promising research field of machine learning, in which the training process of the neural network is equivalent to multi-objective optimization. On one hand, MTL trains all the network weights simultaneously to converge the multi-task loss. On the other hand, multi-objective optimization aims to find the optimum solution, which satisfies the constraints and optimizes the vector of objective functions. Therefore, the performance of MTL is dominated by the computation of the multi-objective solution. This paper proposes a method based on Riemannian optimization to solve the multi-objective optimization in MTL. Firstly, multi-objective optimization is reduced to its Karush-Kuhn-Tucker (KKT) condition as the optimum solution of constrained quadratic optimization. Secondly, by mapping the Euclidean space of the constraint into manifold, the quadratic optimization is transformed to an unconstrained problem. Finally, Riemannian optimization algorithm is used to compute the solution of this problem, which gives a Pareto direction towards the KKT condition. We perform experiments on the MultiMNIST and Fashion MNIST datasets, and the experimental results demonstrate the efficiency of our method.

Keywords: Multi-task learning · Multi-objective optimization · Riemannian optimization

1 Introduction

MTL is a very classic and important machine learning model. There are two noteworthy points about the definition of MTL. 1. Task correlation. Each task needs to have a certain correlation; 2. Task type. In machine learning, learning tasks mainly include supervised learning tasks such as classification and regression, unsupervised learning tasks such as clustering, semi-supervised learning tasks, active learning tasks, reinforcement learning tasks, online learning tasks and multi-perspective learning tasks. Therefore, different learning tasks correspond to different model settings. MTL aims to make each task interact and improve their efficiency respectively [1]. The core of MTL is the knowledge sharing and simultaneous learning among different tasks. In the MTL model, branches

This work was supported by the Fundamental Research Funds for the Central Universities of China under Grant No. 20720190028.

D.-S. Huang et al. (Eds.): ICIC 2021, LNCS 12837, pp. 499–509, 2021.
https://doi.org/10.1007/978-3-030-84529-2_42

share the same trunk, and the parameters in the trunk structure will be affected by all branches when they are updated. The loss of all tasks is expected to converge.

How to combine the loss of each individual task into a single overall loss is worth considering. A classical method which is called linear weighting method has been widely used, which directly assigns the specific weight to the task, so manual adjustment of the weight is also a natural choice [2–4]. These weights determine the performance of an MTL model. However, the cost of finding the best weight by manual adjustment is too expensive. Therefore, we hope to minimize the total loss function automatically within the learning process obtained by the weighted sum of each loss function, which is more complex but more accurate. Fortunately, MTL can be modeled as multi-objective optimization problem, whose Pareto optimal solution is exactly what we need to find for obtaining the best task loss weight. Especially, the multiple gradient descent method (MGDA) [5] is employed in [6], whose core principle is to combine the gradient of each task to decide how to update the shared parameters. And this is also the reason why MGDA is suitable for MTL scenarios.

Our contribution is proposing a method based on Riemannian optimization to solve the multi-objective optimization in MTL. Firstly, multi-objective optimization is reduced to its KKT condition as the optimum solution of constrained quadratic optimization. Secondly, by mapping the Euclidean space of the constraint into manifold, the quadratic optimization is transformed to an unconstrained problem. Finally, Riemannian optimization algorithm is used to compute the solution of this problem, which gives a Pareto direction towards the KKT condition. All the above steps will be elaborated in Sect. 3. Through experiments on MultiMNIST dataset [7], the feasibility and efficiency of this method was verified. In addition, we also conducted experiments on Fashion MNIST dataset [8], and the results show that the performance of our method was much better than that of single task experiment.

2 Related Work

Generally, there are two ways of MTL, hard parameter sharing or soft parameter sharing. In the hard parameter sharing, the upper network structure will produce a set of shared parameters for each task, while other parameters are determined by the characteristics of the task itself. In soft parameter sharing, all parameters are determined by the characteristics w.r.t the corresponding tasks. Recently, a review on MTL is provided in [9]. A basic approach in MTL is fine-tuning [10], which can set weights in advance before the model starts training, and then make small adjustments to make more effective utilization of these tasks. MTL can be widely applied in different fields of machine learning, such as computer vision [3, 11, 12], natural language processing [13], speech recognition [14, 15], etc.

A theoretical analysis of MTL was made in [16], and it believes that each task is equivalent to a learner, and how shared parameters should be updated depends on the meta-algorithm. Every task in MTL was considered to be based on core learning in [17], and each task utilizes multi-objective optimization theory to deal with MTL problems. Multinet [18] proposes a multiple structure network, which can be used for three visual tasks in automatic driving scene. The theory of collaborative multi-objective optimization to multiple perspective learning in multi-task situation was applied in [19]. And an adaptive learning algorithm to calculate the loss weight of multiple tasks using the concept of uncertainty was proposed in [20].

The purpose of multi-objective optimization is to optimize each sub-objective as much as possible by coordinating their weights. Multi-objective optimization was introduced by [21] and [22], respectively. The goal of optimization is to get Pareto optimal solution. A distribution estimation algorithm to calculate the solution set was proposed in [23]. The method of [24] is based on SVM, and the solution set is related to the identification of gene markers. [25] used genetic algorithm to find the Pareto optimal solution of the outer window design problem. Among these methods, the gradient based on multi-objective optimization [26, 27] is theoretically supported.

Manifold optimization [28, 29] is used by transforming constrained optimization problem in Euclidean space into an unconstrained optimization problem on Riemannian manifold. Previous works have shown the potential of Riemannian optimization in machine learning [30, 31].

3   The Method Based on Riemannian Optimization

3.1   Preliminary

Let M denote a topological space, for $\forall m \in M$, there exists an open neighborhood Ω which belongs to m homeomorphism with an open subset in D-dimensional Euclidean space, then M can be called D-dimensional topological manifold, also called D-dimensional manifold. Riemannian manifold is smooth differential manifold with Euclidean inner product defined at the tangent space T_mM of any point on manifold M. This leads to the concept of tangent space, which is a linear space composed of all tangent vectors at a certain point. As is shown in Fig. 1 [28], there is a tangent vector at a point on the sphere, but such tangent vectors exist in countless directions.

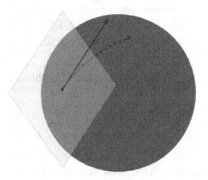

Fig. 1. A simple example of tangent space.

Tangent bundle is an extension of tangent space, which is defined as follows. The tangent bundle of a manifold M is the disjoint union of the tangent spaces of M.

$$TM = \{(x, v) : x \in M, v \in T_xM\} \tag{1}$$

In (1), x represents any point on manifold M, and v represents any tangent vector on tangent space T_mM. Before the Riemannian optimization algorithm can solve the multi-objective optimization, it is prerequisite to map the constraints from Euclidean space to the Riemannian space T_mM, and fix the overall objective function always on the Riemannian manifold, which consequently allows us to only solve the unconstrained problem. Among many mappings which vary from the simplest exponential mapping to other complex methods, and we use the mapping in [32] to accomplish this step.

3.2   From Multi-objective Optimization to Riemannian Optimization

Multi-objective Optimization
MTL can be formulated into the following optimization problem.

$$\min_{\lambda_{sh}, \lambda_1, ..., \lambda_W} \sum_{w=1}^{W} \hat{G}^w(\lambda_{sh}, \lambda_w) \tag{2}$$

μ^w is the task weight to be calculated, λ_{sh} are the shared parameters, λ_w are task-specific parameters. $\hat{G}^w(\lambda_{sh}, \lambda_w)$ is the task-specific loss function. We expand the expression (2) into the problem as shown in (3).

$$\min_{\lambda_{sh}, \lambda_1, ..., \lambda_W} G(\lambda_{sh}, \lambda_1, ..., \lambda_w) = \min_{\lambda_{sh}, \lambda_1, ..., \lambda_W} (\hat{G}^1(\lambda_{sh}, \lambda_1), ..., \hat{G}^W(\lambda_{sh}, \lambda_W))^T \tag{3}$$

In fact, the above process is equivalent to transforming the MTL problem into multi-objective optimization problem.

Reduced Optimization
Then, in order to find the Pareto optimal solution, we employ the idea in [6], which reduced the multi-objective optimization to solving its KKT condition, by using the multiple gradient descent algorithm (MGDA) [5]. KKT condition [33] is the core of MGDA. The KKT condition in MTL is actually defined as follows, 1. There exist $\mu^1, \mu^2, ..., \mu^W \geq 0$, and all μ satisfy: $\sum_{w=1}^{W} \mu^w = 1$ and $\sum_{w=1}^{W} \mu^w \nabla_{\lambda_{sh}} \hat{G}^w(\lambda_{sh}, \lambda_w) = 0$. 2. For all tasks w, $\nabla_{\lambda_w} \hat{G}^w(\lambda_{sh}, \lambda_w) = 0$. Consequently, problem (2) can be reduced to the problem as below.

$$\min_{\mu^1, ..., \mu^w} \left\{ \left\| \sum_{w=1}^{W} \mu^w \nabla_{\lambda_{sh}} \hat{G}^w(\lambda_{sh}, \lambda_w) \right\|_2^2 \middle| \sum_{w=1}^{W} \mu^w = 1, \mu^w \geq 0, \forall w \right\} \tag{4}$$

Because the following properties have been proved in [5]. If the objective function of the above optimization problem is 0, then the KKT condition is satisfied, which means that we have solved the multi-objective optimization. Otherwise, the solution $(\mu^1, \mu^2, ..., \mu^W)$ gives a Pareto direction to improve all the tasks.

Now, let's observe the constraints of the problem (4). This constraint is an n-dimensional plane in Euclidean space, which needs to be mapped to the manifold.

After this step, we will only need to solve the corresponding Riemannian optimization, and subsequently finish the MTL.

Riemannian Optimization

Before searching the solution on the Riemannian manifold, we must define what is gradient on manifold, which is called Riemannian gradient. Let $f: M \rightarrow R$ be a smooth function on a Riemannian manifold M. The Riemannian gradient of f is the vector field grad f on M uniquely defined by these identities,

$$\forall v \in T_x M, Df(x)[v] = \langle v, gradf(x) \rangle_x \tag{5}$$

$\langle \cdot, \cdot \rangle_x$ is the Riemannian metric. It is noteworthy that if x is a local optimum of f, $gradf$ $(x) = 0$. After defining the Riemannian gradient, we have to consider that if there exists $x_k \in M$ and its Riemannian gradient $gradf(x_k)$, how can we obtain $x_{k+1} \in M$? If we use the gradient descent move in Euclidean space as follows (α is the learning rate),

$$x_{k+1} = x_k - \alpha gradf(x_k) \tag{6}$$

then, x_{k+1} above cannot be guaranteed on the manifold, which means that the constraint of the problem (4) is not satisfied. Therefore, we make the gradient descent move always on the Riemannian manifold as follows,

$$x_{k+1} = R_{x_k}(-\alpha_k gradf(x_k)), gradf(x_k) \in T_{x_k} M \tag{7}$$

Here, R represents retraction, which realize the mapping onto Riemannian manifold as described in Sect. 3.1. The retraction is illustrated in Fig. 2 [28]. In fact, different Riemannian manifolds have different retraction operators, and for a certain Riemannian manifold, it can have multiple operators.

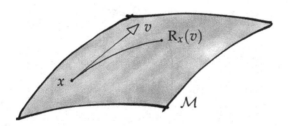

Fig. 2. An example of retraction, which allows to move away from x along $v \in T_x M$ while staying on a Riemannian manifold M

3.3 Solving the Riemannian Optimization

Now we apply Riemannian optimization to solve the Riemannian optimization problem as described in Sect. 3.2. In detail, we use manifold optimization toolbox [29] to accomplish this step. Especially, we map the constraint in problem (4) to the Multinomial Manifold, which is provided in the toolbox.

Because we train the neural network in Python environment while the toolbox is implemented by Matlab, we must call the Matlab engine API in the Python environment. First, after the image is input into the network, the main feature map will be generated through the encoder layer, and the main feature map will branch because of different decoder layers to obtain different branch feature maps. Next, we utilize MGDA [5] to store a large number of gradient values of loss functions, and the gradient is equivalent to the shared parameters mentioned in (2). However, the shapes of them are quite different, so we flattened these tensors into one-dimensional vectors for later computation. Then, we input the number of tasks and gradient vectors into the toolbox.

Next, we formulate the Riemannian optimization problem in form of matrices, which can be further processed in Matlab. The objective function $\sum_{w=1}^{W} \mu^w \nabla_{\lambda^{sh}} G^w (\lambda^{sh}, \lambda^w)$ of the problem (4) can be expanded as below,

$$
\begin{pmatrix}
\mu^1 \nabla_{\lambda^1} G^1 & \mu^1 \nabla_{\lambda^2} G^1 & \cdots & \mu^1 \nabla_{\lambda^{sh}} G^1 \\
\mu^2 \nabla_{\lambda^1} G^2 & \mu^2 \nabla_{\lambda^2} G^2 & \cdots & \mu^2 \nabla_{\lambda^{sh}} G^2 \\
\vdots & \vdots & \ddots & \vdots \\
\mu^W \nabla_{\lambda^1} G^W & \mu^W \nabla_{\lambda^2} G^W & \cdots & \mu^W \nabla_{\lambda^{sh}} G^W
\end{pmatrix}
\tag{8}
$$

Each row number indicates one single task, and column numbers represent different shared parameters (gradient). For convenience, we combine the above expansions in the form of a row vector, which is abbreviated as follows,

$$
\begin{pmatrix} \nabla_1 & \nabla_2 & \cdots & \nabla_{sh} \end{pmatrix}
\tag{9}
$$

If we denote A as gradient vector group and X as weight vector, the cost function of manifold optimization can be defined as follows,

$$
\text{cost}(X) = \|X \cdot A\|_2^2 = \nabla_1^2 + \nabla_2^2 + \cdots + \nabla_{sh}^2
\tag{10}
$$

$$
X \cdot A \Leftrightarrow \begin{pmatrix} \mu^1 & \mu^2 & \cdots & \mu^W \end{pmatrix} \cdot
\begin{pmatrix}
\nabla_{\lambda^1} G^1 & \nabla_{\lambda^2} G^1 & \cdots & \nabla_{\lambda^{sh}} G^1 \\
\nabla_{\lambda^1} G^2 & \nabla_{\lambda^2} G^2 & \cdots & \nabla_{\lambda^{sh}} G^2 \\
\vdots & \vdots & \ddots & \vdots \\
\nabla_{\lambda^1} G^W & \nabla_{\lambda^2} G^W & \cdots & \nabla_{\lambda^{sh}} G^W
\end{pmatrix}
\tag{11}
$$

The Euclidean gradient w.r.t the objective function in reduced problem (4) can be defined as follows,

$$\left(\frac{\partial cost(X)}{\partial \mu^1} \quad \frac{\partial cost(X)}{\partial \mu^2} \quad \cdots \quad \frac{\partial cost(X)}{\partial \mu^W} \right) \tag{12}$$

$$\Rightarrow 2 \cdot \begin{pmatrix} \nabla_1 \cdot \nabla_{\lambda^1} G^1 & \nabla_1 \cdot \nabla_{\lambda^2} G^1 & \cdots & \nabla_1 \cdot \nabla_{\lambda^{sh}} G^1 \\ \nabla_2 \cdot \nabla_{\lambda^1} G^2 & \nabla_2 \cdot \nabla_{\lambda^2} G^2 & \cdots & \nabla_2 \cdot \nabla_{\lambda^{sh}} G^2 \\ \vdots & \vdots & \ddots & \vdots \\ \nabla_{sh} \cdot \nabla_{\lambda^1} G^W & \nabla_{sh} \cdot \nabla_{\lambda^2} G^W & \cdots & \nabla_{sh} \cdot \nabla_{\lambda^{sh}} G^W \end{pmatrix} \tag{13}$$

$$\Rightarrow 2 \cdot (\nabla_1 \quad \nabla_2 \quad \cdots \quad \nabla_{sh}) \cdot \begin{pmatrix} \nabla_{\lambda^1} G^1 & \nabla_{\lambda^1} G^2 & \cdots & \nabla_{\lambda^1} G^W \\ \nabla_{\lambda^2} G^1 & \nabla_{\lambda^2} G^2 & \cdots & \nabla_{\lambda^2} G^W \\ \vdots & \vdots & \ddots & \vdots \\ \nabla_{\lambda^{sh}} G^1 & \nabla_{\lambda^{sh}} G^2 & \cdots & \nabla_{\lambda^{sh}} G^W \end{pmatrix} \tag{14}$$

$$\Rightarrow 2XAA^T \tag{15}$$

After finishing the above operations, we call the *multinomialfactory* function in the toolbox to compute the solution of the Riemannian optimization, and then generate the Pareto direction for scaled back-propagation in MTL. Therefore, all the shared parameters will be improved iteratively and converge efficiently.

4   Experiments

We evaluated our method on two datasets. First, we used MultiMNIST [7], a variant of MNIST [34]. Then, we utilized Fashion MNIST [8]. Although it contains the name of MNIST, its data type is quite different from MNIST. The frameworks we utilize include PyTorch and Matlab. And experiments are run on desktop computer with a single GeForce RTX 2070 SUPER GPU under Ubuntu 16.04.

4.1   MultiMNIST

MultiMNIST is an extension of MNIST. It randomly stacks 10 numbers' images in groups of 2, one in the upper left corner and the other in the lower right corner. Figure 3 shows some examples. Such images provide conditions for MTL. These tasks which will be implemented in the experiment include task-L (classifying the digit on the top-left) and task-R (classifying the digit on the bottom-right). The total number of these images is 60K. The structure of the model based on LeNet [34] is shown in Fig. 4. The network structure is encoder-decoder. The main part of LeNet [34] is used as encoder, and the decoder consists of two fully-connected layers for dual tasks. Our experiments are comparative experiments. Specifically, we compare the following

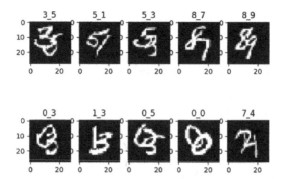

Fig. 3. Some examples from MultiMNIST. Task-L aims to classify the digit on the top-left and task-R aims to classify the digit on the bottom-right.

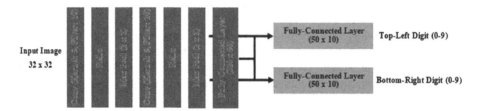

Fig. 4. The architecture of MultiMNIST network.

Table 1. Results on the comparative experiments.

Method	Left digit accuracy (%)	Right digit accuracy (%)
Single task	96.58	95.76
Uniform scaling	96.06	94.35
Euclidean optimization	96.12	95.13
Ours	96.26	94.40

approaches: 1. Single task. 2. Uniform scaling. 3. Euclidean Optimization [6]. 4. Our method. The results are illustrated in Table 1.

We trained 50 epochs in total. For the task-L, all experiments achieve similar accuracy. Our method's accuracy is 0.2% higher than the one that achieves the lowest accuracy, and the accuracy is 0.15% higher than the approach based on computational geometry. For the task-R, the accuracy of our method is slightly lower than that of the computational geometry method, but higher than that of the average weighted method.

4.2 Fashion MNIST

Fashion MNIST contains a large number of images of clothing, taken from a clothing catalogue. Figure 5 includes some samples. The two tasks we implement are classification (classify all images) and reconstruction (compress the image down to a low dimensional representation, then reconstruct the original image). The structure of the

Fig. 5. Samples from Fashion MNIST.

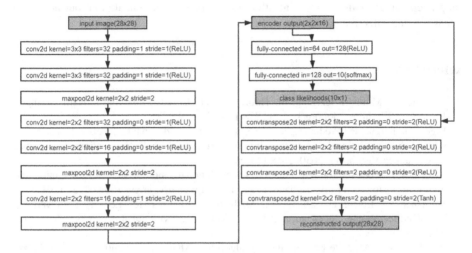

Fig. 6. The encoder and decoders of the model.

Table 2. Results on another comparative experiments.

Method	Classification accuracy (%)	Reconstruction error
Single task	89.7	0.0873
Uniform scaling	89.3	0.0847
Euclidean optimization	93.1	0.0674
Ours	93.2	0.0682

model based on encoder-decoder is shown in Fig. 6. Fashion MNIST experiments' baselines are consistent with MultiMNIST experiments. Table 2 shows final results.

We trained 50 epochs in total. For the task-L, all experiments achieve similar accuracy. Our method's accuracy is 0.2% higher than the one that achieves the lowest accuracy, and the accuracy is 0.15% higher than the approach based on computational geometry. For the task-R, the accuracy of our method is slightly lower than that of the computational geometry method, but higher than that of the average weighted method.

5   Conclusion

In order to solve the problem of multi-objective optimization in MTL, we proposed a method based on Riemannian optimization. KKT condition is the key to solve multi-objective optimization problem. We mapped the constraint from Euclidean space onto the manifold and transformed the quadratic optimization problem into an unconstrained problem. Finally, the optimum task weights under KKT condition were obtained by using Riemannian optimization algorithm. Our experiments indicate that our approach is feasible and effective.

References

1. Caruana, R.: Multitask learning. In: Thrun, S., Pratt, L. (eds.) Learning to learn. Springer, Boston (1998). https://doi.org/10.1007/978-1-4615-5529-2_5
2. Eigen, D., Fergus, R.: Predicting depth, surface normals and semantic labels with a common multi-scale convolutional architecture. In: Proceedings of the IEEE International Conference on Computer Vision, pp. 2650–2658 (2015)
3. Kokkinos, I.: UberNet: training a universal convolutional neural network for low-, mid-, and high-level vision using diverse datasets and limited memory. arXiv:1609.02132 (2017)
4. Sermanet, P., Eigen, D., Zhang, X., Mathieu, M., Fergus, R., LeCun, Y.: OverFeat: integrated recognition, localization and detection using convolutional networks. arXiv:1312.6229 (2013)
5. Désidéri, J.-A.: Multiple-gradient descent algorithm (MGDA) for multiobjective optimization. C.R. Math. **350**(5), 313–318 (2012)
6. Sener, O., Koltun, V.: Multi-task learning as multi-objective optimization. arXiv:1810.04650 (2018)
7. Sabour, S., Frosst, N., Hinton, G.E.: Dynamic routing between capsules. arXiv:1710.09819 (2017)
8. Xiao, H., Rasul, K., Vollgraf, R.: Fashion-MNIST: a novel image dataset for benchmarking machine learning algorithms. arXiv:1708.07747 (2017)
9. Ruder, S.: An overview of multi-task learning in deep neural networks. arXiv:1706.05098 (2017)
10. Oquab, M., Bottou, L., Laptev, I., Sivic, J.: Learning and transferring mid-level image representations using convolutional neural networks. In: Proceedings of the IEEE Conference on Computer Vision and Pattern Recognition, pp. 1717–1724 (2014)
11. Martí, M., Maki A.: A multitask deep learning model for real-time deployment in embedded systems. arXiv:1711.00146 (2017)
12. Bilen, H., Vedaldi, A.: Integrated perception with recurrent multi-task neural networks. arXiv:1606.01735 (2016)
13. Collobert, R., Weston, J.: A unified architecture for natural language processing: deep neural networks with multitask learning. In: Proceedings of the 25th International Conference on Machine Learning, pp. 160–167 (2008)
14. Huang, J.-T., Li, J., Yu, D., Deng, L., Gong, Y.: Cross-language knowledge transfer using multilingual deep neural network with shared hidden layers. In: IEEE International Conference on Acoustics, Speech and Signal Processing, pp. 7304–7308 (2013)

15. Huang, Z., Li, J., Siniscalchi, S.M., Chen, I.-F., Wu, J., Lee, C.-H.: Rapid adaptation for deep neural networks through multi-task learning. In: Proceedings of the Annual Conference of the International Speech Communication Association, pp. 3625–3629 (2015)

16. Baxter, J.: A model of inductive bias learning. J. Artif. Intell. Res. **12**, 149–198 (2000)

17. Li, C., Georgiopoulos, M., Anagnostopoulos, G.C.: Pareto-path multi-task multiple kernel learning. arXiv:1404.3190 (2014)

18. Teichmann, M., Weber, M., Zoellner, M., Cipolla, R., Urtasun, R.: MultiNet: real-time joint semantic reasoning for autonomous driving. arXiv:1612.07695 (2016)

19. Zhou, D., Wang, J., Jiang, B., Guo, H., Li, Y.: Multi-task multi-view learning based on cooperative multi-objective optimization. IEEE Access **6**, 19465–19477 (2017)

20. Kendall, A., Gal, Y., Cipolla, R.: Multi-task learning using uncertainty to weigh losses for scene geometry and semantics. In: Proceedings of the IEEE Conference on Computer Vision and Pattern Recognition, pp. 7482–7491 (2018)

21. Miettinen, K.: Nonlinear Multiobjective Optimization. Springer, New York (2012). https://doi.org/10.1007/978-1-4615-5563-6

22. Ehrgott, M.: Multicriteria Optimization. Springer, Heidelberg (2005). https://doi.org/10.1007/3-540-27659-9

23. Zhou, A., Zhang, Q., Jin, Y.: Approximating the set of pareto-optimal solutions in both the decision and objective spaces by an estimation of distribution algorithm. IEEE Trans. Evol. Comput. **13**(5), 1167–1189 (2009)

24. Mukhopadhyay, A., Bandyopadhyay, S., Maulik, U.: Multi-class clustering of cancer subtypes through SVM based ensemble of pareto-optimal solutions for gene marker identification. PLoS ONE **5**(11), e13803 (2010)

25. Suga, K., Kato, S., Hiyama, K.: Structural analysis of Pareto-optimal solution sets for multi-objective optimization: an application to outer window design problems using multiple objective genetic algorithms. Build. Environ. **45**(5), 1144–1152 (2010)

26. Poirion, F., Mercier, Q., Désidéri, J.-A.: Descent algorithm for nonsmooth stochastic multiobjective optimization. Comput. Optim. Appl. **68**(2), 317–331 (2017). https://doi.org/10.1007/s10589-017-9921-x

27. Peitz, S., Dellnitz, M.: Gradient-based multiobjective optimization with uncertainties. In: Maldonado, Y., Trujillo, L., Schütze, O., Riccardi, A., Vasile, M. (eds.) NEO 2016. SCI, vol. 731, pp. 159–182. Springer, Cham (2018). https://doi.org/10.1007/978-3-319-64063-1_7

28. Boumal, N.: An introduction to optimization on smooth manifolds (2020)

29. Project webpage. https://github.com/NicolasBoumal/manopt

30. Alimisis, F., Orvieto, A., Bécigneul, G., Lucchi, A.: A continuous-time perspective for modeling acceleration in Riemannian optimization. arXiv:1910.10782 (2019)

31. Bonnabel, S.: Stochastic gradient descent on Riemannian manifolds. IEEE Trans. Autom. Control **58**(9), 2217–2229 (2013)

32. Sun, Y., Gao, J., Hong, X., Mishra, B., Yin, B.: Heterogeneous tensor decomposition for clustering via manifold optimization. IEEE Trans. Pattern Anal. Mach. Intell. **38**(3), 476–489 (2015)

33. Kuhn, H.W., Tucker, A.W.: Nonlinear programming. In: Proceedings of the Second Berkeley Symposium on Mathematical Statistics and Probability (1951)

34. LeCun, Y., Bottou, L., Bengio, Y., Haffner, P.: Gradient-based learning applied to document recognition. Proc. IEEE **86**(11), 2278–2324 (1998)

Audio-Visual Salient Object Detection

Shuaiyang Cheng, Liang Song[(⊠)], Jingjing Tang, and Shihui Guo

School of Informatics, Xiamen University, Xiamen 361005, China
songliang@xmu.edu.cn

Abstract. This paper studies audio-visual salient object detection. The task of salient object detection is to detect and mark the objects that are most concerned by people in the visual scene. Traditionally, visual salient object detection uses only images or video frames to detect salient objects, without modeling human multi-modal perception which includes the interaction between vision and hearing. Therefore, in order to improve the visual salient object detection, we incorporate audio modality into the traditional visual salient object detection task by applying a two-stream audio-visual deep learning network. To this end, we also build an audio-visual salient object detection dataset called AVSOD based on the existing dataset. To verify the effectiveness of audio modality in salient object detection, we compare the experimental performance of the deep learning model with and without audio modality. The experimental results demonstrate that audio modality has a good supplementary effect on the task of visual salient object detection, and also verified the effectiveness of the proposed dataset.

Keywords: Salient object detection · Audio-visual modality · Deep learning

1 Introduction

Visual salient object detection uses artificial intelligence algorithms to simulate the human visual system and extract salient objects in images or video frames (that is, objects of interest to human). In cognitive science, due to the bottleneck of information processing ability, human will selectively pay attention to a part of the scene that contains more important information, and relatively ignore other information in the visual scene. This is a powerful attention mechanism [1] in the human visual system, by which human can relatively ignore redundant information that is irrelevant to the current target, reducing the information processing burden on brain. Salient object detection in computer vision aims to train a deep learning model to simulate the attention mechanism so as to focus on useful objects in images or video frames. Furthermore, salient object detection task has played an important role in computer vision, such as image understanding [2], pedestrian recognition [3], video compression [4, 5], and object detection [6, 7], etc., in computer graphics, such as non-photorealistic rendering [8], video summarization [9, 10], etc., and the field of robotics, such as human-robot interaction [11, 12], target discovery.

This work was supported by the Fundamental Research Funds for the Central Universities of China under Grant No. 20720190028, and the National Natural Science Foundation of China under Grant 62072383.

D.-S. Huang et al. (Eds.): ICIC 2021, LNCS 12837, pp. 510–521, 2021.
https://doi.org/10.1007/978-3-030-84529-2_43

However, the task of salient object detection is mainly to extract objects based on images or video frames. Many researchers are dedicated to developing complex deep learning models to improve the detection performance without considering to model the interaction between vision and hearing. For example, [13] builds a triple excitation network, applies ConvLSTM in the third branch, uses optical flow map in the second branch, and takes an optional online excitation strategy during testing phase. Human attention from observation is likely to suffer from distraction due to sound factors. On the other hand, hearing can also assist visual attention. Therefore, integration of multiple modalities such as hearing and vision is able to make deep learning models closer to the real state of human perception and improve the performance of salient object detection.

After understanding the significance of the relevance between audio and visual saliency, we hope to incorporate audio modality into the task of salient object detection. However, to our best knowledge, there is no audio-visual multi-modal salient object detection dataset. Thus, the challenge comes from two aspects, the generation of audio-visual dataset and the construction of audio-visual network. To verify the effect of audio modality on the salient object detection task, this paper makes the following contributions: First, we construct an audio-visual salient object detection dataset that can be applied to audio-visual detection which is based on existing video salient object detection dataset [14, 15]. This dataset contains a variety of simple to complex video and audio scenes, which provided support for our idea of simulating the human visual system as much as possible in the salient object detection task. Secondly, we apply a two-stream audio-visual model [16] into the salient object detection task to verify the importance of audio modality.

2   Related Work

There are two tasks in the field of computer vision that are designed to simulate the attention mechanism of the human visual system, namely, salient object detection [17] and saliency prediction [18]. The difference is that saliency object detection aims to detect specific objects that receive attention, while saliency prediction only cares about a set of saliency points. Many researchers began to pay attention to multi-modal learning [19–21].

The development of salient object detection was surveyed in [22]. [23] combines cross-cognitive psychology, neuroscience, and computer science to propose the lead the salient object detection model. Then, the surging research in this field is due to [24] defining the salient object detection task as a binary segmentation problem, and since then a large number of salient object detection models have been proposed. So far, due to the unprecedented performance that the CNN method can achieve, it has become the mainstream method for the task of salient object detection.

Salient Object Detection. Recently, significant object detection methods based on deep learning have continuously emerged, and many meaningful datasets and models have emerged. In terms of salient object detection models, a large number of models use VGG [25] or ResNet [26] as the backbone network. VGG model has fewer parameters and light model architecture, but the detection performance is relatively worse.

The performance of ResNet network is better but not as lightweight as VGG network. In addition to the model, there are many methods to introduce other features into salient object detection. For example, [13] combines optical flow information with the input image, [27–30] merge the depth information of the objects in the image with the input image, collectively referred to as the RGB-D method. These feature fusion methods usually face the difficulty of the feature collection. For example, the depth information of objects in an image usually requires special equipment to collect and process them.

Although a large number of salient object detection methods based on deep learning have emerged, the datasets and models for salient object detection are still insufficient. Moreover, the existing datasets such as PASCAL-S [31], MSRA-10K [24], SOD [32], etc. are based on pictures, while the salient object detection datasets based on video sequences are relatively rare. A video-level salient object detection dataset is proposed in the paper [14].

Audio-Visual Learning. Incorporating sound features into visual tasks is similar to the way that depth information and optical flow information assist visual tasks. Compared with depth, optical flow information and other features, the sound in the visual scene is easier to be obtained without complicated processing. At the same time, sound is also one of the factors that have an important influence on the human visual system. It plays an important role in assisting human visual observation. [33] uses audio as a preview mechanism to eliminate both short-term and long-term visual redundancies. [34] proposes a method to integrate sound features into the face super-resolution task. In the field of saliency prediction, it is more common to use sound modality, [35] constructed a dataset of audio-visual saliency prediction, and converted audio features into log Mel-spectrogram frames, using a two-stream ResNet based network for predicting. [16] uses SoundNet and ResNet as a two-stream network for audio-visual saliency prediction, and uses a sound source localization to locate audio features when fusing audio and visual features. [36] uses an encoder-decoder network to fuse the audio characteristics with the encoder's results to achieve very good performance.

3   Audio-Visual Dataset and Network

We construct an audio-visual salient object detection dataset for the task of video salient object detection based on an existing video salient object detection dataset DAVSOD [14]. Then, we apply a two-stream audio-visual network architecture [16] to the audio-visual salient object detection task, and verify the role of audio features on the visual salient object detection task.

3.1   Dataset Generation

To create an audio-visual dataset in the field of salient object detection, we collect the original videos of the DAVSOD dataset and then use a multimedia content processing tool FFmpeg [37] to extract the available audio from these videos. A dynamic video salient object detection dataset was proposed in [14], referred to as DAVSOD, which contains 226 videos with 23,938 frames that cover diverse realistic scenes, objects, instances, and motions. The videos in this dataset are mainly derived from DHF1K [15], which is currently the largest dynamic eye-tracking dataset. DHF1K videos are mainly derived from YouTube video websites, which provides us with the possibility of constructing audio-visual salient object detection dataset.

Some data in DAVSOD does not have corresponding audio, so we remove them for these data. Finally, we construct a dataset that can be specifically used for audio-visual salient object detection, which we call audio-visual salient object detection dataset or AVSOD. AVSOD is a subset of the dataset DAVSOD. The dataset contains a variety of ground truth including salient object detection, as well as available audio data, which provides us with the possibility of achieving audio-visual salient object detection. The AVSOD dataset has 172 videos, the corresponding audio is 2-channel stereo, and the sampling rate is 44100Hz.

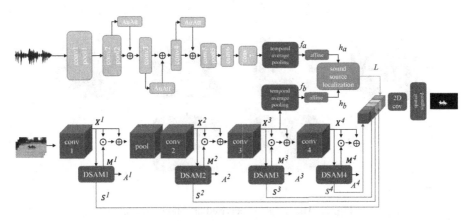

Fig. 1. The overall architecture of the audio-visual model

3.2 Network Architecture

Our purpose is to verify whether the audio modality can assist the task of salient object detection and improve the results of this task. So we use a spatial-temporal audio-visual network, namely Stavis [16], an audio-visual network that has been verified in the task of saliency prediction. The feature fusion strategy of audio and visual modalities adopted by the network has proved to be very effective. Stavis is a two-stream audio-visual network. It uses a vision stream to extract visual features from video frames, an audio feature extraction stream called SoundNet [38] to extract audio features from audio, and then apply a sound source localization module to fuse the audio and one of the visual feature maps, the fusion result of sound source localization module is fused with other visual feature maps, finally, the fused feature maps are predicted to obtain the final result. The network architecture is depicted in Fig. 1. On the other hand, [39] proposed an audio attention module that is used to improve the effect of audio feature extraction, and we also adopt this module. Finally, we migrate the network to the salient object detection task.

Visual Stream. The visual feature extraction stream is based on the 3D extension of the ResNet network, which is composed of four convolution blocks. After each convolution block, the feature maps X_1, X_2, X_3, and X_4 of different scales are output, then each one is fed into an attention block called Deeply Supervised Attention Module (DSAM).

The DSAM module receives the feature map $X_m (m = 1, 2, 3, 4)$ as input. X_m first passes through a 3D average pooling layer and a 3×3 3D convolution layer. After X_m passes through these two layers, it passes through a 2D upsampling layer to obtain S_m, or through a 2D convolution layer to obtain a feature map, this feature map either passes through an upsampling layer to get A_m or passes through a spatial softmax layer to get X_m. The detail of the DSAM module are depicted in Fig. 2.

Deeply Supervised Attention Module(DSAM)

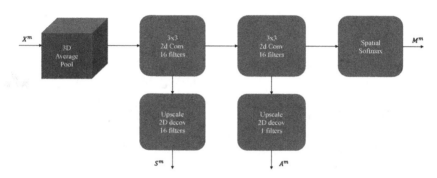

Fig. 2. The deeply supervised attention module

For the feature maps output by the DSAM module, the element-level product of M_m and the input X_m layer is performed and the result of the element-level product is then added to X_m as the element-level, and the result of the element-level sum is input to the next convolutional block. S_m is used to provide multi-level saliency feature maps for our final prediction. DSAM parameters are trained both by the main-path of the visual network and the eye-tracking data through the skip connections of Fig. 1.

Audio Stream. The SoundNet architecture is a 1D fully connected network, which is different from [35] converting sound waveforms into 2D Mel-spectrogram frames and then using a 2D CNN network to extract features, SoundNet works directly on 1D sound waveforms. The original SoundNet architecture has a total of 8 layers. Since our task is pixel-level prediction rather than a binary classification task such as image classification, only the first seven layers of SoundNet are adopted. Since the network only outputs the result of the middle frame in the input video frames, a Hanning window operation is applied to give higher weight to the sound waveforms corresponding to the video frame, so that the model can pay more attention to this clip of sound. The detail of SoundNet is depicted in Fig. 1. In [39], a deep learning module is proposed to optimize the extraction of audio features, we incorporate this module into the audio stream to enhance the learning ability. This module is another work of ours which will be published soon, so we won't describe it too much here.

Sound Source Localization. After the extracted audio features are obtained by the SoundNet network, to assist the salient object detection task, the audio features need to be fused with the feature map output by the visual stream. Here, the third layer of the visual network, X_3 is used to fuse the extracted audio features, because X_3 not only contains rich semantic information but also maintains a considerable resolution. The network performs a temporal average polling operation on the output audio features and visual feature map X_3 respectively to obtain f_a and f_b. However, the dimensions of f_a and f_b are different at this time, so the network uses two affine transformation operations to adjust f_a and f_b, and a spatial tilling operation is performed on f_a, finally, we obtain the adjusted audio and visual features h_a and h_b, $h_a, h_b \in R^{D_h \times N_x \times N_y}$. To learn the correlation between the audio feature h_a and the visual feature h_v, the network uses a sound source localization module to yield either a single or multiple output maps $L^j(x, y) \in R^{N_x \times N_y}, j = 1, 2, 3, \dots N_{out}$. The sound source localization module is a bilinear transformation function of h_a and h_v:

$$L^j(x, y) = \sum_{l=1}^{D_h} \sum_{k=1}^{D_h} M^{j,l,k} \times h_h^l(x, y) \times h_a^k(x, y) + u^j \tag{1}$$

where $M^{j,l,k}$, u^j are learning parameters.

Loss Function. After the sound source localization function, we can get the feature map L after fusing the audio and video features. The network then concatenates the output of the DSAM module s_1, s_2, s_3, s_4 with L, then use a 1×1 2D convolution layer to fuse these features, to obtain the final salient object detection result, a spatial softmax layer is applied. Then we can compute the loss between this result and ground truth. For the loss function, we apply the cross-entropy function:

$$loss_{CE} = -(Y(x, y) \times log(P(x, y)) + (1 - Y(x, y)) \times log(1 - P(x, y))) \tag{2}$$

where Y is ground truth and P is our predicted result. To control the learning process, we set a weight factor w for $loss_{CE}$, so our final loss function is $loss = w \times loss_{CE}$.

4 Experiments

4.1 Experimental Settings

For AVSOD dataset, we employ the strategy in [14] to split videos into three parts for training, validation and test respectively, in the ratio of 4:2:4. The test set is further divided into easy, normal, and difficult subsets. The original resolution of each frame in each video in the dataset is 640×360 pixels. Before feeding it to the network, we resize the image to 256×256 pixels. During training and testing our model, we set 16 frames and 1 step sliding window to sample the data from each video and set a data augment strategy of random horizontal flip with the probability of 0.5, and then feed the corresponding audio part into SoundNet according to the fragment of the current video frames. For the pre-training parameters of the visual stream, we use the training weights on the kinetics 400 database, and for the SoundNet part, we use the pre-training weights trained on Flickr [38], which has 2 million videos. Finally, we employ stochastic gradient descent with momentum 0.9, while we assign a weight decay of $1e-5$ for regularization. We set the batch size of 16 and multistep learning rate, and set the weights of our cross-entropy loss to 0.1, 60 epochs for training.

Table 1. Measure results of the pure visual model and the audio-visual model on three subsets with different complexity.

Metrics	Easy		Normal		Difficult	
	Stavis(STA)	Stavis(ST)	Stavis(STA)	Stavis(ST)	Stavis(STA)	Stavis(ST)
S	**0.632**	0.582	**0.640**	0.538	**0.613**	0.516
adp F	**0.472**	0.394	**0.437**	0.328	**0.356**	0.257
max F	**0.478**	0.416	**0.493**	0.344	**0.398**	0.277
mean F	**0.415**	0.326	**0.436**	0.286	**0.342**	0.222
MAE	**0.137**	0.198	**0.141**	0.224	**0.113**	0.231

4.2   Experimental Results

Due to the diverse scenarios of the AVSOD dataset and the characteristic of salient object detection task, the data in some simple scenes may only have one salient object, and the volume of the object is relatively large and the shape is relatively regular. However, in some complex scenes, there may be multiple objects with different irregular shapes or even overlaps between objects. Therefore, the test set is further divided into three subsets according to the degree of difficulty of the video salient object detection task, which are easy subset, normal subset, and difficult subset. These three subsets contain 34, 24, 20 videos respectively.

There are many measure metrics for the predicted results, and each metric has different considerations and trade-offs for the predicted results. To test the effect of audio on the salient object detection task more comprehensively, we used five different measure metrics to make a more comprehensive measurement of the results, namely the S-measure [40], adp F-measure, max F-measure, mean F-measure, and MAE [41]. S-measure is a novel, efficient and easy metric to evaluate non-binary foreground maps. F-measure [42] is the weighted harmonic average of precision and recall, which is often used to evaluate the classification model. The first four metrics are all positive measurements, which means the higher the calculating value of this metric, the better the model performs. The MAE is a measurement of the error between the predicted result and the ground truth, the smaller the calculating value of this metric, the better the model performs.

According to the split strategy of the test set and different salient object detection metrics, we train and test the audio-visual network respectively to compare the effects of audio features on the salient object detection task. To make a fair measurement, we only test the results of the network with or without audio modality. First, we train and test the audio-visual two-stream network, then perform training and testing by removing the audio stream, that is, before the final prediction, only the s_1, s_2, s_3, and s_4 feature maps output by the DSAM layer are combined to obtain the predicted result, while the sound source localization module and SoundNet network are removed. The visual results and the measure metrics of model's performance are shown in Fig. 3 and the Table 1 respectively.

We provide the visualization results and measure results of the pure visual model and the audio-visual model on the three subsets respectively. By comparing the performance of the audio-visual model and the pure visual model, we can verify the significance of audio on the task of salient object detection.

Measure results can be seen in Table 1 that compared to the pure visual model. The audio-visual model has a significant and comprehensive improvement in the five measurements of the salient object detection task on each subset. And it can be observed that if the complexity of the scene increases, the overall prediction effect of the model will decrease, but the audio-visual model will increase the prediction result more obviously.

Visualization results are shown in Fig. 3, we provide five groups of comparison results for each subset of the test set to analyze the effect of audio features on the salient object detection task. For each group of pictures in these subsets, the video frame, ground truth, audio-visual result and visual-only result are displayed sequentially from top to bottom. In simple scenes, we can see the most salient objects in images are relatively prominent, the shape is relatively clear, and the distinction from the background is relatively obvious. Both the audio-visual model and the pure visual model can roughly predict the position and shape of the salient objects, but the audio-visual model per-forms better. In more complex scenes, the salient object and the background often have similar colors, which is difficult to distinguish, or the salient objects are too small to find out. Although the pure visual model can probably predict the position of the salient objects, the predicted shapes are rather obscure, while the audio-visual model can still distinguish the salient objects from backgrounds. Finally, in difficult scenes, there are smaller and much more salient objects with more similar colors to the background. Now it is difficult for the pure visual model to predict the location and shape of salient objects. Although the prediction of the audio-visual model is not as accurate as the above two categories of scenes, the result is still acceptable.

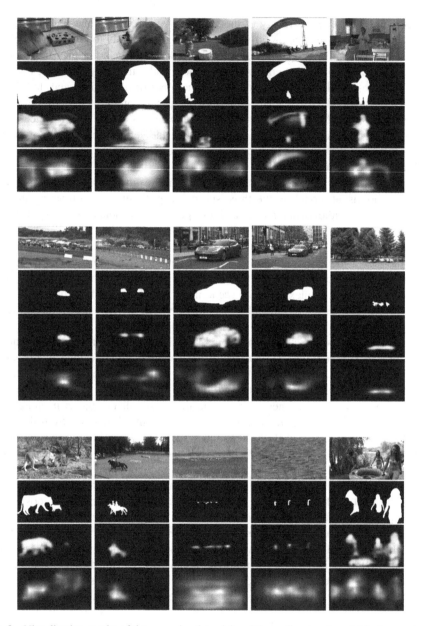

Fig. 3. Visualization results of the pure visual model and the audio-visual model in three subsets of different complexity

5   Conclusion

In this paper, we studied audio-visual salient object detection. To this end, we built an audio-visual salient object detection dataset, which can be used for further research. Then, we applied a two-stream audio-visual deep learning network to verify the effectiveness of audio modality in the salient object detection task. The experimental results demonstrated that the audio features can effectively assist the task of salient object detection. Therefore, this work provides a novel framework to develop the system for salient object detection and improve the performance.

References

1. Koch, C., Ullman, S.: Shifts in selective visual attention: towards the underlying neural circuitry. In: Matters of Intelligence, pp. 115–141. Springer, Cham (1987). https://doi.org/10.1007/978-94-009-3833-5_5
2. Huang, T., Tian, Y., Li, J., Yu, H.: Salient region detection and segmentation for general object recognition and image understanding. Sci. Chin. Inf. Sci. **54**(12), 2461–2470 (2011)
3. Lian, G., Lai, J., Yuan, Y.: Fast pedestrian detection using a modified WLD detector in salient region. In: Proceedings 2011 International Conference on System Science and Engineering, pp. 564–569. IEEE (2011)
4. Guo, C., Zhang, L.: A novel multiresolution spatiotemporal saliency detection model and its applications in image and video compression. IEEE Trans. Image Process. **19**(1), 185–198 (2009)
5. Hadizadeh, H., Bajić, I.V.: Saliency-aware video compression. IEEE Trans. Image Process. **23**(1), 19–33 (2013)
6. Ren, Z., Gao, S., Chia, L.T., Tsang, I.W.H.: Region-based saliency detection and its application in object recognition. IEEE Trans. Circuits Syst. Video Technol. **24**(5), 769–779 (2013)
7. Zhang, D., Meng, D., Zhao, L., Han, J.: Bridging saliency detection to weakly supervised object detection based on self-paced curriculum learning. arXiv:1703.01290 (2017)
8. Kapoor, A., Biswas, K.K., Hanmandlu, M.: An evolutionary learning based fuzzy theoretic approach for salient object detection. Vis. Comput. **33**(5), 665–685 (2016). https://doi.org/10.1007/s00371-016-1216-1
9. Ma, Y.F., Lu, L., Zhang, H.J., Li, M.: A user attention model for video summarization. In: Proceedings of the Tenth ACM International Conference on Multimedia, pp. 533–542 (2002)
10. Simakov, D., Caspi, Y., Shechtman, E., Irani, M.: Summarizing visual data using bidirectional similarity. In: 2008 IEEE Conference on Computer Vision and Pattern Recognition, pp. 1–8. IEEE (2008)
11. Sugano, Y., Matsushita, Y., Sato, Y.: Calibration-free gaze sensing using saliency maps. In: 2010 IEEE Computer Society Conference on Computer Vision and Pattern Recognition, pp. 2667–2674. IEEE (2010)
12. Borji, A., Itti, L.: Defending yarbus: eye movements reveal observers' task. J. Vis. **14**(3), 29 (2014)
13. Ren, S., Han, C., Yang, X., Han, G., He, S.: Tenet: triple excitation network for video salient object detection. In: Vedaldi, A., Bischof, H., Brox, T., Frahm, J.-M. (eds.) ECCV 2020. LNCS, vol. 12350, pp. 212–228. Springer, Cham (2020). https://doi.org/10.1007/978-3-030-58558-7_13

14. Fan, D.P., Wang, W., Cheng, M.M., Shen, J.: Shifting more attention to video salient object detection. In: Proceedings of the IEEE/CVF Conference on Computer Vision and Pattern Recognition, pp. 8554–8564 (2019)

15. Wang, W., Shen, J., Guo, F., Cheng, M.M., Borji, A.: Revisiting video saliency: a large-scale benchmark and a new model. In: Proceedings of the IEEE Conference on Computer Vision and Pattern Recognition, pp. 4894–4903 (2018)

16. Tsiami, A., Koutras, P., Maragos, P.: Stavis: spatio-temporal audiovisual saliency network. In: Proceedings of the IEEE/CVF Conference on Computer Vision and Pattern Recognition, pp. 4766–4776 (2020)

17. Wang, W., Lai, Q., Fu, H., Shen, J., Ling, H., Yang, R.: Salient object detection in the deep learning era: An in-depth survey. IEEE Transactions on Pattern Analysis and Machine Intelligence (2021)

18. Borji, A.: Saliency prediction in the deep learning era: Successes, limitations, and future challenges. arXiv:1810.03716 (2018)

19. Lee, G., Nho, K., Kang, B., Sohn, K.A., Kim, D.: Predicting Alzheimer's disease progression using multi-modal deep learning approach. Sci. Rep. 9(1), 1–12 (2019)

20. Wang, A., Lu, J., Cai, J., Cham, T.J., Wang, G.: Large-margin multi-modal deep learning for RGB-D object recognition. IEEE Trans. Multimedia 17(11), 1887–1898 (2015)

21. Gené-Mola, J., Vilaplana, V., Rosell-Polo, J.R., Morros, J.R., Ruiz-Hidalgo, J., Gregorio, E.: Multi-modal deep learning for Fuji apple detection using RGB-D cameras and their radiometric capabilities. Comput. Electron. Agric. 162, 689–698 (2019)

22. Borji, A., Cheng, M.-M., Hou, Q., Jiang, H., Li, J.: Salient object detection: a survey. Comput. Vis. Media 5(2), 117–150 (2019). https://doi.org/10.1007/s41095-019-0149-9

23. Itti, L., Koch, C., Niebur, E.: A model of saliency-based visual attention for rapid scene analysis. IEEE Trans. Pattern Anal. Mach. Intell. 20(11), 1254–1259 (1998)

24. Liu, T., et al.: Learning to detect a salient object. IEEE Trans. Pattern Anal. Mach. Intell. 33(2), 353–367 (2010)

25. Simonyan, K., Zisserman, A.: Very deep convolutional networks for large-scale image recognition. arXiv:1409.1556 (2014)

26. He, K., Zhang, X., Ren, S., Sun, J.: Deep residual learning for image recognition. In: Proceedings of the IEEE Conference on Computer Vision and Pattern Recognition, pp. 770–778 (2016)

27. Fan, D.-P., Zhai, Y., Borji, A., Yang, J., Shao, L.: BBS-Net: RGB-D salient object detection with a bifurcated backbone strategy network. In: Vedaldi, A., Bischof, H., Brox, T., Frahm, J.-M. (eds.) ECCV 2020. LNCS, vol. 12357, pp. 275–292. Springer, Cham (2020). https://doi.org/10.1007/978-3-030-58610-2_17

28. Luo, A., Li, X., Yang, F., Jiao, Z., Cheng, H., Lyu, S.: Cascade graph neural networks for rgb-d salient object detection. In: Vedaldi, A., Bischof, H., Brox, T., Frahm, J.-M. (eds.) ECCV 2020. LNCS, vol. 12357, pp. 346–364. Springer, Cham (2020). https://doi.org/10.1007/978-3-030-58610-2_21

29. Ling, H.: Cross-modal weighting network for RGB-D salient object detection (2020)

30. Ji, W., Li, J., Zhang, M., Piao, Y., Lu, H.: Accurate rgb-d salient object detection via collaborative learning. arXiv:2007.11782 (2020)

31. Wirth, N.: Pascal-s: a subset and its implementation. Berichte des Instituts fürInformatik, vol. 12 (1975)

32. Movahedi, V., Elder, J.H.: Design and perceptual validation of performance measures for salient object segmentation. In: 2010 IEEE Computer Society Conference on Computer Vision and Pattern Recognition-Workshops, pp. 49–56. IEEE (2010)

33. Gao, R., Oh, T.H., Grauman, K., Torresani, L.: Listen to look: action recognition by previewing audio. In: Proceedings of the IEEE/CVF Conference on Computer Vision and Pattern Recognition, pp. 10457–10467 (2020)

34. Meishvili, G., Jenni, S., Favaro, P.: Learning to have an ear for face super-resolution. In: Proceedings of the IEEE/CVF Conference on Computer Vision and Pattern Recognition, pp. 1364–1374 (2020)

35. Tavakoli, H.R., Borji, A., Rahtu, E., Kannala, J.: Dave: A deep audio-visual embedding for dynamic saliency prediction. arXiv:1905.10693 (2019)

36. Jain, S., Yarlagadda, P., Subramanian, R., Gandhi, V.: Avinet: Diving deep into audio-visual saliency prediction. arXiv:2012.06170 (2020)

37. Tomar, S.: Converting video formats with FFmpeg. Linux J. **2006**(146), 10 (2006)

38. Aytar, Y., Vondrick, C., Torralba, A.: Soundnet: Learning sound representations from unlabeled video. arXiv:1610.09001 (2016)

39. Cheng, S., Gao, X., Song, L., Xiahou, J.: Audio-visual saliency network with audio attention module, unpublished

40. Fan, D., Cheng, M., Liu, Y., Li, T., Borji, A.: A new way to evaluate foreground maps. In: Proceedings of the IEEE Conference on Computer Vision and Pattern Recognition (CVPR), p. 245484557 (2017)

41. Perazzi, F., Krähenbühl, P., Pritch, Y., Hornung, A.: Saliency filters: contrast based filtering for salient region detection. In: 2012 IEEE Conference on Computer Vision and Pattern Recognition, pp. 733–740. IEEE (2012)

42. Achanta, R., Hemami, S., Estrada, F., Susstrunk, S.: Frequency-tuned salient region detection. In: 2009 IEEE Conference on Computer Vision and Pattern Recognition, pp. 1597–1604. IEEE (2009)

Research on Deep Neural Network Model Compression Based on Quantification Pruning and Huffmann Encoding

Cong Wei$^{(\boxtimes)}$, Zhiyong Lu, Zhiyong Lin, and Chong Zhong

Xiamen Road and Bridge Information Co., Ltd., Floor 18, Building A06,
Software Park Phase III, Xiamen 361000, Fujian, China

Abstract. With the rapid development of hardware GPU and the advent of the era of big data, neural networks have developed rapidly and greatly improved the recognition performance in various fields. The application of nerual networks to intelligent mobile embedded military equipment will become the wave of the development of a new generation of deep learning. However, neural networks are both computationally intensive and memory intensive, making them difficult to deploy on embedded systems with limited hardware resources. To address this limitation, we propose a new method called "deep compression", which including three stages: pruning, quantification and Huffmann encoding, to reduce the storage requirement of neural networks without impacting original accuracy. Our method first prunes the network by learning only the important connections. Next, we quantize the weights to enforce weight sharing, finally, weapply Huffmann encoding. We evaluated our method on both MNIST and ImageNet. On the ImageNet dataset, our method reduced the storage of AlexNet by 35× without loss of accuracy and compressed VGG-16 model by 49×, also with no loss of accuracy. Our method is an efficient solution for real-time multi-objective recognition based on lightweight deep neural networks.

Keywords: Model compression · Prune · Quantize · Huffmann coding

1 Introduction

With the rapid development of hardware GPU and the advent of the era of big data, deep learning has developed rapidly and has swept all fields of artificial intelligence, including speech recognition, image recognition, video tracking, natural language processing and other built-in the fields of graphics, text and video. Deep learning technology breaks through traditional technology methods and greatly improves the recognition performance in various fields. The application of deep learning technology to intelligent mobile embedded military equipment will become the wave of the development of a new generation of deep learning.

One of the most typical examples is Alexnet [1]. Alexnet relies on the powerful global feature expression capabilities of convolution neural networks and various network training techniques (data expansion, Dropout, local response standardization, nonlinear ReLU, etc.), and won the first place in the ILSVRC 2012 image recognition competition with nearly ten percentage points higher than traditional methods. Not only

© Springer Nature Switzerland AG 2021
D.-S. Huang et al. (Eds.): ICIC 2021, LNCS 12837, pp. 522–531, 2021.
https://doi.org/10.1007/978-3-030-84529-2_44

that, but with the emergence of Alexnet, various deep neural network models (e.g., VGGNet [2], GoogLeNet [3]}, ResNet [4], etc.) have sprung up in various fields of artificial intelligence, significantly improving recognition performance. With the increase of the performance of the deep neural network model, the depth of the model is getting deeper and deeper. What follows is the drawback of the high calculation and high storage of the deep neural network model, which severely restricts the application environment with limited resources, especially the intelligent mobile embedded devices. For example, AlexNet-8 is equipped with 600,000 network nodes, 61 M network parameters, which requires 240 MB of memory storage and 729 M floating-point operations (FLOP) to classify a color image with a resolution of 224 × 224. As the depth of the model deepens, the storage and calculation overhead will become greater. Similarly, to classify a color image with a resolution of 224 × 224, the VGG-16 is equipped with 1,500,000 network nodes, 144 M network parameters, which requires 528 MB of memory storage and 15 B FLOP. Deepface [4] requires 120 M network parameters to recognize faces. Karpathy et al. [5] converted images into natural speech automatically, which requires 130 M CNN parameters and 100 M RNN parameters. Traditionally relying on the device side to calculate the online model, the server side responsible for model training and calculation method, which is restricted by the network information transmission lag or bandwidth limitation, cannot meet the real-time requirements. In summary, accelerating and compressing the deep neural network model so that it can be applied directly to the intelligent mobile embedded device side will become an effective solution. Therefore, by accelerating and compressing the deep neural network model, we propose a real-time multi-object recognition method based on lightweight deep neural network, and transplant the lightweight deep neural network to an intelligent mobile terminal.

From the point view of power consumption, whether the same level of object recognition tasks can be accomplished with low power consumption to meet the requirements of long-term operation directly affects the performance of the object recognition device. From the perspective of flexibility of use, if it is necessary to update the object recognition algorithm or redefine the object to be recognized, the implementation device of deep neural network needs to have good ability of updating and modification.

At present, the realization of deep neural network is mostly based on the general-purpose computers. These computers are not only large in size, but also limit the overall speed due to the speed of input and output interface. In addition, the designer or user of a deep neural network must have a considerable understanding of the working hardware and software environment of the network before it can be applied to the actual system, which limits the breadth and depth of the application, and also limits the development of deep neural network. Therefore, implementing the controller in the form of a Field Programmable Gate Array (FPGA) chip can not only significantly reduce the size of the hardware, but also have the advantage of fast execution and high flexibility. Although there have been some studies on implementing shallow neural network controllers on FPGA chips, research on implementing the method of deep neural networks is in the ascendant and has begun to show promising prospects. Previous studies have found that the execution speed of hardware after the implementation of FPGA can increase the millisecond speed, originally limited to ordinary computer I/O, to the fast output of microsecond [6–8]. The high processing speed will enable the application of deep neural network in more applications requiring rapid response. Moreover, deep neural networks increase the user's interest due to the small size and low power consumption of FPGA hardware, especially in learning accuracy, which does not vary much from hardware. In addition, deep neural networks will increase the interest of users due to the small size and low power consumption of FPGA hardware, especially in terms of learning accuracy which is not much difference due to hardwareization. Therefore, the hardware architecture design of deep neural network is necessary, and the depth upgrade developed by FPGA hardware implementation is another research topic of this subject.

In recent years, artificial intelligence technology has developed very rapidly. Substantial progress has been made in many applications, including speech recognition, image recognition, machine translation, autonomous driving, etc. In the military field, stimulated by the need for war, the development of artificial intelligence is even more remarkable. In particular, the rapid development of deep learning has made deep learning technology break through traditional technical methods and greatly improve the recognition performance in various fields. Applying deep learning technology to intelligent mobile embedded devices will become the wave of the development of a new generation of deep learning [9]. Designing a lightweight neural network to recognize multiple targets online and in real time, and embedding the lightweight model in mobile embedded devices reflects two requirements: Market demand and Technology requirement.

Through the above analysis, it can be seen that the key scientific problems that have not been solved in real-time multi-objective recognition based on lightweight deep neural network is the lack of a truly effective deep neural network compression and acceleration algorithm for the characteristics of high storage and high consumption of deep neural network models. Driven and challenged by high storage and consumption of deep neural network, we hope to propose a solution for real-time multi-objective recognition based on lightweight deep neural network. In order to achieve this goal, we intend to regard generating adversarial network learning and convolution neural network as the basic structure, to achieve robust small objective recognition technology and solve the contradiction between the high complexity of complex objective recognition model and the limited storage space. We carry out the research of deep neural network model compression methods, and realize multi-level network scale compression by network pruning, weight quantification and sharing, as well as weight entropy coding. Our method can effectively compress the model size without sacrificing model accuracy and implementation efficiency, which make it easy to deploy with limited resources, and support the recognition of multi-objective, especially for small objectives. In this process, the deep neural network model used for objective recognition is compressed and accelerated by using the techniques of pruning, quantification, and Hoffmann encoding.

2   Proposed Method

The model framework, shown in Fig. 1, consists of three modules: The first step, which drives neural network model training through a large number of samples. And then, in the second step, realize weight sharing through parameter quantification and then superimposed with the Hoffmann variable length coding technology in the third step to realize maximum compression of the model.

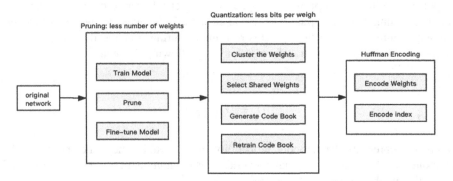

Fig. 1. The framework of deep neural network model based on pruning, quantification, and Hoffmann encoding.

2.1   Parameter Pruning

For a given deep neural network model, the connection between neurons in the model can first be pruned. The specific process of algorithm can be divided into three steps. The first step, driven by a large number of samples, fully train the convolution neural network with given structure using a training algorithm for universal convolution neural network to obtain an available network model. From the point of view of the pruning algorithm, this training process can not only get the value of the connection weight, but also learn which weights are important and which weights can be set to zero through the value of connection weight. Usually those connections with higher weight have a correspondingly greater impact on the final model output. The second step is to prune the connection during the training process. For each connection weight, if its value is below a threshold, then the value is forcibly set to zero. At this time, the gradient of back propagation is also correspondingly set to zero, and the connection is permanently disconnected. In the next third step, the pruned network will be further fine-tuned, the retained weights make up for the loss of the removed connection through learning.

In the pruning process, the choice of parameter global regularization penalty is critical. L1 regularization, linearly penalizing non-zero parameters, tends to give sparse solution, which will bring less loss after pruning. But during the retraining process, it often cannot get better results because the model performance trained with the remaining connection weights is not as good as that trained with L2 regularization. In general, L2 regularization penalty for model training is a better choice.

Dropout is a unique operation in the full connection layer of convolution neural network. In each forward propagation process of the network model, randomly set the output of some neurons to zero. This method has been proved to be effective in preventing overfitting when the neuron scale of the model is large. At the same time, each time the weight of model is set to zero, it can be regarded as a smaller one. During the test process, Dropout is canceled when forward propagation, which can be equivalently regarded as a combined model of several small models for voting to achieve better algorithmic performance. During pruning, because of fewer neuron responses resulting from pruning, Dropout should set a lower zeroing rate when randomly selecting neuron output set to zero to ensure that the network layer after that gets enough feature input. For neural network training during pruning, the corresponding Dropout probability can be transformed into the following formula.

$$D_r = D_o \sqrt{\frac{C_{ir}}{C_{io}}} \tag{1}$$

where D_r represents the Dropout probability of the pruning network and D_o is the Dropout probability of the original network. C_{ir} is the number of input connections and C_{io} is the number of output connections.

The pruning process is carried out layer by layer from front to back, while the threshold for pruning decreases gradually, resulting in a nearly lossless effect compared to the original network.

2.2   Parameter Quantification

Through parameter quantification, the number of bits required to express each weight parameter can be further compressed on the sparsely connected network obtained by pruning. And the number of parameters that need to be stored can be greatly reduced by restricting different neurons to share the same connection weight parameters. Corresponding to each parameter of the pruning network, it can be replaced with an 8-bit index, which can represent data in the range of 0–255, that is, we only need to store the weights of 256 real parameter, and each original parameter location can be represented by an index mapped to 256 shared parameters by sharing these weights. At this point, the network equivalent compression ratio can be calculated by the following formula.

$$r = \frac{Nb}{N \log_2(K) + Kb} \tag{2}$$

where N represents the number of connections to the source network model. Each connection is represented by b bits, and K represents the number of shared parameters.

Specifically, we can determine the value of each shared weight parameter by the most classical k-means clustering method, and for networks that have been fully trained and pruned, the pruning process has greatly reduced the parameter that need to be quantified, which improve the accuracy of quantification. For n parameters of a layer

$W = \{w_1, w_2, ..., w_n\}$, we can cluster them into k cluster centers $C = \{c_1, c_2, ..., c_k\}$ with k-means, where k is much larger than n. The process is to achieve the following optimization goal:

$$\arg \min_C \sum_{i=1}^{k} \sum_{w \subset c_i} \|w - \mu_i\|^2 \tag{3}$$

Unlike the HashNet algorithm, in which parameter sharing is based on a hash function that processes untrained network models, this kind of parameter sharing is performed on a fully trained network, so this parameter quantification sharing method can better approximate the source network model.

After the pruning process completing, the parameter weight of the network becomes bimodal distribution. In the clustering process, the initialization of the clustering center will have an impact on the clustering results and the task performance of the final network. There are three ways to initialize: random initialization, density-based initialization, and linear initialization. The random initialization method initializes the selected sample parameters as the starting cluster center by randomly selecting k parameters. Because of the bimodal distribution of the parameter weights, such clustering results tend to determine the final cluster center around two peaks. Density-based initialization uniformly selects the initial clustering center in the interval of different densities so that the final clustering result is also around two peaks, but will be more scattered than random initialization. The linear initialization method uniformly selects the initial cluster centers in the value space of the weights so that the final cluster center results will have the highest diversity.

In general, parameters with higher weight values are approximately important and have a greater impact on the results. But at the same time, the number of parameters with higher values is low, which makes it less efficient to choose clustering centers around such parameters. In the final clustering results of density-based initialization methods and random initialization methods, clustering centers often rarely have higher absolute values and therefore it will have a bad effect on the approximation effect of the final model. The linear initialization method can avoid this situation, so in the final clustering, initial clustering center by linear initialization is a reasonable choice.

After clustering, 256 shared weights can be selected. During network forward propagation, each network connection can find the corresponding shared weight through its stored index checking table. At the same time, in the back-propagation process of the necessary fine-tuning, for a specific neural network layer with n connections to m clustering centers, W_{ij} representing the j-th connection weight parameter of the i-th neuron. Assuming the loss function is L, the gradient update for the k-th shared parameter can be expressed as.

$$\frac{\partial L}{\partial C_K} = \sum_{i,j} \frac{\partial L}{\partial W_{ij}} \frac{\partial W_{ij}}{\partial C_K} = \sum_{i,j} \frac{\partial L}{\partial W_{ij}} 1(I_{ij} = k) \tag{4}$$

where I_{ij} represents the cluster center index corresponding to W_{ij}.

2.3   Hoffmann Encoding

Hoffmann encoding is a variable length coding technology, usually used for lossless data compression. It uses variable length coding method to encode the source data, and the code length corresponding to the symbol in the source data is determined by the probability of the symbol. The average encoding length of all symbols can be reduced by giving the occurrence of a higher probability of the symbol with a shorter encoding.

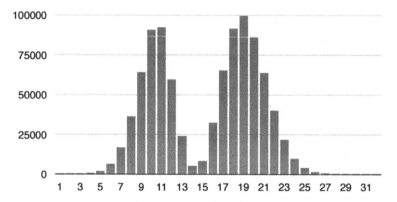

Fig. 2. Distribution of network weight values after parameter pruning.

Table 1. The compression pipeline saves network storage by 35× to 48× across different networks without loss of accuracy.

Network	Top-1 Acc	Parameters	Compress Rate
LeNet-300–100 Ref	98.35%	1080 KB	**40**×
LeNet-300–100 Compressed	98.32%	**27 KB**	
AlexNet Ref	57.32%	240 MB	**35**×
AlexNet Compressed	57.33%	**6.8 MB**	
VGG-16 Ref	68.48%	552 MB	**48**×
VGG-16 Compressed	68.50%	**11.5 MB**	

Figure 2 shows the distribution of network weights after pruning, in which can be seen that the number of weight parameters within each of the different value intervals is very different, i.e. the probability of each weight parameter is very different. Hoffmann encoding technology can effectively compress and encode the weight parameters of this distribution. In addition, because of the lossless nature of Hoffmann encoding, this compression will not result in any loss of network task performance.

3 Experimental Results and Analysis

We pruned, quantized, and Huffman encoded three networks: one on MNIST and two on ImageNet data-sets. The parameters and accuracy of network before and after pruning are shown in Table 1. The compression pipeline saves network storage by $35\times$ to $48\times$ across different networks without loss of accuracy. The total size of AlexNet decreased from 240 MB to 6.8 MB.

Table 2. Compression statistics for LeNet-300–100. P: pruning, Q: quantification, H: Huffmann encoding.

Layer	#Weights	Weight bits (P + Q)	Weight bits (P + Q + H)	Index bits (P + Q)	Index bits (P + Q + H)	Compress rate (P + Q)	Compress rate (P + Q + H)
ip1	236 K	6.0	4.4	5.0	3.7	3.1%	2.3%
ip2	30 K	6.0	4.4	5.0	4.3	3.8%	3.0%
ip3	1 K	6.0	4.3	5.0	3.2	15.7%	12.7%
Total	267 K	6.0	5.1	5.0	3.7	3.1%(32×)	2.5%(40×)

Table 3. Compression statistics for AlexNet. P: pruning, Q: quantification, H: Huffmann encoding.

Layer	#Weights	Weight bits (P + Q)	Weight bits (P + Q + H)	Index bits (P + Q)	Index bits (P + Q + H)	Compress rate (P + Q)	Compress rate (P + Q + H)
conv1	35 K	8.0	6.4	4.0	1.3	32.6%	20.5%
conv2	307 K	8.0	5.6	4.0	2.2	14.4%	9.4%
conv3	885 K	8.0	5.0	4.0	2.5	13.0%	8.4%
conv4	663 K	8.0	5.1	4.0	2.5	14.2%	9.1%
conv5	442 K	8.0	5.5	4.0	2.6	14.0%	9.4%
fc6	38 M	5.0	3.8	4.0	3.1	3.1%	2.4%
fc7	17 M	5.0	3.7	4.0	3.6	3.0%	2.4%
fc8	4 M	5.0	4.0	4.0	3.2	7.3%	5.9%
Total	267 K	5.5	5.1	4.0	3.2	3.6%(27×)	2.89%(35×)

3.1 MNIST

LeNet-300–100 [10] is a fully connected network with two hidden layers, with 300 and 100 neurons each, which achieves 1.6% error rate on MNIST. Table 2 shows the statistics of the compression pipeline. Most of the saving comes from pruning and quantification (compressed $32\times$), while Huffman encoding gives a marginal gain (compressed $40\times$).

3.2   ImageNet

AlexNet. Only two levels of headings should be numbered. We further examine the performance of Deep Compression on the ImageNet dataset, which has 1.2 M training examples and 50 K validation examples. We use the AlexNet, which has 61 million parameters and achieved a top-1 accuracy of 57.32%, to test our method. Table 3 shows that AlexNet can be compressed to 2.89% of its original size without impacting accuracy (see Table 1). There are 256 shared weights in each convolutional layer, which are encoded with 8.0 bits, and 32 shared weights in each FC layer, which are encoded with only 5.0 bits. And the corresponding index is encoded with 4.0 bits.

Table 4. Compression statistics for VGG-16. P: pruning, Q: quantification, H: Huffmann encoding.

Layer	#Weights	Weight bits (P + Q)	Weight bits (P + Q + H)	Index bits (P + Q)	Index bits (P + Q + H)	Compress rate (P + Q)	Compress rate (P + Q + H)
conv1	39 K	8.0	6.6	6.7	2.2	23.5%	16.6%
conv2	222 K	8.0	5.7	5.7	2.3	14.5%	9.2%
conv3	1 M	8.0	4.7	4.7	2.5	18.6%	8.9%
conv4	5 M	8.0	4.5	4.3	2.8	12.7%	6.8%
conv5	6 M	8.0	4.6	4.5	2.4	13.4%	7.6%
fc6	103 M	5.0	3.6	3.6	3.5	1.6%	1.1%
fc7	17 M	5.0	4.0	4.0	4.3	1.5%	1.3%
fc8	4 M	5.0	4.0	4.0	3.4	7.2%	5.2%
Total	138 M	6.4	6.4	4.1	3.2	3.2%(31×)	2.1%(49×)

VGG-16. With promising results on AlexNet, we also looked at a larger one, VGG-16, on the same ImageNet dataset. VGG-16 has far more convolutional layers but still only three fully-connected layers. Following a similar methodology, we aggressively compressed both convolutional and fully-connected layers to realize a significant reduction in the number of weights, which is shown in Table 4. The VGG-16 network has been compressed by 49×. Weights in the convolutional layers are represented with 8.0 bits, and fully-connected layers use 5.0 bits, which does not impact the accuracy.

4   Conclusion

In order to solve the problem of high storage and high complexity of deep neural network model, we explored the existing effective deep neural model compression and acceleration technology, fully combined the characteristics of different types of network compression technology, and gave full play to the effectiveness of various technologies. In this study, we overcame the problem that traditional methods for deep neural model compression cannot greatly improve the trade-off ratio between performance and compression ratio, and converted the high-storage and high-complexity deep neural network model into a lightweight network while maintaining the original model recognition performance.

Acknowledgments. This work is supported by Xiamen Major Science and Technology Projects (No. 3502Z20201017).

References

1. Krizhevsky, A., Sutskever, I., Hinton, G.E.: Imagenet classification with deep convolutional neural networks. Adv. Neural Inform. Process. Syst. **25**, 1097–1105 (2012)
2. Simonyan, K., Zisserman, A.: Very deep convolutional networks for large-scale image recognition. arXiv preprint arXiv:1409.1556 (2014)
3. Szegedy, C., Liu, W., Jia, Y., et al.: Going deeper with convolutions. In: Proceedings of the IEEE Conference on Computer Vision and Pattern Recognition, pp. 1–9 (2015)
4. He, K., Zhang, X., Ren, S., et al.: Deep residual learning for image recognition. In: Proceedings of the IEEE Conference on Computer Vision and Pattern Recognition, pp. 770–778 (2016)
5. Karpathy, A., Fei-Fei, L.: Deep visual-semantic alignments for generating image descriptions. In: Proceedings of the IEEE Conference on Computer Vision and Pattern Recognition, pp. 3128–3137 (2015)
6. Zhang, C., Li, P., Sun, G., Guan, Y., Xiao, B., Cong, J.: Optimizing FPGA-based accelerator design for deep convolutional neural networks. In: Proceedings of the 2015 ACM/SIGDA International Symposium on Field-Programmable Gate Arrays, pp. 161–170 (2015)
7. Motamedi, M., Gysel, P., Akella, V., et al.: Design space exploration of FPGA-based deep convolutional neural networks. In: Design Automation Conference (ASP-DAC), 2016 21st Asia and South Pacific, pp. 575–580. IEEE (2016)
8. Qiu, J., Wang, J., Yao, S., et al.: Going deeper with embedded fpga platform for convolutional neural network. In: Proceedings of the 2016 ACM/SIGDA International Symposium on Field-Programmable Gate Arrays, pp. 26–35 (2016)
9. Dundar, A., Jin, J., Gokhale, V., et al.: Accelerating deep neural networks on mobile processor with embedded programmable logic. In: Neural Information Processing Systems Conference (2013)
10. LeCun, Y., Bottou, L., Bengio, Y., et al.: Gradient-based learning applied to document recognition. Proc. IEEE **86**(11), 2278–2324 (1998)

Extreme Learning Machine Based on Double Kernel Risk-Sensitive Loss for Cancer Samples Classification

Zhen-Xin Niu[1], Liang-Rui Ren[1], Rong Zhu[1], Xiang-Zhen Kong[1],
Ying-Lian Gao[2], and Jin-Xing Liu[1(✉)]

[1] School of Computer Science, Qufu Normal University, Rizhao, China
zhurongsd@qfnu.edu.cn
[2] Qufu Normal University Library, Qufu Normal University, Rizhao, China

Abstract. In recent years, Extreme Learning Machine (ELM) has attracted extensive attention in various research fields. To improve the performance of ELM, we propose an Extreme Learning Machine Based on Double Kernel Risk-Sensitive Loss (DKRSLELM) method in this paper. The Kernel Risk-Sensitive Loss (KRSL) is integrated into the objective function of ELM. This is because KRSL not only can effectively reduce the influence of noise and outliers, but also can eliminate the redundant neurons of ELM. In the experiment, we conduct a classification experiment on cancer integration data-sets. The experimental results indicate that our method can effectively improve the classification performance of ELM.

Keywords: Extreme Learning Machine · Kernel Risk-Sensitive Loss · Robustness · Sparsity · Supervised learning

1 Introduction

Cancer is a common disease that threatens human health, and cancer will develop into malignant tumors in severe cases [1, 2]. Early detection of cancer is very important to reduce cancer mortality [3]. With the development of science and technology, researchers are getting deeper understanding of cancer. The classification of cancer samples is an important method for studying cancer, and it is developing rapidly.

With the development of science and technology, Extreme Learning Machine (ELM) [4–6] has received extensive attention. And ELM is applied for the analysis of actual data in many fields due to the excellent performance, for example, facial recognition [7], time series analysis [8, 9], control and robotics [10], etc.

ELM has received widespread attention in many fields, but its sparsity and robustness need to be improved. Huang et al. proposed the Regularized Extreme Learning Machine (RELM) method, which uses the L_2-norm as the loss function and regularization constraint of ELM [11]. However, the square loss of the L_2-norm would amplify the impact of noise and outliers. In addition, RELM has random hidden layer mapping, which requires a larger number of hidden nodes than other feedforward neural networks to achieve ideal performance [12]. So Li et al. proposed the $L_{2,1}$-norm

© Springer Nature Switzerland AG 2021
D.-S. Huang et al. (Eds.): ICIC 2021, LNCS 12837, pp. 532–539, 2021.
https://doi.org/10.1007/978-3-030-84529-2_45

Based Loss Function and Regularization Extreme Learning Machine (LR21ELM) method, in which a row-sparsity inducing $L_{2,1}$-norm regularization term is integrated into the objective function [13]. These innovations make LR21ELM have better sparsity.

The main contributions of this paper are as follows:

1) A novel method called DKRSLELM is presented to enhance the performance of original ELM. We introduce KRSL into the loss function of ELM to improve the robustness. To reduce the complexity of the ELM network and promote the group sparsity, our method also innovatively uses KRSL to replace the sparse penalty term of original ELM.

2) In the experiment, we first test the classification performance of DKRSLELM on five cancer integration data-sets, and four evaluation indicators including Accuracy, Recall, Precision and F-measure are used to evaluate DKRSLELM method.

2 Related Work

2.1 Extreme Learning Machine

For a data-set $\{\mathbf{X}, \mathbf{Y}\} = \{\boldsymbol{x}_i, \boldsymbol{y}_i\}_{i=1}^{N}$, $\boldsymbol{x}_i \in \mathbb{R}^{n_i}$ and $\boldsymbol{y}_i$ is a n_o-dimensional binary vector with only one entry equals to one for multi-classification tasks, or $\boldsymbol{y}_i \in \mathbb{R}^{n_o}$ is used for regression tasks, where n_i and n_o are the dimensions of input and output, respectively [14]. In this paper, we use $\boldsymbol{h}(\boldsymbol{x}_i) \in \mathbb{R}^{n_h \times n_o}$ to represent the output vector of the hidden layer relative to $\boldsymbol{x}_i$. The output matrix connecting the hidden layer and the output layer is represented by $\boldsymbol{\beta} \in \mathbb{R}^{n_h \times n_o}$. Then the output function of ELM is:

$$\boldsymbol{f}(\boldsymbol{x}_i) = \boldsymbol{h}(\boldsymbol{x}_i)\boldsymbol{\beta}, \quad i = 1, \ldots, N. \tag{1}$$

And the objective function of ELM can be expressed as:

$$\min_{\boldsymbol{\beta}} \frac{1}{2}\|\boldsymbol{\beta}\|^2 + \frac{\xi}{2}\sum_{i=1}^{N}\|\mathbf{e}_i\|^2 \tag{2}$$
$$s.t. \quad \boldsymbol{h}(\boldsymbol{x}_i)\boldsymbol{\beta} = \boldsymbol{y}_i^T - \mathbf{e}_i^T, \quad i = 1, \ldots, N,$$

where $\mathbf{e}_i \in \mathbb{R}^{n_o}$ is the error vector of the i-th sample, and the penalty coefficient of training error is represented by ξ. By bringing constraints into the ELM objective function, we can get its equivalent unconstrained optimization problem:

$$\min_{\boldsymbol{\beta}} \frac{1}{2}\|\boldsymbol{\beta}\|^2 + \frac{\xi}{2}\|\mathbf{Y} - \mathbf{H}\boldsymbol{\beta}\|^2. \tag{3}$$

The number of hidden neurons is represented by L. If the number of training patterns is greater than the number of hidden neurons, that is, $N > L$, β can be obtained by calculating the partial derivative of Eq. (3) and setting it to zero:

$$\beta = (\mathbf{H}^T\mathbf{H} + \frac{\mathbf{I}_{n_h}}{\xi})^{-1}\mathbf{H}^T\mathbf{Y}, \tag{4}$$

where $\mathbf{I}_{n_h}$ is an identity matrix with dimension n_h. If the number of training patterns is less than the number of hidden neurons, that is, $N < L$, β can be calculated as:

$$\beta = \mathbf{H}^T(\mathbf{H}\mathbf{H}^T + \frac{\mathbf{I}_N}{\xi})^{-1}\mathbf{Y}, \tag{5}$$

where $\mathbf{I}_N$ is an identity matrix with dimension N.

2.2 Kernel Risk-Sensitive Loss

The KRSL between two random variables S and Q is defined by:

$$
\begin{aligned}
L_\eta(S, Q) &= \frac{1}{\lambda}\mathrm{E}[\exp(\lambda(1 - k_\sigma(S - Q)))] \\
&= \frac{1}{\lambda}\int \exp(\lambda(1 - k_\sigma(s - q)))dF_{SQ}(s, q),
\end{aligned} \tag{6}
$$

where $\lambda > 0$ represents the risk-sensitive parameter. Equation (6) can also be expressed as:

$$L_\lambda(S, Q) = \frac{1}{\lambda}\mathrm{E}\left[\exp\left(\lambda\left(\frac{1}{2}\|\phi(S) - \phi(Q)\|_{\mathbf{H}}^2\right)\right)\right], \tag{7}$$

which is different from the traditional risk-sensitive loss in the defined space [15, 16]. According to [15], the risk sensitive loss in original space can be defined as:

$$l_\lambda(S, Q) = \frac{1}{\lambda}\mathrm{E}\left(\exp\left(\frac{\lambda}{2}(S - Q)^2\right)\right). \tag{8}$$

In the actual situation, the joint distribution of S and Q is unknown, so the approximate values obtained from a limited number of N samples need to be averaged, which is called empirical KRSL:

$$\hat{L}_\lambda(S, Q) = \frac{1}{N\lambda}\sum_{i=1}^{N}\exp(\lambda(1 - k_\sigma(s(i) - q(i)))). \tag{9}$$

3   The Proposed Method

3.1   The Objective Function of DKRSLELM

The objective function can be expressed as follows:

$$F = \min_{\boldsymbol{\beta}} \frac{C}{N\lambda} \sum_{i=1}^{N} \exp[\lambda(1 - k_{\sigma_1}(\boldsymbol{\beta}))] + \frac{1}{N\lambda} \sum_{i=1}^{N} \exp[\lambda(1 - k_{\sigma_2}(\mathbf{Y} - \mathbf{H}\boldsymbol{\beta}))], \qquad (10)$$

where hidden layer output weight is denoted by $\boldsymbol{\beta}$. C and λ represent the regularization parameter and risk-sensitive parameter, respectively. k_σ denotes a radial kernel function with kernel bandwidth parameter σ. Equation (13) can also be expressed as:

$$F = \min_{\boldsymbol{\beta}} \frac{C}{N\lambda} \sum_{i=1}^{N} \exp\left[\lambda\left(1 - \exp\left(-\frac{\boldsymbol{\beta}^2}{2\sigma_1^2}\right)\right)\right]$$
$$+ \frac{1}{N\lambda} \sum_{i=1}^{N} \exp\left[\lambda\left(1 - \exp\left(-\frac{(\mathbf{Y} - \mathbf{H}\boldsymbol{\beta})^2}{2\sigma_2^2}\right)\right)\right]. \qquad (11)$$

3.2   The Optimization of DKRSLELM

The gradient of $\boldsymbol{\beta}$ can be computed as follows:

$$\frac{\partial F}{\partial \boldsymbol{\beta}} = \frac{C}{N\sigma_1^2} \exp[\lambda(1 - k_{\sigma_1}(\boldsymbol{\beta}))]k_{\sigma_1}(\boldsymbol{\beta})\boldsymbol{\beta}$$
$$- \frac{1}{N\sigma_2^2} \exp[\lambda(1 - k_{\sigma_2}(\mathbf{Y} - \mathbf{H}\boldsymbol{\beta}))]k_{\sigma_2}(\mathbf{Y} - \mathbf{H}\boldsymbol{\beta})\mathbf{H}^T(\mathbf{Y} - \mathbf{H}\boldsymbol{\beta}). \qquad (12)$$

In this paper, we assume $\mathbf{A}_1$ and $\mathbf{A}_2$ represent $\exp[\lambda(1 - k_{\sigma_1}(\boldsymbol{\beta}))]k_{\sigma_1}(\boldsymbol{\beta})$ and $\exp[\lambda(1 - k_{\sigma_2}(\mathbf{Y} - \mathbf{H}\boldsymbol{\beta}))]k_{\sigma_2}(\mathbf{Y} - \mathbf{H}\boldsymbol{\beta})$, respectively. If the number of training patterns is greater than the number of hidden neurons ($\mathbf{H}$ not only has multiple rows and multiple columns, but also has complete column levels) [14], we set Eq. (12) be equal to zero and get:

$$\boldsymbol{\beta} = \left(\lambda^* \mathbf{A}_1 + \mathbf{H}^T \mathbf{A}_2 \mathbf{H}\right)^{-1} \mathbf{H}^T \mathbf{A}_2 \mathbf{Y}, \qquad (13)$$

where λ^* is equal to $C\sigma_2^2/\sigma_1^2$. If the number of training patterns is less than the number of hidden neurons, in this case, $\boldsymbol{\beta}$ will have an infinite number of solutions [14]. So we limit $\boldsymbol{\beta}$ to a linear combination of $\mathbf{H}$ rows: $\lambda^* \mathbf{A}_1 \boldsymbol{\beta} = \mathbf{H}^T \alpha$, and bring in Eq. (12). We can get:

$$\boldsymbol{\beta} = \mathbf{A}_1^{-1} \mathbf{H}^T \left(\lambda^* \mathbf{A}_1 + \mathbf{A}_2 \mathbf{H} \mathbf{A}_1^{-1} \mathbf{H}^T\right)^{-1} \mathbf{A}_2 \mathbf{Y}. \qquad (14)$$

In summary, the solution process of DKRSLELM is described in detail in Algorithm 1.

Algorithm 1: DKRSLELM

Input:

Training data-set $\{S_{train}, Q_{train}\} = \{s_i, q_i\}_{i=1}^{N}$

The regularization C and the risk-sensitive λ

Kernel bandwidth σ_1 and σ_2

The number of hidden neurons L

Output:

The output weight matrix $\boldsymbol{\beta}$

Steps:

1: Initializing the input weight matrix and bias

2: By the sigmoid function, compute the hidden layer output matrix $\mathbf{H}$

3: Randomly generate a hidden layer output weight matrix $\boldsymbol{\beta}_0$

4: Repeat

5: Update the auxiliary matrix:

$$\mathbf{A}_1 = \exp\left[\lambda\left(1 - k_{\sigma_1}(\boldsymbol{\beta})\right)\right] k_{\sigma_1}(\boldsymbol{\beta}),$$

$$\mathbf{A}_2 = \exp\left[\lambda\left(1 - k_{\sigma_2}(\mathbf{Y} - \mathbf{H}\boldsymbol{\beta})\right)\right] k_{\sigma_2}(\mathbf{Y} - \mathbf{H}\boldsymbol{\beta}).$$

6: Calculate the output weight $\boldsymbol{\beta}$ according to Eq. (13) or Eq. (14)

7: Until converge

---

4   Experiments

4.1   Data-Sets

In the experiment, we use the DKRSLELM on the cancer integration data-sets for classification. Details of benchmark data-sets and cancer data-sets are listed in Table 1.

Table 1. The detail of data-sets

Data-sets	Classes	Samples	Training	Testing	Features
CO_ES	2	445	334	111	20502
CO_ES_PA	3	621	446	155	20502
Pan_Cancer	5	801	601	200	16383

4.2 Settings

During this experiment, there are 4 parameters $(C, \lambda, \sigma_1, \sigma_2)$ that need to be adjustment. The selection range of parameter C is $(10^{-4}, 10^{-3}, \ldots, 10^4)$, λ is set as $(1, 2, \ldots, 10)$, σ_1 and σ_2 are set as $(2^{-4.5}, 2^{-3.5}, \ldots, 2^{4.5})$.

4.3 Classification Results on Benchmark Data-Sets and Cancer Integration Data-Sets

To reduce the impact of random initialization on the experimental results, we run each method 20 times. Table 2 respectively records the classification performance of all methods on cancer integration data-sets.

Table 2. The experimental results on cancer based integration data-sets

Data-sets	EVA	COAD_ESCA	COAD_ESCA_PAAD	Pan_Cancer
LSSVM	ACC	0.9342 ± 0.0002	0.9228 ± 0.0013	0.9791 ± 0.0006
	Pre	0.9361 ± 0.0005	0.9275 ± 0.0010	0.9689 ± 0.0003
	Re	0.9575 ± 0.0002	0.9200 ± 0.0008	0.9705 ± 0.0002
	F	0.9439 ± 0.0007	0.9287 ± 0.0016	0.9713 ± 0.0005
RELM	ACC	0.9549 ± 0.0036	0.9348 ± 0.0001	0.9665 ± 0.0001
	Pre	0.9543 ± 0.0001	0.9372 ± 0.0001	0.9687 ± 0.0000
	Re	0.9857 ± 0.0001	0.9340 ± 0.0002	0.9650 ± 0.0001
	F	0.9650 ± 0.0000	0.9352 ± 0.0001	0.9663 ± 0.0000
LR21ELM	ACC	0.9513 ± 0.0003	0.9109 ± 0.0002	0.9725 ± 0.0001
	Pre	0.9512 ± 0.0001	0.9154 ± 0.0002	0.9679 ± 0.0002
	Re	0.9728 ± 0.0003	0.9069 ± 0.0002	0.9609 ± 0.0002
	F	0.9618 ± 0.0000	0.9101 ± 0.0002	0.9638 ± 0.0002
CELM	ACC	0.9315 ± 0.0003	0.8942 ± 0.0003	0.9500 ± 0.0001
	Pre	0.9279 ± 0.0004	0.8962 ± 0.0003	0.9442 ± 0.0001
	Re	0.9671 ± 0.0001	0.8913 ± 0.0003	0.9494 ± 0.0001
	F	0.9469 ± 0.0001	0.8930 ± 0.0003	0.9461 ± 0.0001
ELMKMPE	ACC	0.9117 ± 0.0002	0.9116 ± 0.0002	0.9755 ± 0.0001
	Pre	0.9234 ± 0.0001	0.9027 ± 0.0003	0.9751 ± 0.0002
	Re	0.8499 ± 0.0012	0.9121 ± 0.0001	0.9781 ± 0.0001
	F	0.8821 ± 0.0005	0.9063 ± 0.0002	0.9764 ± 0.0001
DKRSLELM	ACC	$\mathbf{0.9873 \pm 0.0001}$	$\mathbf{0.9445 \pm 0.0000}$	$\mathbf{0.9870 \pm 0.0000}$
	Pre	$\mathbf{0.9845 \pm 0.0000}$	$\mathbf{0.9456 \pm 0.0001}$	$\mathbf{0.9868 \pm 0.0001}$
	Re	$\mathbf{0.9957 \pm 0.0000}$	$\mathbf{0.9434 \pm 0.0001}$	$\mathbf{0.9859 \pm 0.0000}$
	F	$\mathbf{0.9901 \pm 0.0002}$	$\mathbf{0.9411 \pm 0.0000}$	$\mathbf{0.9861 \pm 0.0001}$

Based on Table 2, we can draw the following conclusions:

1) Generally, DKRSLELM method performs better than CELM method. That is to say, the methods based on KRSL are better than that based on correntropy. The reason is that the performance surface of the loss function caused by correntropy is highly non-convex, the slope around the optimal solution is high, the convergence is slow and the accuracy is low. Using KRSL as the loss function of ELM can effectively solve the above problems.

2) It can be seen from Table 2 that compared with the RELM method, the LR21ELM method has not significantly improved classification performance. From the experimental results, we can observe that constraining β with the $L_{2,1}$-norm can't effectively improve the classification performance of ELM. DKRSLELM is significantly better than LR21ELM since KRSL is easier to identify neurons that are not related to the output and improve the model's ability to reduce certain lines to zero, then reduce the complexity of the network.

5   Conclusion

To improve the sparsity and robustness of ELM, Extreme Learning Machine Based on Double Kernel Risk-Sensitive Loss (DKRSLELM) method is proposed. Considering the KRSL is less affected by noise and outliers, the original loss function of ELM is replaced by KRSL. More importantly, we innovatively replaced the sparse penalty term of the original ELM with KRSL to further reduce the complexity of the network structure and improve the classification performance of the model. And to obtain the exact solution of the objective function, an iterative algorithm is proposed to solve the optimization problem of DKRSLELM. We conduct some classification experiments on three cancer integration data-sets to evaluate the performance of DKRSLELM method. The experimental results can strongly illustrate that the DKRSLELM has a competitive advantage in the classification of cancer samples.

Acknowledgment. This work was supported in part by the grants provided by the National Natural Science Foundation of China, Nos. 61872220, and 61572284.

References

1. Jiao, C.-N., Gao, Y.-L., Yu, N., Liu, J.-X., Qi, L.-Y.: Hyper-graph regularized constrained NMF for selecting differentially expressed genes and tumor classification. IEEE J. Biomed. Health Inform. **24**(10), 3002–3011 (2020)
2. Liu, J.-X., Wang, D., Gao, Y.-L., Zheng, C.-H., Xu, Y., Yu, J.: Regularized non-negative matrix factorization for identifying differentially expressed genes and clustering samples: a survey. IEEE/ACM Trans. Comput. Biol. Bioinf. **15**(3), 974–987 (2017)
3. Zhang, J., Liu, Z., Du, B., He, J., Li, G., Chen, D.: Binary tree-like network with two-path fusion attention feature for cervical cell nucleus segmentation. Comput. Biol. Med. **108**, 223–233 (2019)

4. Huang, G.-B., Zhu, Q.-Y., Siew, C.-K.: Extreme learning machine: theory and applications. Neurocomputing **70**(1–3), 489–501 (2006)
5. Huang, G.-B., Chen, L., Siew, C.K.: Universal approximation using incremental constructive feedforward networks with random hidden nodes. Neural Netw. **17**(4), 879–892 (2006)
6. Huang, G.-B., Chen, L.: Convex incremental extreme learning machine. Neurocomputing **70** (16–18), 3056–3062 (2007)
7. He, B., et al.: Fast face recognition via sparse coding and extreme learning machine. Cogn. Comput. **6**(2), 264–277 (2014)
8. Guo, W., Xu, T., Tang, K.: M-estimator-based online sequential extreme learning machine for predicting chaotic time series with outliers. Neural Comput. Appl. **28**(12), 4093–4110 (2016). https://doi.org/10.1007/s00521-016-2301-0
9. Liu, Z., Loo, C.K., Masuyama, N., Pasupa, K.: Recurrent kernel extreme reservoir machine for time series prediction. IEEE Access **6**, 19583–19596 (2018)
10. Yu, Y., Choi, T.-M., Hui, C.-L.: An intelligent quick prediction algorithm with applications in industrial control and loading problems. IEEE Trans. Autom. Sci. Eng. **9**(2), 276–287 (2011)
11. Huang, G.-B., Zhou, H., Ding, X., Zhang, R.: Extreme learning machine for regression and multiclass classification. IEEE Trans. Syst. Man Cybern. Part B (Cybernetics) **42**(2), 513–529 (2011)
12. Miche, Y., Sorjamaa, A., Bas, P., Simula, O., Jutten, C., Lendasse, A.: OP-ELM: optimally pruned extreme learning machine. IEEE Trans. Neural Netw. **21**(1), 158–162 (2009)
13. Li, R., Wang, X., Lei, L., Song, Y.: $L_{2,1}$-norm based loss function and regularization extreme learning machine. IEEE Access **7**, 6575–6586 (2018)
14. Huang, G., Song, S., Gupta, J.N., Wu, C.: Semi-supervised and unsupervised extreme learning machines. IEEE Trans. Cybern. **44**(12), 2405–2417 (2014)
15. Boel, R.K., James, M.R., Petersen, I.R.: Robustness and risk-sensitive filtering. IEEE Trans. Autom. Control **47**(3), 451–461 (2002)
16. Lo, J.T., Wanner, T.: Existence and uniqueness of risk-sensitive estimates. IEEE Trans. Autom. Control **47**(11), 1945–1948 (2002)

Delay to Group in Food Delivery System: A Prediction Approach

Yang Yu[(✉)], Qingte Zhou, Shenglin Yi, Huanyu Zheng,
Shengyao Wang, Jinghua Hao, Renqing He, and Zhizhao Sun

Meituan, Beijing, China
yuyang47@meituan.com

Abstract. Food delivery services are now changing the way people live by connecting customers, restaurants and drivers through online APPs. Dispatching orders is a key issue in food delivery system, which requires effectively matching between drivers and orders. In order to improve driver efficiency, we group orders with similar pick-up and delivery locations. We measure this efficiency by grouping rate, which is the ratio of grouped orders to all orders. On one hand, it is naturally expected that grouping rate increases if the system delays and accumulates more orders in the matching pool. On the other hand, customer waiting time also increases with orders dispatched to drivers at a later time. Considering both driver efficiency and customer experience, we develop a prediction-based approach to decide order delaying decision, combining machine learning and optimization. Machine learning part predicts how much delaying affects grouping possibility and customer experience. Optimization part makes delaying decisions, balancing driver efficiency and customer experience. Through empirical experiments in different cities, we can find our algorithm significantly improves grouping rate, with similar customer experience at the same time.

Keywords: Food delivery · Delaying · Grouping · Optimization with prediction

1 Introduction Background

With convenience with online ordering, food delivery is a fast-growing industry all over the world. Food delivery platforms provide various food choices from nearby restaurants for each customer. When a customer orders online, the platform find the best driver to deliver the order. Because orders are generated randomly, matching with consideration of future orders is a major challenge for the platform.

In this paper, we address future orders problem in a situation called orders grouping. In order to improve driver efficiency, similar orders generated at different time can be grouped before assigned to customers. The similarity is from two aspects. First, orders have similar pick-up and delivery locations. Drivers can save their time waiting at the same restaurant or delivering in the same neighborhood. Second, orders have similar assigned arrival time, otherwise one of customers may receive order late. With similar orders grouped, drivers enjoy earning more money per hour and food

© Springer Nature Switzerland AG 2021
D.-S. Huang et al. (Eds.): ICIC 2021, LNCS 12837, pp. 540–551, 2021.
https://doi.org/10.1007/978-3-030-84529-2_46

delivery platform enjoys serving more orders per hour. Therefore, grouping rate, i.e., the percentage of orders grouped before assigning to drivers, is a good way to measure efficiency of food delivery.

The major challenge of grouping is how to group orders generated at different time. Figure 1 considers delaying decisions for grouping. If orders are dispatched without delay as Fig. 1 (1), at later time, another order may arrive and has to be dispatched to another driver, as shown in Fig. 1 (2). If some orders are held in the order pool as Fig. 1 (3), we may find a new order arriving and grouped with the delayed order, as shown in Fig. 1 (4). With delayed matching considering grouping, one of drivers may benefit from more efficient matching results. Thus, delaying assignment of an order can improve the chance of grouping with a similar order generated later. But at the same time, this customer may receive the order at a later time. Therefore, delaying decision is of great importance, balancing grouping rate and customer experience.

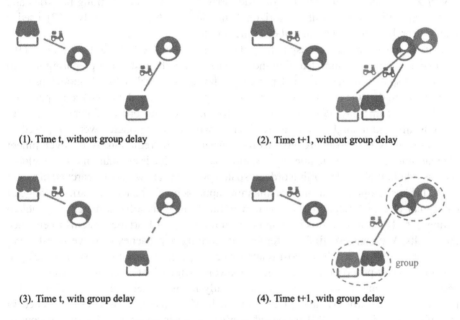

(1). Time t, without group delay   (2). Time t+1, without group delay

(3). Time t, with group delay   (4). Time t+1, with group delay

Fig. 1. Delaying decisions for grouping.

In food delivery system, orders are generated randomly in real time. It is quite challenging to decide delaying time for each order. In this paper, we use supervised learning method to predict whether a similar order will be generated before a certain order is dispatched. At the same time, we use another model to predict how late customer may receive this order. If the possibility of grouping is high and customer does not receive the order late, we tend to hold the order and wait for a group.

In Sect. 2, we present related research. Section 3 describes our problem, while Sect. 4 proposes our approach towards the problem. In Sect. 5, we provide experimental results. Finally, we conclude our work with Sect. 6.

2   Related Work

Food delivery is an extension of last-mile delivery problem [1]. Last-mile delivery problem is considered as the most inefficient part in the whole supply chain process due to the frequent need of specialized delivery service [2]. Traditional last-mile delivery problems are often related to different variants of VRP problems. As for food delivery order dispatching problem, the most relevant VRP variant is the pickup and delivery problem with time windows (PDPTW). In PDPTW each customer order is assigned with a receiving time window. Exact algorithms for PDPTW can only solve small-scale problems [3]. Meta-heuristic [4–7] and construction heuristic [8, 9] are more applicable for large-scale problems. Food delivery problem is extremely hard since it's highly dynamic [10]. Other relevant dynamic VRP variants include dynamic multiperiod routing problem (DMPRP) and vehicle routing problem with release date (VRPRD). DMPRP considers the uncertainty of order arrival within given planning horizons and [11] used a three-phase rolling horizon heuristic to solve it efficiently. [12] used a scatter search with strategic oscillation method to solve VRPRD.

The food delivery order dispatching problem is complex and highly dynamic. [10] used heuristic and machine learning methods to predict the restaurant preparing time to improve dispatching results. [13] presented the deep neural network model used in Elm.me to predict the order fulfillment cycle time. [14] studied a dynamic dispatching problem of same-day delivery service providers where orders enter the system dynamically and a single driver is available. Two dynamic policies were proposed.

Our work is also closely related to order dispatch with prediction. With development of smart phones and mobile internet, location-based service is popular among the globe, such as ride-hailing. Ride-hailing order dispatch problems are widely researched in recent years. [15] developed a learning and planning approach, in which they learn the value of empty car at each spatiotemporal pattern offline, and then solve a matching problem online. [16–18] extended this idea and built reinforcement learning systems to enhance the results. We can also find reinforcement learning approaches to solve ride-hailing problems from [19, 20]. The most related work is [21], in which they studied delayed matching. They proposed a two-stage framework, which first determines delayed time and then matching. In our paper, we specifically focus on delayed time decisions that increases chances for batching. Order dispatch with prediction is also researched in manufacturing industry. [22] proposed reinforcement learning-based adaptive control system for order dispatching in the semiconductor industry. [23] used a machine learning approach for reality awareness and optimization in cloud manufacturing.

3   Problem Description

This section presents problem background and formulates a mathematical model for online order delaying problem.

The decision to delay occurs when an order enters the system. Then, the system matches the order to an optimal driver based on the given optimization objective. Delaying decisions require the system to choose between two actions: to assign it to the driver immediately or to keep it in the system. Delaying decision means to keep the order in the system.

Order delaying decision is to decide when to assign orders. Order delay is a double-edged sword in the system. On one hand, order delay extends order existing time in the system, resulting in higher probability of grouping. On the other hand, order delay increases the chance of late delivery. Figure 2 shows the difference between long delay time and short delay time.

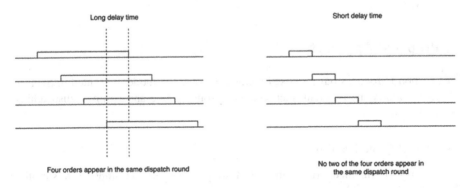

Fig. 2. Difference between long delay time and short delay time.

Food delivery delaying problem can be described by the following. The increase of driver efficiency from grouping is described as grouping rate, i.e., the proportion of orders completed in the form of grouping. As for customer experience, two main objectives are overtime rate and average delivery time. An overtime occurs when order delivery time is later than estimated time of arrival (ETA), which is promised to the customer. Overtime rate is the percentage of overtime orders in all completed orders. Average delivery time is the average time from ordering to receiving an order.

The online order delay problem can be formulated as the following mathematical model. Decision variables are delay time x_i for each order $i \in N$. If the delay time for an order is zero means to dispatch immediately when an order comes. The objective of the problem is to maximize grouping rate. The constraint of the problem is customer experience, including overtime rate and average delivery time. From description above, the online order delay time problem can be formulated as follows:

$$\text{obj.max} \sum_{\forall i \in N} g_i(x_i) \tag{1}$$

$$\text{s.t.} \sum_{\forall i \in N} d_i(x_i) - d_i(0) \leq 0 \tag{2}$$

$$\sum_{\forall i \in N} c_i(x_i) - c_i(0) \leq C_0 \tag{3}$$

$$c_i(x_i) - c_i(0) \leq C_1, \forall i \in N \tag{4}$$

$$x_i \geq 0, \forall i \in N \tag{5}$$

where $g_i(x_i)$ is the group probability function, which is the portion of grouped orders in all orders, $c_i(x_i)$ is the delivery time function, $d_i(x_i)$ is the overtime probability function. The objective is to maximize grouping probability of all orders, which is shown in (1). Equation (2) is a constraint, which guarantees that the overtime rate of all the orders caused by delay does not deteriorate. Equation (3) is another constraint, which limits total delivery time increment. Equation (4) limits the delivery time increment of each order.

4 Proposed Approach

This section introduces our proposed method to solve the model presented in Sect. 3. We first simplify the problem, then estimate grouping rate and optimize dispatching decision.

4.1 Simplify the Problem

In order to solve the model raised in Sect. 3, we propose some assumptions to simplify the problem reasonably.

First, we assume that delivery time increment is linearly related to delay time:

$$c_i(x_i) - c_i(0) = \alpha x_i \tag{6}$$

Here α is the empirical coefficient, which can be obtained from data. For real-world food delivery, this assumption is reasonable. The later an order is assigned to a driver, the later the order is delivered to the customer.

With this assumption, we can simplify constraint (2) in the model, which is overtime rate of each order. Overtime refers to the situation when the driver delivers later than ETA, which is promised by the platform to the customer.

Thus, if we know ETA and estimate delivery time if the order is not delayed, we can find a maximum delay time l_i allowed:

$$l_i = (ETA_i - ETR_i)/\alpha \tag{7}$$

where ETR, estimated time of route, is the estimated delivery time if the order is not delayed. We can estimate pickup and delivery time by a prediction model. An example of model output is shown in Fig. 3.

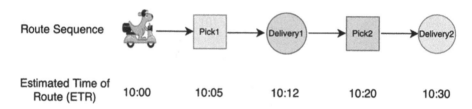

Fig. 3. An example of ETR model output.

Therefore, for a specific order, the system allows a maximum delay. Constraint (2) can be rewritten as follows:

$$0 \leq x_i \leq l_i \tag{8}$$

Second, we assume that grouping possibility is increasing linearly with delay time. For real-world food delivery, this assumption is also reasonable. The later an order is assigned to a driver, the longer the order stays in the system, and has a greater chance to find a similar order. With linearity assumptions, delaying an order is either beneficial or harmful, regardless how longer we delay this order.

Thus, we can simplify decision variable from delaying time to whether to delay. If we choose to delay, then we delay the order to the maximum delay time. Otherwise, we assign the order immediately. In this way, we only need to estimate the grouping probability of an order within the maximum delay time.

Based on the above simplification, we reformulate the original problem as follows:

$$\text{obj.max} \sum_{\forall i \in N} \hat{g}_i(y_i) \tag{9}$$

$$\text{s.t.} \sum_{\forall i \in N} \hat{c}_i(y_i) \leq C_0 \tag{10}$$

$$\hat{c}_i(y_i) \leq C_1, \forall i \in N \tag{11}$$

$$y_i \in \{0, 1\}, \forall i \in N \tag{12}$$

The decision variable for the problem is y_i, which means whether delay the order or not. The objective is to maximize the grouping probability and is estimated by a machine learning model. The constraints are to limit the increment of delivery time, with the following definitions:

$$\hat{c}_i(y_i) = \begin{cases} \alpha l_i & \text{if } y_i = 1 \\ 0 & \text{if } y_i = 0 \end{cases} \tag{13}$$

After simplification, the order delay problem is modeled as a knapsack problem. Constraints are linear under our assumptions. The objective grouping possibility is non-linear. In the next subsection, we introduce how to predict grouping possibility.

4.2 Grouping Possibility Prediction

In this subsection, we predict grouping probability of an order within the maximum delay time.

XgBoost model [24] is used to predict grouping probability. XgBoost has been proved to effective, with great performance in many estimation problems. Another reason we choose XgBoost model is that the features in our model are mostly continuous. We use Area Under Curve (AUC) as an indicator to evaluate the quality of the model. The results show that XgBoost performs well.

Historical data is applied to model training. In our food delivery system, all orders are delayed with a fixed amount of time. Some of them were successfully grouped, while others were not. An order which is successfully grouped is labeled as 1, otherwise 0. The characteristics of these orders are described by some features.

Features consist of three parts. First, order characteristics are applied in the model. This includes basic features, such as delivery distance. Second, time related features can provide information for future order density. In our food-delivery system, the distribution of orders changes significantly over time. Historical statistics are one of the best estimates of future order distribution. Third, spatial features are also very important. Order distribution information provides direct geographic insight for estimating (Fig. 4).

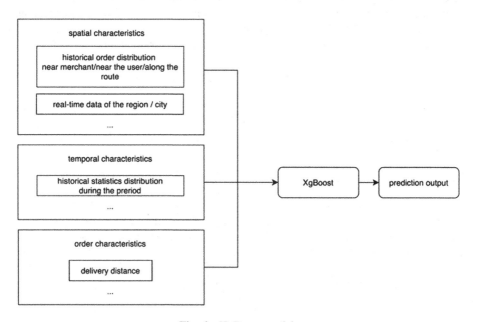

Fig. 4. XgBoost model.

The prediction model performs well, AUC reaches 0.80. In the following Figure, we can see output value of the model perfectly fits actual grouping probability (Fig. 5).

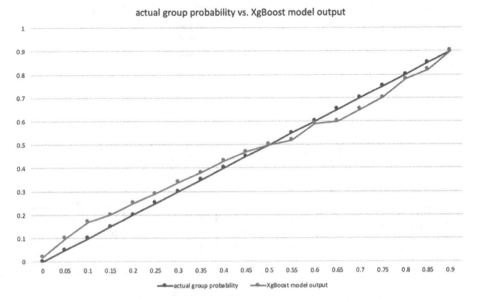

Fig. 5. Performance of XgBoost model.

4.3 Decision Optimization

Decision optimization has two parts: offline optimization and online usage.

First, we introduce offline optimization part. To solve the knapsack problem raised in Sect. 4.1, we use dynamic programming algorithm. We find the optimal solution is to delay if grouping possibility is much higher than increment of delivery time. We define the ratio of grouping possibility to increment of delivery time by:

$$R_i = \frac{\hat{g}_i(y_i)}{\alpha l_i} \tag{14}$$

The optimal solution is described by:

$$R_i > \underline{R} = \min\{R_{y_i=1}\} \tag{15}$$

Here $\underline{R}$ is the minimum ratio allowed for delaying an order.

With offline optimization, we get $\underline{R}$ and use it online to decide whether to delay an order. If an order has a higher ratio of grouping possibility to increment of delivery time, i.e., $R_i > \underline{R}$, then we decide to delay. Otherwise an order is dispatched immediately. At the same time, we need to handle overtime risk by not delaying orders if ETR > ETA. Figure 6 shows this process. At each decision time, we estimate grouping possibility and ETR with separate models. Then by comparing ratio of each order with offline generated minimum ratio, and comparing ETR with ETA, we decide whether to delay an order.

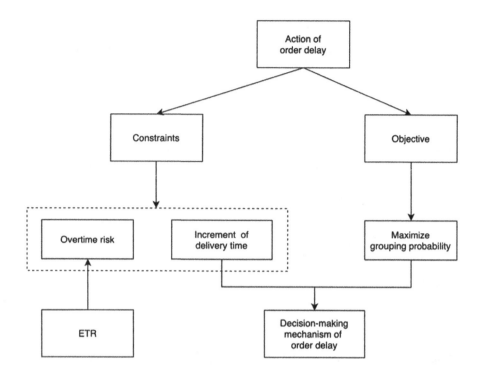

Fig. 6. Order delay decision link of online dispatch system.

5 Experimental Results

In this section, we provide numerical experiments to compare our algorithm with another algorithm, which uses a fixed two minutes delaying strategy.

We use the following metrics to evaluate the proposed method: grouping rate, overtime rate and average delivery time. Our goal is to increase grouping rate, related to driver efficiency. At the same time, customer experience constraints are considered, including overtime rate and average delivery time. For experiment settings, we first divide a city into two similar parts, and randomly select one part as experimental group and the other as control group. The experiment period is one week.

The results are shown in Table 1. From the online AB test results, we can find a significant improvement of grouping rate over all cities. At the same time, overtime rate and average delivery time is similar to the control group, which indicates that customer experience constraints are not violated.

Table 1. Numerical results from AB test in five cities.

City	Grouping rate/pp	Overtime rate/pp	Average delivery time/%
A	3.51	0.11	−0.52
B	1.53	0.14	−0.63
C	5.38	−0.08	−0.21
D	3.44	−0.05	−0.38
E	4.51	0.07	−0.53

In addition, we also examine the relationship between delay and grouping to show the effectiveness of our proposed method. We denote the ratio of orders eventually grouped to the orders we delayed by grouping rate after delay.

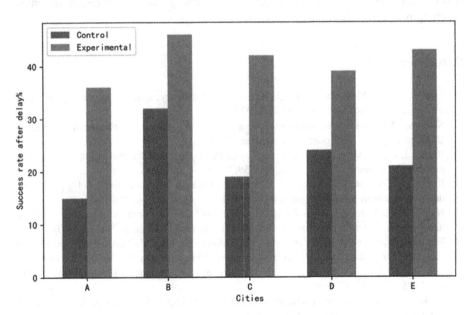

Fig. 7. Comparison of grouping rate after delay from AB test in five experiment cities.

From Fig. 7., it can be found that compared to a fixed two minutes delay, our proposed delay method can achieve a higher success rate. The average grouping rate after delay of experimental and control group is 41.20% and 22.19%, respectively. This indicates that, the proposed method can delay the orders with higher grouping probability in our dispatch system.

6 Conclusion

In food delivery system, delaying orders is a double-edged sword. Delaying orders increase driver efficiency by grouping more orders before dispatching. But at the same time, delaying may harm customer experience by postponing order delivery time. Our work proposes a prediction-based optimization approach for making delaying decisions in food delivery system. The prediction part uses XgBoost machine learning model, in order to predict grouping possibility and customer experience. The optimization part simplifies the optimization problem to a knapsack problem. Then we solve the knapsack problem offline to tune parameters and uses these parameters online to make delaying decisions. With extensive field experiment, we show that our approach can significantly improve driver efficiency by increasing grouping possibility. At the same time, our results show that customer experience is not harmed. This paper provides a novel framework, which combines machine learning and optimization in online matching problem with delaying decisions.

References

1. Liu, S., He, L., Max Shen, Z.J.: On-time last-mile delivery: order assignment with travel-time predictors. Management Science (2020)
2. Slabinac, M.: Innovative solutions for a "last-mile" delivery–a European experience. Business Logistics in Modern Management (2015)
3. Baldacci, R., Bartolini, E., Mingozzi, A.: An exact algorithm for the pickup and delivery problem with time windows. Oper. Res. **59**(2), 414–426 (2011)
4. Ropke, S., Cordeau, J.F., Laporte, G.: Models and branch-and-cut algorithms for pickup and delivery problems with time windows. Netw. Int. J. **49**(4), 258–272 (2007)
5. Nanry, W.P., Barnes, J.W.: Solving the pickup and delivery problem with time windows using reactive tabu search. Transp. Res. Part B: Methodol. **34**(2), 107–121 (2000)
6. Tchoupo, M.N., Yalaoui, A., Amodeo, L., Yalaoui, F., Lutz, F.: Ant colony optimization algorithm for pickup and delivery problem with time windows. In: Sforza, A., Sterle, C. (eds.) ODS 2017. SPMS, vol. 217, pp. 181–191. Springer, Cham (2017). https://doi.org/10.1007/978-3-319-67308-0_19
7. Pankratz, G.: A grouping genetic algorithm for the pickup and delivery problem with time windows. OR Spectr. **27**(1), 21–41 (2005)
8. Lu, Q., Dessouky, M.M.: A new insertion-based construction heuristic for solving the pickup and delivery problem with time windows. Eur. J. Oper. Res. **175**(2), 672–687 (2006)
9. Jiang, Z., Gu, H.Y., Xi, Y.G.: Quick heuristic algorithm for solving PDPTW problem. Control Eng. Chin. **13**(5), 413–415 (2006)
10 Ameddah, M.A., Moussab, D.: Multi-delivery dispatching system based on heuristic and machine learning mechanisms
11. Wen, M., Cordeau, J.F., Laporte, G., Larsen, J.: The dynamic multi-period vehicle routing problem. Comput. Oper. Res. **37**(9), 1615–1623 (2010)
12. Soman, J.T., Patil, R.J.: A scatter search method for heterogeneous fleet vehicle routing problem with release dates under lateness dependent tardiness costs. Expert Syst. Appl. **150**, 113302 (2020)

13. Zhu, L., et al.: Order fulfillment cycle time estimation for on-demand food delivery. In: Proceedings of the 26th ACM SIGKDD International Conference on Knowledge Discovery Data Mining, pp. 2571–2580 (2020)

14. Klapp, M.A., Erera, A.L., Toriello, A.: The one-dimensional dynamic dispatch waves problem. Transp. Sci. **52**(2), 402–415 (2018)

15. Xu, J., Rahmatizadeh, R., Bölöni, L., Turgut, D.: Taxi dispatch planning via demand and destination modeling. In: 2018 IEEE 43rd Conference on Local Computer Networks (LCN), pp. 377–384. IEEE (2018)

16. Wang, Z., Qin, Z., Tang, X., Ye, J., Zhu, H.: Deep reinforcement learning with knowledge transfer for online rides order dispatching. In: 2018 IEEE International Conference on Data Mining (ICDM), pp. 617–626. IEEE (2018)

17. Li, M., et al.: Efficient ridesharing order dispatching with mean field multi-agent reinforcement learning. In: The World Wide Web Conference, pp. 983–994 (2019)

18. Tang, X., et al.: A deep value-network based approach for multi-driver order dispatching. In: Proceedings of the 25th ACM SIGKDD International Conference on Knowledge Discovery & Data Mining, pp. 1780–1790 (2019)

19. Al-Abbasi, A.O., Ghosh, A., Aggarwal, V.: Deeppool: Distributed model-free algorithm for ride-sharing using deep reinforcement learning. IEEE Trans. Intell. Transp. Syst. **20**(12), 4714–4727 (2019)

20. Shi, J., Gao, Y., Wang, W., Yu, N., Ioannou, P.A.: Operating electric vehicle fleet for ride-hailing services with reinforcement learning. IEEE Trans. Intell. Transp. Syst. **21**(11), 4822–4834 (2019)

21. Ke, J., Yang, H., Ye, J.: Learning to delay in ride-sourcing systems: a multi-agent deep reinforcement learning framework. IEEE Transactions on Knowledge and Data Engineering (2020)

22. Stricker, N., Kuhnle, A., Sturm, R., Friess, S.: Reinforcement learning for adaptive order dispatching in the semiconductor industry. CIRP Ann. **67**(1), 511–514 (2018)

23. Morariu, C., Morariu, O., Răileanu, S., Borangiu, T.: Machine learning for predictive scheduling and resource allocation in large scale manufacturing systems. Comput. Ind. **120**, 103244 (2020)

24. Chen, T., Guestrin, C.: Xgboost: a scalable tree boosting system. In: Proceedings of the 22nd ACM SIGKDD International Conference on Knowledge Discovery Data Mining, pp. 785–794 (2016)

Variational EM Algorithm for Student-*t* Mixtures of Gaussian Processes

Xiangyang Guo, Xiaoyan Li, and Jinwen Ma$^{(\boxtimes)}$

Department of Information and Computational Sciences, School of Mathematical
Sciences and LMAM, Peking University, Beijing 100871, China
jwma@math.pku.edu.cn

Abstract. Student-*t* mixture of Gaussian processes (TMGP) extends the conventional mixture of Gaussian processes (MGP) by using Student-*t* mixture as the input distribution instead of Gaussian mixture for robust regression and the un *v*-Hardcut EM algorithm has already been established for Student-*t* mixtures of Gaussian processes. However, this Hardcut EM algorithm takes an approximation of the *Q*-function with the maximum a posteriori estimate of the hidden variables in the E-step, and separately optimizes the degrees of freedom of Student-*t* distributions and the hyperparameters of the covariance functions in the M-step. In order to get rid of these problems, we propose a variational EM (VEM) algorithm for Student-*t* mixtures of Gaussian processes from the viewpoint of variational inference. It is demonstrated by the experimental results on three datasets that our proposed VEM algorithm is effective and outperforms the un *v*-Hardcut EM algorithm.

Keywords: Gaussian process · Student-*t* mixtures of gaussian processes · EM algorithm · Variational inference

1 Introduction

Gaussian process (GP) is an effective and popular tool for Bayesian nonlinear nonparametric regression and classification, e.g., modeling the inverse dynamics of a robot arm and classifying the images of handwritten digits [1]. However, it cannot deal with data with multi-modality and suffers from an expensive training time complexity scaling as $\mathcal{O}(N^3)$, where N is the number of training samples. To overcome these limitations, the mixture of Gaussian processes (MGP) [2] was proposed and then many researches have been devoted to learning its parameters, analyzing its performance, and applying it to the real-world problems [3–11].

Unfortunately, the MGP model may suffer from strong overlaps between GP components, which frequently appear due to the fact that the true input distributions often have long or heavy tails. As Gaussian distributions have relatively short or light tails, the conventional MGP model cannot perform well on the datasets where the actual components are strongly overlapped. In order to tackle this problem, Li et al. [11] established the Student-*t* mixture of Gaussian processes (TMGP) model by using Student-*t* distributions as the input distributions instead of Gaussian distributions.

© Springer Nature Switzerland AG 2021
D.-S. Huang et al. (Eds.): ICIC 2021, LNCS 12837, pp. 552–563, 2021.
https://doi.org/10.1007/978-3-030-84529-2_47

In fact, student-t distribution has longer tail than a Gaussian and therefore it is much less sensitive than Gaussian distribution when there appear some outliers [11, 12].

Moreover, Li et al. [11] proposed an efficient un v-Hardcut EM algorithm to train the TMGP model, and the experimental results demonstrated the robustness of the TMGP model. However, the un v-Hardcut EM algorithm has two major drawbacks. Firstly, in the E-step, it just uses the maximum a posteriori estimate of the latent indicators to calculate an approximation of the Q-function, which may cause large inaccuracy. Secondly, the degrees of freedom of Student-t distributions and the hyperparameters of the covariance functions are optimized separately in the M-step.

Taking the success of the variational EM algorithm in training MGP models into consideration, we propose a variational EM (VEM) algorithm for TMGPs to avoid those limitations in this paper. Firstly, a more accurate approximation of the Q-function is calculated by replacing the true posterior with a variational posterior. Secondly, we optimize the approximation with respect to the degrees of freedom of Student-t distributions and the hyperparameters of the covariance functions together using gradient ascent method. The dependency between observation values is the main obstacle to conducting the variational inference, and we kick away it by reconstructing the standard GP through the linear GP model in the way of sparse GP models [4, 5, 9, 13–15]. We test the proposed VEM algorithm on three datasets and the experimental results demonstrate that it is effective and outperforms the un v-Hardcut EM algorithm.

The rest of this paper is organized as follows. Section 2 briefly introduces the GP, MGP and TMGP models. In Sect. 3, we present the VEM algorithm in detail. The experimental results are further summarized in Sect. 4. Finally, we conclude in Sect. 5.

2 Related Models

First we briefly introduce the GP model. GP is a classic stochastic process in which any finite number of states are subject to a joint Gaussian distribution [1]. In order to specify a GP, $\{f(\mathbf{x})|\mathbf{x} \in \mathbb{X}\}$, we only need to define a mean function $m(\mathbf{x})$ and a covariance function $c(\mathbf{x}, \mathbf{x}')$, where

$$m(\mathbf{x}) = \mathbb{E}[f(\mathbf{x})], c(\mathbf{x}, \mathbf{x}') = \mathbb{E}[(f(\mathbf{x}) - m(\mathbf{x}))(f(\mathbf{x}') - m(\mathbf{x}'))]. \qquad (1)$$

Here, we simply assume $m(\mathbf{x})$ to be zero and utilize the squared-exponential covariance function which is defined by

$$c(\mathbf{x}, \mathbf{x}'; \boldsymbol{\theta}) = \theta_0^2 \exp\left\{ -\frac{1}{2} \sum_{i=1}^{d} \frac{(x_i - x_i')^2}{\theta_i^2} \right\}, \qquad (2)$$

where d is the dimensionality of $\mathbf{x}$ and $\theta_i, i = 0, 1, \ldots, d$ are positive hyperparameters. In this way, a GP can be denoted as

$$f(\mathbf{x}) \sim \mathcal{GP}(0, c(\mathbf{x}, \mathbf{x}'; \boldsymbol{\theta})). \tag{3}$$

Suppose that we have a training dataset $\mathcal{D} = \{(\mathbf{x}_n, y_n)\}_{n=1}^N$, in which y_n is the noisy observation of $f_n = f(\mathbf{x}_n)$, i.e. $y_n = f_n + \epsilon_n$ with $\epsilon_n \sim \mathcal{N}(0, r^{-1})$, where r^{-1} denotes the noise level. Let $\mathbf{X}$ and $\mathbf{y}$ represent all inputs and all outputs, respectively. The marginal distribution is given by

$$\mathbf{y}|\mathbf{X} \sim \mathcal{N}(0, c(\mathbf{X}, \mathbf{X}) + r^{-1}\mathbf{I}_N), \tag{4}$$

where $c(\mathbf{X}, \mathbf{X})$ is an $N \times N$ covariance matrix consisting of covariance functions between training samples and $\mathbf{I}_N$ is an $N \times N$ identity matrix.

Next, a short description of the MGP model is given. Suppose that there are K GPs in a MGP model, having hyperparameters $\boldsymbol{\theta}_k$, noise level r_k^{-1} and mixing proportions π_k, where $\pi_k > 0$ and $\sum_{k=1}^K \pi_k = 1$, respectively. An indicator z_n is introduced for each $\mathbf{x}_n$, and we have

$$p(z_n = k) = \pi_k, \tag{5}$$

$$\mathbf{x}_n|z_n = k \sim \mathcal{N}(\boldsymbol{\mu}_k, \boldsymbol{\Lambda}_k^{-1}), \tag{6}$$

and

$$y_n|z_n = k \sim \mathcal{GP}(0, c(\mathbf{x}, \mathbf{x}'; \boldsymbol{\theta}_k) + r_k^{-1}\delta(\mathbf{x}, \mathbf{x}')), \tag{7}$$

where $\delta(\mathbf{x}, \mathbf{x}')$ is the Kronecker delta function.

Compared to MGP models, the input distributions in a TMGP model are replaced with Student-t distributions St $(\boldsymbol{\mu}_k, \boldsymbol{\Lambda}_k, v_k), k = 1, 2, \ldots, K$. For simplicity, we introduce an latent variable τ_n for each $\mathbf{x}_n$. If

$$\tau_n|z_n = k \sim \Gamma(v_k/2, v_k/2) \text{ and } \mathbf{x}_n|\tau_n, z_n = k \sim \mathcal{N}\left(\boldsymbol{\mu}_k, (\tau_n \boldsymbol{\Lambda}_k)^{-1}\right), \tag{8}$$

then it is obtained that $\mathbf{x}_n|z_n = k \sim \text{St}(\boldsymbol{\mu}_k, \boldsymbol{\Lambda}_k, v_k)$ by integrating out τ_n.

3 Variational EM Algorithm

3.1 Linear GP Model

From Eq. (4), we see that there is dependency between $y_n, n = 1, 2, \ldots, N$, which makes variational inference infeasible. In order to eliminate the dependency, the linear GP model is introduced to reconstruct a standard GP [4, 5, 9].

It is assumed that each GP has a support set $\mathcal{D}_k$ of size $M \ll N$ chosen from $\mathcal{D}$ according to certain criteria to be discussed in Sect. 3.4. We introduce K latent random vectors $\gamma_k, k = 1, 2, \ldots, K$ with

$$\gamma_k \sim \mathcal{N}\left(0, \mathbf{C}_k^{-1}\right), \tag{9}$$

where $\mathbf{C}_k$ is an $M \times M$ covariance matrix consisting of covariance functions between training samples in $\mathcal{D}_k$. Suppose there is a sample $(\mathbf{x}_n, y_n)$ from kth GP. Let

$$p\left(y_n | z_n = k, \mathbf{x}_n, \gamma_k, r_k^{-1}\right) = \mathcal{N}\left(\mathbf{c}_k(\mathbf{x}_n)^T \gamma_k, r_k^{-1}\right), \tag{10}$$

where $\mathbf{c}_k(\mathbf{x}_n)$ is a column vector of size M consisting of covariance functions between $(\mathbf{x}_n, y_n)$ and training samples in $\mathcal{D}_k$. We see that outputs are independent of each other given γ_k for k th GP. Assuming that $K = 1$ and $M = N$, we obtain that

$$p(\mathbf{y}|\mathbf{X}) = \mathcal{N}\left(0, c(\mathbf{X}, \mathbf{X}) + r^{-1}\mathbf{I}_N\right), \tag{11}$$

the same as Eq. (4), which demonstrates the rationality of linear GP model.

3.2 Variational Inference

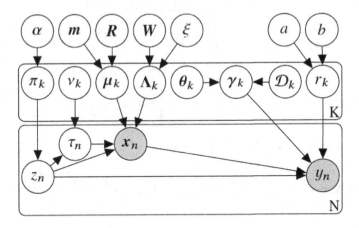

Fig. 1. The graphical model representation for the proposed model.

Figure 1 shows the graphical model representation for the proposed model. In the E-step of our proposed VEM algorithm, a variational posterior is calculated to approximate the true one. The priors of π, μ_k, Λ_k, γ_k and r_k are Dir (α), $\mathcal{N}\left(\mathbf{m}, \mathbf{R}^{-1}\right)$, Wishart $(\mathbf{W}, \xi)$, $\mathcal{N}\left(0, \mathbf{C}_k^{-1}\right)$ and $\Gamma(a, b)$, respectively.

Let $\Omega = \{\pi, \mu, \Lambda, \gamma, \mathbf{r}, \mathbf{z}, \tau\}$ represent all latent variables. The complete data log-likelihood is given by

$$\begin{aligned}
\ln p(\Omega, \mathbf{X}, \mathbf{y}) &= \ln p(\pi) + \ln p(\mu) + \ln p(\Lambda) + \ln p(\gamma) + \ln p(\mathbf{r}) \\
&+ \ln p(\mathbf{z}|\pi) + \ln p(\tau|\mathbf{z}) + \ln p(\mathbf{X}|\mathbf{z}, \tau, \mu, \Lambda) + \ln p(\mathbf{y}|\mathbf{z}, \mathbf{X}, \gamma, \mathbf{r}).
\end{aligned} \tag{12}$$

Omitting terms independent of variables of interest, we have

$$\ln p(\pi) = \sum_{k=1}^{K} (\alpha_k - 1) \ln \pi_k, \tag{13}$$

$$\ln p(\mu) = -\frac{1}{2} \sum_{k=1}^{K} (\mu_k - \mathbf{m})^T \mathbf{R}(\mu_k - \mathbf{m}), \tag{14}$$

$$\ln p(\Lambda) = \frac{\xi - d - 1}{2} \sum_{k=1}^{K} \ln|\Lambda_k| - \frac{1}{2} \sum_{k=1}^{K} \mathrm{tr}(\mathbf{W}^{-1}\Lambda_k), \tag{15}$$

$$\ln p(\gamma) = \frac{1}{2} \sum_{k=1}^{K} \ln|\mathbf{C}_k| - \frac{1}{2} \sum_{k=1}^{K} \gamma_k^T \mathbf{C}_k \gamma_k, \tag{16}$$

$$\ln p(\mathbf{r}) = (a - 1) \sum_{k=1}^{K} \ln r_k - b \sum_{k=1}^{K} r_k, \tag{17}$$

$$\ln p(\mathbf{z}|\pi) = \sum_{n=1}^{N} \sum_{k=1}^{K} 1_{\{z_n=k\}} \ln \pi_k, \tag{18}$$

$$\ln p(\tau|\mathbf{z}) = \sum_{n=1}^{N} \sum_{k=1}^{K} 1_{\{z_n=k\}} \left\{ \frac{v_k}{2} \ln \frac{v_k}{2} - \ln \Gamma\left(\frac{v_k}{2}\right) + \left(\frac{v_k}{2} - 1\right) \ln \tau_n - \frac{v_k}{2} \tau_n \right\}, \tag{19}$$

$$\ln p(\mathbf{X}|\mathbf{z}, \tau, \mu, \Lambda) = \sum_{n=1}^{N} \sum_{k=1}^{K} 1_{\{z_n=k\}} \left\{ \frac{1}{2} \ln|\Lambda_k| + \frac{d}{2} \ln \tau_n - \frac{\tau_n \omega_{kn}}{2} \right\}, \tag{20}$$

where $\omega_{kn} = (\mathbf{x}_n - \mu_k)^T \Lambda_k (\mathbf{x}_n - \mu_k)$, and

$$\ln p(\mathbf{y}|\mathbf{z}, \mathbf{X}, \gamma, \mathbf{r}) = \sum_{n=1}^{N} \sum_{k=1}^{K} 1_{\{z_n=k\}} \left\{ \frac{1}{2} \ln r_k - \frac{r_k (y_n - \mathbf{c}_k(\mathbf{x}_n)^T \gamma_k)^2}{2} \right\}. \tag{21}$$

Based on the widely used mean field theory, we assume the variational distribution takes the form of

$$q(\Omega) = q(\pi) \prod_{k=1}^{K} \{q(\mu_k)q(\Lambda_k)q(\gamma_k)q(r_k)\} \prod_{n=1}^{N} \{q(z_n)q(\tau_n)\}. \tag{22}$$

Because the process of calculating the factors is analogous to that in Sun et al. [5], we omit the details and just present the results here.

$q(\boldsymbol{\pi}) \sim \mathrm{Dir}(\boldsymbol{\beta})$, where

$$\beta_k = \alpha_k + \sum\nolimits_{n=1}^{N} q_{nk}(q_{nk} = q(z_n = k)). \tag{23}$$

$q(\boldsymbol{\mu}_k) \sim \mathcal{N}(\mathbf{u}_k, \mathbf{S}_k^{-1})$, where

$$\mathbf{S}_k = \mathbf{R} + \sum\nolimits_{n=1}^{N} q_{nk}\mathbb{E}_q\tau_n\mathbb{E}_q\Lambda_k, \mathbf{u}_k = \mathbf{S}_k^{-1}\left(\mathbf{Rm} + \mathbb{E}_q\Lambda_k \sum_{n=1}^{N} q_{nk}\mathbb{E}_q\tau_n\mathbf{x}_n\right). \tag{24}$$

$q(\boldsymbol{\Lambda}_k) \sim \mathrm{Wishart}(\mathbf{Q}_k, \eta_k)$, where

$$\mathbf{Q}_k = \left(\mathbf{W}^{-1} + \sum\nolimits_{n=1}^{N} q_{nk}\mathbb{E}_q\tau_n\mathbb{E}_q\left((\mathbf{x}_n - \boldsymbol{\mu}_k)(\mathbf{x}_n - \boldsymbol{\mu}_k)^T\right)\right)^{-1}, \eta_k = \xi + \sum\nolimits_{n=1}^{N} q_{nk}. \tag{25}$$

$q(\boldsymbol{\gamma}_k) \sim \mathcal{N}(\mathbf{v}_k, \mathbf{T}_k^{-1})$, where

$$\mathbf{T}_k = \mathbf{C}_k + \mathbb{E}_q r_k \sum\nolimits_{n=1}^{N} q_{nk}\mathbf{c}_k(\mathbf{x}_n)\mathbf{c}_k(\mathbf{x}_n)^T, \mathbf{v}_k = \mathbb{E}_q r_k \mathbf{T}_k^{-1} \sum\nolimits_{n=1}^{N} q_{nk}y_n\mathbf{c}_k(\mathbf{x}_n). \tag{26}$$

$q(r_k) \sim \Gamma(\phi_k, \zeta_k)$, where

$$\phi_k = a + \frac{1}{2}\sum\nolimits_{n=1}^{N} q_{nk}, \zeta_k = b + \frac{1}{2}\sum\nolimits_{n=1}^{N} q_{nk}\mathbb{E}_q\left(y_n - \mathbf{c}_k(\mathbf{x}_n)^T\boldsymbol{\gamma}_k\right)^2. \tag{27}$$

$q(z_n = k) \propto e^{\lambda_{nk}}$, where

$$\begin{aligned}
\lambda_{nk} = {} & \mathbb{E}_q \ln \pi_k + \frac{v_k}{2}\ln\frac{v_k}{2} - \ln\Gamma\left(\frac{v_k}{2}\right) + \left(\frac{v_k + d}{2} - 1\right)\mathbb{E}_q \ln \tau_n \\
& - \frac{1}{2}\mathbb{E}_q\tau_n\left(v_k + \mathrm{tr}\left(\mathbb{E}_q\left((\mathbf{x}_n - \boldsymbol{\mu}_k)(\mathbf{x}_n - \boldsymbol{\mu}_k)^T\right)\mathbb{E}_q\Lambda_k\right)\right) + \frac{1}{2}\mathbb{E}_q \ln|\Lambda_k| \\
& + \frac{1}{2}\mathbb{E}_q \ln r_k - \frac{1}{2}\mathbb{E}_q r_k \mathbb{E}_q\left(y_n - \mathbf{c}_k(\mathbf{x}_n)^T\boldsymbol{\gamma}_k\right)^2.
\end{aligned} \tag{28}$$

$q(\tau_n) \sim \Gamma(h_n, s_n)$, where

$$\begin{aligned}
h_n &= \tfrac{1}{2}\sum\nolimits_{k=1}^{K} q_{nk}v_k + \tfrac{d}{2}, \\
s_n &= \tfrac{1}{2}\sum\nolimits_{k=1}^{K} q_{nk}\left(v_k + \mathrm{tr}\left(\mathbb{E}_q\left((\mathbf{x}_n - \boldsymbol{\mu}_k)(\mathbf{x}_n - \boldsymbol{\mu}_k)^T\right)\mathbb{E}_q\Lambda_k\right)\right)
\end{aligned} \tag{29}$$

3.3   Approximation of Q-Function and Its Maximization

Let Θ_t be the current hyperparameters, and $q_t(\Omega)$ the current variational posterior. We obtain that

$$
\hat{Q}(\Theta|\Theta_t) = \mathbb{E}_{q_t} \ln p(\Omega, \mathbf{X}, \mathbf{y}) = \frac{1}{2} \sum_{k=1}^{K} \left(\ln|\mathbf{C}_k| - \text{tr}\left(\mathbf{C}_k \mathbb{E}_{q_t}\left(\boldsymbol{\gamma}_k \boldsymbol{\gamma}_k^T\right)\right) \right)
$$
$$
+ \sum_{n=1}^{N} \sum_{k=1}^{K} q_{tnk} \left(\frac{v_k}{2} \ln \frac{v_k}{2} - \ln \Gamma\left(\frac{v_k}{2}\right) + \left(\frac{v_k}{2} - 1\right) \mathbb{E}_{q_t} \ln \tau_n - \frac{v_k}{2} \mathbb{E}_{q_t} \tau_n \right) \quad (30)
$$
$$
- \frac{1}{2} \sum_{n=1}^{N} \sum_{k=1}^{K} q_{tnk} \mathbb{E}_{q_t} r_k \mathbb{E}_{q_t} \left(y_n - \mathbf{c}_k(\mathbf{x}_n)^T \boldsymbol{\gamma}_k \right)^2 + \text{const},
$$

where the terms independent of Θ are absorbed into the constant. We utilize gradient ascent method to find better parameters in the M-step and the derivatives of $\hat{Q}(\Theta|\Theta_t)$ with respect to Θ are presented as follows.

$$
\frac{\partial \hat{Q}}{\partial v_k} = \frac{1}{2} \sum_{n=1}^{N} q_{tnk} \left(\ln \frac{v_k}{2} + 1 - \psi\left(\frac{v_k}{2}\right) + \mathbb{E}_{q_t} \ln \tau_n - \mathbb{E}_{q_t} \tau_n \right), \quad (31)
$$

where $\psi(\cdot)$ is the digamma function.

$$
\frac{\partial \hat{Q}}{\partial \theta_{kj}} = \text{tr}\left(\frac{\partial \hat{Q}}{\partial \mathbf{C}_k} \frac{\partial \mathbf{C}_k}{\partial \theta_{kj}} \right) + \sum_{n=1}^{N} \left(\frac{\partial \hat{Q}}{\partial \mathbf{c}_k(\mathbf{x}_n)} \right)^T \frac{\partial \mathbf{c}_k(\mathbf{x}_n)}{\partial \theta_{kj}}
$$
$$
= \frac{1}{2} \text{tr}\left(\left(\mathbf{C}_k^{-1} - \mathbb{E}_{q_t}\left(\boldsymbol{\gamma}_k \boldsymbol{\gamma}_k^T\right)\right) \frac{\partial \mathbf{C}_k}{\partial \theta_{kj}} \right) \quad (32)
$$
$$
+ \mathbb{E}_{q_t} r_k \sum_{n=1}^{N} q_{tnk} \left(y_n \mathbb{E}_{q_t} \boldsymbol{\gamma}_k - \mathbb{E}_{q_t}\left(\boldsymbol{\gamma}_k \boldsymbol{\gamma}_k^T\right) \mathbf{c}_k(\mathbf{x}_n) \right)^T \frac{\partial \mathbf{c}_k(\mathbf{x}_n)}{\partial \theta_{kj}}.
$$

3.4   Support Set Selection

At the beginning of the proposed VEM algorithm, K support sets are randomly chosen form $\mathcal{D}$. To improve the training result, we update $\mathcal{D}_k$ according to certain criteria after completing the VEM algorithm, and then run it again, which is repeated until appropriate hyperparameters are obtained.

Regarding $\mathbf{T}_k$ in Eq. (26) as a function of $\mathcal{D}_k$ with hyperparameters and variational posteriors of other hidden variables fixed, to obtain a better support set $\mathcal{D}_k$ for the k th GP, we maximize the variational probability density $q(\boldsymbol{\gamma}_k)$ at its mean [4, 5, 9], i.e. maximize the determinant

$$
|\mathbf{T}_k| = \left| \mathbf{C}_k + \mathbb{E}_q r_k \sum_{n=1}^{N} q_{nk} \mathbf{c}_k(\mathbf{x}_n) \mathbf{c}_k(\mathbf{x}_n)^T \right|. \quad (33)
$$

Apparently, determining the best support sets is an NP problem, because there are $\binom{N}{M}$ candidate sets for each $\mathcal{D}_k$. Thus, a greedy algorithm, seen in Algorithm 1, is used to alleviate the computational burden.

Algorithm 1: Greedy Algorithm for Support Set Selection

Require: Training set $\mathcal{D}$, Support set size M, Candidate set size P.
  Initialize $\mathcal{D}_k = \emptyset$.
  Repeat
    Let $\mathcal{C}_k$ be the set of P candidates randomly chosen from $\mathcal{D} \setminus \mathcal{D}_k$.
    Find $s_k = \text{argmax}_{\{s'\} \cup \mathcal{D}_k, s' \in \mathcal{C}_k} |\mathbf{T}_k|$.
    $\mathcal{D}_k = \{s_k\} \cup \mathcal{D}_k$.
  Until $|\mathcal{D}_k| = M$.
  Output: $\mathcal{D}_k, k = 1, 2, \dots, K$.

3.5 Predictive Distribution

For a new input x^*, the predictive distribution is given by

$$
\begin{aligned}
p(y^*|\mathcal{D}, \mathbf{x}^*) &= \int p(y^*|\mathbf{x}^*, \boldsymbol{\Omega}) p(\boldsymbol{\Omega}|\mathcal{D}) d\boldsymbol{\Omega} \\
&\approx \int p(y^*|\mathbf{x}^*, \boldsymbol{\Omega}) q(\boldsymbol{\Omega}) d\boldsymbol{\Omega} \\
&\approx p(y^*|\mathbf{x}^*, \mathbb{E}_q \boldsymbol{\Omega}) \\
&= \sum_{k=1}^{K} p(y^*, z^* = k|\mathbf{x}^*, \mathbb{E}_q \boldsymbol{\Omega}) \\
&= \sum_{k=1}^{K} p(z^* = k|\mathbf{x}^*, \mathbb{E}_q \boldsymbol{\Omega}) p(y^*|z^* = k, \mathbf{x}^*, \mathbb{E}_q \boldsymbol{\Omega}) \\
&= \sum_{k=1}^{K} \frac{p(z^* = k|\mathbb{E}_q \boldsymbol{\Omega}) p(\mathbf{x}^*|z^* = k, \mathbb{E}_q \boldsymbol{\Omega})}{\sum_{i=1}^{K} p(z^* = i|\mathbb{E}_q \boldsymbol{\Omega}) p(\mathbf{x}^*|z^* = i, \mathbb{E}_q \boldsymbol{\Omega})} p(y^*|z^* = k, \mathbf{x}^*, \mathbb{E}_q \boldsymbol{\Omega}).
\end{aligned}
\tag{34}
$$

There are two approximations in Eq. (34) to reduce the computation complexity. Firstly, the variational posterior is substituted for the true one. Secondly, the value of $p(y^*|\mathbf{x}^*, \boldsymbol{\Omega})$ at $\boldsymbol{\Omega} = \mathbb{E}_q \boldsymbol{\Omega}$ is used to approximate its expectation with respect to $q(\boldsymbol{\Omega})$.

4 Experimental Results

In this section, we perform experiments on three datasets. Given a test dataset $\{(\mathbf{x}_l, y_l)|l = 1, \dots, L\}$ and corresponding prediction values $\{\hat{y}_l|l = 1, \dots, L\}$, the prediction accuracy is measured by the root mean squareds error (RMSE), defined as

$$
\text{RMSE} = \sqrt[2]{\frac{1}{L} \sum_{l=1}^{L} (\hat{y}_l - y_l)^2}.
\tag{35}
$$

In the experiments, following Yuan et al. [9], we set $\boldsymbol{\alpha} = (1, \dots, 1)^T$, $\mathbf{W} = 100 \mathbf{I}_d$, $\xi = d$, $a = 0.01$ and $b = 0.0001$. Then, $\mathbf{m}$ and $\mathbf{R}$ are initialized as the mean and the precision matrix of training inputs, respectively. We set $\mathbf{v} = (5, \dots, 5)^T$ to make initial variances of Student-t distributions not very large. The hyperparameters of covariance functions are all initialized as 1. We conduct all experiments on a personal computer (Intel(R) Core(TM) i7-8750H CPU 2.20 GHz, 8G RAM). From the results to be discussed, we can see that our proposed method outperforms the un v-Hardcut EM algorithm, and it is proven again that the TMGP model performs better than MGPs.

4.1   On Synthetic Dataset

The synthetic dataset, shown in Fig. 2, is drawn from a TMGP model composed of three GP components with a prior distribution $\pi = (0.2, 0.3, 0.5)^T$. The three GP components are used to model the following three functions, respectively:

$$f_0(x) = -2.6 \sin \frac{2}{5} \pi x, x \in [-7.5, -2.5],$$

$$f_1(x) = 2.0 \sin \frac{2}{5} \pi x, x \in [-2.5, 2.5], \tag{36}$$

$$f_2(x) = -1.2 \sin \frac{2}{5} \pi x, x \in [2.5, 7.5].$$

Gaussian noises with standard deviation std = 0.3 are added to all training points and all test points. From Fig. 2 (left), overlaps appear around −2.5 and 2.5.

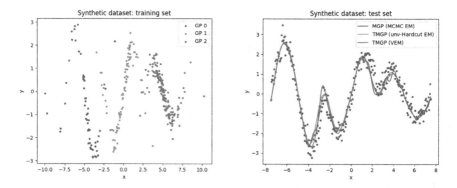

Fig. 2. Synthetic dataset

This dataset consists of 300 training points and 300 test points evenly distributed in $[-7.5, 7.5]$. The numbers of training samples from three GP components are 60, 90 and 150, respectively. M and P need not to be very large [4, 5, 9]. Here, M and P are set to be 30 and 40, respectively. The experimental result is presented in Fig. 2 (right) and Table 1. The code, also used in experiment 2 and 3, for training the MGP model via MCMC EM algorithm and training the TMGP model via un v-Hardcut EM algorithm is offered by Wu et al. [6] and Li et al. [11].

Table 1. RMSEs on three datasets

	Synthetic dataset	Coal gas dataset	Motorcycle dataset
MGP (MCMC EM)	0.4344	0.6011	24.20
TMGP (un v-Hardcut EM)	0.4222	0.5939	23.89
TMGP (VEM)	**0.3819**	**0.5854**	**21.61**

4.2   On Coal Gas Concentration Dataset

We also applied our VEM algorithm to a coal gas concentration dataset [11]. The dataset is divided into a training set, consisting of 200 points, and a test set, containing 113 points. The original data, where all observations are positive, is shown in the upper left of Fig. 3. It is necessary to make the observations distributed around 0.0 and change the scales of both inputs and observations, which are presented in the upper right. The transformed training points are plotted on the lower left. Finally, the three predicted curves are drawn on the lower right and corresponding RMSEs are described in Table 1. Here, we set K, M and P to be 4, due to the fact that the dataset apparently has 4 different parts, 20 and 20, respectively.

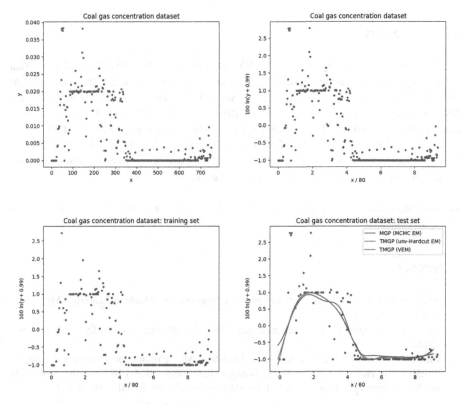

Fig. 3. Coal gas concentration dataset

4.3   On Motorcycle Dataset

Our proposed VEM algorithm was also applied to the public motorcycle dataset [6], which consists of 133 samples. Following Wu et al. [6], we use the whole dataset for both training and testing the models. Here, we set K, M and P to be 2, 3 0 and 3 0, respectively. We show the dataset and the experimental result in Fig. 4 and Table 1. The prediction result of the MGP model can be found in Wu et al. [6].

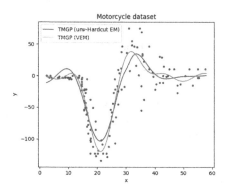

Fig. 4. Motorcycle dataset

5   Conclusion

We have proposed a new variational EM algorithm for the TMGP model. Similar complexity analyses can be found in Yuan et al. [9]. Compared to the un ν-Hardcut EM algorithm [11], a better posterior over latent variables can be obtained in the E-step through variational inference. It is demonstrated by the experimental results on three datasets that our proposed algorithm performs better than the un ν-Hardcut EM algorithm. However, in the process of training a TMGP model, the number of GP components cannot be determined automatically. Thus, a model selection algorithm [10] should be established for TMGP models in the future.

Acknowledgement. This work is supported by the National Key R & D Program of China (2018YFC0808305).

References

1. Rasmussen, C.E., Williams, C.K.I.: Gaussian Processes for Machine Learning. MIT Press, Cambridge (2006)
2. Tresp, V.: Mixtures of gaussian processes. Adv. Neural. Inf. Process. Syst. **13**, 654–660 (2000)
3. Chen, Z., Ma, J., Zhou, Y.: A precise hard-cut EM algorithm for mixtures of gaussian processes. In: Huang, D.-S., Jo, K.-H., Wang, L. (eds.) ICIC 2014. LNCS (LNAI), vol. 8589, pp. 68–75. Springer, Cham (2014). https://doi.org/10.1007/978-3-319-09339-0_7

4. Luo, C., Sun, S.: Variational mixtures of Gaussian processes for classification. In: Proceedings of the 26th International Joint Conference on Artificial Intelligence, vol. 26, pp. 4603–4609 (2017)
5. Sun, S., Xu, X.: Variational inference for infinite mixtures of Gaussian processes with applications to traffic flow prediction. IEEE Trans. Intell. Transp. Syst. **12**(2), 466–475 (2011)
6. Wu, D., Chen, Z., Ma, J.: An MCMC based EM algorithm for mixtures of gaussian processes. In: Hu, X., Xia, Y., Zhang, Y., Zhao, D. (eds.) ISNN 2015. LNCS, vol. 9377, pp. 327–334. Springer, Cham (2015). https://doi.org/10.1007/978-3-319-25393-0_36
7. Wu, D., Ma, J.: A two-layer mixture model of Gaussian processes functional regressions and its MCMC EM algorithm. In: IEEE Transactions on Neural Networks and Learning Systems, vol. 29 (2018)
8. Yang, Y., Ma, J.: An efficient EM approach to parameter learning of the mixture of Gaussian processes. In: Liu, D., Zhang, H., Polycarpou, M., Alippi, C., He, H. (eds.) ISNN 2011. LNCS, vol. 6676, pp. 165–174. Springer, Heidelberg (2011). https://doi.org/10.1007/978-3-642-21090-7_20
9. Yuan, C., Neubauer, C.: Variational mixture of Gaussian process experts. Adv. Neural. Inf. Process. Syst. **21**, 1897–1904 (2009)
10. Qiang, Z., Ma, J.: Automatic model selection of the mixtures of Gaussian processes for regression. In: Hu, X., Xia, Y., Zhang, Y., Zhao, D. (eds.) ISNN 2015. LNCS, vol. 9377, pp. 335–344. Springer, Cham (2015). https://doi.org/10.1007/978-3-319-25393-0_37
11. Li, X., Li, T., Ma, J.: The un -hardcut EM algorithm for non-central Student- mixtures of Gaussian processes. In: 15th IEEE International Conference on Signal Processing (ICSP), pp. 289–294 (2020)
12. Bishop, C.M.: Pattern Recognition and Machine Learning. Springer (2006)
13. Gibbs, M., Mackay, D.J.C.: Efficient implementation of Gaussian processes. Technical report, Cavendish Laboratory, Cambridge, UK (1997)
14. Smola, A.J., Bartlett, P.: Sparse greedy Gaussian process regression. Adv. Neural Inform. Proc. Syst. **13**, 619–625 (2001)
15. Snelson, E., Ghahramani, Z.: Sparse Gaussian processes using pseudo-inputs. Adv. Neural. Inf. Process. Syst. **18**, 1257–1264 (2006)

Ensemble Learning with Resampling for Imbalanced Data

Firuz Kamalov[1]($\boxtimes$), Ashraf Elnagar[2], and Ho Hon Leung[3]

[1] Faculty of Electrical Engineering, Canadian University Dubai, Dubai, UAE
firuz@cud.ac.ae
[2] Department of Computer Science, University of Sharjah, Sharjah, UAE
ashraf@sharjah.ac.ae
[3] Department of Mathematics, UAE University, Al Ain, UAE
hohon.leung@uaeu.ac.ae

Abstract. Imbalanced class distribution is an issue that appears in various applications. In this paper, we undertake a comprehensive study of the effects of sampling on the performance of bootstrap aggregating in the context of imbalanced data. Concretely, we carry out a comparison of sampling methods applied to single and ensemble classifiers. The experiments are conducted on simulated and real-life data using a range of sampling methods. The contributions of the paper are twofold: i) demonstrate the effectiveness of ensemble techniques based on resampled data over a single base classifier and ii) compare the effectiveness of different resampling techniques when used during the bagging stage for ensemble classifiers. The results reveal that ensemble methods overwhelmingly outperform single classifiers based on resampled data. In addition, we discover that NearMiss and random oversampling (ROS) are the optimal sampling algorithms for ensemble learning.

Keywords: Imbalanced data · Undersampling · Oversampling · Ensemble method · Data preprocessing sampling

1 Introduction

Imbalanced class distribution refers to a situation where one class considerably outnumbers another class. It appears in a variety of contexts including text classification, medical diagnostics, fraud detection and many others involving rare events. Skewed class distribution causes bias against the minority class in learning models. In particular, the prediction accuracy is often higher on the majority class relative to the minority class [23]. There exists a variety of approaches to deal with imbalanced data including feature selection, cost-sensitive learning, one-class learning, and others. One of the most popular such approaches is sampling the data to balance the class distributions. However, the use of sampling alone may result in a high variance classifier. To reduce the variance researchers have employed ensemble methods. In an ensemble method, the data is repeatedly sampled to obtain a collection of balanced datasets which are used to train base classifiers (weak learners). Then an ensemble rule is applied to aggregate the predictions of the base classifiers into a single response. In a

© Springer Nature Switzerland AG 2021
D.-S. Huang et al. (Eds.): ICIC 2021, LNCS 12837, pp. 564–578, 2021.
https://doi.org/10.1007/978-3-030-84529-2_48

basic ensemble method for imbalanced class distribution, the majority class is repeatedly undersampled to match the size of the minority data and a corresponding decision tree is constructed based on the balanced bootstrap sample. The process is carried multiple times depending on the user preference resulting in a collection of decision trees. Then the predictions of the resulting ensemble method are based on the majority or the mode of predictions of the constituent decision trees. The use of multiple tree to make predictions reduces the variance of the classifier. Since decision trees are very efficient algorithms ensemble methods do not experience any significant deterioration in execution time.

Ensemble learning with sampling for imbalanced data has been an active area of research with several authors proposing their own methods to combine ensemble and sampling techniques to improve classification performance. The goal of this paper is to carry out a comprehensive study of the effects of ensemble learning in regards to sampling for imbalanced data. Concretely, for each sampling technique, we compare classification performance between individual and ensemble tree classifiers. We consider a range of undersampling and oversampling techniques including random undersampling (RUS), NearMiss, random oversampling (ROS), synthetic minority oversampling technique (SMOTE), and ADASYN. To obtain broadly applicable results we carry out multiple numerical experiments using both simulated and real-life data. The real life-data covers a range of applications including astronomy, social science, medical diagnostics, and image recognition.

It is well known that bootstrap aggregating methods outperform single decision trees on balanced data. However, there does not exist an extensive study on bootstrap methods in the context of imbalanced data. The performance of a classifier on a balanced dataset is measured by its overall accuracy rate. By aggregating the predictions from a collection of classifiers an ensemble method reduces the variance of the predictions. As a result, it achieves improved accuracy on the testing set. On the other hand, the performance of a classifier on an imbalanced set is measured by AUC and F1-score. It is not immediately clear that the gains obtained by an ensemble classifier in accuracy rate will also materialize in AUC and F1-score. Therefore, a separate study is required to investigate the effects of bootstrap aggregating on the AUC and F1-score in the context of imbalanced data.

To fill the gap in the literature we test a range of sampling methods applied to a collection of simulated and real-life data. The results of the experiments show that ensemble methods consistently outperform single classifiers. In particular, we find that ensemble classifier outperforms single classifier on all tested datasets when using undersampling techniques. The ensemble classifier similarly produces better results when using oversampling techniques on the simulated data and 4 out 5 real-life datasets.

The paper is organized follows. In Sect. 2, we briefly review the current literature on ensemble methods and imbalanced data. In Sect. 3, we discuss the methodology and the results of the numerical experiments. Section 4 concludes the paper with a summary of our findings.

2 Literature Review

There exists a range of approaches in the literature to combat imbalance class distribution [17]. One of the approaches is based on selecting the optimal feature subset for identifying the minority class points [2, 3]. Another approach is based on balancing the class distribution in the dataset. Balancing the class distribution through resampling is arguably the most popular approach to dealing with imbalanced data. Resampling methods can be divided into two groups: undersampling and oversampling. Undersampling techniques consist of selecting a subset of the majority class that is of the same size as the minority class. The RUS algorithm is the simplest undersampling technique whereby a subset of the majority class is selected with uniform probability. In more intelligent approach called NearMiss the points of the majority class that are close to the border with the minority class are more likely to be selected [20]. Oversampling techniques consist of creating new minority points based on the existing minority points. A popular oversampling technique called SMOTE generates new points by linear interpolation between the existing minority points [4, 9]. An extension of the SMOTE algorithm called ADASYN generates new points in the same fashion as SMOTE but with greater emphasis on the minority points that lie in regions with high concentration of the majority class points [12]. In a more sophisticated approach, the authors in [14] employ kernel density estimation to estimate the underlying distribution of the existing minority points. The estimated distribution is used to generate the new minority points. Fusion approaches that combine multiple methods have also been used to deal with imbalanced data [25, 26]. As an alternative to resampling, other approaches such as weighted misclassification penalty and features selection can be employed to combat imbalanced data [16].

It has been widely accepted that ensemble approach reduces the variance of an estimator by introducing randomization into its construction procedure and then aggregating individual estimators. Ensemble methods such as bagging and random forest have been successfully modified to fit imbalanced data [19]. Each method applies a particular sampling technique during the bagging/boosting stage to balance the data. The most popular approach to constructing a bagging classifier is by random undersampling of the majority class to obtain a balanced subset which is used to train an estimator. An ensemble of estimators is created by repeating the sampling and training procedure. There exists several extensions of the random undersampling ensemble method. The authors in [10], propose a variation of the underbagging ensemble method by ap- plying evolutionary undersampling on the majority class. According to the proposed method new subsets of the majority class are sampled using evolutionary approach and base classifiers are constructed. The resulting ensemble method performs well on highly imbalanced data. Hido et al. [13] introduced Roughly Balanced Bagging (RBB), where the numbers of instances for both classes are determined in different ways. The number of minority points in each bootstrap set equals the size of the minority class while the number of majority points is determined according to the negative binomial distribution. RBB has been a popular ensemble learning tool that has been applied to various contexts [18]. Diez-Pastor et al. [6] proposed an ensemble method based on bootstrap sets with random class ratios. Each base classifier in the ensemble is trained on a dataset with arbitrarily chosen class ratio. The proposed method aims to increase the resulting AUC. The authors in [5] propose an ensemble learning technique based on the threshold moving technique which applies a threshold to the continuous output of a model. The proposed method preserves the natural class distribution of the data resulting in well-calibrated posterior probabilities. The authors in [22] consider adjusting the existing ensemble rules to account for data imbalance. In [24], the authors explore the relationship between the diversity of the base classifiers in an ensemble and the performance of the final ensemble. Investigation of three ensemble approaches based on undersampling, oversampling and SMOTE reveal a positive relationship between the diversity and recall rates on the minority class.

In [27], the authors used one-vs-one (OVO) approach together with ensemble methods to tackle imbalanced distribution for multi-class data. An empirical study of various ensemble methods indicates the high effectiveness of ensemble learning with OVO scheme in dealing with the multi-class imbalance classification problems. Concretely, the authors find that decision tree-based ensemble classifier SMOTEBoost achieves average accuracy of 0.7676 while neural network-based classifier SMOTE + AdaBoost achieves average accuracy of 0.7915. The authors in [1], build on the previous work by considering different base classifiers on each subset of the OVO decomposition. Thus, a different base model is selected for each data subset in the ensemble classifier. In addition, the authors replace the OVO strategy with Error Correcting Output Codes. The test results show that the proposed method produces a higher F1-score compared to the other 17 benchmark algorithms. Concretely, the average F1-score for the proposed method is 0.7750 while the accuracy is 0.9323.

3 Sampling Techniques

In this section we present the sampling algorithms used to balance data with skewed class distribution. There are two types of sampling methods: undersampling and oversampling. In undersampling, a subset of the majority class, of the same size as the minority class, is selected (Fig. 1, top). There exists a number of undersampling techniques in the literature. The simplest undersampling technique - the RUS algorithm - consists of randomly selecting a subset of the majority class with or without replacement. The main advantage of RUS is its simplicity. However, it fails to take full advantage of the data by underutilizing the majority class subset. In a more sophisticated algorithm called NearMiss, the majority class instances that are closest to the minority class are more likely to be selected [20]. Concretely, pairwise distances between members of the majority and minority classes are calculated. Then for each point in the majority class p^+, the average distance to the closest k points of the minority class $\overline{d}_{p+}$ is calculated. The points p^+ with the corresponding smallest values of $\overline{d}_{p+}$ are selected as the majority class sample. The motivation for the NearMiss algorithm is that given more points in the border region a classifier will be more likely to separate the two classes along the border. Despite its intelligent approach to undersampling it similar to RUS - also fails to take full advantage of the data.

In oversampling, the existing minority class points are used to generate the new minority points (Fig. 1, bottom). The simplest oversampling approach - the ROS algorithm - consists of randomly selecting (with replacement) from the existing minority points. The main advantage of ROS is its simplicity. However, it leads to overfitting by generating several minority points in the same location. In a more creative approach called SMOTE the new points are created synthetically by interpolating between the existing minority points [4]. Concretely, for each point in the minority class its k nearest neighbors in the minority class are determined. Given a minority point and one of its neighbors, the difference between the two is calculated and multiplied by a random number between 0 and 1. The new minority class instance is obtained by adding the preceding result to the minority point:

$$p_{new}^- = p^- + t\left(p^- - p_k^-\right), \tag{1}$$

where p_{new}^- denotes the new minority class point, p^- is the minority point under consideration, p_k^- is a kth nearest neighbor, and t is a random number between 0 and 1. Although SMOTE generates minority points in new locations it is bound to only linear paths between neighboring minority points which restricts the range of points that can be generated. To combat the issue of restricted linear generation other methods employing KDE and Gamma distribution have been proposed recently [15]. An extension of SMOTE called Adaptive Synthetic (ADASYN) attempts to generate the new minority points around the minority points that are harder to learn [12]. Concretely, ADASYN employs the ratio of the majority to minority points in the neighborhood of an existing minority point to determine the number of new minority points to generate in that neighborhood. ADASYN is motivated by the logic that the minority points that lie in regions of high concentration of majority points are more likely to be ignored by a classifier.

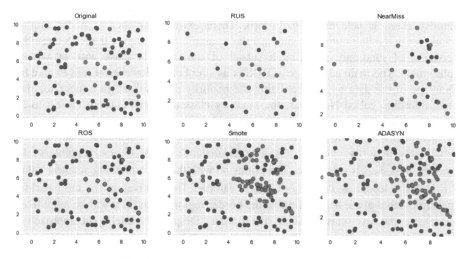

Fig. 1. Illustration of the sampling algorithms.

4   Numerical Experiments

In this section, we present the results of the numerical experiments that were carried out to compare single and ensemble classifiers in the context of imbalanced class distribution. In our experiments we use both simulated and real-life data. The simulated data contains 10,000 instances with 100 features while the real-life data is comparatively smaller.

4.1   Experimental Design

Constructing a single decision tree classifier for imbalanced data is straightforward. First, the original imbalanced data is resampled using an appropriate sampling technique in order to obtain a balanced dataset. Then a decision tree classifier is trained on the balanced data (Fig. 2).

Fig. 2. Construction of a balanced decision tree classifier.

The ensemble classifier construction process for imbalanced data is illustrated in Fig. 3. To construct an ensemble classifier the original imbalanced data is resampled 50 times via an appropriate sampling technique. The resampling procedure produces a set of balanced datasets. We train a decision tree estimator on each balanced dataset. We obtain 50 individual decision tree estimators that are used to construct an ensemble classifier using the majority rule. Given new data, we make predictions using the individual estimators and select the most popular (majority) prediction as the final output of the ensemble classifier. During the experiments the data is divided into training and testing sets according to 75/25 ratio. The experiments are carried out in Python using machine learning libraries sklearn [21] and imblearn [19].

Traditional measures such as accuracy and error rate do not adequately capture the performance of a classifier on the minority class in the context of imbalanced data. Therefore, we use the F1-score to obtain a more unbiased measure of classifier performance. The F1-score is a balanced metric that combines precision and recall into a single value. Recall represents the fraction of positive instances that were correctly labeled as such. It is given by the equation

$$recall = \frac{tp}{tp+fn},\tag{2}$$

where tp and fn denote the number of true positives and false negatives, respectively. Precision represents the fraction of truly positive instances from the total positively labeled instances,

$$precision = \frac{tp}{tp+fp}, \tag{3}$$

where fp denotes the number of false positives. The F1-score is the harmonic mean of precision and recall. It is given by the equation

$$F1 = 2\frac{precision.recall}{precision+recall}, \tag{4}$$

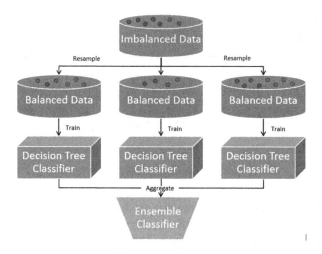

Fig. 3. Construction of a balanced ensemble decision tree classifier.

4.2 Simulated Data Experiments

We generate a random binary classification problem using the *make_classification* function from sklearn library. The dataset consists of a total of 100 features of which 15 are informative, 20 are redundant, 15 are repeated and the remaining are generated as random noise. The data is generated by placing clusters of points normally distributed (std = 1) about vertices of a 15-dimensional hypercube with sides of length 2 and assigning 2 clusters to each class. The redundant features are created as linear combinations of the informative features, while the repeated features are drawn randomly from the in- formative and redundant features [11]. The number of samples in the simulated dataset is 10,000. The ratio of minority to majority class points is 5/95.

We use the simulated dataset to examine the performance of ensemble and single decision tree classifiers in the context of various sampling algorithms. In particular, we investigate 2 undersampling and 2 oversampling algorithms: RUS, NearMiss, ROS, and SMOTE. The simulated data is balanced according to each sampling algorithm. We train ensemble and single decision tree classifiers on each balanced dataset. The performance of the classifiers is measured via the F1-score, recall, precision, and accuracy rates. As shown in Fig. 4, the ensemble classifier outperforms a single decision tree classifier in precision but underperforms it in recall. However, on balance

- as reflected by the F1-score - the ensemble classifier outperforms the single decision tree classifier with all 4 sampling algorithms. We also note that the ensemble classifier produces higher accuracy rate in almost all the tested sampling algorithms. Although accuracy rate is not the primary measure of performance for imbalanced data, it further supports the superiority of the ensemble approach. Observe that in the simulated data the average accuracy of ensemble classifier is 0.9619 which is higher than the benchmark score of 0.95 using the naive approach of classifying all instances as positive. Finally, we note that the ensemble approach produces better results with both undersampling (RUS, NearMiss) and oversampling (ROS, SMOTE) algorithms. The better performance in both the F1-score and accuracy indicates that the ensemble method is effective at identifying the minority as well as the majority class points. Identifying the minority (positive) instances is particularly important in many applications such as medical diagnostics. Thus, given an imbalanced data with a coherent underlying structure a sampling technique coupled with an ensemble method seems to be an effective solution.

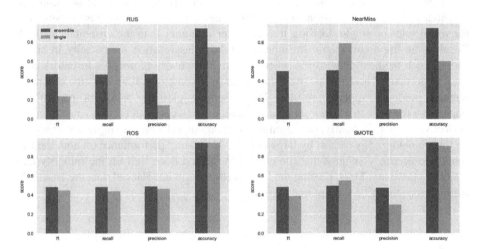

Fig. 4. Performance comparison of the ensemble and single decision tree classifier on simulated data consisting of 10,000 samples with 100 features.

Table 1. Experimental datasets.

Name	Repository & Target	Ratio	#S	#F
1 ecoli	UCI target: imU	8.6:1	336	7
2 spectometer	UCI, target: > = 44	11:1	531	93
3 us crime	UCI, target: >0.65	12:1	1,994	100
4 libras move	UCI, target: 1	14:1	360	90
5 letter_img	UCI, target: Z	26:1	20,000	16
6 mammography	UCI, target: minority	42:1	11,183	6

4.3   Real-Life Data Experiments

We use a number of real-life datasets to compare the performance of ensemble and single tree classifiers in imbalanced class setting. The data is fetched through imblearn library where it is appropriately preprocessed. Alternatively, it can be obtained directly from the UCI repository [7]. The details of the datasets are presented in Table 1. We use a diverse set of datasets in order to obtain comprehensive results. The datasets are chosen from a range of fields including astronomy, image recognition, medical diagnostics, and social science. The class ratio of the datasets ranges from 8.6:1 to 42:1, the number of samples ranges from 336 to 20,000, and the number of features ranges from 6 to 100.

We begin our experiments with the RUS algorithm. Recall that RUS operates by randomly selecting a subset of the majority class to balance with the minority class. We apply RUS to balance each dataset given in Table 1. Then, we train an ensemble and single decision tree classifiers on the balanced sets as described in Sect. 4.1. The performance of the classifiers is measured by the F1-score, recall, precision, and accuracy rates. The results of the experiments are presented in Fig. 5. As can be observed from the figure, the ensemble classifier outperforms the single decision tree classifier in F1-score, precision, and accuracy rates for all the tested datasets. Note that the recall rate for single decision tree classifier is higher than the ensemble method. Recall is an important metric in imbalanced data classification - especially in applications such medical diagnostics, where effective discovery of positive instances is of great importance. Nevertheless, on balance - as reflected by the F1-score - the ensemble method produces superior results. The ensemble classifier achieves particularly impressive results in the *letter_img* dataset with the F1-score and accuracy rate near the perfect value of 1. Similarly, the ensemble method outperforms the single tree classifier on the *mammography* dataset by large margin. The strong performance on both the F1-score and accuracy suggest that the ensemble method does well on the minority and majority class data. Thus, if using the RUS algorithm on imbalanced data the ensemble method appears to be the optimal approach.

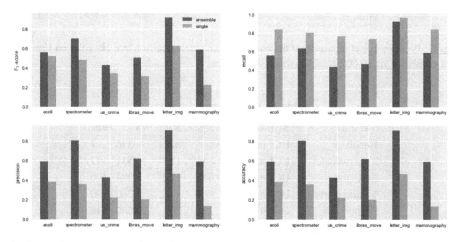

Fig. 5. Performance comparison of the ensemble and single decision tree classifier on simulated data consisting of 10,000 samples with 100 features.

Next, we compare the classification performance using the NearMiss algorithm. The NearMiss algorithm selects the samples of the majority class that are near the minority class points. As shown in Fig. 6, the ensemble classifier outperforms the single decision tree classifier on all the tested datasets. Concretely, the F1-score, precision, and accuracy of the ensemble classifier are substantially higher than a single tree classifier for all the tested datasets. The difference in the margin of the scores is particularly large in the case of spectrometer, letter_img, and mammography datasets. The results for the NearMiss algorithm are consistent with the RUS sampling algorithm.

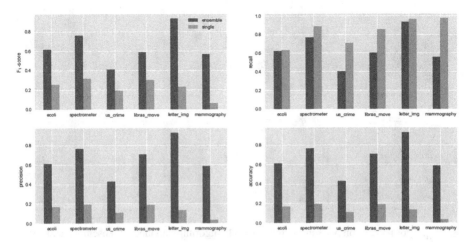

Fig. 6. Performance comparison of the NearMiss algorithm applied to ensemble and single decision tree classifiers.

We move on to investigate the performance of the ensemble and single tree classifiers using oversampling techniques. SMOTE is a popular oversampling algorithm that creates new minority samples through linear interpolation between existing minority points. As shown in Fig. 7, the ensemble classifier again outperforms the single decision tree classifier albeit in a slightly different manner than previously. In particular, the F1-score and accuracy of the ensemble method are higher on 5 out of 6 datasets. The results indicate that the ensemble method is effective in identifying the minority class instances while simultaneously producing strong outcomes on majority class data. Delving deeper into the results we see that the ensemble method produces better precision scores in all but one dataset. Unlike the case with the oversampling algorithms above, SMOTE-based ensemble classifier produces an even performance in recall rates. In particular, the ensemble classifier yields higher recall scores on 2 out 6 datasets. Given the overall results, as shown in Fig. 7, we conclude that the ensemble classifier is superior to the single decision tree classifier on SMOTE-sampled data.

Our final experiment is based on ADSYN sampled data. The ADASYN algorithm resembles SMOTE in the way it crates the new minority points through linear interpolation. The main difference is that ADASYN creates the new minority points around the existing minority points with high concentration of the majority class points. In this way, ADASYN tries to account for the learning difficulty on each minority point. The performance of the ensemble and single decision tree classifier on ADASYN-sampled data is very similar to that of SMOTE-sampled data. In particular, as shown in Fig. 8, the ensemble classifier produces higher F1-score, precision, and accuracy rates on 5 out of 6 tested datasets. The recall performance of the ensemble classifier is even with the single decision tree classifier. Thus, on balance the ensemble classifier produces significantly better results.

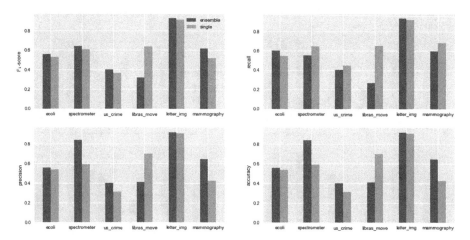

Fig. 7. Performance comparison of the SMOTE algorithm applied to ensemble and single decision tree classifiers.

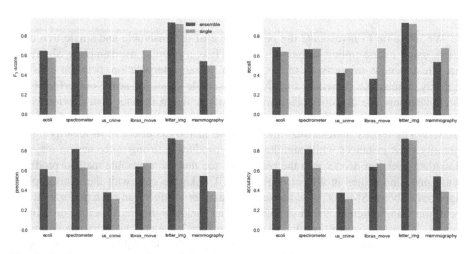

Fig. 8. Performance comparison of the ADASYN algorithm applied to ensemble and single decision tree classifiers.

Since they provide a way to reduce overfitting, bagging methods work best with strong and complex models such as fully developed decision trees. However, other base classifiers such as the k-nearest neighbors algorithm or logistic regression can also be employed as base classifiers. In Table 2, we present the average accuracy for ensemble and single classifier methods using 3 different base classifiers. The averages are calculated based on the results over the 6 datasets given in Table 1. It can be seen from the table that the ensemble classifiers outperform single classifiers for all 3 base classifiers. In addition, we observe that ROS and NearMiss achieve higher accuracy than other sampling methods in ensemble classification.

Table 2. Average accuracy using different base classifiers.

Model	ROS	SMOTE	ADASYN	RUS	NearMiss
Ensemble DT	0.9582	0.9532	0.9597	0.9577	0.9621
Single DT	0.9483	0.9367	0.9383	0.8397	0.5720
Ensemble KNN	0.9546	0.9483	0.9460	0.9474	0.9479
Single KNN	0.9436	0.9322	0.9342	0.8770	0.7534
Ensemble LR	0.9536	0.9473	0.9490	0.9494	0.9469
Single LR	0.9203	0.9235	0.9077	0.8852	0.7411

Similarly, in Table 3, we present the average F1-score for ensemble and single classifier methods. Although the results of the F1-scores are not exactly the same as the ac- curacy scores, they are generally consistent with our previous observations. Concretely, note that ensemble classifiers are generally better than single classifiers and that ROS and NearMiss are better sampling techniques in ensemble learning.

Table 3. Average F1-score using different base classifiers.

Model	ROS	SMOTE	ADASYN	RUS	NearMiss
Ensemble DT	0.6316	0.5770	0.6150	0.6181	0.6441
Single DT	0.6193	0.5953	0.6102	0.4204	0.2244
Ensemble KNN	0.6461	0.6303	0.6187	0.5944	0.6150
Single KNN	0.6991	0.6742	0.6635	0.4904	0.3608
Ensemble LR	0.6461	0.6303	0.6187	0.5944	0.6150
Single LR	0.5824	0.5951	0.5654	0.5041	0.3821

5 Conclusion

Imbalanced data is a widespread issue in a number of fields including medical diagnostics, text classification, fraud detection, and many others. Standard classifiers often struggle with imbalanced data by favoring the majority class data. However, the minority class data is often of more importance. For instance, it is more critical to identify fraudulent transactions than regular ones within credit card or insurance data.

One of the popular approaches to deal with imbalanced data is resampling whereby the original data is balanced prior to training a classifier. Resampling is naturally associated with ensemble classifiers as individual estimators of an ensemble can be trained on different iterations of resampled data. Therefore, we postulate that given a sampling procedure ensemble classifier would outperform single classifiers. To this end, we compared the performance of a single decision tree classifier to the performance an ensemble of decision trees in the context of applying a sampling procedure to imbalanced data. Concretely, investigated several undersampling and oversampling techniques and the performance of the classifiers on sampled data.

In order to obtain comprehensive results, we used simulated and real-life data. The simulated data (Sect. 4.2) consisted of 10,000 samples and 100 features of which only 15 were relevant. The results, as shown in Fig. 4, demonstrate the superiority of the ensemble classifier. The ensemble classifier outperformed the single classifier with respect to every sampling technique. In particular, the F1-score of the ensemble classifier exceeds that of the single decision tree classifier on every tested sampling technique. The results from the simulated data suggest that ensemble classifiers perform well on structured data in the context of resampled data.

We also investigated the performance single and ensemble classifiers on real-life resampled data. We used data from a range of applications to obtain a robust analysis. The experimental results, as presented in Sect. 4.3, show the superiority of the ensemble classifier. Using undersampling techniques the ensemble classifier yielded better F1- score on all 6 tested datasets. Using oversampling techniques, the ensemble classifier yielded better F1-score on 5 out of 6 tested datasets. Given the diversity of the tested datasets and sampling techniques, and the consistency of the results, we conclude an ensemble classifier is more suitable than a single classifier in the context of resampled data.

It is important to note that ensemble classifiers have a theoretically greater of algorithmic complexity. However, in practice the added training time is negligible because the underlying decision tree estimators are very efficient. Thus, in the light of the above discussion ensemble classifiers offer significantly better performance with little added cost.

As part of future work, our study can be extended to include multi-label imbalanced data. Using OVO or OVA approach multi-label data can be decomposed into a set of binary problems. Then an resampling-based ensemble approach can be applied to each binary problem. The effects of different resampling techniques on the performance of the corresponding ensemble methods would be interesting to study. In particular, applying these methods on high-dimensional big data can be a valuable addition to the existing literature.

References

1. Bi, J., Zhang, C.: An empirical comparison on state-of-the-art multi-class imbalance learning algorithms and a new diversified ensemble learning scheme. Knowl.-Based Syst. **158**, 81–93 (2018)

2. Bolon-Canedo, V., Sanchez-Marono, N., Alonso-Betanzos, A.: A review of feature selection methods on synthetic data. Knowl. Inf. Syst. **34**(3), 483–519 (2013)
3. Chandrashekar, G., Sahin, F.: A survey on feature selection methods. Comput. Electr. Eng. **40**(1), 16–28 (2014)
4. Chawla, N.V., Bowyer, K.W., Hall, L.O., Kegelmeyer, W.P.: SMOTE: synthetic minority over-sampling technique. J. Artif. Intell. Res. **16**, 321–357 (2002)
5. Collell, G., Prelec, D., Patil, K.R.: A simple plug-in bagging ensemble based on threshold-moving for classifying binary and multiclass imbalanced data. Neurocomputing **275**, 330–340 (2018)
6. Diez-Pastor, J.F., Rodrıguez, J.J., Garcıa-Osorio, C., Kuncheva, L.I.: Random balance: ensembles of variable priors classifiers for imbalanced data. Knowl. Based Syst. **85**, 96–111 (2015)
7. Dua, D., Graff, C.: UCI Machine Learning Repository. University of California, School of Information and Computer Science, Irvine, CA (2019). http://archive.ics.uci.edu/ml
8. Fawcett, T.: An introduction to ROC analysis. Pattern Recogn. Lett. **27**(8), 861–874 (2006)
9. Fernandez, A., Garcia, S., Herrera, F., Chawla, N.V.: Smote for learning from im- balanced data: progress and challenges, marking the 15-year anniversary. J. Artif. Intell. Res. **61**, 863–905 (2018)
10. Galar, M., Fernández, A., Barrenechea, E., Herrera, F.: EUSBoost: enhancing ensembles for highly imbalanced datasets by evolutionary undersampling. Pattern Recogn. **46**(12), 3460–3471 (2013)
11. Guyon, I., Gunn, S., Hur, A.B., Dror, G.: Design and analysis of the NIPS2003 challenge. In: Guyon, I., Nikravesh, M., Gunn, S., Zadeh, L.A. (eds.) Feature Extraction. Studies in Fuzziness and Soft Computing, vol. 207, pp. 237–263. Springer, Heidelberg (2006). https://doi.org/10.1007/978-3-540-35488-8_10
12. He, H., Bai, Y., Garcia, E.A., Li, S.: ADASYN: adaptive synthetic sampling approach for imbalanced learning. In: 2008 IEEE International Joint Conference on Neural Networks (IEEE World Congress on Computational Intelligence), pp. 1322–1328. IEEE (2008)
13. Hido, S., Kashima, H., Takahashi, Y.: Roughly balanced bagging for imbalanced data. Stat. Anal. Data Min. ASA Data Sci. J. **2**(5–6), 412–426 (2009)
14. Kamalov, F.: Kernel density estimation based sampling for imbalanced class distribution. Inf. Sci. **512**, 1192–1201 (2020)
15. Kamalov, F., Denisov, D.: Gamma distribution-based sampling for imbalanced data. Knowl. Based Syst. **207**, 106368 (2020)
16. Kamalov, F.: Sensitivity analysis for feature selection. In: 2018 17th IEEE International Conference on Machine Learning and Applications (ICMLA), pp. 1466–1470. IEEE (2018)
17. Krawczyk, B.: Learning from imbalanced data: open challenges and future directions. Progress Artif. Intell. **5**(4), 221–232 (2016). https://doi.org/10.1007/s13748-016-0094-0
18. Lango, M., Stefanowski, J.: Multi-class and feature selection extensions of roughly balanced bagging for imbalanced data. J. Intell. Inf. Syst. **50**(1), 97–127 (2018)
19. Lemaitre, G., Nogueira, F., Aridas, C.K.: Imbalanced-learn: a python toolbox to tackle the curse of imbalanced datasets in machine learning. J. Mach. Learn. Res. **18**(1), 559–563 (2017)
20. Mani, I., Zhang, I.: kNN approach to unbalanced data distributions: a case study involving information extraction. In: Proceedings of Workshop on Learning from Imbalanced Datasets, vol. 126 (2003)
21. Pedregosa, F., et al.: Scikit-learn: machine learning in Python. J. Mach. Learn. Res. **12**, 2825–2830 (2011)
22. Sun, Z., Song, Q., Zhu, X., Sun, H., Xu, B., Zhou, Y.: A novel ensemble method for classifying imbalanced data. Pattern Recogn. **48**(5), 1623–1637 (2015)

23. Thabtah, F., Hammoud, S., Kamalov, F., Gonsalves, A.: Data imbalance in classification: experimental evaluation. Inf. Sci. **513**, 429–441 (2020)
24. Wang, S., Yao, X.: Diversity analysis on imbalanced data sets by using ensemble models. In: 2009 IEEE Symposium on Computational Intelligence and Data Mining, pp. 324–331. IEEE (2009)
25. Yang, P., Liu, W., Zhou, B.B., Chawla, S., Zomaya, A.Y.: Ensemble-based wrapper methods for feature selection and class imbalance learning. In: Pei, J., Tseng, V.S., Cao, L., Motoda, H., Xu, G. (eds.) PAKDD 2013. LNCS (LNAI), vol. 7818, pp. 544–555. Springer, Heidelberg (2013). https://doi.org/10.1007/978-3-642-37453-1_45
26. Li, Y., Guo, H., Liu, X., Li, Y., Li, J.: Adapted ensemble classification algorithm based on multiple classifier system and feature selection for classifying multi-class imbalanced data. Knowl. Based Syst. **94**, 88–104 (2016)
27. Zhang, Z., Krawczyk, B., Garcia, S., Rosales-Perez, A., Herrera, F.: Empowering one-vs-one decomposition with ensemble learning for multi-class imbalanced data. Knowl. Based Syst. **106**, 251–263 (2016)

Dual-Channel Recalibration and Feature Fusion Method for Liver Image Classification

Tingting Niu[1,2,3(✉)], Xiaolong Zhang[1,2,3], Chunhua Deng[1,2,3], and Ruoqin Chen[3,4]

[1] School of Computer Science and Technology,
Wuhan University of Science and Technology, Wuhan, China
{xiaolong.zhang,dengchunhua}@wust.edu.cn
[2] Institute of Big Data Science and Engineering,
Wuhan University of Science and Technology, Wuhan, China
[3] Hubei Key Laboratory of Intelligent Information Processing and Real-Time
Industrial System, Wuhan University of Science and Technology, Wuhan, China
[4] Tianyou Hospital, Wuhan University of Science and Technology,
Wuhan, China

Abstract. Aiming at the problem of poor medical image recognition accuracy caused by the large individual differences in liver pathological images and the large single feature size, a liver image classification method based on dual-channel recalibration and feature fusion is proposed. Firstly, we construct a dual-channel recalibration model based on attention mechanism, suppress useless features and calculate different feature channel weights, and embed them into the Inception_ResNet_V2 network structure. Secondly, we design fully connected layers on the basis of two-dimensional and three-dimensional convolutional neural networks, then add feature fusion layers to obtain deep semantic information in different dimensions. Finally, we use pre-training models to initialize the network structure and input the fused features into the XGBoost classifier for classification prediction. Experiments have been carried out with liver cancer race imaging datasets and a hospital patient dataset. Experimental results show that this method is superior to previous ones.

Keywords: Liver cancer image classification · Deep learning · Channel recalibration · Feature fusion · Transfer learning

1 Introduction

Early detection of hepatocellular carcinoma and the diagnosis and treatment at early lesions [1] can greatly contribute to the treatment and recovery of patients. In actual clinical application, due to the complex and indistinguishable texture structure of liver lesions and their distribution patterns, purely manual film reading can lead to problems such as misdiagnosis and missed diagnosis. In recent years, with the development of computer visualization, image processing, pattern recognition and other related fields of technology, computer-aided diagnosis [2] has gradually become a development trend of future medical diagnosis as an auxiliary tool.

© Springer Nature Switzerland AG 2021
D.-S. Huang et al. (Eds.): ICIC 2021, LNCS 12837, pp. 579–591, 2021.
https://doi.org/10.1007/978-3-030-84529-2_49

Research on liver image classification has gone through manual feature extraction stage and deep learning stage. The algorithm based on traditional manual extraction of texture features is the main theoretical basis for early recognition of lesions. Liu et al. [3] classified liver pathology images by capturing macroscopic local binary pattern (LBP, Local Binary Pattern) texture information, and the recognition accuracy reached 70.3%. Since this algorithm only covers local areas within a fixed radius and doesn't satisfy the needs of different sizes and frequencies of textures, it is difficult to improve its recognition accuracy. Based on this, Kirubakaran et al. [4] improved the single feature extraction model by combining the multiscale features of Gabor transform and local features obtained by local binary pattern algorithm, and feature fusion can improve the accuracy to 88.23%. Although this method obtains multi-scale feature information by cascading, it still fundamentally fails to break through the limitations of manually extracted features. In contrast, the high-level semantic features extracted by neural networks [5] can precisely compensate for the lack of expressiveness of manually extracted features with better abstraction and interference resistance. Zhu et al. [6] extracted features such as angular co-moments, inverse differential moments and entropy of grayscale coeval matrices from liver pathology images as feature vectors into BP neural networks for classification and recognition, and from the classification of normal and fatty liver could reach 85.71% and 88.89% accuracy can be seen from the fact that the recognition is significantly improved by using a simple neural network algorithm. Romero [7] et al. implemented an end-to-end deep learning approach by combining the effective features extracted in InceptionV3 and the weights obtained by training in ImageNet, and the accuracy of lesion types can reach 96%. The Inception [8] structure of this network implements convolution and pooling operations in parallel to obtain more potential features of the image. However, the above-mentioned research methods don't differentiate the expressiveness of features, and therefore, problems such as network degradation occur when the number of network layers increases.

The traditional manual feature extraction methods are limited to low-level texture semantic information [9]. Moreover, neural network extraction feature methods often face problems such as single feature channel, restricted image information, and network structure degradation. To overcome these problems, the thesis proposes a liver image classification model with dual-channel recalibration and multidimensional feature fusion, and is used to liver image classification. A dual-channel calibration model is embedded in the fully connected layer of a two-dimensional convolutional network [10], the effective features are enhanced while the invalid features are suppressed by changing feature channel weights [11], and the liver image domain and spatial domain information are extracted and loaded into the residual network structure at the same time, in which more key feature information is extracted while ensuring the stability of the network structure.The proposed method constructs a channel recalibration model based on the attention module to achieve feature importance distinction; then, bulid convolutional neural networks with different dimensions to increase the diversity and complementarity of features by combining image neighboring semantic information; finally, input the features to the XGBoost classifier to obtain classification results with the residual network structure as the main structure.

2 Liver Cancer Imaging Classification Method

Firstly, the attention mechanism based on Squeeze-and-Excitation (SE) model is investigated and improved, and the fused dual-channel weights are obtained using the average pooling and maximum pooling structures. Then, the two-channel rescaled model is embedded into the Inception_ResNet_V2 structure to extract image feature information together with the TC3D network. Finally, the depth features of different dimensions are input to the classifier for recognition. The overall structure of liver image classification is shown in Fig. 1.

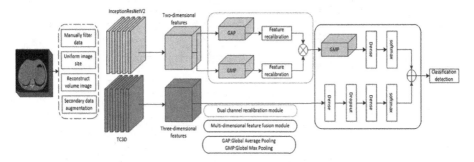

Fig. 1. Overall structure of liver image classification.

2.1 Dual-Channel Recalibration Mechanism

The channel recalibration mechanism is an attention mechanism that works on the feature channels. A feature recalibration strategy is used to differentiate the feature importance by assigning different weights to the convolutional channels [12]. It was first proposed in the framework of SENet [13]. The SENet structure is centered on learning feature weights and highlighting feature differences based on the loss values of the training network model. And it can be combined with networks such as Inception, ResNet, and GoogleNet. Models using the SENet structure have significant performance improvements.

Since the initial channel recalibration mechanism only recalibrates for a single channel, the extracted key feature information is restricted and there is more room for improvement in feature differentiation. For this reason, a dual-channel recalibration mechanism model is proposed in this paper: (1) the model's input is changed from a single channel to a dual channel, on the basis of the original structure, introduce the maximum pooling layers and retain the most significant feature information in the feature map, fuse the different features obtained in multiple channels, then obtain the final channel weights. (2) Neighbor global pooling layers, two BatchNormalization structures and a fully connected layer. Which is not only to realize the fast convergence of the model, reduce the probability of overfitting phenomenon,gradient disappearance problem, but also make the learned features easier to fit the liver image. The optimized structure contains more layers of liver region information in the output features while ensuring as few operations as possible, and expands the differences between different features and weakens the interference of invalid features. The different channel recalibration mechanisms are shown in Fig. 2. On the left is the single-channel recalibration mechanism SE, while on the right is the dual-channel recalibration mechanism SSE.

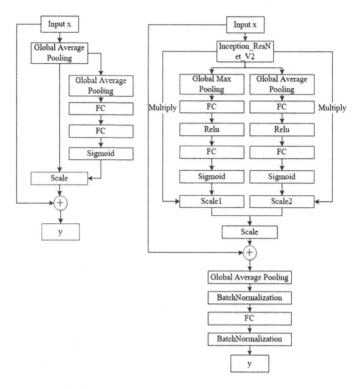

Fig. 2. Different channel recalibration mechanisms.

2.2 Maintaining the Integrity of the Specifications

The proposed multidimensional feature fusion structure is based on Inception_ResNet_V2 and TC3D networks. Among them, the Inception_ResNet_V2 network structure combines the advantages of Inception and ResNet modules, where Inception is ease to acquire sparse as well as non-sparse features at the same layer, and ResNet can improve the training speed and prevent gradient dispersion, which makes it have improved performance.

The 3D convolutional neural network structure consists of multiple layers of nonlinear neurons that can obtain representative high-level abstract semantic information from images. Since 3D convolutional neural networks are highly expressive of features, for non-large scale datasets, the classification performance of the model may decrease as the depth of the 3D network increases [14], but limiting the network depth can also cause incomplete feature extraction. Therefore, based on the structure of

VGGNet, we designed a C3D-like network architecture and named it TC3D network, which is essentially a 3D CNN capable of processing consecutive multi-slice images. The input size is an array of n × 224 × 224 superimposed images, where n is the extracted consecutive liver slices and 224 × 224 is the height and width of each liver slice image. The network architecture contains ten 3D convolutional layers, five 3D maximum pooling layers, five activation functions and two fully connected layers. Firstly, combine the convolution layers with the activation functions to produce more nonlinear operations and match the image features better. Then, add a pooling layer after each two convolutional layers to form a fixed double convolutional pooling unit which samples in both time and space dimensions, making the number of parameters smaller and the network structure more streamlined. Nextly, Use the MaxPooling technique for pooling operation to ensure feature invariance and dimensionality reduction operation, the pooling model converges more easily. Finally, insert two fully connected layers and flattening functions, which is aiming at the transition from convolutional to fully connected layers and the one-dimensionalization of multidimensional inputs. Each double convolution pooling unit is followed by an attached Dropout operation to prevent the overfitting phenomenon caused by the model parameters.

It is proved that 3 × 3 × 3 is an effective convolution kernel for processing 3D spatial features, therefore, the size of the convolution kernel of the 3D convolution structure designed in this paper is fixed to 3 × 3 × 3. The model takes 3D liver data as input, and the designed neural network model is trained in an end-to-end manner without complicated pre-processing and post-processing. With the same amount of data, the TC3D network used in this paper can effectively improve the accuracy and efficiency of classification with fewer parameters than other neural network models. Figure 3 shows the model structure of the TC3D designed in this paper.

The strategy of TC3D network structure to extract spatial features is: extract 9 sliced images with a size of 224 × 224 from the original volume image, reconstruct the image data with a size of (224, 224, 3, 3) after grayscale processing. After the nonlinear transformation of the convolutional layer, the 3D tensor is generated to complete the extraction of spatial features.

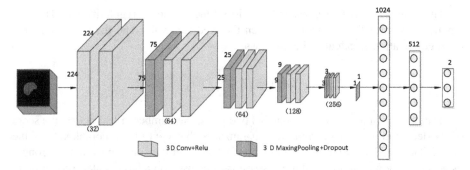

Fig. 3. TC3D model structure diagram.

In the field of computer vision [15], feature fusion is a common strategy to combine features from different perspectives for information acquisition and compensation. To construct the feature fusion layer [16], this paper designs a fully connected layer with uniform dimension after the convolutional neural network and performs the fusion of the same layers to facilitate the effective use of object information at different levels. The feature matrix splicing method is also used to enhance the feature information in the horizontal direction to obtain the final feature fusion results, which are input to the XGBoost classifier for classification prediction.

2.3   Maintaining the Integrity of the Specifications

The XGBoost [17] algorithm is an integrated learning algorithm, which has the advantages of efficiency, flexibility, and robustness comparing with other Boosting algorithms. Its essence is to form a strong classifier by continuously iterating through multiple weak classifiers. The base learners of this algorithm are classification trees and regression trees, which continuously split features and add new tree structures to update functions and reduce errors, thus achieving the goal of improving training accuracy. Therefore, the final result of classifier learning is the definition of the objective function, which includes training loss and regularization. The former illustrates the accuracy of the model prediction and the latter illustrates the complexity of the model operation. The overall operation process is referred to the literature [17]. Prediction output results are as follows:

$$\widehat{y}_k = \sum_{k=1}^{K} f_k(x_i) \tag{1}$$

Among them, representing the predicted cumulative output value of the i-th leaf node of k trees, represents the classification result of the i-th node of the k-th tree. Define the objective function as:

$$Obj = \sum_{i=1}^{n} L(y_i, \widehat{y}_i) + \sum_{k=1}^{K} \Omega(f_k) \tag{2}$$

The objective function consisted of a loss function and a regularization. The loss function measures the difference between the predicted value and the true value, and the regularization is calculated as follows:

$$\Omega(f) = \gamma T + \frac{1}{2}\lambda \|w\|^2 \tag{3}$$

The regularization term also contains two parts: the number of leaf nodes, the score of the leaf nodes, both of which are parameters that control the complexity of the model. The smaller the regularization value is, the lower the complexity is, the stronger its generalization ability is, and the easier it is to be used on a large scale. According to formula (1) and formula (3), substitute it into the objective function formula (2), expand according to Taylor formula to obtain the first-order partial derivative and the second-order partial derivative, and define $H_j = \sum_{i \in I_j} h_i$, $G_j = \sum_{i \in I_j} g_j$, remove the terms that don't affect the optimization of the objective function to obtain the optimal solution of the objective function.

$$Obj = \frac{1}{2} \sum_{j=1}^{T} \frac{G_j^2}{H_j + \lambda} + \gamma T \qquad (4)$$

At this point, Obj is the structure score of the tree, which minimizes the objective function as a result of the guidance of model learning and prediction. The XGBoost model usually has the advantages of high accuracy, high generalization, and not easy to overfit. It is currently widely used in image classification, data mining, financial risk control and so on.

3   Experimental Results and Analysis

3.1   Experimental Platform

This experiment is carried out on a high-performance workstation, the machine is running on Linux 16.04 operating system, the GPU graphics card model is GeForce GTX 1080 Ti, the video memory is 11 GB, and the disk space is 233 GB. This experiment is based on the Keras2.2.4 framework for deep learning. The framework is highly modular and extensible, which greatly improves the efficiency of users.

3.2   Datasets

We conducts experiments on an open liver cancer imaging dataset and a hospital imaging dataset. The former is the dataset taken for the competition question of "Big Data Healthcare - AI Diagnosis of Liver Cancer Imaging" in 2019, and there are always 1,397 participating teams, the optimal classification result with F1_score of 91.56 is obtained by the best team, which has a total of 7,398 liver sequences;the latter are real patient liver sequences collected in a hospital from 2014 to 2018,and expanded up to 9436 liver sequences after secondary data enhancement. The image format is Dicom file format, the background area and overall structure are similar, the differences are: (1) the open liver cancer image dataset is CT scan sequence images with a fixed image size of 512 × 512 pixels; the hospital image dataset is CT scan sequence images and MRI sequence images with a fixed image size of 456 × 456 pixels. (2) The category of open liver cancer image dataset is divided into benign/malignant. The malignant lesions were a series of irregular hollow masses; the categories of hospital imaging dataset were normal/abnormal. The abnormal categories include fatty liver, cirrhosis, liver tumor, etc. Figure 4 shows an example of liver data.

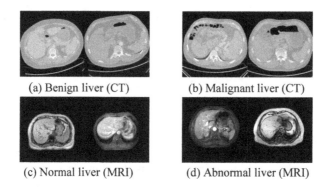

(a) Benign liver (CT)   (b) Malignant liver (CT)

(c) Normal liver (MRI)   (d) Abnormal liver (MRI)

Fig. 4. Example of liver data.

In this paper, we name the liver cancer image dataset as DataSet_I for benign/malignant classification of liver cancer and the hospital image dataset as DataSet_II for normal/abnormal classification of liver to identify and classify the liver. In order to ensure the independence and fairness of the training and test sets during the experiment, the training and test sets are randomly divided according to the ratio of 5:1 in this paper. The distribution of the datasets is shown in Table 1.

Table 1. Comparison of datasets.

Datasets	Training sets	Test sets	Total
DataSet_I	6165	1233	7398
DataSet_II	7864	1572	9436

3.3   Evaluation for Training

To fully quantify the classification performance of the model, Acc (Accuracy), Pre (Precision), Rec (Recall) and F1 value (F1_score) are used as evaluation indexes in this paper, the formulae are shown in Eqs. (5), (6), (7), and (8).

$$Acc = \frac{TP + TN}{TP + TN + FP + FN} \tag{5}$$

$$Pre = \frac{TP}{TP + FP} \tag{6}$$

$$Rec = \frac{TP}{TP + FN} \tag{7}$$

$$F1 = 2 \times \frac{Pre \times Rec}{Pre + Rec} \tag{8}$$

In the formula, where *TP* is the true case, *TN* is the true negative case, *FP* is the false positive case, and *FN* is the false negative case. From this, it can be seen that the accuracy rate is the percentage of category-predicted correct results to the total liver samples; the precision rate refers to the probability that all liver samples predicted to be positively labeled are actually positive; the recall rate is the percentage of predicted positive results among all liver samples that are actually positive; and the F1 value is the summed average of the precision rate and recall rate.

3.4   Experimental Process and Parameter Setting

Firstly, the dataset is pre-processed to generate uniform 2D images and 3D stereo data; then the dataset is randomly divided into training set and test set according to the scale; then input the data to the pre-trained Inception_ResNet_V2 network and the original TC3D model, fine-tune the network parameters to fuse image plane and spatial features; finally, the fused features are input to the XGBoost classifier for classification prediction, and the final classification results are obtained.

In the operation of the preprocessing method, due to the large differences in the quality, size and category of liver image samples, the following treatments were done to ensure the accuracy and validity of the experimental results. (1) Manual screening of data to eliminate obvious liver slice defects and confusion of label sequences Waiting for image interference. (2) Unify the image size and adjust the image size selected in the experiment to 224 × 224. (3) Reconstruct the volume image. After grayscaling the original image to a single-channel image, it is converted to an image array for super-imposition and used as input for the 3D network structure. (4) Secondary data enhancement. In order to reduce the gap in the number of image classes, the primary enhancement of the source data is mirrored, rotated and randomly cropped. To ensure that the images have better robustness and generalization capability, the ImageDataGenerator [18] method is used for secondary augmentation, and the data secondary augmentation parameters are shown in Table 2.

Table 2. Data secondary expansion parameter setting.

Parameter	Value
rotation_range	90
width_shift_range	0.5
height_shift_range	0.5
shear_range	0.2
horizontal_flip	True
vertical_flip	True

For the selection of model training parameters, the batch size was specified as 32 and the number of training sessions was 100. And use the SGD optimizer, set the learning rate to 0.0001, the momentum to 0.9, and the Dropout node hiding rate to 0.5. Cross entropy

[19] was used as the loss function for liver classification, and the network parameters were updated according to the loss values.The formula is described:

$$Loss = \frac{1}{N} \sum_i -[y_i \log(p_i) + (1 - y_i) \log(1 - p_i)] \tag{9}$$

Loss is the cross-entropy loss function, N is the number of samples,y_i represents the actual label value corresponding to sample i, and p_i represents the probability value that sample i is predicted to be positive. The quality of the model is judged by calculating the actual label value and the predicted label value to get the current loss function value. Then the network parameters are updated according to the back-propagation algorithm to train the network model, and then the difference between the actual label value of the sample and the predicted label value is reduced so that the extracted features can better fit the data samples. The classification result is the average of five randomly divided data sets for the experiment.

3.5 Experimental Result and Analysis

In order to verify the classification effectiveness of the proposed method on liver images, a large number of comparison experiments are conducted on the basis of the existing dataset, and the accuracy Acc (Accuracy) and F1 value (F1_score) were used as evaluation metrics to evaluate the classification results. Firstly, the experiments compare the classification metrics of the current mainstream models in the dataset DataSet_I. The selected 2D convolutional neural network models are:Vgg19 [20], ResNet18, InceptionV3 and Inception_ResNet_V2. Each sub-network is trained independently with the same loss function and parameter settings as the whole experiment. The performance comparison of each classification is shown in Table 3. It can be seen that the Inception_ResNet_V2 network model combining the Inception structure and the ResNet structure has the most significant performance in the benign/malignant classification results of liver cancer, and has incomparable advantages compared with other two-dimensional network models.

Table 3. Classification performance comparison of mainstream models.

Models	Acc	F1
VGG19	85.99	83.72
ResNet18	88.14	84.65
InceptionV3	89.48	88.73
Inception_ResNet_V2	92.87	91.12

Table 4 is the experiment results of comparsion of the improved dual-channel recalibration model with other models. The network structure with the addition of the single-channel recalibration mechanism SE module shows a significant improvement in accuracy. In DataSet_I, compared with the single-channel recalibration mechanism SE module, the average accuracy and F1 value of the improved SSE module improved by

0.17 and 1.45% points; in DataSet_II, compared with the single-channel recalibration mechanism SE module, the average accuracy and F1 value also improved by 3.82% and 5.13%, which further proves the effectiveness of the dual-channel recalibration mechanism for the recognition of CT liver slices. It also has a strong generalization ability on MRI liver slices.

Table 4. Performance comparison of different channel recalibration mechanism models.

Method	DataSet_I		DataSet_II	
	Acc	F1	Acc	F1
Inception_ResNet_V2	92.87	91.12	73.09	71.62
Inception_ResNet_V2 + SE	93.94	92.32	76.91	76.65
Inception_ResNet_V2 + SSE	94.11	93.77	77.36	81.08

The single-dimensional model's are compared with the multidimensional model after fusing the features for experiments. The experimental results are shown in Table 5. In the DataSet_I dataset, CNN is the highest F1 value obtained by the champion team "FDU-Zero Krypton Technology" in the "Big Data Healthcare-Liver Cancer Imaging AI Diagnosis" competition in 2019, and the F1 value of this paper is 3.28% points higher than this result, which shows the superiority of this method. This shows the superiority of the present method. Compared with the single 2-D model embedded with a two-channel calibration mechanism, the average accuracy and F1 values show an improvement of 2.95% and 1.07%, and compared with the single 3-D model, the average accuracy and F1 values show an improvement of 1.22% and 0.59%. It also has good classification performance in DataSet_II, indicating that the fused multidimensional model has stronger feature representation capability than the single model for different liver types of datasets.

Table 5. Performance comparison of different dimensional models.

Method	DataSet_I		DataSet_II	
	Acc	F1	Acc	F1
CNN	–	91.56	–	–
Inception_ResNet_V2 + SE	93.94	92.32	76.91	76.65
Inception_ResNet_V2 + SSE	94.11	93.77	77.36	81.08
TC3D	95.84	94.25	76.12	78.99
The proposed method	97.06	94.84	80.53	82.41

4 Conclusion

In this paper, we propose a classification algorithm for liver cancer images based on dual-channel recalibration and multidimensional feature fusion to solve the problems of large individual differences and single feature dimension in liver medical images. The algorithm embeds the improved dual-channel calibration mechanism into a two-dimensional convolutional neural network to achieve the differentiation of key feature information. Then, unites different dimensional convolutional neural networks to enhance their feature expression capability to obtain more information at different levels. Nextly, designs a feature fusion structure to fuse multidimensional features to make full use of image domain and spatial domain features. Finally, uses XGBoost classifier to classify liver pathology images prediction. The experimental results show that the method in this paper has advantage over the existing ones, and also has good classification effect and generalization ability on the hospital image dataset, which verifies the effectiveness of the method in this paper.

References

1. Hessinger Née Reimann, C.: Dielectric contrast between normal and tumor ex-vivo human liver tissue. IEEE Access 164113–164119(2019)
2. Khosravan, N., Celik, H., Turkbey, B.: A collaborative computer aided diagnosis (C-CAD) system with eye-tracking, sparse attentional model, and deep learning. Med. Image Anal. **51**, 101–115 (2019)
3. Liu, L., Long, Y., Fieguth, P.W.: BRINT: binary rotation invariant and noise tolerant texture classification. IEEE Trans Image Process **23**(7), 3071–3084 (2014)
4. Ramamoorthy, S., Kirubakaran, R., Subramanian, R.S.: Texture feature extraction using mgrlbp method for medical image classification. In: Suresh, L.P., Dash, S.S., Panigrahi, B.K. (eds.) Artificial Intelligence and Evolutionary Algorithms in Engineering Systems. AISC, vol. 324, pp. 747–753. Springer, New Delhi (2015). https://doi.org/10.1007/978-81-322-2126-5_80
5. Bacciu, D., Crecchi, F.: Augmenting recurrent neural networks resilience by dropout. IEEE Trans. Neural Netw. Learn. Syst. **31**, 345–351 (2020)
6. Zhu, F., Zhu, B., Li, P., Wang, Z., Wei, L.: Quantitative analysis and identification of liver B-scan ultrasonic image based on BP neural network. In: 2013 International Conference on Optoelectronics and Microelectronics (ICOM), pp. 62–66 (2013)
7. Romero, F.P.: End-to-end discriminative deep network for liver lesion classification. In: 2019 IEEE 16th International Symposium on Biomedical Imaging (ISBI 2019), pp. 1243–1246(2019)
8. Szegedy, C., Vanhoucke, V., Ioffe, S.: Rethinking the Inception Architecture for Computer Vision. 32, 2818–2826 (2016)
9. Xie, X., Cai, X., Zhou, J., Cao, N.: A semantic-based method for visualizing large image collections. IEEE Trans. Vis. Comput. Graph. **25**, 2362–2377 (2019)
10. Chen, S., Sun, W., Huang, L.: Compressing fully connected layers using Kronecker tensor decomposition. In: 2019 IEEE 7th International Conference on Computer Science and Network Technology (ICCSNT), pp. 308–312 (2019)
11. Zheng, J., Wu, Y., Song, W.: Multi-scale feature channel attention generative adversarial network for face sketch synthesis. IEEE Access **8**, 146754–146769 (2020)

12. Zhang, D.: clcNet: improving the efficiency of convolutional neural network using channel local convolutions. In: 2018 IEEE/CVF Conference on Computer Vision and Pattern Recognition, pp. 7912–7919 (2018)
13. Hu, J., Shen, L., Sun, G.: Squeeze-and-excitation networks. In: 2018 IEEE/CVF Conference on Computer Vision and Pattern Recognition, pp. 7132–7141 (2018)
14. Hara, K., Kataoka, H., Satoh, Y.: Can spatiotemporal 3D CNNs Retrace the History of 2D CNNs and ImageNet? In: 2018 IEEE/CVF Conference on Computer Vision and Pattern Recognition, pp. 6546–6555 (2018)
15. Chiu, M.T.: Agriculture-vision: a large aerial image database for agricultural pattern analysis. In: 2020 IEEE/CVF Conference on Computer Vision and Pattern Recognition (CVPR), pp. 2825–2835(2020)
16. Amin, S.U., Alsulaiman, M., Muhammad, G.A.M.: Multilevel weighted feature fusion using convolutional neural networks for EEG motor imagery classification. IEEE Access 7, 18940–18950 (2019)
17. Gong, H., Zhang, H., Zhou, L.: An interpretable artificial intelligence model of chinese medicine treatment based on XGBoost algorithm. In: 2020 IEEE International Conference on Bioinformatics and Biomedicine (BIBM), pp. 1550–1554 (2020)
18. Gu, K., Tao, D., Qiao, J.: Learning a no-reference quality assessment model of enhanced images with big data. IEEE Trans. Neural Netw. Learn. Syst. 29, 1301–1313 (2018)
19. Udhayakumar, R.K., Karmakar, C., Palaniswami, M.: Cross entropy profiling to test pattern synchrony in short-term signals. In: 2019 41st Annual International Conference of the IEEE Engineering in Medicine and Biology Society (EMBC), pp. 737–740 (2019)
20. Xia, Y., Cai, M., Ni, C.: A switch state recognition method based on improved VGG19 network. In: 2019 IEEE 4th Advanced Information Technology, Electronic and Automation Control Conference (IAEAC), pp. 1658–1662 (2019)

Research on Path Planning Algorithm for Mobile Robot Based on Improved Reinforcement Learning

Junwei Liu, Aihua Zhang$^{(\boxtimes)}$, and Yang Zhang

School of Control Science and Engineering, Bohai University,
Jinzhou 121013, China

Abstract. This paper proposes an improved Q-learning algorithm to solve the problem that the traditional Q-learning algorithm is applied to the path planning of mobile robot in complex environment, the convergence speed is slow due to large number of iterations, and even the actual reward signal is sparse so that the agent cannot get the optimal path. The improved algorithm reduces the number of iterative runs of the agent in the path planning process to improve the convergence speed by further updating the iterative formula of the Q-value. At the same time, adding sparse reward algorithm leads to finding the optimal path successfully. In order to verify the effectiveness of the algorithm, simulation experiments are carried out in two groups of environments: the simple and the complex. The final simulation results show that the improved algorithm can avoid obstacles effectively and find the optimal path to the target position after less iterations, which proves that the performance of the improved algorithm is better than the traditional Q-learning in the path planning.

Keywords: Reinforcement learning · Markov decision process · Path planning · Sparse reward · Q-learning · Reward shaping

1 Introduction

In recent years, with the continuous development of artificial intelligence, unmanned driving technology has gradually emerged, and path planning technology has got more and more attention. Path planning has a widely range of applications in many fields. Applied in new high-tech fields including: autonomous collision-free path exploration of robots; obstacle avoidance flight of UAV; and applied in daily life including: GPS navigation; planning and navigation of urban road networks, etc. Therefore, the research on path planning of mobile robots is of great significance.

The purpose of path planning is that the agent can avoid obstacles automatically from the current position to the target position, the path of the agent should be as short as possible, and the smoothness of the path should meet the dynamics requirements. According to the known degree of environmental information in the path planning process, path planning can be divided into global path planning and local path planning [1]. A variety of ways can realize the path planning of mobile robot. The traditional path planning methods include artificial potential field method [2], dynamic window

© Springer Nature Switzerland AG 2021
D.-S. Huang et al. (Eds.): ICIC 2021, LNCS 12837, pp. 592–604, 2021.
https://doi.org/10.1007/978-3-030-84529-2_50

method [3], graphic method [4], A-start algorithm [5], etc.; and artificial intelligence methods of path planning include genetic algorithm [6], neural network [7], ant colony algorithm [8], etc. However, each type of path planning algorithm has its best usage scenarios and its own limitations. Traditional path planning methods are easier to fall into traps and cannot escape in a complex environment, so robot cannot reach the target position with a high probability; while the artificial intelligence algorithms have a huge amount of calculations and requires a large number of iterations to successfully plan the route, even the requirements for hardware are highly. But as a new intelligent algorithm, reinforcement learning imitates human thinking, the agent uses trial-and-error to learn to continuously interact with the environment. At the same time, the agent selects and executes the next action to find the optimal strategy according to the feedback information of the environment. It is currently shown outstanding performance in the path planning of mobile robot [9, 10].

Reinforcement learning is an algorithm based on reward and punishment mechanism. In order to improve the ability of the agent to complete specific tasks, corresponding rewards and punishments will be set. In the path planning process, the agent will be rewarded by avoiding obstacles autonomously and reaching the target point. And the agent also will be punished when it collides with the obstacle or cannot reach the target point. The greedy thought of exploration and utilization is well reflected in the path planning with unknown environments, the greater the probability of exploration, the better the final path will be obtained. Theoretically, it can complete path planning in many complex environments. It also has strongly development capabilities and research significance [11, 12].

Q-learning is an efficient off-policy algorithm based on the Markov decision process in reinforcement learning [13, 14]. However, the traditional Q-learning initialization process is randomly. Due to the unknown environment information and the random planning of the initial path, efficiency of the path planning is low, the number of iterations is so many, and the calculation time is so long [15]. Therefore, this paper proposes an improved Q-learning algorithm for the problem that the agent cannot get the optimal path due to the large number of iterations of the Q-learning algorithm, the slow convergence speed and the sparse reward signal. This algorithm reduces the number of iterative runs of the agent in the path planning process by updating the iterative formula of the Q-value, and accelerates the convergence speed. At the same time, adding the sparse reward algorithm improves the problem that the agent is difficult to learn the optimal strategy from the interaction with the environment.

2   Reinforcement Learning

As a branch of the machine learning field, reinforcement learning is different from task-driven supervised learning algorithm and data-driven unsupervised learning algorithm [16]. Both supervised learning and unsupervised learning require corresponding data, but the biggest feature of reinforcement learning is that the agent can learn through its interaction with the environment autonomously without given label training data. Therefore, reinforcement learning algorithm is also one of the core technologies to improve the intelligent level of robots. As one of the hot issues in the field of machine learning in recent years, the research and development of reinforcement learning has great and far-reaching significance, and it has had rich achievements in theoretical research and practical applications.

2.1 Principles of Reinforcement Learning

The structure of reinforcement learning is shown in Fig. 1. The agent first obtains the initial state information from the environment, and then selects and executes the action corresponding to the current state of the agent according to the strategy function. After the action is executed, the environment state of the agent will be changed, and then it will get a reward to measure the quality of the action performed. The agent constantly adjusts its behavior according to the reward it received, and learns through its interaction with the environment, and then obtains the optimal strategy gradually. In the entire reinforcement learning process, the goal of the agent is to maximize the cumulative reward.

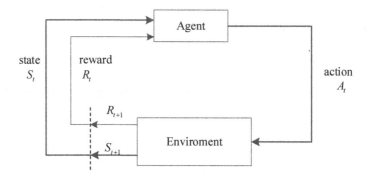

Fig. 1. The structure of reinforcement learning flow chart

The Markov decision process provides a clear theoretical framework for reinforcement learning. The inertial order decision problem handled by reinforcement learning can generally be formalized as an MDP model, which can be described by a four-tuple $\{S, A, R, P\}$, where:

S is the environment state space of the agent;
A is the action space performed by the agent;
R is the reward function, which is the evaluation of the degree of performance of the agent;
P is the state transition probability function. For each state $s \in S$ and each action $a \in A$, $P(s'|s, a)$ is the probability distribution of the next state transition to s'.

The agent obtains environmental state information $s_t \in S$ by observing at time t. According to the current state of the environment, the agent will adopt a corresponding strategy to select an appropriate action a_t from the action space A, which contains all possible actions, and the agent executes a_t. Then the agent will get a new state s_{t+1} and a reward r_t. Special attention is paid to the fact that the new state s_{t+1} only depends on the current state s_t and action a_t, and has no connection with the previous historical state and action.

For the traditional reinforcement learning algorithm, the reward function G_t is defined as the discount sum of the rewards that the agent obtains after the time step t:

$$G_t = \sum_{j=0}^{\infty} \gamma^j r_{t+j} = r_t + \gamma G_{t+1} \tag{1}$$

where $\gamma \in [0, 1)$ and it is the discount factor, which is used to indicate the degree of influence of future rewards on actions performed at the current state. In reinforcement learning, the goal of the agent is to find an optimal strategy π^* to maximize the expectation of the reward function.

2.2 Q-Learning Algorithm

The proposal of the timing differential control algorithm under the off-orbit strategy is an important breakthrough in the early stage of reinforcement learning, this algorithm is called Q-learning. Q-learning was cited by Watkins in 1989 [17], it is an efficient off-policy algorithm in reinforcement learning, and it is a typical model-free algorithm. Firstly, Q-learning algorithm establishes an initialized Q-value table, and then it updates the Q-value table continuously through the interaction between the agent and the environment. However, the renewal of the Q-table is different from the strategy of selecting actions. That is to say, when the Q-table is updated, it calculates the maximum reward value of the next state. And if the maximum value is taken, the corresponding action does not depend on the current strategy. With the continuously interaction with the environment, the agent changes its strategy constantly so that the actions tend to the optimal set of actions eventually. Q-learning uses a method to obtain the optimal strategy, which is iteration of the state-action pair $Q(s, a)$. The algorithm needs to examine the value of each state-action pair in each iteration. The basic form of the Q-learning algorithm is given in:

$$Q(s_t, a_t) = Q(s_t, a_t) + \alpha[r_t + \gamma \max_a Q(s_{t+1}, a) - Q(s_t, a_t)] \tag{2}$$

Where s_t is the state of the agent at time t. When the action a_t is executed in the state s_t, the state of the agent changes to s_{t+1} and the reward value r_t is obtained at the same time; r_t is the evaluation of the state a_t, which means that the agent obtains reward value from the current state to the next state with the action a_t; α is the learning rate; $\max_a Q(s_{t+1}, a)$ means that an action a is selected from the action space A to maximize the value of $Q(s_{t+1}, a)$.

The characteristic of the Q-learning algorithm is that it can compare the expected rewards of all alternative actions without considering the environmental model, and it can deal with random conversions and rewards without any modification. Theoretical research shows that it can effectively learn an optimal strategy for any MDP problem with limited state space and action space. However, in the research of mobile robot path planning, if the robot faces a complex environment with dense obstacles, the Q-learning algorithm will require a large number of iterative operations, so the convergence speed will be slow. And it is time-consuming, even it is difficult to find the optimal path due to the sparse reward problem. Therefore, Q-learning is difficult to apply to the path planning of mobile robots in complex state spaces. The corresponding improvements made in this paper based on the Q-learning algorithm will be discussed in the next chapter.

3   Algorithm Improvement Based on Q-Learning

3.1   Sparse Reward Algorithm

Reinforcement learning is widely used in the field of artificial intelligence and has obtained the remarkable result, setting the reward can lead the agent to learn in reinforcement learning algorithm. However, the problem of sparse reward signals exists in many researches on practical problems. Due to the lack of reward information, it is difficult for agent to learn the optimal strategy from the interaction with the environment, which is the sparse reward problem [18, 19].

Not only the sparse reward signals lead to learn slowly, due to the small number of reward samples, the variance of the value function estimate is large, and the sparse reward problem will also cause the model training to converge difficultly. The current algorithms to solve the sparse reward problem mainly include: reward shaping, imitation learning, course learning, replay of after-the-fact experience, curiosity-driven, and division into reinforcement learning, etc. [18]. In order to speed up the iterative speed of path planning and make the mobile robot to find the optimal path faster, this paper uses reward shaping algorithm combined with Q-learning, because it is easy to implement and the difficulty of the model is simple.

Reward shaping refers to design additional rewards based on prior knowledge artificially in order to guide the agent to complete the expected task successfully. Research shows that designing an appropriate additional reward function can effectively improve the sparse reward problem and accelerate the learning speed of the agent [20, 21]. Generally, if $r(s, a)$ represents the reward function of the original MDP, then $f(s, a)$ represents the additional reward function. The reward function of the new MDP after reward shaping is:

$$r'(s, a) = r(s, a) + f(s, a) \tag{3}$$

However, the optimal strategy learned in the new MDP problem may be not the optimal strategy of the original MDP process, it may lead the agent to learning non-theoretical optimal strategy after reward shaping. Existing research shows that the optimal strategy could be guaranteed to remain unchanged unless the additional reward function was expressed as the differential form of the potential energy function $\Phi(s)$. Therefore, setting an additional reward function directly in the path planning process may cause the robot to obtain a non-optimal path [22]. Existing research has proved that we can express the additional reward function as the difference form of the potential based function, it can guarantee that the optimal strategy remains unchanged. In order to ensure that the mobile robot finds the optimal path quickly, the additional reward function is set as the differential form of the potential based function $\Phi(s)$, the additional reward function is given in:

$$f(s, a) = \gamma\Phi(s_{t+1}) - \Phi(s) \tag{4}$$

Among them, γ is the discount factor in the original MDP, s and s_{t+1} are the state of the agent at the current moment and the state at the next moment. Respectively, the potential energy function $\Phi(s)$ is the mapping of the state of the state to the real number.

When setting the reward function in the robot path planning process, this paper adopts the method of reward shaping and introduces the additional reward function, which improves the problem of sparse reward and speeds up the learning process significantly.

3.2 Update Q-Value Iteration Formula

The convergence of the previous state depends on the next state and has no connection with the initial value in the Q-learning algorithm, and the convergence can be guaranteed without knowing the environment model. Therefore, the Eq. (2) is iterated once, and then:

$$Q(s_t, a_t) = (1 - \alpha)Q(s_t, a_t) + \alpha[r_t + \gamma \max_a Q(s_{t+1}, a)] \tag{5}$$

It can be seen that the necessary condition for the stability of the Q-value of the state s_t is that the state s_{t+1} corresponding to the $\max_a Q(s_{t+1}, a)$ is constant, otherwise the Q-value of the previous state will change with the change of the Q-value in the next state, because the whole is a backtracking process, and the states of all the previous actions cannot reach a stable value.

Therefore, the second iteration (denoted $Q_t = Q(s_t, a_t)$, $Q_{t+1} = \max_a Q(s_{t+1}, a)$) is performed on the basis of the formula (5), and the formula (6) is obtained:

$$Q_t = (1 - \alpha)[(1 - \alpha)Q_t + \alpha(r_t + \gamma Q_{t+1})] + \alpha[r_t + \gamma Q_{t+1}] \tag{6}$$

which is:

$$Q_t = (1 - \alpha)^2 Q_t + [1 - (1 - \alpha)^2][r_t + \gamma Q_{t+1}] \tag{7}$$

Perform the third iteration on the basis of the formula (7) to obtain the formula (8):

$$Q_t = (1 - \alpha)[(1 - \alpha)^2 Q_t + (1 - (1 - \alpha)^2)(r_t + \gamma Q_{t+1})] + \alpha[r_t + \gamma Q_{t+1}] \tag{8}$$

which is:

$$Q_t = (1 - \alpha)^3 Q_t + [1 - (1 - \alpha)^3][r_t + \gamma Q_{t+1}] \tag{9}$$

After iterating n times in the above way, and we can get the formula (10):

$$Q_t = (1 - \alpha)^n Q_t + [1 - (1 - \alpha)^n][r_t + \gamma Q_{t+1}] \tag{10}$$

Because $0 < \alpha < 1$, so $0 < (1 - \alpha) < 1$, when $n \to \infty$, $(1 - \alpha)^n \to 0$. Then $Q_t \to r_t + \gamma Q_{t+1}$, at this time $Q(s_t, a_t)$ converges.

By updating the formula (2), the entire Q-value expression will eventually be given in:

$$Q(s_t, a_t) = r_t + \gamma \max_a Q(s_{t+1}, a) \tag{11}$$

3.3   Path Planning Algorithm Flow

The improved algorithm is applied to the path planning of the mobile robot. The specific algorithm flow is shown in the Fig. 2:

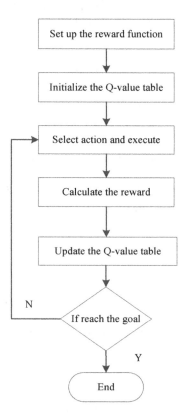

Fig. 2. Improved algorithm flowchart

4 Simulation Experiment and Analysis

In order to verify the effectiveness of the improved algorithm proposed in this paper in the mobile robot path planning, two sets of simulation experiments were carried out, which were verified in a simple map environment and a complex map environment. Then we set the action space contains 8 actions, which respectively are: move up, move down, move left, move right, and move up left, move down left, move up right, move down right. Among them, move up left, move down left, move up right, and move down right are the diagonal movement. Since the higher the learning rate and discount factor, the faster the Q-value converges, the easier it is to produce oscillations. However, in order to compare the difference between the two algorithms, the selection of the discount factor and the learning rate can be randomly. In these two groups of simulation experiments, we conducted many experiments with different α and γ according to the actual situation, and finally we found that the learning rate α is 0.6 and the discount factor γ is 0.9, the agent performs better in the environment we create.

4.1 Simple Map Environment Simulation

As shown in Fig. 3, a simple map simulation environment is established. The model size of the map is a two-dimensional grid environment of 20 * 20. The starting point coordinate of the robot in the map is (1, 1), and the target point coordinate is (20, 20). In the map environment, the robot is regarded as a movable mass point. The minimum movement unit is 1, and each grid represents a state of the robot. Among them, the black represents obstacles, the gray represents feasible areas, and the white represents starting point and target point.

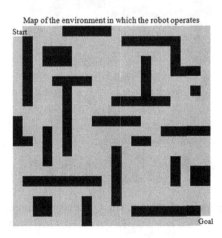

Fig. 3. Simple map simulation environment

Figure 4 shows the optimal path of the robot in the simple map environment with the improved algorithm, and Fig. 5 shows the optimal path display with traditional Q-learning in simple environment. Compared with the running result of the traditional Q-learning algorithm, we can see that the agent can choose diagonal actions quickly after improving algorithm, so that the optimal path has a smoother trajectory and fewer steps in Fig. 4.

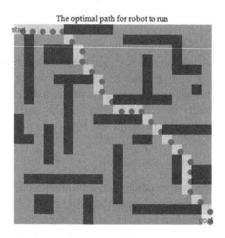

Fig. 4. The optimal path display after the improved algorithm in simple environment

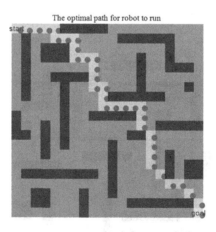

Fig. 5. The optimal path display with Q-learning in simple environment

In the simple environment, the simulation experiment data shows that the optimal path step length of the traditional Q-algorithm is 35, and the optimal path step size of the improved algorithm is 30.2, which is reduced by 13.7%.

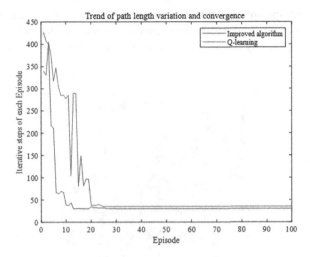

Fig. 6. The change curve of the step length and the number of iterations to the target point

Figure 6 shows the curve of the change in the number of iterations of each episode when the two algorithms are running in the simple map environment. It can be seen from the Fig. 6 that the improved algorithm has a smaller oscillation amplitude and the convergence speed is better than the traditional Q-learning algorithm.

4.2   Complex Map Environment Simulation

In order to verify that the improved algorithm still has better performance in complex environment maps, a more complex environment map is designed as shown in Fig. 7. The starting point coordinate of the robot is (4, 5) and the end point coordinate is (27, 27). The model size of the map is 30 * 30.

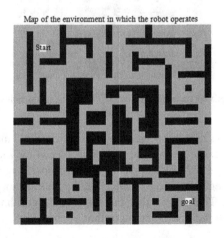

Fig. 7. Complex map simulation environment

Figure 8 shows the optimal path of the robot in the complex map environment after the improved algorithm, and Fig. 9 represents the running result of the optimal path with traditional Q-learning algorithm in the complex map environment.

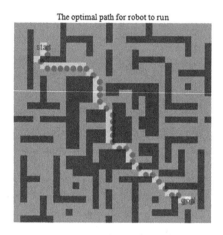

Fig. 8. The optimal path after improved algorithm in complex environment

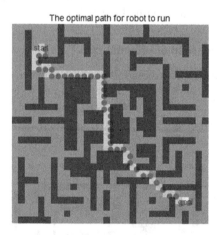

Fig. 9. The optimal path display with Q-learning in complex environment

From Fig. 9, we can see that the agent chooses the diagonal motion after a period of learning, and we can clearly see that the path trajectory in the second half is smoother than that in the first half. But in Fig. 8, the agent can choose diagonal motion according to the situation at the beginning, so that the overall path trajectory is smoother and the path length is shorter. This shows that the improved algorithm makes the learning speed faster compare to the traditional algorithm.

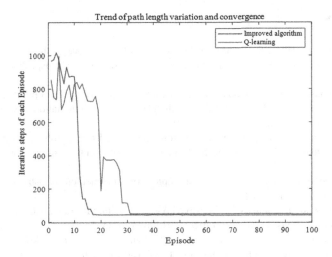

Fig.10. The change curve of the step length and the number of iterations to the target point

The simulation experiment data shows that in a complex environment, the optimal path step length of the traditional Q-algorithm is 50.2, and the optimal path step length of the improved algorithm is 42.2, which is reduced by 15.9%, and the optimal path are smoother. It can be seen the change curve of the path length and the number of iterations from Fig. 10, compare with traditional Q-learning, the convergence speed of the improved algorithm is increased obviously, and change of the curve is smooth.

5 Conclusion

In the research on path planning of mobile robot, the traditional Q-learning algorithm has a slower convergence speed and a large number of iterations. It is difficult to obtain the optimal path of robot movement, and it is not suitable for complex environments with large state spaces. The improved algorithm on the basis of the traditional Q-learning algorithm proposed in this paper, combines the sparse reward algorithm in the field of reinforcement learning, sets the reward function in the form of reward shaping, and updates the Q-value iteration formula. Simulation experiments show that the performance of the improved algorithm in the path planning of mobile robot is significantly better than the traditional Q-learning algorithm. And the improved algorithm has better path smoothness, faster convergence speed, and it performs well in a complex map environment. Therefore, the application of this improved algorithm in the path planning of mobile robot deserves further research.

Acknowledgments. This works is partly supported by the Natural Science Foundation of Liaoning, China under Grant 2019MS008, Education Committee Project of Liaoning, China under Grant LJ2019003.

References

1. Marin-Plaza, P., Hussein, A., Martin, D., et al.: Global and local path planning study in a ROS-based research platform for autonomous vehicles. J. Adv. Transp. **2018**(PT.1), 1–10 (2018)
2. Li, G., Yamashita, A., Asama, H., et al.: An efficient improved artificial potential field based regression search method for robot path planning. In: International Conference on Mechatronics & Automation. IEEE (2012)
3. Chang, L., Shan, L., Jiang, C., et al.: Reinforcement based mobile robot path planning with improved dynamic window approach in unknown environment. Auton. Robot. **45**(2), 51–76 (2020)
4. Liu, G., Lao, S.Y., Tan, D.F., et al.: Fast graphic converse method for path planning of anti-ship missile. J. Ballist. (2011)
5. Tang, X.R., Zhu, Y., Jiang, X.X.: Improved A-star algorithm for robot path planning in static environment. J. Phys. Conf. Ser. **1792**(1), 012067 (2021). 8p.
6. Shen, Z., Hao, Y., Li, K.: Application research of an Adaptive Genetic Algorithms based on information entropy in path planning. Int. J. Inf. **13**(6) (2010)
7. Wang, G., Zhou, J.: Dynamic robot path planning system using neural network. J. Intell. Fuzzy Syst. **40**(2), 3055–3063 (2021)
8. Zhang, C., You, X.: Improved quantum ant colony algorithm of path planning for mobile robot based on grid model. Electron. Sci. Technol. (2016)
9. Gaskett, C., Fletcher, L., Zelinsky, A.: Reinforcement learning for a vision based mobile robot. IEEE (2018)
10. Kamalapurkar, R., Walters, P., Rosenfeld, J., et al.: Model-Based Reinforcement Learning for Approximate Optimal Control (2018)
11. Polydoros, A.S., Nalpantidis, L.: Survey of model-based reinforcement learning: applications on robotics. J. Intell. Rob. Syst. **86**(2), 1–21 (2017)
12. Tong, L., Wang, J.: Application of reinforcement learning in robot path planning. Comput. Simul. **30**(012), 351–355 (2013)
13. Huang, C., Sheng, Z., Jie, X., et al.: Markov Decision Process (2018)
14. Surya, S., Rakesh, N.: Traffic Congestion Prediction and Intelligent Signaling Based on Markov Decision Process and Reinforcement Learning (2018)
15. Chu, P., Vu, H., Yeo, D., Lee, B., Um, K., Cho, K.: Robot reinforcement learning for automatically avoiding a dynamic obstacle in a virtual environment. In: Park, J.J.H., Chao, H.-C., Arabnia, H., Yen, N.Y. (eds.) Advanced Multimedia and Ubiquitous Engineering. LNEE, vol. 352, pp. 157–164. Springer, Heidelberg (2015). https://doi.org/10.1007/978-3-662-47487-7_24
16. Chen, L.P., Mohri, M., Rostamizadeh, A., Talwalkar, A.: Foundations of machine learning, second edition. Statistical Papers 60 (2019)
17. Watkins, C., Dayan, P.: Q-learning. Machine Learning (1992)
18. Yang, R., Yan, J.P., Li, X.: Research on sparse reward algorithm for reinforcement learning: theory and experiment . J. Intell. Syst. **15**(05), 888–899 (2020)
19. Riedmiller, M., Hafner, R., Lampe, T., et al.: Learning by Playing-Solving Sparse Reward Tasks from Scratch (2018)
20. Gullapalli, V., Barto, A.G.: Shaping as a Method for Accelerating Reinforcement Learning. IEEE (2002)
21. Wiewiora, E.: Reward Shaping. Springer, Heidelberg (2011)
22. Ng, A.Y., Harada, D., Russell, S.: Policy invariance under reward transformations: theory and application to reward shaping. In: ICML, vol. 99, pp. 278–287 (1999)

OnSum: Extractive Single Document Summarization Using Ordered Neuron LSTM

Xue Han[1], Qing Wang[2], Zhanheng Chen[3], Lun Hu[4],
and Pengwei Hu[5(✉)]

[1] IBM Research, Beijing, China
[2] Institute of Computing Technology, Chinese Academy of Science,
Beijing, China
[3] Shenzhen University, Shenzhen, China
[4] Xinjiang Technical Institute of Physics and Chemistry,
Chinese Academy of Sciences, Beijing, China
[5] Merck China Innovation Hub, Shanghai, China

Abstract. A growing trend of extractive summarization research is to take the document structure into account which has been shown to correlate with the important contents in a text. However, building complex document structures such as Rhetorical Structure Theory (RST) is time-consuming and requires a large effort to prepare labeled training data. Therefore, how to effectively learn a document structure for summarization remains an open question. Recent findings in the language model area show that the syntactic distance of the basic semantic unit could be used to induce the syntactic structure without any extra labeled data. Inspired by these findings, we propose to extend the basic semantic units from words to sentences and extract the syntactic distance of each sentence in the document by building an ON-LSTM (Ordered Neuron LSTM) based model with the document-level language model objective. We then leverage these syntactic distances to evaluate whether a sentence could be extracted to the summary. Our model achieves state-of-the-art performance in terms of Rouge-1 (48.67) and Rouge-2 (26.32) on the CNN/Daily Mail data set.

Keywords: Extractive summarization · Ordered neuron LSTM · Document structure

1 Introduction

Document summarization is an important NLP (Natural Language Processing) task due to its potential for various applications. Summarization aims to produce a shorter version of a text while preserving its principal information content. Existing proposed summarization methods can be divided into two categories: extractive and abstractive. The objective of extractive summarization methods is to select and subsequently concatenating the most important text snippets (usually are sentences) from the document, while abstractive summarization techniques aim to concisely paraphrase the information contained in the text. Extractive techniques are very attractive as they are more reliable to yield semantically and grammatically correct sentences [1]. Extractive

© Springer Nature Switzerland AG 2021
D.-S. Huang et al. (Eds.): ICIC 2021, LNCS 12837, pp. 605–615, 2021.
https://doi.org/10.1007/978-3-030-84529-2_51

summarization could also meet the requirements of many application scenarios, that require the generated summarization do not introduce irrelevant or wrong information, such as dialogue generation [2]. Therefore, we focus on extractive summarization in this work.

A vast majority of the literature on document summarization is devoted to extractive summarization. A growing trend is to take the structure of the document into account which has been shown to correlate with what readers perceive as important in a text [3]. On the basis of the document structure information, recurrent neural networks are trained to predict a label for each sentence specifying whether it should be included in the summary [4]. There are various methods proposed differently from the ways of obtaining the document structure. Most of them [7] try to train a parser out of discourse annotated corpora such as Rhetorical Structure Theory (RST) [8]. The reliance on the parser which is both expensive to obtain (since it must be trained on labeled data) and error-prone, presents a major obstacle to their widespread use. Other methods include imposing structural constraints on the basic attention mechanism [13], formulating structure learning as a reinforcement learning problem [12]. Despite their success, how to effectively learn a document structure for summarization remains an open question [14].

In this paper, we are aiming at learning the structure of a document by capturing the syntactic distance of sentences without extra human-labeled knowledge. Recent findings in the language model area show that the syntactic distances of basic semantic units (e.g. words in a sentence) could be used to induce the syntactic structure of a sentence [26] without any labeled data. The syntactic distances are higher when there is a higher level of the constituent boundary between the corresponding pair of the basic semantic units. Inspired by these findings, we propose to extend the basic semantic units from words to sentences and extract the syntactic distance of each sentence in the document by building a model with the document-level language model objective. Our proposal is based on the wide agreement that text units such as words, clauses, or sentences are usually not isolated. Instead, they correlate with each other to form coherent and meaningful discourse together [27]. We then leverage these syntactic distances to evaluate whether a sentence could be extracted to the summary. The main contributions of this work are:

- We propose to train a document-level language model on the basis of the state-of-the-art distance-based model: Ordered Neurons LSTM (ON-LSTM) [17]. We retrieve the syntactic distance information from the trained model and joint train an MLP (Multilayer Perceptron) to predict a label for each sentence specifying whether it should be included in the summary.
- Our proposed ON-LSTM based summarization method (OnSum) learns to identify the relative importance of sentences without recourse to any expensive discourse parser or additional annotations.
- We conduct experiments on the benchmark dataset CNN/Daily Mail. Our model achieves state-of-the-art performance in terms of Rouge-1 (48.67) and Rouge-2 (26.32).

2   Related Work

A very popular algorithm for extractive single-document summarization is TextRank [15]. It represents document sentences as nodes in a graph with undirected edges whose weights are computed based on sentence similarity. In order to decide which sentence to include in the summary, a node's centrality is often measured using graph-based ranking algorithms such as PageRank [28].

Furthermore, a growing trend is to frame this sentence selection process as a sequential binary labeling problem, where binary inclusion/exclusion labels are chosen for sentences one at a time, starting from the beginning of the document, and decisions about later sentences may be conditioned on decisions about earlier sentences. Recurrent neural networks may be trained with stochastic gradient ascent to maximize the likelihood of a set of ground-truth binary label sequences [5, 6]. However, this approach has two well-recognized disadvantages. Firstly, it suffers from exposure bias, a form of mismatch between training and testing data distributions which can hurt performance [29]. Secondly, extractive labels must be generated by a heuristic, as summarization datasets do not generally include ground-truth extractive labels.

Recent approaches to (single-document) extractive summarization frame the task as a sequence labeling problem taking advantage of the success of neural network architectures. The idea is to predict a label for each sentence specifying whether it should be included in the summary. Existing systems mostly rely on recurrent neural networks to model the document and obtain a vector representation for each sentence [5, 30].

Inter-sentential relations are captured in a sequential manner, without taking the structure of the document into account, although the latter has been shown to correlate with what readers perceive as important in a text [3]. The summarization literature offers examples of models which exploit the structure of the underlying document, inspired by existing theories of discourse such as Rhetorical Structure Theory (RST). Most approaches produce summaries based on tree-like document representations obtained by a parser trained on discourse annotated corpora [31]. For instance, [3] argues that a good summary can be generated by traversing the RST discourse tree structure top-down, following nucleus nodes (discourse units in RST are characterized regarding their text importance; nuclei denote central units, whereas satellites denote peripheral ones). Other work [35] extends this idea by transforming RST trees into dependency trees and generating summaries by tree trimming. [33] summarize product reviews; their system aggregates RST trees representing individual reviews into a graph, from which an abstractive summary is generated. Despite the intuitive appeal of discourse structure for the summarization task, the reliance on a parser which is both expensive to obtain (since it must be trained on labeled data) and error prone, presents a major obstacle to its widespread use.

Recognizing the merits of structure-aware representations for various NLP tasks, recent efforts have focused on learning latent structures (e.g., parse trees) while optimizing a neural network model for a down-stream task. Various methods impose structural constraints on the basic attention mechanism [32], formulate structure learning as a reinforcement learning problem [36], or sparsify the set of possible structures [34]. Although latent structures are mostly induced for individual sentences, [32] induce dependency-like structures for entire documents.

However, despite their success, the reliance on a parser which is both expensive to obtain (since it must be trained on labeled data) and error-prone, becomes a major obstacle.

3   Methodology

We define the extractive summarization as a binary classification task with labels indicating whether a sentence should be extracted to represent the important meanings of a document. Given a document $Doc = (S_1, S_2, \ldots, S_l)$ containing L sentences, where S_l is the l_{th} sentence in the document. Each sentence S_l in Doc is represented as $S_l = [w_1^l, w_2^l, \ldots, w_t^l]$, where w_t^l means the t_{th} word in l_{th} sentence. An extractive summarization task is to assign a label $y_l \in \{0, 1\}$ to each S_l, indicating whether the sentence should be included in the summary.

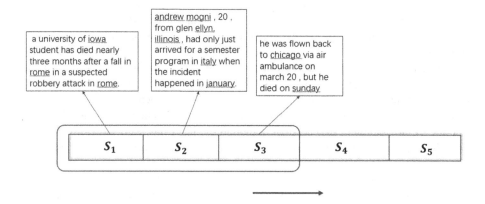

Fig. 1.   Example of document-level language model.

We propose to extract the syntactic distance of each sentence in the document by building a model that leverages ON-LSTM [17] with the document-level language model objective. The encoder is a hierarchical model that consists of two Bi-LSTMs (bidirectional LSTM) to represent the meaning of the context by word-sentence-document hierarchy. The decoder is an ON-LSTM network which is trained with document-level language model objective without any extra human-labeled knowledge. We leverage previous sentences to predict next sentence like language model predicting upcoming words from prior word context [37]. As shown Fig. 1, we use previous sentences sequence (S_1, S_2, S_3) to predict S_4. Finally, the syntactic distance of the sentence which is the hidden vector of the decoder could be input into an MLP to predict the label of each sentence. The overall architecture of the proposed approach is as Fig. 2, where H_t^e and H_t^d represent all the hidden vectors in encoder and decoder separately.

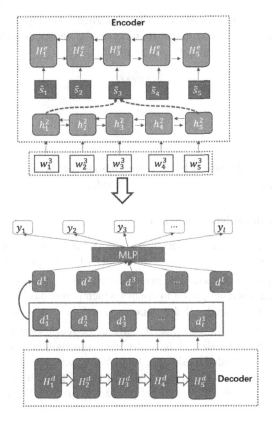

Fig. 2. Architecture of Onsum

3.1 Document Encoder

We firstly introduce the document encoder component. We follow previous work [18] in modeling documents hierarchically by first obtaining representations for sentences and then composing those into a document representation. The encoder consists of a sentence level Bi-LSTM network to process the words and a document level Bi-LSTM network to process the sentences.

Sentence level encoder takes the word embedding u_t^l for each word w_t^l in sentence S_l as input. Then process the sentences with Bi-lSTM as below:

$$h_t^l = BiLSTM(h_{t-1}^l, u_t^l) \tag{1}$$

Then we average each word hidden representation as to the sentence representation:

$$\tilde{s}_l = \frac{1}{n_l}\sum_{t=1}^{n_l} h_t^l \tag{2}$$

where n_l is the number of word in the l_{th} sentence.

Document level encoder is built upon the sentence level representation to derive a global representation of the document. It processes the sentences with another BiLSTM that takes the sentence representation $\tilde{s}_l$ as input:

$$o_l = BiLSTM(o_{l-1}, \tilde{s}_l) \tag{3}$$

$$c = \frac{1}{L}\sum_{j=1}^{L} o_j \tag{4}$$

Where o_l is the document-level representation of S_l and c is the document embedding for sentence l that fed in to the decoder.

3.2   Decoder with Ordered Neurons

We use an ON-LSTM based decoder to train a document-level language model as [25]. Compared with standard LSTM architecture, ON-LSTM introduces novel ordered neuron rules to update cell state. The master forget gate $\tilde{f}_t$ and the master input gate $\tilde{i}_t$ are newly introduced as a split point to control which part should be updated or retained and the cell state is ordered, all the neurons that follow it in the ordering are also updated. The product of the two master gates a_t represents the overlap of $\tilde{f}_t$ and $\tilde{i}_t$. Whenever the overlap exists $(\exists k, a_{tk} > 0)$, the corresponding segment of neurons encodes are further controlled by the standard LSTM gates f_t and i_t. ON-LSTM further introduced a new activation function CUMSUM to find the splitting point d. Based on this activation function, the master forget gate are defined as Eq. (5):

$$\tilde{f}_t = CU_f(x_t, h_{t-1}) \tag{5}$$

where x_t is the current input and h_{t-1} is the hidden state of previous step. CU_f is an individual activation function with its own trainable parameters.

To be specific, the decoder predicts the next sentence in the document word by word given the contextual document information and the previous sentence information. Each word probability could be calculated by:

$$\bar{h}_t^l = ONLSTM_{decoder}(o_{l-1}; w_{t-1}^l; \bar{h}_{t-1}^l) \tag{6}$$

$$\widehat{h}_l = [\bar{h}_t^l; c] \tag{7}$$

$$\tilde{h}_t^l = \tanh(W_p.\widehat{h}_t^l) \tag{8}$$

$$P(w_t|w_{1:t-1}^l, S_{1:l-1}) = \frac{\exp\left(\tilde{h}_t^l.w_t^l\right)}{\sum_{j=1}^{|V|} \exp\left(\tilde{h}_t^l.w_j\right)} \tag{9}$$

Where W_p is a projection matrix to transform the decoder hidden representation into the word embedding space. $|V|$ is the vocabulary size.

The objective of the decoder is to maximize the log-likelihood of the target sentences:

$$L = \sum_t \log P(w_t^l | w_{1:t-1}^l, S_{1:l-1}) \tag{10}$$

After training the ON-LSTM based document-level language model, the syntactic distance of each sentence in the document could be inferred. Specifically, given a document as input, each layer of trained ON-LSTM based decoder infers a value $d_l = (d_1^l, d_2^l, \ldots, d_t^l)$ that measures the syntactic distance between S_{l-1} and S_l. An estimate of the t_{th} element of d_l is calculated in Eq. (11):

$$\hat{d}_t^l = E(d_t^l) = D_m - \sum_{k=1}^{D_m} \tilde{f}_{tk}^l \tag{11}$$

where D_m is the size of the hidden state. $\tilde{f}_{tk}^l$ refers to the k_{th} element in the master forget gate $\tilde{f}_t^l$ as in Eq. (5).

3.3 Summarization Classifier

Finally, we need to label the sentence included in the summary according to the syntactic distance d_l of each sentence. The final output layer is an MLP based classifier:

$$\hat{y}_t = MLP(\hat{d}_1^l, \hat{d}_2^l, \ldots, \hat{d}_t^l) \tag{12}$$

Where $\left(\hat{d}_1^l, \hat{d}_2^l, \ldots, \hat{d}_t^l\right)$ is the vector representation of d_l from the hidden states of ON-LSTM decoder. The loss of the classifier is the binary classification entropy of prediction $\hat{y}_l$ against gold label y_l.

4 Experiments

4.1 Data Sets

We perform experiments on two divergent mainstream data sets to verify the effectiveness of our proposed method as follows.

CNN/DailyMail [19]: we use extractive summarization CNN/Daily Mail corpus created by (Liu, 2019b) as training set in our experiments. In addition to the traditional CNN/DailyMail which contains news articles and associated highlights, this corpus including the labels for extractive summarization which are determined using a rule-based system.

XSum [12]: we use XSum as test data set which is a one-sentence summary data set to answer the question What is the article about?.

4.2   Experimental Setup

We perform experiments on the mainstream data set CNN/DailyMail [19] to verify the effectiveness of our proposed method. We followed the conventional split of 287,226/13,368/11,490 samples for training, validation, and testing, respectively [6]. For the encoder, it is found that stacking BiLSTMs works better than a one-layer BiLSTM [21]. Therefore, we used 3 layers for the Stacked BiLSTMs to encode both the sentence-level and document-level inputs. For each input word, we used word embeddings with pre-trained BERT model "bert-base-uncased"[1]. The word embedding dimension is 768. The vocabulary is the default vocabulary provided by the BERT model. The size of the vocabulary is 30522 tokens. For the decoder, we used three-layer ON-LSTM model[2]. We only modified the objective function which is defined as Eq. (10). The hidden size of ON-LSTM is 768. We trained our whole model using Adam optimizer with learning rate 0.003 and batch size 32. Embedding dropouts for both encoder and decoder inputs are 0.1. We also followed the same hyper-parameters from the original ON-LSM [17]. We used the syntactic distance of the second ON-LSTM layer which is found to perform best. We trained the summarization classifier using an MLP network in which the number of neural units is the same with the maximum document length.

4.3   Evaluation Results

Table 1. Results on CNN/DM test set.

Models	Rouge-1	Rouge-2	Rouge-L
Lead3 Baseline [23]	40.34	17.70	36.57
NeuSum [18]	41.59	19.01	37.98
BERTSUM [19]	43.25	20.24	39.63
MATCHSUM [24]	44.41	20.86	40.55
SUMO [13]	41.0	18.4	37.2
DISCOBERT [7]	43.77	20.85	40.67
OnSum	48.67	26.32	39.64

Following previous work, we used ROUGE-1, ROUGE- 2 and ROUGE-L of ROUGE F [22] to evaluate our experimental results. We used version 1.5.5 of the ROUGE script to evaluate the summary output.

Results on CNN/Daily Mail

Table 1 illustrates the results on two kinds of state-of-the-art models. The first five models (Lead3 baseline, NeuSum, BERTSUM and MATCHSUM) are neural network-based extractive models including models basing on similar hierarchical encoders.

---

[1] https://huggingface.co/transformers/pretrained_models.html.

[2] https://github.com/bojone/on-lstm.

OnSum outperforms the strong baseline MATCHSUM in terms of Rouge-1 and Rouge-2 by 4.26 and 5.46 points separately. OnSum also shows comparative performance on Rouge-L scores. We also compared with the state-of-the-art models (SUMO and DISCOBERT) which utilize the discourse information. OnSum also outperforms the strong baseline DISCOBERT in terms of Rouge-1 and Rouge-2 separately while saving the effort to learn the document structure.

Results on XSum
We also test the CNN/Mail Daily trained model on XSum data set. We also split the XSum data set which contains 226,711 news articles as 90:10 into training and test sets. We just used the test sets to evaluate our model trained on CNN/Mail Daily. We shuffled the XSum data set and get 10 test sets randomly to evaluate. Table 2) illustrates the two base lines methods (BERTSUM and MATCHSUM) which are trained on the training set of XSum. OnSum shows strong performance on Rouge-2 score. Results in terms of Rouge-1 and Rouge-L are also promising. It demonstrated that OnSum is robust on different data corpus.

Table 2. Results on XSum test set.

Models	Rouge-1	Rouge-2	Rouge-L
BERTSUM [19] (Liu, 2019b)	38.76	16.33	39.63
MATCHSUM [24] (Zhong, 2020)	24.86	14.66	18.41
OnSum	33.72	19.57	26.87

5 Conclusion

We propose a novel sentence unit syntactic distance based framework to select the sentences in summarization. We did experiments on two of the popular benchmark data sets, CNN/Mail Daily and XSum. Experimental results show that OnSum outperforms or achieves competitive results compared with the current state-of-the-art extractive model on popular benchmark data sets.

References

1. Dong, Y., Shen, Y., Crawford, E., van Hoof, H., Cheung, J.C.K.: Banditsum: extractive summarization as a contextual bandit. In: Proceedings of the 2018 Conference on Empirical Methods in Natural Language Processing. Brussels, Belgium (2018)
2. Su, H., et al.: Diversifying dialogue generation with non-conversational text. In: Proceedings of the 58th Annual Meeting of the Association for Computational Linguistics (2020)
3. Marcu, D.: A decision-based approach to rhetorical parsing. SIGIR (1999)
4. Nallapati, R., Zhai, F., Zhou, B.: SummaRuNNer: a recurrent neural network based sequence model for extractive summarization of documents. In: Proceedings of the Thirty-First AAAI Conference on Artificial Intelligence. San Francisco, California USA (2017)

5. Cheng, J., Lapata, M.: Neural summarization by extracting sentences and words. In: Proceedings of the 54th Annual Meeting of the Association for Computational Linguistics, Berlin, Germany (2016)

6. Nallapati, R., Zhou, B., Santos, C., Gulcehre, C., Bing, X.: Abstractive text summarization using sequence-to-sequence RNNs and beyond. In: Proceedings of the 54th Annual Meeting of the Association for Computational Linguistics, Berlin, Germany (2016)

7. Xu, J., Gan, Z., Cheng, Y., Liu, J.: Discourse-aware neural extractive text summarization. In: Proceedings of the 58th Annual Meeting of the Association for Computational Linguistics (2020)

8. Mann, W.C., Thompson, S.A.: Rhetorical structure theory: toward a functional theory of text organization. Text Interdisc. J. Study Discours. **8** (1998)

9. Carlson, L., Marcu, D., Okurowski, M.E.: Building a discourse-tagged corpus in the framework of rhetorical structure theory. In: van Kuppevelt, J., Smith, R.W. (eds.) Current and New Directions in Discourse and Dialogue. Text, Speech and Language Technology, vol. 22. Springer, Dordrecht (2003). https://doi.org/10.1007/978-94-010-0019-2_5

10. Prasad, R., et al.: The penn discourse treebank 2.0. In: LREC, Marrakech, Morocco (2008)

11. Zhao, K., Huang, L.: Joint syntacto discourse parsing and the syntacto-discourse treebank. In: Proceedings of the 2017 Conference on Empirical Methods in Natural Language Processing. Copenhagen, Denmark (2017)

12. Narayan, S., Cohen, S.B., Lapata, M.: Ranking sentences for extractive summarization with reinforcement learning. In: Proceedings of the 2018 Conference of the North American Chapter of the Association for Computational Linguistics, New Orleans, Louisiana USA (2018)

13. Yang, L., Titov, I., Lapata, M.: Single document summarization as tree induction. In: Proceedings of the 2019 Conference of the North American Chapter of the Association for Computational Linguistics, Minneapolis, Minnesota USA (2019)

14. Wang, D., Liu, P., Zheng, Y., Qiu, X., Huang, X.: Heterogeneous graph neural networks for extractive document summarization. In: Proceedings of the 58th Annual Meeting of the Association for Computational Linguistics (2020)

15. Mihalcea, R., Tarau, P.: Textrank: bringing order into text. In: Proceedings of the 2004 Conference on Empirical Methods in Natural Language Processing, Barcelona, Spain (2004)

16. Brin, S., Page, M.: Anatomy of a large-scale hypertextual Web search engine. In: Proceedings of the 7th Conference on World Wide Web, pp. 107–117 (1999)

17. Shen, Y., Tan, S., Sordoni, A., Courville, A.: Ordered neurons: integrating tree structures into recurrent neural networks. In: Proceedings of Seventh International Conference on Learning Representations (2019)

18. Zhou, Q., Yang, N., Wei, F., Huang, S., Zhou, M., Zhao, T.: Neural document summarization by jointly learning to score and select sentences. In: Proceedings of the 56th Annual Meeting of the Association for Computational Linguistics, Melbourne, Australia (2018)

19. Liu, Y., Lapata, M.: Text summarization with pretrained encoders. In: Proceedings of the 2019 Conference on Empirical Methods in Natural Language Processing and the 9th International Joint Conference on Natural Language Processing, Hong Kong, China (2019)

20. Narayan, S., Cohen, S.B., Lapata, M.: Dont give me the details, just the summary! topic-aware convolutional neural networks for extreme summarization. In: Proceedings of the 2018 Conference on Empirical Methods in Natural Language Processing, Brussels, Belgium (2018)

21. Chen, D.: Neural reading comprehension and beyond. Stanford University (2018)

22. Lin, C.-Y.: ROUGE: a package for automatic evaluation of summaries. In: Text summarization branches out: Proceedings of the ACL-04 Workshop, vol. 8 (2004)

23. See, A., Liu, P.J., Manning, C.D.: Get to the point: summarization with pointer-generator networks. In: Proceedings of the 55th Annual Meeting of the Association for Computational Linguistics, Vancouver, Canada (2017)

24. Zhong, M., Liu, P., Chen, Y., Wang, D., Qiu, X., Huang, X.: Extractive summarization as text matching. In: Proceedings of the 58th Annual Meeting of the Association for Computational Linguistics (2020)

25. Kang, D., Hovy, E.: Linguistic versus latent relations for modeling coherent flow in paragraphs. In: Proceedings of the 2019 Conference on Empirical Methods in Natural Language Processing and the 9th International Joint Conference on Natural Language Processing, Hong Kong, China (2019)

26. Du, W., Lin, Z., Shen, Y., O'Donnell, T.J., Bengio, Y., Zhang, Y.: Exploiting syntactic structure for better language modeling: a syntactic distance approach. In: Proceedings of the 58th Annual Meeting of the Association for Computational Linguistics (2020)

27. Wang, Y., Li, S., Yang, C.-Y., Sun, X., Wang, H.: Tag-enhanced tree-structured neural networks for implicit discourse relation classification. In: Proceedings of the Eighth International Joint Conference on Natural Language Processing, Taipei, Taiwan (2017)

28. Page, L., Brin, S., Motwani, R., Winograd, T.: The PageRank Citation Ranking: Bringing Order to the Web (1999)

29. Paulus, R., Xiong, C., Socher, R.: A deep reinforced model for abstractive sum- marization. In: Proceedings of the 6th International Conference on Learning Representations, Vancouver, BC, Canada (2018)

30. Nallapati, R., Zhai, F., Zhou, B.: SummaRuNNer: A recurrent neural network based sequence model for extractive summarization of documents. In Proceedings of the 31st AAAI Conference on Artificial Intelligence (2017)

31. Carlson, L., Marcu, D., Okurowski, M.E.: RST Discourse Treebank (RST– DT) LDC2002T07. Linguistic Data Consortium, Philadelphia (2002)

32. Liu, J., Cohen, S.B., Lapata, M.: Discourse representation structure parsing. In: 429 Proceedings of the 56th Annual Meeting of the Association for Computational Linguistics (Long Papers), Melbourne, Australia, 15–20 July, pp. 429–439 (2018)

33. Gerani, S., Mehdad, Y., Carenini, G., Ng, R.T., Nejat, B.: Abstractive summarization of product reviews using discourse structure. In: Proceedings of the Conference on Empirical Methods in Natural Language Processing (EMNLP) (2014)

34. Niculae, V., Martins, A.F.T., Cardie, C.: Towards dynamic computation graphs via sparse latent structure. In: Proceedings of EMNLP (2018)

35. Hirao, T., Yoshida, Y., Nishino, M., Yasuda, N., Nagata, M.: Single-document summarization as a tree knapsack problem. In: Proceedings of the Conference on Empirical Methods in Natural Language Processing (EMNLP) (2013)

36. Yogatama, D., Blunsom, P., Dyer, C., Grefenstette, E., Ling, W.: Learning to compose words into sentences with reinforcement learning. In Proceedings of ICLR (2017)

37. Jurafsky, D., Martin, J.H.: Speech and Language Processing, Third Edition draft (2020). https://web.stanford.edu/~jurafsky/slp3/

Diagnosing COVID-19 on Limited Data: A Comparative Study of Machine Learning Methods

Rita Zgheib[1] , Firuz Kamalov[1] , Ghazar Chahbandarian[2] ,
and Osman El Labban[3]( )

[1] School of Computer Engineering, Canadian University Dubai, Dubai, UAE
{rita.zgheib,Firuz}@cud.ac.ae
[2] Institut de Recherche en Informatique de Toulouse, Toulouse, France
[3] Head of Family Medicine Department, Al Zahra Hospital, Dubai, UAE
osman.labban@azhd.ae

Abstract. Given the enormous impact of COVID-19, effective and early detection of the virus is a crucial research question. In this paper, we compare the effectiveness of several machine learning algorithms in detecting COVID-19 virus based on patient's age, gender, and nationality. The results of the experiments show that neural networks, support vector machines, and gradient boosting decision tree models achieve an 89% accuracy, and the random forest model produces an 87% accuracy in the identification of the COVID-19 cases.

Keywords: COVID-19 detection · Neural network · SVM · Random forest · GBDT

1 Introduction

COVID-19 is a type of virus known to cause respiratory infections in humans. It attacks the respiratory system and causes illnesses such as cough, fever, fatigue, and breathlessness[1]. It has rapidly expanded worldwide and is severely posing enormous health, economic, environmental, and social challenges to the entire human population. According to the World Health Organization report (WHO as of April 14, 2021), the current outbreak of COVID-19 has affected over 130,000,000 people and killed around 3,000,000 people in more than 200 countries throughout the world. Till now, there is no report of any clinically approved antiviral drugs. Many vaccine solutions have been proposed by Pfizer, Moderna, Sinopharm, Aztrazenica, Sputnik, etc. As of the day of writing, many countries have started to vaccinate the elderly population and health staff. With a wealth of research demonstrating vaccines' efficiency and safety simultaneously, many arguments against vaccines have been distinguished. To summarize the situation, after more than two months of vaccination, governments are still putting more confinements, expecting more lockdowns and no immediate complete solution.

---

[1] https://www.who.int/emergencies/diseases/novel-coronavirus-2019/question-and-answers-hub/q-a-detail/coronavirus-disease-COVID-19#:∼:text=symptoms.

D.-S. Huang et al. (Eds.): ICIC 2021, LNCS 12837, pp. 616–627, 2021.
https://doi.org/10.1007/978-3-030-84529-2_52

Decision support models have been widely adopted in the medical field and diagnose illnesses such as diabetes [7,13]. Many studies have been done to predict diseases and epidemics, such as in [19] to the extent that today's humans take advantage of decision support models and intelligent methods to predict. Research and tech companies have issued a call for global artificial intelligence (AI) researchers to develop novel techniques to assist COVID-19-related research. The large-scale data of COVID-19 patients can be integrated and analyzed by advanced machine learning algorithms to understand the viral spread pattern better. Further, it can improve diagnostic speed and accuracy, develop novel effective therapeutic approaches, and potentially identify the most susceptible people based on personalized genetic and physiological characteristics.

As a response to the call, research solutions have proposed the support of AI, the Internet of Things (IoT), Big Data, and Machine Learning (ML) to fight, go back to everyday life, and look forward against new diseases. For instance, in [2] the authors proposed a Recovered-Deceased model with a recursive approach using Neural Network, and SIR-based formulation guiding government actions mainly in two fundamental aspects: real-time data assessment and dynamic predictions of COVID-19 curves. Most of these projects converge on predicting and forecasting the number of confirmed and recovered cases of COVID-19 in the upcoming days, like in [4]. This quantitative approach was based on a Gaussian spreading hypothesis that emerged from imposed measures in a simple dynamical infection model. Hence, an NN-based curve fitting model has been proposed to predict COVID-19 spread and death cases in India, the USA, the UK, and France based on the data trend up to the first week of May 2020 [12].

In previous work [18], we proposed an NN model combined with IoT to predict if a patient has COVID-19 or not in a public area like universities based on criteria like age, gender, nationality, primary diagnosis, and secondary diagnosis. This approach aims to reduce healthcare workers' burden by reducing the daily unnecessary PCR tests required to go to work. Healthcare workers can focus then on helping patients with higher risk. Moreover, the proposed model allows for regaining everyday life by implementing it in public places like the university. Students and professors could go back to their regular professional activity safely. This approach works well with patients that present symptoms. However, asymptomatic patient's data needs to be studied further.

Motivated by the need to develop a full solution to aid in the battle against the COVID-19 pandemic, we present in this paper a comparative study of the performance of four ML algorithms Random Forest (RF), Neural Networks (NN), Support Vector Machine (SVM), and Gradient Boosting Decision Tree (GBDT), in identifying COVID-19. Each of the algorithms is implemented to predict whether a person has caught COVID-19 disease or not based on three factors only: age, gender, and nationality. One of the main strengths of the paper is the real dataset from a local hospital in Dubai, Al Zahra hospital Dubai[2], that is analyzed and studied.

The rest of the paper is organized as follows: First, we present in Sect. 2 the different algorithms that have been used in this study. Section 3 describes the dataset

---

[2] https://azhd.ae/.

and characteristics of 9416 patients who run a PCR test et al. Zahra hospital. In Sect. 4, the methodology adopted in this research is presented. Section 5 presents and discusses the results of experiments conducted to evaluate the algorithm's efficacy in both a quantitative and qualitative manner. Finally, conclusions are drawn, and future directions are discussed in Sect. 6.

2   Related Work

Currently the main reliable method to detect COVID-19 is the Polymerase Chain Reaction (PCR) tests. In the UAE alone around 240 thousand tests are performed daily[3] and the number is increasing overtime. Researchers trying to reduce the number of the PCR tests, in particular the tests that can be determined using other methods such as AI. Therefore, the unnecessary overload over the labs and hospitals can be reduced.

There is considerable research in the literature on the use of AI algorithms to detect COVID-19. In this paper four main algorithms are used: Random Forest, Neural Networks SVM, and GBDT. Therefore, in this section the main contributions of these algorithms are presented.

2.1   Random Forest (RF)

The Random Forest is a well known Machine Learning classification algorithm, it combines multiple decision trees to create a \forest" that has more accurate and stable predictions.

One of the interesting works using Random Forest is by [17]. The paper investigated in estimating the near future case numbers for 190 countries in the world and it is mapped in comparison with actual confirmed cases results.

Another work proposes a fine-tuned Random Forest model boosted by the Ada-Boost algorithm. [6] The model uses the COVID-19 patient's geographical, travel, health, and demographic data to predict the severity of the case and the possible outcome, recovery, or death. The model has an accuracy of 94% and a F1 Score of 0.86 on the dataset used.

A random forest (RF) algorithm was used to predict the prognoses of COVID-19 patients and identify the optimal diagnostic predictors for patients' clinical prognoses. The proposed method achieved an accuracy of 97.24% [15].

2.2   Neural Networks (NN)

Neural Network (NN) is a famous Machine learning algorithm that inspires from real biological neural systems to build a data processing system consisting of a large number of simple highly interconnected processing elements capable of solving intelligent and complex problems. In the literature NN is one of the most used algorithms in the COVID-19 detection area. NN is effective to extract information from images, such as

---

[3] https://ourworldindata.org/.

predicting COVID-19 cases using X-ray images [14]. NN is also used to reveal how the number of infected persons will change in the upcoming days and how many people will recover from COVID-19. In Saudi Arabia for example the forecast predicted the confirmed cases will increase [5]. Similarly, using NN curve fitting techniques in prediction and forecasting of the COVID-19 cases and death cases in India, USA, France, and UK, considering the progressive trends of China and South Korea [12].

2.3 Support Vector Machine (SVM)

SVM is another attractive Machine Learning algorithm used a lot in the literature. Basically, SVM finds a hyper-plane that creates a boundary between the classes of data. This hyper-plane is simply a line in 2 dimensional space. Few papers used SVM to detect COVID-19 compared to the other Machine learning algorithm. One of the interesting papers that used SVM to produce real-time forecasts and to investigate the Corona Virus Disease 2019 (COVID-19) prediction of confirmed, deceased and recovered cases. [11] Moreover, a study used SVM to build COVID-19 severeness detection model, in order to estimate whether the COVID-19 would develop severe symptoms [16].

In some papers SVM and deep feature is used to detect COVID-19 infected patients using X-ray images [9].

Finally, one paper used the Internet of Things (IoT) in smart hospitals and different ML models to diagnose patients with COVID-19. Three ML models were used, Naive Bayes (NB), Random Forest (RF), and Support Vector Machine (SVM). SVM model obtained the best diagnosis performance [1].

2.4 Gradient Boosting Decision Trees

Gradient boosting Decision Trees GBDT, is a well know ML model that uses an ensemble of weak learners (decision trees) to improve the performance of a machine learning model. The weak learners work sequentially. Each model tries to improve on the error from the previous model in order to have a better model. One paper in Ontario, Canada used GBDT to predict risk of hospitalization due to COVID-19 severity from routinely collected health records of the entire adult population [3]. Another paper in India used GBDT to explore the association between the coronavirus disease 2019 (COVID-19) transmission rates and meteorological parameters [10].

3   Dataset

Data is the core of any AI algorithm. This study makes the foundation strong by using a real data set of patients from a local hospital in Dubai named Al Zahra Hospital because it is extremely difficult to obtain patient data due to privacy concerns. The original dataset includes for each patient who run a PCR test his date of birth, gender, nationality, result, primary diagnosis and secondary diagnosis. For some patients, the vital signs were also included. However, Many patients did not present any symptoms or records for their vitals signs. In this dataset, we focus on four attributes (age, gender, nationality, and result) of 9416 patients who run a PCR test at the hospital and we exclude the diagnosis and vitals since the patients in study are asymptomatic. The dataset is distributed as 3427 Females and 5989 Male, from all age categories (Fig. 1) and more than 100 nationalities (Fig. 2). Dubai is a multinational city where more than 50% of the population are expatriates and frequent travelers, which increases the possibility of spreading the virus. When positive results show a high percentage of the same nationality, this information will help the authorities notify the original country and take the measurement and monitor the airport traffic.

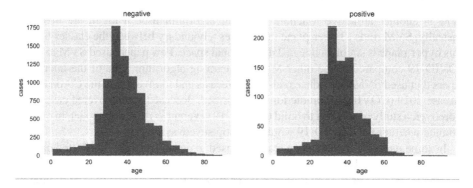

Fig. 1. Age distribution

Input/Output Variables. The input variables to the neural network are the personal information taken about the patient (age, gender, nation). These variables are presented in Table 1.

Two examples of patients are presented in Table 2. The first example with negative PCR test is a male patient from Afghanistan 49 years old. The second example with positive PCR test is a female patient from Egypt 51 years old. The output variable represents whether a person is suspected of having contracted COVID-19 or not (Sick, Healthy).

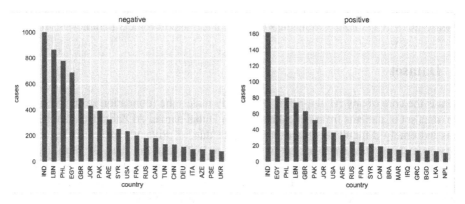

Fig. 2. Top 20 nations distribution

Table 1. Attributes of the dataset

Feature description	
1 Age	The age of the person generated from the date of birth
2 Gender	It takes 2 values F for Female and M for Male
3 Nation	Nationality of the patient
4 Result	PCR test result showing if COVID-19 is detected or not detected

Table 2. Samples of one patient detected positive one patient detected negative in the data set

Age	Gender	Nation	Result
49	M	AFG	NOT DETECTED
51	F	EGY	DETECTED

Concerning the population distribution in the used dataset. The Fig. 2 shows top 20 nations in the dataset. Most of the patients were from India with around 1160 patients, 160 of them had positive COVID-19 cases. Second most patients were from Lebanon with around 900 patients, 75 of them had positive COVID-19 cases. The third most patients were from the Philippines with around 800 patients, 80 of them had positive COVID-19 cases.

The 20 top nations in the dataset that performed COVID-19 PCR tests are in order India, Lebanon, Philippines, Egypt, Great Britain, Jordan, Pakistan, ARE, Syria, USA, France, Russia, Canada, Tunis, China, Germany, Italy Azerbaijan, Palestine and Ukraine.

Concerning the age distribution of the used data set. Most of the patients were around 25 to 40 years old, from whom the patients that have negative cases were around 35 years old, they were around 1750 patients. The patients with positive cases were around 30 years old with about 250 cases.

30 years old with about 250 cases. On the other hand, the fewer data were available in the case of children below 18 years old and seniors over 60 years old. Children below 18 years old were around 400 cases from whom less than 100 positive cases. Seniors over 60 years old were around 300 cases from whom less than 50 with positive cases. As shown in Fig. 2.

4 Methodology

The methodology consists of three steps:

- **Pre-processing:** Data pre-processing is a fundamental step to most ma-chine learning algorithms. ML algorithms require quality datasets in a specific format in order to generate quality predictions. If the dataset contains incomplete values, the algorithm will not use it. If the dataset contains inconsistent or noisy values, the predictions of the algorithm will be inaccurate. Data pre-processing mainly includes data cleaning, data reduction as well as data transformation to improve the dataset quality [20].

The used dataset is extracted from the database of Al Zahra Hospital. The base for each record/instance in the dataset is a medical visit of a patient who did a PCR test for COVID detection. In the database, each visit has age, gender, nation, vitals, primary symptom and many secondary symptoms. Ina previous research work [18] a neural network model has been propose during age, gender, nation, primary, and secondary symptoms. The objective of this paper is to compare different ML algorithms with minimal number of features. Hence, the data pre-processing includes the following steps:

- **Feature selection**: In this paper, the features presented in the Table1 are selected manually; other features such as BPType, Systole, Dias-tole, Heart rate, primary diagnosis, and secondary diagnosis have been discarded. The selected features have not any medical information, the objective is to experiment if non medical features can detect COVID-19, such a model can be interesting to use in scenarios when no medical information is available or when it is difficult to get medical equipment, such as in universities when students do not show necessarily any medical symptoms. In future work, all of the available features will be considered to address the situations when medical information is easy to collect.
- **Incomplete data**: The dataset with the selected features contains few missing values less than 5%, and they are completely at random places. Therefore ignoring the missing values is the chosen strategy.
- **Data transformation**: Gender and Nation are categorical features that have too many values. For example, the nation has over 200 values. The Neural Network, SVM, and RF algorithms work best when the features have binary values. Therefore, the remaining features have been trans-formed into simple features with binary values using the "One hot encoding" method [8].

- Training: The prepared medical dataset is split into 80% of the Training set and 20% Testing set. The Training set is used as an input to the AI model. The output is a trained AI model.
- Testing: The Testing set split from the previous step is used to predict the classifications. Those predictions are compared to the true classifications to evaluate the trained model (Fig. 3).

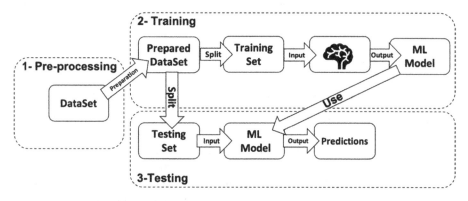

Fig. 3. Methodology

5 Experiments

In our experiments, we employed four machine learning algorithms: feed-forward neural network, Random Forest (RF), SVM, and Gradient Boosting Decision Trees (GBDT). All the experiments were carried in Python using sklearn and Keras libraies. The ML methods used in the paper are selected due to their simplicity and off-the-shelf availability which makes it easy for a non-expert to employ in practice. Many doctors and healthcare professionals do not have expert knowledge of advanced ML methods so we restrict our attention to only the accessible ML methods. In addition, the data does not have a complicated structure with only a few features. Therefore, there is no need to employ overly complicated models especially given that MLP, SVM, and RF are already very effective methods in their own right.

To evaluate the algorithms, a 5-fold cross-validation is used for training and testing purposes. After training a model, the accuracy of the trained model is measured on the test set. The experiments are carried out using two versions of the dataset: full and top 20 nations. Since the full dataset contains many nations with only a few instances it is informative to experiment with the dataset of the top 20 nations. The inputs for the models are the individual's age, gender and nationality. The output is positive/negative determination of COVID-19 infection.

5.1 Neural Networks

In our experiment, we used multilayer perceptron (MLP) neural networks with 1, 2, and 3 hidden layers. In the models with 1 and 2 hidden layers, each fully connected layer is followed by a dropout layer with the dropout rate of 0.2. There are two versions of the model with 3 hidden layers: with and without the dropout layers. The reason MLP was chosen was because the dataset contains only three features and MLP is powerful enough to handle such a dataset. It is a commonly used neural network architecture used in many applications.

Furthermore, we utilized different sized layers depending on the dataset. For the experiment with the full dataset, we used layers with 100, 50, and 25 nodes. For the experiment with the dataset of the top 20 nations, we used layers with sizes 20, 10, and 5. All the hidden layers utilize the Rectified Linear Unit (Relu) function for activation. We used binary crossentropy as the loss function. We used RMSprop for loss function optimization. The models are implemented in Keras with the default hyperparameters unless specified above.

Table 3. The accuracy results of the experiments with neural network models based on 5-fold cross-validation.

Model	MLP-1	MLP-2	MLP-3a	MLP-3b
All nations	0.8964	0.8965	0.8966	0.8959
Top 20 nations	0.895	0.8976	0.8977	0.8978

The details of the neural networks experiments are presented in Table 3 and Table 4. Table 3 shows the accuracy results of experiments based on 5-fold cross-validation. The accuracy of the models is nearly identical for the different configurations. The average accuracy is 89.6% for the full and top 20 datasets.

Table 4 presents the AUC measurements of the neural networks experiments. The average result is 0.53 for the full and top 20 datasets. The model with three hidden layers and no dropout optimization yields a slightly inferior performance.

Table 4. The AUC results of the experiments with neural network models based on 5-fold cross-validation.

Model	MLP-1	MLP-2	MLP-3a	MLP-3b
All nations	0.51	0.52	0.54	0.58
Top 20 nations	0.50	0.53	0.55	0.57

5.2   Random Forest

Random forest is a well-known ensemble classifier. It operates by training a set of decision tree classifiers and then determining the final label based on the majority vote among the base classifiers. In our experiments, we used the random forest classifier with various numbers of estimators. Concretely, we employed 5, 10, 20, and 30 base tree classifiers. The maximum depth of the tree classifiers was set to 20. The results of the experiments are presented in Table 5 and Table 6. The results are similar across different model configurations. The average accuracy is 87.75% and 87.7% for the full and top 20 datasets respectively. The AUC average is 0.62 and 0.60 for the full and top 20 datasets respectively. The configuration with 10 base estimators is slightly better than the rest.

Table 5. The accuracy results of the experiments with random forest models using different number of the base estimators based on 5-fold cross-validation.

Estimators	5	10	20	30
All nations	0.8718	0.8799	0.8809	0.8805
Top 20 nations	0.8766	0.8793	0.8781	0.8793

Table 6. The AUC results of the experiments with random forest models using different number of the base estimators based on 5-fold cross-validation.

Estimator	5	10	20	30
All nations	0.622	0.6265	0.6293	0.6306
Top 20 nations	0.6046	0.6103	0.6097	0.6097

5.3   SVM

The SVM classifier is a powerful algorithm that can efficiently estimate nonlinear patterns. In our experiment, we employed an SVM classifier with the radial basis function kernel. We tested the performance of the model under different regularization parameter settings. The results are similar across different model configurations (Table 7).

Table 7. The accuracy results of the experiments with SVM models using different regularization level based on 5-fold cross-validation.

C	0.2	0.5	1	2	5	10
All nations	0.8966	0.8966	0.897	0.8966	0.895	0.8927
Top 20 nations	0.897	0.897	0.897	0.8793	0.8966	0.8954

The average accuracy is 89.5% and 89.6% for the full and top 20 datasets respectively. Note that the strength of the regularization is inversely proportional to C. The models with the lower values of C appear to perform slightly better. The average AUC value is 0.55 and 0.52 for the full and top 20 datasets respectively (Table 8).

Table 8. The AUC results of the experiments with SVM models using different regularization level based on 5-fold cross-validation.

C	0.2	0.5	1	2	5	10
All nations	0.5621	0.5585	0.5601	0.5579	0.5581	0.5637
Top 20 nations	0.545	0.5283	0.5406	0.515	0.5216	0.5237

5.4   Gradient Boosting Decision Trees

Gradient boosting decision trees (GBDT) is an ensemble classifier that uses decision stumps - decision tree with a single node - to build strong classifiers. GBDT is built incrementally whereby at each iteration a new stump is added in order to correct for the prior errors. In our experiments, we used the GBDT classifier with various number of estimators - the number of boosting stages to perform. Gradient boosting is fairly robust to overfitting so a large number usually results in better performance. Concretely, we experimented with 50, 100, 200, and 300 estimators.

The results of the experiments are presented in Table 9. The results are similar across different model configurations. The average accuracy is 89.66% and 89.75% for the full and top 20 datasets respectively. As shown in Table 10, the AUC values range between 0.5507 and 0.5566. Using only the top 20 nations with 300 estimators achieves the maximum accuracy.

Table 9. The accuracy results of the experiments with GBDT models using different number of the base estimators based on 5-fold cross-validation.

Estimators	50	100	200	300
All nations	0.8964	0.8965	0.8966	0.8966
Top 20 nations	0.871	0.8974	0.8976	0.8979

Table 10. The AUC results of the experiments with GBDT models using different number of the base estimators based on 5-fold cross-validation.

Estimator	50	100	200	300
All nations	0.5564	0.5561	0.5566	0.5554
Top 20 nations	0.5511	0.5503	0.5512	0.5507

5.5   Analysis

The results of the experiments show that neural network and GBDT models produce better accuracy than random forest in identifying the classifying cases of COVID-19. The top 20 nation sets produce better average results than the full dataset. The results are similar across different model configurations.

The best models are MLP-2, and MLP-3 as well as GBDT with estimator 300. From the other side, GBDT and random forest present better AUC values for the full and top 20 datasets.

6   Conclusion

The coronavirus COVID-19 pandemic defines the global health crisis of our time and the most significant challenge in our society. It has become paramount to design and implement an effective predictive architecture for COVID-19 detection. This paper has studied real data sets of asymptomatic patients who run PCR tests et al. Zahra Hospital. We have also fed this data set to the different machine learning algorithms SVM, Neural Networks, and Random Forest for training and finally used a 5-fold cross-validation. The results of numerical experiments show that The results of the experiments show that Neural Networks and SVM models have an 89% accuracy, and the random forest model shows an 87% accuracy in identifying the COVID-19 cases. The presented algorithms have been tested with only three attributes age, gender, and nationality of 9416 patients who run PCR tests et al. Zahra Hospital. In the future, more attributes will be used, such as vitals, symptoms, and historical diagnoses. We will also investigate the blood types, heart rate, or something else for the asymptomatic patients.

Acknowledgment. The authors would like to thank Al Zahra hospital for their cooperation, support, and sharing knowledge.

References

1. Abdulkareem, K.H., et al.: Realizing an effective covid-19 diagnosis system based on machine learning and IoT in smart hospital environment. IEEE Internet Things J., 1 (2021)
2. Amaral, F., Casaca, W., Oishi, C.M., Cuminato, J.A.: Towards providing effective data-driven responses to predict the covid-19 in Sao Paulo and Brazil. Sensors **21**(2), 540 (2021)
3. Gutierrez, J.M., Volkovs, M., Poutanen, T., Watson, T., Rosella, L.: Development of a multivariable model for covid-19 risk stratification based on gradient boosting decision trees. medRxiv (2020)
4. Hamadneh, N.N., Khan, W.A., Ashraf, W., Atawneh, S.H., Khan, I., Hamadneh, B.N.: Artificial neural networks for prediction of covid-19 in Saudi Arabia. Comput. Mater. Contin. **66**, 2787–2796 (2021)
5. Hamadneh, N., Khan, W., Ashraf, W., Atawneh, S., Khan, I., Hamadneh, B.: Artificial neural networks for prediction of covid-19 in Saudi Arabia. Comput. Mater. Cont. **66**(3) (2021)
6. Iwendi, C., et al.: Covid-19 patient health prediction using boosted random forest algorithm. Front. Pub. Health **8**, 357 (2020)
7. World Organization, et al.: Definition, diagnosis and classification of diabetes mellitus and its complications: report of a who consultation. part 1, diagnosis and classification of diabetes mellitus. Technical report, World Health Organization (1999)
8. Rodríguez, P., Bautista, M.A., Gonzalez, J., Escalera, S.: Beyond one-hot en-coding: lower dimensional target embedding. Image Vis. Comput. **75**, 21–31 (2018)
9. Sethy, P., Santi, K., Behera, Kumar, P., Biswas, P.: Detection of coronavirus disease(covid-19) based on deep features and support vector machine, pp. 643–651 (2020)
10. Shrivastav, L.K., Jha, S.K.: A gradient boosting machine learning approach in modeling the impact of temperature and humidity on the transmission rate of covid-19 in India. Appl. Intell., 1–13 (2020). https://europepmc.org/articles/PMC7609380
11. Singh, V., et al.: Prediction of covid-19 corona virus pandemic based on time series data using support vector machine. J. Discrete Math. Sci. Crypt. **23**(8), 1583–1597 (2020)
12. Tamang, S., Singh, P., Datta, B.: Forecasting of covid-19 cases based on prediction using artificial neural network curve fitting technique. Global J. Environ. Sci. Manag. **6**(Special Issue (Covid-19)), 53–64 (2020)
13. Temurtas, H., Yumusak, N., Temurtas, F.: A comparative study on diabetes disease diagnosis using neural networks. Expert Syst. Appl. **36**(4), 8610–8615 (2009)
14. Toraman, S., Alakus, T.B., Turkoglu, I.: Convolutional caps net: a novel artificial neural network approach to detect covid-19 disease from x-ray images using capsule networks. Chaos Solitons Fract. **140**, 110122 (2020)
15. Wang, J., et al.: A descriptive study of random forest algorithm for predicting covid-19 patients outcome. PeerJ **8**, e9945 (2020)
16. Yao, H., et al.: Severity detection for the coronavirus disease 2019 (covid-19) patients using a machine learning model based on the blood and urine tests. Front. Cell Dev. Biol. **8**, 683 (2020)
17. Ye Silkanat, C.M.: Spatio-temporal estimation of the daily cases of covid-19 in worldwide sing random forest machine learning algorithm. Chaos Solitons Fract. **140**, 110210 (2020)
18. Zgheib, R., Chahbandarian, G., Firuz, K., Osman, A.L.: Neural networks architecture for covid-19 early detection. J. Ambient Intell. Human. Comput., 1–19 (2020)
19. Zgheib, R., Kristiansen, S., Conchon, E., Plageman, T., Goebel, V., Bastide, R.: A scalable semantic framework for IoT healthcare applications. J. Ambient Intell. Human. Comput., 1–19 (2020)
20. Zhang, S., Zhang, C., Yang, Q.: Data preparation for data mining. Appl. Artif. Intell. **17**(5–6), 375–381 (2003)

An Inverse QSAR Method Based on Decision Tree and Integer Programming

Kouki Tanaka[1], Jianshen Zhu[1], Naveed Ahmed Azam[1(✉)] [iD],
Kazuya Haraguchi[1] [iD], Liang Zhao[2] [iD], Hiroshi Nagamochi[1] [iD],
and Tatsuya Akutsu[3] [iD]

[1] Department of Applied Mathematics and Physics,
Kyoto University, Kyoto 606-8501, Japan
{ktanaka, zhujs, azam, haraguchi,
nag}@amp.i.kyoto-u.ac.jp
[2] Graduate School of Advanced Integrated Studies in Human Survavibility
(Shishu-Kan), Kyoto University, Kyoto 606-8306, Japan
liang@gsais.kyoto-u.ac.jp
[3] Bioinformatics Center, Institute for Chemical Research,
Kyoto University, Uji 611-0011, Japan
takutsu@kuicr.kyoto-u.ac.jp

Abstract. Recently a novel framework has been proposed for designing the molecular structure of chemical compounds using both artificial neural networks (ANNs) and mixed integer linear programming (MILP). In the framework, we first define a feature vector $f(\mathbb{C})$ of a chemical graph $\mathbb{C}$ and construct an ANN that maps $x = f(\mathbb{C})$ to a predicted value $\eta(x)$ of a chemical property π to $\mathbb{C}$. After this, we formulate an MILP that simulates the computation process of $f(\mathbb{C})$ from $\mathbb{C}$ and that of $\eta(x)$ from x. Given a target value y^* of the chemical property π, we infer a chemical graph $\mathbb{C}^\dagger$ such that $\eta(f(\mathbb{C}^\dagger)) = y^*$ by solving the MILP. In this paper, in pursuit of alternative learning models, we design an original decision tree that is based on a set of separating hyperplanes in the feature space which is then used to construct a prediction function η instead of ANNs. For this, we derive an MILP formulation that simulates the computation process of the proposed decision tree. The results of computational experiments suggest our method can infer chemical graphs with around up to 50 non-hydrogen atoms and that the prediction function based on a decision tree outperforms that with ANNs for several chemical properties.

Keywords: Machine learning · Decision tree · Integer programming · Cheminformatics · Materials informatics · QSAR/QSPR · Molecular design

1 Introduction

Background. Analysis of chemical compounds is one of the important applications of intelligent computing. Indeed, various machine learning methods have been applied to the prediction of chemical activities from their structural data, where such a problem is

© Springer Nature Switzerland AG 2021
D.-S. Huang et al. (Eds.): ICIC 2021, LNCS 12837, pp. 628–644, 2021.
https://doi.org/10.1007/978-3-030-84529-2_53

often referred to as *quantitative structure activity relationship* (QSAR) [1, 2]. Recently, neural networks and deep-learning technologies have extensively been applied to QSAR [3].

In addition to QSAR, extensive studies have been done on inverse quantitative structure activity relationship (inverse QSAR), which seeks for chemical structures having desired chemical activities under some constraints. Since it is difficult to directly handle chemical structures in both QSAR and inverse QSAR, chemical compounds are usually represented as vectors of real or integer numbers, which are often called *descriptors* in chemoinformatics and correspond to *feature vectors* in machine learning. One major approach in inverse QSAR is to infer feature vectors from given chemical activities and constraints and then reconstruct chemical structures from these feature vectors [4–6], where chemical structures are usually treated as undirected graphs. However, the reconstruction itself is a challenging task because the number of possible chemical graphs is huge. For example, chemical graphs with up to 30 atoms (vertices) C, N, O, and S may exceed 10^{60} [7]. Indeed, it is NP-hard to infer a chemical graph from a given feature vector except for some simple cases [8]. Due to this inherent difficulty, most existing methods for inverse QSAR do not guarantee optimal or exact solutions.

As a new approach, extensive studies have recently been done for inverse QSAR using *artificial neural networks* (ANNs), especially using graph convolutional networks [9]. For example, recurrent neural networks [11, 12], variational autoencoders [10], grammar variational autoencoders [13], generative adversarial networks [14], and invertible flow models [15, 16] have been applied. These ANN-based methods try to output feature vectors so that the corresponding chemical graphs are optimal in the sense that their property values are closer to the desired value. These methods use some statistical and heuristic approaches, and therefore do not guarantee optimality, i.e., there may exist some other feature vectors that satisfy the given constraints and have the desired property closer to the required property. Furthermore these methods have tendency to output invalid feature vectors that do not correspond to any chemical graphs.

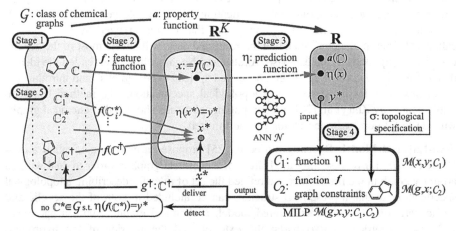

Fig. 1. An illustration of a framework for inferring a set of chemical graphs $\mathbb{C}^*$.

Framework. Akutsu and Nagamochi [17] proved that the computation process of a given ANN can be simulated with a mixed integer linear programming (MILP). Based on this, a novel framework for inferring chemical graphs has been developed [18, 19], as illustrated in Fig. 1. It constructs a prediction function in the first phase and infers a chemical graph in the second phase. The first phase of the framework consists of three stages. In Stage 1, we choose a chemical property π and a class $\mathcal{G}$ of graphs, where a property function a is defined so that $a(\mathbb{C})$ is the value of π for a compound $\mathbb{C} \in \mathcal{G}$, and collect a data set D_π of chemical graphs in G such that $a(\mathbb{C})$ is available for any $\mathbb{C} \in D_\pi$. In Stage 2, we introduce a feature function $f: \mathcal{G} \to \mathbb{R}^K$ for a positive integer K. In Stage 3, we construct a prediction function η with an ANN $\mathcal{N}$ that, given a vector $x \in \mathbb{R}^K$, returns a value $y = \eta(x) \in \mathbb{R}$ so that $\eta(f(\mathbb{C}))$ serves as a predicted value to the real value of π for each $\mathbb{C} \in D_\pi$. Given a target chemical value y^*, the second phase infers chemical graphs $\mathbb{C}^*$ with $\eta(f(\mathbb{C}^*)) = y^*$ in the next two stages. We have obtained a feature function f and a prediction function η and call an additional constraint on the substructures of target chemical graphs a *topological specification*. In Stage 4, we prepare the following two MILP formulations:

- MILP $\mathcal{M}(x, y; \mathcal{C}_1)$ with a set $\mathcal{C}_1$ of linear constraints on variables x and y (and some other auxiliary variables) simulates the process of computing $y := \eta(x)$ from a vector x; and
- MILP $\mathcal{M}(g, x; \mathcal{C}_2)$ with a set $\mathcal{C}_2$ of linear constraints on variable x and a variable vector g that represents a chemical graph $\mathbb{C}$ (and some other auxiliary variables) simulates the process of computing $x := f(\mathbb{C})$ from a chemical graph $\mathbb{C}$ and chooses a chemical graph $\mathbb{C}$ that satisfies the given topological specification σ.

Given a target value y^*, we solve the combined MILP $\mathcal{M}(g, x, y; \mathcal{C}_1, \mathcal{C}_2)$ to find a feature vector x^* and a chemical graph $\mathbb{C}^\dagger$ with the specification σ such that $f(\mathbb{C}^\dagger) = x^*$ and $\eta(x^*) = y^*$ (where if the MILP instance is infeasible then this suggests that there does not exist such a desired chemical graph). In Stage 5, we generate other chemical graphs $\mathbb{C}^*$ such that $\eta(f(\mathbb{C}^*)) = y^*$ based on the output chemical graph $\mathbb{C}^\dagger$.

MILP formulations required in Stage 4 have been designed for chemical compounds with cycle index 0 (i.e., acyclic) [19, 20], cycle index 1 [21] and cycle index 2 [22], where no sophisticated topological specification was available yet. Azam et al. [20] introduced a restricted class of acyclic graphs that is characterized by an integer ρ, called a "branch-parameter" such that the restricted class still covers most of the acyclic chemical compounds in the database. Akutsu and Nagamochi [23] extended the idea to define a restricted class of cyclic graphs, called "ρ-lean cyclic graphs" and introduced a set of flexible rules for describing a topological specification. The method has been implemented by Zhu et al. [24] and computational results showed that chemical graphs with up to 50 non-hydrogen atoms can be inferred.

Two-Layered Model. Recently Shi et al. [25] proposed a new model, called a *two-layered model* for representing the feature of a chemical graph in order to deal with an arbitrary graph in the framework and refined the set of rules for describing a topological specification so that a prescribed structure can be included in both of the acyclic and cyclic parts of $\mathbb{C}$. In the two-layered model, a chemical graph $\mathbb{C}$ with a parameter $\rho \geq 1$ is regarded as two parts: the exterior and the interior of the hydrogen-suppressed chemical graph $\langle \mathbb{C} \rangle$ obtained from $\mathbb{C}$ by removing hydrogen. The exterior consists of maximal acyclic induced subgraphs with height at most ρ in $\langle \mathbb{C} \rangle$ and the interior is the connected subgraph of $\langle \mathbb{C} \rangle$ obtained by ignoring the exterior. Shi et al. [25] defined a feature vector $f(\mathbb{C})$ of a chemical graph $\mathbb{C}$ to be a combination of the frequency of adjacent atom pairs in the interior and the frequency of chemical acyclic graphs among the set of chemical rooted trees T_u rooted at interior-vertices u.

Contribution. In this paper, we first make a modification to the two-layered model to improve the feature vector $f(\mathbb{C})$. Since a chemical rooted tree T_u used in the feature vector by Shi et al. [25] does not include any hydrogens adjacent to the interior-vertex u, we use the frequency of a chemical acyclic graph that is obtained from T_u by putting back the hydrogens originally attached T_u in $\mathbb{C}$. This modification allows us to treat more variety of chemical rooted trees to represent the feature of the exterior.

Note that the quality of a prediction function η constructed in Stage 3 is one of the most important factors in the framework. We used scikit-learn ANNs and scikit-learn decision trees to construct predication functions and observed that decision trees perform better than ANNs for some chemical properties; see Sect. 5 for details. It is also pointed out that overfitting is a major issue in ANN-based approaches for QSAR because ANNs have many parameters to be optimized [3]. Furthermore we have also designed a new algorithm to construct decision tree based on hyperplanes in the feature space in Stage 3 and demonstrated that the proposed algorithm has better performance as compared to that of scikit-learn ANNs and scikit-learn decision trees for some properties. In this paper, we next used our new decision tree algorithm for constructing a prediction function η in Stage 3 instead of ANNs. For this, we derive a new MILP formulation $\mathcal{M}(x, y; \mathcal{C}_1)$ that simulates the computation process of a prediction function by the proposed decision tree algorithm in Stage 4.

For an MILP formulation $\mathcal{M}(x, y; \mathcal{C}_2)$ that represents a feature function f and a specification σ in Stage 4, we can use the same formulation proposed by Shi et al. [25] with a slight modification and omit a full description of the MILP in this paper. To generate target chemical graphs $\mathbb{C}^*$ in Stage 5, we can also use the dynamic programming algorithm due to Shi et al. [25] with a slight modification and omit the details in this paper.

We implemented the refined two-layered model based on a decision tree. The results of computational experiments suggest that there are some chemical properties for which a decision tree is more effective than an ANN and that the proposed method can infer chemical graphs with up to 50 non-hydrogen atoms.

The paper is organized as follows. Section 2 introduces some notions on graphs, a modeling of chemical compounds and a choice of descriptors. Section 3 describes our modification to the two-layered model. Section 4 reviews the idea of decision trees in a general setting based on hyperplanes in the feature space. Section 5 reports the results on some computational experiments conducted for chemical properties such as energy of highest occupied molecular orbital (HOMO), energy of lowest unoccupied molecular orbital (LUMO), the energy difference between the HOMO and LUMO, critical pressure, critical temperature, experimental amorphous density, biological half life, dissociation constants and refractive index. Section 6 makes some concluding remarks.

2 Preliminary

This section introduces some notions and terminologies on graphs, modeling of chemical compounds and our choice of descriptors.

Let $\mathbb{R}$, $\mathbb{Z}$ and $\mathbb{Z}_+$ denote the sets of reals, integers and non-negative integers, respectively. For two integers a and b, let $[a, b]$ denote the set of integers i with $a \leq i \leq b$.

Graph. Given a graph G, let $V(G)$ and $E(G)$ denote the sets of vertices and edges, respectively. For a subset $V' \subseteq V(G)$ (resp., $E' \subseteq E(G)$) of a graph G, let $G - V'$ (resp., $G - E'$) denote the graph obtained from G by removing the vertices in V' (resp., the edges in E'). The *rank* $r(G)$ of a graph G is defined to be the minimum $|F|$ of an edge subset $F \subseteq E(G)$ such that $G - F$ is acyclic. A path with two end-vertices u and v is called a u,v-*path*. An edge $e = u_1u_2$ in a connected graph G is called a *bridge* if the graph $G - e$ obtained from G by removing edge e is not connected. For a cyclic graph G, an edge e is called a *core-edge* if it is in a cycle of G or is a bridge $e = u_1u_2$ such that each of the connected graphs G_i, $i = 1, 2$ of $G - e$ contains a cycle. A vertex incident to a core-edge is called a *core-vertex* of G. The chemical graph obtained by all core-edges of G is called *core* of G.

For a graph G, let $V_{\text{leaf}}(G)$ denote the set of vertices of degree 1 in G, and define graphs G_i, $i \in \mathbb{Z}_+$ obtained from G by removing the set of leaf-vertices i times so that $G_0 := G$; $G_{i+1} := G_i - V_{\text{leaf}}(G_i)$, where we call a vertex $v \in V_{\text{leaf}}(G_k)$ a *leaf k-branch* and we say that a vertex $v \in V_{\text{leaf}}(G_k)$ has *height* $\text{ht}(v) = k$ in G. The *height* $\text{ht}(T)$ of a rooted tree T is defined to be the maximum of $\text{ht}(v)$ of a vertex $v \in V(T)$.

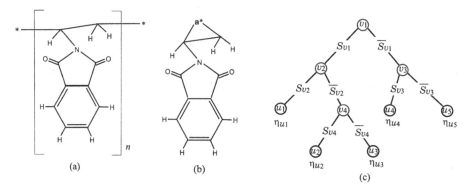

Fig. 2. (a) An example of polymer: N-vinylPhthalimide, where the asterisks indicate the two connecting edges; (b) A monomer form of N-vinylPhthalimide with a fictitious chemical element a^*; (c) An illustration of a decision tree $\mathcal{DT} = (T, \mathcal{S}, \eta)$, where $V_{\text{left}}(u_1) = \{v_1, v_2\}$, $V_{\text{right}}(u_1) = \emptyset$, $V_{\text{left}}(u_2) = \{v_1, v_4\}$, Vright(u2) = \{v2\}, Vleft(u3) = \{v1\}, Vright(u3) = \{v2, v4\}, Vleft (u4) = \{v3\}, Vright(u4) = \{v1\}, Vleft(u5) = \emptyset and Vright(u5) = \{v1, v3\}.

Chemical Graph. To represent a chemical compound, we introduce a set Λ of chemical elements such as H (hydrogen), C (carbon), O (oxygen), N (nitrogen) and so on, where we assume that each chemical element a has a fixed valence $\text{val}(a) \in [1, 4]$.

For each chemical element $a \in \Lambda$, let mass(a) and val(a) denote the mass and valence of a, respectively. A chemical compound is represented by a tuple $C = (H, \alpha, \beta)$ of a simple, connected undirected graph H and functions $\alpha: V(H) \to \Lambda$ and $\beta: E(H) \to [1, 3]$. The set of atoms and the set of bonds in the compound are represented by the vertex set $V(H)$ and the edge set $E(H)$, respectively. The chemical element assigned to a vertex $v \in V(H)$ is represented by $\alpha(v)$ and the bond-multiplicity between two adjacent vertices $u, v \in V(H)$ is represented by $\beta(e)$ of the edge $e = uv \in E(H)$.

Polymer. In this paper, we treat a polymer which is a sequential concatenation of a chemical compound with two connecting-edges, as illustrated in Fig. 2(a). To treat such a polymer as a monomer, we introduce a fictitious chemical element a^* with val(a^*) = 2 and mass(a^*) = 0 into a set Λ of chemical elements, and join the two connecting-edges of a polymer with a single atom a^*, as illustrated in Fig. 2(b). In what follows, we treat a polymer as a monomer with a fictitious chemical element a^* of degree 2.

3 Two-Layered Model

This section reviews the two-layered model and describes our modification to the model.

Let $\mathbb{C} = (H, \alpha, \beta)$ be a chemical graph and $\rho \geq 1$ be an integer, which we call a *branch parameter*. A *two-layered model* of $\mathbb{C}$ is a partition of the hydrogen-suppressed chemical graph $\langle \mathbb{C} \rangle$ into an "interior" and an "exterior" in the following way. We call a vertex $v \in V(\langle \mathbb{C} \rangle)$ (resp., an edge $e \in E(\langle \mathbb{C} \rangle)$) of G an *exterior-vertex* (resp., *exterior-edge*) if ht(v) $< \rho$ (resp., e is incident to an exterior-vertex) and denote the sets of exterior-vertices and exterior-edges by $V^{ex}(\mathbb{C})$ and $E^{ex}(\mathbb{C})$, respectively and denote $V^{int}(\mathbb{C}) = V(\langle \mathbb{C} \rangle) \backslash V^{ex}(\mathbb{C})$ and $E^{int}(\mathbb{C}) = E(\langle \mathbb{C} \rangle) \backslash E^{ex}(\mathbb{C})$, respectively. We call a vertex in $V^{int}(\mathbb{C})$ (resp., an edge in $E^{int}(\mathbb{C})$) an *interior-vertex* (resp., *interior-edge*). The set $E^{ex}(\mathbb{C})$ of exterior-edges forms a collection of connected graphs each of which is regarded as a rooted tree T rooted at the vertex $v \in V(T)$ with the maximum ht(v). For $\langle \mathbb{C} \rangle$, we denote the subgraph $(V(E^{int}(\mathbb{C}))$ by $\mathcal{T}^{ex}(\langle \mathbb{C} \rangle)$. The *interior* of $\mathbb{C}$ is defined to be the subgraph $(V^{int}(\mathbb{C}), E^{int}(\mathbb{C}))$ of $\langle \mathbb{C} \rangle$.

Figure 3 illustrates an example of a hydrogen-suppressed chemical graph $\langle \mathbb{C} \rangle$. For a branch parameter $\rho = 2$, the interior of the chemical graph $\langle \mathbb{C} \rangle$ in Fig. 3 is obtained by removing the set of vertices with degree 1 $\rho = 2$ times; i.e., first remove the set $V_1 = \{w_1, w_2, \ldots, w_{14}\}$ of vertices of degree 1 in $\langle \mathbb{C} \rangle$ and then remove the set $V_2 = \{w_{15}, w_{16}, \ldots, w_{19}\}$ of vertices of degree 1 in $\langle \mathbb{C} \rangle - V_1$, where the removed vertices become the exterior-vertices of $\langle \mathbb{C} \rangle$.

For each interior-vertex $u \in V^{int}(\mathbb{C})$, we denote by T_u the connected graph in $\mathcal{T}^{ex}(\langle \mathbb{C} \rangle)$ which is regarded as chemical tree rooted at u (where possibly T_u consists of a single vertex) and define the *ρ-fringe-tree* $\mathbb{C}[u]$ to be the chemical rooted tree obtained from T_u by putting back the hydrogens originally attached T_u in $\mathbb{C}$. Let $\mathcal{T}(\mathbb{C})$ denote the set of ρ-fringe-trees $\mathbb{C}[u]$, $u \in V^{int}(\mathbb{C})$. Figure 4 illustrates the set $\mathcal{T}(\mathbb{C}) = \{\mathbb{C}[u_i] | i \in [1, 28]\}$ of the 2-fringe-trees of the example $\mathbb{C}$ in Fig. 3.

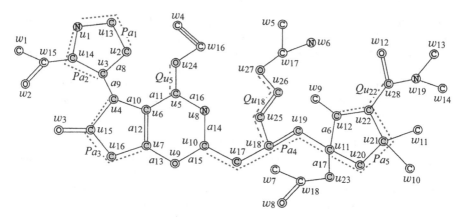

Fig. 3. An illustration of a hydrogen-suppressed chemical graph $\langle \mathbb{C} \rangle$ obtained from a chemical graph $\mathbb{C}$ by removing all the hydrogens, where for $\rho = 2$, $V^{ex}(\mathbb{C}) = \{w_i \mid i \in [1, 19]\}$ and $V^{int}(\mathbb{C}) = \{u_i \mid i \in [1, 28]\}$.

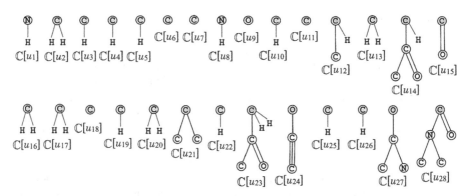

Fig. 4. The set $\mathcal{T}(\mathbb{C})$ of 2-fringe-trees $\mathbb{C}[u_i]$, $I \in [1, 28]$ of the example $\mathbb{C}$ in Fig. 3, where the root of each tree is depicted with a gray circle and the hydrogens attached to non-root vertices are omitted in the figure

Feature Function. The feature of an interior-edge $e = uv \in E^{int}(\mathbb{C})$ such that $\alpha(u) =$ a, $\deg_{\langle \mathbb{C} \rangle}(u) = d$, $\alpha(v) = $ b, $\deg_{\langle \mathbb{C} \rangle}(v) = d'$ and $\beta(e) = m$ is represented by a tuple (ad, bd', m), which is called the *edge-configuration* of the edge e.

For an integer K, a feature vector $f(\mathbb{C})$ of a chemical graph $\mathbb{C}$ is defined by a *feature function f* that consists of K descriptors. We call $\mathbb{R}^K$ the *feature space*.

Shi et al. [25] defined a feature vector $f(\mathbb{C}) \in \mathbb{R}^K$ to be a combination of the frequency of edge-configurations of the interior-edges and the frequency of chemical rooted trees among the set of chemical rooted trees T_u over all interior-vertices u.

In this paper, we modify the feature vector $f(\mathbb{C})$ by including the hydrogens attached to chemical rooted trees T_u of interior-vertices; i.e., by using the frequency of chemical rooted trees among the set of chemical rooted trees $\mathbb{C}[u]$.

Topological Specification. Shi et al. [25] also introduced a set of rules for describing a topological specification in the following way:

(i)   A *seed graph* G_C as an abstract form of a target chemical graph $\mathbb{C}$;
(ii) A set $\mathcal{F}$ of chemical rooted trees with no hydrogens as candidates for trees in the exterior of $\mathbb{C}$; and
(iii) Lower and upper bounds on the number of components in a target chemical graph such as chemical elements, double/triple bounds and the interior-vertices in $\mathbb{C}$.

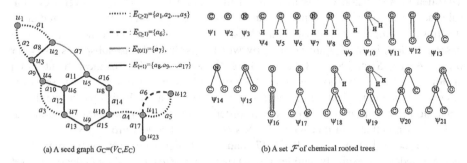

(a) A seed graph $G_C=(V_C, E_C)$ (b) A set $\mathcal{F}$ of chemical rooted trees

Fig. 5. (a) An illustration of a seed graph G_C where the vertices in V_C are depicted with gray circles, the edges in $E_{(\geq 2)}$ are depicted with dotted lines, the edges in $E_{(\geq 1)}$ are depicted with dashed lines, the edges in $E_{(0/1)}$ are depicted with gray bold lines and the edges in $E_{(=1)}$ are depicted with black solid lines; (b) A set $\mathcal{F} = \{\psi_1, \psi_2, \ldots, \psi_{21}\} \subseteq \mathcal{F}(D_\pi)$ of 21 chemical rooted trees ψ_i, $i \in [1, 21]$, where the root of each tree is depicted with a gray circle, where the hydrogens attached to non-root vertices are omitted in the figure.

According to our modification on featuring the exterior mentioned above, this paper defines $\mathcal{F}$ in rule (ii) to be a set of chemical rooted trees with hydrogens. Figure 5(a) and (b) illustrate examples of a seed graph G_C and a set $\mathcal{F}$ of chemical rooted trees, respectively. Given a seed graph G_C, the interior of a target chemical graph $\mathbb{C}$ is constructed from G_C by replacing some edges $a = uv$ with paths P_a between the end-vertices u and v and by attaching new paths Q_v to some vertices v. For example, a chemical graph $\mathbb{C}$ in Fig. 3 is constructed from the seed graph G_C in Fig. 5(a) as follows.

- First replace five edges $a_1 = u_1u_2$, $a_2 = u_1u_3$, $a_3 = u_4u_7$, $a_4 = u_{10}u_{11}$ and $a_5 = u_{11}u_{12}$ in G_C with new paths $P_{a1} = (u_1, u_{13}, u_2)$, $P_{a2} = (u_1, u_{14}, u_3)$, $P_{a3} = (u_4, u_{15}, u_{16}, u_7)$, $P_{a4} = (u_{10}, u_{17}, u_{18}, u_{19}, u_{11})$ and $P_{a5} = (u_{11}, u_{20}, u_{21}, u_{22}, u_{12})$, respectively to obtain a subgraph G_1 of $\langle \mathbb{C} \rangle$.
- Next attach to this graph G1 three new paths $Q_{u5} = (u_5, u_{24})$, $Q_{u18} = (u_{18}, u_{25}, u_{26}, u_{27})$ and $Q_{u22} = (u_{22}, u_{28})$ to obtain the interior of $\langle \mathbb{C} \rangle$ in Fig. 3.
- Finally attach to the interior 28 trees selected from the set $\mathcal{F}$ and assign chemical elements and bond-multiplicities in the interior to obtain a chemical graph $\mathbb{C}$ in Fig. 3. In Fig. 4, $\psi_1 \in \mathcal{F}$ is selected for $\mathbb{C}[u_i]$, $i \in \{6, 7, 11, 18\}$. Similarly ψ_2 for $\mathbb{C}[u_9]$, ψ_4 for $\mathbb{C}[u_i]$, $i \in \{3, 4, 5, 10, 19, 22, 25, 26\}$, ψ_5 for $\mathbb{C}[u_i]$, $i \in \{2, 13, 16,$

17, 20}, ψ_7 for $\mathbb{C}[u_i]$, $i \in \{1, 8\}$, ψ_9 for $\mathbb{C}[u_{12}]$, ψ_{11} for $\mathbb{C}[u_{15}]$, ψ_{13} for $\mathbb{C}[u_{21}]$, ψ_{16} for $\mathbb{C}[u_{24}]$, ψ_{17} for $\mathbb{C}[u_{27}]$, ψ_{18} for $\mathbb{C}[u_{14}]$, ψ_{19} for $\mathbb{C}[u_{23}]$ and ψ_{21} for $\mathbb{C}[u_{28}]$.

Our definition of a topological specification is analogous with the one by Shi et al. [25] except for a necessary modification due to the introduction of hydrogen in fringe-configurations.

4 Decision Trees

The idea of a decision tree is to partition the feature space $\mathbb{R}^K$ into several disjoint subspaces each of which we prepare an output value (or we can prepare a prediction function). A simple way of dividing a subspace S of $\mathbb{R}^K$ is to use a threshold $\alpha_k \in \mathbb{R}$ of a descriptor dcp_k in the feature vector f by which S is divided into subspaces $S+ = \{x(k) \geq \alpha_k \mid x \in S\}$ and $S- = \{x(k) < \alpha_k \mid x \in S\}$, where $x(k)$ denotes the k-th entry of a vector $x \in \mathbb{R}^K$. This method has a merit for us to observe which descriptor is sensitive or crucial in the classification or regression of a given data set. In this paper, we define a decision tree in a more general way in order to construct a prediction function that may perform better than that with an ANN for some chemical properties.

For a rooted binary tree T, let $V_{inl}(T)$ and $V_{leaf}(T)$ denote the sets of internal nodes and leaf nodes, respectively, and let $V_{left}(u)$ (resp., $V_{right}(u)$), $u \in V_{leaf}(T)$ denote the set of internal nodes $v \in V_{inl}(T)$ such that u is a descendant of the left (resp., right) child of v.

Let K be a positive integer. A decision tree $\mathcal{DT}$ in the space $\mathbb{R}^K$ is defined to be a tuple $\mathcal{DT} = (T, \mathcal{S}, \eta)$ of a rooted binary tree T, a set $\mathcal{S}$ of subspaces $S_v \subseteq \mathbb{R}^K$, $v \in V_{inl}(T)$ and a set η of prediction functions $\eta_u: \mathbb{R}^K \to \mathbb{R}$, $u \in V_{leaf}(T)$. Let $\overline{S}_v$ denote the subspace $\mathbb{R}^K \backslash S_v$. Figure 2(c) illustrates a decision tree $\mathcal{DT}$ with five leaf nodes.

A prediction function $\eta_{\mathcal{DT}}: \mathbb{R}^K \to \mathbb{R}$ is defined so that, for a given vector $x \in \mathbb{R}^K$, $\eta_{\mathcal{DT}}(x) := \eta_u(x)$ for the leaf node $u \in V_{leaf}(T)$ such that $x \in S_v, \forall v \in V_{left}(u)$ and $x \in \overline{S}_v, \forall v \in V_{right}(u)$.

In this paper, we derive an MILP formulation $\mathcal{M}(x, y; \mathcal{C}_1)$ that simulates the computation process of our decision tree in order to solve the inverse problem to a decision tree in Stage 4. We also propose a prediction function based on a set of hyperplanes in the feature space and an algorithm for constructing a decision tree by using the prediction function.

5 Computational Results

We implemented our method of Stages 1 to 5 for inferring chemical graphs under a given topological specification and conducted experiments to evaluate the computational efficiency. We executed the experiments on a PC with Processor: Core i7-9700 (3.0 GHz; 4.7 GHz at the maximum) Memory: 16 GB RAM DDR4.

Results on Phase 1. To conduct experiments for Phase 1, we selected nine chemical properties π: energy of highest occupied molecular orbital (HOMO), energy of lowest unoccupied molecular orbital (LUMO), the energy difference between the HOMO and LUMO (GAP), critical pressure for acyclic chemical compounds (CP$_{(a)}$), critical temperature for acyclic chemical compounds (CT$_{(a)}$), experimental amorphous density for polymers (AD$_{(p)}$), biological half life (BHL), dissociation constants (DC) and refractive index for polymers (RF$_{(p)}$). We used data sets provided by HSDB from PubChem [26] for CP$_{(a)}$, CT$_{(a)}$, BHL and DC; MoleculeNet [27] for HOMO, LUMO and GAP; and J. Bicerano [28] for AD$_{(p)}$ and RF$_{(p)}$. HOMO, LUMO and GAP share the original data set in common. The actual data set is so massive that it contains more than 130,000 graphs. We use the first 1,000 graphs (in terms of the index) for the data set.

We implemented Stages 1, 2 and 3 in Phase 1 as follows.

Stage 1. We set a graph class $\mathcal{G}$ to be the set of all chemical graphs with any graph structure, and set a branch-parameter ρ to be 2.

For each of the nine properties, we first select a set Λ of chemical elements and then collect a data set D_π on chemical graphs over the set Λ of chemical elements. To construct the data set D_π, we eliminated chemical compounds that are disconnected or have at most three carbon atoms or contain a charged element such as N^+ or an element $a \in \Lambda$ whose valence is different from our setting of valence function val.

Table 1. Data sets for Stage 1 in Phase 1.

π	$\Lambda \setminus \{H\}$	$|D_\pi|$	$\Gamma^{int}(D_\pi)$	$|\mathcal{F}(D_\pi)|$	$[n, \bar{n}]$	$[a, \bar{a}]$
HOMO	C, O, N	686	24	184	[4, 8]	[-0.343, -0.163]
LUMO	C, O, N	686	24	184	[4, 8]	[-0.121, 0.108]
GAP	C, O, N	686	24	184	[4, 8]	[0.149, 0.411]
CP$_{(a)}$	C, O, N	124	8	74	[4, 63]	[4.7×10^{-6}, 5.52]
CT$_{(a)}$	C, O, N	124	8	75	[4, 63]	[56.1, 3607.5]
AD$_{(p)}$	C, O, N, a*	86	21	26	[5, 46]	[0.84, 1.34]
BHL	C, O, N	282	19	68	[5, 36]	[0.03, 732.99]
DC	C, O, N	126	20	56	[5, 44]	[0.50, 17.11]
RF$_{(p)}$	C, O, N, a*	92	20	36	[5, 30]	[0.49, 1.68]

Table 1 shows the size and range of data sets that we prepared for each chemical property in Stage 1, where we denote the following: $\Lambda \setminus \{H\}$: the set of selected non-hydrogen chemical elements, where a* denotes a fictitious chemical element introduced to join the two connecting-edges in a polymer; $|D_\pi|$: the size of data set D_π over Λ for the property π; $|\Gamma^{int}(D_\pi)|$: the number of different edge-configurations of interior-edges over the compounds in D_π; $|\mathcal{F}(D_\pi)|$: the number of non-isomorphic chemical rooted trees in the set of all 2-fringe-trees in the compounds in D_π; $[n, \bar{n}]$: the minimum and maximum values of the number $n(\mathbb{C})$ of non-hydrogen atoms in compounds $\mathbb{C}$ in D_π; and $[a, \bar{a}]$: the minimum and maximum values of $a(\mathbb{C})$ for π over compounds $\mathbb{C}$ in D_π.

Stage 2. We used the new feature function defined in our chemical model without suppressing hydrogen.

Stage 3. We conducted an experiment in Stage 3 based on cross-validation by using ANN and decision tree to demonstrate that the later can perform better than the former. For a property π, an execution of a *cross-validation* consists of five rounds of constructing a prediction function as follows. First partition the data set D_π into five subsets $D_\pi^{(i)}, i \in [1, 5]$ randomly; for each $i \in [1, 5]$, the i-round constructs a prediction function $\eta(i)$ (by selecting an ANN or a decision tree using the set $D_\pi \backslash D_\pi^{(i)}$ as a training data set). When we iterate an execution of a cross-validation p times, let $\eta(j, i)$ denote the prediction function $\eta(i)$ of the i-th round in the j-th iteration for each $j \in [1, p]$. Let $R^2(\eta(j, i), D_\pi^{(i)})$ denote the coefficient of determination of $\eta(j, i)$ for the test data set $D_\pi^{(i)}$. For each property π, let t-$R^2_{cv}(j)$ denote the average of $R^2(\eta(j, i), D_\pi^{(i)})$ over all $i \in [1, 5]$ in the j-th cross-validation.

We constructed prediction functions η as follows:

(I)   For each of the nine properties, construct an ANN $\mathcal{N}$ using scikit-learn as used by Shi et al. [25] to obtain a prediction function $\eta_\mathcal{N}$ based on $\mathcal{N}$;

(II) For each of the nine properties, construct a decision tree $\mathcal{DT}_{sc}$ using scikit-learn to obtain a prediction function $\eta_{\mathcal{DT}_{sc}}$ based on $\mathcal{DT}_{sc}$;

(III) For each property $\pi \in \{B_{HL}, D_C, R_{F(p)}\}$, construct a decision tree $\mathcal{DT}_{hp}$ using our algorithm on hyperplanes to obtain a prediction function $\eta_{\mathcal{DT}_{hp}}$ based on $\mathcal{DT}_{hp}$.

More precisely, (I) to (III) are conducted in the following way.

(I)   We prepare ten architectures $A_j, j \in [1, 10]$ with one or two hidden layers. For a property π, we iterate an execution of a cross-validation $p = 10$ times, where we use the architecture A_j in the j-th iteration for each $j \in [1, 10]$. We use scikit-learn version 0.23.2 with Python 3.8.5, MLPRegressor and ReLU activation function to construct an ANN $\mathcal{N}(j, i)$ based on the architecture A_j and the training data set $D_\pi \backslash D_\pi^{(i)}$ of the i-th round. During an iterative process of constructing an ANN $\mathcal{N}(j, i)$, we choose the best iteration that maximizes the coefficient of determination to the test data $D_\pi^{(i)}$. Let $\eta_{\mathcal{N}(j,i)}$ denote the resulting prediction function obtained by the ANN $\mathcal{N}(j, i)$.

(II) Construct a decision tree $\mathcal{DT}_{sc}$ by using scikit-learn of version 0.23.2 with Python 3.8.5. We can bound from above the height of a decision tree with a prescribed integer b, called a *height bound*, where $b = \inf$ means that the height of a decision tree is not bounded. Choose a set $\{1, 2, 3, 4, 5, 10, 20, \inf\}$ of height bounds. For a property π, we iterate a cross-validation $p = 10$ times, where we change a partition of a data set D_π into five subsets $D_\pi^{(i)}, i \in [1, 5]$ for each iteration. In each round $i \in [1, 5]$ of an execution of a cross-validation, we construct eight decision trees $\mathcal{DT}_b^{(i)}, b \in \{1, 2, 3, 4, 5, 10, 20, \inf\}$ for each training data set $D_\pi \backslash D_\pi^{(i)}$ and choose a decision tree $\mathcal{DT}_b^{(i)}$ with the maximum test coefficient of determination $R^2(\eta(j, i), D_\pi^{(i)})$ as a decision tree $\mathcal{DT}_{sc}(i)$ by scikit-learn for the i-th round. Let $DT_{sc}^{(j,i)}$ denote the decision tree $DT_{sc}(i)$ of the i-th round in the j-th iteration and $\eta_{DT_{sc}^{(j,i)}}$ denote the resulting prediction function obtained by $DT_{sc}^{(j,i)}$. Note that #ℓ in DT_{sc} means the number of thresholds (i.e., hyperplanes).

(III)   To construct a decision tree DT_{hp}, we design an algorithm DT-Algorithm. Analogously with (II), we iterate a cross-validation $p = 10$ times for a property π, where we change a partition of a data set D_π into five subsets $D_\pi^{(i)}, i \in [1, 5]$ for each iteration. To construct a decision tree $DT_{hp}^{(j,i)}$ of the i-th round in the j-th cross-validation, we execute DT-Algorithm $(D_{train}, D_{test}, h, s)$ for $D_{train} := D_\pi \backslash D_\pi^{(i)}$, $D_{test} := D_\pi^{(i)}$, $h := 15$ and $s := 20$. Let $\eta_{DT_{hp}^{(j,i)}}$ denote the resulting prediction function obtained by $DT_{hp}^{(j,i)}$.

Table 2. Results of Stages 2 and 3 in Phase 1 for the properties $\pi \in \{$HOMO, LUMO, GAP, CP$_{(a)}$, CT$_{(a)}$, AD$_{(p)}\}$

π	ANN $\mathcal{N}$				Decision Tree DT_{sc}				
	L-time	t-R$^2_{cv}$	t-R$^2_{max}$	Arch.	L-time	t-R$^2_{cv}$	t-R$^2_{max}$	#ℓ	$\frac{\#\ell}{|D_\pi|}$
HOMO	0.12	0.383	0.513	(227,30,20,1)	3.2×10^{-3}	0.831	0.866	213	0.31
LUMO	0.11	0.616	0.703	(227,30,20,1)	3.9×10^{-3}	0.775	0.858	505	0.74
GAP	0.12	0.633	0.701	(227,10,10,1)	3.7×10^{-3}	0.766	0.840	235	0.34
CP$_{(a)}$	0.20	0.328	0.712	(101,20,10,1)	5.7×10^{-4}	0.536	0.945	95	0.77
CT$_{(a)}$	1.26	0.510	0.895	(102,40,1)	6.4×10^{-4}	0.734	0.996	41	0.33
AD$_{(p)}$	0.07	0.616	0.790	(67,30,20,1)	5.7×10^{-4}	0.888	0.975	67	0.78

Table 3. Results of Stages 2 and 3 in Phase 1 for the properties $\pi \in \{$BHL, DC, RF$_{(p)}\}$

π	ANN $\mathcal{N}$				Decision Tree DT_{sc}					Decision Tree DT_{hp}					
	L-time	t-R$^2_{cv}$	t-R$^2_{max}$	Arch.	L-time	t-R$^2_{cv}$	t-R$^2_{max}$	#ℓ	$\frac{\#\ell}{|D_\pi|}$	L-time	t-R$^2_{cv}$	t-R$^2_{max}$	#ℓ	#h	$\frac{\#h}{|D_\pi|}$
BHL	1.02	0.417	0.774	(106,30,20,1)	1.2×10^{-3}	0.475	0.855	16	0.06	10.7	0.659	0.920	4	32	0.11
DC	0.25	0.291	0.841	(95,50,1)	6.3×10^{-4}	0.518	0.918	69	0.55	2.02	0.605	0.945	1	16	0.13
RF$_{(p)}$	0.11	-3.837	-0.014	(76,50,1)	5.4×10^{-4}	0.703	0.955	20	0.22	2.96	0.421	0.756	2	21	0.23

Table 2 (resp., Table 3) shows the results on Stages 2 and 3 for properties $\pi \in \{$HOMO, LUMO, GAP, CP$_{(a)}$, CT$_{(a)}$, AD$_{(p)}\}$ (resp., $\pi \in \{$BHL, DC, RF$_{(p)}\}$), where we denote the following: L-time: the average time (sec.) to construct a single ANN $\mathcal{N}$ (resp., a single decision tree DT_{sc} or DT_{hp}) over all $10 \times 5 = 50$ constructions; t-R$^2_{cv}$: the best value of t-R$^2_{cv}(j)$ over all ten cross-validations for ANNs $\mathcal{N}$; (resp., decision trees $\mathcal{DT}_{sc}$ and $\mathcal{DT}_{hp}$); t-R²$_{max}$: the maximum of $R^2(\eta(j,i), D_\pi^{(i)})$ (resp., $R^2\left(\eta_{\mathcal{DT}_{sc}^{(j,i)}}, D_\pi^{(i)}\right)$ and $R^2\left(\eta_{\mathcal{DT}_{hp}^{(j,i)}}, D_\pi^{(i)}\right)$ over all 50 prediction functions $\eta_{\mathcal{N}(j,i)}$ (resp., $\eta_{\mathcal{DT}_{sc}^{(j,i)}}$ and $\eta_{\mathcal{DT}_{hp}^{(j,i)}}$, $j \in [1, 5]i \in$; Arch.: The architecture A_j, $j \in [1, 10]$ that attains t-R²$_{max}$. An architecture $(K, p, 1)$ (resp., $(K, p_1, p_2, 1)$) consists of an input layer with K nodes, a hidden layer with p nodes (resp., two hidden layers with p_1 and p_2 nodes, respectively) and an output layer with a single node, where K is equal to the number of descriptors in the feature vector; #ℓ: the number of leaf nodes in the decision tree $\mathcal{DT}_{sc}$ or $\mathcal{DT}_{hp}$ that attains t-R²$_{max}$; and $\frac{\#\ell}{|D_\pi|}$: the number of leaf nodes in the decision tree $\mathcal{DT}_{sc}$ divided by the number of graphs in the data set D_π. Note that this indicates the simplicity of the decision tree $\mathcal{DT}_{sc}$; #h: the total number of hyperplanes used in a decision tree $\mathcal{DT}_{hp} = (T, \mathcal{S}, \eta)$ constructed by our algorithm; i.e., #h = #$\ell + \sum\{|W_u| \mid$ leaf nodes $u\}$, where W_u denotes the separator set used to define the prediction function η_u of the leaf node u; and $\frac{\#h}{|D_\pi|}$: the total number of hyperplanes used in the decision tree $\mathcal{DT}_{hp}$ divided by the number of graphs in the data set D_π; Note that this indicates the simplicity of the decision tree $\mathcal{DT}_{hp}$.

From Tables 2 and 3, we see that the prediction functions constructed by decision trees perform considerably better than those by ANNs, and all of t-R²$_{max}$ attained by decision trees are around 0.75 to 0.99. For property RF$_{(p)}$, the best performance of a prediction function among 50 ANNs is apparently much worse than that by the both decision trees. The number of leaf nodes in $\mathcal{DT}_{hp}$ is much smaller than that in $\mathcal{DT}_{sc}$ since we use a prediction function based on hyperplanes at each leaf node. We observe that the simplicity measure #$\ell/|D_\pi|$ of $\mathcal{DT}_{sc}$ is 0.06 to 0.78 and #h/$|D_\pi|$ of $\mathcal{DT}_{hp}$ is 0.11 to 0.23.

Results on Phase 2. To execute Stages 4 and 5 in Phase 2, we used the same set of seven instances I_a, I_b^i, $i \in [1, 4]$, I_c and I_d prepared by Shi et al. [25]. We here present their seed graphs G_C. The seed graph G_C of instance I_a is given by the graph in Fig. 5 (a). The seed graph G_C' of instance I_b^i for each $i \in [1, 4]$ is illustrated in Fig. 6.

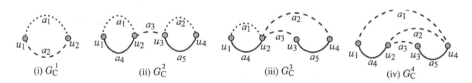

(i) G_C^1 (ii) G_C^2 (iii) G_C^3 (iv) G_C^4

Fig. 6. (i) A seed graph G_C^1 of rank-1 for instances I_b^1 and I_d; (ii) A seed graph G_C^2 of rank-2 for instance I_b^2; (iii) A seed graph G_C^3 of rank-2 for instance I_b^3; (iv) A seed graph G_C^4 of rank-2 for instance I_b^4.

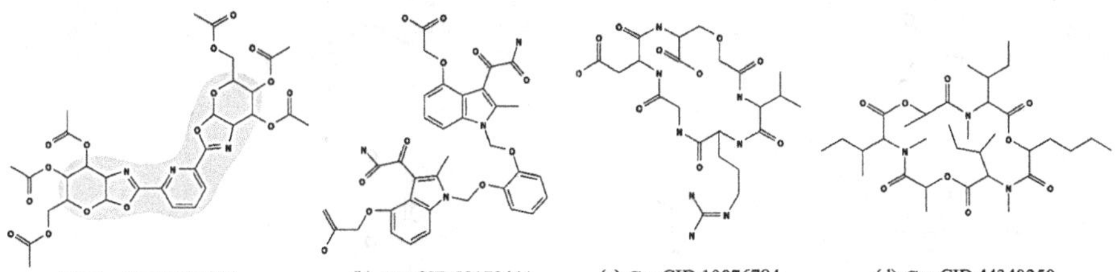

(a) $\mathbb{C}_A$: CID 24822711   (b) $\mathbb{C}_B$: CID 59170444   (c) $\mathbb{C}_A$: CID 10076784   (d) $\mathbb{C}_B$: CID 44340250

Fig. 7. An illustration of chemical compounds for instances I_c and I_d: (a) $\mathbb{C}_A$: CID 24822711; (b) $\mathbb{C}_B$: CID 59170444; (c) $\mathbb{C}_A$: CID 10076784; (d) $\mathbb{C}_B$: CID 44340250, where hydrogens are omitted.

Instance I_c has been introduced in order to infer a chemical graph $\mathbb{C}^\dagger$ such that the core of $\mathbb{C}^\dagger$ is equal to the core of chemical graph $\mathbb{C}_A$: CID 24822711 in Fig. 7(a) and the frequency of each edge-configuration in the non-core of $\mathbb{C}^\dagger$ is equal to that of chemical graph $\mathbb{C}_B$: CID 59170444 in Fig. 7(b). This means that the seed graph G_C of I_c is the core of $\mathbb{C}_A$ which is indicated by a shaded area in Fig. 7(a).

Instance I_d has been introduced in order to infer a chemical monocyclic graph $\mathbb{C}^\dagger$ such that the frequency vector of edge-configurations in $\mathbb{C}^\dagger$ is a vector obtained by merging those of chemical graphs $\mathbb{C}_A$: CID 10076784 and $\mathbb{C}_B$: CID 44340250 in Fig. 7(c) and (d), respectively. The seed graph G_C of I_d is given by G^1_C in Fig. 6(i).

Table 4. Results of Stages 4 and 5 for BHL and DC.

inst.	BHL									DC								
	y^*	#v	#c	I-time	n	n^{int}	D-time	C-LB	#C	y^*	#v	#c	I-time	n	n^{int}	D-time	C-LB	#C
I_a	0.05	7918	8957	2.7	41	27	0.171	3456	100	8.5	7852	8875	2.2	41	24	0.171	576	100
I_b^1	0.05	9516	6291	1.6	38	12	0.123	2.5×10^7	100	8.5	9450	6212	3.9	38	12	0.169	2.9×10^4	100
I_b^2	0.05	11158	8633	11.5	50	30	5.16	2.7×10^8	100	8.5	11092	8554	13.1	50	30	0.197	3.4×10^8	100
I_b^3	0.05	10972	8636	8.8	50	30	1.81	3.1×10^{11}	100	8.5	10906	8557	9.4	50	30	42.3	12	12
I_b^4	0.05	10786	8639	13.1	50	30	6.88	2.7×10^{10}	100	8.5	10720	8560	23.7	50	30	0.93	2.6×10^8	100
I_c	0.05	67697	1411	1.6	50	34	0.059	32	32	4.7	6703	7063	0.7	50	33	0.058	32	32
I_d	0.05	53966	651	3.5	44	23	8.11	1.9×10^9	100	8.5	5330	6573	8.4	40	23	1.29	8.2×10^7	100

Stage 4. We executed Stage 4 for the above seven instances with each of properties $\pi \in \{\text{BHL}, \text{DC}\}$. For the MILP formulation $\mathcal{M}(x, y; \mathcal{C}_1)$, we use the prediction function $\eta_{\mathcal{DT}_{hp}}$ based on $\mathcal{DT}_{hp}$ that attained t-R^2_{max} in Table 3. To solve an MILP in Stage 4, we used CPLEX version 12.10. Table 4 shows the results of Stage 4, where we denote the following: y^*: a target value for a property π; #v (resp., #c): the number of variables (resp., constraints) in the MILP in Stage 4; I-time: the time (sec.) to solve the MILP in Stage 4; n: the number $n(\mathbb{C}^\dagger)$ of non-hydrogen atoms in the chemical graph $\mathbb{C}^\dagger$ inferred in Stage 4; and n^{int}: the number $n^{int}(\mathbb{C}^\dagger)$ of interior-vertices in the chemical graph $\mathbb{C}^\dagger$ inferred in Stage 4.

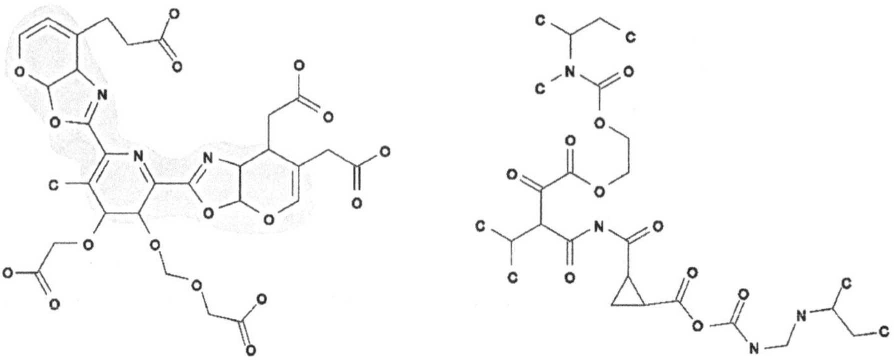

(a) $\mathbb{C}^{\dagger}$ to I_c (b) $\mathbb{C}^{\dagger}$ to I_d

Fig. 8. (a) $\mathbb{C}^{\dagger}$ inferred from I_c with $y^* = 0.05$ of BHL; (b) $\mathbb{C}^{\dagger}$ inferred from I_d with $y^* = 8.5$ of Dc, where hydrogens are omitted.

Figure 8(a) (resp., (b)) illustrates the chemical graph $\mathbb{C}^{\dagger}$ inferred from instance I_c with $y^* = 0.05$ of BHL (resp., I_d with $y^* = 8.5$ of Dc) in Table 4.

We see that the running time for solving an MILP instance with $n = 50$ is 0.7 to 23.7 (s), which is generally faster than the running time of 8.5 to 93.9 (s) to solve a similar set of MILP instances with $n = 50$ in the experimental result for the previous method with ANNs by Shi et al. [25].

Stage 5. We executed Stage 5 to generate a more number of target chemical graphs $\mathbb{C}^*$, where we call a chemical graph $\mathbb{C}^*$ a *chemical isomer* of a target chemical graph $\mathbb{C}^{\dagger}$ of isomers $\mathbb{C}^*$ of each target chemical graph $\mathbb{C}^{\dagger}$ inferred in Stage 4. We execute an algorithm for generating chemical isomers of $\mathbb{C}^{\dagger}$ up to 100 when the number of all chemical isomers exceeds 100. Such an algorithm can be obtained from the dynamic programming proposed by Shi et al. [25] with a slight modification. The algorithm can evaluate a lower bound on the total number of all chemical isomers $\mathbb{C}^{\dagger}$ without generating all of them.

Table 4 shows the computational results of the experiment, where we denote the following: D-time: the running time (sec.) to execute the dynamic programming algorithm in Stage 5 to compute a lower bound on the number of all chemical isomers $\mathbb{C}^*$ of $\mathbb{C}^{\dagger}$ and generate all (or up to 100) chemical isomers $\mathbb{C}^*$; C-LB: a lower bound on the number of all chemical isomers $\mathbb{C}^*$ of $\mathbb{C}^{\dagger}$; and #C: the number of all (or up to 100) chemical isomers $\mathbb{C}^*$ of $\mathbb{C}^{\dagger}$ generated in Stage 5.

From Table 4, we observe that the running time for generating up to 100 target chemical graphs in Stage 5 is not considerably larger than that in Stage 4.

6 Concluding Remarks

The framework of inferring chemical graphs was based on the result on ANNs and MILP by Akutsu and Nagamochi [17]. For some of various types of chemical properties, a prediction function based on a decision tree performs better than that by ANNs. In this paper, we used a decision tree to construct a prediction function in the framework for the first time. For this, we derived a new MILP formulation that simulates the computation process of a decision tree. We also designed a decision tree based on a set of separating hyperplanes in the feature space selected by solving an adequately introduced LP. From the experimental results on constructing a prediction function in Stage 3, we observed that our decision tree method outperforms the previous method with ANNs [25] for chemical properties such as energy of highest occupied molecular orbital, energy of lowest unoccupied molecular orbital, the energy difference between the HOMO and LUMO, critical pressure, critical temperature, experimental amorphous density, biological half life, dissociation constants and refractive index. The experimental result on Stage 4 suggests that our MILP formulation still can infer chemical graphs with up to 50 non-hydrogen atoms.

It is left as a future work to use other learning methods such as random forest, graph convolution and an ensemble method in Stages 3 and 4 in the framework.

References

1. Lo, Y.-C., Rensi, S.E., Torng, W., Altman, R.B.: Machine learning in chemoinformatics and drug discovery. Drug Discov. Today **23**, 1538–1546 (2018)
2. Tetko, I.V., Engkvist, O.: From Big Data to Artificial Intelligence: chemoinformatics meets new challenges. J. Cheminformat. **12**, 74 (2020)
3. Ghasemi, F., Mehridehnavi, A., Pérez-Garrido, A., Pérez-Sánchez, H.: Neural network and deep-learning algorithms used in QSAR studies: merits and drawbacks. Drug Discov. Today **23**(10), 1784–1790 (2018)
4. Miyao, T., Kaneko, H., Funatsu, K.: Inverse QSPR/QSAR analysis for chemical structure generation (from y to x). J. Chem. Inf. Model. **56**, 286–299 (2016)
5. Ikebata, H., Hongo, K., Isomura, T., Maezono, R., Yoshida, R.: Bayesian molecular design with a chemical language model. J. Comput. Aided Mol. Des. **31**(4), 379–391 (2017)
6. Rupakheti, C., Virshup, A., Yang, W., Beratan, D.N.: Strategy to discover diverse optimal molecules in the small molecule universe. J. Chem. Inf. Model. **55**, 529–537 (2015)
7. Bohacek, R.S., McMartin, C., Guida, W.C.: The art and practice of structure-based drug design: a molecular modeling perspective. Med. Res. Rev. **16**, 3–50 (1996)
8. Akutsu, T., Fukagawa, D., Jansson, J., Sadakane, K.: Inferring a graph from path frequency. Discrete Appl. Math. **160**(10–11), 1416–1428 (2012)
9. Kipf, T.N., Welling, M.: Semi-supervised classification with graph convolutional networks (2016). arXiv:1609.02907
10. Gomez-Bombarelli, R., et al.: Automatic chemical design using a data-driven continuous representation of molecules. ACS Cent. Sci. **4**, 268–276 (2018)
11. Segler, M.H.S., Kogej, T., Tyrchan, C., Waller, M.P.: Generating focused molecule libraries for drug discovery with recurrent neural networks. ACS Cent. Sci. **4**, 120–131 (2017)

12. Yang, X., Zhang, J., Yoshizoe, K., Terayama, K., Tsuda, K.: ChemTS: an efficient python library for de novo molecular generation. STAM **18**, 972–976 (2017)
13. Kusner, M.J., Paige, B., Hernandez-Lobato, J.M.: Grammar variational autoencoder. In: Proceedings of the 34th International Conference on Machine Learning, vol. 70, pp. 1945–1954 (2017)
14. De Cao, N., Kipf, T.: MolGAN: an implicit generative model for small molecular graphs (2018). arXiv:1805.11973
15. Madhawa, K., Ishiguro, K., Nakago, K., Abe, M.: GraphNVP: an invertible flow model for generating molecular graphs (2019). arXiv:1905.11600
16. Shi, C., Xu, M., Zhu, Z., Zhang, W., Zhang, M., Tang, J.: GraphAF: a flow-based autoregressive model for molecular graph generation (2020). arXiv:2001.09382
17. Akutsu, T., Nagamochi, H.: A mixed integer linear programming formulation to artificial neural networks. In: Proceedings of the 2nd International Conference on Information Science and Systems, pp. 215–220 (2019)
18. Pereira, G.: In: Schweiger, G. (ed.) Poverty, Inequality and the Critical Theory of Recognition. PP, vol. 3, pp. 83–106. Springer, Cham (2020). https://doi.org/10.1007/978-3-030-45795-2_4
19. Zhang, F., Zhu, J., Chiewvanichakorn, R., Shurbevski, A., Nagamochi, H., Akutsu, T.: A new integer linear programming formulation to the inverse QSAR/QSPR for acyclic chemical compounds using skeleton trees. In: Fujita, H., Fournier-Viger, P., Ali, M., Sasaki, J. (eds.) IEA/AIE 2020. LNCS (LNAI), vol. 12144, pp. 433–444. Springer, Cham (2020). https://doi.org/10.1007/978-3-030-55789-8_38
20. Azam, N.A., et al.: A novel method for inference of acyclic chemical compounds with bounded branch-height based on artificial neural networks and integer programming. To appear in Algorithms for Molecular Biology (2021)
21. Ito, R., Azam, N.A., Wang, C., Shurbevski, A., Nagamochi, H., Akutsu, T.: A novel method for the inverse QSAR/QSPR to monocyclic chemical compounds based on artificial neural networks and integer programming. In: Proceedings of the BIOCOMP2020, Las Vegas, Nevada, USA, 27–30 July (2020)
22. Zhu, J., Wang, C., Shurbevski, A., Nagamochi, H., Akutsu, T.: A novel method for inference of chemical compounds of cycle index two with desired properties based on artificial neural networks and integer programming. Algorithms **13**(5), 124 (2020)
23. Akutsu, T., Nagamochi, H.: A novel method for inference of chemical compounds with prescribed topological substructures based on integer programming (2020). arXiv:2010.09203
24. Zhu, J., et al.: A novel method for inferring of chemical compounds with prescribed topological substructures based on integer programming (submitted)
25. Shi, Y., et al.: An inverse QSAR method based on a two-layered model and integer programming. Int. J. Mol. Sci. **22**, 2847 (2021)
26. Annotations from HSDB (on pubchem): https://pubchem.ncbi.nlm.nih.gov/
27. QM9 @ MoleculeNet: http://moleculenet.ai
28. Bicerano, J.: Prediction of Polymer Properties. 3rd Edn, Revised and Expanded. CRC Press, Boca Raton (2002)

A Link-Based Ensemble Cluster Approach for Identification of Cell Types

Xinguo Lu$^{(\boxtimes)}$, Yan Gao, Daoxu Tang, and Yue Yuan

College of Computer Science and Electronic Engineering, Hunan University, Changsha, China

Abstract. Clustering is a necessary step in analyzing single-cell RNA-seq (sRNA-seq) data to illuminate the complexity of the tissue, including the number of cell types and transcriptome characteristics of each cell type. However, the clustering results obtained from different single-cell clustering methods are often different, and sometimes even contradictory conclusions are drawn. Biologists often cannot obtain the correct clustering results. To overcome this challenge, researchers have developed an integrated learning strategy that can effectively solve this problem. Here, we propose a new unsupervised ensemble clustering method LE2CT. First, we obtained five clustering results that have been used in the scRNA-seq data clustering method. Second, we construct a similarity consensus matrix based on multiple clustering solutions. Finally, hierarchical clustering is used as a consensus function to generate final data partitions. We identified cell clusters on twelve scRNA-seq benchmark data sets and used the adjusted RAND index (ARI) and normalized mutual information (NMI) to measure the accuracy of clustering. The experimental results are encouraging. Compared with the classic single clustering method, LE2CT has higher clustering accuracy and stronger robustness in various data sets, which shows that LE2CT has a competitive advantage compared with existing methods.

Keywords: Single-cell · Clustering · Heterogeneity

1 Introduction

Although multicellular organisms have similar genome sequences at the DNA level, cells in different tissues differ in gene expression and maintain the functions of corresponding tissues or organs through differential expression at the mRNA level. However, researchers have also found that even cell populations derived from similar morphology or even the same cell clone have differences in gene expression and stress response [1–4]. This cell heterogeneity has been observed in many studies, such as the heterogeneity and development of ureteral progenitor cells, the heterogeneity of human-induced pluripotent stem cells (hiPS), the developmental map of Arabidopsis root cells, etc. [5, 6]. In the same organism, cell populations in different organs or different tissues are heterogeneous at the genetic and epigenetic levels. Genetic differences include variations in ploidy number, such as haploid gametes (sperm, egg) and polyploid cells. This cellular heterogeneity has been detected in many organisms and

© Springer Nature Switzerland AG 2021
D.-S. Huang et al. (Eds.): ICIC 2021, LNCS 12837, pp. 645–654, 2021.
https://doi.org/10.1007/978-3-030-84529-2_54

developmental environments and plays an important role in tissue biology and disease states (such as cancer) [7–9]. It is also often related to the heterogeneity of expression of phenotype-related genes. For example, expression heterogeneity between embryonic stem cells can change differentiation characteristics. Besides, the expression heterogeneity between pathogenic bacteria cells or cancer cells may be related to human diseases. The transcriptional variation in the population of the single-celled malaria pathogen Plasmodium falciparum has potentially important implications for the adaptive response of the host organism [10, 11]. Cancer cells collected from the same tumor have significant heterogeneity in morphology and gene expression, and these differences are related to treatment and disease progression [12, 13]. Single-cell RNA sequencing (scRNA-seq) technology can measure the morphological gene expression of the whole transcriptome in a single cell and reveal the heterogeneity of the expression level between cells. This is essential for identifying clusters of cell types, inferring the arrangement of cell populations based on trajectory topology, highlighting the clonal structure of somatic cells, and characterizing cell heterogeneity in complex diseases [14, 15].

Bulk sequencing can only provide the average value of gene expression or apparent marker signals in the whole tissue. Therefore, Bulk sequencing is not suitable for solving differences in gene expression or epigenetic marks between cells [16, 17]. The increased resolution of single-cell RNA sequencing data and the statistical power of ever-growing data sets have brought exciting opportunities. Cluster analysis is a favorable calculation method for studying and analyzing the heterogeneity of these huge single-cell data. In the past few years, many clustering algorithms have been proposed to analyze scRNA-seq data. However, due to the existence of problems such as high noise and high dimension in scRNA-seq data [18–20], it increases the time and complexity of cell-type identification and makes systematic data analysis more challenging. Therefore, the algorithm needs to be constantly adjusted.

Several single-cell clustering algorithms have been proposed to deal with these problems. Yau et al. [21] proposed a hierarchical clustering method based on a cell hierarchy tree. The method combines dimensionality reduction of principal component analysis with hierarchical clustering. Firstly, the scRNA-seq data were projected onto the structure after principal component analysis, and initial k-means clustering was performed in the subspace to obtain the initial cluster allocation. Then, these clusters were iteratively merged according to the probability density function of each pair of clusters. Xu et al. [22] proposed a graph-based clustering method. They proposed a new similarity evaluation standard based on the concept of shared nearest neighbors to measure the similarity between cells, called the main similarity. The concept considers the influence of surrounding neighbor data points to process high-dimensional data. They obtained the cell similarity matrix based on Euclidean distance and listed the k nearest neighbors of each data point. At the same time, they proposed a second-level similarity measure, which calculates two similarities based on the shared neighborhood of two data points. The similarity between. Moreover, an SNN graph is constructed based on the connectivity between data points. Then, the graph-based clustering method is applied to the SNN graph, where nodes and weighted edges represent the similarity between data points and data points, respectively. Kiselev et al. [23]

proposed a consensus clustering method. This method deletes unrelated genes and uses three distance measures to estimate the similarity between cells. Next, use principal component analysis (PCA) or Laplacian to transform the similarity matrix, and perform k-means clustering based on the transformed similarity matrix. Then construct a consensus matrix. Finally, the clustering results are obtained through hierarchical clustering. However, the methods proposed above are all single clustering methods, which cannot provide an accurate and comprehensive overview of cell clustering. To solve the challenging problem of choosing the best method when the true cell type is unknown, combining information from multiple individual methods is an interesting solution.

Here, we designed a single-cell clustering based on the link ensemble method LE2CT (a link-based ensemble clustering method for identifying cell types). This method is based on a clustering ensemble strategy to cluster high-dimensional data. First, take the scRNA-seq data gene expression matrix as input, and obtain five different clustering solutions based on five published single-cell clustering analysis methods. Then, in order to construct a consensus matrix of the five clustering solutions, they are integrated with a link similarity evaluation method. Finally, perform hierarchical clustering on the consensus matrix to cluster units.

2 Materials and Method

2.1 Dataset Description

In this section, in order to evaluate the performance of the proposed single-cell clustering algorithm, we compare the ARI and NMI between LE2CT and five clustering methods on 12 published datasets. These data sets were collected from different tissues using different sequencing techniques, where the number of cells and the number of cell types were completely different. (1) Baise [24]: The dataset was sequenced using Smart-Seq protocol for the transcriptome information of mouse embryonic cells at each developmental stage: zygote, 2-cell stage, 4-cell stage, and primitive stage. (2) Zheng [25]: The Zheng data set contains 500 human peripheral blood mononuclear cells (PBMC) sequenced using the GemCode platform, including three cell types: CD56 + natural killer cells, CD19 + B cells and CD4 + / CD25 + regulatory T cells. (3) Yan [26]: This data set includes developmental transcriptome information for each mouse embryonic cell: oocyte, zygote, stage 2 cell, stage 4 cell, stage 8 cell, stage 16 cell, and late-stage blastocyst cell in the hatching stage. (4) Goolam [27]: The transcriptomes of all blastomeres of 28 embryos at the 2-cell stage, 4-cell stage, and 8-cell stages were determined, and the embryonic cells at 16 stages and 32 stages were sequenced. (5) Darmanis [28]: The dataset sequenced 466 individual cells from adult and fetal human brains, dividing the individual cells into all the major neuronal, glial, and vascular cell types in the brain. (6) Xin [29]: The authors used scRNA-seq to determine the transcriptomes of 1,492 human pancreatic α-, β-, δ- and PP cells from non-diabetic and type 2 diabetic organ donors. (7) Pollen [30]: The dataset contained 249 cells with 11 clusters. These include skin cells, pluripotent stem cells, blood cells, and nerve cells. (8) Kolodziejczyk [31]: The dataset contained 704 cells with three clusters obtained from mouse embryonic stem cells under different culture conditions.

(9) Usoskin [32]: The dataset consisted of 622 mouse neurons with four clusters: peptide-nociceptors, non-peptide-nociceptors, nerve filaments, and tyrosine hydroxylase cells. (10) Tasic [33]: The author classifies cells in the cortical area of adult mice. 49 transcriptome cell types were identified, including 23 GABAergic, 19 glutamatergic, and 7 non-neuronal types. (11) Muraro [34]: Sequencing data of 2126 human pancreatic cells, including 9 labeled cell types acinar, alpha, beta, delta, ductal, colorectal cancer, epsilon, gamma, mesenchymal, and some unknown cell types, were collected. (30) Zeisel [35]: The data were sequenced from the CA1 region of the somatosensory cortex and hippocampus of mice. Cell types included: interneurons, microglia, oligodendrocytic cells, pyramidal CA1, astrocytes, endothelial cells, and pyramidal SS neurons. The number and type of cells are summarized in Table 1.

Table 1. Summary of the Single-cell datasets in this study.

Dataset	Samples	Cluster	Data source
Biase	49	3	GSE57249
Zheng	500	3	10X Chromium
Yan	90	7	GSE36552
Goolam	124	5	E-MTAB-3321
Darmains	466	9	GSE67835
Xin	1600	8	GSE81608
Pollen	301	11	SRP041736
Kolodziejczyk	704	3	E-MTAB-2600
Usoskin	622	8	GSE59739
Tasic	1679	18	GSE71585
Muraro	2126	10	GSE85241
Zeisel	3005	9	GSE60361

2.2 Method Overview

In this section, we will introduce the workflow of LE2CT. This method designs a clustering ensemble strategy to cluster high-dimensional data. First, take the scRNA-seq data gene expression matrix as input, and based on the published single-cell clustering analysis methods, namely Seurat [36], SC3 [23], SIMLR [37], CIDR [38] and t-SNE + k-means [39] obtained multiple different clustering results. Secondly, based on the link similarity evaluation method as a consensus function, integrate information from multiple clustering results, and finally, the layer uses sub-clustering to generate the final data partition.

2.3   Method of Generating Base Clustering

(1) Seurat [36] is a method that combines the dimensionality reduction process and graph-based clustering. First, perform cell filtering and gene filtering on the input gene expression matrix. Then a k-nearest neighbor graph with Euclidean distance is constructed in the PCA space, and edges are drawn between similar cells. The weights of these edges are refined according to the Jaccard distance, which measures the difference between local neighborhoods. Finally, use graph clustering to classify cells.

(2) SC3 [23]. First, delete genes with unqualified expression values, and estimate multiple similarities between cells through Euclidean distance, Pearson correlation and cosine similarity. Then, the PCA dimensionality reduction method and the normalized Laplacian are used to transform the similarity matrix, and a consensus matrix is constructed based on the transformed similarity matrix. Finally, the clustering results are obtained through hierarchical clustering.

(3) SIMLR [37]. The model first flexibly estimates the similarity between elements by learning several Gaussian cores. Secondly, t-SNE is used to reduce the dimension of the matrix and k-means clustering is performed. When SIMLR processes big data, in order to reduce the cost of model calculation, knn is used instead of t-SNE to approximate the similarity matrix. The subsequent clustering steps are based on spectral clustering.

(4) CIDR [38]. This method uses implicit calculation to reduce missing events in monocyte RNA sequencing data. The author uses PCoA to achieve dimensionality reduction, after reducing zero-expansion noise, calculate the degree of dissimilarity between units, and obtain the final clustering result through hierarchical clustering based on the Calinski-Harabasz index.

(5) t-SNE + k-means [39]. The method is simple and effective. t-SNE algorithm is used to reduce the dimension of high-dimensional scRNA-seq data to a low-dimensional space, and then the data is clustered into n clusters based on k-means.

2.4   Link-Based Similarity Learning

The link-based similarity evaluation algorithm extends the range of similarity estimation beyond the local context of neighboring neighbors, assuming that neighboring neighbors are also similar [40].

The consensus matrix is calculated based on the cell labels inferred from the five scRNA-seq clustering methods mentioned in this article. Therefore, we obtained a single-cell network diagram based on the consensus matrix, which reflects the integrated similarity between cells. Essentially, given a cluster ensemble L, a graph $G = (V, W)$ can be constructed, where V is the set of vertices representing the data points and clusters in the set, and W is the set of edges between the data points and the clusters to which they are assigned.

Let $SRS(a, b)$ be the item in the SRS matrix, and represent the similarity between any pair of data points in the ensemble, or the similarity between any two clusters in the ensemble, is defined as follows:

$$SRS(a, b) = \frac{dc}{|N_a||N_b|} \sum_{a' \in N_a} \sum_{b' \in N_b} SRS(a', b') \tag{1}$$

where dc is constant decay factor within the interval (0, 1], $N_a \subset V$ denotes the set of vertices connecting to $a \subset V$. Accordingly, the similarity between data points x_i and x_j is the average similarity between the clusters to which they belong, and likewise, the similarity between clusters is the average similarity between their members.

2.5   Evaluation of Clustering Performance

Clustering algorithms often use an adjusted Rand index (ARI) to quantify how well our clustering results match another given set of labels. ARI ranges from [0–1]. Let $X = \{x_1, x_2, \ldots, x_m\}$ represents the true cell type label in m clusters, $Y = \{y_1, y_2, \ldots, y_n\}$ represents the predicted cell type in the n cluster. k is the total number of cells. The adjusted Rand index is defined as:

$$ARI(X, Y) = \frac{\sum_{i=1}^{m} \sum_{j=1}^{n} \binom{k_{ij}}{2} - \left[\sum_{i}^{m} \binom{a_i}{2} \sum_{j=1}^{n} \binom{b_j}{2}\right] / \binom{k}{2}}{\frac{1}{2}\left[\sum_{i=1}^{m} \binom{a_i}{2} + \sum_{j=1}^{n} \binom{b_i}{2}\right] - \left[\sum_{i}^{m} \binom{a_i}{2} \sum_{j=1}^{n} \binom{b_j}{2}\right] / \binom{k}{2}} \tag{2}$$

where k_{ij} represents the number of overlap cells between x_i and y_j $(k_{ij} = |x_i \cap y_j|)$, a_i is the sum of the i-th row in the contingency table, b_j is the sum of the j-th column in the contingency table.

The mutual information between X and Y is defined as:

$$I(X, Y) = \sum_{i=1}^{m} \sum_{j=1}^{n} \frac{x_i \cap y_j}{n} \log \frac{n|x_i \cap y_j|}{|x_i| \times |y_j|} \tag{3}$$

$H(X)$ and $H(Y)$ indicates the entry X and Y respectively. It is defined as follows:

$$H(X) = -\sum_{i=1}^{m} \frac{x_i}{k} \log \frac{x_i}{k} \tag{4}$$

$$H(Y) = -\sum_{j=1}^{n} \frac{y_i}{k} \log \frac{y_i}{k} \tag{5}$$

The normalized mutual information is given by:

$$NMI(X, Y) = \frac{2I(X, Y)}{H(X) + H(Y)} \tag{6}$$

NMI ranges from [0–1]. The higher NMI represents the better clustering.

3   Results

In this section, we benchmarked our clustering LE2CT method and 6 methods based on two evaluation indicators of ARI and NMI on 12 published data sets. These methods include the ensemble clustering method and the individual clustering method. These data sets cover different sequence technologies, species, tissues, number of single cells, and number of cell types.

Table 2. The ARI for single-cell clustering algorithm.

Dataset	SC3	Seurat	t-SNE + k-means	CIDR	SIMLR	SAME	LE2CT
Biase	0.948	0.947	0.948	1.000	1.000	0.948	1.000
Zheng	0.869	0.988	0.994	0.632	0.842	0.983	0.994
Yan	0.621	0.650	0.621	0.718	0.542	0.621	0.873
Goolam	0.629	0.582	0.828	0.606	0.513	0.582	0.892
Darmains	0.562	0.642	0.674	0.555	0.649	0.682	0.720
Xin	0.284	0.464	0.516	0.475	0.340	0.484	0.598
Pollen	0.934	0.853	0.775	0.923	0.921	0.923	0.928
Kolodziejczyk	0.703	0.489	0.710	0.398	0.656	0.711	0.597
Usoskin	0.814	0.415	0.366	0.517	0.407	0.775	0.558
Tasic	0.860	0.414	0.778	0.597	0.342	0.850	0.865
Muraro	0.643	0.646	0.610	0.242	0.621	0.704	0.621
Zeisel	0.407	0.387	0.772	0.489	0.410	0.602	0.848
Avg	0.690	0.623	0.716	0.596	0.604	0.739	0.791

Table 1 and Table 2 summarize the clustering results of all methods on twelve data sets. Taking ARI and NMI as the standard, LE2CT clustering outperforms all other methods in 66% data performance. Among the 12 scRNA-seq datasets, LE2CT produced the best results in 8 datasets (Biase, Zheng, Yan, Goolam, Domains, Xin, Tasic, Zeisel). In particular, the ARI and NMI scores of LE2CT on the two data Biase and Zheng are the same as the results of some clustering methods. For Biase, LE2CT, and CIDR, SIMLR correctly identified all cell clusters and cell types, with ARI and NMI scores of 1. The ARI and NMI scores for Zheng, LE2CT, and t-SNE + k-means are the same, 0.994 and 0.989, respectively. The results show that LE2CT is superior to all other methods (Table 3).

Table 3. The NMI for single-cell clustering algorithm.

Dataset	SC3	Seurat	t-SNE + k-means	CIDR	SIMLR	SAME	LE2CT
Biase	0.9294	0.9226	0.9294	1.000	1.000	0.9294	1.000
Zheng	0.905	0.980	0.989	0.771	0.847	0.973	0.989
Yan	0.726	0.752	0.719	0.844	0.726	0.726	0.874
Goolam	0.695	0.760	0.828	0.817	0.828	0.760	0.881
Darmains	0.725	0.690	0.750	0.701	0.742	0.74	0.773
Xin	0.580	0.676	0.568	0.556	0.418	0.649	0.659
Pollen	0.945	0.93	0.900	0.952	0.964	0.952	0.958
Kolodziejczyk	0.753	0.729	0.750	0.449	0.735	0.756	0.624
Usoskin	0.823	0.668	0.503	0.6333	0.618	0.793	0.678
Tasic	0.878	0.851	0.857	0.734	0.625	0.868	0.885
Muraro	0.791	0.802	0.794	0.449	0.744	0.825	0.793
Zeisel	0.649	0.658	0.738	0.581	0.645	0.724	0.800
Avg	0.783	0.785	0.777	0.707	0.741	0.808	0.826

4 Conclusion

Increasing sequence data and clustering algorithms have helped solve many research problems. However, the experimental results show that the existing single-based clustering methods are not robust on multiple data sets, and even perform poorly on some complex data sets. Because the scRNA-seq data onto different technologies or sequencers is unlabeled and has limited additional information, it is difficult to evaluate the best clustering method. To solve this problem, we propose a novel clustering framework LE2CT, a link-based ensemble cluster approach for the identification of cell types. Based on the five published methods of single-cell clustering analysis, five different clustering solutions are obtained, and a refined cluster association matrix is created based on a link integration method of link similarity evaluation. Finally, hierarchical clustering was used as a consensus function to generate the final data partition. Through experiments on single-cell data sets, using ARI and NMI to compare the experimental results with the most advanced single-cell clustering methods that have been published, the feasibility of the method proposed in this paper is verified.

Acknowledgements. This work was supported by Natural Science Foundation of China (Grant No. 61972141) and Natural Science Foundation of Hunan Province, China (Grant No. 2018JJ2053).

References

1. Toyooka, Y., Shimosato, D., Murakami, K., Takahashi, K., Niwa, H.: Identification and characterization of subpopulations in undifferentiated ES cell culture. Development **135**, 909–918 (2008)

2. Bumgarner, S.L., et al.: Single-cell analysis reveals that noncoding RNAs contribute to clonal heterogeneity by modulating transcription factor recruitment. Mol. Cell **45**, 470–482 (2012)
3. Chang, H.H., Hemberg, M., Barahona, M., Ingber, D.E., Huang, S.: Transcriptome-wide noise controls lineage choice in mammalian progenitor cells. Nature **453**, 544–547 (2008)
4. Shalek, A.K., et al.: Single-cell RNA-seq reveals dynamic paracrine control of cellular variation. Nature **510**, 363–369 (2014)
5. Zhang, T.Q., Xu, Z.G., Shang, G.D., Wang, J.W.: A single-cell RNA sequencing profiles the developmental landscape of arabidopsis root. Mol. Plant **12**, 648–660 (2019)
6. Nguyen, Q.H., et al.: cRNA-seq of human induced pluripotent stem cells reveals cellular heterogeneity and cell state transitions between subpopulations. Genome Res. **28**, 1053–1066 (2018)
7. Zhao, Q., et al.: Single-cell transcriptome analyses reveal endothelial cell heterogeneity in tumors and changes following antiangiogenic treatment. Cancer Res. **78**, 2370–2382 (2018)
8. Calbo, J., et al.: A functional role for tumor cell heterogeneity in a mouse model of small cell lung cancer. Cancer Cell **19**, 244–256 (2011)
9. Tellez-Gabriel, M., Ory, B., Lamoureux, F., Heymann, M.F., Heymann, D.: Tumour heterogeneity: the key advantages of single-cell analysis. Int. J. Mol. Sci. **17**, 2142 (2016)
10. Walzer, K.A., Fradin, H., Emerson, L.Y., Corcoran, D.L., Chi, J.T.: Latent transcriptional variations of individual Plasmodium falciparum uncovered by single-cell RNA-seq and fluorescence imaging. PLoS Genet. **15**, e1008506 (2019)
11. Wen, L., Tang, F.: Single-cell sequencing in stem cell biology. Genome Biol. **17**, 71 (2016)
12. Dagogo-Jack, I., Shaw, A.T.: Tumour heterogeneity and resistance to cancer therapies. Nat. Rev. Clin. Oncol. **15**, 81–94 (2018)
13. Saliba, A.-E., Westermann, A.J., Gorski, S.A., Vogel, J.: Single-cell RNA-seq: advances and future challenges. Nucleic Acids Res. **42**(14), 8845–8860 (2014)
14. Haque, A., Engel, J., Teichmann, S.A., Lönnberg, T.: A practical guide to single-cell RNA-sequencing for biomedical research and clinical applications. Genome Med. **9**(1), 75 (2017)
15. Chung, W., et al.: Single-cell RNA-seq enables comprehensive tumour and immune cell profiling in primary breast cancer. Nat. Commun. **8**, 15081 (2017)
16. Li, W.V., Li, J.J.: An accurate and robust imputation method scimpute for single-cell rna-seq data. Nat. Commun. **9**(1), 997 (2018)
17. Stegle, O., Teichmann, S.A., Marioni, J.C.: Computational and analytical challenges in single-cell transcriptomics. Nat. Rev. Genet. **16**(3), 133 (2015)
18. Soneson, C., Robinson, M.D.: Bias, robustness and scalability in single-cell differential expression analysis. Nat. Methods **15**, 255–261 (2018)
19. Andrews, T.S., Hemberg, M.: M3Drop: dropout-based feature selection for scRNASeq. Bioinformatics **35**, 2865–2867 (2019)
20. Grun, D.: Revealing dynamics of gene expression variability in cell state space. Nat. Methods **17**, 45–49 (2020)
21. Yau, C., et al.: pcareduce: hierarchical clustering of single cell transcriptional profiles. BMC Bioinforma **17**(1), 140 (2016). https://doi.org/10.1186/s12859-016-0984-y
22. Xu, C., Su, Z.: Identification of cell types from single-cell transcriptomes using a novel clustering method. Bioinformatics **31**(12), 1974–1980 (2015)
23. Kiselev, V.Y., et al.: Sc3: consensus clustering of single-cell rna-seq data. Nat. Methods **14**(5), 483 (2017)
24. Biase, F.H., Cao, X., Zhong, S.: Cell fate inclination within 2-cell and 4-cell mouse embryos revealed by single-cell RNA sequencing. Genome Res. **24**, 1787–1796 (2014)
25. Zheng, G., Terry, J.M., Belgrader, P., Ryvkin, P., Bent, Z.W., Wilson, R., et al.: Massively parallel digital transcriptional profiling of single cells. Nat. Commun. **8**, 14049 (2017)

26. Yang, M., Guo, H., Yang, L., Wu, J., Li, R., et al.: Single-cell RNA-seq profiling of human preimplantation embryos and embryonic stem cells. Nat. Struct. Mol. Biol. **20**, 1131–1139 (2013)

27. Goolam, M., Scialdone, A., Graham, S.J., Macaulay, I.C., Jedrusik, A., Hupalowska, A., et al.: Heterogeneity in Oct4 and Sox2 targets biases cell fate in 4-cell mouse embryos. Cell **165**, 61–74 (2016)

28. Darmanis, S., Sloan, S.A., Zhang, Y., Enge, M., Caneda, C., Shuer, L.M., et al.: A survey of human brain transcriptome diversity at the single cell level. Proc. Natl. Acad. Sci. U.S.A. **112**, 7285–7290 (2015)

29. Xin, Y., et al.: RNA sequencing of single human Islet cells reveals type 2 diabetes genes. Cell Metab. **24**, 608–615 (2016)

30. Pollen, A.A., Nowakowski, T.J., Shuga, J., et al.: Low-coverage single-cell mRNA sequencing reveals cellular heterogeneity and activated signaling pathways in developing cerebral cortex. Nat. Biotechnol. **32**(10), 1053–1058 (2014)

31. Kolodziejczyk, A.A., et al.: Single cell RNA-sequencing of pluripotent states unlocks modular transcriptional variation. Cell Stem Cell **17**, 471–485 (2015)

32. Usoskin, D., et al.: Unbiased classification of sensory neuron types by large-scale single-cell RNA sequencing. Nat. Neurosci. **18**, 145–153 (2015)

33. Tasic, B., et al.: Adult mouse cortical cell taxonomy revealed by single cell transcriptomics. Nat. Neurosci. **19**, 335–346 (2016)

34. Muraro, M.J., et al.: A single-cell transcriptome atlas of the human pancreas. Cell Syst. **3**, 385-394.e3 (2016)

35. Zeisel, A., et al.: Brain structure. Cell types in the mouse cortex and hippocampus revealed by single-cell RNA-seq. Science **347**, 1138–1142 (2015)

36. Macosko, E.Z., et al.: Highly parallel genome-wide expression profiling of individual cells using nanoliter droplets. Cell **161**(5), 1202–1214 (2015)

37. Wang, B., Zhu, J., Pierson, E., Ramazzotti, D., Batzoglou, S.: Visualization and analysis of single-cell RNA-seq data by kernel-based similarity learning. Nat. Methods **14**(4), 414–416 (2017)

38. Lin, P., Troup, M., Ho, J.W.K.: CIDR: ultrafast and accurate clustering through imputation for single-cell RNA-seq data. Genome Biol. **18**(1), 59 (2017)

39. Grün, D., Lyubimova, A., Kester, L., Wiebrands, K., Basak, O., Sasaki, N., et al.: Single-cell messenger RNA sequencing reveals rare intestinal cell types. Nature **525**(7568), 251–255 (2015)

40. Jeh, G., Widom, J.: SimRank: a measure of structural-context similarity. In: The Eighth ACM SIGKDD International Conference ACM, pp. 538–543 (2002)

A Defect Detection Method for Diverse Texture Fabric Based on CenterNet

Wenjing Kong⬚, Huanhuan Zhang$^{(\boxtimes)}$⬚, Junfeng Jing⬚,
and Mingyang Shi⬚

School of Electronics and Information,
Xi'an Polytechnic University, Xi'an 710048, China

Abstract. Fabric defect detection is a crucial step for the fabric production process. However, existing traditional methods of fabric defect detection only pay attention to the simplest plain and twill fabrics, while ignoring the complicated diverse texture fabric. In this paper, we proposed a robust defect detection method based on CenterNet for diverse texture fabric. Firstly, we used ResNet-50 backbone network to extract the fabric image features. Then we applied up-sampling of three-fold deconvolution to obtain the high-resolution feature images. Finally, defects in the diverse texture were located and recognized. Experimental results showed that the proposed method can successfully detect the defects and achieve a high mAP. Compared with the current state-of-the-art methods, such as Faster-RCNN, EffcientDet and Yolo-V4, the proposed method outperforms these methods in terms of accuracy and robustness.

Keywords: Fabric defect detection · CenterNet · Convolution operation · Deconvolutions

1 Introduction

For modern textile enterprises, textile appearance quality inspection is an indispensable process, since defects have a direct impact on the quality and price of textile end products [1, 2]. However, most textile enterprises still rely on manual inspection method to detect fabric defects. Traditional manual inspection method is easily affected by personal subjective factors. Especially, for the defects with complex texture can hardly be recognized by human eyes. It is urgent to develop an automatic defect detection method to replace traditional manual inspection method. In recent years, various defect detection methods based on machine vision have attracted extensive attention. Some achievements have been made in fabric defect detection of grey cloth, checkers, stripes, dots and stars with relatively simple texture structure and single color, but the adaptability of complex texture fabrics with variable colors and various patterns are insufficient [3, 4].

In order to solve the problem of fabric defects with complex texture and pattern characteristics, Pan [5] proposed a defect detection method based on cross-correlation, which realized the automatic detection of pattern offset, color difference and other defects of printed fabric. Kuo [6] proposed a defect detection method for printed fabrics based on RGB cumulative mean method, and effectively detected the warp break, weft

© Springer Nature Switzerland AG 2021
D.-S. Huang et al. (Eds.): ICIC 2021, LNCS 12837, pp. 655–664, 2021.
https://doi.org/10.1007/978-3-030-84529-2_55

break and other defects. However, the RGB space calculation is large, the hardware requirements are high, and the aperiodic printed fabrics are not discussed. Liu [7] used the multi-level Gan network to expand the samples of a small number of printed fabrics, improved the accuracy of the segmentation network, and successfully completed the defect detection of periodic printed fabrics. Jing [8] used mobile UNET network model to detect various types of fabric defects, and the results were excellent. Li [9] applied Gaussian mixture model to complete the detection of printing fabric defects such as wrong pattern, color difference and pattern skew, but the detection target of this method is still periodic printing fabric, and it needs the comparison and matching of positive samples.

Diverse texture fabric has many feature, including changeable color, various pattern, and low contrast between defect object and background. The existing traditional detection based on machine vision cannot be effectively detected. In our work, a robust defect detection method for diverse texture fabric based on CenterNet was proposed.

2   Based on the CenterNet Network Fabric Defect Detection Method

2.1   CenterNet Network Structure

Compared with YOLO and Faster-RCNN, CenterNet [10] is an anchor-free target detection network, which has advantages in speed and accuracy [11]. In addition, the peak center point was extracted by CenterNet convolution according to the heat map, which can effectively describe the defect feature of the complex background texture fabric.

The CenterNet network consists of three parts: the backbone network, the upsampling network and the prediction network. The framework of the CenterNet is shown in Fig. 1.

The backbone network of CenterNet is ResNet50. ResNet50 network consists of several Bottleneck modules. Each Bottleneck module has two residual blocks which are named Conv Block and Identity Block, as shown in Fig. 2. The function of Conv Block is to change the dimension of the network. Identity Block has the same input and output dimensions and can be connected in series to deepen the network. ResNet50 contains 49 convolutional layers and 1 fully connected layer, the structure is shown in Fig. 3.

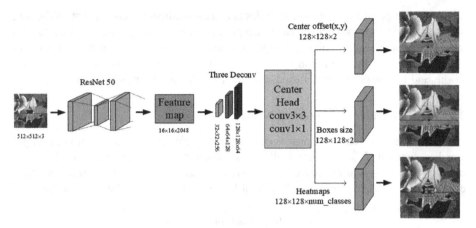

Fig. 1. The framework of CenterNet

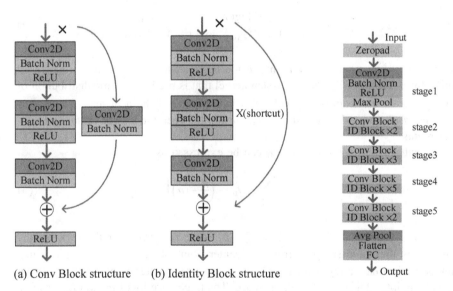

(a) Conv Block structure   (b) Identity Block structure

Fig. 2. Bottleneck structure chart

Fig. 3. The framework of ResNet50

In our work, we obtained a preliminary feature layer for the next step of processing using the backbone network which shape is $16 \times 16 \times 2048$, and then we discarded the last layer of the ResNet full connection layer FC, and replaced it with CenterNet using three rewinding to get a higher resolution output. In order to save the amount of calculation, the number of output channels of these 3 deconvolutions are 256, 128, and 64 respectively. Therefore, after three deconvolution and upsampling, the height and width of the feature layer we obtained are 128×128, and the number of channels is 64. Finally, we use the effective feature layer to obtain the final prediction result.

CenterNet prediction network includes heat map prediction, center point prediction, and width and height prediction. Heat map prediction is used to predict whether there is an object at each heating point and to predict the type of object. Center point prediction is used to predict the deviation of the center of each object from the heating point. Width and height prediction is used to predict the width and height of each object.

2.2   Loss Function of CenterNet Network

The CenterNet loss function consists of three parts: center point loss function L_k, target size loss function L_{size} and center offset loss function L_{off}. λ_{size} and λ_{off} are hyperparameter to modify the influence of off loss and size loss. Each loss function has a corresponding weight parameter. The total loss function L_{det} can be expressed as:

$$L_{det} = L_k + \lambda_{size}L_{size} + \lambda_{off}L_{off} \tag{1}$$

In total loss function, the center point loss function L_k can be expressed as:

$$L_k = -\frac{1}{N}\sum_{xyc}\begin{cases}(1 - \hat{Y}_{xyc})^\alpha \log(\hat{Y}_{xyc}) & , Y_{xyc} = 1 \\ (1 - Y_{xyc})^\beta(\hat{Y}_{xyc})^\alpha \log(1 - \hat{Y}_{xyc}), \text{otherwise}\end{cases} \tag{2}$$

Where, N is the number of key points in the image, which is used to normalize all positive focal loss to 1. Y_{xyc} is a Gaussian kernel that is used to find the distribution of the keypoints in the feature map [11]. $\hat{Y}_{xyc}$ is the value of the heat map. The hyperparameters α, β are used to adjust the relationship between the loss of center and non-center points.

The center offset loss function L_{off} can be expressed as:

$$L_{off} = \frac{1}{N}\sum_{p}\left|\hat{O}_{\hat{p}} - \left(\frac{p}{R} - \tilde{p}\right)\right| \tag{3}$$

Where, $\hat{O}_{\hat{p}}$ is the predicted local offset. $\left(\frac{p}{R} - \tilde{p}\right)$ is the real error calculated before the training. $\frac{p}{R}$ is the value sampled from the center point of the original image, which contains the decimal part. p is the coordinate of the target center point, $\tilde{p}$ is the position of the key point obtained in the output space. The target size loss function L_{size} can be expressed as:

$$L_{size} = \frac{1}{N}\sum_{k=1}^{N}\left|\hat{S}_{pk} - s_k\right|s_k \tag{4}$$

Where, $\hat{S}_{pk}$ is the size that our model predicts, s_k is a truth size calculated by using the value after downsampling the location of the top-left and bottom-right of the dataset.

3   Results and Discussion

In our work, we use the PyTorch to build the CenterNet network. The specific environment is followed: GPU is NVIDIA GeForce RTX 2080Ti, CPU is Intel (R) Xeon (R) Gold5118 @ 2.30 GHz. The operating system is Windows 10.0, the development environments are Python 3.7, CUDA10.1, Torch1.5.0 and some necessary dependent libraries.

3.1   Dataset

The experimental samples are derived from the texture fabric defect of the actual factory which are classified according to the category of the defect. Image feature details are clear and all have been confirmed by experienced inspectors. To verify the wide applicability of our method, diverse textures and defects are adopted, texture including simple plaid background texture and complex printed background textures, defects including pilling, wrinkle, beltyarn, leakink, filament and pass. The part of samples are shown in Fig. 4.

Large-scale data sets are the prerequisite for the successful application of deep convolution neural networks. In our work, we use image enhancement techniques to randomly rotate the training sample with a parameter of 180°, the horizontal and vertical translation ratio is 30%, the zoom change range ratio is 10%, horizontal and vertical random flips, cropping and adjustment of brightness and color. The number of samples before and after enhancement is shown in Table 1.

(a) pilling   (b) filament   (c) wrinkle

(d) beltyarn   (e) leakink   (f) pass

Fig. 4. Six types of defects

Table 1. Number of samples before and after expansion

Type of defect	Number of data before augment	Number of data after augment
pilling	50	106
filament	100	368
wrinkle	100	300
beltyarn	100	295
leakink	100	304
pass	100	300
Total	550	1673

There are 1673 defect data images after enhancement, which were labeled with LabelImg. The labeled data set was divided into training set and test set with the ratio of 8:2. 1205 images were selected from 1339 training sets for training, and 134 images were selected as the validation set. The specific data are shown in Table 2.

Table 2. Test set and training set

Data type	Training set	Validation set	Test set	Total
pilling	76	9	21	106
filament	265	30	73	368
wrinkle	216	24	60	300
beltyarn	213	23	59	295
leakink	219	24	61	304
pass	216	24	60	300
Total	1205	134	334	1673

3.2   Parameter Settings and Evaluation Metrics

In the training stage, the speed of backbone feature extraction is general, freezing training can speed up the training speed and prevent the weights from being destroyed. Therefore, we freezes the backbone feature extraction network with a pre-trained model. Some experimental parameters are shown in Table 3.

We chose precision P (precision), R (Recall), mAP (mean average precision), and FPS to test the performance of our method [16].

Table 3. Some experimental parameters

Parameters	Lr (freeze/unfreeze)	Epoch	Batch size	Init epoch	Freeze epoch
Parameter value	1e−3/1e−4	100	8	0	50

3.3 Results and Discussion

Figure 5 is the detection result of fabric defect. The red box represents the pilling defect, the light blue box represents the beltyarn defect, the yellow box represents the filament defect, the green box represents the wrinkle defect, the blue box represents the leakink defect, the rose red box represents the pass defect.

Fig. 5. Test results of different defects (Color figure online)

The statistical graphs of mAP and log-average miss rate are shown in Fig. 6. It can be seen that the our method can accurately identify and locate the defect location.

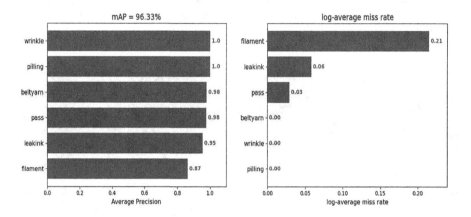

Fig. 6. The statistical graphs of mAP and log-average miss rate

In order to further verify the effectiveness of the proposed method, we compared our method with other existing defect detection methods. They are EfficientDet [12], YOLO-V4 [13, 14] and Faster-RCNN [15], and tested on various types of fabric defect data. Table 4 is the detection results of various types of defects by different methods.

Table 4. Detection results of various types of defects by different models

	YoloV4		Fast-RCNN		EfficientDet		CenterNet (ours)	
	P (%)	R (%)	P (%)	R (%)	P (%)	R (%)	P (%)	R (%)
Pilling	37.5	70.59	100	100	100	100	**100**	94.1
Filament	90.72	**100**	61.16	84.09	**98.67**	84.09	98.21	62.50
Wrinkle	93.94	95.38	36.31	93.85	0	0	**100**	95.38
Pass	88.33	89.83	72.84	100	0	0	**98.25**	94.92
Beltyarn	79.22	**93.85**	83.12	98.46	**100**	1.54	96.6	87.69
Leakink	94.74	94.74	73.91	89.47	94.74	94.74	**96.49**	**96.49**

From Table 4, the CenterNet model has an accuracy of 100% for pilling. The accuracy of detecting defects such as filament is 98.21%, although it is lower than EfficientDet and ranked second, but there is not much difference. The detection accuracy and recall rate of wrinkle, pass, leakink defects are the highest, and the detection accuracy of beltyarn defects is ranked second, but very close.

The CenterNet and other methods are compared with the average accuracy rate (mAP) and FPS. The experimental results are shown in Fig. 7.

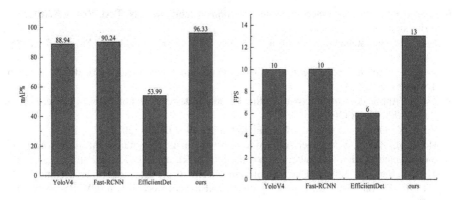

Fig. 7. The experimental results of mAP and FPS

From Fig. 7, it can be seen that mAP and FPS of CenterNet are higher than other methods, it achieved a good balance between detection accuracy and detection efficiency. Experimental results showed that the CenterNet performs well with higher accuracy, faster detection speed, and better real-time performance.

4 Conclusion

A robust fabric defect detection method based on CenterNet was proposed. The experimental results showed that our method has higher average detection accuracy and faster detection speed than the three comparison methods of Yolo-V4, Faster-RCNN and EffcientDet. Our method can accurately and effectively detect the defects of various complex texture fabrics.

Acknowledgment. This work was supported by financial support received from the National Natural Science Foundation of China (No. 61902302). Shaanxi Provincial Key R&D Program Project (2021GY-261), Shaanxi Innovation Ability Support Program (2021TD-29).

References

1. Jing, J.F.: A coarse-to-fine method for glass fiber fabric surface defect detection. J. Text. Inst. **112**(3), 388–397 (2021)
2. Li, F.: Bag of tricks for fabric defect detection based on Cascade R-CNN. Text. Res. J. **91**(5–6), 599–612 (2021)
3. Zhou, T.: A series of efficient defect detectors for fabric quality inspection. Measurement **172**, 108885 (2021)
4. Li, C., Liu, C., Gao, G., Liu, Z., Wang, Y.: Robust low-rank decomposition of multi-channel feature matrices for fabric defect detection. Multimed. Tools Appl. **78**(6), 7321–7339 (2018). https://doi.org/10.1007/s11042-018-6483-6
5. Pan, R.: Defect detection of printed fabrics using normalized cross correlation. J. Text. Res. **31**(12), 134–138 (2010)

6. Kuo, C.F.: Automatic detection system for printed fabric defects. Text. Res. J. **82**(6), 591–601 (2012)
7. Liu, J.H.: Multistage GAN for fabric defect detection. IEEE Trans. Image Process. **29**, 3388–3400 (2019)
8. Jing, J.F.: Mobile-Unet: an efficient convolutional neural network for fabric defect detection. Text. Res. J. (2020)
9. Li, M.: Application of Gaussian mixture model on defect detection of print fabric. J. Text. Res. **36**(8), 94–98 (2015)
10. Zhou, X.: Objects as points. arXiv preprint arXiv:1904.07850 (2019)
11. He, K.M., Zhang, X.Y.: Deep residual learning for image recognition. In: IEEE Conference on Computer Vision and Pattern Recognition (CVPR), pp. 770–778 (2016)
12. Tan, M., Pang, R.: EfficientDet: scalable and efficient object detection. In: Proceedings of the IEEE/CVF Conference on Computer Vision and Pattern Recognition (CVPR), pp. 10778–10787 (2020)
13. Jing, J.: Fabric defect detection using the improved YOLOv3 model. J. Eng. Fibers Fabr. **15**(1), 1–10 (2020)
14. Bochkovskiy, A.: YOLOv4: optimal speed and accuracy of object detection (2020)
15. Liu, Z., Guo, Z.: Research on texture defect detection based on faster-RCNN and feature fusion. In: Proceedings of the 11th International Conference on Machine Learning and Computing, Zhuhai, China, 22–23 February 2019, pp. 429–433 (2019)
16. Liu, S., Ma, H.: Combined attention mechanism and CenterNet pedestrian detection algorithm. In: 2021 IEEE 5th Advanced Information Technology, Electronic and Automation Control Conference (IAEAC), pp. 1978–1982 (2021). https://doi.org/10.1109/IAEAC50856.2021.9391037

Accelerating Deep Reinforcement Learning via Hierarchical State Encoding with ELMs

Tao Tang⑩, Qiang Fang⑩, Xin Xu$^{(\boxtimes)}$⑩, and Yujun Zeng⑩

National University of Defense Technology, Changsha 410073, China
{tangtao16,nxu}@nudt.edu.cn

Abstract. Image-based deep reinforcement learning has made great break-through and achievements in recent years, while unavoidably facing with the requirement of a large amount of interaction data and the problem of low training efficiency. In order to refine this problem, we propose a new method to accelerate the learning process of deep reinforcement learning by combining hierarchical encoder network with an actor-critic RL algorithm. Through making use of stacked extreme learning machines (ELMs) and training it in a supervised way, we are able to convert the high-dimensional raw image into meaningful states and send to the cascaded state-based RL algorithm. To deal with continuous state and action spaces we adopt the Cerebellar Model Articulation Controller (CMAC) network to be the critic, and take advantage of recursive least-squares TD (RLS-TD) learning method to improve the learning efficiency. Our algorithm has provided a novel useful mechanism to solve image-based end-to-end continuous control problems. Simulations on the typical cart-pole task show that our method can solve such problem efficiently and significantly improve the learning effectiveness, compared with previous deep reinforcement algorithms such as DDPG and PPO.

Keyword: Deep reinforcement learning · State encoding · ELMs

1 Introduction

As an important branch of machine learning, reinforcement learning (RL) has received plenty of attention in recent years, and many notable achievements have been made [1–3]. The combination with function approximator (especially the deep neural network) makes it possible for RL algorithms to deal with high-dimensional or continuous control problems, which derives another important branch—deep reinforcement learning (DRL). The function approximator is usually used to approximate the state value function, state-action value function or policy, in which way the iterative value calculation in traditional tabular RL algorithms (such as Q-learning, SARSA) is transformed into the learning process of the approximator's weight parameters.

Currently, most DRL algorithms are designed for specific task, such as playing Chess [4], StarCraft [5], Atari games [6], autonomous driving [7], robotic manipulation [8] and so on. In order to obtain an optimal policy, a large amount of interactive data is required to be prepared and the agent is usually trained from scratch. From the perspective of algorithm input, such interactive data generally appears in two forms: states

© Springer Nature Switzerland AG 2021
D.-S. Huang et al. (Eds.): ICIC 2021, LNCS 12837, pp. 665–680, 2021.
https://doi.org/10.1007/978-3-030-84529-2_56

and image. Therefore, RL algorithms can also be divided into two types: state-based and image-based. The state-based algorithms are relatively straightforward because the input data is state value that closely relates to the policy making, such as the location, speed and posture data in the classic continuous control task based on Mujoco physics engine [9]. However, this method is difficult for us to obtain accurate and comprehensive state data in the real world, which makes it infeasible to provide solutions for practical tasks.

In contrast, environment images are taken as input for image-based algorithms, such as playing Atari games, and then the agent is usually trained in an end-to-end manner. The advantage of this method is that it is easy to obtain generous image data. However, it inevitably brings the problem of excessively high input data dimensions, which causes additional burdens on computation and storage resources, and the network converts the image into a feature vector without any physical meaning. Moreover, training an agent based on images usually requires a great number of image data. For example, model-free RL algorithms on Atari [6] and DeepMind Control (DMC) [10] have cost tens of millions of steps to reach the stable stage. The low online learning efficiency and unstable training process of image-based DRL algorithms are still a great barrier to application.

As for the image-based RL algorithms, a number of researchers have done related work and proposed many useful approaches, like self-supervised learning, contrastive learning, auxiliary task and so on. Autoencoder is the simplest auxiliary task among of them which aims to accomplish learning representation with pixel reconstruction objective. Most prior work take the two-step training procedure, where a state representation autoencoder is trained first and then the control policy is trained on the top of the fixed representation [11–13]. However, this kind of training method are both data and time consuming. Another form of training procedure was developed in recent years, represented by CURL [14], which trains the encoder module along with cascaded RL algorithm by constructing the contrastive loss and has achieved great performance in both Atari games and DMC after taking a long time and computation to train. In addition, model learning methods on images [15, 16] is proved to be more successful and sample efficiency but brittle to hyperparameter settings and difficult to reproduce.

In this paper, in order to overcome these drawbacks described above, we propose a new end-to-end DRL algorithm that integrates a hierarchical encoding network that trained in supervised way in the first stage, which encodes the raw image into low-dimensional states. The hierarchical encoding network is constructed by stacked extreme learning machines (ELMs) [17], which is guaranteed to extract states effectively in theory while greatly improve the learning speed. We cascade a learning control algorithm in an actor-critic framework based on the pre-trained hierarchical encoding network. In order to deal with continuous state and action spaces, we take the Cerebellar Model Articulation Controller (CMAC) [18] in the critic network and use the recursive least-squares TD (RLS-TD) method for policy evaluation. Simulations taken in the benchmark continuous control task, cart-pole, shows that the proposed approach outperforms PPO [19] and DDPG [20] and demonstrates higher learning efficiency and effectiveness.

The motivation of our method is to provide solution for some scenes where the task requires high learning efficiency and accurate control performance while states data of controlled object is inaccessible in certain stage. That's when we have to provide another form of input information, like image, to compensate the absence of states. Taking the automatic driving as an example, if one or more sensors on the vehicle are out of work and cannot provide corresponding state data, it will be quite dangerous in such situation and might cause serious damage to both life and property. It will be different if we pre-train an encoding network based on labeled images and apply it into the above scenes, we can make up for the lack of sensors and ensure the normal running of system.

The primary contributions of this paper are: (1) proposing a cascaded DRL algorithm, which encodes states first using hierarchical encoding network, and then executes online learning control. (2) Providing a fast end-to-end learning architecture with high learning efficiency and effectiveness.

The rest of this paper is organized as follows. Section 2 introduces some preliminaries about our approach. And the general architecture and implementation detail of our method will be described in Sect. 3. Furthermore, we validate our method in cartpole task and analyze the experiment results in Sect. 4.

2 Preliminaries

2.1 Markov Decision Processes (MDPs)

Many learning control problems can be abstracted as Markov Decision Processes (MDPs), which is a mathematically idealized form of the reinforcement learning problem. MDPs corresponds to the process where the agent obtains the environment's state $s_t \in \mathcal{S}$ at each time step t, and selects an action $A_t \in A(s)$. One time step later, in part as a consequence of its action, the agent receives a numerical reward $R_{t+1} \in \mathcal{R}$, and changes into a new state s_{t+1}. This process can be described as Fig. 1.

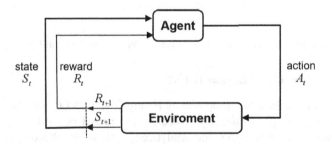

Fig. 1. The Markov decision processes.

When we talking about finite MDPs, it means that the sets of states, actions, and rewards (S, A, and R) all have a finite number of elements. Thus, the MDPs can be represented as a tuple (S, A, P, R, γ), in which S is a finite set of states, A is a finite set of actions, P is a state transition probability matrix, R is a reward function and γ is the discount factor, which is used to balance the immediate and future reward [21].

The final goal of agent training is to maximize the cumulative reward G_t it receives in the long run, that is

$$G_t = R_{t+1} + \gamma R_{t+2} + \gamma R_{t+3} + \cdots = \sum_{k=0}^{\infty} \gamma^k R_{t+k+1} \tag{1}$$

In MDPs, policy π maps states into probabilities of each action. And in order to improve policy, value function $v_\pi(s)$ and state-value function $q_\pi(s,a)$ are used to estimate the given state s and action a with respect to the current policy.

$$v_\pi(s) = E_\pi[G_t|S_t = s] = E_\pi\left[\sum_{k=0}^{\infty} \gamma^k R_{t+k+1}|S_t = s\right] \tag{2}$$

$$q_\pi(s,a) = E_\pi[G_t|S_t = s, A_t = a] = E_\pi\left[\sum_{k=0}^{\infty} \gamma^k R_{t+k+1}|S_t = s, A_t = a\right] \tag{3}$$

Optimal policy is the one that leads to the most cumulative reward, denotes as π_*. And the corresponding optimal value v_* function and state-value q_* function are denoted as

$$v_*(s) = \max_\pi v_\pi(s) \tag{4}$$

$$q_*(s,a) = \max_\pi q_\pi(s,a) \tag{5}$$

According to Bellman optimality equation, the value of a state under an optimal policy must equal the expected return for the best action from that state, that is

$$v_*(s) = \max_{a \in A} q_*(s,a) \tag{6}$$

The training process ends when the optimal value function and policy are found.

2.2   Extreme Learning Machine (ELM)

Extreme learning machine was proposed by Huang et al. in 2006, who creatively came up with a kind of single-hidden layer feedforward neural networks (SLFNs) which randomly chooses hidden nodes and analytically determines the output weights of SLFNs. Their experiments on regression and classification problems had demonstrated great accuracy and efficiency. In theory, ELM tends to provide good generalization performance at extremely fast learning speed.

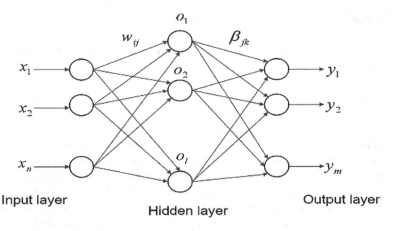

Fig. 2. The structure of extreme learning machine.

For SLFNs demonstrates in Fig. 2, the input is N distinct samples (x_i, t_i), where $x_i = [x_{i1}, x_{i2}, \cdots, x_{in}]^T \in R^n$, $t_i = [t_{i1}, t_{i2}, \cdots, t_{im}]^T \in R^m$. The output of standard SLFNs can be modeled as

$$y_j = \sum_{i=1}^{L} \beta_i g_i(x_j) = \sum_{i=1}^{L} \beta_i g(w_i x_j + b_i) \tag{7}$$

where L is the number of hidden nodes and $g(x)$ is activation function.

Huang et al. had proved that the output matrix H of hidden layer is invertible for w_i and b_i generated in any intervals of R_n and R respectively [17]. Then $\|H\beta - T\| = 0$ and the parameters β can be computed in a least-squares way of the linear system $H\beta = T$, that is

$$\left| H(w_1, \cdots, w_L, b_1, \cdots, b_L)\hat{\beta} - T \right| = \min_{\beta} |H(w_1, \cdots, w_L, b_1, \cdots, b_L)\beta - T| \tag{8}$$

If we choose distinct training samples N that makes it equals to the number of hidden nodes L, $L = N$, the matrix H will be square and invertible and parameter β can be computed by

$$\hat{\beta} = H^{-1}T \tag{9}$$

However, in most cases, L is much less than N, $L \ll N$, which makes H a nonsquare matrix. It becomes difficult to find correct parameters that satisfy the linear condition $H\beta = T$, in which situation β can only be computed by the following form

$$\hat{\beta} = H^{\dagger}T \tag{10}$$

where $H^{\dagger}$ is the Moore-Penrose generalized inverse of matrix H.

3   Methods

In the section we will introduce the main architecture of our method in detail and describe its training process.

3.1   The Cascaded DRL Framework

The general framework of our approach is shown in Fig. 3. It is composed of two cascaded modules: the feature encoding module and the RL algorithm module. Firstly, the feature encoding module is pre-trained and responsible for converting the input image into a low-dimensional state vector s_t, then the cascaded RL algorithm module generates action based on its policy, which acts on the environment and receive reward r_t. We take this two-step training procedure so as to improve the encoding accuracy and the control stability. In this framework, we can choose any suitable RL algorithm to act as agent according to the task properties, such as discrete or continuous state space and action space.

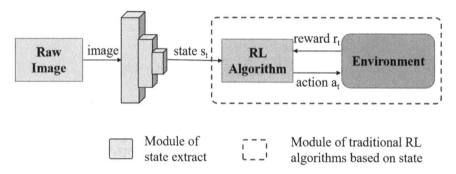

Fig. 3. The general framework of our approach.

3.2   Hierarchical Encoding Network

The hierarchical encoding network is based on ELM and consists of a stacked autoencoder and a regressor, as shown in Fig. 4. The input is raw image and the output is encoded states vector, while the middle part of the encoding network maps the high-dimensional image into low-dimensional states vector.

The stacked autoencoder is made up of multiple ELMs. In theory, the multi-layer ELMs has better learning representation and feature extraction capabilities, which makes it suitable for high-dimensional signal, such as image. In our method, the stacked autoencoder is used to map the raw high-dimensional image data into low-dimensional encoded features. And the role of regressor is to convert these low-dimensional encoded features into states for the learning of RL algorithms. In this way we convert the image-based control problems into state-based ones while the encoded states vector corresponds to real meaningful states, which we found to be great helpful for reaching our control aim.

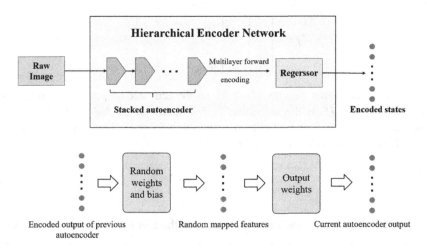

Fig. 4. The architecture of hierarchical encoder network in our approach.

The training process of the Hierarchical Encoder Network contains two separate stages: the unsupervised hierarchical feature learning of stacked autoencoder and the supervised feature regression of regressor. The stacked autoencoder is built up by stacking multiple unsupervised trained autoencoder. As the number of the autoencoder being stacked increases, the encoded features will become more compact and meaningful. The lower part of Fig. 4 shows the flowchart of single autoencoder.

The objective function of the train process is

$$O = \underset{\beta}{\arg\min}\left\{|H\beta - Y| + |\beta|_{l_1}\right\} \tag{11}$$

where H stands for the random mapping output of the hidden layer, Y is the original input data that could be the raw images input or the output of the previous autoencoder T, and β denotes the output layer weights. Besides, the l_1-penalty is conducted to make the autoencoder discover more compact and meaningful latent features.

$$T_i = f(T_{i-1} \cdot \beta) \quad i = 2, 3, \ldots, k \tag{12}$$

where $f(\cdot)$ is the activation function, k is the total number of ELM autoencoders. If $i = 1$ the input will be the image.

In terms of the regressor, which is a typical SLFN, the algorithm in ELM is used for the supervised learning process, where the objective function is formulated as

$$O = \underset{\beta_r}{\arg\min}\left\{\frac{1}{2}C|H\beta_r - L|^2 + \frac{1}{2}|\beta_r|^2\right\} \tag{13}$$

$$L = \begin{bmatrix} l_1^T \\ \vdots \\ l_N^T \end{bmatrix} = \begin{bmatrix} l_{11} & \cdots & l_{1m} \\ \vdots & \ddots & \vdots \\ l_{N1} & \cdots & l_{Nm} \end{bmatrix} \tag{14}$$

where β_r is the output weights which relates the activated output of the hidden layer to the output layer, H_r is the activation output matrix of the hidden layer. C is the regularization factor, L is the label matrix shown in formula (14). According to ELM theory, β_r can be calculated as follows

$$\beta_r = H_r^T \left(\frac{1}{C} + H_r H_r^T \right)^{-1} L \tag{15}$$

Algorithm 1 illustrates the training process of the hierarchical encoder network.

Algorithm 1 Training of Hierarchical Encoder Networks

Require: N_k:number of hidden nodes in the k-th autoencoder,$k = 1, 2, \ldots, m$
        N_{m+1}: number of hidden nodes in regressor
        C: regularization parameter
        $\{I_i, l_i\}^m$: image samples with label
1: Set k to 1
2: **repeat**
3:     Randomly set the input weights w_k and biases b_k for the k-th autoencoder
4:     Compute the hidden activation output matrix H_k
5:     Compute the output weights β_k
6:     Set $k \leftarrow k+1$
7: **until** $k = m$
8: Randomly set the input weights w and biases b for the regressor
9: Compute the hidden activation output matrix H
10: Compute the output weights β
11: **return** β_k and β

3.3 Linear Value Function Approximation Based on Encoded States

Our approach can be applied to both discrete and continuous state and action spaces. For the discrete case, the typical value based RL algorithms (such as SARSA, DQN) can be utilized in the control process straightly. In this subsection, we will mainly pay attention to the case of continuous action space, and introduce the actor-critic algorithm with CMAC.

The cascaded RL algorithm in our method take the actor-critic framework, in which the critic network uses RLS-TD [22] to estimate the value function and the actor network updates the policy by using the estimated policy gradient. We combine CMAC networks with linear approximation structures and apply RLS-TD learning method to improve the learning efficiency.

CMAC is an associative network with local reception properties for continuous inputs inspired from the human brain which usually has three layers [18]. The first layer maps the continuous input space S into a high-dimensional address space including $A_1, A_2, \ldots, A_n$ with an overlapped quantization encoding structure. The second layer converts the high-dimensional address space into a one-dimensional feature space F using Hash encoding method, to reduce the storage space. The third layer computes linearly weighted output-based feature $F(s)$ and weight W. The structure of CMAC is shown in Fig. 5.

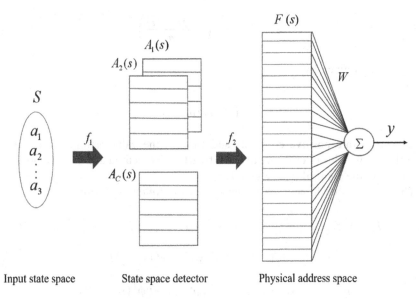

Fig. 5. The CMAC network used in our approach.

In the CMAC network, every input has C state detectors and M partitions. For the i-th element of an input s, the detector output is computed as

$$A(s) = \sum_{i=1}^{4} \left[a(i) + M^{i-1} \right] \tag{16}$$

$$F(s) = A(s) \bmod K \tag{17}$$

where K represents the physical memory size and $F(s)$ represents the corresponding address.

The last layer of CMAC computes the linear weighted sum in the active physical address unit, which is

$$y(s) = w^T F(s) \tag{18}$$

The above introduced CMAC network is used as the value function approximator in the critic. And we adopt the RLS-TD algorithm to accelerate the learning process. As for the actor network, the action selection policy is defined as

$$\widehat{y}_t = f(u, s_t) \tag{19}$$

where $u = [u_1, u_2, \ldots, u_m]$ is weight vector of actor network and s_t is the state encoded by hierarchical encoder network described in Sect. 3.2.

The actor outputs can be approximated by the Gaussian probabilistic distribution

$$p_r(y_t) = \exp\left(-\frac{(y_t - \widehat{y}_t)}{\sigma_t^2}\right) \tag{20}$$

where σ_t is given by

$$\sigma_t = \frac{m_1}{1 + \exp(m_2 V(s_t))} \tag{21}$$

m_1 and m_2 are positive constants and $V(s_t)$ is the value estimation of the critic network. The weights of critic are updated by RLS-TD method and the weights of actor is updated in the form of policy gradients.

The update of critic network weights takes the following form:

$$P_{t+1} = P_t - P_t \overrightarrow{z_t} \left[1 + \left(\phi^T(s_t) - \gamma\phi^T(s_{t+1})\right)P_t\overrightarrow{z_t}\right]^{-1}\left(\phi^T(s_t) - \gamma\phi^T(s_{t+1})\right)P_t \tag{22}$$

$$K_{t+1} = P_t\overrightarrow{z_t}/\left(1 + \left(\phi^T(s_t) - \gamma\phi^T(s_{t+1})\right)P_t\overrightarrow{z_t}\right) \tag{23}$$

$$Wc_{t+1} = Wc_t + K_{t+1}\left(r_t - \left(\phi^T(s_t) - \gamma\phi^T(s_{t+1})\right)Wc_t\right) \tag{24}$$

where s_t and z_t are states and eligibility trace at step t, and P_t, K_t, Wc_t are the variance matrix, auxiliary variables and weights parameter of critic respectively.

And the update of actor network weights is calculated by

$$Wa_{t+1} = Wa_t + \beta_t\frac{\partial J_\pi}{\partial a_t} \tag{25}$$

where β_t is the learning rate of actor and J_π is the objective function.

The training process of our approach is listed as Algorithm 2.

Algorithm 2 Training process of the online Actor-Critic with Encoded States

---

Require: critic network C, actor network A, trained encoder network E, stop criterion ϵ, and learning rate of actor network β

1: **repeat**
2:   receive the image input I_t
3:   convert I_t into s_t using E
4:   calculate the output of actor network y_t
5:   calculate the probability distribution $p_r(s_t)$ and sample action a_t
6:   execute a_t and obtain the next state s_{t+1} and reward r_{t+1}
7:   update the weights of critic network W_c by equation (22), (23) and (24)
8:   update the weights of actor network W_a by equation (25)
9: **until** the criterion is satisfied
10: **return** W_a and W_c

---

4   Experiments

4.1   Experiment Design

We have conducted experiments in the benchmark continuous action control task, cart-pole, to demonstrate the effectiveness of our approach, and compare it with both state-based and image-based DDPG and PPO algorithms.

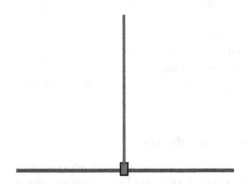

Fig. 6. The created cart-pole environment.

The cart-pole environment interacted in our simulations is shown in Fig. 6. The dynamical model of the cart-pole system is described by the following equations

$$\ddot{\theta} = \frac{(m+M)g\sin\theta - \cos\theta\left[F + ml\theta^2\sin\theta - \mu_c sgn(\dot{x})\right] - \frac{\mu_p(m+M)\dot{\theta}}{ml}}{\frac{4}{3}(m+M)l - ml\cos^2\theta} \tag{26}$$

$$\ddot{x} = \frac{F + ml\left(\dot{\theta}^2\sin\theta - \ddot{\theta}con\theta\right) - \mu_c sgn(\dot{x})}{m + M} \tag{27}$$

Where $g = -9.8$ m/s, x denotes the distance between the cart and the track and F is the force applied to the cart, which is limited between $[-10$ N, 10 N$]$. The masses of the cart and the pole are 1.0 kg and 0.1 kg respectively. The length of the half-pole is $l = 0.5$ m, and $\mu_c = 0.0005$ and $\mu_p = 0.000002$ both represent the friction coefficient, the former is that of the cart on the track and the latter is the pole on the cart. The position and angle are bounded in $[-1.2$ m, 1.2 m$]$ and $[-12°, 12°]$ respectively.

We treat the cart-pole task as an episodic task, which means that when the position of cart or the angle of pole exceed the above range, the corresponding episode will end and randomly starts at the initial state between $[-0.05, 0.05]$. At each step, the reward $r_t = 0$ if this episode does not over otherwise $r_t = -1$. The hyperparameters of our algorithm is listed in the Table 1.

For DDPG and PPO algorithms, the parameters are default. Through a number of experiments, we found that DDPG and PPO were hard to converge in the sparse reward situation after more than 10000 episodes, so we change the reward for these two algorithms into $r_t = 1$ for not fall and $r_t = 0$ for failure, which is easier to learn compared with the sparse case.

Table 1. The hyperparameters of our method

Hyperparameter	Value
Earning rate of actor	0.5
Discount factor	0.96
Lambda	0.6
Forgetting rate	1
The number of tiles	4
The number of partitions	6
Physical memory sizes	30 for critic and 60 for actor

4.2   Effectiveness and Generalization

Firstly, we have trained the hierarchical encoder network in a supervised way with collected real states data at each step, and Fig. 7 demonstrates the encoding precision. It's clear that our hierarchical encoder network has achieved a high accuracy to encode the image into meaningful states with supervised data, which are position x and angle θ, from which we can derive the velocity v and angle rate w. Based on the precise states input, the cascaded actor-critic algorithm also displays excellent control performance.

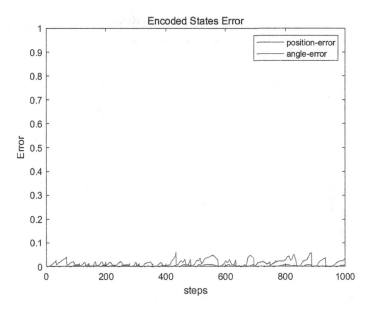

Fig. 7. The encoded states error of hierarchical encoder network.

We have examined our method, DDPG, and PPO three algorithms in the above cart-pole task independently. DDPG and PPO take both states and image as input, which we named DDPG-state, PPO-state, DDPG-image, PPO-image respectively. The total training episodes is 500 and the maximum number of steps in each episode is 200, at which we think the controller has mastered the task completely. Figure 8 shows the training results.

From the results curve recorded in the figure, we can clearly see that our approach possesses the capacity to control cart-pole quickly, only after about 40 episodes, while both DDPG-state and PPO-state converge at 460 and 320 episodes respectively. Our method surpasses them at least 8 times in terms of the learning efficiency on this task. In addition, it's difficult for DDPG-image and PPO-image to learn to control this task in such short time without straight states information. In our simulations we find that it usually takes at least 10000 episodes to converge for both DDPG-image and PPO-image algorithms.

We also carry out a series of experiments to test the generalization of our method. On the one hand, we change the initial position of cart-pole in each episode, at −1 m and +1 m respectively, which named Position-left and Position-right. The results in Fig. 9 validates that our method generates into these two situations well. On the other hand, we change the length of pole with different values, 0.4 m, 0.45 m, 0.55 m, 0.6 m respectively, and train them in the same way with normal condition. Except the 0.4 m case, other cases of pole length all success in controlling the cart-pole, though with different speed.

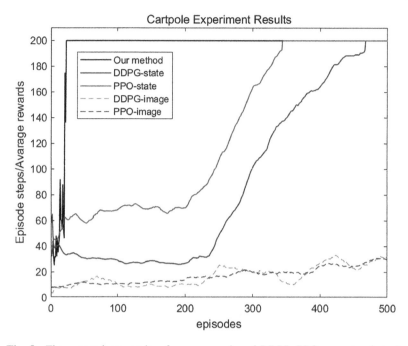

Fig. 8. The comparison results of our approach and DDPP, PPO on cart-pole task.

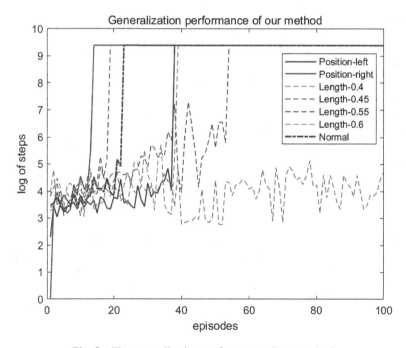

Fig. 9. The generalization performance of our method.

The excellent performance of our method comes from several aspects. The main advantage should be owed to the hierarchical encoder network, which provides precise states about the current MDP situation. And with the help of CMAC network and RLS-TD learning method, the learning speed is improved a lot. The slow convergency and poor performance of DDPG-state and PPO-state methods is mostly limited by their complex network architecture and learning mechanism.

5   Conclusion

In this paper, we have proposed a novel method to accelerate the deep reinforcement learning by using ELM-based hierarchical state encoding, and describe its principle and training process in detail. The experiment results carried on the cartpole control task demonstrate the great effectiveness and efficiency of our method compared with DDPG and PPO. The advantage of our method is that provides a useful solution to the image-based scene where other RL algorithms are data-consuming and time-wasting. And it can be applied to many other scenes such as robotic control and manipulation.

Acknowledgements. This work was supported by the Nation Natural Science Foundation of China under Grant 61825305, and the National Key R&D Program of China 2018YFB1305105. And this work is completed with the help of many persons, they are Xin Xu, Qiang Fang, Junkai Ren, Yixin Lan, and Yichuan Zhang. Their suggestion and help have greatly encouraged me to do more rigorous and meaningful work.

References

1. Schrittwieser, J., Antonoglou, I., Hubert, T.: Mastering Atari, Go, chess and shogi by planning with a learned model. Nature **588**(7839), 604–609 (2020)
2. Mnih, V., Kavukcuoglu, K., Silver, D.: Human-level control through deep reinforcement learning. Nature **518**(7540), 529–533 (2015)
3. Neftci, E.O., Averbeck, B.: Reinforcement learning in artificial and biological systems. Nat. Mach. Intell. **1**(3), 133–143 (2019)
4. Silver, D., Huang, A., Maddison, C.: Mastering the game of go with deep neural networks and tree search. Nature **529**(7587), 484–489 (2016)
5. Vinyals, O., Babuschkin, I., Czarnecki, W.M.: Grandmaster level in StarCraft II using multi-agent reinforcement learning. Nature **575**(7782), 350–354 (2019)
6. Mnih, V., Kavukcuoglu, K., Silver, D.: Playing Atari with deep reinforcement learning. Computer Science (2013)
7. Sallab, A.E.L., Abdou, M., Perot, E.: Deep reinforcement learning framework for autonomous driving. Electron. Imaging **2017**(19), 70–76 (2017)
8. Gu, S., Holly, E., Lillicrap, T.: Deep reinforcement learning for robotic manipulation with asynchronous off-policy updates. In: IEEE International Conference on Robotics and Automation (ICRA), pp. 3389–3396. IEEE, Singapore (2017)
9. Todorov, E., Erez, T., Tassa, Y.: MuJoCo: a physics engine for model-based control. In: IEEE/RSJ International Conference on Intelligent Robots and Systems, Portugal, pp. 5026–5033. IEEE (2012)
10. Tassa, Y., Doron, Y.: Deepmind control suite. arXiv preprint arXiv:1801.00690 (2018)

11. Lange, S., Riedmiller, M.: Deep auto-encoder neural networks in reinforcement learning. In: International Joint Conference on Neural Networks, Spain, pp. 1–8. IEEE (2010)

12. Mattner, J., Lange, S., Riedmiller, M.: Learn to swing up and balance a real pole based on raw visual input data. In: Huang, T., Zeng, Z., Li, C., Leung, C.S. (eds.) ICONIP 2012. LNCS, vol. 7667, pp. 126–133. Springer, Heidelberg (2012). https://doi.org/10.1007/978-3-642-34500-5_16

13. Dwibedi, D., Tompson, J., Lynch, C.: Learning actionable representations from visual observations. In: IEEE/RSJ International Conference on Intelligent Robots and Systems (IROS), pp. 1577–1584. IEEE (2019)

14. Srinivas, A., Laskin, M., Abbeel, P.: CURL: contrastive unsupervised representations for reinforcement learning. arXiv e-prints arXiv:2004.04136 (2020)

15. Ha, D., Schmidhuber, J.: Recurrent world models facilitate policy evolution. arXiv e-prints arXiv:1809.01999 (2018)

16. Lee, A.X., Nagabandi, A., Abbeel, P.: Stochastic latent actor-critic: deep reinforcement learning with a latent variable model. arXiv e-prints arXiv:1907.00953 (2019)

17. Huang, G.B., Zhu, Q.Y., Siew, C.K.: Extreme learning machine: theory and applications. Neurocomputing 70(1–3), 489–501 (2006)

18. Albus, J.S.: A new approach to manipulator control: the Cerebellar Model Articulation Controller (CMAC). Trans. ASME J. Dyn. Syst. 97, 220–227 (1975)

19. Schulman, J., Wolski, F., Dhariwal, P.: Proximal policy optimization algorithms. arXiv e-prints arXiv:1707.06347 (2017)

20. Lillicrap, T.P., Hunt, J., Pritzel, A.: Continuous control with deep reinforcement learning. arXiv preprint arXiv:1509.02971(2015)

21. Sutton, R.S., Barto, A.G.: Reinforcement Learning: An Introduction. MIT Press, Cambridge (2018)

22. Xu, X., He, H., Hu, D.: Efficient reinforcement learning using recursive least-squares methods. J. Artif. Intell. Res. 16, 259–292 (2002)

Mal_PCASVM: Malonylation Residues Classification with Principal Component Analysis Support Vector Machine

Tong Meng[1], Yuehui Chen[2], Wenzheng Bao[3(✉)], and Yi Cao[4]

[1] School of Information Science and Engineering,
University of Jinan, Jinan, China
[2] School of Artificial Intelligence Institute and Information Science
and Engineering, University of Jinan, Jinan, China
[3] School of Information School of Information Engineering (School of Big
Data), Xuzhou University of Technology, Xuzhou, China
baowz55555@126.com
[4] Shandong Provincial Key Laboratory of Network Based Intelligent
Computing, School of Information Science and Engineering),
University of Jinan, Jinan, China

Abstract. Post-translational modification (PTM) is considered a significant biological process with a tremendous impact on the function of proteins in both eukaryotes, and prokaryotes cells. Malonylation of lysine is a newly discovered post-translational modification, which is associated with many diseases, such as type 2 diabetes and different types of cancer. In addition, compared with the experimental identification of propionylation sites, the calculation method can save time and reduce cost. In this paper, we combine principal component analysis with support vector machine (SVM) to propose a new computational model - Mal-PCASVM (malonylation prediction). Firstly, the one-hot encoding, physicochemical properties and the composition of k-spacer acid pairs were used to extract sequence features. Secondly, we preprocess the data, select the best feature subset by principal component analysis (PCA), and predict the malonylation sites by SVM. And then, we do a five-fold cross validation, and the results show that compared with other methods, Mal-PCASVM can get better prediction performance. In the 10-fold cross validation of independent data sets, AUC (area under receiver operating characteristic curve) analysis has reached 96.39%. Mal-PCASVM is used to identify the malonylation sites in the protein sequence, which is a computationally reliable method. It is superior to the existing prediction tools that found in the literature and can be used as a useful tool for identifying and discovering novel malonylation sites in human proteins.

Keywords: Post-translational modification · Malonylation · One-hot encoding · Principal component analysis · Support vector machine

© Springer Nature Switzerland AG 2021
D.-S. Huang et al. (Eds.): ICIC 2021, LNCS 12837, pp. 681–695, 2021.
https://doi.org/10.1007/978-3-030-84529-2_57

1 Introduction

Protein post-translational modification refers to a covalent process of protein during or after translation, that is, to change the properties of protein by adding modification groups to one or several amino acid residues or cutting off groups by protein hydrolysis. As we all know, PTM is very important in the prevention and treatment of some diseases. Malonylation of lysine is a newly discovered post-translational modification, in which a negatively charged malonyl group is added to the positively charged Lysine side chain by chemical modification. And it was initially detected by mass spectrometry and widely exists in eukaryotes and prokaryotes [1]. For example, kmal is rich in key signaling molecules in mouse liver [2], plant cells [3], and Gram-positive bacteria Saccharopolyspora spinosa [4, 5]. Although many efforts have been made to study the cellular mechanism of Kmal, its biological significance is unknown. In the substrate, malonylation sites'recognition is an important step in elucidating the molecular mechanism of malonylation.

So far, there have developed various computational tools to predict malonylation sites in protein sequences [9–14].For example, Wang et al. [10] established a predictor named MaloPred, which considered five characteristics, including amino acid binary coding (BINA), k-nearest neighbor feature(KNN), encoding based on grouped weight (EBGW),amino acid composition(AAC) and position specific scoring matrix(PSSM). Then evaluate their information gain (IG) to select the most meaningful and important features. Xu et al. [9] constructed a prediction tool called Mal-Lys by integrating the sequence information of residual sequences, position specific amino acid tendency and physicochemical properties of each peptide using the minimum Redundant Maximum (mMRM) correlation model. Chen et al. [12] proposed an integrated malonylation predictor based on LSTM, which combines the long-term and short-term memory algorithm with word embedding, and combines the random forest algorithm with new amino acid content enhancement coding. Hasan and Kurata [11] put forward a prediction tool called identification of Lysine Malonylation Site (iLMS), which encodes the fragment using the composition of map based k-spacer amino acid pair, dipeptide amino acid composition (DC) and amino acid index attribute (AAindex). Furthermore, Bao et al. [14] developed the IMKPse model, which uses the general PseAAC as the classification feature and the flexible neural tree as the classification model. Obviously, many achievements have been made in the prediction of malonylation sites, but the performance prediction can still be greatly improved.

In this study, we used the dimensionality reduction algorithm principal component analysis to see if it is helpful to predict malonylation sites. Here we want to know whether the integration of sequence features can produce better prediction accuracy, which is a problem. Based on our results, Mal-SVMPCA is significantly superior to the existing predictors, and shows that PCA can improve the accuracy of prediction with three sequence features, one is one-hot coding, one is physicochemical property (AAindex) and one is k-interval amino acid pair composition (cksaap). Therefore, Mal-SVMPCA can be used as a powerful tool to identify malonylation sites in proteins.

2 Methods

2.1 Data Sets and Preprocessing

In this study, we collected 1768 sequences from 934 human proteins. We used CD-HIT to remove redundant sequences [18] whose similarity is equal to or greater than 40% in order to reduce redundancy and avoid human error. Then the treated sequence was truncated into a long sequence of 17 residues with Lysine (k) as the center. Each peptide fragment is defined as follows:

$$P = R - \eta R - \eta + 1 \ldots R - 1KR1R2 \ldots R\varepsilon \tag{1}$$

Here, the length of the fragment is $\eta + \varepsilon + 1$. $R - \eta$ represents the η-th upstream sequence fragment, $R\varepsilon$ represents the ε-th downstream peptide of center K, and so on. Because there may be less amino acids around the center K and the downstream peptide of the center K is less than ε, we can use X to fill these residues. Therefore, our data set consists of 20 natural amino acids and virtual code X. Different lengths of malonylated peptides need to be analyzed under different research conditions. In this project, we set $\varepsilon = 8$ and $\eta = 8$, and the length of peptide is 17. It can be seen that the complete sequence fragment describing lysine belongs to any of two categories ($\delta 1$, $\delta 2$), The expression is as follows:

$$P \in (\delta 1, \delta 2)T (\delta 1, \delta 2) \in (0, 1)T \tag{2}$$

While lysine is malonylation site, $\delta 1 = 0$, otherwise $\delta 2 = 1$. Thus, we selected 1735 sequences from 931 human proteins as the positive dataset. The sequence fragments surrounding lysine (abbreviated as Lys or K) not included in the positive data set are combined to form a negative dataset. So we got 45,607 negative samples. Since unbalanced data sets may lead to incorrect predictions, the down sampling method was used to construct balanced data sets [19]. As a result, we have 3470 positive and negative data sets in which half of the data sets are balanced. For validating the performance of the predictor, 20% of the dataset was divided into independent dataset (695), and the rest are training dataset (2775).

2.2 Flowchart of the Mal-PCASVM

Figure 1 shows the flowchart of the malonylation site prediction method in this study. And the steps of the Mal-SVMPCA are described as follows:

1. Data collection and preprocessing. We collected data sets by searching literature and NCBI website. Then the peptide with length of 17 was selected by sliding window, and lysine was in the center. The down-sampling method was used to construct positive and negative data sets equally.
2. Feature representation. We chose CKSAAP, AAindex, and One-hot coding method as features to indicate each peptide segment.
3. Dimension reduction. In this study, we used PCA to reduce dimensions, because high-dimensional data sets are likely to lead to dimension disaster, and get the most appropriate dimension through experiments.

4. Classifier.SVM is chosen as the classification algorithm of data set, because compared with other classical classification algorithms, SVM has the best performance.
5. Performance evaluation of the model. In order to avoid over fitting and find the appropriate penalty parameters and kernel radius, we use 10 fold cross validation algorithm and classical metrics, such as ACC and Sen, to evaluate the performance of the algorithm.

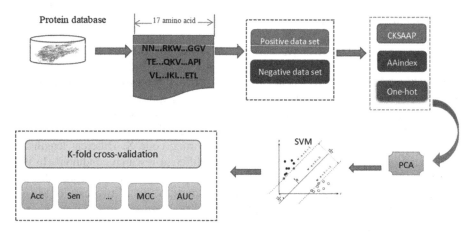

Fig. 1. Schematic illustration of the Mal-SVMPCA method from protein data selection to k-fold cross-validation

2.3　Feature Extraction

i. *Composition of K-spaced amino acid pairs (CKSAAP)*
   CKSAAP has been successfully applied to many PTM prediction, which reflects the composition of K-spaced amino acid pairs and was used to calculate the frequency of k-spacer amino acid pairs. Specifically, 21 * 21 = 441 amino acid pairs could be formed because there are 20 amino acids and 1 pseudo amino acid x. The probability of these residues appearing in these 441 amino acid pairs can be counted, so we extracted amino acids from K (k = 0, 1, 2, ...). Therefore, a 441 dimensional eigenvector was generated. Let's take an example of the peptide "RVFEDESGKHWSKSVMD". The length was set to 17, and when k = 0, 17 amino acid pairs {RV, VF, FE, ..., VM, MD} can be extracted, that is to say, each amino acid combines with its next adjacent amino acid to form a pair. Hence, we used NRV to calculate the number of times RV appears. The record is as follows:

$$NRV = occurrences(RV) \tag{3}$$

Then the probability of these residues appearing in 441 amino acid pairs was counted as follow:

$$(NRV, NVF, NFE, ...)441 \qquad (4)$$

ii. *One-hot encoding*

One-hot coding is a binary vector representation of classification variables, which can transform amino acids into orthogonal digital vectors, and has been used in many protein sequence analysis. Each peptide sequence can be expressed as a 21 dimensional vector, because there are 21 amino acids (20 conventional amino acids and 1 Pseudo amino acid X). Take the protein sequence "ACDEF-GHIKLMNPORSTVWYX" as an example, in which alanine (A) encodes "100000000000000000000". Particularly, the pseudo amino acid X is encoded as "000000000000000000001". Assuming that the peptide sequence is "RVFE-DESGKHWSKSVMD", we can obtain a 17 * 21 dimensional vector encoding the peptide.

iii. *Physiochemical properties*

AAindex is a digital index database, which represents various physicochemical and biochemical properties of amino acids. The database consists of two parts: AAindex1 and AAindex2. AAindex1 is the collection of published amino acid index and AAindex2 is the collection of published amino acid mutation matrix. An amino acid index is a set of 20 values representing various physicochemical and biochemical properties of amino acids. At present, AAindex has collected 544 amino acid indexes. AAindex (version 9.2) adds AAindex3 with protein contact potential. Nine kinds of physical and chemical properties, namely hydrophilic value, average polarity, isoelectric point, refractive index, average flexibility index, average volume of buried residue, electron ion interaction potential, surface transfer free energy and consistent normalized hydrophobicity, were used in this paper. The length of each peptide is 17, so the physicochemical property is 17 * 9 dimensional vectors.

2.4   Algorithm and SVM

i. *Support vector machine*

This paper used support vector machine (SVM) as classifier. SVM is a kind of generalized linear classifier which classifies data by supervised learning and has good generalization ability. It can classify data nonlinearly by kernel method, which is one of the common kernel learning methods. The basic principle of SVM is to map the samples in the input space to the high-dimensional feature space through the kernel function, so as to obtain the optimal classification hyperplane of the low dimensional vector space in the high-dimensional kernel space, that is, to solve the separation hyperplane which can correctly divide the training data set and has the largest geometric interval. Of course, it has achieved good performance in protein secondary structure prediction, cancer classification and subtype, biomarker/feature discovery, protein-protein interaction, cancer driving gene discovery, cancer therapeutic drug discovery and many other fields.

ii. *Principal component analysis*

PCA (principal component analysis) is a common way of data analysis [40], which is often used to reduce the dimension of high-dimensional data, and can be used to extract the main feature components of data. PCA is of great significance for feature extraction in this study.

2.5   Performance Measures

In this paper, we conducted model selection through 10-fold cross-validation, so it can effectively avoid over learning and lack of learning, and the result is more convincing. In the 10-fold cross validation, the whole training dataset is randomly divided into 10 subsets of approximately equal size. Each subset was sequentially used as the test set, and the remaining 9 subsets were used to train the classifier. Besides, in order to measure the prediction quality more intuitively and easily, we used the following five indicators to evaluate the prediction performance: Sensitivity (Sen), Specificity (Spec), F1 score, Accuracy (ACC) and Matthews correlation coefficient (MCC). The selected performance has been demonstrated in Eqs. (5)–(9).

$$\text{Sen} = \frac{\text{TP}}{\text{TP} + \text{FN}} \tag{5}$$

$$\text{Spec} = \frac{TN}{TN + \text{FP}} \tag{6}$$

$$\text{F1} = 2 \times \frac{SN \times \text{PPV}}{SN + \text{PPV}} \tag{7}$$

$$\text{ACC} = \frac{TP + TN}{TP + FP + TN + FN} \tag{8}$$

$$MCC = \frac{(\text{TP} \times \text{TN}) - (\text{FP} \times \text{FN})}{(TP + \text{FN}) \times (\text{TN} + \text{FP}) \times (\text{TP} + \text{FP}) \times (\text{TN} + \text{FN})} \tag{9}$$

Here, TP, TN, FP and FN represent the number of true positive, true negative, false positive and false negative respectively. In addition, according to Sen and spec, the receiver operating characteristic curve is drawn with different threshold, and their area under ROC (AUC) value is calculated based on trapezoidal approximation.

3   Results and Discussion

3.1   Determination of CKSAAP Features

Although the CKSAAP features were used by many methods to predict PTM sites, most methods only used the CKSAAP features generated by a single K value, and the best K value were not determined to construct the CKSAAP features. We analyzed the performance of different combinations of CKSAAP features to obtain effective

CKSAAP features. Particularly, we not only analyzed the CKSAAP characteristics obtained by a single K value from 0 to 6, but also analyzed their combined effects. Principal component analysis (PCA) was used to determine the size of all data sets as 100. The LIBSVM tool we used is provided by https://www.csie.ntu.edu.tw/~cjlin/libsvmtools/. Two important parameters of SVM, c and g, are optimized by grid search method, which are penalty parameter and kernel parameter in SVM algorithm. At last, we set c = 10 and g = 2 in SVM model, and used radial basis function as kernel function. 50 times of 5-fold cross validation were performed to optimize the parameters in the training model. Results as shown in supplementary Table 1, in which ACC, F1 and MCC showed little change under different K values. For example, the Sen value changed from 81.46% to 98.63%, the Spec value changed from 65.72% to 82.24%, and the F1 score value changed from 81.49% to 84.77%.

Therefore, finding out which is more suitable is very difficult. Since then, we compare by combining all the features (CKSAAP, one-hot encoding, AAindex). The parameters c and g of SVM were set to 1.9 and 0.07 by grid search, separately. The supplementary Table 2 showed the performance, from which we can see that the performance of the proposed method does not change much when k is set to 0 to 6. ACC, Sen, spec, F1 and MCC changed from 88.58% to 89.87%, 89.01% to 90.38%, 87.53% to 89.77%, 88.65% to 89.87%, 79.79% to 81.81% respectively. When combining the feature vectors calculated by different K values, the results have a certain rule, as shown in Fig. 2, according to which we can see that when the first four CKSAAP features were combined together, the accuracy reaches the best score, which is also true in the other four measures. Therefore, in this paper, K is set to 0, 1, 2 and 3, and the CKSAAP eigenvector is 441 * 4 = 1764 dimensions.

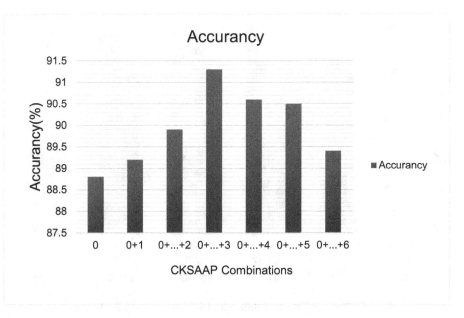

Fig. 2. Comparison of accuracy of different CKSAAP feature combinations

3.2   Effectiveness of PCA

For determining the dimension of principal component analysis suitable for our prediction, we run the training model when the dimensions are equal to 50, 100, 150, 200, 250 and 300 respectively. We perform 50 times of 5-fold cross validation optimize the parameters. The results are shown in Table 1.

Table 1. 5-fold cross-validation results of different dimensions.

Dimensions	Acc (%)	Sen (%)	Spec (%)	F1 (%)	MCC (%)
50	85.91	86.09	85.68	85.94	75.77
100	**91.24**	**91.71**	**90.83**	**91.18**	**84.03**
150	90.20	91.00	89.48	90.30	82.31
200	88.11	89.53	86.65	88.39	79.02
250	84.61	86.15	83.10	84.73	73.91
300	82.31	84.34	80.36	82.27	70.89

In Table1, we can see that when the dimension is equal to 100, the proposed method performs best, and the average ACC, Sen, spec, F1 and MCC can reach 91.24, 91.71, 90.83, 91.18 and 84.03% respectively. When the size is equal to 100, the maximum accuracy is 91.24%. When the size value is greater than 100, the larger the size, the lower the accuracy.

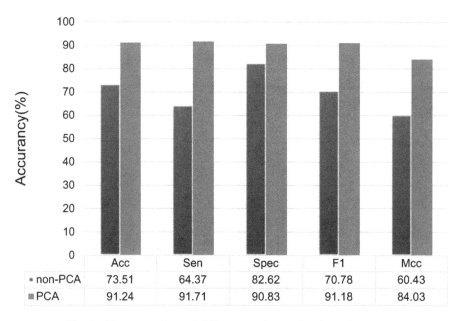

Fig. 3. The comparison of different metrics result using PCA or not

In order to analyze the role of principal component analysis in our proposed method, we applied the same process as our proposed method without principal component analysis. Parameters c and g were set to 2 and 0.1 by grid search. For analysis, Fig. 3 shows the comparison of different measurement results with or without PCA. Principal component analysis is used to reduce the size to 100, and non principal component analysis is not used. By analyzing the data in Fig. 3, it is obvious that compared with the proposed method without PCA, the average ACC, Sen, spec, F1 and MCC of the proposed method with PCA can increase by 17.73%, 27.34%, 8.21%, 20.4% and 19.6% respectively. It can be seen that principal component analysis (PCA) can improve the performance of the algorithm effectively.

3.3   Performance Comparison of Different Feature Combination

We analyzed the performance of each feature and multiple feature combinations to further determine the role of various features. As shown in supplementary Table 3, the performance comparison of each single feature showed that CKSAAP is superior to the other two features, especially in ACC, Sen, Spec and F1, which were almost 20%–30% higher than the other two features. At the same time, the supplementary Table 4 showed the performance comparison of multiple features, and it shows the performance of different feature combinations. AAindex (exclude) stands for excluding AAindex from three features, that is, the combination of CKSAAP and One-hot. CKSAAP (exclude) and One-hot (exclude) also have the same meaning. The combination of the three features is represented by all of these. Obviously, supplementary Table 5 shows that the combination of AAindex and One-hot performs best among all the two features. Interestingly, we know that CKSAAP performs best in comparing individual features. Therefore, to sum up this phenomenon with a Chinese proverb is that three cobblers add up to one genius. That is to say, the combination of the three is the best. From supplementary Fig. 1, we can know that ECKSAAP is defined as excluding CKSAAP from the three features, as are AAindex and One-hot. It is not difficult to see that the proposed method combining all features achieves the best in all metrics. Of course, ECKSAAP ranked second in ACC, spec, F1 and MCC.

Through the above analysis, Mal-SVMPCA can achieve better performance by combining four attributes of CKSAAP, One-hot coding and nine attributes of AAindex, and then reducing the dimension to 100 with PCA.

3.4   Comparison of Five Classical Algorithms

The performance of five classical classifiers is compared on the training data set, including our Mal-SVMPCA (SVM), Ensemble of decision tree, k-nearest neighbor (KNN), Random Forest (RF) and Naive Bayes (NB) [15–17]. We used the Euclidean distance in KNN, and set the number of its neighbors to 2. In RF and Ensemble, we set the number of decision trees to 20 and 50 respectively. Then 50 times 5-cross validation was performed on each of them, and the Table 2 shows their performance comparison.

As we all know, compared with other classical classifiers, ensemble classifier has better accuracy and robustness. Table 2 shows that Mal-SVMPCA model performs better than other classical classifiers in all metrics. It can be concluded that different data sets need different models.

Table 2. The performance comparisons of different classical classifiers

Classifier	Acc (%)	Sen (%)	Spec (%)	F1 (%)	MCC (%)
KNN	59.68	26.73	92.18	34.34	34.98
NB	83.24	84.17	82.36	83.39	72.11
RF	68.25	62.41	74.04	66.27	56.36
Ensemble	64.11	60.11	68.20	62.50	53.69
Mal-SVMPCA (SVM)	**91.24**	**91.71**	**90.83**	**91.18**	**84.03**

3.5 Comparison of the State-of-the-Art Approaches

Some of the most advanced methods for predicting malonylation sites were compared with our proposed methods, including MaloPred, LEMP, SPRINT-mal, Mal-Lys, iLMS. There are two reasons for the good performance of our method. On the one hand, principal component analysis (PCA) is used to extract features. Principal component analysis can extract more effective feature information, and it is a dimension reduction method. On the other hand, SVM classifier is used to classify. The support vector machine classifier combining principal component analysis and three features (pseaac, one hot, cksaap) is more suitable for predicting malonylation sites than the existing methods.

3.6 Performance on Independent Data Set

In order to make an objective performance comparison, the independent data set which is blind to the training data set was used to evaluate the performance of the method. As shown in Table 3, the best method is the recommended method, including ACC, Sen, spec, F1 and MCC values of 90.65%, 89.71%, 91.59%, 90.62 and 83.04%,separately. In Fig. 4, ROC curves for different feature combinations on independent data sets are shown. It is obvious that the AUC value of the proposed method (all features) is 90.72% on the independent data set, ECKSAAP ranks second, followed by EOne-hot,

Table 3. Performance of different feature combinations on the independent data set

Features	Acc (%)	Sen (%)	Spec (%)	F1 (%)	MCC (%)
CKSAAP	77.55	77.14	77.97	77.59	65.18
AAindex	64.73	65.43	57.97	63.26	52.61
One-hot	58.71	61.43	55.94	59.97	51.44
CKSAAP (exclude)	86.19	86.86	85.51	86.36	76.19
AAindex (exclude)	79.42	80.57	78.26	79.77	67.30
One-hot (exclude)	71.08	76.29	65.80	72.65	58.65
ALL	90.65	89.71	91.59	90.62	83.04

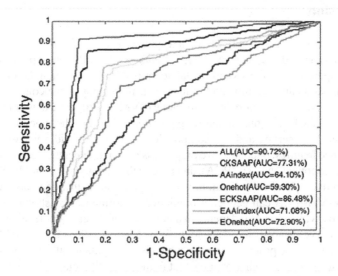

Fig. 4. ROC curves performed by different feature combinations on the independent data set

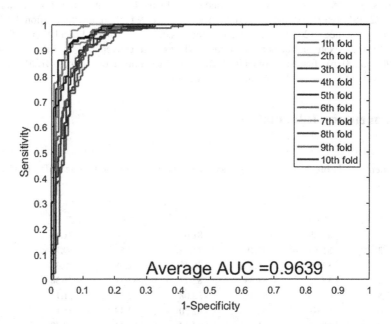

Fig. 5. ROC curves performed by different feature combinations on the independent data set

EAAindex, AAindex and One-hot, which are the same as the results on the test data set. All kinds of evidences show that the Mal-SVMPCA model has a good effect after combining these three characteristics and PCA.

4 Conclusion

Malonylation of lysine is a newly discovered post-translational modification, which is associated with many diseases, such as type 2 diabetes and different types of cancer. In addition, compared with the experimental identification of propionylation sites, the calculation method can save time and reduce cost. In this paper, we combine principal component analysis with support vector machine (SVM) to propose a new computational model - Mal-SVMPCA (malonylation prediction). Firstly, the one-hot encoding, physicochemical properties and the composition of k-spacer acid pairs were used to extract sequence features. Secondly, we preprocess the data, select the best feature subset by principal component analysis (PCA), and predict the malonylation sites by SVM. And then, we do a five-fold cross validation, and the results show that compared with other methods, Mal-SVMPCA can get better prediction performance. As shown in Fig. 5, in the 10-fold cross validation of independent data sets, AUC (area under receiver operating characteristic curve) analysis has reached 96.39%. Mal-SVMPCA is used to identify the malonylation sites in the protein sequence, which is a computationally reliable method. It is superior to the existing prediction tools that found in the literature and can be used as a useful tool for identifying and discovering novel malonylation sites in human proteins.

Acknowledgments. This work was supported in part by the University Innovation Team Project of Jinan (2019GXRC015), and in part by Key Science &Technology Innovation Project of Shandong Province (2019JZZY010324), the Natural Science Foundation of China (No. 61902337), Natural Science Fund for Colleges and Universities in Jiangsu Prov-ince (No. 19KJB520016), Jiangsu Provincial Natural Science Foundation (No. SBK2019040953), Young talents of science and technology in Jiangsu.

Supplementary Information

Supplementary Table 1. The performance of the proposed method using different CKSAAP features.

K	Acc (%)	Sen (%)	Spec (%)	F1 (%)	MCC (%)
0	82.63	86.78	78.21	83.62	71.09
1	**82.85**	84.07	81.69	83.00	**71.48**
2	81.59	87.19	75.92	82.66	69.76
3	82.02	85.45	78.43	82.68	70.33
4	82.74	82.84	82.72	82.71	71.36
5	82.70	83.32	82.08	82.77	71.11
6	82.63	83.11	82.20	82.49	71.11
0 + 1	82.05	81.86	**82.24**	82.02	70.29
0 + 1 + 2	81.66	81.46	81.37	81.49	69.48
0 +... + 3	82.67	89.68	75.36	83.63	70.59
0 +... + 4	82.56	95.88	69.02	84.58	70.15
0 +... + 5	82.41	98.23	66.54	84.70	69.50
0 +... + 6	82.23	**98.63**	65.72	**84.77**	69.19
Mean	82.36 ± 0.42	87.58 ± 6.17	77.04 ± 6.20	83.16 ± 1.03	70.42 ± 0.78

Supplementary Table 2. Performance of the proposed method using different CKSAAP combinations

K	Acc (%)	Sen (%)	Spec (%)	F1 (%)	MCC (%)
0	88.76	89.92	87.53	89.11	80.01
1	88.94	90.27	87.64	89.19	80.30
2	89.30	89.79	88.78	89.42	80.89
3	89.87	89.99	89.77	89.87	81.81
4	88.58	89.01	88.16	88.65	79.79
5	89.37	90.09	88.68	89.40	81.00
6	89.41	90.38	88.31	89.58	81.02
0 + 1	89.19	89.58	88.84	89.22	80.73
0 + 1 + 2	89.91	90.46	89.31	89.93	81.87
0 + ... + 3	**91.24**	**91.71**	**90.83**	**91.18**	**84.03**
0 +... + 4	90.63	91.41	89.97	90.63	83.02
0 +... + 5	90.49	91.41	89.50	90.64	82.76
0 +... + 6	89.44	91.25	87.64	89.60	81.12

Supplementary Table 3. The performance of 5-fold cross-validation without PCA

K	Acc (%)	Sen (%)	Spec (%)	F1 (%)	MCC (%)
1	74.59	64.60	85.61	72.73	61.40
2	70.63	61.85	78.95	67.20	57.81
3	72.43	62.22	82.11	68.71	59.16
4	74.23	67.13	81.78	72.87	61.40
5	75.68	66.04	84.67	72.39	62.40
Mean	73.51 ± 1.99	64.37 ± 2.32	82.62 ± 2.63	70.78 ± 2.64	60.43 ± 1.89

Supplementary Table 4. The performance comparison of different single feature.

Feature	Acc (%)	Sen (%)	Spec (%)	F1 (%)	MCC (%)
CKSAAP	**82.38**	**91.66**	**74.16**	**83.69**	**69.83**
AAindex	59.78	63.87	55.81	61.27	51.76
One-hot	59.75	65.52	53.86	62.14	51.62

Supplementary Table 5. The performance comparison of different feature combination.

Features	Acc (%)	Sen (%)	Spec (%)	F1 (%)	MCC (%)
CKSAAP (exclude)	85.44	85.82	85.14	85.52	75.09
AAindex (exclude)	78.27	79.81	76.65	78.66	65.98
One-hot (exclude)	69.77	73.61	65.92	70.96	57.64
ALL	**91.24**	**91.71**	**90.83**	**91.18**	**84.03**

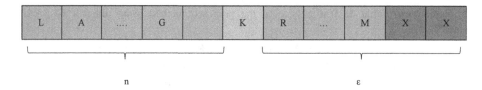

Supplementary Fig. 1. The comparison of different feature combinations.

References

1. Molinie, B., Giallourakis, C.C.: Genome-wide location analyses of N6-Methyladenosine modifications (m(6)A-Seq). Methods Mol Biol **1562**, 45–53 (2017)
2. Nye, T.M., van Gijtenbeek, L.A., Stevens, A.G., et al.: Methyltransferase DnmA is responsible for genome-wide N6-methyladenosine modifications at non-palindromic recognition sites in Bacillus subtilis. Nucleic Acids Res. **48**, 5332–5348 (2020)
3. O'Brown, Z.K., Greer, E.L.: N6-methyladenine: a conserved and dynamic DNA mark. In: Jeltsch, A., Jurkowska, R.Z. (eds.) DNA Methyltransferases - Role and Function. AEMB, vol. 945, pp. 213–246. Springer, Cham (2016). https://doi.org/10.1007/978-3-319-43624-1_10
4. Zhang, G., et al.: N6-methyladenine DNA modification in drosophila. Cell **161**(4), 893–906 (2015)
5. Janulaitis, A., et al.: Cytosine modification in DNA by BCNI methylase yields N4-methylcytosine. FEBS Lett. **161**, 131–134 (1983)
6. Unger, G., Venner, H.: Remarks on minor bases in spermatic desoxyribonucleic acid. Hoppe-Seylers Z. Physiol. Chem. **344**, 280–283 (1966)
7. Fu, Y., et al.: N6-methyldeoxyadenosine marks active transcription start sites in Chlamydomonas. Cell **161**, 879–892 (2015)
8. Greer, E.L., et al.: DNA methylation on N6-adenine in C. Elegans. Cell **161**, 868–878 (2015)
9. Zhang, G., et al.: N6-methyladenine DNA modification in Drosophila. Cell **161**, 893–906 (2015)
10. Wu, T.P., et al.: DNA methylation on N6-adenine in mammalian embryonic stem cells. Nature **532**, 329–333 (2016)
11. Xiao, C.L., et al.: N-methyladenine DNA modification in the human genome. Mol. Cell **71**, 306–318 (2018)
12. Zhou, C., et al.: Identification and analysis of adenine N6-methylation sites in the rice genome. Nat. Plants **4**, 554–563 (2018)
13. Chen, W., et al.: i6mA-Pred: identifying DNA N6-methyladenine sites in the rice genome. Bioinformatics **35**, 2796–2800 (2019)
14. Almagor, H.A.: A Markov analysis of DNA sequences. J. Theor. Biol. **104**, 633–645 (1983)
15. Borodovsky, M., et al.: Detection of new genes in a bacterial genome using Markov models for three gene classes. Nucleic Acids Res. **17**, 3554–3562 (1995)
16. Durbin, R., et al.: Biological Sequence Analysis Probabilistic Models of Proteins and Nucleic Acids. Cambridge University Press, Cambridge (1998)
17. Ohler, U., et al.: Interpolated Markov chains for Eukaryotic promoter recognition. Bioinformatics **15**, 362–369 (1999)
18. Reese, M., et al.: Improved splice site detection in genie. J. Comput. Biol. **4**, 311–323 (1997)

19. Wren, J.D., et al.: Markov model recognition and classification of DNA/protein sequences within large text databases. Bioinformatics **21**, 4046–4053 (2005)
20. Yakhnenko, O., et al.: Discriminatively trained Markov model for sequence classification. In: IEEE International Conference on Data Mining (2005)
21. Matthews, B.W.: Comparison of the predicted and observed secondary structure of t4 phage lysozyme. Biochimica et Biophysica Acta (BBA)- Protein Structure **405**(2), 442–451 (1975)

Theoretical Computational Intelligence
and Applications

The Influence of Sliding Windows Based on MM-6mAPred to Identify DNA N6-Methyladenine

Wenzhen Fu[1], Yixin Zhong[2], Wenzheng Bao[3(✉)], and Yi Cao[4]

[1] School of Information Science and Engineering, University of Jinan,
Jinan, China
[2] Artificial Intelligence Institute, School of Information Science
and Engineering, University of Jinan, Jinan, China
[3] School of Information and Electrical Engineering,
Xuzhou University of Technology, Xuzhou, China
baowz55555@126.com
[4] Shandong Provincial Key Laboratory of Network Based Intelligent
Computing, School of Information Science and Engineering, University of Jinan,
Jinan, China

Abstract. Methylation of DNA N6-methyladenine (6mA) is a type of epigenetic modification that plays an essential role in eukaryotic organisms. It is of great importance to discriminate precisely N6-methyladenine (6mA) in genomes. Based on this, we can recognize its biological functions. As a result, finding a stable method which can identify quickly and precisely is indispensable. In this paper, we have changed the number of sliding windows in the MM-6mAPred, which based on the Markov model. The model identifies 6mA sites through the transition probability between adjacent nucleotides. The results show that the transition probability among adjacent nucleotides is benefit to capture more discriminant sequence information.

Keywords: Methylation of DNA N6-methyladenine · Sliding windows · Machine learning

1 Introduction

As we all know, epigenetic modification is a crucial modification which can regulate gene expression without altering DNA sequences. Nowadays, the definitive epigenetic phenomenon includes DNA methylation, RNA methylation, genomic imprinting, gene silencing, RNA editing, maternal effect, transposon activation and so on. Among them, DNA N6-methyladenine (6mA) plays a pivotal role in controlling tissue-specific gene expression, transcript synthesis, gene imprinting X-chromosome inactivation and positioning and stability of nucleosome [1, 2]. It is a significant epigenetic modification of nucleic acid, which exist in diverse genomes such as bacteria, eukaryotes, and archaea [3, 4].

DNA 6mA was found in Escherichia coli at the beginning. Then, it was identified in several other species such as bacteria [5]. In the past years, researchers maintained

D.-S. Huang et al. (Eds.): ICIC 2021, LNCS 12837, pp. 699–708, 2021.
https://doi.org/10.1007/978-3-030-84529-2_58

that 6mA existed nothing but in prokaryotes and single-cell organisms. Until 1966, although Unger and Venner (1996) [6] found the existence of 6mA in bovine and human sperm cells, we could not duplicate the results as well as possible. As biotechnology developed, researchers identified 6mA and its function in Chlamydomonas, Nematodes and Drosophila in 2015 [7–9].

In 2016, there was a paper which is about the study of 6mA modification of mouse embryos published in Nature [10]. The 6mA of Chinese DNA was detected for the first time by a team from the Sun Yat-sen University through sequencing technologies [11]. Then, the distribution of 6mA sites in the rice genome was analyzed by Zhou et al. [12]. They used multiple sequencing methods, such as 6mA-IP-Seq, liquid chromatograph-mass spectrometer (LC-MS/MS) and single-molecule and real-time sequencing (SMRT). Although DNA 6mA can not express greatly, the modification is bound up with gene transcription regulation in current researches.

Recently, we usually use SMRT which is the mainstream experimental technique to detect 6mA sites. However, we can not identify 6mA sites from the whole genome through the technology. What's more, the technology is high-cost and labor-intense. Taking into account, developing efficient algorithms for detecting DNA 6mA sites computationally is of high priority. With regard to predict methylation modification of 5-methylcytosine, there are numerous methods. But there are extremely few methods for 6mA methylation modification. A method called i6mA-Pred for identifying DNA 6mA sites was proposed by Chen et al. in 2019 [13]. They used the method which is employed by a support vector machine (SVM) classifier on the basis of the chemical features of nucleotides and position-specific nucleotide frequencies. Nonetheless, the study ignored the relationship of nucleotides near 6mA sites. Our study has shown that the transition probabilities among neighboring bases in 6mA sequences are greatly different from the non-6mA sequences. As a result, we develop a novel method named MM-6mAPred. It identifies DNA 6mA sites on the basis of Markov model (MM). As a matter of fact, Markov chain models have long been used to model neighboring dependency among biological sequences [14–20]. Some basic biological and chemical features of nucleic acids is closely related to the frequencies of dinucleotides in recent studies [14]. Nevertheless, for what we know, MM-6mAPred which overcame the above-mentioned issues of using this neighboring dependency information is the first tool to identify 6mA sites. The results validate that the proposed model produces better scores than existing models.

2 Materials and Methods

2.1 Dataset

In order to compare with the existing method called i6mA-Pred and MM-6mAPred, we used the same dataset in our study. It is used by the model called i6mA-Pred (Chen et al., 2019; Zhou et al., 2018) [12, 13]. We downloaded this dataset from http://lingroup.cn/server/i6mAPred/data. The DNA sequences are 41 bp long both in positive and in negative samples. In summary, 880 positive sequences contain 6mA sites and 880 negative sequences contain no 6mA sites from the rice genome.

2.2 Construction of MM

As we all know, the Markov Model (MM) is a stochastic process. In the MM, the next variable only depends on the most recent variable(s) instead of all the previous variables. In DNA sequence, we usually use the first-order Markov chain where the next nucleotide will solely depend on the current nucleotide. In our study, we build a model about the sequence of 6mA sites through a first-order Markov chain. In particular, the MM assumes that $N_t \in (A, G, C, T)$ be the random variable at the t-th location of the sequence of length L. While t = 2, 3,..., L, $P(N_t|N_{t-1}, N_{t-2}, \ldots, N_1) = P(N_t|N_{t-1})$. The flow chart for identifying sequences containing 6mA sites by MM-6mAPred is showed in Fig. 1. In addition, from the 6mA sequences in the training dataset, let $P^P_{N_1}$ be the probability vector, the nucleotide N_1 be the initial position, and $T^1_P, T^2_P, \ldots, T^{L-1}_P$ be the transition probability matrices of $N_1 \rightarrow N_2, N_2 \rightarrow N_3, \ldots, N_{L-1} \rightarrow N_L$, respectively. In the same case, from the non-6mA sequences in the training dataset, we estimate $P^N_{N_1}$, which is initial distribution of nucleotides, $T^1_N, T^2_N, \ldots, T^{L-1}_N$ are the corresponding transition probability. As a result, we use both the 6mA sequences and the non-6mA sequences in the training dataset to train two MMs which is called $O_P = (P^P_{N_1}, T^1_P, T^2_P, \ldots, T^{L-1}_P)$ and $O_N = (P^N_{N_1}, T^1_N, T^2_N, \ldots, T^{L-1}_N)$. In our study, we predict a test sequence 'Seq = GTAT...AA' of 41 nucleotides. Based on generating the sequence 'Seq' under the model O_p and O_N, we get the probabilities $P(seq|O_p)$ and $P(seq|O_N)$. Then, we use the ratio of $P(seq|O_p)$ to $P(seq|O_N)$, which can determine whether 'Seq' is a 6mA or non-6mA sequence. Here, $P(seq|O_P) = P^P_G \times P^{P_1}_{GT} \times P^{P_2}_{TA} \times P^{P_3}_{AT} \times \ldots \times P^{P_{40}}_{AA}$ and $P(seq|O_N) = P^N_G \times P^{N_1}_{GT} \times P^{N_2}_{TA} \times P^{N_3}_{AT} \times \ldots \times P^{N_{40}}_{AA}$. Provided that $Ratio = P(seq|O_P)/P(seq|O_N) > 1$, 'Seq' is a 6mA sequence, and vice 'Seq' is a non-6mA sequence. The likelihood ratio can be any non-negative value. We can use the log-likelihood ratio for symmetricity. Its natural threshold is zero instead of one.

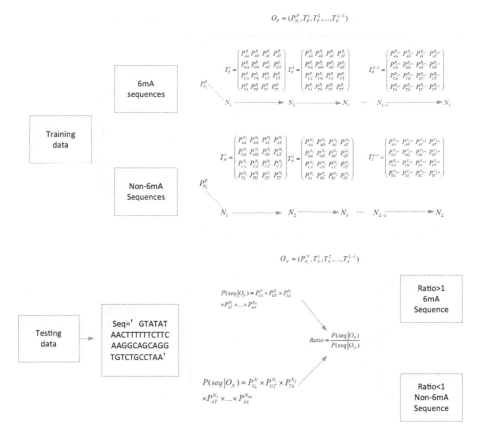

Fig. 1. The flow chart for identifying sequences containing 6mA sites by MM-6mAPred.

2.3   Prediction Accuracy Assessment

We use the common performance evaluation, including the total prediction accuracy (ACC), Specificity (Sp), Sensitivity (Sn) and the Mathew's correlation coefficient (MCC) (Matthews, 1975):

The metric sensitivity is also known as True Positive Rate (TPR). It measures the model's ratio of positive samples accepted as positive. Sensitivity can be expressed as:

$$S_n = \frac{T_P}{T_P + T_N} \tag{1}$$

The metric specificity is also known as True Negative Rate (TNR). It measures the model's ratio of negative samples known as negative. Specificity can be expressed as:

$$S_P = \frac{T_N}{T_N + F_P} \tag{2}$$

Accuracy is the correct predictions to the test data. It can be defined as:

$$Accuracy = \frac{T_P + T_N}{T_P + T_N + F_P + F_N} \times 100\% \tag{3}$$

The Matthews correlation coefficient (MCC) reflects the model's output by a binary classifier [21]. It measures the correlation between the real and binary classifications ranging between -1 and $+1$. The $+1$ denotes a perfect binary classification, while -1 denotes complete deviation between the real and predicted labels. It can be defined as:

$$MCC = \frac{T_P \times T_N - F_p \times F_N}{\sqrt{(T_P + F_P) \times (T_N + F_N) \times (T_P + F_N) \times (T_N + F_P)}} \tag{4}$$

where TP is the number of real 6mA sequences identified correctly, FN is the number of 6mA sequences classified incorrectly, TN is the number of non-6mA sequences correctly identified and FP is the number of non-6mA sequences classified incorrectly.

3 Result and Discussion

3.1 Analysis of Transition Probability Between Adjacent Nucleotides

In our study, we have found that transition probabilities between adjacent nucleotides of 6mA and non-6mA sequences are different from each other. Based on this, we can prove that 6mA sequence recognition on the basis of MM is reasonable. It is obvious that there are significant differences in the transition probabilities of A to A and C to G. As a result, it is available for us to identify sequences containing 6mA sites through the first-order Markov chain. Due to the distinctness of transition probabilities in different nucleotide pairs, it is crucial to determine an appropriate region for classification.

The values of all $P^{P_i}_{N_i N_{i+1}}/P^{N_i}_{N_i N_{i+1}}, i = 1, 2, \ldots, 40$ are shown in Fig. 2. It presents the simple calculation of ratio and the detail discriminant information. We set $\{1.09, 1.00, 0.91, 0.97\}$ for the first position of the 41 nucleotides in the input dataset as the vector $P^P_{N_1}/P^{N_1 N}$ for the four nucleotides $\{A, G, C, T\}$. When the value of ratio is close to 1, the corresponding transition probabilities of 6mA is similar to the corresponding transition probabilities of non-6mA. In the Fig. 3, we can see the significant difference in the transition probabilities among the regions from [22–23] to [27–28].

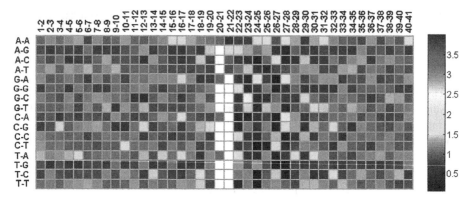

Fig. 2. The visualization of $P^{P_i}_{N_iN_{i+1}} / P^{N_i}_{N_iN_{i+1}}$.

The variable names along the vertical axis represent the nucleotide $N_i - N_{i+1}$ $(N_i, N_{i+1} \in (A, G, C, T))$.The variable names along the horizontal axis represent the location pair i − i + 1(i = 1, 2,..., 40). The reason why white squares in the figure represent undefined is that the 21st position of the input sequences is all A.

3.2   Comparison with the Different Number of Sliding Windows

To date, i6mA-Pred and MM-6mAPred are the existing identification algorithms for 6mA sequences. The method named i6mA-Pred used an SVM classifier on the basis of 164 features. The features include nucleotide chemical properties and nucleotide frequencies. In our study, we found that the number of sliding windows has a great influence on the identification of 6mA. Based on this, we have changed the number of sliding windows in the MM-6mAPred which based on the Markov model. In order to compare these algorithms more conveniently, we used the same dataset which is provided by the i6mA-Pred paper. The dataset which has 880 6mA sequences and 880 non-6mA sequences comes from the rice genome. For the ease of researching the different number of sliding windows in the MM-6mAPred, we showed the performance of i6mA-Pred and MM-6mAPred in Table 1.

Table 1. The performance of i6mA-Pred and the different number of sliding windows in the MM-6mAPred based on the same dataset

Method	Sn (%)	Sp (%)	ACC (%)	MCC
i6mA-Pred	82.95	83.30	83.13	0.662
[3,39]	90.13	89.31	89.72	0.816
[3,40]	89.73	88.58	89.15	0.807
[1,39]	89.92	88.41	89.15	0.807
[3,30]	89.16	88.80	88.98	0.804
[1,40]	89.43	88.54	88.98	0.804
[1,16]	51.90	50.93	51.25	0.470
[3,25]	49.86	49.31	49.77	0.371

The S_n, S_p and ACC of MM-6mAPred which is at the region [3,39] are 90.13%, 89.31% and 89.72%, respectively. Our ACC is 6.59% higher than that of i6mA-Pred (83.13%). Our MCC is 0.154 higher than that of i6mA-Pred (0.662). In addition, the Fig. 4 shows the identification of DNA 6mA sites by using the MM-6mAPred at the region [3,39]. The results show that the transition probabilities between adjacent nucleotides can reflect better discriminant information (Fig. 5).

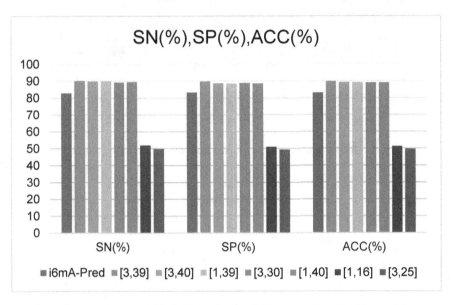

Fig. 3. Comparison with i6mA-Pred and the different number of sliding windows

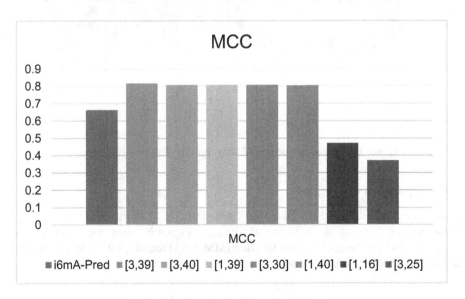

Fig. 4. The MCC value of different slide

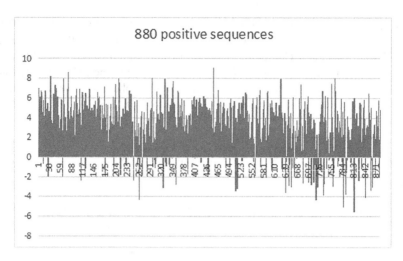

Fig. 5. The histograms of Log (Ratio) from the 880 negative sequence

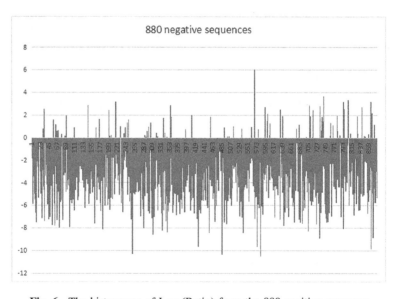

Fig. 6. The histograms of Log (Ratio) from the 880 positive sequence

3.3 Application of Program

In our study, we identified the DNA 6mA sites from 41 nt short sequences by developing a Matlab function (MM-6mAPred(Seq)). Figure 6 when we input a short sequence of 41 nt, the results of the function (MM-6mAPred(Seq)) include the Location and Ratio of DNA 6mA site. For example:

Seq1 = 'AATTGGATAGGGAGAAGCCGATGTAGCTGATTCTAGCAAGA'.
[Ratio, Location, Class] = MM6mAPred(Seq1);
Output: Ratio = 33.2297; Location = 21, Class = 6mA.

4 Conclusion

With regard to explore organisms' biological functions, it is a must for us to identify precisely DNA 6mA. In this study, we have changed the number of sliding windows based on the MM-6mAPred. The MM is on the basis of the Markov model. In our study, we make full use of the relationship of nucleotides near 6mA sites. We use the 10-fold cross-validation to evaluate the classification performance. The MM performs better than the existing algorithm i6mA-Pred. The results show that the transition probability among adjacent nucleotides is benefit to capture more discriminant sequence information. However, in the i6mA-Pred used by the SVM, this neighboring dependency information did not be fully utilized. In this paper, we get a more stable model by comparing different sliding windows. By and large, it proves the great advantages in the 6mA sites.

In the future, we can improve the MM in numerous aspects. For the sequence longer than 41 nt, we can select the best subsequence region by considering a wider window. We can also choose a discontinuous region such as two or more discontinuous areas. We can revise the region selection criterion in order to balance between the classification accuracy and the model complexity. In addition, the MM of the first-order Markov chain can be extended two or more orders to capture more dependency information.

Acknowledgments. This work was supported in part by the University Innovation Team Project of Jinan (2019GXRC015), and in part by Key Science & Technology Innovation Project of Shandong Province (2019JZZY010324), the Natural Science Foundation of China (No. 61902337), Natural Science Fund for Colleges and Universities in Jiangsu Province (No. 19KJB520016), Jiangsu Provincial Natural Science Foundation (No. SBK2019040953), Young talents of science and technology in Jiangsu.

References

1. Molinie, B., Giallourakis, C.C.: Genome-wide location analyses of N6-Methyladenosine modifications (m(6)A-Seq). Methods Mol. Biol. **1562**, 45–53 (2017)
2. Nye, T.M., van Gijtenbeek, L.A., Stevens, A.G., et al.: Methyltransferase DnmA is responsible for genome-wide N6-methyladenosine modifications at non-palindromic recognition sites in Bacillus subtilis. Nucleic Acids Res. **48**, 5332–5348 (2020)
3. O'Brown, Z.K., Greer, E.L.: N6-Methyladenine: a conserved and dynamic DNA mark. In: Jeltsch, A., Jurkowska, R. (eds.) DNA Methyltransferases - Role and Function. Advances in Experimental Medicine and Biology, vol. 945. Springer, Cham (2016). https://doi.org/10.1007/978-3-319-43624-1_10
4. Zhang, G., et al.: N6-methyladenine dna modification in drosophila. Cell **161**(4), 893–906 (2015)

5. Janulaitis, A., et al.: Cytosine modification in DNA by BCNI methylase yields N4-methylcytosine. FEBS Lett. **161**, 131–134 (1983)

6. Unger, G., Venner, H.: Remarks on minor bases in spermatic desoxyribonucleic acid. Hoppe-Seylers Zeitschrift physiologische Chemie **344**, 280–283 (1966)

7. Fu, Y., et al.: N6-methyldeoxyadenosine marks active transcription start sites in Chlamydomonas. Cell **161**, 879–892 (2015)

8. Greer, E.L., et al.: DNA methylation on N6-adenine in C. elegans. Cell **161**, 868–878 (2015)

9. Zhang, G., et al.: N6-methyladenine DNA modification in Drosophila. Cell **161**, 893–906 (2015)

10. Wu, T.P., et al.: DNA methylation on N6-adenine in mammalian embryonic stem cells. Nature **532**, 329–333 (2016)

11. Xiao, C.L., et al.: N-methyladenine DNA modification in the human genome. Mol. Cell **71**, 306–318 (2018)

12. Zhou, C., et al.: Identification and analysis of adenine N6-methylation sites in the rice genome. Nat. Plants **4**, 554–563 (2018)

13. Chen, W., et al.: i6mA-Pred: identifying DNA N6-methyladenine sites in the rice genome. Bioinformatics **35**, 2796–2800 (2019)

14. Almagor, H.A.: A Markov analysis of DNA sequences. J. Theor. Biol. **104**, 633–645 (1983)

15. Borodovsky, M., et al.: Detection of new genes in a bacterial genome using Markov models for three gene classes. Nucleic Acids Res. **17**, 3554–3562 (1995)

16. Durbin, R., et al.: Biological Sequence Analysis Probabilistic Models of Proteins and Nucleic Acids. Cambridge University Press, Cambridge (1998)

17. Ohler, U., et al.: Interpolated Markov chains for Eukaryotic promoter recognition. Bioinformatics **15**, 362–369 (1999)

18. Reese, M., et al.: Improved splice site detection in genie. J. Comput. Biol. **4**, 311–323 (1997)

19. Wren, J.D., et al.: Markov model recognition and classification of DNA/protein sequences within large text databases. Bioinformatics **21**, 4046–4053 (2005)

20. Yakhnenko, O., et al.: Discriminatively trained Markov model for sequence classification. In: IEEE International Conference on Data Mining (2005)

21. Matthews, B.W.: Comparison of the predicted and observed secondary structure of t4 phage lysozyme. Biochimica et Biophysica Acta (BBA)-Protein Struct. **405**(2), 442–451 (1975)

RF_Bert: A Classification Model of Golgi Apparatus Based on TAPE_BERT Extraction Features

Qingyu Cui[1], Wenzheng Bao[2(✉)], Yi Cao[1], Bin Yang[3], and Yuehui Chen[1]

[1] School of Information Science and Engineering, University of Jinan, Jinan 250024, China
[2] School of Information and Electrical Engineering, Xuzhou University of Technology, Xuzhou 221018, China
baowz55555@126.com
[3] School of Information Science and and Engineering, Zaozhuang University, Zaozhuang 277160, China

Abstract. Golgi is an important eukaryotic organelle. Golgi plays a key role in protein synthesis in eukaryotic cells, and its dysfunction will lead to various genetic and neurodegenerative diseases. In order to better develop drugs to treat diseases, one of the key problems is to identify the protein category of Golgi apparatus. In the past, the physical and chemical properties of Golgi proteins have often been used as feature extraction methods, but more accurate sub-Golgi protein identification is still challenged by existing methods. In this article, we use the Tape-Bert model to extract the features of Golgi body. To create a balanced dataset from an unbalanced Golgi dataset, we used the SMOTE over-sampling method. In addition, we screened out the important eigenvalues of 300 dimensions to identify the types of Golgi proteins. In 10-fold cross validation and independent test set test, the accuracy rate reached 90.6% and 95.31%.

Keywords: Golgi apparatus · Feature extraction · SMOTE · Deep learning

1 Introduction

Golgi is a kind of organelle in eukaryotic cells. Its main function is to store, package and classify proteins [1]. The Golgi apparatus consists mainly of two parts: cis Golgi network and trans Golgi network [2]. The main task of cis Golgi is to accept proteins, while the trans Golgi is to release synthesized proteins. Studies have shown that the functional defects of Golgi apparatus in cells can lead to the occurrence of some diseases, such as diabetes [3], Parkinson's disease [4], Alzheimer's disease [5] and some cancers [6]. There are treatments to treat these diseases, but they only relieve them, not cure them permanently. Correct identification of Golgi protein types can help medical and researchers better solve the occurrence of these diseases [7].

As machine learning technologies continue to mature, a large number of machine learning models have been applied to protein localization [8–13]. However, only a few models have been designed to study Golgi protein localization. In the past few years,

© Springer Nature Switzerland AG 2021
D.-S. Huang et al. (Eds.): ICIC 2021, LNCS 12837, pp. 709–717, 2021.
https://doi.org/10.1007/978-3-030-84529-2_59

Van Dijk et al. [14] proposed a method to predict type-ii membrane protein types. It uses linear kernel support vector machine as classifier. Ding et al. [15] used PseAAC and a custom Markov discriminant to identify Golgi protein types with an overall accuracy of 74.7%. They then used an improved spacer dipeptide composition method to achieve a prediction accuracy of 85.4% [16]. Jiao and Du et al. [17] used the position specific physico-chemical properties (PSPCP) of amino acid residues to extract features, and selected important eigenvalues through variance analysis, which improved the prediction accuracy of the model to 86.9%. In their subsequent experiments, they combined PSPCP with the pseudo amino acid composition of Chou [18]. Lv et al. [19] used SAAC and 2gDPC methods to extract sequence features, and then used ANOVA to reduce feature dimensions, and the prediction accuracy reached 90.5%. Zhao et al. [20] used PseAAC and FUNDES to extract features, and the prediction accuracy reached 78.4%. Yang et al. [21] used 3gDPC, CSP-PSSM-DC, CSP-Bigram-PSSM and CSP-ED-PSSM fusion feature vectors, and SMOTE algorithm and random forest recursive feature elimination are used to search the optimal feature. Finally, the prediction accuracy of the model reached 88.5%.

In order to improve the prediction accuracy of the model, Zhao et al. and Yang et al. all chose to make efforts on feature extraction, because feature extraction often plays an important role in protein sequence analysis, and they often used fusion feature method to increase the ability of feature representation to achieve a good prediction accuracy of the model. With the continuous development of deep learning, more and more pre-trained deep learning networks can extract features from new data, and achieve good results in natural language processing and image recognition [22–28].

In this experiment, the pre-training model Bert was used to extract the sequence feature of Golgi proteins. To create a balanced dataset from an unbalanced Golgi dataset, we used the SMOTE oversampling method. Then, a sub-Golgi protein locator model was developed based RF model. The flow chart of the whole work is shown in Fig. 1. Based on the data set constructed by Yang. The 10-fold cross-validation score was ACC 90.6%, Sn 91.32%, Sp 89.328%, MCC 0.8154, and the independent test set score was ACC 95.31%, Sn 84.62%, Sp 98.04%, MCC 0.852. The experimental results show that this method can improve the prediction accuracy of sub-Golgi protein.

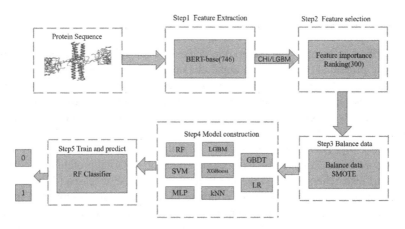

Fig. 1. Work flow chart

2   Data Sets and Feature Extraction

The data set constructed by Yang et al., among them, there were 304 Golgi body protein sequences in the training set, including 217 negative samples and 87 positive samples, in order to avoid overfitting of the model, a separate test set was used to verify the capability of the model. The test set contained 64 Golgi protein sequences, with a positive/negative sample ratio of about 1:5.

Feature extraction method is a particularly important step for Golgi protein classification judgment, and the selection of appropriate feature extraction method will greatly improve the prediction accuracy. In the previous model, the feature extraction is mainly based on amino acid sequence and physical and chemical properties, such as: AAC, SAAC, PSSM, PsePSSM, N-gram and other methods. With the continuous development of deep learning, some pre-trained deep learning models are used for feature extraction. In this paper, Bert model in the pre-trained Tape model proposed by Rao et al. [29] in 2019 is used for feature extraction.

3   Classifiers

In this article, we use the eight popular machine learning classifier to do experiments, they are random forests, LightGBM, SVM, MLP, XGBoost, KNN, LR, GBDT, some of the classifier is to use scikit-learn [30] tools to build, The purpose of building these 8 classifiers is to pick out the best model.

4   Evaluation Metrics and Methods

In order to objectively and justly evaluate and test our method, k-fold cross validation and independent test set methods are adopted in this paper to test and evaluate the model. Selecting appropriate evaluation indexes to evaluate the performance of the model is a necessary step in the task of protein classification. Accuracy (ACC), sensitivity (SN), specificity (SP), Matthews correlation coefficient (MCC), F-score and AUC (Area Under Roc Curve) were used in this paper [31–37]. They are calculated as follows:

$$Sp = \frac{TN}{TN + FP} \qquad (4-1)$$

$$Sn = \frac{TP}{TP + FN} \qquad (4-2)$$

$$Acc = \frac{TP + TN}{TP + FN + TN + FP} \qquad (4-3)$$

$$F1 = \frac{2 * TP}{2TP + FN + FP} \qquad (4-4)$$

$$MCC = \frac{TP \times TN - FP \times FN}{\sqrt{(TP+FP) \times (TP+FN) \times (TN+FN) \times (TN+FP)}} \qquad (4-5)$$

Where, TP is the correct number of positive samples, FP is the wrong number of negative samples, TN is the number of negative samples that are predicted to be positive, and FN is the number of positive samples that are predicted to be negative. Sn is the percentage of correct predictions for the positive cases, while Sp is the percentage of correct predictions for the negative cases. ACC reflects the overall accuracy of the predictor. When the data set is not balanced, ACC can't truly evaluate the quality of the classification results. In this case, MCC can still effectively measure the overall quality of the model. An F1 score, also known as a balanced F-score, can be interpreted as a weighted average of accuracy and recall rates. In addition, ROC is a curve based on sensitivity and specificity, and AUC is the area under the ROC curve. As an indicator to measure the robustness of the prediction model, the closer the AUC value is to 1, the better the prediction performance of the model will be.

5   Experimental Method and Result Analysis

5.1   Bert Extraction Evalution

First, we used the pre-training model Bert to extract the features of protein sequences, this feature extraction method extracted is a 304 * 768 feature matrix. Then, without any other processing, the feature vectors were directly put into the machine learning classifiers. Table 1 shows the performance of the eight classifiers. Among them, the accuracy of RF model is the highest, reaching 74.032%, and its AUROC value is 0.6902. The ROC curve of 10-fold cross-validation is shown in Fig. 2.

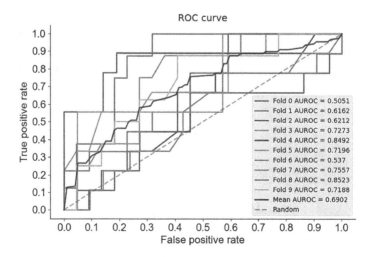

Figure 2. RF_ROC curve

Table 1. Only Bert extraction classifiers evaluation

Model	Sn	Sp	Pre	Acc	MCC	F1	ROC
RF	30.0	91.644	63.273	74.032	0.2876	0.3875	0.6902
LightGBM	36.944	87.445	54.665	73.022	0.2765	0.4147	0.7092
SVM	19.304	94.349	61.786	73.022	0.221	0.2675	0.6889
MLP	52.222	75.475	46.166	68.773	0.2714	0.4738	0.6814
XGBoost	38.332	87.987	57.509	73.678	0.3039	0.446	0.7009
KNN	51.111	79.243	46.722	71.107	0.2981	0.4872	0.6618
GBDT	36.943	88.462	55.837	73.678	0.291	0.434	0.6877
LR	53.195	78.355	52.074	71.108	0.3199	0.5159	0.6737

5.2 Bert Extraction and SMOTE

Because the ratio of positive and negative samples in the data set is 1:3, the positive and negative samples are unbalanced, which may lead to the deviation of the classification prediction results to the categories with more samples. Therefore, we choose to over-sample the data to alleviate the imbalance of positive and negative samples. Table 2 shows that after the proposed features are over-sampled and put into the classifier, the model effect is greatly improved. The accuracy of RF classifier is increased from 74.032 to 88.744, the ROC value is increased from 0.6902 to 0.9251, and the MCC value is increased from 0.2876 to 0.7843.

Table 2. Bert Extraction + SMOTE

Model	Sn	Sp	Pre	Acc	MCC	F1	ROC
RF	89.934	87.532	88.375	88.744	0.7843	0.886	0.9251
LightGBM	90.888	84.243	86.733	87.569	0.7706	0.8754	0.9241
SVM	79.308	71.861	73.794	75.622	0.5182	0.7618	0.7966
MLP	91.819	74.567	79.165	83.218	0.6846	0.8452	0.8967
XGBoost	88.593	81.969	83.818	85.286	0.7221	0.8524	0.9181
KNN	90.867	52.815	66.699	71.897	0.4862	0.7643	0.8364
GBDT	89.978	80.583	82.949	85.286	0.7217	0.8558	0.9159
LR	90.865	71.84	77.532	81.391	0.6486	0.8313	0.8634

5.3 Bert Extraction + SMOTE + Feature Selection

Since the feature dimension of training set is larger than the number of test sets, feature selection is carried out for them. During feature selection, feature importance is ranked first, and then feature values ranked top 300 are selected. In terms of feature selection, we chose Chi-square test and LGBM feature importance ranking. Table 3 and Table 4 respectively show the model effect after Chi-square test and LGBM feature selection.

Table 3. Bert Extraction + SMOTE + CHI-Square

Model	Feature selection	Sn	Sp	Pre	Acc	MCC	F1	ROC
RF	CHI square	89.48	87.532	88.194	88.521	0.7774	0.8841	0.9259
LightGBM		91.362	82.856	85.131	87.113	0.7595	0.8723	0.9262
SVM		76.538	73.68	74.196	75.153	0.507	0.7507	0.8073
MLP		91.363	72.813	78.055	82.087	0.6622	0.8371	0.9061
XGBoost		90.433	80.584	83.455	85.508	0.729	0.86	0.9188
KNN		90.412	54.265	66.752	72.389	0.4897	0.7652	0.8307
GBDT		89.978	81.47	83.567	85.746	0.7314	0.8584	0.918
LR		87.661	71.493	76.174	79.576	0.6062	0.8111	0.8541

Table 4. Bert Extraction + SMOTE + LightGBM

Model	Feature selection	Sn	Sp	Pre	Acc	MCC	F1	ROC
RF	LightGBM	91.32	89.328	90.239	90.349	0.8154	0.9028	0.9349
LightGBM		91.774	87.033	88.255	89.423	0.7991	0.8941	0.9363
SVM		81.127	72.792	74.702	77.004	0.5463	0.7753	0.8335
MLP		93.636	80.604	83.899	87.141	0.7578	0.8804	0.9131
XGBoost		90.866	83.832	85.466	87.352	0.7591	0.8752	0.9351
KNN		92.684	63.031	72.118	77.882	0.5924	0.808	0.8656
GBDT		89.069	84.286	86.364	86.671	0.7526	0.8635	0.9274
LR		88.571	78.268	80.909	83.456	0.6818	0.8411	0.885

5.4   Comparison with Other Methods

In order to prove the validity of the proposed model, we compare its prediction accuracy with that of other models on the basis of 10-fold cross validation and independent test set validation. Table 5 shows the comparison results of the four methods.

Table 5. Comparison with other methods

	ACC	Sn	Sp	MCC
Ding et al.	85.4	90.5	73.8	0.65
Jiao et al.	86.9	92.6	73.8	0.68
Yang et al.	88.5	88.9	88.0	0.68
This work	90.3	91.3	89.3	0.82

6 Conclusion

This paper proposes a new method to extract the features of Golgi body protein based on deep learning. In this experiment, the pre-training model BERT was used to extract the sequence features of Golgi body protein, and then a few over-samples were synthesized to alleviate the imbalance between the positive and negative samples of the training set and the test set. Subsequently, LightGBM feature selection method was adopted. A sub-Golgi protein localization classifier based on RF was developed. Finally, the 10-fold cross-validation score of this model is ACC 90.6%, Sn 91.32%, Sp 89.3%, MCC 0.82, independent test set score is ACC 95.3%, Sn 84.6%, Sp 98.0%, MCC 0.85. By comparing with the existing algorithm, this method can improve the prediction accuracy of sub-Golgi protein. In the next step, the Golgi protein data set will be expanded to enrich the deep learning features of protein sequences.

Acknowledgments. This work was supported in part by the University Innovation Team Project of Jinan (2019GXRC015), and in part by Key Science & Technology Innovation Project of Shandong Province (2019JZZY010324), the Natural Science Foundation of China (No. 61902337), the talent project of "Qingtan scholar" of Zaozhuang University, Natural Science Fund for Colleges and Universities in Jiangsu Province (No. 19KJB520016), Jiangsu Provincial Natural Science Foundation (No. SBK2019040953), Young talents of science and technology in Jiangsu.

References

1. Fujita, Y., et al.: Fragmentation of Golgi apparatus of nigral neurons with α-synuclein-positive inclusions in patients with Parkinson's disease. Acta Neuropathol. **112**(3), 261–265 (2006)
2. Hoyer, S.: Is sporadic Alzheimer disease the brain type of non-insulin dependent diabetes mellitus? A challenging hypothesis. J. Neural Transm. **105**(4–5), 415–422 (1998)
3. Rose, D.R.: Structure, mechanism and inhibition of Golgi α-mannosidase II. Curr. Opin. Struct. Biol. **22**(5), 558–562 (2012)
4. Gonatas, N.K., Gonatas, J.O., Stieber, A.: The involvement of the Golgi apparatus in the pathogenesis of amyotrophic lateral sclerosis, Alzheimer's disease, and ricin intoxication. Histochem. Cell Biol. **109**(5–6), 591–600 (1998)
5. Yang, W., et al.: A brief survey of machine learning methods in protein sub-Golgi localization. Curr. Bioinform. **14**(3), 234–240 (2019)
6. Wang, Z., Ding, H., Zou, Q.: Identifying cell types to interpret scRNA-seq data: how, why and more possibilities. Brief. Funct. Genomics **19**(4), 286–291 (2020)
7. Yuan, L., Guo, F., Wang, L., Zou, Q.: Prediction of tumor metastasis from sequencing data in the era of genome sequencing. Brief. Funct. Genomics **18**(6), 412–418 (2019)
8. Hummer, B.H., Maslar, D., Gutierrez, M.S., de Leeuw, N.F., Asensio, C.S.: Differential sorting behavior for soluble and transmembrane cargoes at the trans-Golgi network in endocrine cells. Mol. Biol. Cell **31**(3), 157–166 (2020)
9. Deng, S., Liu, H., Qiu, K., You, H., Lei, Q., Lu, W.: Role of the Golgi apparatus in the blood-brain barrier: Golgi protection may be a targeted therapy for neurological diseases. Mol. Neurobiol. **55**(6), 4788–4801 (2018)

10. Villeneuve, J., Duran, J., Scarpa, M., Bassaganyas, L., Van Galen, J., Malhotra, V.: Golgi enzymes do not cycle through the endoplasmic reticulum during protein secretion or mitosis. Mol. Biol. Cell **28**(1), 141–151 (2017)

11. Hou, Y., Dai, J., He, J., Niemi, A.J., Peng, X., Ilieva, N.: Intrinsic protein geometry with application to non-proline cis peptide planes. J. Math. Chem. **57**(1), 263–279 (2019)

12. Wei, L., Xing, P., Tang, J., Zou, Q.: PhosPred-RF: a novel sequence-based predictor for phosphorylation sites using sequential information only. IEEE Trans. Nanobiosci. **16**(4), 240–247 (2017)

13. Du, X., et al.: DeepPPI: boosting prediction of protein–protein interactions with deep neural networks. J. Chem. Inf. Model. **57**(6), 1499–1510 (2017)

14. van Dijk, A.D.J., et al.: Predicting sub-Golgi localization of type II membrane proteins. Bioinformatics **24**(16), 1779–1786 (2008)

15. Ding, H., et al.: Identify Golgi protein types with modified mahalanobis discriminant algorithm and pseudo amino acid composition. Protein Pept. Lett. **18**(1), 58–63 (2011)

16. Ding, H., et al.: Prediction of Golgi-resident protein types by using feature selection technique. Chemom. Intell. Lab. Syst. **124**, 9–13 (2013)

17. Jiao, Y.-S., Pu-Feng, D.: Predicting Golgi-resident protein types using pseudo amino acid compositions: approaches with positional specific physicochemical properties. J. Theor. Biol. **391**, 35–42 (2016)

18. Jiao, Y.-S., Pu-Feng, D.: Prediction of Golgi-resident protein types using general form of Chou's pseudo-amino acid compositions: approaches with minimal redundancy maximal relevance feature selection. J. Theor. Biol. **402**, 38–44 (2016)

19. Lv, Z., et al.: A random forest sub-Golgi protein classifier optimized via dipeptide and amino acid composition features. Front. Bioeng. Biotechnol. **7**, 215 (2019)

20. Zhao, W., et al.: Predicting protein sub-Golgi locations by combining functional domain enrichment scores with pseudo-amino acid compositions. J. Theor. Biol. **473**, 38–43 (2019)

21. Yang, R., Zhang, C., Gao, R., Zhang, L.: A novel feature extraction method with feature selection to identify Golgi–resident protein types from imbalanced data. Int. J. Mol. Sci. **17**(2), 218 (2016)

22. Jia, J., Liu, Z., Xiao, X., Liu, B., Chou, K.-C.: IPPBS-Opt: a sequence based ensemble classifier for identifying protein–protein binding sites by optimizing imbalanced training datasets. Molecules **21**(1), 95 (2016)

23. Jia, J., Liu, Z., Xiao, X., Liu, B., Chou, K.-C.: IPPI-Esml: an ensemble classifier for identifying the interactions of proteins by incorporating their physicochemical properties and wavelet transforms into PseAAC. J. Theor. Biol. **377**, 47–56 (2015)

24. Liu, B., Fang, L., Wang, S., Wang, X., Li, H., Chou, K.-C.: Identification of microRNA precursor with the degenerate K-tuple or Kmer strategy. J. Theor. Biol. **385**, 153–159 (2015)

25. Liu, B., Long, R., Chou, K.-C.: IDHS-EL: Identifying DNase I hyper sensitive sites by fusing three different modes of pseudo nucleotide composition into an ensemble learning framework. Bioinformatics **32**(16), 2411–2418 (2016)

26. Ding, H., et al.: ICTX-type: A sequence–based predictor for identifying the types of conotoxins in targeting ion channels. Biomed. Res. Int. **2014**, 1–10 (2014)

27. Liu, B., Gao, X., Zhang, H.: BioSeq–Analysis2.0: An updated platform for analyzing DNA, RNA and protein sequences at sequence level and residue level based on machine learning approaches. Nucleic Acids Res. **47**(20), e127 (2019)

28. Chen, W., Feng, P., Liu, T., Jin, D.: Recent advances in machine learning methods for predicting heat shock proteins. Curr. Drug Metab. **20**(3), 224–228 (2019)

29. Rao, R., et al.: Evaluating protein transfer learning with tape. Adv. Neural Inf. Process. Syst. **32**, 9689 (2019)

30. Pedregosa, F., et al.: Scikit-learn: machine learning in Python. J. Mach. Learn. Res. **12**, 2825–2830 (2011)
31. Zeng, X., Lin, W., Guo, M., Zou, Q.: A comprehensive overview and evaluation of circular RNA detection tools. PLoS Comput. Biol. **13**(6), e1005420 (2017)
32. Wei, L., Xing, P., Su, R., Shi, G., Ma, Z.S., Zou, Q.: CPPred–RF: a sequence-based predictor for identifying cell–penetrating peptides and their uptake efficiency. J. Proteome Res. **16**(5), 2044–2053 (2017)
33. Wei, L., Xing, P., Zeng, J., Chen, J., Su, R., Guo, F.: Improved prediction of protein–protein interactions using novel negative samples, features, and an ensemble classifier. Artif. Intell. Med. **83**, 67–74 (2017)
34. Hu, Y., Zhao, T., Zhang, N., Zang, T., Zhang, J., Cheng, L.: Identifying diseases-related metabolites using random walk. BMC Bioinf. **19**(S5), 116 (2018)
35. Zhang, M., et al.: MULTiPly: a novel multi-layer predictor for discovering general and specific types of promoters. Bioinformatics **35**(17), 2957–2965 (2019)
36. Song, T., Rodriguez-Paton, A., Zheng, P., Zeng, X.: Spiking neural P systems with colored spikes. IEEE Trans. Cogn. Dev. Syst. **10**(4), 1106–1115 (2018)
37. Lin, X., Quan, Z., Wang, Z.-J., Huang, H., Zeng, X.: A novel molecular representation with BiGRU neural networks for learning atom. Briefings Bioinf. Art. no. bbz125 (2019)

PointPAVGG: An Incremental Algorithm for Extraction of Points' Positional Feature Using VGG on Point Clouds

Yanzhao Shi, Chongyu Zhang, Xiaohui Zhang, Kai Wang[✉],
Yumeng Zhang, and Xiuyang Zhao

Shandong Provincial Key Laboratory of Network Based Intelligent Computing,
University of Jinan, Jinan 250022, People's Republic of China
ise_wangk@ujn.edu.cn

Abstract. Many works have been devoted to improving the accuracy of point cloud classification and segmentation, which are essential problems in computer vision. Although these works have achieved excellent performance using advanced feature extraction methods, it is still a challenging task to extract high-level features from disordered data in the form of point clouds. To tackle this issue, we propose a VGG-based network, called Point Positional Attention VGG (PointPAVGG) for 3D point cloud feature extracting and processing, which is inspired by the classical VGG network. Concretely, in order to combine global and local features, we extract the local point cloud geometric information by every sphere domain and analyze its global position score by our point attention (PA) module. This novel network, namely, PointPAVGG, with graph structure point cloud feature extraction and PA, is mainly presented and applied in point cloud classification as well as segmentation tasks. Comprehensive experiments carried out on ShapeNet and modelNet, which demonstrate that our methods deliver superior performance, showing state-of-the-art results in classification and segmentation tasks.

Keywords: Point cloud · Classification · Segmentation

1 Introduction

Learning and analyzing the point clouds are challenging works and how to reduce the huge number of parameters is also a problem that needs to be addressed urgently. In this paper, we proposed a special VGG-based architecture which has a smaller number of parameters to learn high-level features from the irregular and disordered 3D point clouds.

In the field of computer vision, convolutional neural networks are designed to learn the features in regular grid data. In recent years, the architectures of classical convolutional neural networks are used in point clouds and some CNN-based architectures [14, 24] have made significant progress in analyzing 3D point clouds. To process irregular point cloud, several works [11, 16, 17, 21, 25, 28] make efforts to replicate the remarkable performance on the analysis of image [12, 24]. In the early explorations, one common method is to transform point cloud data into regular voxels [3, 18, 22], but

D.-S. Huang et al. (Eds.): ICIC 2021, LNCS 12837, pp. 718–731, 2021.
https://doi.org/10.1007/978-3-030-84529-2_60

this method may cause the loss of rich 3D geometric information to a large extent and cause high complexity. Another way is using multi-view images [2, 5, 13] to analyze point clouds, but the same problem arises.

To solve all these problems, PointNet [26] is a pioneer to learn directly from irregular point cloud, which learns on each point from the point cloud independently and then gathers the final features for a global representation. Several works [6, 7] use other strategies such as grouping and multilayer perceptron (MLP) to get valid features. Although these strategies can make some progress theoretically, it is time-costing for grouping and sampling.

Based on the classical convolution operation in images [10, 11], some works [4, 27] have applied convolutional methods to point clouds. DGCNN [15] is an effective measure to get enhanced local geometric features in a local region and the key to DGCNN is updating the graph based on k-nearest neighbors dynamically. Therefore, it has this capability to capture local geometric structure. To capture the further relationship among each point and its neighbors, a hierarchical architecture called RS-CNN [16] is presented to analyze point cloud by contextual shape-aware learning. In addition, KPConv [27] has designed a deformable kernel to extract patterns. These above-mentioned researches have captured the information of geometric relations in the local regions and those works have dramatically improved shape awareness and robustness.

However, the limited receptive field of MLPs may lead to non-negligible loss of complex shape information. The works [16, 29, 30] with MLPs may be difficult to learn the geometric relations between each central point and its neighbors. Therefore, PointVGG [9] magnifies receptive fields, trying to capture the relations between neighbor points. Based on classic VGG [24] architecture, PointVGG [9] redefines the convolution and pooling method on point cloud and pays attention to the relations between each point and its neighbors.

Although PointVGG [9] achieves excellent results in object classification, the local semantic features of point clouds lack the connection with its positional information in global structure, which means that the information of each sphere domain contains more about the its local geometric structure but less associated with the positional information of central point. Thus, when it comes to extremely irregular-shaped objects, PointVGG [9] has difficulty measuring the relation between edge and center regions. To remedy this problem and extract more high-reliable features in each local structure, we propose a point attention unit (PA) to extract the position feature of each central point and combine normalized position feature with local geometric features. To make feature processing more focused on global position information, we regard the normalized position scores of all central points in point cloud as a matrix to regulate the result of convolution. Experiments prove that our work achieves state-of-the-art results.

The main contributions of our work are listed as follows:

1) We presented a new feature extraction method (PA) to combine the position features and local geometric features, which enabled the spherical domain to get local geometric features accurately while simultaneously cascading their corresponding global location feature. It can extract high-level features in the central points from the original point clouds more effectively.

2) We proposed PAconv to regulate the convolution by combining it with the normalized position features, which enabled PointPAVGG to learn high-level features efficiently and focused more on global positional information.

3) We named our novel network PointPAVGG, which can effectively process the features extracted by PA with the regulation of convolutional operation. We use PAconv and Ptcldpool methods based on the convolution and pooling methods of traditional VGG network to process the positional features and other geometric information.

4) PointPAVGG achieves state-of-the-art performance for classification and segmentation tasks on Modelnet40 and ShapeNet dataset.

2   Related Works

2.1   Classical Architecture of Point Clouds Processing

To process the irregular points from point clouds, one early method is to transform point cloud data into regular voxels [3, 18, 22], but it may lead to high complexity and loss the rich 3D geometric information. Then, PointNet [26], the pioneer of deep learning method used for point clouds, is aimed to learn directly from irregular point cloud, which gathers the features of each point for a global representation. However, it ignores the local structures, which is proven to be a crucial part of CNN, especially for accuracy. To get the features of all the related neighborhood, it uses a hierarchical architecture by MLPs, aiming to learn geometric relations among local subsets. More works [13, 15, 20] have already shown that using MLPs has achieved excellent result in getting representing features of multiscale 2D images. Besides, max pooling is also an effective measure to learn the local and regional feature. What's more, it is meaningful to learn the contextual representation efficiently [22–24]. However, MLP also has defects. For example, MLPs only concentrate on the relations between central point and its neighbors sampled by k-NN, and that results in causing loss of features involved in the local region.

To solve this problem, DGCNN [15] focuses on enhanced local geometric features in a local region and updates the graph based on k-nearest neighbors dynamically. Thus, it captures local geometric structure effectively. However, the limited receptive field of MLPs may cause non-negligible loss of complex shape information. Therefore, PointVGG [9] magnifies receptive fields and try to capture the high-level relations between neighbor points. It also redefines the convolution and pooling method on point cloud and pays attention to the relations between each point and its neighbors.

However, the local semantic features of point clouds have little connection with global position information, which means that although each sphere domain contains more information about the local structure, it lacks the information of its position in all sphere domains. By calculating the Eulerian distance of one central point and other N-1 points in point clouds, and then sum the number of N-1 distances can we get the original feature of its position. After normalizing the position feature of each central point, we aggregate this normalized position feature with local feature and train them with the regulation of normalized position metrics.

2.2   Point Convolutional Networks

Applying convolutional networks on point clouds is an effective way to learn high-level features. RS-CNN [16] using convolution methods to aggregates the high-level relations between each central point and its neighbors in order to represent the spatial layout of the point cloud locally and inductively. Dynamic Graph CNN for Learning on Point Clouds (DGCNN) [15] captures the information of similar local shapes by learning point relation in a high-dimensional feature space, which could be not reliable in some cases. However, the works [9, 16, 29, 30] mentioned above lack the representation of the positional information about each local structure in global structure, which is also a high-reliable feature of point clouds. Other works [19, 20] also have such problem.

To remedy this problem, in this study, we proposed a hierarchical architecture named PointPAVGG to extract this position feature and aggregate it with local feature, and then, training them by VGG-based convolutional architecture. We designed a point attention unit (PA) to aggregate the position features and other local geometric features. And we also regard all position scores of central points as positional matrix to regulate the result of convolution in order to let feature processing focuses more on global position information.

3   Method

Our work is an extension of PointVGG [9] with the added learning of positional feature and the regulation of convolution operation. We review the work of PointVGG (Sect. 3.1) at first, and then introduce our basic extension of PointVGG with the added extraction of positional feature to learn more high-reliable features in each local structure (Sect. 3.2). Next step, we introduce our regulation method of convolution (Sect. 3.3). Finally, we present our PointPAVGG which is able to process high-level features and focus more on positional information (Sect. 3.4).

3.1   Review of PointVGG [9]: Graph Convolutional Network with Progressive Aggregating Features on Point Clouds

Given $P = \{p_1, ..., p_n\} \subseteq R^F$ to indicate **n** points in point clouds, and F denotes the dimension $p_i = (x_i, y_i, z_i)$, it uses $N(p_i)$ to indicate the nearest neighbors of p_i sampled by farthest nearest sampling. The graph is represented as $G = \{V, E\}$, with $V = \{1, ..., n\}$ to indicate the vertices and $\varepsilon \in V \times V$ to indicate the edges respectively.

Feature Extraction. PointVGG applied k-NN algorithm to find k nearest neighbors $N(p_i)$, and sorted them by their Euclidean distance. PointVGG calculates edge features in local structure and aggregates them as

$$e_{ij} = \left(\sqrt{(x_i - x_j)^2 + (y_i - y_j)^2 + (z_i - z_j)^2}, x_i - x_j, y_i - y_j, z_i - z_j, \right.$$
$$\left. x_i y_i, z_i, x_j, y_j, z_j \right) \tag{1}$$

which contains some important geometric information.

Feature Learning Method. PointVGG processes the extracted features by VGG-based architecture, which contains the Pconv operation and Ppool operation.

Pconv is an essential convolution part on point cloud, which is defined as.

$$f_{N(p_i)} = \psi(f_{p_j}), \ p_j \in N \ (p_i) \tag{2}$$

In this equation, p_j is a neighbor point of center point p_i, f_{pj} is the feature vector of p_j, $f_{N(pi)}$ is the feature vector captured by ψ function. ψ is the Pconv operation to get abstract high-level features from the original features. In order to obtain additional abstract features and acquire globalization ability, PointVGG magnifies the receptive fields in its layer-by-layer architecture gradually.

Ppool is a special pooling method to process features in point cloud, which is defined as

$$f_{\varphi} = \varphi(f_{N(p_i)}), \tag{3}$$

$f_{N(pi)}$ indicates the result in Pconv operation, f_{φ} is the result of Ppool. φ is Ppool function to converge features progressively, which is similar to the max pooing method in 2D images. Ppool uses **1xN** max pooling to process the feature vector, it can *half* the last dimension (the number of neighbors) to converge progressively.

Base Architecture. PointPAVGG stacks Pconv and Ppool layer by layer to gradually magnify the receptive fields and converge the number of neighbors. Afterwards, it uses two fully connected layers to learn global features.

3.2 Aggregating Global Positional Feature and Local Geometric Feature

The edge features in local structure generated by PointVGG contains

$$\sqrt{\left(x_i - x_j\right)^2 + \left(y_i - y_j\right)^2 + \left(z_i - z_j\right)^2}, x_i - x_j, y_i - y_j, z_i - z_j, x_i y_i, z_i, x_j, y_j, z_j \tag{4}$$

These features can remedy the lack of local geometric features. However, in this way the relationship among all local domains is hard to represent adequately. Our work is aimed to relate every single local structure to other local domains. We find that all central points lack the representation of the relation to other central points in point cloud (see in Fig. 1).

To get this relation, we present the attention method PA to calculate the relation score, here we name this specific relation as positional feature. We define one central point as p_i, and $P = \{p_1, p_2, ..., p_n\}$ is the set of all points in point cloud. We calculate the distance between p_i, and $p_1 \in P$ and named it dis_1, and then calculate the distance between p_i and $p_2 \in P$ as dis_2. Repeat this way to calculate the distance between p_i and all central points and sum the n distances and define it as Dis. Every central point has its Dis_i, we record all of them as $Dis = \{dis_1, dis_2, ..., dis_n\}$. In our original experiment, we use dis_i as pi's positional feature, but this may cause the mismatch in magnitude of features, which means that we need to normalize this feature. We use softmax to normalize Dis as

$$a_{j,k} = softmax_j(dis_j, dis_{j,k}) = \frac{exp(dis_j^T \cdot dis_{j,k})}{\sum_{k \in \sqcup(p_j)} exp(dis_j^T \cdot dis_{j,k})} \tag{5}$$

where a_i is the position score of the spherical domain with p_i as its center point. In this way, we can find that edge points in point cloud usually have larger positional score than internal points. Thus, we regard this feature as a sign to distinguish edge points from intermediate points.

We combine local features with positional feature to learn high-reliable feature. The whole features contain $\sqrt{(x_i - x_j)^2 + (y_i - y_j)^2 + (z_i - z_j)^2}, x_i - x_j, y_i - y_j, z_i - z_j,$ $x_i y_i, z_i, x_j, y_j, z_j, a_i$.

In this way we can not only capture local geometric features, but also consider the relations between current local structure and others. Here, we consider this positional feature as a kind of attention feature to describe the position of each local region.

Although the similar attention feature is extracted in SVGA-Net [1], our combination method is an innovation to get high-level feature and learn the domain's position relations in the global simultaneously. The aggregation of positional feature and local geometric feature is paired with a subsequent convolution and pooling operation to achieve state of art.

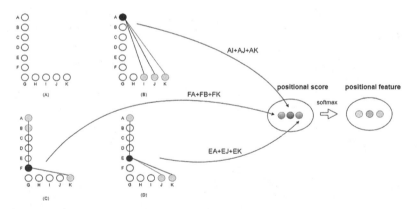

Fig. 1. The **method of getting positional feature.** PointPAVGG learns positional feature to distinguish edge points and intermediate points. (A) represents the point cloud, (B)(C)(D) represent the conditions where the center points are A, F, and E, respectively. Here, for one center point, we picked three neighbors for example. As it shown in (B), we picked point I, J, K as the neighbors of A, then calculate the sum of Euclidean distance of A_i, A_j, A_k as the original positional feature. We can see that A's positional score is higher than E and F, while A is the furthest point from the center of point cloud and represents the edge points. And the distance from E to the center of point cloud is farther than that of F, which highly proves that our positional feature has the capability to distinguish internal points from edge points.

3.3 The Regulation of Convolution

PointVGG applied convolutional operation to learn high-level features. It redefined the traditional convolution method as Pconv, which applied $1 \times n$ convolution kernel to get additional local geometric features. However, Pconv improved the generalization ability by magnifying the receptive fields step by step, which may lack the ability to perform effective for model to control global relationship among all local structures. In other words, if we add a regulation about positional relationship to convolution, it may be more effectual to enhance the generalization ability. Our method is shown in Fig. 2, where we illustrate the differences between our novel work and PointNet.

We have calculated the normalized positional feature $A = \{a_1, a_2, a_3, ..., a_n\}$, where a_n is the positional score of central point p_n. A is the set of all points' positional score and represents the global positional relationship among all local structures. We use A to regulate the result of convolution and we define this method as PAconv.

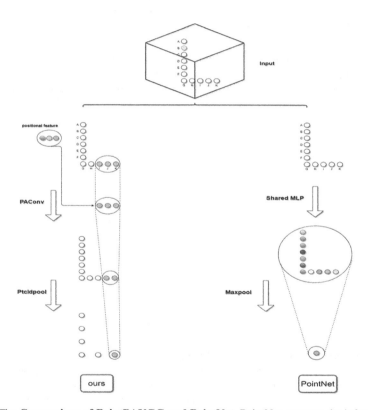

Fig. 2. The **Comparison of PointPAVGG and PointNet.** PointNet captures the information of points individually, so it lacks the learning of local structure. However, PointPAVGG can not only learn additional features of local structure but also pays attention to global positional relations by introducing *PAconv*. *Ptcldpool* is also applied to get detailed features in each local region by converging points gradually.

Ptcldpool. Ptcldpool can be defined as: $f_{N(pi)} = \xi(F_{pj}) \times A$, where ξ is the MLP architecture to learn high-level information from the extracted features. A is the metric of normalized positional feature, which is applied to add the representation of positional relations between each local spherical domain and other local structures. $F_{N(pi)}$ is the result of multiplication, which is simultaneously rich of local geometric information and global positional information. Experiment proves that our novel method PAconv has excellent performance, which achieves the high generalization ability and pays attention to global positional relationship simultaneously.

3.4 Hierarchical Point Set Feature Learning by PointPAVGG

The hierarchical architecture of PointPAVGG is shown in Fig. 3. We construct our VGG-based architecture by applying PAconv, Ptcldpool and Pconv. PAconv uses 1×3 convolutional kernel size instead of 3×3, because 3×3 convolution may cause the lack of local geometric information. PAconv also considered the representation of positional relations which is efficient to distinguish edge points from intermediate points. Ptcldpool applies 2×2 max pooling, which halves the number of

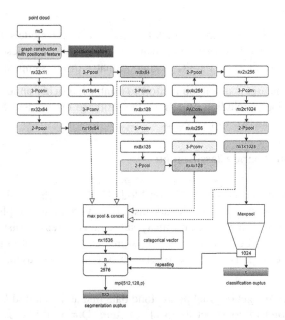

Fig. 3. Architecture **of PointPAVGG to achieve point cloud classification (right branch) and segmentation (left branch).** PointPAVGG is designed to implementing classification tasks and segmentation tasks. Based on the architecture of PointVGG, we added positional features to achieve better model performance. In this architecture, we constructed the graph feature with positional feature in order to add relations among all center points in point cloud. Then we applied a stack of Pconv and Ptcldpool to process the features. As the green block shown in the middle of figure, we used PAconv to regulate the result of convolution. Then, after two FC-layers, we used two kinds of architecture as the bottom of figure shown to us in order to solve classification tasks and segmentation tasks, respectively (Color figure online).

surroundings to minimize the feature maps. And in this strategy, we avoid using too large pooling window to capture enough geometric features. Pconv is an operation that proposed in PointVGG [9], which uses 1×3 convolutional kernel size as well. 1×1 Pconv is also applied in our architecture, which is aimed to capture high-level features while remaining the original size of feature metric.

Our PointPAVGG has a stack of PAconv, Ptcldpool and Pconv, then we add two fully connected layers. All these mentioned layers are equipped with ReLU and BN.

4 Experiments

In this section, we empirically demonstrate the efficacy and efficiency of our Point-PAVGG by conducting essential experiments with different benchmark tasks (Sect. 4.2 and 4.3), namely classification and segmentation. All the experiments are implemented in pyTorch and are run on NVIDIA TITAN V GPU card with 32G RAM. We also provide details to study PointPAVGG thoroughly (Sect. 4.4).

4.1 Datasets

ModelNet40. This dataset consists of 9,843 meshed CAD models for training and 2,468 meshed CAD models for testing, which contains 40 categories. We discard the original meshes and only use the (x, y, z) coordinates of the sampled points.

ShapeNet. This dataset is an ongoing effort to establish a richly-annotated, large-scale dataset of 3D shapes, including 16,881 shapes from 16 classes and 50 parts in total, where each shape has two to five parts.

4.2 Point Cloud Shape Classification

Model. We evaluate our PointPAVGG on ModelNet40, which is a highly competitive computer vision classification benchmark task. For every point sampled from a CAD model, we divide its neighbours into many subregions based on distances; the point cloud is established to build a graph which is predicted to feed into the convolutional network. The batch-size is set to be 16. The loss function we used, SFLoss, can be referred to [8].

Implementation. We get original point cloud data first and then find k neighbor points of each center point to construct its local structure. During this process, we sort the neighbors by Euclidean distance to solve the disorder of inputs. Then, we construct the graph of local features as

$$e_{ij} = \left(\sqrt{\left(x_i - x_j\right)^2 + \left(y_i - y_j\right)^2 + \left(z_i - z_j\right)^2}, x_i - x_j, y_i - y_j, z_i - z_j, x_i y_i, z_i, x_j, y_j, z_j, a_i \right)$$

$$(6)$$

a_i is the global positional score of each local structure, which contains the positional relationship among all local structures. After constructing the feature graph, we process the extracted feature by PointPAVGG network. The size of extracted feature is $b \times c \times n \times k$, where b is batch size, c is the channels of feature, n is the points we extracted from the point clouds and k is the number of neighbors in each local structure. We use *3-Pconv* to process it, the result is $b \times 2c \times n \times k$. Afterwards we half the k by *Ptcldpool* and the result is $b \times 2c \times n \times k/2$. In our experiment, the size of original extracted features is $16 \times 11 \times 1024 \times 32$, we use one *3-Pconv* and *Ptcldpool* to get the outputs $16 \times 64 \times 1024 \times 16$. Then we use a stack of *3-Pconv* and *Ptcldpool* repeatedly until the size of channel reached 256. We use *PAConv* to regulate the convolution by global positional features, which is an effective measure to pay attention to positional relationship among all local structures. Then, we use Ptcldpool, 3-Pconv and Ptcldpool, respectively. We already get the feature size 16×1024 1024×1, and then we use fully connected layer to learn global feature. The output is 16×46, and we use this result as classification output score.

Results. Table 2 gives a breakdown of our work and PointVGG in detail. Although our improvement is not quite large, still it occupies the highest position currently. Without using normal vector and voting, it can achieve the classification accuracy of RSCNN, which uses voting. Our method also outperforms several methods that use additional standard data and dense points 340 (5k) (Table 1).

Table 1. Classification results on **ModelNet40.** mA: mean per-class accuracy. OA: Overall accuracy. Comparison of PointPAVGG(ours) method by using Model B. Best results are denoted in boldface. (-: unknown)

method	input	points	mA	OA
PointNet[24]	point	1k	86.2	89.2
PointNet ++ [25]	point	1k	-	90.7
DGCNN[36]	point	1k	90.2	92.9
PointCNN[16]	point	1k	88.1	92.2
A-CNN[12]	point	1k	90.3	92.6
RSCNN[20]	point	1k	-	92.9
DensePoint[19]	point	1k	-	92.1
Point2Sequence[18]	point	1k	90.4	92.6
Point2Node[6]	point	1k	-	93
PointPAVGG(ours)	point	1k	**90.9**	**93.6**
SO-Net[15]	point	2k	-	90.9
PVR[42]	view	-	-	91.7
DGCNN[36]	point	5k	90.7	93.5
Pointconv[38]	point,normal	1k	-	92.5
PointNet ++ [25]	point,normal	5k	-	91.9
SO-Net[15]	point,normal	5k	90.8	93.4
RSCNN(with voting)[20]	point	1k	-	93.6
DensePoint(with voting)[29]	point	1k	-	93.2

Table 2. Comparison of PointPAVGG (ours) and PointVGG on classification results

method	mA	OA
PointPAVGG(ours)	90.94	**93.63**
PointVGG[10]	**91.07**	93.60

4.3   Point Cloud Shape Part Segmentation

Model. It is proved that part segmentation is a challenging fine-grained 3D recognition task. The network architecture and parameter setting are quite similar to the shape classification and the structure of which is presented in Fig. 3. In part segmentation network, 512 points are selected randomly as inputs and after convolution and pooling, we obtain a global feature vector. This vector is fused with the multilevel middle features produced by the convolution and pooling. Then, the fused features are concatenated with the one-hot encoding of the object label to form a congregated feature vector. With different numbers of neighbors, e.g., k = 32 and k = 16, we can acquire multilevel congregated feature vectors. The multilevel congregated features are then fed into fully connected layers for segmentation. Figure 4. shows the details of our method. For comparison, we use intersection over union (IoU) on points to evaluate the accuracy of our model and other benchmark methods.

Implementation. We use a stack of Pconv, Ptcldpool and PAConv which is same as classification task. Differently, after Pconv and Ptcldpool operation, we get global feature vector, which is fused with multilevel middle features. We concatenated the fused features with the one-hot encoding of object label and used different number of neighbors, e.g. k = 32, k = 64 to get multilevel congregated feature vectors. Then, we applied FC layer to get n × p segmentation score.

Results. Table 3 illustrates the segmentation results. Our PointPAVGG model outperforms some of the remarkable networks, which used 2048 points for part segmentation. We compare our results with PointVGG, DGCNN, and SO-Net. PointPAVGG achieves a remarkable performance and ranks first in three categories.

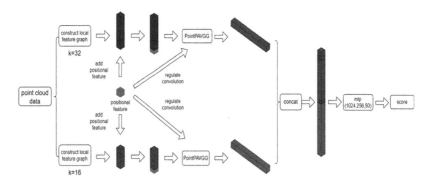

Fig. 4. Application of multiscale approach in segmentation.

Table 3. The result of part segmentation on ShapeNet

method	instance mIOU	air plane	bag	cap	car	chair	ear phone	guitar	knife	lamp	laptop	motor bike	mug	pistol	rocket	skate board	table
PointNet[24]	83.7	83.4	78.7	82.5	74.9	89.6	73.0	91.5	85.9	80.8	95.3	65.2	93.0	81.2	57.9	72.8	80.6
SO-Net[44]	84.6	81.9	83.5	84.8	78.1	90.8	72.2	90.1	83.6	82.3	95.2	69.3	94.2	80.0	51.6	72.1	82.6
PointNet++[34]	85.1	82.4	79.0	87.7	77.3	90.8	71.8	91.0	85.9	83.7	95.3	71.6	94.1	81.3	58.7	76.4	82.6
DGCNN[36]	85.2	84.0	83.4	86.7	77.8	90.6	74.7	91.2	87.5	82.8	95.7	66.3	94.9	81.1	63.5	74.5	82.6
Point2Sequence[45]	85.2	82.6	81.8	87.5	77.3	90.8	77.1	91.1	86.9	83.9	95.7	70.8	94.6	79.3	58.1	75.2	82.8
PointVGG[10]	86.1	83.3	82.7	87.5	77.9	90.7	79.6	91.3	87.0	84.9	95.6	71.1	95.1	81.7	56.4	76.3	85.0
PointPAVGG(ours)	85.6	81.4	80.4	87.4	79.1	90.5	76.5	91.7	80.7	84.4	95.4	71	94.5	82.3	57.5	74.6	83.7

4.4 PointPAVGG Design Analysis

In this section, we present the experimental results of our proposed network with diverse structures and parameters.

Table 4. The results of edge feature at different levels on ModelNet40

model	edge feature	mA	OA
a	(Euclidean Metric)	73.0	80.0
b	$(p_i\text{-}p_j)$	88.3	93.0
c	$(p_i\text{-}p_j,\ p_i)$	90.7	93.5
d	(Euclidean Metric, $p_i\text{-}p_j$, p_i, p_j)	90.9	93.6

Edge Feature at Different Levels. Determining how to design edge features is also significant. By representing the underlying shape information, the edge feature, e_{ij} can be defined flexibly. The experimental results are summarized in Table 4. We find that if we only use the Euclidean metric, PointPAVGG can achieve 80.0% accuracy; but when the edge feature is abundant, the model performance can be enhanced and can achieve 93.6% accuracy. In summary, the effectiveness of relation learning in Point-PAVGG is verified, and the experimental results can achieve higher accuracy when the edge feature contains information.

4.5 Poitrines Framework Design on ModelNet40

ResNet is a full of influence network structure building a deep CNN through skip connection. As a result, we attempt to build PointPAResNet in accordance with our experience in building PointPAVGG. To apply the similar structure of ResNet to point clouds, We replace ordinary image convolution with PAconv and downsampling with Ptcldpool. Due to the limitation of video memory, other deep network experiments are not conducted. In addition, we find that when the network is deeper and deeper, the performance degenerates dramatically probably due to the lack of rich local features and the small amount of training dataset.

5 Conclusion

We proposed PointPAVGG, a convolutional network with PAconv, Ptcldpool, and a graph structure used for point clouds, which is an extension of PointVGG [9] with the added learning of positional feature and the regulation of convolution operation. This network can learn the underlying geometric relations between each point and its neighbors. Moreover, it can abstract substantial discriminative features of point clouds. We evaluate our method on tasks, such as point cloud *shape classification* and *shape part segmentation*, on several benchmark datasets. The excellent results illustrates that our method outperforms other SOTA methods in terms of classification and achieves the better performance in part segmentation. Meanwhile, our regulation method of convolution is effective. Besides, our experimental results can convincingly validate that adding extraction of positional feature and additional discriminative features, which refine the results are able to learn more high-reliable features in each local structure when they are fed into the network.

Acknowledgements. This research was supported by Natural Science Foundation of Shandong province (No. ZR2019MF 013), Project of Jinan Scientific Research Leader's Laboratory (No. 2018GXRC023).

References

1. He, Q., Wang, Z., Zeng, H., et al.: SVGA-net: sparse voxel-graph attention network for 3D object detection from point clouds (2020)
2. Feng, Y., Zhang, Z., Zhao, X., Ji, R., Gao, Y.: GVCNN: group-view convolutional neural net-works for 3D shape recognition. In: CVPR, pp. 264–272 (2018). 1,2,5
3. Huang, X., Mei, G., Zhang, J.: Feature-metric registration: a fast semi-supervised approach for robust point cloud registration without correspondences (2020)
4. Guerrero, P., Kleiman, Y., Ovsjanikov, M., Mitra, N.J.: PCPNET learning local shape properties from raw point clouds. In: Computer Graphics Forum, no. 2, pp. 75–85. Wiley (2018)
5. Guo, H., Wang, J., Gao, Y., Li, J., Lu, H.: Multi-view 3D object retrieval with deep embedding network. IEEE Trans. Image Process. **25**(12), 5526–5537 (2016)
6. Hermosilla, P., et al.: Monte Carlo convolution for learning on non-uniformly sampled point clouds. ACM Trans. Graph. (TOG) **37**(6), 1–12 (2018)
7. Hua, B.S., Tran, M.K., Yeung, S.K.: Pointwise convolutional neural net-works. In: Proceedings of CVPR, pp. 984–993 (2018)
8. Huang, F., Xu, C., Tu, X., Li, S.: Weight loss for point clouds classification. In: Journal of Physics: Conference Series, vol. 1229, p. 012045 (2019)
9. Li, R., Zhang, Y., Niu, D., et al.: PointVGG: graph convolutional network with progressive aggregating features on point clouds. Neurocomputing **429**, 187–198 (2021). https://doi.org/10.1016/j.neucom.2020.10.086
10. Kipf, T.N., Welling, M.: Semi-supervised classification with graph convolutional networks. In: Proceedings of ICLR, pp. 1–14 (2017)
11. Wang, C., Samari, B., Siddiqi, K.: Local spectral graph convolution for point set feature learning. In: Ferrari, V., Hebert, M., Sminchisescu, C., Weiss, Y. (eds.) ECCV 2018. LNCS, vol. 11208, pp. 56–71. Springer, Cham (2018). https://doi.org/10.1007/978-3-030-01225-0_4
12. Krizhevsky, A., Sutskever, I., Hinton, G.E: ImageNet classification with deep convolutional neural net-works. In: NeurIPS, pp. 1106–1114 (2012). 1,4
13. Su, H., Maji, S., Kalogerakis, E., et al.: Multi-view convolutional neural networks for 3D shape recognition (2015)
14. Krizhevsky, A., Sutskever, I., Hinton, G.E.: ImageNet classification with deep convolutional neural networks. In: Advances in Neural Information Processing Systems, pp. 1097–1105 (2012)

15. Vaswani, A., et al.: Attention is all you need. In: Advances in Neural Information Processing Systems, pp. 5998–6008 (2017)
16. Liu, Y., Fan, B., Xiang, S., Pan, C.: Relation-shape convolutional neural network for point cloud analysis. In: Proceedings of CVPR, pp. 8895–8904 (2019)
17. Yin, K., Huang, H., Cohen-Or, D., Zhang, H.: P2P-NET: bidirectional point displacement net for shape transform. ACM Trans. Graph. 37(4), 152:1-152:13 (2018)
18. Wu, Z., et al.: 3D ShapeNets: a deep representation for volumetric shapes. In: CVPR, pp. 1912–1920 (2015). 1,2,5,7
19. Qi, C.R., Su, H., Mo, K., Guibas, L.J.: PointNet: deep learning on point sets for 3D classification and segmentation. In: Proceedings of CVPR, pp. 652–660 (2017)
20. Li, R., Li, X., Heng, P.A., et al.: PointAugment: an auto-augmentation framework for point cloud classification. In: 2020 IEEE/CVF Conference on Computer Vision and Pattern Recognition (CVPR). IEEE (2020)
21. Jiang, M., Wu, Y., Lu, C.: PointSIFT: a SIFT-like network module for 3D point cloud semantic segmentation. arXiv preprint arXiv:1807.00652 (2018)
22. Maturana, D., Scherer, S.: VoxNet: a 3D convolutional neural network for real-time object recognition. In: IROS, pp. 922–928 (2015)
23. Shen, Y., Feng, C., Yang, Y., Tian, D.: Mining point cloud local structures by kernel correlation and graph pooling. In: CVPR, pp. 4548–4557 (2018). 1,2,5,6,8
24. Simonyan, K., Zisserman, A.: Very deep convolutional networks for large-scale image recognition. In: Proceedings of ICLR, pp. 1–14 (2015)
25. Ravanbakhsh, S., Schneider, J., Poczos, B.: Deep learning with sets and point clouds. In: ICLR, pp. 1–12 (2017)
26. Qi, C.R., Su, H., Mo, K., Guibas, L.J.: PointNet: deep learning on point sets for 3D classification and segmentation. In: CVPR, pp. 77–85 (2016). 1,2,5,6,8
27. Thomas, H., Qi, C.R., Deschaud, J.E., et al.: KPConv: flexible and deformable convolution for point clouds (2019)
28. Qi, C.R., Yi, L., Su, H., Guibas, L.J.: Pointnet++: deep hierarchical feature learning on point sets in a metric space. In: Advances in Neural Information Processing Systems, pp. 5099–5108 (2017)
29. Wang, Y., Sun, Y., Liu, Z., Sarma, S.E., Bronstein, M.M., Solomon, J.M.: Dynamic graph CNN for learning on point clouds. ACM Trans. Graph. (TOG) 38(5), 1–12 (2019)
30. Wu, W., Qi, Z., Fuxin, L.: PointConv: deep convolutional networks on 3D point clouds. In: Proceedings of CVPR, pp. 9621–9630 (2019)

Predicting Course Score for Undergrade Students Using Neural Networks

Ming Liu[1], Zhuohui Li[2], Runyuan Sun[1], and Na Zhang[1(✉)]

[1] Shandong Provincial Key Laboratory of Network
Based Intelligent Computing, University of Jinan, Jinan 250022, China
zhangn@ujn.edu.cn
[2] Digital Department, Shandong Branch of China Unicom, Jinan 250021, China

Abstract. The rapid development of education big data has accumulated valuable data resources for the modernization of education. The teaching mode has gradually developed from the traditional "experience" teaching mode to a brand new "data" teaching model. The applications of education big data increasingly diversified, among them the research on education data is called Education Data Mining (EDM). In order to improve the academic performance of students and the teaching quality of universities, this paper proposed a score prediction model based on multi-layer feedforward neural network. During the optimization of this model, the accuracy of the experiment results improving gradually. Experimental results showed that the score prediction model can supply different levels of risking warning based on students score, and it also can help teachers to take early interventions to make sure students graduate successfully.

Keywords: Educational Data Mining · Neural network · Academic pre-warning

1 Introduction

In the information age, the educational datasets of universities are more and more diversity. How to take fully advantage of the valuable datasets became the important way to construct high-level university. In the teaching model, the student's score is the key factor to evaluate its learning situation in each semester. And the performance of the courses in the new semester often depends on the performance of the students in the historical stage [1–3].

At present, student achievement and dropout rates in higher education have also received attention [4, 5]. People are starting to use data mining techniques to solve some of the problems in education. Most people regard online learning behavior as a dynamic data source, which is the most mainstream basis for course score prediction [6–9]. Most of the data sources for their studies were from online learning platforms, including massive open platforms (MOOC) and various online education systems [10, 11]. These data are unreliable and untimely. First, platforms cannot distinguish whether students are actually engaged in the course based on attributes such as the length of time they have been online and the degree of completion of the video. Second, the findings depend

© Springer Nature Switzerland AG 2021
D.-S. Huang et al. (Eds.): ICIC 2021, LNCS 12837, pp. 732–744, 2021.
https://doi.org/10.1007/978-3-030-84529-2_61

on the student's performance in the current phase to predict the final performance in the current phase, so there is a lag [1].

In fact, the study of traditional educational data is more relevant to the reform of university education. Someone has experimented for student's performance based on historical student datasets, and they believed that neural network models can effectively capture the learning process of student and are suitable for the prediction of student performance [12–15]. So far, the above studies have achieved some success, there are still some common problems that have not been solved. First, the different majors of students lead to a small amount of data for similar samples that cannot be widely used for more courses [1]. Second, there are no uniform data selection criteria in the field to determine which attributes of students are associated with scores.

Therefore, in this paper we also used neural networks as a framework for our study and improved the method for the problems mentioned above. Firstly, we designed a data processing scheme in which students' history semester scores in all required courses were used as input criteria. Beside that, GPA, mean score and standard deviation for the same course taught by different teachers are considered to solve the small data problems. Secondly, for the loss function of the model we selected the *weighted mse loss* (*wmse*) function, and designed the weight term according to the target scores of students, which makes the model focus on high-risk students in particular, and ensures the reliability of the pre-warning.

The rest of the paper is organized as follows. Section 2 describes the development of the field of educational data mining and the main work in the field. Section 3 presents the model we designed. The experimental procedure and results are described in detail in Sect. 4 and Sect. 4.3. Section 5 provides a conclusion and outlook.

2 Related Works

Educational Data Mining (EDM) is a subject that intersects with the field of big data and artificial intelligence. In the field of EDM, the term "at-risk" is often used to refer to groups of students who are at risk of learning, may fail, or even drop out of school [16]. According to studies, prediction of students' course scores has become one of the most important research field [17–20].

The traditional algorithms such as decision trees, KNN and SVM were used in predicting final performance of student's examinations. Chiheb et al. [21] used decision trees and the J48 algorithm to predict the final performance of students. Hasnawi et al. [22] used CART and KNN algorithms to predicted the performance of 55 students and achieved an accuracy of 65.5% and 60.5%. Although there are many existing improvement methods, such as Ma [23] optimizing decision tree and SVM through a grid search algorithm to improve the accuracy of the algorithm. However, there are still have large room for progress in machine learning field such as increase prediction accuracy and solve big sample datasets problem.

In order to improve the shortcomings of machine learning, the use of neural networks to predict student performance has been proposed in recent years [24–28]. Rifat [29] proposed a DNN based model to predict the CGPA of undergraduate business students, demonstrating that the model was capable of consistent learning from the transcript data and showed promising predictive performance. Rodolfo [30] implemented several deep neural networks to predict students' future grades using their online activity logs as input attributes. Sikder [31] addressed early prediction of student risk by training a number of neural networks to predict which students are likely to submit assignments on time, and although the proposed neural network models have relatively good accuracy, they cannot be said to be applicable to all courses.

3 Methodology

3.1 Problem Statement

Assume that the set of student features is S. S is composed of the student' history semester course information and the teacher' information set of the target course. Therefore, S is defined as:

$$S = \{feature, teacher\}$$

The student's history course information includes grades in required history semester courses, GPA, and failure rate in required courses. In particular, the student's required course attributes include all required courses offered by the student's college. In the set we use m to represent the total number of required courses in the history semester. The number of history semesters directly affects the value of m, and thus the dimensions of the model input change accordingly. The input set $feature$ is defined as:

$$feature = \{score_1, score_2, \cdots, score_m, gpa, fail\_rate\}$$

Teacher' information includes teacher name, class mean and standard deviation of the course taught by the teacher. It is worth noting that the average score and standard deviation in the teacher information are based on the class as a unit. For example, a teacher teaches the same course in class A and class B, but the average score and standard deviation of the final statistical are different. The $teacher$ set is as follows:

$$teacher = \{t\_name, t\_avg, t\_std\}$$

Assuming that the original data is divided into p semesters, the total number of courses in each semester is t_p. The model can predict the grades of the kth course in j semesters based on the relevant information of the previous i semesters. That is, the target set l can be defined as:

$$l = \{l_{jk}|i+1 \leq j \leq p, 1 \leq k \leq t_j\}$$

Since there is no course input before the first semester, it can only be used as a feature, not as a prediction target, so $i \geq 1$. As shown in Fig. 1, S is used as model input and l is used as model output. It should be noted that the compulsory courses of l are the courses that students will take in the future.

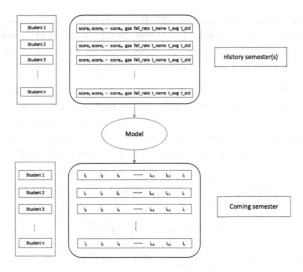

Fig. 1. Problem statement

3.2 Course Score Prediction Model

Model Architecture. To improve the accuracy of performance prediction, the model adopted a single-output feedforward neural network with two hidden layers, and the model architecture is shown in as Fig. 2. Firstly, we adopted a network structure with two hidden layers and used LeakyReLU as the activation function. Besides, in order to avoid data bias in the training process, batch normalization layers are added to maintain the same distribution of inputs on each layer, which reduces the difficulty of model learning. Finally, a dropout layer is added to prevent overfitting and make the model more generalizable.

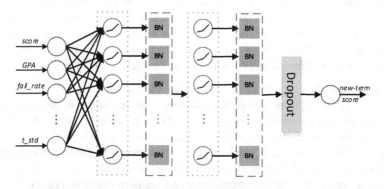

Fig. 2. Model architecture

Optimizer. AdamW adds a regular term to the loss function of the Adam optimization algorithm, and the overall convergence of the model is faster and reduces the over-fitting phenomenon that occurs during model training [32]. Therefore, AdamW is chosen as the optimizer for this experiment in this paper.

Loss Function. The final purpose of the course score prediction model we study is to find out the students who may have risk in the future course score. Therefore, the *weighted mse loss* function can be a useful solution to our needs, and the formula is shown in (1). Where the weighted term w depends on the score of the target course, $w \in [0, 1]$. For example, a student's target course score is 100, then $w = 0$.

$$wmse = \frac{1}{n}\sum_{i=0}^{n} w_i * (Ypred_i - Ytarg_i)^2 \tag{1}$$

$$w_i = |(Ytarg_i - 100)/100| \tag{2}$$

where *Ypred*, *Ytarg* and *w* are vectors with the same dimension; *Ypred* represents the predicted value; *Ytarg* represents the target value, and i is the subscript.

4   Experiments and Discussions

4.1   Data Collection and Processing

Data Collection. The data involved in this study were obtained from information on all undergraduate courses related to a college from 2014–2016. In addition, since there are few courses in the third and fourth grades of the university, and the professionalism is strong, we only select the student data of the first and second years of the university as the sample space. Total of 51,498 students' course data and 190,986 teachers' data are collected.

Unifying Feature. After data preprocessing, we obtained a sample set for each student, and the course attributes in the sample set included all required courses set by the student's college. But in reality, the course plans set by different majors in the same college are different. Therefore, we use 0 to replace students' untaken course grades, which solves the problem of inconsistent course characteristics of students.

Data Description. Take a college as an example, the raw data was processed to obtain a total of 1740 students information. Each student's information is in the form of an n-dimensional row vector of elements. And we filtered the dataset from it based on the presence or absence of the target course scores. In this paper, we used different semester courses as an example, such as the courses shown in Table 1. Given that our proposed course score prediction model is designed to identify at-risk students for the course, we are more interested in whether students are performing at a pass or fail level. For this purpose, we counted data on the total number of students in the course, the number of students who passed and the number of students who failed, and so on.

Table 1. Data description

Course-name	Course-total	Course-pass	Course-fail	Failure rate
Probability theory	1731	1537	194	11.21%
English 4	1725	1626	99	5.74%
Data structures	1078	960	118	10.95%
Computer principles	843	743	100	11.86%
Math 2	1728	1496	232	13.43%

According to the statistics, we see that the failure rate is more than 10% for all the courses except for the course English 4. This shows that if we can pre-warning by way of course score prediction, it has significant implications for reducing the failure rate of the course.

Standardization. The scores between students are discrete, so the attribute values will vary greatly. High or low scores will affect the reasonable training of the model to a certain extent. Therefore, we use the StandardScaler method of Eq. (3) to normalize the standard deviation of each feature. The processed data conform to the normal distribution, with an average value of 0 and a variance of 1.

$$Z = \frac{x - \bar{x}}{S} \tag{3}$$

where x is the current value, $\bar{x}$ is the average value, and s is the standard deviation of the sample.

4.2 Experimental Environment

The experimental environment is based on the PyTorch 1.8.1 deep learning framework, and the values of the parameters of the model in the experiment are as follows:

a. *split ratio.* We split the total sample into the train set and test set at a ratio of 7: 3.
b. *learning rate.* The Adamw optimization algorithm is more flexible in the way the learning rate is changed, no artificial adjustment of the learning rate is needed, and the optimal value can be obtained quickly using the default parameters.
c. *weight decay.* In this paper, the weights of the nodes are decayed by introducing a weight decay penalty term, which reduces the fitting problem of the model to some extent.
d. *batch size.* The sample set is randomly selected by the value of batch size for each training, and an appropriate batch size can make the model training better. In this paper, we set the size of the batch size to 128.
e. *iteration.* The network model is iterated 1000 times to learn more implicit information on the basis of avoiding underfitting or overfitting.

4.3 Model Evaluation

Regression Model Evaluation Index. For the course score prediction model, the evaluation indicators we use are mean square error (MSE), mean absolute error (MAE), and R2 coefficient of determination. The formula for each evaluation index is as follows. MSE is used to measure the degree of variation in the data, while MAE calculates the mean of the absolute error, which better reflects the true picture of the error; the coefficient of determination of R2 reflects the ability of the independent variable to explain the dependent variable. The value is between 0–1. In theory, when the value is 1, a perfect fit should be achieved.

$$MSE = \frac{1}{m}\sum_{i=1}^{m}(p_i - y_i)^2 \tag{4}$$

$$MAE = \frac{1}{m}\sum_{i=1}^{m}|p_i - y_i|^2 \tag{5}$$

$$R2 = 1 - \frac{\sum_{i=1}^{m}(p_i - y_i)^2}{\sum_{i=1}^{m}(y_i - \hat{y})^2} \tag{6}$$

where y_i is the actual score, p_i is the predicted score, and $\hat{y}$ is the mean of the actual score.

Classification Model Evaluation Index. The confusion matrix is usually used to summarize the prediction results of the classification model. According to the decision criteria of the real category and the decision criteria of the predicted category, the records in the data set are summarized into a matrix form. Where the rows of the matrix indicate whether the students' real performance are passing or not, and the columns of the matrix indicate whether the model is labeled with risk, as shown in Table 2.

Table 2. Confusion matrix

Confusion matrix		Risk marker	
		Risky = 0	Risk-free = 1
Real performance	Qualified = 0	TP	FN
	Unqualified = 1	FP	TN

The real score is scored as a decision criterion value of 60 to determine whether the score is qualified, and the risk prediction is based on whether the predicted score is less than the set criterion value (usually less than 60 points) to mark whether the students' performance is threatened. Here is the concept of a decision criterion value, which is a criterion to judge whether a student's performance is qualified (risky) or unqualified (risk-free).

Therefore, we can calculate the accuracy, precision, and recall rate based on the confusion matrix to judge the classification and prediction capabilities of the model.

The accuracy rate is calculated as the ratio of the correct number of model risk labels to the total sample size.

$$Accuracy = (TP + TN)/(TP + FN + FP + TN) \tag{8}$$

The precision of the model is calculated as the ratio of the number of samples that are correctly labeled as risky to the number of samples that are all labeled as risky.

$$Precision = TP/(TP + FP) \tag{9}$$

The recall rate is calculated as the ratio of the number of samples that are correctly predicted to be at risk to the number of samples that are truly at risk.

$$Recall = TP/(TP + FN) \tag{10}$$

In the next section, we will judge the prediction (pre-warning) capabilities of the model in different scenarios based on the above evaluation indicators.

4.4 Experimental Results and Analysis

Regression Evaluation. To begin with, we tested the predictive power of the model using various evaluation metrics of the regression task. We carried out experiments with required courses from several colleges of a university. Next, we still used a college as an example to predict the courses for different semesters, and the prediction results are shown in Table 3.

Table 3. Regression model evaluation results

Course-name	Number of test samples	MSE	MAE	R2
Probability theory	520	114.85	8.33	0.54
English 4	517	30.32	4.19	0.47
Data structure	324	98.50	7.85	0.4
Computer principles	253	117.30	8.61	0.48
Math 2	519	125.83	8.93	0.54

From the table, we can see that the MAE value can be controlled within 10, which indicates that the course score prediction model proposed in this paper is effective. While the value of R2 is not very satisfactory, the reason is that due to our use of the *wmse* loss function, the model will pay particular attention to risky students, so the overall fitting ability of the model is affected to some extent.

Since the number of test samples was too large to visualize clearly, we randomly selected some test samples to visualize the prediction results for the Math 2 course in the second semester and the English 4 course in the fourth semester, as shown in Fig. 3 and Fig. 4.

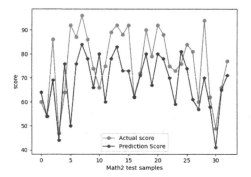

Fig. 3. Fitting curve of Math2

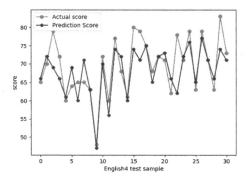

Fig. 4. Fitting curve of English4

Since the loss function and focused on at-risk students, we can see that the fit under the image is better than the top one, providing assurance of pre-warning. The final results of the pre-warning will be reflected in the next section.

Classification Evaluation. The ultimate goal of our model is to take personalized pre-warning for students who are at risk in the course, so we convert the results of regression prediction into classification results and use the evaluation indicators of the classification task to judge the pre-warning ability of the model.

In order to ensure the reliability and fault tolerance of the model, we divide the model into a high-risk model and a low-risk model according to the decision boundary, and their focus is different. The purpose of the high-risk model is to more accurately mark the students who are at high risk. This part of the students is a small number but their risks are certain; the goal of the low-risk model is to mark all students who may be at risk. The number is large but their risk is uncertain.

In this experiment, for the high-risk model, we set the decision boundary value as 50 points, and the predicted value lower than the decision boundary is the high-risk warning (the results are shown in Table 4 below). The accuracy ensures the reliability of the prediction results.

Table 4. High-risk models predict outcomes

Course-name	Number of test samples	Actual number of risks	Mark the number of risks	Mark the correct number	Accuracy	Precision
Probability theory	520	58	15	13	0.91	0.87
English 4	517	33	6	5	0.94	0.83
Data structure	324	32	4	4	0.91	1
Computer principles	253	36	9	8	0.89	0.88
Math 2	519	70	34	31	0.92	0.91

After that, for the low-risk model, if we increase the decision boundary value to 60 points or higher, the model flags more students who are likely to be at risk, and the recall rate best reflects the competence of the low-risk model. (See Table 5).

Table 5. Low-risk models predict outcomes

Course-name	Number of test samples	Actual number of risks	Mark the number of risks	Mark the correct number	Accuracy	Recall
Probability theory	520	58	64	36	0.87	0.62
English 4	517	33	56	19	0.90	0.58
Data structure	324	32	25	15	0.92	0.47
Computer principles	253	36	38	20	0.87	0.56
Math 2	519	70	99	51	0.87	0.73

From the calculation results, both the high-risk model and the low-risk model guarantee more than 85% classification accuracy. The high-risk model can determine which students are at risk by more than 80% and can assist schools in the scientific intervention so that students can notice the risks and take measures in advance.

Although the low-risk model will "accidental injury" some students who are not at risk, it will not actually hurt these students. On the contrary, it can mark the students who are really at risk in the future to a greater extent and make them more vigilant.

5   Conclusions and Outlook

The undergraduate score prediction model proposed in this paper can predict course scores for a new semester based on students' previous academic performance, it also can supply early risk-warning in specific course for high risk students who maybe fail in this course. This model can help university improve the student's quality and decrease the student's dropout rate, what's more it can change the management mode of the university.

In the future, we will enrich the data dimensions, including ways to increase the basic attributes of students, online characteristics, and the degree of correlation between courses, so as to improve the accuracy of the course score prediction model. Finally, it is packaged as an application and integrated on the school's smart campus platform to achieve the purpose of assisting teaching.

References

1. Ma, Y., Cui, C., Nie, X., Yang, G., Shaheed, K., Yin, Y.: Pre-course student performance prediction with multi-instance multi-label learning. Sci. China Inf. Sci. **62**(2), 1–3 (2018). https://doi.org/10.1007/s11432-017-9371-y
2. Sweeney, M., Lester, J., Rangwala, H.: Next-term student grade prediction. In: 2015 IEEE International Conference on Big Data, pp. 970–975 (2015)
3. Gull, H., Saqib, M., Iqbal, S.Z., et al.: Improving learning experience of students by early prediction of student performance using machine learning. In: 2020 IEEE International Conference for Innovation in Technology, pp. 1–4 (2020)
4. Polyzou, A., Karypis, G.: Feature extraction for next-term prediction of poor student performance. IEEE Trans. **12**, 237–248 (2019)
5. Pongpaichet, S., Jankapor, S., Janchai, S., et al.: Early detection at-risk students using machine learning. In: 2020 International Conference on Information and Communication Technology Convergence, pp. 283–287 (2020)
6. Chen, X., Vorvoreanu, M., Madhavan, K.: Mining social media data for understanding students' learning experiences. IEEE Trans. Learn. Technol. **7**(3), 246–259 (2014)
7. Pardo, A., Han, F., Ellis, R.A.: Combining university student self-regulated learning indicators and engagement with online learning events to predict academic performance. IEEE Trans. **10**(1), 82–92 (2017)
8. Jiang, J., Zeng, L., Zhi, L.: Research on the process data mining of online wisdom learning. In: 2019 IEEE 4th International Conference on Big Data Analytics, pp. 135–139 (2019)
9. Monllaó Olivé, D., Huynh, D.Q., Reynolds, M., et al.: A quest for a one-size-fits-all neural network: early prediction of students at risk in online courses. IEEE Trans. **12**(2), 171–183 (2019)
10. Voghoei, S.N., Yazdansepas, D., et al.: University online courses: correlation between students' participation rate and academic performance. In: 2019 International Conference on Computational Science and Computational Intelligence, pp. 772–777 (2019)

11. Jain, H., Anika: Applying data mining techniques for generating MOOCs recommendations on the basis of learners online activity. In: 2018 IEEE 6th International Conference on MOOCs, Innovation and Technology in Education (MITE), pp. 6–13 (2018)
12. Alkhasawneh, R., Hobson, R.: Modeling student retention in science and engineering disciplines using neural networks. In: 2011 IEEE Global Engineering Education Conference (EDUCON), pp. 660–663 (2011)
13. Huang, S., Fang, N.: Work in progress—prediction of students' academic performance in an introductory engineering course. In: 2011 Frontiers in Education Conference (FIE), pp. S4D-1–S4D-3 (2011)
14. Liu, Q., Huang, Z., et al.: EKT: exercise-aware knowledge tracing for student performance prediction. IEEE Trans. **33**(1), 100–115 (2019)
15. Patil, A.P., Ganesan, K., Kanavalli, A.: Effective deep learning model to predict student grade point averages. In: 2017 IEEE International Conference on Computational Intelligence and Computing Research (ICCIC), pp. 1–6 (2017)
16. Ho, L.C., Shim, K.: Data mining approach to the identification of at-risk students. In: 2018 IEEE International Conference on Big Data (Big Data), pp. 5333–5335 (2018)
17. Baneres, D.M., Gonzalez, E., Serra, M.: An early feedback prediction system for learners at-risk within a first-year higher education course. IEEE Trans. **12**(2), 249–263 (2019)
18. Wan, H., Liu, K., Yu, Q., et al.: Pedagogical intervention practices: improving learning engagement based on early prediction. IEEE Trans. **12**(2), 278–289 (2019)
19. Hung, J., Shelton, B.E., Yang, J., et al.: Improving predictive modeling for at-risk student identification: a multistage approach. IEEE Trans. **12**(2), 148–157 (2019)
20. Cano, A., Leonard, J.D.: Interpretable multiview early warning system adapted to underrepresented student populations. IEEE Trans. **12**(2), 198–211 (2019)
21. Chiheb, F., Boumahdi, F., Bouarfa, H.: Predicting students performance using decision trees: case of an Algerian University. In: 2017 International Conference on Mathematics and Information Technology (ICMIT), pp. 113–121 (2017)
22. Hasnawi, M., Kurniati, N., Mansyur, S.H., et al.: Combination of case based reasoning with nearest neighbor and decision tree for early warning system of student achievement. In: 2018 2nd East Indonesia Conference on Computer and Information Technology (EIConCIT), pp. 78–81 (2018)
23. Ma, X., Zhou, Z.: Student pass rates prediction using optimized support vector machine and decision tree. In: 2018 IEEE 8th Annual Computing and Communication Workshop and Conference (CCWC), pp. 209–215 (2018)
24. Hossen, M.A., Alamgir, R.B., Alam, A.U., et al.: A web based four-tier architecture using reduced feature based neural network approach for prediction of student performance. In: 2021 2nd International Conference on Robotics, Electrical and Signal Processing Techniques (ICREST), pp. 269–273 (2021)
25. Cazarez, R.L., Martin, C.: Neural networks for predicting student performance in online education. IEEE Lat. Am. Trans. **16**(7), 2053–2060 (2018)
26. Deuja, R., Kusatha, R.: Data-driven predictive analysis of student performance in college using neural networks. In: 2018 IEEE 3rd International Conference on Computing, Communication and Security (ICCCS), pp. 77–81 (2018)
27. Yang, C.T., Brinton, C.G., Chiang, M.: Behavior-based grade prediction for MOOCs via time series neural networks. IEEE J. Sel. Top. Signal Process. **11**(5), 716–728 (2017)
28. Monllaó Olivé, D.Q., Huynh, M., et al.: A quest for a one-size-fits-all neural network: early prediction of students at risk in online courses. IEEE Trans. **12**(2), 171–183 (2019)
29. Islam Rifat, M.R., Badrudduza, A.S.: EduNet: a deep neural network approach for predicting CGPA of undergraduate students. In: 2019 1st International Conference on Advances in Science, Engineering and Robotics Technology (ICASERT), pp. 1–6 (2019)

30. Raga, R.C., Raga, J.D.: Early prediction of student performance in blended learning courses using deep neural networks. In: 2019 International Symposium on Educational Technology (ISET), pp. 39–43 (2019)
31. Sikder, M.F., Uddin, M.J., Halder, S.: Predicting students yearly performance using neural network: A case study of BSMRSTU. In: 2016 5th International Conference on Informatics, Electronics and Vision (ICIEV), pp. 524–529 (2016)
32. Loshchilov, I., Hutter, F.: Decoupled weight decay regularization. In: International Conference on Learning Representations (2017)

Classification of Heart Sounds Using MFCC and CNN

Kai Wang[✉] and Kang Chen

School of Information Science and Engineering, University of Jinan,
Jinan 250022, People's Republic of China
ise_wangk@ujn.edu.cn

Abstract. An excellent heart sound classification system can be used as a good method to complete the daily heart sound detection under the condition of low cost and high efficiency, which is convenient to detect problems in the early stage of heart disease, and at the same time can alleviate the problem of medical staff shortage. In this paper, the data set of 3000 heart sound samples was used to accomplish the system. The noise reduction and feature extraction of cardiac sound signals, as well as the construction and training of network models was used to describe in detail. In order to improve the accuracy of the system, contrast the coif5 wavelet and db6 wavelet noise reduction effect, without segmentation of one mind under the condition of the second heart sounds, compared the LPCC, MFCC and the effect of pure heart sound wave input, used for feature extraction has advantages of convolution neural network classification task, and the loss function and optimizer batches for cross training, finally realizes a in 3000 samples with 98% accuracy of heart sound classification system.

Keywords: Heart sound · MFCC · Convolution neural network

1 Introduction

Heart sound is the sound produced by the vibration of heart organs and cardiovascular system as a whole. It is one of the physiological signals with great significance and value in human body. The heart sound in pathological changes contains a lot of pathological information such as atrium, ventricle and cardiovascular system, which has very important time-saving and guiding significance for self examination of related diseases and preliminary diagnosis of doctors.

As a link in the whole diagnosis chain, heart sound classification is of great significance in the rapid auxiliary diagnosis of heart disease, especially when the number of patients is large, or the medical staff is lack of auscultation experience, so as to quickly screen out patients with obvious related diseases and seize the treatment time.

The analysis and prediction of heart sound signal mainly includes three parts: noise reduction, feature extraction and classification [1].

There are many effective algorithms for signal de-noising, such as DB wavelet, coif wavelet and so on. For example, at the end of last century, Liang H proposed a heart sound processing algorithm based on the envelope diagram of heart sound (mainly the

© Springer Nature Switzerland AG 2021
D.-S. Huang et al. (Eds.): ICIC 2021, LNCS 12837, pp. 745–756, 2021.
https://doi.org/10.1007/978-3-030-84529-2_62

connection of wave peaks) [2]. Wavelet analysis was carried out to calculate the Shannon energy envelope, judge the threshold and find the peak, The first heart sound and the second heart sound are distinguished by wavelet transform. And then through the neural network method to denoise heart sound [3], can achieve an ideal effect.

In the aspect of feature extraction of heart sound signal, many different methods are used, such as over wavelet analysis, EMD [4, 5], LPCC [6], MFCC [7–9] and so on. MFCC feature is a widely used feature extraction method (Mel frequency cepstrum coefficient). Mel frequency and the characteristics of animal hearing fit very well, and it belongs to nonlinear mapping with frequency. The main steps are pre emphasis, short frame, Hamming window, fast Fourier transform, filtering by Mel filter (22–26), DCT (discrete cosine transform) and so on. In recent years, MFCC feature parameters have been widely used in heart sound feature.

In the recognition and classification of heart sound signals, most of the algorithms related to machine learning are used in the early stage, such as K-Neighborhood algorithm (KNN) [10, 11], Gaussian hybrid model (GMM) [12], random forest (RF) [13], support vector machine (SVM) [14], etc., which have been applied to some extent in the recognition of heart sound signals, and have achieved certain results. Recently, with the rise of deep neural network, the paper analyzes the characteristics of the new algorithm, there are also many DNN classification algorithms [15].

Because the first heart sound and the second heart sound are more significant, the relevant analysis methods are relatively mature.

The closure of the mitral and tricuspid valves produces a first heart sound. According to statistics, only a few children can be observed and monitored for division. When the first heart sounds appear abnormal (such as "wiring", "blurring, tearing") it is often caused by abnormal aortic valve activity, ventricular septal defect, regurgitation of atrioventricular valve, unclosed artery catheter and obstruction of outflow channel of right ventricle.

The second heart sounds are produced by the closure of pulmonary artery and aortic valve. The pathological condition is mainly caused by Division (usually divided, fixed, abnormal) [16], which may be caused by severe hypertension, atrial septal defect, left bundle branch block or aortic stenosis.

2   Heart Sound Signal Preprocessing

2.1   Noise Reduction of Heart Sound Signal

Heart sound is the vibration produced by the contraction and relaxation of myocardium, the opening and closing of valves, and the contact between blood and ventricular wall and blood vessels in a single cardiac cycle. Heart sounds fluctuate periodically and regularly with heart beat, producing the first, second, third and fourth heart sounds (S1/S2/S3/S4). Generally, the first and second heart sounds are obvious, while the third and fourth heart sounds are weak.

Heart sound signal is relatively weak, and is usually collected in relatively strong noise environment, which is easily interfered by many environmental factors, resulting in a lot of noise. Before the classification of heart sound signal, it is necessary to de noise it.

At present, the common noise reduction methods are spectral subtraction, wavelet and so on. Effective noise reduction methods can improve the recognition rate and robustness of the classification model. In order to improve the accuracy of heart sound classification, DB6 wavelet is used to decompose heart sound signal into five layers.

Wavelet transform is usually used as an advanced tool for time-frequency analysis [17]. The idea of wavelet originates from Fourier transform and short-time Fourier transform, which can be improved while preserving the advantages of the former two. Compared with Fourier transform, wavelet transform decomposes the signal into several trigonometric function curves, while the former decomposes and transforms the signal into a series of wavelets, which not only solves the problem of locality, but also obtains the time-frequency diagram by wavelet transform, which is also of great value in signal analysis.

Wavelet formula:

$$\Psi_{(a,\tau)} = \frac{1}{\sqrt{a}} \Psi\left(\frac{t-\tau}{a}\right) \tag{1}$$

The formula of wavelet transform is as follows

$$WT(a,\tau) = \frac{1}{\sqrt{a}} \int_{-\infty}^{\infty} f(t) * \Psi\left(\frac{t-\tau}{a}\right) dt \tag{2}$$

The wavelet depends on the parameter (a, τ). From wavelet generating function Ψ(x) Generated continuous wavelet function, change a can stretch the function, change a τ you can translate this function. A is inversely proportional to the frequency, τ it is positively proportional to time.

The wavelet de-noising method in this paper uses decomposition and reconstruction method.

Noise reduction steps:

① The preprocessed heart sound data is decomposed by wavelet transform to get the components (coefficients) at different levels.
② The components are thresholded.
③ Wavelet reconstruction, the components are reconstructed to form the de-noising signal.

In the realization of wavelet de-noising, the determination of wavelet base, the selection of vanishing moment, the selection of decomposition level, the selection of threshold and threshold processing algorithm are all very important to the final de-noising effect of heart sound.

In this paper, coif wavelet and DB wavelet are tried in the simulation, and finally DB wavelet is used for noise reduction.

In the process of wavelet de-noising, the larger the number of decomposition layers, the greater the difference between the effective information and the invalid information, which is good for separating the noise. However, the larger the number of decomposition layers is, the time delay after decomposition will become unstable, and the stability in time domain will become worse. The new signal formed after reconstruction is more likely to be distorted, and the signal itself may be filtered, which will obviously affect the final performance of noise reduction. According to Chen Tianhua's research on filtering measures of heart sound signal, this paper chooses 5-layer decomposition.

Daubechies created compactly supported orthogonal wavelets, or Daubechies wavelets, abbreviated as DBN, where N is the order of wavelets. 2N − 1 is their support domain, and with the increase of order N, the smoothness will become better [18]. The better the ability to extract local features in frequency domain, the better the result of subband decomposition, but it will cause some uncontrollable problems in time domain, so the effect will become worse. In this paper, the value of N is set to 6.

Daubechies wavelet has the following characteristics.

(1) In the time domain, it is finite supported, i.e. $\Psi(t)$ The length is limited.
(2) When $\omega = 0$, $\Psi(\omega)$ has zero point of n-th order.
(3) We can use $\Phi(t)$ to derive the wavelet function $\Psi(t)$。 $\Phi(t)$ is low pass filter function, $t \in (0\text{--}2N - 1)$.

The threshold value is not a fixed value, and the rules for selecting the threshold value are: unbiased likelihood estimation, threshold value of extreme value multiplying proportion and fixed threshold estimation. After the threshold is selected, the wavelet needs to be filtered by the threshold function, and the commonly used threshold function is the soft and hard threshold.

Hard threshold denoising method: when the absolute value of wavelet coefficient is less than the determined threshold, it will be taken as zero, and it will not change when it is greater than the given threshold. Soft threshold denoising method: when the absolute value of the wavelet coefficient is less than the determined threshold, it will return to zero; when it is greater than the threshold, it will subtract the threshold from the wavelet coefficient (Figs. 1 and 2).

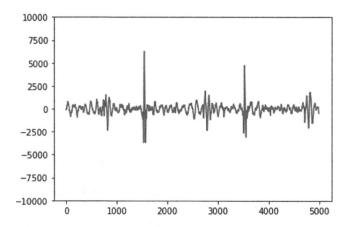

Fig. 1. Original heart sound signal

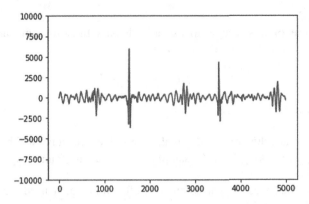

Fig. 2. Heart sound signal after noise reduction

2.2 Feature Extraction of Heart Sound Signal

The heart sound signals used in this paper come from 2016 PhysioNet/Computing in Cardiology (CinC) Challenge data set [19].

As a special sound signal, heart sound has many differences from ordinary signals, such as periodicity, which can be analyzed only by using a few seconds of signal.

MFCC method is chosen in this paper.

Mel frequency cepstral coefficients (MFCC) was proposed by Davis and mermelstein in 1980 after studying the mechanism of human hearing. It consists of a series of band-pass filters from small to large in different frequency bands according to different bandwidth [20]. Each filter is used to filter the input data, the output data is the MFCC coefficient, and MFCC has good tolerance for noise, low requirements for the input signal, and it is very consistent with the human auditory model. Therefore, compared with LPCC, MFCC has better stability and inclusiveness, and is more in line with animal characteristics. It has analysis value and excellent recognition ability in different application scenarios.

MFCC formula (Fig. 3):

$$\text{Mel}(f) = 2595 * \lg\left(1 + \frac{f}{700}\right) \tag{3}$$

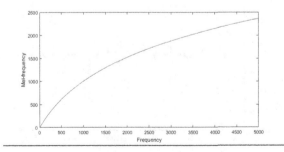

Fig. 3. Frequency to mel-frequency curve

(1) Preemphasis

The purpose of this step is to use a high pass filter to compensate the high frequency part and flatten the spectrum.

$$H(Z) = 1 - \mu z^{-1} \tag{4}$$

In this formula, $\mu = 0.97$.

(2) Framing

In order to make the processed signal more stable, the signal will be processed in frames, usually taking 2^8 or 2^9 sampling points as the frame length.

(3) Add windows

In order to make the signal continuous after framing, it needs to overlap in a certain proportion.

$$W(n) = 0.54 - 0.42 * \cos((2 * pi * n)/(n - 1)) \tag{5}$$

(4) Fast Fourier transform (FFT)

At this time, the signal already belongs to the short-time stationary signal in the time domain, and each frame can be converted into the frequency domain signal for processing.

(5) Triangular bandpass filter banks

24 different Mel triangular filters are used to filter the frequency domain signal obtained in step (4) of the fourth step, and the subband energy is calculated. This step is completed by librosa library.

(6) Discrete cosine transform

The obtained subband energy is substituted into the discrete cosine transform to obtain the MFCC coefficients:

$$C(n) = \sum_{m=0}^{N-1} s(m)\cos\left(\frac{\pi n(m - 0.5)}{M}\right) \tag{6}$$

A heart sound audio file with a sampling rate of 2000 and wav format is sliced, denoised and MFCC extracted to get the preprocessed result (Figs. 4 and 5).

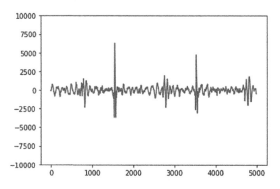

Fig. 4. Input audio waveform

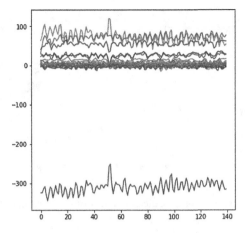

Fig. 5. Output MFCC feature array

3   Heart Sound Classification Algorithm

Common heart sound classification algorithms include KNN, RF, GMM and so on. KNN takes a long time to calculate when there is a large amount of data, and its robustness is poor when there is an error signal. The accuracy of RF is always lower than 80%. GMM is obviously not as good as neural network in fitting nonlinear characteristics. After several commonly used algorithms are implemented and run for many times, convolution neural network is selected to complete the classification after comparison, and the training is carried out without extracting the wave crest and cutting the first heart sound and the second heart sound in advance, which greatly shortens the preprocessing time and improves the simplicity of the code (Table 1 and Fig. 6).

Table 1. Comparison of classification effect of different algorithms (%)

GMM	KNN	RF	SVM	CNN
66	73	72	66	77
27	68	79	70	76
69	65	71	74	74
32	74	75	73	81
31	70	69	72	78
24	78	76	75	75
73	76	74	65	79
69	73	80	68	84
34	72	77	59	77
71	69	75	70	81

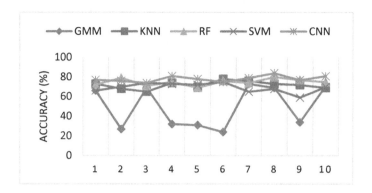

Fig. 6. Comparison of classification effect of different algorithms (%)

3.1 Convolutional Neural Network

Convolutional neural network (CNN) is a multi-layer feedforward network. Its main structure is input layer, output layer, convolutional layer, excitation layer, pooling layer and fully connected layer. CNN has a strong ability of feature recognition [21].

In convolution network, usually * denotes convolution operation:

$$s(t) = (x * w)(t) \tag{7}$$

The function w is called kernel or filer, which is a probability density function (integral is 1 in the domain of definition), so the output will be a weighted average. X is the input.

Usually, in the implementation, it is discrete data, and discrete convolution can be defined:

$$s(t) = (x * w)(t) = \sum_{a=-\infty}^{\infty} x(a)w(t-a) \tag{8}$$

3.2 Training of Convolutional Neural Network

In the training of convolution neural network, we should pay attention to the following points.

One is that the input format is determined, but the heart sound signal belongs to the data of variable length, but different from the voice data, the heart sound data can show a complete state in just a few seconds, so in the preprocessing, only the first few seconds are intercepted, and the data with insufficient length is padded to make the input fixed.

The second is the loss (cost) function, which needs to find a loss function to define the gap between the predicted and the actual results, or the closeness between the actual output and the expected results. The common loss functions are logarithmic loss function, absolute value loss function, 0–1 loss function, square loss function, etc., and the appropriate loss function is very important for the accuracy of the model.

The cross entropy loss function can well represent the probability distribution, and is often used in two or more classification problems.

The formula of cross entropy loss is as follows:

$$Loss(Y, f(x)) = -\frac{1}{n} \sum_x [Ylna + (1 - y)\ln(1 - a)] \tag{9}$$

Here y is the label, X is the input, a is the predicted value, and N is the total sample size.

Through optimizers such as MBGD, SGD, bgd, Adam, adadelta and adagrad, the weight of the minimum error is found. Bgd uses all the data as a batch to calculate the gradient, so it will cause the calculation process to be very slow. SGD calculates every sample once. MBGD, as the name suggests, uses Mini batch. Small batch samples are used as a batch to calculate the gradient, so the convergence is more stable and fast. In the face of nonconvex functions, these three kinds of optimizers may be limited to local optima (or oscillate out) (Table 2).

Table 2. CNN

	Input dimension	Convolution window size (filter)	Output
Convolution layer 1	(n, 1, 24, 24)	5	(n, 10, 20, 20)
Pooling layer 1	(n, 10, 20, 20)	NA	(n,10, 10, 10)
Convolution layer 2	(n,10, 10, 10)	3	(n, 20, 8, 8)
Pooling layer 2	(n, 20, 8, 8)	NA	(n,20, 4, 4)

In order to get the best model, test and compare the effects of different loss functions and optimizers. After the analysis, the loss function chooses to test cross entropy and mean square error, and the optimizer uses SGD and Adam for comparative test (Fig. 7 and Table 3).

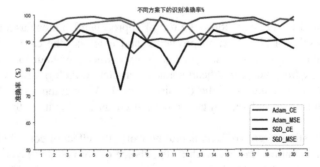

Fig. 7. Performance of different loss functions and optimizers

Table 3. Performance of different loss functions and optimizer

Adam+CE	Adam+MSE	SGD+CE	SGD+MSE
97.6	90.4	90.4	79.4
96.5	91.2	95.8	89
98.6	92.8	90.4	88.8
99	91.8	96.4	94.2
99.2	92	97	92.6
98.4	92.6	97.6	91
98.8	90.6	98	72.2
97	85.6	95.8	93.4
90.4	90.4	98.2	90
98.9	91	97.8	87.4

4   Conclusion

The purpose of this paper is to complete an end-to-end automatic heart sound classi-fication system, which can help patients or suspected patients quickly diagnose heart condition, make patients realize that their body has sent out discomfort signals, can timely seek medical treatment, can help doctors save time, so as to have time for in-depth analysis and diagnosis and treatment of patients in need. In the process of realizing the system, the noise reduction of heart sound, the feature extraction of heart sound and the realization of convolution neural network are mainly discussed.

(1) Noise reduction of heart sound signal data: heart sound signal usually passes through relatively large noise environment and is relatively weak. It is easy to be interfered by many environmental factors and many noises are recorded. The purpose of noise reduction of heart sound signal is to filter out the noise that interferes with heart sound. After comparing the performance of different wave-lets, this system chooses DB6 wavelet for five level decomposition to reduce the noise of heart sound signal, so as to improve the accuracy of heart sound classification.

(2) Heart sound feature extraction: the use of appropriate features has a great impact on the results of classification problems, and it is also the case in heart sound classification problems. In the process of experiment, without locating and seg-menting the first and second heart sounds, complete waveform, LPCC and MFCC are used as the features of the denoised signal. After comparison, MFCC with good robustness and recognition rate is used as the input of the back neural network.

(3) Implementation of convolution neural network: the effect of convolution network is related to various factors, such as the choice of convolution layers, the size of convolution window, the choice of excitation function, the choice of loss function, and the choice of optimizer. After analyzing the data complexity, the network is simplified from five layer convolution, and two-layer convolution is adopted, the

first layer uses ten 5 * 5 convolution kernels, and the second layer uses twenty 3 * 3 convolution kernels. Relu is selected as the excitation layer, cross entropy loss as the loss function, Adam optimizer and python are used to complete the architecture, completed a very simple and well targeted network.

Finally, the system saves the model with the highest accuracy and completes a heart sound classification system with an accuracy of 98%. It can predict a segment of input heart sound data and get two results of "normal heart sound" or "abnormal heart sound".

References

1. Reed, T.R., Reed, N.E., Fritzson, P.: Heart sound analysis for symptom detection and computer-aided diagnosis. Simul. Model. Pract. Theory. 12(2), 129–146 (2004)
2. Liang, H., Lukkarinen, S., Hartimo, I.: Heart sound segmentation algorithm based on heart sound envelogram. In: Computers in Cardiology 1997, pp. 105–108. IEEE (1997)
3. Hadi, H.M., Mashor, M.Y., Suboh, M.Z., Mohamed, M.S.: Classification of heart sound based on s-transform and neural network. In: 10th International Conference on Information Science, Signal Processing and Their Applications (ISSPA 2010), pp. 189–192 (2010)
4. Salman, A.H., Ahmadi, N., Mengko, R., et al.: Empirical mode decomposition (EMD) based denoising method for heart sound signal and its performance analysis. Int. J. Electr. Comput. Eng. 6(5), 2197 (2016)
5. Boutana, D., Benidir, M., Barkat, B.: On the selection of intrinsic mode function in EMD method: application on heart sound signal. In: 2010 3rd International Symposium on Applied Sciences in Biomedical and Communication Technologies (ISABEL 2010), pp. 1–5 (2010)
6. Ye-wei, T., Xia, S., Hui-xiang, Z., Wei, W.: A biometric identification system based on heart sound signal. In: 3rd International Conference on Human System Interaction, pp. 67–75 (2010)
7. Fu, W., Yang, X., Wang, Y.: Heart sound diagnosis based on DTW and MFCC. In: 2010 3rd International Congress on Image and Signal Processing, pp. 2920–2923 (2010)
8. Anusuya, M., Katti, S.: Comparison of different speech feature extraction techniques with and without wavelet transform to kannada speech recognition. Int. J. Comput. Appl. 26, 19–24 (2011)
9. Turner, C., Joseph, A.: A wavelet packet and mel-frequency cepstral coefficients-based feature extraction method for speaker identification. Procedia Comput. Sci. 61, 416–421 (2015)
10. Juniati, D., Khotimah, C., Wardani, D., et al.: Fractal dimension to classify the heart sound recordings with KNN and fuzzy c-mean clustering methods. J. Phys. Conf. Ser. 953, 012202 (2018)
11. Singh, S.A., Majumder, S.: Classification of unsegmented heart sound recording using KNN classifier. J. Mech. Med. Biol. 19(04), 1950025 (2019)
12. Zhang, W.Y., Guo, X.M., Weng, J.: Application of improved GMM in classification and recognition of heart sound. J. Vib. Shock 33(6), 29–34 (2014)
13. Esmail, M.Y., Ahmed, D.H., Eltayeb, M.: Classification system for heart sounds based on random forests. J. Clin. Eng. 44(2), 76–80 (2019)

14. Springer, D.B., Tarassenko, L., Clifford, G.D.: Support vector machine hidden semi-Markov model-based heart sound segmentation. In: Computing in Cardiology 2014, vol. 41, pp. 625–628 (2014)
15. Chen, T.-E., et al.: S1 and S2 heart sound recognition using deep neural networks. IEEE Trans. Biomed. Eng. **64**(2), 372–380 (2017)
16. Wang, K., Cheng, X., et al.: Heart sound model based on cascaded and lossless acoustic tubes. J. Mech. Med. Biol. **19**(05), 1950031 (2019)
17. Daubechies: The wavelet transform, time-frequency localization and signal analysis. IEEE Trans. Inf. Theory **36**(5), 961–1005(1990)
18. Sanamdikar, S.T., Hamde, S.T., Asutkar, V.G.: Extraction of different features of ECG signal for detection of cardiac arrhythmias by using wavelet transformation Db 6. In: 2017 International Conference on Energy, Communication, Data Analytics and Soft Computing (ICECDS), pp. 2407–2412 (2017)
19. Liu, C., Springer, D.B., et al.: An open access database for the evaluation of heart sound algorithms. Physiol. Meas. **37**(12), 2181–2213 (2016)
20. Muda, L., Begam, M., Elamvazuthi, I.: Voice recognition algorithms using mel frequency cepstral coefficient (MFCC) and dynamic time warping (DTW) techniques. J. Comput. **2**(3), 138–143 (2010)
21. Ketkar, N., Moolayil, J.: Deep Learning with Python. Apress, Berkeley (2021)

Author Index

Printed in the United States
by Baker & Taylor Publisher Services